Common Formulas

Distance

$$d = rt$$

d = distance traveled
t = time
r = rate

Temperature

$$F = \frac{9}{5}C + 32$$

F = degrees Fahrenheit
C = degrees Celsius

Simple Interest

$$I = Prt$$

I = interest
P = principal
r = annual interest rate
t = time in years

Compound Interest

$$A = P\left(1 + \frac{r}{n}\right)^{nt}$$

A = balance
P = principal
r = annual interest rate
n = compoundings per year
t = time in years

Coordinate Plane: Midpoint Formula

Midpoint of line segment joining (x_1, y_1) and (x_2, y_2)

$$\left(\frac{x_1 + x_2}{2}, \frac{y_1 + y_2}{2}\right)$$

Coordinate Plane: Distance Formula

d = distance between points (x_1, y_1) and (x_2, y_2)

$$d = \sqrt{(x_2 - x_1)^2 + (y_2 - y_1)^2}$$

Quadratic Formula

Solutions of $ax^2 + bx + c = 0$

$$x = \frac{-b \pm \sqrt{b^2 - 4ac}}{2a}$$

Rules of Exponents

(Assume $a \neq 0$ and $b \neq 0$.)

$$a^0 = 1 \qquad a^m \cdot a^n = a^{m+n}$$

$$(ab)^m = a^m \cdot b^m \qquad (a^m)^n = a^{mn}$$

$$\frac{a^m}{a^n} = a^{m-n} \qquad \left(\frac{a}{b}\right)^m = \frac{a^m}{b^m}$$

$$a^{-n} = \frac{1}{a^n} \qquad \left(\frac{a}{b}\right)^{-n} = \frac{b^n}{a^n}$$

Basic Rules of Algebra

Commutative Property of Addition

$$a + b = b + a$$

Commutative Property of Multiplication

$$ab = ba$$

Associative Property of Addition

$$(a + b) + c = a + (b + c)$$

Associative Property of Multiplication

$$(ab)c = a(bc)$$

Left Distributive Property

$$a(b + c) = ab + ac$$

Right Distributive Property

$$(a + b)c = ac + bc$$

Additive Identity Property

$$a + 0 = 0 + a = a$$

Multiplicative Identity Property

$$a \cdot 1 = 1 \cdot a = a$$

Additive Inverse Property

$$a + (-a) = 0$$

Multiplicative Inverse Property

$$a \cdot \frac{1}{a} = 1, \quad a \neq 0$$

Properties of Equality

Addition Property of Equality

If $a = b$, then $a + c = b + c$.

Multiplication Property of Equality

If $a = b$, then $ac = bc$.

Cancellation Property of Addition

If $a + c = b + c$, then $a = b$.

Cancellation Property of Multiplication

If $ac = bc$, and $c \neq 0$, then $a = b$.

Zero Factor Property

If $ab = 0$, then $a = 0$ or $b = 0$.

Elementary Algebra

FOURTH EDITION

Instructor's
Annotated
Edition

Elementary Algebra

FOURTH EDITION

• **Ron Larson**
The Pennsylvania State University
The Behrend College

• **Robert P. Hostetler**
The Pennsylvania State University
The Behrend College

With the assistance of
Patrick M. Kelly
Mercyhurst College

Houghton Mifflin Company   Boston   New York

Vice President and Publisher: Jack Shira
Associate Sponsoring Editor: Cathy Cantin
Development Manager: Maureen Ross
Associate Editor: Marika Hoe
Assistant Editor: James Cohen
Supervising Editor: Karen Carter
Senior Project Editor: Patty Bergin
Editorial Assistant: Allison Seymour
Production Technology Supervisor: Gary Crespo
Executive Marketing Manager: Michael Busnach
Senior Marketing Manager: Ben Rivera
Marketing Assistant: Lisa Lawler
Senior Manufacturing Coordinator: Priscilla Bailey
Composition and Art: Meridian Creative Group
Cover Design Manager: Diana Coe

Cover art © by Dale Chihuly
20,000 Pounds of Ice and Neon, detail
1992
Honolulu Academy of Arts, Honolulu, Hawaii
Photo: Linny Morris Cunningham

We have included examples and exercises that use real-life data as well as technology output from a variety of software. This would not have been possible without the help of many people and organizations. Our wholehearted thanks go to all their time and effort.

Trademark acknowledgement: TI is a registered trademark of Texas Instruments, Inc.

Printed in the U.S.A.

Library of Congress Catalog Card Number: 2003107480

ISBN: 0-618-38818-4

123456789–DOW–08 07 06 05 04

Contents

CONTENTS

*Appendices B and C are available on the textbook website and the instructor ClassPrep CD-ROM. To access the appendices online, go to math.college.hmco.com/instructors and link to **Elementary Algebra**, Fourth Edition.*

A Word from the Authors

Welcome to *Elementary Algebra*, Fourth Edition. In this revision, we have continued to focus on developing students' proficiency and conceptual understanding of algebra. We hope you enjoy the Fourth Edition.

In response to suggestions from elementary and intermediate algebra instructors, we have revised and reorganized the coverage of topics for the Fourth Edition. To give more emphasis to operations with integers, Chapter 1 now contains two separate sections: Section 1.2, "Adding and Subtracting Integers," and Section 1.3, "Multiplying and Dividing Integers." We have consolidated the discussion of complex fractions and placed it as its own section in Chapter 7. In addition, we have added a new section to Chapter 8 on "Systems of Linear Inequalities" and a new section to Chapter 10 on "Complex Numbers." To improve the flow of the material, we have relocated the section on "Negative Exponents and Scientific Notation," including a discussion on the rules of exponents, to the beginning of Chapter 5. Chapter 7, "Rational Expressions and Equations," now appears before "Systems of Linear Inequalities" in Chapter 8. Also, our introduction to functions has been moved to the end of Chapter 10, where it can be easily incorporated or omitted by instructors.

In order to address the diverse needs and abilities of students, we offer a straightforward approach to the presentation of difficult concepts. In the Fourth Edition, the emphasis is on helping students learn a variety of techniques—symbolic, numeric, and visual—for solving problems. We are committed to providing students with a successful and meaningful course of study.

Our approach begins with *Motivating the Chapter,* a feature that introduces each chapter. These multipart problems are designed to show students the relevance of algebra to the world around them. Each *Motivating the Chapter* feature is a real-life application that requires students to apply the concepts of the chapter in order to solve each part of the problem. Problem-solving and critical thinking skills are emphasized here and throughout the text in applications that appear in the examples and exercise sets.

To improve the usefulness of the text as a study tool, we have added two new, paired features to the beginning of each section: *What You Should Learn* lists the main objectives that students will encounter throughout the section, and *Why You Should Learn It* provides a motivational explanation for learning the given objectives. To help keep students focused as they read the section, each objective presented in *What You Should Learn* is restated in the margin at the point where the concept is introduced.

In this edition, the *Study Tip, Technology: Tip,* and *Technology: Discovery* features have been revised. *Study Tip* features provide hints, cautionary notes, and words of advice for students as they learn the material. *Technology: Tip* features provide point-of-use instruction for using a graphing calculator, whereas *Technology: Discovery* features encourage students to explore mathematical concepts using their graphing or scientific calculators. All technology features are highlighted and can easily be omitted without loss of continuity in coverage of material.

The new chapter summary feature *What Did You Learn?* highlights important mathematical vocabulary (*Key Terms*) and primary concepts (*Key Concepts*) from the chapter. For easy reference, the *Key Terms* are correlated to the chapter by page number and the *Key Concepts* by section number.

As students proceed through each chapter, they have many opportunities to assess their understanding and practice skills. A set of *Exercises*, located at the end of each section, correlates to the *Examples* found within the section. *Mid-Chapter Quizzes* and *Chapter Tests* offer students self-assessment tools halfway through and at the conclusion of each chapter. *Review Exercises*, organized by section, restate the *What You Should Learn* objectives so that students may refer back to the appropriate topic discussion when working through the exercises. In addition, the *Integrated Review* exercises that precede each exercise set, and the *Cumulative Tests* that follow Chapters 3, 6, and 9, give students more opportunities to revisit and review previously learned concepts.

To show students the practical uses of algebra, we highlight the connections between the mathematical concepts and the real world in the multitude of applications found throughout the text. We believe that students can overcome their difficulties in mathematics if they are encouraged and supported throughout the learning process. Too often, students become frustrated and lose interest in the material when they cannot follow the text. With this in mind, every effort has been made to write a readable text that can be understood by every student. We hope that your students find our approach engaging and effective.

Ron Larson

Robert P. Hostetler

Features

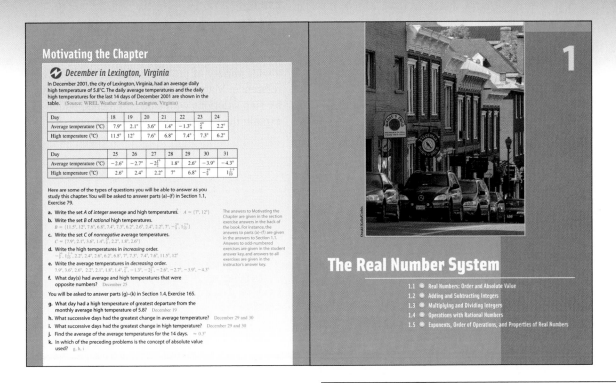

Motivating the Chapter

🌀 December in Lexington, Virginia

In December 2001, the city of Lexington, Virginia, had an average daily high temperature of 5.8°C. The daily average temperatures and the daily high temperatures for the last 14 days of December 2001 are shown in the table. (Source: WREL Weather Station, Lexington, Virginia)

Day	18	19	20	21	22	23	24
Average temperature (°C)	7.9°	2.1°	3.6°	1.4°	−1.3°	$\frac{8}{9}°$	2.2°
High temperature (°C)	11.5°	12°	7.6°	6.8°	7.4°	7.3°	6.2°

Day	25	26	27	28	29	30	31
Average temperature (°C)	−2.6°	−2.7°	$-2\frac{1}{2}°$	1.8°	2.6°	−3.9°	−4.3°
High temperature (°C)	2.6°	2.4°	2.2°	7°	6.8°	$-\frac{20}{9}°$	$1\frac{7}{10}°$

Here are some of the types of questions you will be able to answer as you study this chapter. You will be asked to answer parts (a)–(f) in Section 1.1, Exercise 79.

a. Write the set *A* of *integer* average and high temperatures. $A = \{7°, 12°\}$
b. Write the set *B* of *rational* high temperatures.
$B = \{11.5°, 12°, 7.6°, 6.8°, 7.4°, 7.3°, 6.2°, 2.6°, 2.4°, 2.2°, 7°, -\frac{20}{9}°, 1\frac{7}{10}°\}$
c. Write the set *C* of *nonnegative* average temperatures.
$C = \{7.9°, 2.1°, 3.6°, 1.4°, \frac{8}{9}°, 2.2°, 1.8°, 2.6°\}$
d. Write the high temperatures in *increasing* order.
$-\frac{20}{9}°, 1\frac{7}{10}°, 2.2°, 2.4°, 2.6°, 6.2°, 6.8°, 7°, 7.3°, 7.4°, 7.6°, 11.5°, 12°$
e. Write the average temperatures in *decreasing* order.
$7.9°, 3.6°, 2.6°, 2.2°, 2.1°, 1.8°, 1.4°, \frac{8}{9}°, -1.3°, -2\frac{1}{2}°, -2.6°, -2.7°, -3.9°, -4.3°$
f. What day(s) had average and high temperatures that were opposite numbers? December 25

You will be asked to answer parts (g)–(k) in Section 1.4, Exercise 165.

g. What day had a high temperature of greatest departure from the monthly average high temperature of 5.8? December 19
h. What successive days had the greatest change in average temperature? December 29 and 30
i. What successive days had the greatest change in high temperature? December 29 and 30
j. Find the average of the average temperatures for the 14 days. ≈ 0.3°
k. In which of the preceding problems is the concept of absolute value used? g, h, i

The answers to Motivating the Chapter are given in the section exercise answers in the back of the book. For instance, the answers to parts (a)–(f) are given in the answers to Section 1.1. Answers to odd-numbered exercises are given in the student answer key, and answers to all exercises are given in the instructor's answer key.

1

The Real Number System

Chapter Opener

Every chapter opens with *Motivating the Chapter.* These multipart problems use concepts discussed in the chapter and present them in the context of a single real-world application. *Motivating the Chapter* problems are correlated to specific sections and can be assigned as part of an exercise set or as an individual or group project. The icon 🌀 identifies an exercise that relates back to *Motivating the Chapter.* Answers to these problems are found in place within the Instructor's Annotated Edition.

Section Opener   *New*

Every section begins with a list of learning objectives called *What You Should Learn.* Each objective is restated in the margin at the point where it is covered. *Why You Should Learn It* provides a motivational explanation for learning the given objectives.

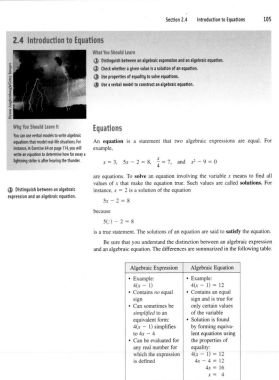

2.4 Introduction to Equations

What You Should Learn
① Distinguish between an algebraic expression and an algebraic equation.
② Check whether a given value is a solution of an equation.
③ Use properties of equality to solve equations.
④ Use a verbal model to construct an algebraic equation.

Why You Should Learn It
You can use verbal models to write algebraic equations that model real-life situations. For instance, in Exercise 64 on page 114, you will write an equation to determine how far away a lightning strike is after hearing the thunder.

① Distinguish between an algebraic expression and an algebraic equation.

Equations

An **equation** is a statement that two algebraic expressions are equal. For example,

$$x = 3, \quad 5x - 2 = 8, \quad \frac{x}{4} = 7, \quad \text{and} \quad x^2 - 9 = 0$$

are equations. To **solve** an equation involving the variable x means to find all values of x that make the equation true. Such values are called **solutions.** For instance, $x = 2$ is a solution of the equation

$$5x - 2 = 8$$

because

$$5(2) - 2 = 8$$

is a true statement. The solutions of an equation are said to **satisfy** the equation.

Be sure that you understand the distinction between an algebraic expression and an algebraic equation. The differences are summarized in the following table.

Algebraic Expression	Algebraic Equation
• Example: $4(x - 1)$	• Example: $4(x - 1) = 12$
• Contains *no* equal sign	• Contains an equal sign and is true for only certain values of the variable
• Can sometimes be *simplified* to an equivalent form: $4(x - 1)$ simplifies to $4x - 4$	• Solution is found by forming equivalent equations using the properties of equality: $4(x-1) = 12$ $4x - 4 = 12$ $4x = 16$ $x = 4$
• Can be evaluated for any real number for which the expression is defined	

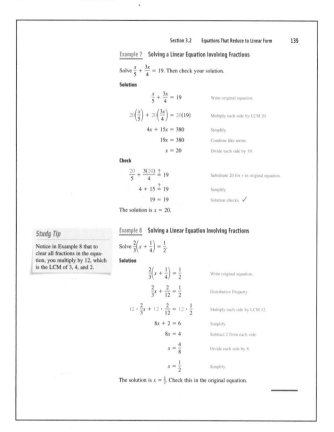

Section 3.2   Equations That Reduce to Linear Form   139

Example 7   Solving a Linear Equation Involving Fractions

Solve $\frac{x}{5} + \frac{3x}{4} = 19$. Then check your solution.

Solution

$$\frac{x}{5} + \frac{3x}{4} = 19 \qquad \text{Write original equation.}$$

$$20\left(\frac{x}{5}\right) + 20\left(\frac{3x}{4}\right) = 20(19) \qquad \text{Multiply each side by LCM 20.}$$

$$4x + 15x = 380 \qquad \text{Simplify.}$$

$$19x = 380 \qquad \text{Combine like terms.}$$

$$x = 20 \qquad \text{Divide each side by 19.}$$

Check

$$\frac{20}{5} + \frac{3(20)}{4} \stackrel{?}{=} 19 \qquad \text{Substitute 20 for } x \text{ in original equation.}$$

$$4 + 15 \stackrel{?}{=} 19 \qquad \text{Simplify.}$$

$$19 = 19 \qquad \text{Solution checks. } \checkmark$$

The solution is $x = 20$.

Study Tip

Notice in Example 8 that to clear all fractions in the equation, you multiply by 12, which is the LCM of 3, 4, and 2.

Example 8   Solving a Linear Equation Involving Fractions

Solve $\frac{2}{3}\left(x + \frac{1}{4}\right) = \frac{1}{2}$.

Solution

$$\frac{2}{3}\left(x + \frac{1}{4}\right) = \frac{1}{2} \qquad \text{Write original equation.}$$

$$\frac{2}{3}x + \frac{2}{12} = \frac{1}{2} \qquad \text{Distributive Property}$$

$$12 \cdot \frac{2}{3}x + 12 \cdot \frac{2}{12} = 12 \cdot \frac{1}{2} \qquad \text{Multiply each side by LCM 12.}$$

$$8x + 2 = 6 \qquad \text{Simplify.}$$

$$8x = 4 \qquad \text{Subtract 2 from each side.}$$

$$x = \frac{4}{8} \qquad \text{Divide each side by 8.}$$

$$x = \frac{1}{2} \qquad \text{Simplify.}$$

The solution is $x = \frac{1}{2}$. Check this in the original equation.

Examples

Each example has been carefully chosen to illustrate a particular mathematical concept or problem-solving technique. The examples cover a wide variety of problems and are titled for easy reference. Many examples include detailed, step-by-step solutions with side comments, which explain the key steps of the solution process.

Applications

A wide variety of real-life applications are integrated throughout the text in examples and exercises. These applications demonstrate the relevance of algebra in the real world. Many of the applications use current, real data. The icon indicates an example involving a real-life application.

Section 3.4   Ratios and Proportions   163

Example 8   Gasoline Cost

You are driving from New York to Phoenix, a trip of 2450 miles. You begin the trip with a full tank of gas and after traveling 424 miles, you refill the tank for $24.00. How much should you plan to spend on gasoline for the entire trip?

Solution

Verbal Model:
$$\frac{\text{Cost for trip}}{\text{Cost for tank}} = \frac{\text{Miles for trip}}{}$$

Labels:
Cost of gas for entire trip = x (dollars)
Cost of gas for tank = 24   (dollars)
Miles for entire trip = 2450   (miles)
Miles for tank = 424   (miles)

Proportion:
$$\frac{x}{24} = \frac{2450}{424} \qquad \text{Original proportion}$$

$$x = 24\left(\frac{2450}{424}\right) \qquad \text{Multiply each side by 24.}$$

$$x \approx 138.68 \qquad \text{Simplify.}$$

You should plan to spend approximately $138.68 for gasoline on the trip. Check this in the original statement of the problem.

In examples such as Example 8, you might point out that an "approximate" answer will not check "exactly" in the original statement of the problem. However, the process of checking solutions is still important.

④ Solve application problems using the Consumer Price Index.

The Consumer Price Index

The rate of inflation is important to all of us. Simply stated, *inflation* is an economic condition in which the price of a fixed amount of goods or services increases. So, a fixed amount of money buys less in a given year than in previous years.

The most widely used measurement of inflation in the United States is the *Consumer Price Index* (CPI), often called the *Cost-of-Living Index*. The table below shows the "All Items" or general index for the years 1970 to 2001. (Source: Bureau of Labor Statistics)

Year	CPI	Year	CPI	Year	CPI	Year	CPI
1970	38.8	1978	65.2	1986	109.6	1994	148.2
1971	40.5	1979	72.6	1987	113.6	1995	152.4
1972	41.8	1980	82.4	1988	118.6	1996	156.9
1973	44.4	1981	90.9	1989	124.0	1997	160.5
1974	49.3	1982	96.5	1990	130.7	1998	163.0
1975	53.8	1983	99.6	1991	136.2	1999	166.6
1976	56.9	1984	103.9	1992	140.3	2000	172.2
1977	60.6	1985	107.6	1993	144.5	2001	177.1

Example panel 1

Section 3.3   Problem Solving with Percents   149

Example 6 Course Grade

You missed an A in your chemistry course by only three points. Your point total for the course is 402. How many points were possible in the course? (Assume that you needed 90% of the course total for an A.)

Solution

Verbal Model:
$$\frac{\text{Your}}{\text{points}} + \frac{3}{\text{points}} = \frac{\text{Percent}}{\text{(in decimal form)}} \cdot \frac{\text{Total}}{\text{points}}$$

Labels:
Your points = 402   (points)
Percent = 90% = 0.9   (in decimal form)
Total points for course = b   (points)

Equation:
$402 + 3 = 0.9b$ Original equation
$405 = 0.9b$ Add.
$\frac{405}{0.9} = b$ Divide each side by 0.9.
$450 = b$ Simplify.

So, there were 450 total points for the course. You can check your solution as follows.

$402 + 3 = 0.9b$ Write original equation.
$402 + 3 \stackrel{?}{=} 0.9(450)$ Substitute 450 for b.
$405 = 405$ Solution checks. ✓

Example 7 Percent Increase

The monthly basic cable TV rate was $7.69 in 1980 and $30.08 in 2000. Find the percent increase in the monthly basic cable TV rate from 1980 to 2000. (Source: Paul Kagan Associates, Inc.)

Solution

Verbal Model:
$$\frac{2000}{\text{price}} = \frac{1980}{\text{price}} \cdot \frac{\text{Percent increase}}{\text{(in decimal form)}} + \frac{1980}{\text{price}}$$

Labels:
2000 price = 30.08   (dollars)
Percent increase = p   (in decimal form)
1980 price = 7.69   (dollars)

Equation:
$30.08 = 7.69p + 7.69$ Original equation
$22.39 = 7.69p$ Subtract 7.69 from each side.
$2.91 \approx p$ Divide each side by 7.69.

So, the percent increase in the monthly basic cable TV rate from 1980 to 2000 is approximately 291%. Check this in the original statement of the problem.

Problem Solving

This text provides many opportunities for students to sharpen their problem-solving skills. In both the examples and the exercises, students are asked to apply verbal, numerical, analytical, and graphical approaches to problem solving. In the spirit of the AMATYC and NCTM standards, students are taught a five-step strategy for solving applied problems, which begins with constructing a verbal model and ends with checking the answer.

Geometry

The Fourth Edition continues to provide coverage and integration of geometry in examples and exercises. The icon ▲ indicates an exercise involving geometry.

Example panel 2

Section 9.4   Radical Equations and Applications   527

Pendulum Length In Exercises 79 and 80, the time t (in seconds) for a pendulum of length L (in feet) to go through one complete cycle, both forward and back (its period), is given by

$$t = 2\pi \sqrt{\frac{L}{32}}.$$

79. How long is the pendulum of a grandfather clock with a period of 2 seconds (see figure)? *3.24 feet*

80. How long is the pendulum of a mantel clock with a period of 0.8 second? *0.52 foot*

81. ▲ *Geometry* A ladder is 15 feet long, and the bottom of the ladder is 3 feet from the side of a house (see figure). How far does the ladder reach up the side of the house? *6√6 ≈ 14.70 feet*

82. ▲ *Geometry* A 39-foot guy wire on a sailboat is attached to the top of the mast and to the deck 15 feet from the base of the mast (see figure). How tall is the mast? *36 feet*

83. ▲ *Geometry* A 12-foot plank is used to brace a basement wall during construction of a home. The plank is nailed to the wall 4 feet above the floor (see figure). Find the slope of the plank. $-\frac{\sqrt{2}}{4}$

84. ▲ *Geometry* A volleyball court is 30 feet wide and 60 feet long. Find the length of the diagonal of the court. *30√5 ≈ 67.08 feet*

85. ▲ *Geometry* A baseball diamond is a square that is 90 feet on a side (see figure). Determine the distance between first base and third base. *90√2 ≈ 127.28 feet*

86. ▲ *Geometry* The distance between Memphis and New Orleans is 410 miles. The distance between Memphis and Chattanooga is 317 miles (see figure). Approximate the distance between Chattanooga and New Orleans. *√268,589 ≈ 518.3 miles*

Example panel 3

162   Chapter 3   Linear Equations and Problem Solving

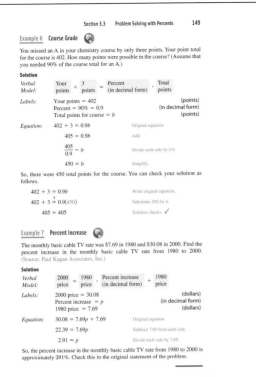

Triangular lot

Sketch
Figure 3.2

Example 6 Geometry: Similar Triangles

A triangular lot has perpendicular sides of lengths 100 feet and 210 feet. You are to make a proportional sketch of this lot using 8 inches as the length of the shorter side. How long should you make the other side?

Solution

This is a case of similar triangles in which the ratios of the corresponding sides are equal. The triangles are shown in Figure 3.2.

$$\frac{\text{Shorter side of lot}}{\text{Longer side of lot}} = \frac{\text{Shorter side of sketch}}{\text{Longer side of sketch}}$$ Proportion for similar triangles

$\frac{100}{210} = \frac{8}{x}$ Substitute.
$x \cdot 100 = 210 \cdot 8$ Cross-multiply.
$x = \frac{1680}{100} = 16.8$ Divide each side by 100.

So, the length of the longer side of the sketch should be 16.8 inches.

Example 7 Resizing a Picture

You have a 7-by-8-inch picture of a graph that you want to paste into a research paper, but you have only a 6-by-6-inch space in which to put it. You go to the copier that has five options for resizing your graph: 64%, 78%, 100%, 121%, and 129%.

a. Which option should you choose?

b. What are the measurements of the resized picture?

Solution

a. Because the longest side must be reduced from 8 inches to no more than 6 inches, consider the proportion

$$\frac{\text{New length}}{\text{Old length}} = \frac{\text{New percent}}{\text{Old percent}}$$ Original proportion

$\frac{6}{8} = \frac{x}{100}$ Substitute.
$\frac{6}{8} \cdot 100 = x$ Multiply each side by 100.
$75 = x.$ Simplify.

To guarantee a fit, you should choose the 64% option, because 78% is greater than the required 75%.

b. To find the measurements of the resized picture, multiply by 64% or 0.64.

Length = 0.64(8) = 5.12 inches   Width = 0.64(7) = 4.48 inches

The size of the reduced picture is 5.12 inches by 4.48 inches.

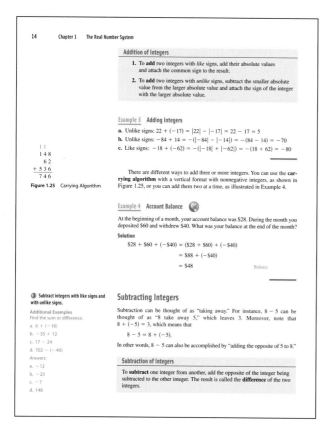

14   Chapter 1   The Real Number System

Addition of Integers

1. To **add** two integers with *like* signs, add their absolute values and attach the common sign to the result.
2. To **add** two integers with *unlike* signs, subtract the smaller absolute value from the larger absolute value and attach the sign of the integer with the larger absolute value.

Example 3   Adding Integers

a. Unlike signs: $22 + (-17) = |22| - |-17| = 22 - 17 = 5$
b. Unlike signs: $-84 + 14 = -(|-84| - |-14|) = -(84 - 14) = -70$
c. Like signs: $-18 + (-62) = -(|-18| + |-62|) = -(18 + 62) = -80$

There are different ways to add three or more integers. You can use the **carrying algorithm** with a vertical format with nonnegative integers, as shown in Figure 1.25, or you can add them two at a time, as illustrated in Example 4.

$$
\begin{array}{r}
1\ 1 \\
1\ 4\ 8 \\
6\ 2 \\
+\ 5\ 3\ 6 \\
\hline
7\ 4\ 6
\end{array}
$$

Figure 1.25   Carrying Algorithm

Example 4   Account Balance

At the beginning of a month, your account balance was $28. During the month you deposited $60 and withdrew $40. What was your balance at the end of the month?

Solution

$$\$28 + \$60 + (-\$40) = (\$28 + \$60) + (-\$40)$$
$$= \$88 + (-\$40)$$
$$= \$48 \qquad \text{Balance}$$

③ Subtract integers with like signs and with unlike signs.

Additional Examples
Find the sum or difference.
a. $6 + (-18)$
b. $-35 + 12$
c. $17 - 24$
d. $102 - (-46)$
Answers:
a. -12
b. -23
c. -7
d. 148

Subtracting Integers

Subtraction can be thought of as "taking away." For instance, $8 - 5$ can be thought of as "8 take away 5," which leaves 3. Moreover, note that $8 + (-5) = 3$, which means that

$$8 - 5 = 8 + (-5).$$

In other words, $8 - 5$ can also be accomplished by "adding the opposite of 5 to 8."

Subtraction of Integers

To **subtract** one integer from another, add the opposite of the integer being subtracted to the other integer. The result is called the **difference** of the two integers.

Definitions and Rules

All important definitions, rules, formulas, properties, and summaries of solution methods are highlighted for emphasis. Each of these features is also titled for easy reference.

Study Tips

Study Tips offer students specific point-of-use suggestions for studying algebra, as well as pointing out common errors and discussing alternative solution methods. They appear in the margins.

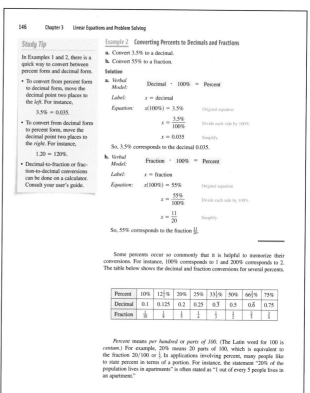

146   Chapter 3   Linear Equations and Problem Solving

Study Tip

In Examples 1 and 2, there is a quick way to convert between percent form and decimal form.

• To convert from percent form to decimal form, move the decimal point two places to the *left*. For instance,

$$3.5\% = 0.035.$$

• To convert from decimal form to percent form, move the decimal point two places to the *right*. For instance,

$$1.20 = 120\%.$$

• Decimal-to-fraction or fraction-to-decimal conversions can be done on a calculator. Consult your user's guide.

Example 2   Converting Percents to Decimals and Fractions

a. Convert 3.5% to a decimal.
b. Convert 55% to a fraction.

Solution

a. *Verbal Model:*   Decimal $\cdot$ 100% = Percent
   Label: x = decimal
   Equation: $x(100\%) = 3.5\%$ Original equation
   $$x = \frac{3.5\%}{100\%}$$ Divide each side by 100%.
   $$x = 0.035$$ Simplify.

So, 3.5% corresponds to the decimal 0.035.

b. *Verbal Model:*   Fraction $\cdot$ 100% = Percent
   Label: x = fraction
   Equation: $x(100\%) = 55\%$ Original equation
   $$x = \frac{55\%}{100\%}$$ Divide each side by 100%.
   $$x = \frac{11}{20}$$ Simplify.

So, 55% corresponds to the fraction $\frac{11}{20}$.

Some percents occur so commonly that it is helpful to memorize their conversions. For instance, 100% corresponds to 1 and 200% corresponds to 2. The table below shows the decimal and fraction conversions for several percents.

Percent	10%	$12\frac{1}{2}\%$	20%	25%	$33\frac{1}{3}\%$	50%	$66\frac{2}{3}\%$	75%
Decimal	0.1	0.125	0.2	0.25	$0.\overline{3}$	0.5	$0.\overline{6}$	0.75
Fraction	$\frac{1}{10}$	$\frac{1}{8}$	$\frac{1}{5}$	$\frac{1}{4}$	$\frac{1}{3}$	$\frac{1}{2}$	$\frac{2}{3}$	$\frac{3}{4}$

Percent means *per hundred* or *parts of 100*. (The Latin word for 100 is *centum*.) For example, 20% means 20 parts of 100, which is equivalent to the fraction 20/100 or $\frac{1}{5}$. In applications involving percent, many people like to state percent in terms of a portion. For instance, the statement "20% of the population lives in apartments" is often stated as "1 out of every 5 people lives in an apartment."

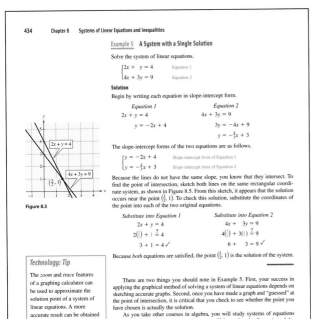

Example 5 A System with a Single Solution

Solve the system of linear equations.

$$\begin{cases} 2x + y = 4 & \text{Equation 1} \\ 4x + 3y = 9 & \text{Equation 2} \end{cases}$$

Solution

Begin by writing each equation in slope-intercept form.

Equation 1	Equation 2
$2x + y = 4$	$4x + 3y = 9$
$y = -2x + 4$	$3y = -4x + 9$
	$y = -\frac{4}{3}x + 3$

The slope-intercept forms of the two equations are as follows.

$$\begin{cases} y = -2x + 4 & \text{Slope-intercept form of Equation 1} \\ y = -\frac{4}{3}x + 3 & \text{Slope-intercept form of Equation 2} \end{cases}$$

Because the lines do not have the same slope, you know that they intersect. To find the point of intersection, sketch both lines on the same rectangular coordinate system, as shown in Figure 8.5. From this sketch, it appears that the solution occurs near the point $(\frac{3}{2}, 1)$. To check this solution, substitute the coordinates of the point into each of the two original equations.

Substitute into Equation 1	Substitute into Equation 2
$2x + y = 4$	$4x + 3y = 9$
$2(\frac{3}{2}) + 1 \stackrel{?}{=} 4$	$4(\frac{3}{2}) + 3(1) \stackrel{?}{=} 9$
$3 + 1 = 4 \checkmark$	$6 + 3 = 9 \checkmark$

Because *both* equations are satisfied, the point $(\frac{3}{2}, 1)$ is the solution of the system.

Figure 8.5

Technology: Tip

The *zoom* and *trace* features of a graphing calculator can be used to approximate the solution point of a system of linear equations. A more accurate result can be obtained by using the *intersect* feature of a graphing calculator. Consult the user's guide of your graphing calculator for the steps in using this feature. Then, use the *intersect* feature to find the solution point in Example 5.

See Technology Answers.

There are two things you should note in Example 5. First, your success in applying the graphical method of solving a system of linear equations depends on sketching accurate graphs. Second, once you have made a graph and "guessed" at the point of intersection, it is critical that you check to see whether the point you have chosen is actually the solution.

As you take other courses in algebra, you will study systems of equations that are *not* linear. When you do that, you will learn that the discussion of the number of solutions on page 432 applies only to systems of *linear* equations. A nonlinear system such as

$$\begin{cases} y = 2x + 3 & \text{Equation 1} \\ y = x^2 & \text{Equation 2} \end{cases}$$

does not have to have zero, one, or infinitely many solutions. Try sketching a graph of this system. How many solutions does it have? Try estimating the solutions from the graphs and then check your estimates in each equation.

Graphics

Visualization is a critical problem-solving skill. To encourage the development of this skill, students are shown how to use graphs to reinforce algebraic and numeric solutions and to interpret data. The numerous figures in examples and exercises throughout the text were computer-generated for accuracy.

Technology: Tips

Point-of-use instructions for using graphing calculators appear in the margins. These features encourage the use of graphing technology as a tool for visualization of mathematical concepts, for verification of other solution methods, and for facilitation of computations. The *Technology: Tips* can easily be omitted without loss of continuity in coverage. Answers to questions posed within these features are located in the back of the Instructor's Annotated Edition.

Technology: Discovery

Technology: Discovery features invite students to engage in active exploration of mathematical concepts and discovery of mathematical relationships through the use of scientific or graphing calculators. These activities encourage students to utilize their critical thinking skills and help them develop an intuitive understanding of theoretical concepts. *Technology: Discovery* features can easily be omitted without loss of continuity in coverage. Answers to questions posed within these features are located in the back of the Instructor's Annotated Edition.

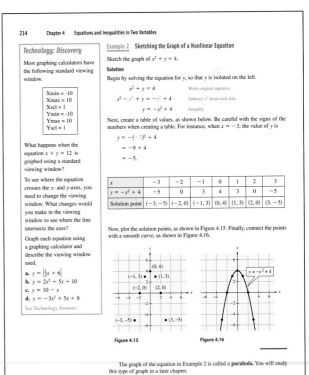

Technology: Discovery

Most graphing calculators have the following standard viewing window.

```
Xmin = -10
Xmax = 10
Xscl = 1
Ymin = -10
Ymax = 10
Yscl = 1
```

What happens when the equation $x + y = 12$ is graphed using a standard viewing window?

To see where the equation crosses the x- and y-axes, you need to change the viewing window. What changes would you make in the viewing window to see where the line intersects the axes?

Graph each equation using a graphing calculator and describe the viewing window used.

a. $y = |\frac{1}{2}x + 6|$
b. $y = 2x^2 + 5x + 10$
c. $y = 10 - x$
d. $y = -3x^3 + 5x + 8$

See Technology Answers.

Example 2 Sketching the Graph of a Nonlinear Equation

Sketch the graph of $x^2 + y = 4$.

Solution

Begin by solving the equation for y, so that y is isolated on the left.

$$x^2 + y = 4 \qquad \text{Write original equation.}$$
$$x^2 - x^2 + y = -x^2 + 4 \qquad \text{Subtract } x^2 \text{ from each side.}$$
$$y = -x^2 + 4 \qquad \text{Simplify.}$$

Next, create a table of values, as shown below. Be careful with the signs of the numbers when creating a table. For instance, when $x = -3$, the value of y is

$$y = -(-3)^2 + 4$$
$$= -9 + 4$$
$$= -5.$$

x	-3	-2	-1	0	1	2	3
$y = -x^2 + 4$	-5	0	3	4	3	0	-5
Solution point	$(-3, -5)$	$(-2, 0)$	$(-1, 3)$	$(0, 4)$	$(1, 3)$	$(2, 0)$	$(3, -5)$

Now, plot the solution points, as shown in Figure 4.15. Finally, connect the points with a smooth curve, as shown in Figure 4.16.

Figure 4.15 **Figure 4.16**

The graph of the equation in Example 2 is called a **parabola.** You will study this type of graph in a later chapter.

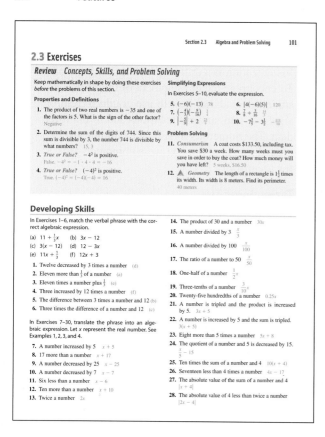

Review: Concepts, Skills, and Problem Solving

Each exercise set (except in Chapter 1) is preceded by these review exercises that are designed to help students keep up with concepts and skills learned in previous chapters. Answers to all *Review: Concepts, Skills, and Problem Solving* exercises are given in the back of the student text and are located in place in the Instructor's Annotated Edition.

Exercises

The exercise sets are grouped into three categories: *Developing Skills, Solving Problems,* and *Explaining Concepts.* The exercise sets offer a diverse variety of computational, conceptual, and applied problems to accommodate many teaching and learning styles. Designed to build competence, skill, and understanding, each exercise set is graded in difficulty to allow students to gain confidence as they progress. Detailed solutions to all odd-numbered exercises are given in the *Student Solutions Guide*, and answers to all odd-numbered exercises are given in the back of the student text. Answers are located in place in the Instructor's Annotated Edition.

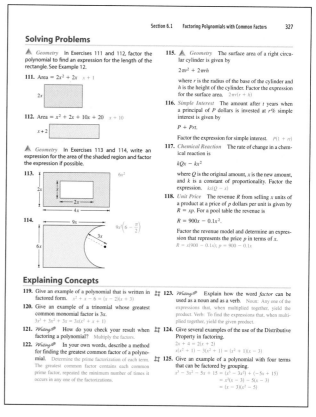

What Did You Learn? (Chapter Summary)

Located at the end of every chapter, *What Did You Learn?* summarizes the *Key Terms* (referenced by page) and the *Key Concepts* (referenced by section) presented in the chapter. This effective study tool aids students as they review concepts and prepare for exams.

Review Exercises

The *Review Exercises* at the end of each chapter have been reorganized in the Fourth Edition. All skill-building and application exercises are first ordered by section, then grouped according to the objectives stated within *What You Should Learn*. This organization allows students to easily identify the appropriate sections and concepts for study and review.

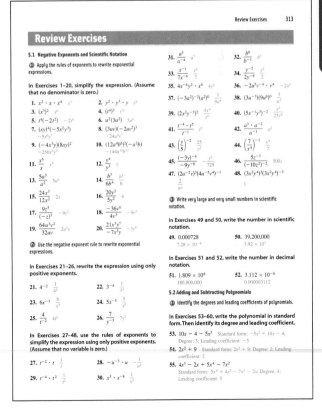

Mid-Chapter Quiz

Take this quiz as you would take a quiz in class. After you are done, check your work against the answers in the back of the book.

In Exercises 1–10, solve the equation.

1. $74 - 12x = 2$ 6
2. $10(y - 8) = 0$ 8
3. $3x + 1 = x + 20$ $\frac{19}{2}$
4. $6x + 8 = 8 - 2x$ 0
5. $-10x + \frac{2}{3} = \frac{7}{3} - 5x$ $-\frac{1}{3}$
6. $\frac{x}{5} + \frac{x}{8} = 1$ $\frac{40}{13}$
7. $\frac{9 + x}{3} = 15$ 36
8. $7 - 2(5 - x) = -7$ -2
9. $\frac{x + 3}{6} = \frac{4}{3}$ 5
10. $\frac{x + 7}{5} = \frac{x + 9}{7}$ -2

In Exercises 11 and 12, solve the equation. Round your answer to two decimal places. In your own words, explain how to check the solution.

11. $32.86 - 10.5x = 11.25$ 2.06
12. $\frac{x}{5.45} + 3.2 = 12.6$ 51.23

Substitute 2.06 for x. After simplifying, the equation should be an identity.

Substitute 51.23 for x. After simplifying, the equation should be an identity.

13. What number is 62% of 25?   15.5
14. What number is $\frac{1}{2}$% of 8400?   42
15. 300 is what percent of 150?   200%
16. 145.6 is 32% of what number?   455
17. The perimeter of a rectangle is 60 meters. The length is $1\frac{1}{2}$ times its width. Find the dimensions of the rectangle.   12 meters × 18 meters
18. You have two jobs. In the first job, you work 40 hours a week at a candy store and earn $7.50 per hour. In the second job, you earn $6.00 per hour baby-sitting and can work as many hours as you want. You want to earn $360 a week. How many hours must you work at the second job?   10 hours
19. A region has an area of 42 square meters. It must be divided into three subregions so that the second has twice the area of the first, and the third has twice the area of the second. Find the area of each subregion.   6 square meters, 12 square meters, 24 square meters
20. To get an A in a psychology course, you must have an average of at least 90 points for three tests of 100 points each. For the first two tests, your scores are 84 and 93. What must you score on the third test to earn a 90% average for the course?   93
21. The price of a television set is approximately 108% of what it was 2 years ago. The current price is $535. What was the approximate price 2 years ago? $495.37
22. The circle graph at the left shows the number of endangered wildlife and plant species for the year 2001. What percent of the total endangered wildlife and plant species were birds?   (Source: U.S. Fish and Wildlife Service)   17%
23. Two people can paint a room in t hours, where t must satisfy the equation $t/4 + t/12 = 1$. How long will it take for the two people to paint the room?   3 hours

Endangered Wildlife and Plant Species

Plants 593
Mammals 314
Other 169
Birds 253
Reptiles 78
Fishes 81

Figure for 22

Chapter Test

Take this test as you would take a test in class. After you are done, check your work against the answers in the back of the book.

In Exercises 1–10, factor the polynomial completely .

1. $7x^2 - 14x^3$
$7x^2(1 - 2x)$
2. $z(z + 7) - 3(z + 7)$
$(z + 7)(z - 3)$
3. $t^2 - 4t - 5$ $(t - 5)(t + 1)$
4. $6x^2 - 11x + 4$ $(3x - 4)(2x - 1)$
5. $3y^3 + 72y^2 - 75y$
$3y(y - 1)(y + 25)$
6. $4 - 25v^2$
$(2 + 5v)(2 - 5v)$
7. $4x^2 - 20x + 25$
$(2x - 5)^2$
8. $16 - (z + 9)^2$
$(-z - 5)(z + 13)$
9. $x^3 + 2x^2 - 9x - 18$
$(x + 2)(x + 3)(x - 3)$
10. $16 - z^4$
$(4 + z^2)(2 + z)(2 - z)$
11. Fill in the missing factor: $\frac{2}{5}x - \frac{3}{5} = \frac{1}{5}(\quad)$. $2x - 3$
12. Find all integers b such that $x^2 + bx + 5$ can be factored. ± 6
13. Find a real number c such that $x^2 + 12x + c$ is a perfect square trinomial.   36
14. Explain why $(x + 1)(3x - 6)$ is not a complete factorization of $3x^2 - 3x - 6$. $3x^2 - 3x - 6 = 3(x + 1)(x - 2)$

In Exercises 15–18, solve the equation.

15. $(x + 4)(2x - 3) = 0$ $-4, \frac{3}{2}$
16. $3x^2 + 7x - 6 = 0$ $-3, \frac{2}{3}$
17. $y(2y - 1) = 6$ $-\frac{3}{2}, 2$
18. $2x^2 - 3x = 8 + 3x$ $-1, 4$
19. The suitcase shown below has a height of x and a width of $x + 2$. The volume of the suitcase is $x^3 + 6x^2 + 8x$. Find the length l of the suitcase. $x + 4$

20. The width of a rectangle is 5 inches less than its length. The area of the rectangle is 84 square inches. Find the dimensions of the rectangle.   7 inches × 12 inches
21. An object is thrown upward from the top of the AON Center in Chicago, with an initial velocity of 14 feet per second at a height of 1136 feet. The height h (in feet) of the object is modeled by the position equation

$$h = -16t^2 + 14t + 1136$$

where t is the time measured in seconds. How long will it take for the object to reach the ground? How long will it take the object to fall to a height of 806 feet?   8.875 seconds; 5 seconds
22. Find two consecutive positive even integers whose product is 624.   24, 26

Cumulative Test: Chapters 4–6

Take this test as you would take a test in class. After you are done, check your work against the answers in the back of the book.

1. Describe how to identify the quadrants in which the points $(-2, y)$ must be located. (y is a real number.)

1. Because $x = -2$, the point must lie in Quadrant II or Quadrant III.

2. Determine whether the ordered pairs are solution points of the equation $9x - 4y + 36 = 0$.
(a) $(-1, -1)$ (b) $(8, 27)$ (c) $(-4, 0)$ (d) $(3, -2)$
Not a solution   Solution   Solution   Not a solution

In Exercises 3 and 4, sketch the graph of the equation and determine any intercepts of the graph.   See Additional Answers.

3. $y = 2 - |x|$ $(-2, 0), (2, 0), (0, 2)$
4. $x + 2y = 8$ $(8, 0), (0, 4)$

5. The slope of a line is $-\frac{1}{4}$ and a point on the line is $(2, 1)$. Find the coordinates of a second point on the line. Explain why there are many correct answers. $(-2, 2)$; There are infinitely many points on a line.
6. Find an equation of the line through $(0, -\frac{2}{3})$ with slope $m = \frac{4}{5}$. $y = \frac{4}{5}x - \frac{2}{3}$

In Exercises 7 and 8, sketch the lines and determine whether they are parallel, perpendicular, or neither.   See Additional Answers.

7. $y_1 = \frac{2}{3}x - 3, y_2 = -\frac{3}{2}x + 1$ Perpendicular
8. $y_1 = 2 - 0.4x, y_2 = -\frac{2}{5}x$ Parallel

9. Subtract: $(x^3 - 3x^2) - (x^3 + 2x^2 - 5)$. $-5x^2 + 5$
10. Multiply: $(6z)(-7z)(z^2)$. $-42z^4$
11. Multiply: $(3x + 5)(x - 4)$. $3x^2 - 7x - 20$
12. Multiply: $(5x - 3)(5x + 3)$. $25x^2 - 9$
13. Expand: $(5x + 6)^2$. $25x^2 + 60x + 36$
14. Divide: $(6x^2 + 72x) \div 6x$. $x + 12$
15. Divide: $\frac{x^2 - 3x - 2}{x - 4}$. $x + 1 + \frac{2}{x - 4}$
16. Simplify: $\frac{(3xy^2)^{-2}}{6x^{-3}}$. $\frac{x}{54y^4}$
17. Factor: $2u^2 - 6u$. $2u(u - 3)$
18. Factor and simplify: $(x - 2)^2 - 16$. $(x + 2)(x - 6)$
19. Factor completely: $x^3 + 8x^2 + 16x$. $x(x + 4)^2$
20. Factor completely: $x^3 + 2x^2 - 4x - 8$. $(x + 2)^2(x - 2)$
21. Solve: $u(u - 12) = 0$.   0, 12
22. Solve: $5x^2 - 12x - 9 = 0$. $-\frac{3}{5}, 3$
23. Rewrite the expression $\left(\frac{x}{2}\right)^{-2}$ using only positive exponents. $\frac{4}{x^2}$
24. A sales representative is reimbursed $125 per day for lodging and meals, plus $0.35 per mile driven. Write a linear equation giving the daily cost C to the company in terms of x, the number of miles driven. Explain the reasoning you used to write the model. Find the cost for a day when the representative drives 70 miles. $C = 125 + 0.35x$; $149.50
25. The cost of operating a pizza delivery car is $0.70 per mile after an initial investment of $9000. What mileage on the car will keep the cost at or below $36,400?   39,142 miles

---

Mid-Chapter Quiz

Each chapter contains a *Mid-Chapter Quiz* midway through the chapter. Answers to all questions in the *Mid-Chapter Quiz* are given in the back of the student text and are located in place in the Instructor's Annotated Edition.

Chapter Test

Each chapter ends with a *Chapter Test*. Answers to all questions in the *Chapter Test* are given in the back of the student text and are located in place in the Instructor's Annotated Edition.

Cumulative Test

The *Cumulative Tests* that follow Chapters 3, 6, and 9 provide a comprehensive self-assessment tool that helps students check their mastery of previously covered material. Answers to all questions in the *Cumulative Tests* are given in the back of the student text and are located in place in the Instructor's Annotated Edition.

Supplements

Elementary Algebra, Fourth Edition, by Larson and Hostetler is accompanied by a comprehensive supplements package, which includes resources for both students and instructors. All items are keyed to the text.

Printed Resources

For Students

Student Solutions Guide by Carolyn F. Neptune, Johnson County Community College
(0-618-38819-2)

- Detailed, step-by-step solutions to all Review: Concepts, Skills, and Problem Solving exercises and to all odd-numbered exercises in the section exercise sets and in the review exercises
- Detailed, step-by-step solutions to all Mid-Chapter Quiz, Chapter Test, and Cumulative Test questions

For Instructors

Instructor's Annotated Edition
(0-618-38818-4)

- Includes answers in place for Exercise sets, Review Exercises, Mid-Chapter Quizzes, Chapter Tests, and Cumulative Tests
- Additional Answers section in the back of the text lists those answers that contain large graphics or lengthy exposition
- Answers to the Technology: Tip and Technology: Discovery questions are provided in the back of the book
- Annotations at point of use that offer strategies and suggestions for teaching the course and point out common student errors

Instructor's Resource Guide by Carolyn F. Neptune, Johnson County Community College
(0-618-38820-6)

- Chapter and Final Exam test forms with answer key
- Individual test items and answers for Chapters 1–10
- Notes to the instructor including tips and strategies on student assessment, cooperative learning, classroom management, study skills, and problem solving

SUPPLEMENTS

Technology Resources

For Students

HM mathSpace™ Student CD-ROM   (0-618-38825-7)

Website (http://math.college.hmco.com/students)

Houghton Mifflin Instructional Videos and DVDs by Dana Mosely
(Video ISBN: 0-618-38822-2; DVD ISBN: 0-618-38823-0)

SMARTHINKING™ Live, Online Tutoring   Houghton Mifflin has partnered with SMARTHINKING to provide an easy-to-use and effective online tutorial service. *Whiteboard Simulations* and *Practice Area* promote real-time visual interaction. Three levels of service are offered.

- **Text-Specific Tutoring** provides real-time, one-on-one instruction with a specially qualified "e-structor."

- *Questions Any Time* allows students to submit questions to the tutor outside the scheduled hours and receive a reply within 24 hours.

- *Independent Study Resources* connect students with around-the-clock access to additional educational services, including interactive websites, diagnostic tests, and Frequently Asked Questions posed to SMARTHINKING e-structors.

For Instructors

HMClassPrep™ with HM Testing   (0-618-38824-9)
Website (http://math.college.hmco.com/instructors)

Acknowledgments

We would like to thank the many people who have helped us revise the various editions of this text. Their encouragement, criticisms, and suggestions have been invaluable to us.

Reviewers

Mary Kay Best, Coastal Bend College; Patricia K. Bezona, Valdosta State University; Connie L. Buller, Metropolitan Community College; Mistye R. Canoy, Holmes Community College; Maggie W. Flint, Northeast State Technical Community College; William Hoard, Front Range Community College; Andrew J. Kaim, DePaul University; Jennifer L. Laveglia, Bellevue Community College; Aaron Montgomery, Purdue University North Central; William Naegele, South Suburban College; Jeanette O'Rourke, Middlesex County College; Judith Pranger, Binghamton University; Kent Sandefer, Mohave Community College; Robert L. Sartain, Howard Payne University; Jon W. Scott, Montgomery College; John Seims, Mesa Community College; Ralph Selensky, Eastern Arizona College; Charles I. Sherrill, Community College of Aurora; Kay Stroope, Phillips Community College of the University of Arkansas; Bettie Truitt, Black Hawk College; Betsey S. Whitman, Framingham State College; George J. Witt, Glendale Community College.

We would also like to thank the staff of Larson Texts, Inc. and the staff of Meridian Creative Group, who assisted in preparing the manuscript, rendering the art package, and typesetting and proofreading the pages and the supplements.

On a personal level, we are grateful to our wives, Deanna Gilbert Larson and Eloise Hostetler, for their love, patience, and support. Also, a special thanks goes to R. Scott O'Neil.

If you have suggestions for improving this text, please feel free to write to us. Over the past two decades we have received many useful comments from both instructors and students, and we value these comments very much.

Ron Larson
Robert P. Hostetler

ACKNOWLEDGMENTS

How to Study Algebra

Your success in algebra depends on your active participation both in class and outside of class. Because the material you learn each day builds on the material you learned previously, it is important that you keep up with the course work every day and develop a clear plan of study. To help you learn how to study algebra, we have prepared a set of guidelines that highlight key study strategies.

Preparing for Class

The syllabus your instructor provides is an invaluable resource that outlines the major topics to be covered in the course. Use it to help you prepare. As a general rule, you should set aside two to four hours of study time for each hour spent in class. Being prepared is the first step toward success in algebra. Before class,

- Review your notes from the previous class.

- Read the portion of the text that will be covered in class.

- Use the *What You Should Learn* objectives listed at the beginning of each section to keep you focused on the main ideas of the section.

- Pay special attention to the definitions, rules, and concepts highlighted in boxes. Also, be sure you understand the meanings of mathematical symbols and of terms written in boldface type. Keep a vocabulary journal for easy reference.

- Read through the solved examples. Use the side comments that accompany the solution steps to help you follow the solution process. Also, read the *Study Tips* given in the margins.

- Make notes of anything you do not understand as you read through the text. If you still do not understand after your instructor covers the topic in question, ask questions before your instructor moves on to a new topic.

- If you are using technology in this course, read the *Technology*: *Tips* and try the *Technology*: *Discovery* exercises.

Keeping Up

Another important step toward success in algebra involves your ability to keep up with the work. It is very easy to fall behind, especially if you miss a class. To keep up with the course work, be sure to

- Attend every class. Bring your text, a notebook, and a pen or pencil. If you miss a class, get the notes from a classmate as soon as possible and review them carefully.

- Take notes in class. After class, read through your notes and add explanations so that your notes make sense to *you*.

- Reread the portion of the text that was covered in class. This time, work each example *before* reading through the solution.

- Do your homework as soon as possible, while concepts are still fresh in your mind.

Use your notes from class, the text discussion, the examples, and the *Study Tips* as you do your homework. Many exercises are keyed to specific examples in the text for easy reference.

Getting Extra Help

It can be very frustrating when you do not understand concepts and are unable to complete homework assignments. However, there are many resources available to help you with your study of algebra.

- Your instructor may have office hours. If you are feeling overwhelmed and need help, make an appointment to discuss your difficulties with your instructor.

- Find a study partner or a study group. Sometimes it helps to work through problems with another person.

- Arrange to get regular assistance from a tutor. Many colleges have a math resource center available on campus as well.

- Consult one of the many ancillaries available with this text: the *Student Solutions Guide,* HM mathSpace™ Student CD-ROM, videotapes, DVDs, and additional study resources available at our website at *http://math.college.hmco.com/students.*

Preparing for an Exam

The last step toward success in algebra lies in how you prepare for and complete exams. If you have followed the suggestions given above, then you are almost ready for exams. Do not assume that you can cram for the exam the night before—this seldom works. As a final preparation for the exam,

- Read the *What Did You Learn?* chapter summary, which is keyed to each section, and review the concepts and terms.

- Work through the *Review Exercises* if you need extra practice on material from a particular section.

- Take the *Mid-Chapter Quiz* and the *Chapter Test* as if you were in class. You should set aside at least one hour per test. Check your answers against the answers given in the back of the book.

- Review your notes and the portion of the text that will be covered on the exam.

- Avoid studying up until the last minute. This will only make you anxious.

- Once the exam begins, read through the directions and the entire exam before beginning. Work the problems that you know how to do first to avoid spending too much time on any one problem. Time management is extremely important when taking an exam.

- If you finish early, use the remaining exam time to go over your work.

- When you get an exam back, review it carefully and go over your errors. Rework the problems you answered incorrectly. Discovering the mistakes you made will help you improve your test-taking ability.

STUDY PLAN

Motivating the Chapter

 December in Lexington, Virginia

In December 2001, the city of Lexington, Virginia, had an average daily high temperature of 5.8°C. The daily average temperatures and the daily high temperatures for the last 14 days of December 2001 are shown in the table.   (Source: WREL Weather Station, Lexington, Virginia)

Day	18	19	20	21	22	23	24
Average temperature (°C)	7.9°	2.1°	3.6°	1.4°	−1.3°	$\frac{5}{9}°$	2.2°
High temperature (°C)	11.5°	12°	7.6°	6.8°	7.4°	7.3°	6.2°

Day	25	26	27	28	29	30	31
Average temperature (°C)	−2.6°	−2.7°	$-2\frac{1}{2}°$	1.8°	2.6°	−3.9°	−4.3°
High temperature (°C)	2.6°	2.4°	2.2°	7°	6.8°	$-\frac{2}{9}°$	$1\frac{7}{10}°$

Here are some of the types of questions you will be able to answer as you study this chapter. You will be asked to answer parts (a)–(f) in Section 1.1, Exercise 79.

a. Write the set A of *integer* average and high temperatures. $A = \{7°, 12°\}$

b. Write the set B of *rational* high temperatures.
   $B = \{11.5°, 12°, 7.6°, 6.8°, 7.4°, 7.3°, 6.2°, 2.6°, 2.4°, 2.2°, 7°, -\frac{2}{9}°, 1\frac{7}{10}°\}$

c. Write the set C of *nonnegative* average temperatures.
   $C = \{7.9°, 2.1°, 3.6°, 1.4°, \frac{5}{9}°, 2.2°, 1.8°, 2.6°\}$

d. Write the high temperatures in *increasing* order.
   $-\frac{2}{9}°, 1\frac{7}{10}°, 2.2°, 2.4°, 2.6°, 6.2°, 6.8°, 7°, 7.3°, 7.4°, 7.6°, 11.5°, 12°$

e. Write the average temperatures in *decreasing* order.
   $7.9°, 3.6°, 2.6°, 2.2°, 2.1°, 1.8°, 1.4°, \frac{5}{9}°, -1.3°, -2\frac{1}{2}°, -2.6°, -2.7°, -3.9°, -4.3°$

f. What day(s) had average and high temperatures that were opposite numbers?   December 25

You will be asked to answer parts (g)–(k) in Section 1.4, Exercise 165.

g. What day had a high temperature of greatest departure from the monthly average high temperature of 5.8?   December 19

h. What successive days had the greatest change in average temperature?   December 29 and 30

i. What successive days had the greatest change in high temperature?   December 29 and 30

j. Find the average of the average temperatures for the 14 days.   ≈ 0.3°

k. In which of the preceding problems is the concept of absolute value used?   g, h, i

The answers to Motivating the Chapter are given in the section exercise answers in the back of the book. For instance, the answers to parts (a)–(f) are given in the answers to Section 1.1. Answers to odd-numbered exercises are given in the student answer key, and answers to all exercises are given in the instructor's answer key.

Owaki-Kulla/Corbis

The Real Number System

1.1 Real Numbers: Order and Absolute Value

© George B. Diebold/Corbis

What You Should Learn

1. Define sets and use them to classify numbers as natural, integer, rational, or irrational.
2. Plot numbers on the real number line.
3. Use the real number line and inequality symbols to order real numbers.
4. Find the absolute value of a number.

Why You Should Learn It

Understanding sets and subsets of real numbers will help you to analyze real-life situations accurately.

1 Define sets and use them to classify numbers as natural, integer, rational, or irrational.

Sets and Real Numbers

The ability to communicate precisely is an essential part of a modern society, and it is the primary goal of this text. Specifically, this section introduces the language used to communicate numerical concepts.

The formal term that is used in mathematics to talk about a collection of objects is the word **set.** For instance, the set $\{1, 2, 3\}$ contains the three numbers 1, 2, and 3. Note that a pair of braces $\{\ \}$ is used to list the members of the set. Parentheses $(\ )$ and brackets $[\ ]$ are used to represent other ideas.

The set of numbers that is used in arithmetic is called the set of **real numbers.** The term *real* distinguishes real numbers from *imaginary* numbers—a type of number that is used in some mathematics courses. You will not study imaginary numbers in Elementary Algebra.

If each member of a set A is also a member of a set B, then A is called a **subset** of B. The set of real numbers has many important subsets, each with a special name. For instance, the set

$$\{1, 2, 3, 4, \ldots\} \qquad \text{A subset of the set of real numbers}$$

is the set of **natural numbers** or **positive integers.** Note that the three dots indicate that the pattern continues. For instance, the set also contains the numbers 5, 6, 7, and so on. Every positive integer is a real number, but there are many real numbers that are not positive integers. For example, the numbers -2, 0, and $\frac{1}{2}$ are real numbers, but they are not positive integers.

Positive integers can be used to describe many things that you encounter in everyday life. For instance, you might be taking four classes this term, or you might be paying $180 a month for rent. But even in everyday life, positive integers cannot describe some concepts accurately. For instance, you could have a zero balance in your checking account, or the temperature could be $-5°F$. To describe such quantities you need to expand the set of positive integers to include **zero** and the **negative integers.** The expanded set is called the set of **integers.**

$$\underbrace{\{\ldots, -3, -2, -1,}_{\text{Negative integers}} \overset{\text{Zero}}{0}, \underbrace{1, 2, 3, \ldots\}}_{\text{Positive integers}} \qquad \text{Set of integers}$$

The set of integers is also a subset of the set of real numbers.

Even with the set of integers, there are still many quantities in everyday life that you cannot describe accurately. The costs of many items are not in whole-dollar amounts, but in parts of dollars, such as $1.19 or $39.98. You might work $8\frac{1}{2}$ hours, or you might miss the first half of a movie. To describe such quantities, you can expand the set of integers to include **fractions.** The expanded set is called the set of **rational numbers.** In the formal language of mathematics, a real number is **rational** if it can be written as a ratio of two integers. So, $\frac{3}{4}$ is a rational number; so is 0.5 $\left(\text{it can be written as } \frac{1}{2}\right)$; and so is every integer. A real number that is not rational is called **irrational** and cannot be written as the ratio of two integers. One example of an irrational number is $\sqrt{2}$, which is read as the positive square root of 2. Another example is π (the Greek letter pi), which represents the ratio of the circumference of a circle to its diameter. Each of the sets of numbers mentioned—natural numbers, integers, rational numbers, and irrational numbers—is a subset of the set of real numbers, as shown in Figure 1.1.

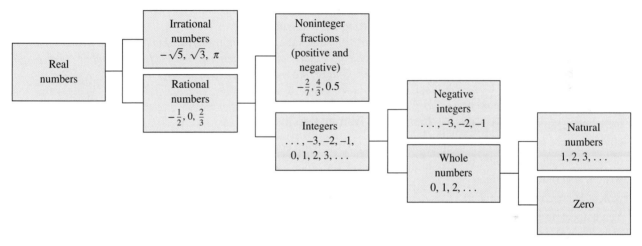

Figure 1.1   Subsets of Real Numbers

In *decimal form,* you can recognize rational numbers as decimals that terminate

$$\frac{1}{2} = 0.5 \quad \text{or} \quad \frac{3}{8} = 0.375$$

or repeat

$$\frac{4}{3} = 1.\overline{3} \quad \text{or} \quad \frac{2}{11} = 0.\overline{18}.$$

Irrational numbers are represented by decimals that neither terminate nor repeat, as in

$$\sqrt{2} = 1.414213562\ldots$$

or

$$\pi = 3.141592653\ldots\ldots$$

Example 1   Classifying Real Numbers

Which of the numbers in the following set are (a) natural numbers, (b) integers, (c) rational numbers, and (d) irrational numbers?

$$\left\{\frac{1}{2}, -1, 0, 4, -\frac{5}{8}, \frac{4}{2}, -\frac{3}{1}, 0.86, \sqrt{2}, \sqrt{9}\right\}$$

Solution

a. Natural numbers: $\left\{4, \frac{4}{2} = 2, \sqrt{9} = 3\right\}$

b. Integers: $\left\{-1, 0, 4, \frac{4}{2} = 2, -\frac{3}{1} = -3, \sqrt{9} = 3\right\}$

c. Rational numbers: $\left\{\frac{1}{2}, -1, 0, 4, -\frac{5}{8}, \frac{4}{2} = 2, -\frac{3}{1} = -3, 0.86, \sqrt{9} = 3\right\}$

d. Irrational number: $\left\{\sqrt{2}\right\}$

The Real Number Line

The diagram used to represent the real numbers is called the **real number line.** It consists of a horizontal line with a point (the **origin**) labeled 0. Numbers to the left of 0 are **negative** and numbers to the right of 0 are **positive**, as shown in Figure 1.2. The real number zero is neither positive nor negative. So, the term **nonnegative** implies that a number may be positive or zero.

Figure 1.2   The Real Number Line

Drawing the point on the real number line that corresponds to a real number is called **plotting** the real number.

Example 2 illustrates the following principle. *Each point on the real number line corresponds to exactly one real number, and each real number corresponds to exactly one point on the real number line.*

Technology: Tip

The Greek letter pi, denoted by the symbol π, is the ratio of the circumference of a circle to its diameter. Because π cannot be written as a ratio of two integers, it is an irrational number. You can get an approximation of π on a scientific or graphing calculator by using the following keystroke.

Keystroke	Display
$\boxed{\pi}$	3.141592654

Between which two integers would you plot π on the real number line?

See Technology Answers.

Example 2   Plotting Real Numbers

a. In Figure 1.3, the point corresponds to the real number $-\frac{1}{2}$.

b. In Figure 1.4, the point corresponds to the real number 2.

c. In Figure 1.5, the point corresponds to the real number $-\frac{3}{2}$.

d. In Figure 1.6, the point corresponds to the real number 1.

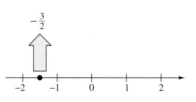

Figure 1.3

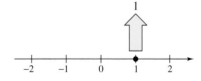

Figure 1.4

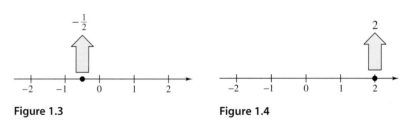

Figure 1.5

Figure 1.6

Ordering Real Numbers

The real number line provides you with a way of comparing any two real numbers. For instance, if you choose any two (different) numbers on the real number line, one of the numbers must be to the left of the other number. The number to the left is **less than** the number to the right. Similarly, the number to the right is **greater than** the number to the left. For example, from Figure 1.7 you can see that -3 is less than 2 because -3 lies to the left of 2 on the number line. A "less than" comparison is denoted by the **inequality symbol** <. For instance, "-3 is less than 2" is denoted by $-3 < 2$.

Similarly, the inequality symbol > is used to denote a "greater than" comparison. For instance, "2 is greater than -3" is denoted by $2 > -3$. The inequality symbol ≤ means **less than or equal to,** and the inequality symbol ≥ means **greater than or equal to.**

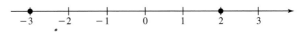

Figure 1.7 -3 lies to the left of 2.

When you are asked to **order** two numbers, you are simply being asked to say which of the two numbers is greater.

Example 3   Ordering Integers

Place the correct inequality symbol (< or >) between each pair of numbers.

a. 3   5   **b.** -3 -5 **c.** 4   0

d. -2 2   **e.** 1 -4

Solution

a. $3 < 5$, because 3 lies to the *left* of 5.   See Figure 1.8.

b. $-3 > -5$, because -3 lies to the *right* of -5.   See Figure 1.9.

c. $4 > 0$, because 4 lies to the *right* of 0.   See Figure 1.10.

d. $-2 < 2$, because -2 lies to the *left* of 2.   See Figure 1.11.

e. $1 > -4$, because 1 lies to the *right* of -4.   See Figure 1.12.

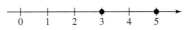

Figure 1.8

Figure 1.9

Figure 1.10

Figure 1.11

Figure 1.12

There are two ways to order fractions: you can write both fractions with the same denominator, or you can rewrite both fractions in decimal form. Here are two examples.

$$\frac{1}{3} = \frac{4}{12} \quad \text{and} \quad \frac{1}{4} = \frac{3}{12} \quad \Longrightarrow \quad \frac{1}{3} > \frac{1}{4}$$

$$\frac{11}{131} \approx 0.084 \quad \text{and} \quad \frac{19}{209} \approx 0.091 \quad \Longrightarrow \quad \frac{11}{131} < \frac{19}{209}$$

The symbol $\approx$ means "is approximately equal to."

Example 4　Ordering Fractions

Place the correct inequality symbol ($<$ or $>$) between each pair of numbers.

a. $\dfrac{1}{3}$ ▢ $\dfrac{1}{5}$　　**b.** $-\dfrac{3}{2}$ ▢ $\dfrac{1}{2}$

Solution

a. $\frac{1}{3} > \frac{1}{5}$, because $\frac{1}{3} = \frac{5}{15}$ lies to the *right* of $\frac{1}{5} = \frac{3}{15}$ (see Figure 1.13).

b. $-\frac{3}{2} < \frac{1}{2}$, because $-\frac{3}{2}$ lies to the *left* of $\frac{1}{2}$ (see Figure 1.14).

Figure 1.13

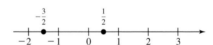

Figure 1.14

Example 5　Ordering Decimals

Place the correct inequality symbol ($<$ or $>$) between each pair of numbers.

a. -3.1 ▢ 2.8　　**b.** -1.09 ▢ -1.90

Solution

a. $-3.1 < 2.8$, because -3.1 lies to the *left* of 2.8 (see Figure 1.15).

b. $-1.09 > -1.90$, because -1.09 lies to the *right* of -1.90 (see Figure 1.16).

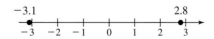

Figure 1.15

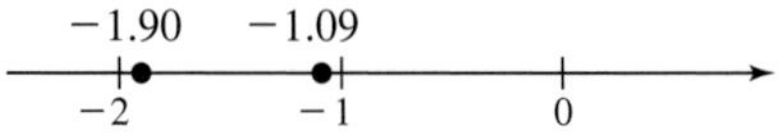

Figure 1.16

④ Find the absolute value of a number.

Absolute Value

Two real numbers are **opposites** of each other if they lie the same distance from, but on opposite sides of, zero. For example, -2 is the opposite of 2, and 4 is the opposite of -4, as shown in Figure 1.17.

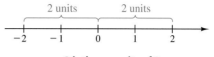

−2 is the opposite of 2.

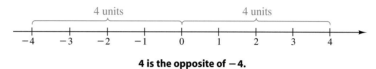

4 is the opposite of −4.

Figure 1.17

Parentheses are useful for denoting the opposite of a negative number. For example, $-(-3)$ means the opposite of -3, which you know to be 3. That is,

$$-(-3) = 3.$$ The opposite of -3 is 3.

For any real number, its distance from zero on the real number line is its **absolute value.** A pair of vertical bars, $|\ \ |$, is used to denote absolute value. Here are two examples.

$$|5| = \text{"distance between 5 and 0"} = 5$$

$$|-8| = \text{"distance between } -8 \text{ and 0"} = 8$$ See Figure 1.18.

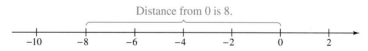

Figure 1.18

Because opposite numbers lie the same distance from zero on the real number line, they have the same absolute value. So, $|5| = 5$ and $|-5| = 5$ (see Figure 1.19).

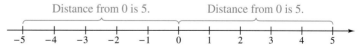

Figure 1.19

You can write this more simply as $|5| = |-5| = 5$.

Definition of Absolute Value

If a is a real number, then the **absolute value** of a is

$$|a| = \begin{cases} a, & \text{if } a \geq 0 \\ -a, & \text{if } a < 0 \end{cases}.$$

The absolute value of a real number is either positive or zero (never negative). For instance, by definition, $|-3| = -(-3) = 3$. Moreover, zero is the only real number whose absolute value is 0. That is, $|0| = 0$.

The word **expression** means a collection of numbers and symbols such as $3 + 5$ or $|-4|$. When asked to **evaluate** an expression, you are to find the *number* that is equal to the expression.

The concept of absolute value may be difficult for some students. (The formal definition of absolute value is given in Section 1.3.)

Example 6   Evaluating Absolute Values

Evaluate each expression.

a. $|-10|$

b. $\left|\dfrac{3}{4}\right|$

c. $|-3.2|$

d. $-|-6|$

Solution

a. $|-10| = 10$, because the distance between -10 and 0 is 10.

b. $\left|\dfrac{3}{4}\right| = \dfrac{3}{4}$, because the distance between $\dfrac{3}{4}$ and 0 is $\dfrac{3}{4}$.

c. $|-3.2| = 3.2$, because the distance between -3.2 and 0 is 3.2.

d. $-|-6| = -(6) = -6$

Note in Example 6(d) that $-|-6| = -6$ does not contradict the fact that the absolute value of a real number cannot be negative. The expression $-|-6|$ calls for the *opposite* of an absolute value and so it must be negative.

Example 7   Comparing Absolute Values

Place the correct symbol ($<$, $>$, or $=$) between each pair of numbers.

a. $|-9|$ ___ $|9|$

b. $|-3|$ ___ 5

c. 0 ___ $|-7|$

d. -4 ___ $-|-4|$

e. $|12|$ ___ $|-15|$

f. 2 ___ $-|-2|$

Solution

a. $|-9| = |9|$, because $|-9| = 9$ and $|9| = 9$.

b. $|-3| < 5$, because $|-3| = 3$ and 3 is less than 5.

c. $0 < |-7|$, because $|-7| = 7$ and 0 is less than 7.

d. $-4 = -|-4|$, because $-|-4| = -4$ and -4 is equal to -4.

e. $|12| < |-15|$, because $|12| = 12$ and $|-15| = 15$, and 12 is less than 15.

f. $2 > -|-2|$, because $-|-2| = -2$ and 2 is greater than -2.

1.1 Exercises

Developing Skills

In Exercises 1–4, determine which of the numbers in the set are (a) natural numbers, (b) integers, and (c) rational numbers. See Example 1.

1. $\left\{-3, 20, -\frac{3}{2}, \frac{9}{3}, 4.5\right\}$

   (a) $20, \frac{9}{3}$ (b) $-3, 20, \frac{9}{3}$ (c) $-3, 20, -\frac{3}{2}, \frac{9}{3}, 4.5$

2. $\left\{10, -82, -\frac{24}{3}, -8.2, \frac{1}{5}\right\}$

   (a) 10 (b) $10, -82, -\frac{24}{3}$ (c) $10, -82, -\frac{24}{3}, -8.2, \frac{1}{5}$

3. $\left\{-\frac{5}{2}, 6.5, -4.5, \frac{8}{4}, \frac{3}{4}\right\}$

   (a) $\frac{8}{4}$ (b) $\frac{8}{4}$ (c) $-\frac{5}{2}, 6.5, -4.5, \frac{8}{4}, \frac{3}{4}$

4. $\left\{8, -1, \frac{4}{3}, -3.25, -\frac{10}{2}\right\}$

   (a) 8 (b) $8, -1, -\frac{10}{2}$ (c) $8, -1, \frac{4}{3}, -3.25, -\frac{10}{2}$

In Exercises 5–8, plot the numbers on the real number line. See Example 2.   See Additional Answers.

5. $-7, 1.5$

6. $4, -3.2$

7. $\frac{1}{4}, 0, -2$

8. $-\frac{3}{2}, 5, 1$

In Exercises 9–18, plot each real number as a point on the real number line and place the correct inequality symbol ($<$ or $>$) between the pair of real numbers. See Examples 3 and 4.   See Additional Answers.

9. 3 $>$ -4

10. 6 $>$ -2

11. 4 $>$ $-\frac{7}{2}$

12. 2 $>$ $\frac{3}{2}$

13. 0 $>$ $-\frac{7}{16}$

14. $-\frac{7}{3}$ $>$ $-\frac{7}{2}$

15. -4.6 $<$ 1.5

16. 28.60 $>$ -3.75

17. $\frac{7}{16}$ $<$ $\frac{5}{8}$

18. $-\frac{3}{8}$ $>$ $-\frac{5}{8}$

In Exercises 19–22, find the distance between a and zero on the real number line.

19. $a = 2$ 2

20. $a = 5$ 5

21. $a = -4$ 4

22. $a = -10$ 10

In Exercises 23–28, find the opposite of the number. Plot the number and its opposite on the real number line. What is the distance of each from 0?

See Additional Answers.

23. 5 -5; Distance: 5 **24.** 2 -2; Distance: 2

25. -3.8 3.8; Distance: 3.8 **26.** -7.5 7.5; Distance: 7.5

27. $-\frac{5}{2}$ $\frac{5}{2}$; Distance: $\frac{5}{2}$ **28.** $-\frac{3}{4}$ $\frac{3}{4}$; Distance: $\frac{3}{4}$

In Exercises 29–32, find the absolute value of the real number and its distance from 0.

29.

$\frac{5}{2}, \frac{5}{2}$

30.

2.4

$2.4, 2.4$

31.

$3, 3$

32.

$-\frac{4}{3}$

$\frac{4}{3}, \frac{4}{3}$

In Exercises 33–46, evaluate the expression. See Example 6.

33. $|7|$ 7 **34.** $|6|$ 6

35. $|-11|$ 11 **36.** $|-15|$ 15

37. $|-3.4|$ 3.4 **38.** $|-16.2|$ 16.2

39. $\left|-\frac{7}{2}\right|$ $\frac{7}{2}$ **40.** $\left|-\frac{9}{16}\right|$ $\frac{9}{16}$

41. $-|4.09|$ -4.09 **42.** $-|91.3|$ -91.3

43. $-|-23.6|$ -23.6 **44.** $-|-43.8|$ -43.8

45. $|0|$ 0 **46.** $|\pi|$ π

In Exercises 47–58, place the correct symbol ($<$, $>$, or $=$) between the pair of real numbers. See Example 7.

47. $|-15|$ $=$ $|15|$

48. $|525|$ $=$ $|-525|$

49. $|-4|$ $>$ $|3|$

50. $|16|$ $<$ $|-25|$

51. $|32|$ < $|-50|$

52. $|-1026|$ > $|800|$

53. $\left|\frac{3}{16}\right|$ < $\left|\frac{3}{2}\right|$

54. $\left|\frac{7}{8}\right|$ < $\left|\frac{4}{3}\right|$

55. $-|-48.5|$ < $|-48.5|$

56. $-|-64|$ < $|-64|$

57. $|-\pi|$ > $-|-2\pi|$

58. $-|-4.9|$ < $|-10.2|$

In Exercises 59–62, plot the numbers on the real number line.   See Additional Answers.

59. $\frac{5}{2}, \pi, -2, -|-3|$

60. $3.7, \frac{16}{3}, -|-1.9|, -\frac{1}{2}$

61. $-4, \frac{7}{3}, |-3|, 0, -|4.5|$

62. $|-2.3|, 3.2, -2.3, -|3.2|$

In Exercises 63–68, find all real numbers whose distance from a is given by d.

63. $a = 8, d = 12$
$-4, 20$

64. $a = 6, d = 7$
$-1, 13$

65. $a = 21.3, d = 6$
$15.3, 27.3$

66. $a = 42.5, d = 7$
$35.5, 49.5$

67. $a = -2, d = 3.5$
$-5.5, 1.5$

68. $a = -7, d = 7.2$
$-14.2, 0.2$

Solving Problems

In Exercises 69–78, give three examples of numbers that satisfy the given conditions.

69. A real number that is a negative integer
Sample answers: $-3, -100, -\frac{4}{1}$

70. A real number that is a whole number
Sample answers: $7, 1032, 15$

71. A real number that is not a rational number
Sample answers: $\sqrt{2}, \pi, -3\sqrt{3}$

72. A real number that is not an irrational number
Sample answers: $\frac{2}{3}, 201, 3.\overline{3}$

73. An integer that is a rational number
Sample answers: $-7, 1, 341$

74. A rational number that is not an integer
Sample answers: $\frac{3}{4}, 1\frac{1}{2}, 0.1\overline{6}$

75. A rational number that is not a negative number
Sample answers: $\frac{1}{2}, 10, 20\frac{1}{5}$

76. A real number that is not a positive rational number
Sample answers: $-\frac{1}{2}, \pi, -\sqrt{2}$

77. A real number that is not an integer
Sample answers: $\frac{1}{7}, 0.25, 10\frac{1}{2}$

78. An integer that is not a whole number
Sample answers: $-1, -10, -100$

Explaining Concepts

79. ⚡ Answer parts (a)–(f) of Motivating the Chapter.

80. *Writing* ✎ Explain why $\frac{8}{4}$ is a natural number, but $\frac{7}{4}$ is not. $\frac{8}{4} = 2, \frac{7}{4} = 1.75$

81. *Writing* ✎ How many numbers are three units from 0 on the real number line? Explain your answer.
Two. They are -3 and 3.

82. *Writing* ✎ Explain why the absolute value of every real number is positive.   The absolute value of every real number is a distance from zero on the real number line. Distance is always positive.

83. *Writing* ✎ Which real number lies farther from 0 on the real number line?

(a) -25 (b) 10

Explain your answer. $-25; |-25| > |10|$

84. *Writing* ✎ Which real number lies farther from -7 on the real number line?

(a) 3 (b) -10

Explain your answer.   3. -10 is 3 units from -7 and 3 is 10 units from -7.

The symbol ⚡ indicates an exercise in which you are asked to answer parts of the Motivating the Chapter problem found on the Chapter Opener pages.

85. *Writing* ✐   Explain how to determine the smaller of two different real numbers.   The smaller number is located to the left of the larger number on the real number line.

86. *Writing* ✐   Select the smaller real number and explain your answer.

(a) $\frac{3}{8}$ (b) 0.35

$\frac{3}{8} = 0.375$, so 0.35 is the smaller number.

True or False?   In Exercises 87–96, decide whether the statement is true or false. Justify your answer.

87. $-5 > -13$

True. $-5 > -13$ because -5 lies to the right of -13.

88. $-10 > -2$

False. $-10 < -2$ because -10 lies to the left of -2.

89. $6 < -17$

False. $6 > -17$ because 6 lies to the right of -17.

90. $4 < -9$

False. $4 > -9$ because 4 lies to the right of -9.

91. The absolute value of any real number is always positive.   False. $|0| = 0$

92. The absolute value of a number is equal to the absolute value of its opposite.

True. For $a \geq 0$, $|a| = a$ and $|-a| = a$.

For $a < 0$, $|a| = -a$, and $|-a| = -a$.

93. The absolute value of a rational number is a rational number.   True. For example, $\left|\frac{2}{3}\right| = \frac{2}{3}$.

94. A given real number corresponds to exactly one point on the real number line.

True. Definition of real number line.

95. The opposite of a positive number is a negative number.   True. Definition of opposite.

96. Every rational number is an integer.

False. $\frac{1}{2}$ is not an integer.

1.2 Adding and Subtracting Integers

© Chuck Savage/Corbis

What You Should Learn

1. Add integers using a number line.
2. Add integers with like signs and with unlike signs.
3. Subtract integers with like signs and with unlike signs.

Why You Should Learn It

Real numbers are used to represent many real-life quantities. For instance, in Exercise 101 on page 19, you will use real numbers to find the increase in enrollment at private and public schools in the United States.

1. Add integers using a number line.

Adding Integers Using a Number Line

In this and the next section, you will study the four operations of arithmetic (addition, subtraction, multiplication, and division) on the set of integers. There are many examples of these operations in real life. For example, your business had a gain of $550 during one week and a loss of $600 the next week. Over the two-week period, your business had a combined profit of

$$550 + (-600) = -50$$

which means you had an overall loss of $50.

The number line is a good visual model for demonstrating addition of integers. To add two integers, $a + b$, using a number line, start at 0. Then move left or right a units depending on whether a is positive or negative. From that position, move left or right b units depending on whether b is positive or negative. The final position is called the **sum.**

Example 1   Adding Integers with Like Signs Using a Number Line

Find each sum.

a. $5 + 2$ **b.** $-3 + (-5)$

Solution

a. Start at zero and move five units to the right. Then move two more units to the right, as shown in Figure 1.20. So, $5 + 2 = 7$.

b. Start at zero and move three units to the left. Then move five more units to the left, as shown in Figure 1.21. So, $-3 + (-5) = -8$.

Figure 1.20   **Figure 1.21**

Example 2   Adding Integers with Unlike Signs Using a Number Line

Find each sum.

a. $-5 + 2$ **b.** $7 + (-3)$ **c.** $-4 + 4$

Solution

a. Start at zero and move five units to the left. Then move two units to the right, as shown in Figure 1.22.

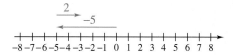

Figure 1.22

So, $-5 + 2 = -3$.

b. Start at zero and move seven units to the right. Then move three units to the left, as shown in Figure 1.23.

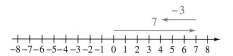

Figure 1.23

So, $7 + (-3) = 4$.

c. Start at zero and move four units to the left. Then move four units to the right, as shown in Figure 1.24.

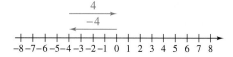

Figure 1.24

So, $-4 + 4 = 0$.

In Example 2(c), notice that the sum of -4 and 4 is 0. Two numbers whose sum is zero are called **opposites** (or **additive inverses**) of each other, So, -4 is the opposite of 4 and 4 is the opposite of -4.

2 Add integers with like signs and with unlike signs.

Adding Integers

Examples 1 and 2 illustrated a *graphical* approach to adding integers. It is more common to use an *algebraic* approach to adding integers, as summarized in the following rules.

Addition of Integers

1. To **add** two integers with *like* signs, add their absolute values and attach the common sign to the result.

2. To **add** two integers with *unlike* signs, subtract the smaller absolute value from the larger absolute value and attach the sign of the integer with the larger absolute value.

Example 3   Adding Integers

a. Unlike signs: $22 + (-17) = |22| - |-17| = 22 - 17 = 5$

b. Unlike signs: $-84 + 14 = -(|-84| - |-14|) = -(84 - 14) = -70$

c. Like signs: $-18 + (-62) = -(|-18| + |-62|) = -(18 + 62) = -80$

$$
\begin{array}{r}
1\ 1 \\
1\ 4\ 8 \\
6\ 2 \\
+\ 5\ 3\ 6 \\
\hline
7\ 4\ 6
\end{array}
$$

Figure 1.25   Carrying Algorithm

There are different ways to add three or more integers. You can use the **carrying algorithm** with a vertical format with nonnegative integers, as shown in Figure 1.25, or you can add them two at a time, as illustrated in Example 4.

Example 4   Account Balance

At the beginning of a month, your account balance was \$28. During the month you deposited \$60 and withdrew \$40. What was your balance at the end of the month?

Solution

$$
\begin{aligned}
\$28 + \$60 + (-\$40) &= (\$28 + \$60) + (-\$40) \\
&= \$88 + (-\$40) \\
&= \$48 \qquad\qquad \text{Balance}
\end{aligned}
$$

③ Subtract integers with like signs and with unlike signs.

Subtracting Integers

Subtraction can be thought of as "taking away." For instance, $8 - 5$ can be thought of as "8 take away 5," which leaves 3. Moreover, note that $8 + (-5) = 3$, which means that

$$8 - 5 = 8 + (-5).$$

In other words, $8 - 5$ can also be accomplished by "adding the opposite of 5 to 8."

Subtraction of Integers

To **subtract** one integer from another, add the opposite of the integer being subtracted to the other integer. The result is called the **difference** of the two integers.

Example 5   Subtracting Integers

a. $3 - 8 = 3 + (-8) = -5$ Add opposite of 8.

b. $10 - (-13) = 10 + 13 = 23$ Add opposite of -13.

c. $-5 - 12 = -5 + (-12) = -17$ Add opposite of 12.

d. $-4 - (-17) - 23 = -4 + 17 + (-23) = -10$ Add opposite of -17 and opposite of 23.

$$
\begin{array}{r}
3\ 10\ 15 \\
\cancel{4}\ \cancel{1}\ \cancel{5} \\
-\ 2\ \ 7\ \ 6 \\
\hline
1\ \ 3\ \ 9
\end{array}
$$

Figure 1.26   Borrowing Algorithm

Be sure you understand that the terminology involving subtraction is not the same as that used for negative numbers. For instance, -5 is read as "negative 5," but $8 - 5$ is read as "8 subtract 5." It is important to distinguish between the operation and the signs of the numbers involved. For instance, in $-3 - 5$ the operation is subtraction and the numbers are -3 and 5.

For subtraction problems involving only two nonnegative integers, you can use the **borrowing algorithm** shown in Figure 1.26.

Example 6   Subtracting Integers

a. Subtract 10 from -4 means: $-4 - 10 = -4 + (-10) = -14$.

b. -3 subtract -8 means: $-3 - (-8) = -3 + 8 = 5$.

To evaluate expressions that contain a series of additions and subtractions, write the subtractions as equivalent additions and simplify from left to right, as shown in Example 7.

Example 7   Evaluating Expressions

Evaluate each expression.

a. $-13 - 7 + 11 - (-4)$ **b.** $5 - (-9) - 12 + 2$

c. $-1 - 3 - 4 + 6$ **d.** $5 - 1 - 8 + 3 + 4 - (-10)$

Solution

a. $-13 - 7 + 11 - (-4) = -13 + (-7) + 11 + 4$ Add opposites.

$= -20 + 15$ Add two numbers at a time.

$= -5$ Add.

b. $5 - (-9) - 12 + 2 = 5 + 9 + (-12) + 2$ Add opposites.

$= 14 + (-10)$ Add two numbers at a time.

$= 4$ Add.

c. $-1 - 3 - 4 + 6 = -1 + (-3) + (-4) + 6$ Add opposites.

$= -4 + 2$ Add two numbers at a time.

$= -2$ Add.

d. $5 - 1 - 8 + 3 + 4 - (-10) = 5 + (-1) + (-8) + 3 + 4 + 10$

$= 4 + (-5) + 14 = 13$

Example 8   **Temperature Change**

The temperature in Minneapolis, Minnesota at 4 P.M. was 15°F. By midnight, the temperature had decreased by 18°. What was the temperature in Minneapolis at midnight?

Solution

To find the temperature at midnight, subtract 18 from 15.

$$15 - 18 = 15 + (-18)$$
$$= -3$$

The temperature in Minneapolis at midnight was $-3°$F.

___

This text includes several examples and exercises that use a calculator. As each new calculator application is encountered, you will be given general instructions for using a calculator. These instructions, however, may not agree precisely with the steps required by *your* calculator, so be sure you are familiar with the use of the keys on your own calculator.

For each of the calculator examples in the text, two possible keystroke sequences are given: one for a standard *scientific* calculator and one for a *graphing* calculator.

Example 9   **Evaluating Expressions with a Calculator**

Evaluate each expression with a calculator.

a. $-4 - 5$ **b.** $2 - (3 - 9)$

Keystrokes	*Display*	
a. 4 [+/−] [−] 5 [=]	−9	Scientific
[(−)] 4 [−] 5 [ENTER]	−9	Graphing

Keystrokes	*Display*	
b. 2 [−] [(] 3 [−] 9 [)] [=]	8	Scientific
2 [−] [(] 3 [−] 9 [)] [ENTER]	8	Graphing

___

Technology: Tip

The keys [+/−] and [(−)] change a number to its opposite and [−] is the subtraction key. For instance, the keystrokes [−] 4 [−] 5 [ENTER] will not produce the result shown in Example 9(a).

1.2 Exercises

Developing Skills

In Exercises 1–8, find the sum and demonstrate the addition on the real number line. See Examples 1 and 2. See Additional Answers.

1. $2 + 7$ 9

2. $3 + 9$ 12

3. $10 + (-3)$ 7

4. $14 + (-8)$ 6

5. $-6 + 4$ -2

6. $-12 + 5$ -7

7. $(-8) + (-3)$ -11

8. $(-4) + (-7)$ -11

In Exercises 9–42, find the sum. See Example 3.

9. $6 + 10$ 16

10. $8 + 3$ 11

11. $14 + (-14)$ 0

12. $10 + (-10)$ 0

13. $-45 + 45$ 0

14. $-23 + 23$ 0

15. $14 + 13$ 27

16. $20 + 19$ 39

17. $-23 + (-4)$ -27

18. $-32 + (-16)$ -48

19. $18 + (-12)$ 6

20. $34 + (-16)$ 18

21. $75 + 100$ 175

22. $54 + 68$ 122

23. $9 + (-14)$ -5

24. $18 + (-26)$ -8

25. $10 - 6 + 34$ 38

26. $7 - 4 + 1$ 4

27. $-15 + (-3) + 8$ -10

28. $-82 + (-36) + 82$ -36

29. $8 + 16 + (-3)$ 21

30. $2 + (-51) + 13$ -36

31. $17 + (-2) + 5$ 20

32. $24 + 1 + (-19)$ 6

33. $-13 + 12 + 4$ 3

34. $-31 + 20 + 15$ 4

35. $15 + (-75) + (-75)$ -135

36. $32 + (-32) + (-16)$ -16

37. $104 + 203 + (-613) + (-214)$ -520

38. $4365 + (-2145) + (-1873) + 40,084$ 40,431

39. $312 + (-564) + (-100)$ -352

40. $1200 + (-1300) + (-275)$ -375

41. $-890 + 90 + (-82)$ -882

42. $-770 + (-383) + 492$ -661

In Exercises 43–76, find the difference. See Example 5.

43. $12 - 9$ 3

44. $55 - 20$ 35

45. $39 - 13$ 26

46. $45 - 35$ 10

47. $4 - (-1)$ 5

48. $9 - (-6)$ 15

49. $18 - (-7)$ 25

50. $27 - (-12)$ 39

51. $32 - (-4)$ 36

52. $47 - (-43)$ 90

53. $19 - (-31)$ 50

54. $12 - (-5)$ 17

55. $27 - 57$ -30

56. $18 - 32$ -14

57. $61 - 85$ -24

58. $53 - 74$ -21

59. $22 - 131$ -109

60. $48 - 222$ -174

61. $2 - 11$ -9

62. $3 - 15$ -12

63. $13 - 24$ -11

64. $26 - 34$ -8

65. $-135 - (-114)$ -21

66. $-63 - (-8)$ -55

67. $-4 - (-4)$ 0

68. $-942 - (-942)$ 0

69. $-10 - (-4)$ -6

70. $-12 - (-7)$ -5

71. $-71 - 32$ -103

72. $-84 - 106$ -190

73. $-210 - 400$ -610

74. $-120 - 142$ -262

75. $-110 - (-30)$ -80

76. $-2500 - (-600)$ -1900

77. Subtract 15 from -6. -21
78. Subtract 24 from -17. -41
79. Subtract -120 from 380. 500
80. Subtract -80 from 140. 220
81. *Think About It*   What number must be added to 10 to obtain -5? -15
82. *Think About It*   What number must be added to 36 to obtain -12? -48
83. *Think About It*   What number must be subtracted from -12 to obtain 24? -36
84. *Think About It*   What number must be subtracted from -20 to obtain 15? -35

In Exercises 85–90, evaluate the expression. See Example 7.

85. $-1 + 3 - (-4) + 10$ 16
86. $12 - 6 + 3 - (-8)$ 17
87. $6 + 7 - 12 - 5$ -4
88. $-3 + 2 - 20 + 9$ -12
89. $-(-5) + 7 - 18 + 4$ -2
90. $-15 - (-2) + 4 - 6$ -15

Solving Problems

91. *Temperature Change*   The temperature at 6 A.M. was $-10°$F. By noon, the temperature had increased by 22°F. What was the temperature at noon? $12°$F

92. *Account Balance*   A credit card owner charged $142 worth of goods on her account. Find the balance after a payment of $87 was made. $55

93. *Sports*   A hiker hiked 847 meters down the Grand Canyon. He climbed back up 385 meters and then rested. Find his distance down the canyon where he rested.   462 meters

94. *Sports*   A fisherman dropped his line 27 meters below the surface of the water. Because the fish were not biting there, he decided to raise his line by 8 meters. How far below the surface of the water was his line?   19 meters

95. *Profit*   A telephone company lost $650,000 during the first 6 months of the year. By the end of the year, the company had an overall profit of $362,000. What was the company's profit during the second 6 months of the year? $1,012,000

96. *Altitude*   An airplane flying at an altitude of 31,000 feet is instructed to descend to an altitude of 24,000 feet. How many feet must the airplane descend?   7000 feet

97. *Account Balance*   At the beginning of a month, your account balance was $2750. During the month you withdrew $350 and $500, deposited $450, and earned interest of $6.42. What was your balance at the end of the month? $2356.42

98. *Account Balance*   At the beginning of a month, your account balance was $1204. During the month, you withdrew $725 and $821, deposited $150 and $80, and earned interest of $2.02. What was your balance at the end of the month? $-109.98

99. *Temperature Change*   When you left for class in the morning, the temperature was 25°C. By the time class ended, the temperature had increased by 4°. While you studied, the temperature increased by 3°. During your soccer practice, the temperature decreased by 9°. What was the temperature after your soccer practice?   23°

100. *Temperature Change*   When you left for class in the morning, the temperature was 40°F. By the time class ended, the temperature had increased by 13°. While you studied, the temperature decreased by 5°. During your club meeting, the temperature decreased by 6°. What was the temperature after your club meeting?   42°

101. *Education*   The bar graph shows the total enrollment (in millions) at public and private schools in
the United States for the years 1995 to 2001.
(Source: U.S. National Center for Education
Statistics)

(a) Find the increase in enrollment from 1996 to
2001.   2.7 million

(b) Find the increase in enrollment from 1999 to
2001.   0.8 million

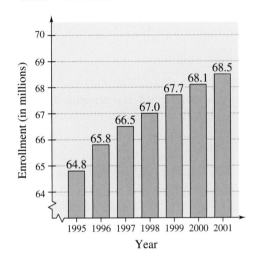

102. *Retail Price*   The bar graph shows the average
retail price of a half-gallon of ice cream in the
United States for the years 1996 to 2000.   (Source:
U.S. Bureau of Labor Statistics)

(a) Find the increase in the average retail price of
ice cream from 1997 to 1998. $0.28

(b) Find the increase in the average retail price of
ice cream from 1998 to 1999. $0.10

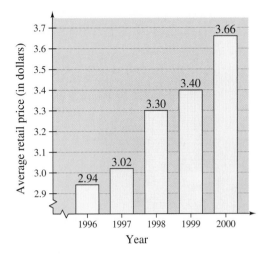

Figure for 102

In Exercises 103 and 104, an addition problem is shown
visually on the real number line. (a) Write the addition
problem and find the sum. (b) State the rule for the
addition of integers demonstrated.

103.

(a) $3 + 2 = 5$

(b) Adding two integers with like signs

104.

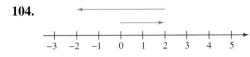

(a) $2 + (-4) = -2$

(b) Adding two integers with unlike signs

Explaining Concepts

105. *Writing*   Explain why the sum of two negative
integers is a negative integer.   To add two negative
integers, add their absolute values and attach the negative
sign.

106. *Writing*   In your own words, write the rule for
adding two integers of opposite signs. How do you
determine the sign of the sum?   Subtract the smaller
absolute value from the larger absolute value and attach
the sign of the integer with the larger absolute value.

107. Write an expression that illustrates 8 subtracted
from 5. $5 - 8$

108. Write an expression that illustrates -6 subtracted
from -4. $-4 - (-6)$

109. Write an expression using addition that can be used
to subtract 12 from 9. $9 + (-12)$

110. Write a simplified expression that can be used to
evaluate $-9 - (-15)$. $-9 + 15$

1.3 Multiplying and Dividing Integers

Joe Sohm/The Image Works

What You Should Learn

1. Multiply integers with like signs and with unlike signs.
2. Divide integers with like signs and with unlike signs.
3. Find factors and prime factors of an integer.
4. Represent the definitions and rules of arithmetic symbolically.

Why You Should Learn It

You can multiply integers to solve real-life problems. For instance, in Exercise 107 on page 31, you will multiply integers to find the area of a football field.

1. **Multiply integers with like signs and with unlike signs.**

Multiplying Integers

Multiplication of two integers can be described as repeated addition or subtraction. The result of multiplying one number by another is called a **product.** Here are three examples.

Multiplication	*Repeated Addition or Subtraction*
$3 \times 5 = 15$	$\underbrace{5 + 5 + 5}_{} = 15$
	Add 5 three times.
$4 \times (-2) = -8$	$\underbrace{(-2) + (-2) + (-2) + (-2)}_{} = -8$
	Add -2 four times.
$(-3) \times (-4) = 12$	$\underbrace{-(-4) - (-4) - (-4)}_{} = 12$
	Subtract -4 three times.

Multiplication is denoted in a variety of ways. For instance,

$$7 \times 3, \quad 7 \cdot 3, \quad 7(3), \quad (7)3, \quad \text{and} \quad (7)(3)$$

all denote the product of "7 times 3," which is 21.

Rules for Multiplying Integers

1. The product of an integer and zero is 0.
2. The product of two integers with *like* signs is *positive.*
3. The product of two integers with *different* signs is *negative.*

As you move through this section, be sure your students understand the relationship between multiplication and division. This will help them as they learn to solve equations.

To find the product of more than two numbers, first find the product of their absolute values. If there is an *even* number of negative factors, then the product is positive. If there is an *odd* number of negative factors, then the product is negative. For instance,

$$5(-3)(-4)(7) = 420. \qquad \text{Even number of negative factors}$$

Example 1   Multiplying Integers

a. $4(10) = 40$ (Positive) · (positive) = positive

b. $-6 \cdot 9 = -54$ (Negative) · (positive) = negative

c. $(-5)(-7) = 35$ (Negative) · (negative) = positive

d. $3(-12) = -36$ (Positive) · (negative) = negative

e. $-12 \cdot 0 = 0$ (Negative) · (zero) = zero

f. $(-2)(8)(-3)(-1) = -(2 \cdot 8 \cdot 3 \cdot 1)$ Odd number of negative factors

$$= -48$$ Answer is negative.

Be careful to distinguish properly between expressions such as $3(-5)$ and $3 - 5$ or $-3(-5)$ and $-3 - 5$. The first of each pair is a *multiplication* problem, whereas the second is a *subtraction* problem.

Multiplication	*Subtraction*
$3(-5) = -15$	$3 - 5 = -2$
$-3(-5) = 15$	$-3 - 5 = -8$

To multiply two integers having two or more digits, we suggest the **vertical multiplication algorithm** demonstrated in Figure 1.27. The sign of the product is determined by the usual multiplication rule.

$$
\begin{array}{r}
47 \\
\times \quad 23 \\
\hline
141 \\
94 \\
\hline
1081
\end{array}
$$

141 ⟸ Multiply 3 times 47.

94 ⟸ Multiply 2 times 47.

1081 ⟸ Add columns.

Figure 1.27   Vertical Multiplication Algorithm

Example 2   Geometry: Volume of a Box

Find the volume of the rectangular box shown in Figure 1.28.

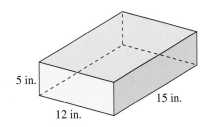

5 in.

15 in.

12 in.

Figure 1.28

Solution

To find the volume, multiply the length, width, and height of the box.

$$\text{Volume} = (\text{Length}) \cdot (\text{Width}) \cdot (\text{Height})$$

$$= (15 \text{ inches}) \cdot (12 \text{ inches}) \cdot (5 \text{ inches})$$

$$= 900 \text{ cubic inches}$$

So, the box has a volume of 900 cubic inches.

② Divide integers with like signs and with unlike signs.

Dividing Integers

Just as subtraction can be expressed in terms of addition, you can express division in terms of multiplication. Here are some examples.

Division		Related Multiplication
$15 \div 3 = 5$	because	$15 = 5 \cdot 3$
$-15 \div 3 = -5$	because	$-15 = -5 \cdot 3$
$15 \div (-3) = -5$	because	$15 = (-5) \cdot (-3)$
$-15 \div (-3) = 5$	because	$-15 = 5 \cdot (-3)$

The result of dividing one integer by another is called the **quotient** of the integers. **Division** is denoted by the symbol $\div$, or by $/$, or by a horizontal line. For example,

$$30 \div 6, \quad 30/6, \quad \text{and} \quad \frac{30}{6}$$

all denote the quotient of 30 and 6, which is 5. Using the form $30 \div 6$, 30 is called the **dividend** and 6 is the **divisor.** In the forms $30/6$ and $\frac{30}{6}$, 30 is the **numerator** and 6 is the **denominator.**

It is important to know how to use 0 in a division problem. Zero divided by a nonzero integer is always 0. For instance,

$$\frac{0}{13} = 0 \quad \text{because} \quad 0 = 0 \cdot 13.$$

On the other hand, division by zero, $13 \div 0$, is *undefined.*

Because division can be described in terms of multiplication, the rules for dividing two integers with like or unlike signs are the same as those for multiplying such integers.

Technology: Discovery

Does $\frac{1}{0} = 0$? Does $\frac{2}{0} = 0$? Write each division above in terms of multiplication. What does this tell you about division by zero? What does your calculator display when you perform the division?

See Technology Answers.

You may want to emphasize the important distinction between division *of* zero by a nonzero number and division *by* zero. For example,

$\frac{0}{4}$ and $\frac{0}{-23}$ are equal to 0.

$\frac{-1}{0}$ and $\frac{8}{0}$ are undefined.

Additional Examples
Find the product or quotient.

a. $5(-12)$

b. $(-6)(-11)$

c. $64 \div (-4)$

d. $81 \div 0$

Answers:

a. -60

b. 66

c. -16

d. Undefined

Rules for Dividing Integers

1. Zero divided by a nonzero integer is 0, whereas a nonzero integer divided by zero is *undefined.*

2. The quotient of two nonzero integers with *like* signs is *positive.*

3. The quotient of two nonzero integers with *different* signs is *negative.*

Example 3   Dividing Integers

a. $\frac{-42}{-6} = 7$ because $-42 = 7(-6)$.

b. $36 \div (-9) = -4$ because $(-4)(-9) = 36$.

c. $0 \div (-13) = 0$ because $(0)(-13) = 0$.

d. $-105 \div 7 = -15$ because $(-15)(7) = -105$.

e. $-97 \div 0$ is undefined.

$$
\begin{array}{r}
27 \\
13\overline{)351} \\
\underline{26} \\
91 \\
\underline{91}
\end{array}
$$

Figure 1.29 Long Division Algorithm

When dividing large numbers, the **long division algorithm** can be used. For instance, the long division algorithm shown in Figure 1.29 shows that

$$351 \div 13 = 27.$$

Remember that division can be checked by multiplying the answer by the divisor. So it is true that

$$351 \div 13 = 27 \quad \text{because} \quad 27(13) = 351.$$

All four operations on integers (addition, subtraction, multiplication, and division) are used in the following real-life example.

Example 4   Stock Purchase

On Monday you bought $500 worth of stock in a company. During the rest of the week, you recorded the gains and losses in your stock's value as shown in the table.

Tuesday	Wednesday	Thursday	Friday
Gained $15	Lost $18	Lost $23	Gained $10

a. What was the value of the stock at the close of Wednesday?

b. What was the value of the stock at the end of the week?

c. What would the total loss have been if Thursday's loss had occurred each of the four days?

d. What was the average daily gain (or loss) for the four days recorded?

Solution

a. The value at the close of Wednesday was

$$500 + 15 - 18 = \$497.$$

b. The value of the stock at the end of the week was

$$500 + 15 - 18 - 23 + 10 = \$484.$$

c. The loss on Thursday was $23. If this loss had occurred each day, the total loss would have been

$$4(23) = \$92.$$

d. To find the average daily gain (or loss), add the gains and losses of the four days and divide by 4. So, the average is

$$\text{Average} = \frac{15 + (-18) + (-23) + 10}{4} = \frac{-16}{4} = -4.$$

This means that during the four days, the stock had an average loss of $4 per day.

Factors and Prime Numbers

The set of positive integers

$$\{1, 2, 3, \ldots\}$$

is one subset of the real numbers that has intrigued mathematicians for many centuries.

Historically, an important number concept has been *factors* of positive integers. From experience, you know that in a multiplication problem such as $3 \cdot 7 = 21$, the numbers 3 and 7 are called *factors* of 21.

$$3 \cdot 7 = 21$$

Factors   Product

It is also correct to call the numbers 3 and 7 *divisors* of 21, because 3 and 7 each divide evenly into 21.

Definition of Factor (or Divisor)

If a and b are positive integers, then a is a **factor** (or **divisor**) of b if and only if there is a positive integer c such that $a \cdot c = b$.

The concept of factors allows you to classify positive integers into three groups: *prime* numbers, *composite* numbers, and the number 1.

Definitions of Prime and Composite Numbers

1. A positive integer greater than 1 with no factors other than itself and 1 is called a **prime number,** or simply a **prime.**

2. A positive integer greater than 1 with more than two factors is called a **composite number,** or simply a **composite.**

The numbers 2, 3, 5, 7, and 11 are primes because they have only themselves and 1 as factors. The numbers 4, 6, 8, 9, and 10 are composites because each has more than two factors. The number 1 is neither prime nor composite because 1 is its only factor.

Every composite number can be expressed as a *unique* product of prime factors. Here are some examples.

$$6 = 2 \cdot 3, \ 15 = 3 \cdot 5, \ 18 = 2 \cdot 3 \cdot 3, \ 42 = 2 \cdot 3 \cdot 7, \ 124 = 2 \cdot 2 \cdot 31$$

According to the definition of a prime number, is it possible for any negative number to be prime? Consider the number -2. Is it prime? Are its only factors one and itself? No, because

$$-2 = 1(-2),$$
$$-2 = (-1)(2),$$
or $-2 = (-1)(1)(2).$

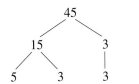

45

15   3

5   3   3

Figure 1.30   Tree Diagram

One strategy for factoring a composite number into prime numbers is to begin by finding the smallest prime number that is a factor of the composite number. Dividing this factor into the number yields a *companion* factor. For instance, 3 is the smallest prime number that is a factor of 45 and its companion factor is $15 = 45 \div 3$. Because 15 is also a composite number, continue hunting for factors and companion factors until each factor is prime. As shown in Figure 1.30, a *tree diagram* is a nice way to record your work. From the tree diagram, you can see that the prime factorization of 45 is

$$45 = 3 \cdot 3 \cdot 5.$$

Example 5   Prime Factorization

Write the prime factorization for each number.

a. 84   **b.** 78   **c.** 133   **d.** 43

Solution

a. 2 is a recognized divisor of 84. So, $84 = 2 \cdot 42 = 2 \cdot 2 \cdot 21 = 2 \cdot 2 \cdot 3 \cdot 7$.

b. 2 is a recognized divisor of 78. So, $78 = 2 \cdot 39 = 2 \cdot 3 \cdot 13$.

c. If you do not recognize a divisor of 133, you can start by dividing any of the prime numbers 2, 3, 5, 7, 11, 13, etc., into 133. You will find 7 to be the first prime to divide 133. So, $133 = 7 \cdot 19$ (19 is prime).

d. In this case, none of the primes less than 43 divides 43. So, 43 is prime.

Other aids to finding prime factors of a number n include the following divisibility tests.

Divisibility Tests

	Example
1. A number is divisible by 2 if it is *even*.	364 is divisible by 2 because it is even.
2. A number is divisible by 3 if the sum of its digits is divisible by 3.	261 is divisible by 3 because $2 + 6 + 1 = 9$.
3. A number is divisible by 9 if the sum of its digits is divisible by 9.	738 is divisible by 9 because $7 + 3 + 8 = 18$.
4. A number is divisible by 5 if its units digit is 0 or 5.	325 is divisible by 5 because its units digit is 5.
5. A number is divisible by 10 if its units digit is 0.	120 is divisible by 10 because its units digit is 0.

When a number is **divisible** by 2, it means that 2 divides into the number without leaving a remainder.

Summary of Definitions and Rules

So far in this chapter, rules and procedures have been described more with words than with symbols. For instance, subtraction is verbally defined as "adding the opposite of the number being subtracted." As you move to higher and higher levels of mathematics, it becomes more and more convenient to use symbols to describe rules and procedures. For instance, subtraction is symbolically defined as

$$a - b = a + (-b).$$

At its simplest level, algebra is a symbolic form of arithmetic. This arithmetic–algebra connection can be illustrated in the following way.

The transition from verbal and numeric descriptions to symbolic descriptions is an important step in a student's progression from arithmetic to algebra.

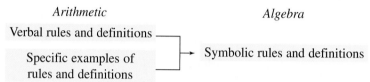

An illustration of this connection is shown in Example 6.

Example 6   Writing a Rule of Arithmetic in Symbolic Form

Write an example and an algebraic description of the arithmetic rule: *The product of two integers with unlike signs is negative.*

Solution

Example

For the integers -3 and 7,

$$(-3) \cdot 7 = 3 \cdot (-7)$$
$$= -(3 \cdot 7)$$
$$= -21.$$

Algebraic Description

If a and b are positive integers, then

$$(-a) \cdot b = a \cdot (-b) = -(a \cdot b).$$

Unlike signs   Unlike signs   Negative product

The list on the following page summarizes the algebraic versions of important definitions and rules of arithmetic. In each case a specific example is included for clarification.

Arithmetic Summary

Definitions: Let a, b, and c be integers.

Encourage students to read and study the definitions and rules on the left and compare them with the examples on the right. Most students need practice in "reading mathematics."

Definition

Example

1. Subtraction:

$$a - b = a + (-b)$$

$$5 - 7 = 5 + (-7)$$

2. Multiplication: (a is a positive integer)

$$a \cdot b = \underbrace{b + b + \cdots + b}_{a \text{ terms}}$$

$$3 \cdot 5 = 5 + 5 + 5$$

3. Division: ($b \neq 0$)

$$a \div b = c, \text{ if and only if } a = c \cdot b.$$

$$12 \div 4 = 3 \text{ because } 12 = 3 \cdot 4.$$

4. Less than:

$a < b$ if there is a positive real number c such that $a + c = b$.

$-2 < 1$ because $-2 + 3 = 1$.

5. Absolute value: $|a| = \begin{cases} a, & \text{if } a \geq 0 \\ -a, & \text{if } a < 0 \end{cases}$

$$|-3| = -(-3) = 3$$

6. Divisor:

a is a divisor of b if and only if there is an integer c such that $a \cdot c = b$.

7 is a divisor of 21 because $7 \cdot 3 = 21$.

Rules: Let a and b be integers.

Rule

Example

1. Addition:
   (a) To add two integers with *like* signs, add their absolute values and attach the common sign to the result.
   (b) To add two integers with *unlike* signs, subtract the smaller absolute value from the larger absolute value and attach the sign of the integer with the larger absolute value.

$$3 + 7 = |3| + |7| = 10$$

$$-5 + 8 = |8| - |-5|$$
$$= 8 - 5$$
$$= 3$$

2. Multiplication:
   (a) $a \cdot 0 = 0 = 0 \cdot a$
   (b) Like signs: $a \cdot b > 0$
   (c) Different signs: $a \cdot b < 0$

$$3 \cdot 0 = 0 = 0 \cdot 3$$
$$(-2)(-5) = 10$$
$$(2)(-5) = -10$$

3. Division:
   (a) $\dfrac{0}{a} = 0$

   (b) $\dfrac{a}{0}$ is undefined.

   (c) Like signs: $\dfrac{a}{b} > 0$

   (d) Different signs: $\dfrac{a}{b} < 0$

$$\frac{0}{4} = 0$$

$$\frac{6}{0} \text{ is undefined.}$$

$$\frac{-2}{-3} = \frac{2}{3}$$

$$\frac{-5}{7} = -\frac{5}{7}$$

Example 7   Using Definitions and Rules

a. Use the definition of subtraction to complete the statement.

$$4 - 9 = \boxed{}$$

b. Use the definition of multiplication to complete the statement.

$$6 + 6 + 6 + 6 = \boxed{}$$

c. Use the definition of absolute value to complete the statement.

$$|-9| = \boxed{}$$

d. Use the rule for adding integers with unlike signs to complete the statement.

$$-7 + 3 = \boxed{}$$

e. Use the rule for multiplying integers with unlike signs to complete the statement.

$$-9 \times 2 = \boxed{}$$

Solution

a. $4 - 9 = 4 + (-9) = -5$

b. $6 + 6 + 6 + 6 = 4 \cdot 6 = 24$

c. $|-9| = -(-9) = 9$

d. $-7 + 3 = -(|-7| - |3|) = -4$

e. $-9 \times 2 = -18$

Example 8   Finding a Pattern

Complete each pattern. Decide which rules the patterns demonstrate.

a. $3 \cdot (3) = 9$ $\qquad$ **b.** $-3 \cdot (3) = -9$
$\quad 3 \cdot (2) = 6$ $\qquad\qquad -3 \cdot (2) = -6$
$\quad 3 \cdot (1) = 3$ $\qquad\qquad -3 \cdot (1) = -3$
$\quad 3 \cdot (0) = 0$ $\qquad\qquad -3 \cdot (0) = 0$
$\quad 3 \cdot (-1) = \boxed{}$ $\qquad -3 \cdot (-1) = \boxed{}$
$\quad 3 \cdot (-2) = \boxed{}$ $\qquad -3 \cdot (-2) = \boxed{}$
$\quad 3 \cdot (-3) = \boxed{}$ $\qquad -3 \cdot (-3) = \boxed{}$

Solution

a. $3 \cdot (-1) = -3$ $\qquad$ **b.** $-3 \cdot (-1) = 3$
$\quad 3 \cdot (-2) = -6$ $\qquad\qquad -3 \cdot (-2) = 6$
$\quad 3 \cdot (-3) = -9$ $\qquad\qquad -3 \cdot (-3) = 9$

The product of integers with unlike signs is negative and the product of integers with like signs is positive.

1.3 Exercises

Developing Skills

In Exercises 1–4, write each multiplication as repeated addition or subtraction and find the product.

1. $3 \cdot 2$ $2 + 2 + 2 = 6$
2. 4×5 $5 + 5 + 5 + 5 = 20$
3. $5 \times (-3)$
   $(-3) + (-3) + (-3) + (-3) + (-3) = -15$
4. $(-6)(-2)$
   $-(-2) - (-2) - (-2) - (-2) - (-2) - (-2) = 12$

In Exercises 5–30, find the product. See Example 1.

5. 7×3 21
6. 6×4 24
7. $0 \cdot 2$ 0
8. $13 \cdot 0$ 0
9. $4(-8)$ -32
10. $10(-5)$ -50
11. $(310)(-3)$ -930
12. $(125)(-4)$ -500
13. $-7(5)$ -35
14. $-9(3)$ -27
15. $(-6)(-12)$ 72
16. $(-20)(-8)$ 160
17. $(-500)(-6)$ 3000
18. $(-350)(-4)$ 1400
19. $5(-3)(-6)$ 90
20. $6(-2)(-4)$ 48
21. $-7(3)(-1)$ 21
22. $-2(5)(-3)$ 30
23. $(-2)(-3)(-5)$ -30
24. $(-10)(-4)(-2)$ -80
25. $|(-3)4|$ 12
26. $|8(-9)|$ 72
27. $|3(-5)(6)|$ 90
28. $|8(-3)(5)|$ 120
29. $|6(20)(4)|$ 480
30. $|9(12)(2)|$ 216

In Exercises 31–40, use the vertical multiplication algorithm to find the product.

31. 26×13 338
32. 14×9 126
33. $(-14) \times 24$ -336
34. $(-8) \times 30$ -240
35. $75(-63)$ -4725
36. $(-72)(866)$ $-62,352$
37. $(-13)(-20)$ 260
38. $(-11)(-24)$ 264
39. $(-21)(-429)$ 9009
40. $(-14)(-585)$ 8190

In Exercises 41–60, perform the division, if possible. If not possible, state the reason. See Example 3.

41. $27 \div 9$ 3
42. $35 \div 7$ 5
43. $72 \div (-12)$ -6
44. $54 \div (-9)$ -6
45. $(-28) \div 4$ -7
46. $(-108) \div 9$ -12
47. $-35 \div (-5)$ 7
48. $(-24) \div (-4)$ 6
49. $\frac{8}{0}$ Division by zero is undefined.
50. $\frac{17}{0}$ Division by zero is undefined.
51. $\frac{0}{8}$ 0
52. $\frac{0}{17}$ 0
53. $\frac{-81}{-3}$ 27
54. $\frac{-125}{-25}$ 5
55. $\frac{6}{-1}$ -6
56. $\frac{-33}{1}$ -33
57. $\frac{-28}{4}$ -7

58. $\dfrac{72}{-12}$ -6

59. $(-27) \div (-27)$ 1

60. $(-83) \div (-83)$ 1

In Exercises 61–70, use the long division algorithm to find the quotient.

61. $1440 \div 45$ 32

62. $936 \div 52$ 18

63. $1440 \div (-45)$ -32

64. $936 \div (-52)$ -18

65. $-1312 \div 16$ -82

66. $-5152 \div 23$ -224

67. $2750 \div 25$ 110

68. $22{,}010 \div 71$ 310

69. $-9268 \div (-28)$ 331

70. $-6804 \div (-36)$ 189

In Exercises 71–74, use a calculator to perform the specified operation(s).

71. $\dfrac{44{,}290}{515}$ 86

72. $\dfrac{33{,}511}{47}$ 713

73. $\dfrac{169{,}290}{162}$ 1045

74. $\dfrac{1{,}027{,}500}{250}$ 4110

Mental Math In Exercises 75–78, find the product mentally. Explain your strategy.

75. $72(8)(25)$ $14{,}400$

76. $64(5)(20)$ 6400

77. $(-2)(532)(500)$ $-532{,}000$

78. $(-4)(262)(50)$ $-52{,}400$

In Exercises 79–88, decide whether the number is prime or composite.

79. 240 Composite

80. 533 Composite

81. 643 Prime

82. 257 Prime

83. 3911 Prime

84. 1321 Prime

85. 1281 Composite

86. 1323 Composite

87. 3555 Composite

88. 8324 Composite

In Exercises 89–98, write the prime factorization of the number. See Example 5.

89. 12 $2 \cdot 2 \cdot 3$

90. 52 $2 \cdot 2 \cdot 13$

91. 561 $3 \cdot 11 \cdot 17$

92. 245 $5 \cdot 7 \cdot 7$

93. 210 $2 \cdot 3 \cdot 5 \cdot 7$

94. 525 $3 \cdot 5 \cdot 5 \cdot 7$

95. 2535 $3 \cdot 5 \cdot 13 \cdot 13$

96. 1521 $3 \cdot 3 \cdot 13 \cdot 13$

97. 192 $2 \cdot 2 \cdot 2 \cdot 2 \cdot 2 \cdot 2 \cdot 3$

98. 264 $2 \cdot 2 \cdot 2 \cdot 3 \cdot 11$

In Exercises 99–102, complete the statement using the indicated definition or rule. See Example 7.

99. Definition of division: $12 \div 4 = \boxed{3}$

100. Definition of absolute value: $|8| = \boxed{8}$

101. Rule for multiplying integers by 0:

$6 \cdot 0 = \boxed{0} = 0 \cdot 6$

102. Rule for dividing integers with unlike signs:

$\dfrac{30}{-10} = \boxed{-3}$

Solving Problems

103. *Temperature Change*   The temperature measured by a weather balloon is decreasing approximately 3° for each 1000-foot increase in altitude. The balloon rises 8000 feet. What is the total temperature change? $-24°$

104. *Stock Price*   The Dow Jones average loses 11 points on each of four consecutive days. What is the cumulative loss during the four days?   44 points

105. *Savings Plan*   After you save $50 per month for 10 years, what is the total amount you have saved? $6000

106. *Loss Leaders*   To attract customers, a grocery store runs a sale on bananas. The bananas are *loss leaders,* which means the store loses money on the bananas but hopes to make it up on other items. The store sells 800 pounds at a loss of 26 cents per pound. What is the total loss? $208

107. ▲ *Geometry*   Find the area of the football field.

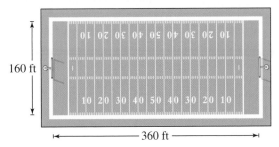

160 ft

360 ft

57,600 square feet

108. ▲ *Geometry*   Find the area of the garden.

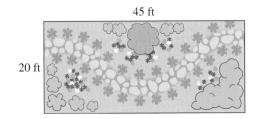

45 ft

20 ft

900 square feet

109. *Average Speed*   A commuter train travels a distance of 195 miles between two cities in 3 hours. What is the average speed of the train in miles per hour?   65 miles per hour

110. *Average Speed*   A jogger runs a race that is 6 miles long in 54 minutes. What is the average speed of the jogger in minutes per mile?   9 minutes per mile

111. *Exam Scores*   A student has a total of 328 points after four 100-point exams.

(a) What is the average number of points scored per exam?   82

(b) The scores on the four exams are 87, 73, 77, and 91. Plot each of the scores and the average score on the real number line.   See Additional Answers.

(c) Find the difference between each score and the average score. Find the sum of these distances and give a possible explanation of the result.
5, -9, -5, and 9; Sum is 0; Explanations will vary.

112. *Sports*   A football team gains a total of 20 yards after four downs.

(a) What is the average number of yards gained per down?   5 yards

(b) The gains on the four downs are 8 yards, 4 yards, 2 yards, and 6 yards. Plot each of the gains and the average gain on the real number line.
See Additional Answers.

(c) Find the difference between each gain and the average gain. Find the sum of these distances and give a possible explanation of the result.
3 yards, -1 yard, -3 yards, and 1 yard; Sum is 0; Explanations will vary.

▲ *Geometry*   In Exercises 113 and 114, find the volume of the rectangular solid. The volume is found by multiplying the length, width, and height of the solid. See Example 2.

113.

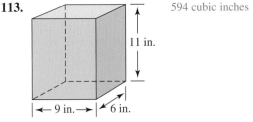

594 cubic inches

11 in.

9 in.   6 in.

114.

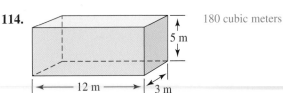

180 cubic meters

5 m

12 m   3 m

Explaining Concepts

115. *Writing* What is the only even prime number? Explain why there are no other even prime numbers. 2; it is divisible only by 1 and itself. Any other even number is divisible by 1, itself, and 2.

116. *Investigation* Twin primes are prime numbers that differ by 2. For instance, 3 and 5 are twin primes. How many other twin primes are less than 100? There are seven other twin primes. They are: 5, 7; 11, 13; 17, 19; 29, 31; 41, 43; 59, 61; 71, 73.

117. *Writing* The number 1997 is not divisible by a prime number that is less than 45. Explain why this implies that 1997 is a prime number. $\sqrt{1997} < 45$

118. *Think About It* If a negative number is used as a factor 25 times, what is the sign of the product? Negative

119. *Think About It* If a negative number is used as a factor 16 times, what is the sign of the product? Positive

120. *Writing* Write a verbal description of what is meant by $3(-5)$. The sum of three terms each equal to -5

121. *Writing* In your own words, write the rules for determining the sign of the product or quotient of real numbers. The product (or quotient) of two nonzero real numbers of like signs is positive. The product (or quotient) of two nonzero real numbers of unlike signs is negative.

122. *The Sieve of Eratosthenes* Write the integers from 1 through 100 in 10 lines of 10 numbers each.

(a) Cross out the number 1. Cross out all multiples of 2 other than 2 itself. Do the same for 3, 5, and 7.

(b) Of what type are the remaining numbers? Explain why this is the only type of number left. Prime; Explanations will vary.

123. *Writing* Explain why the product of an even integer and any other integer is even. What can you conclude about the product of two odd integers? $(2m)n = 2(mn)$. The product of two odd integers is odd.

124. *Writing* Explain how to check the result of a division problem. Multiply the divisor and quotient to obtain the dividend.

125. *Think About It* An integer n is divided by 2 and the quotient is an even integer. What does this tell you about n? Give an example. n is a multiple of 4. $\frac{12}{2} = 6$

126. Which of the following is (are) undefined: $\frac{1}{1}, \frac{0}{1}, \frac{1}{0}? \quad \frac{1}{0}$

127. *Investigation* The **proper factors** of a number are all its factors less than the number itself. A number is **perfect** if the sum of its proper factors is equal to the number. A number is **abundant** if the sum of its proper factors is greater than the number. Which numbers less than 25 are perfect? Which are abundant? Try to find the first perfect number greater than 25.
Perfect (< 25): 6; Abundant (< 25): 12, 18, 20, 24; First perfect greater than 25 is 28.

122. (a)
```
 1   2   3   4   5   6   7   8   9  10
11  12  13  14  15  16  17  18  19  20
21  22  23  24  25  26  27  28  29  30
31  32  33  34  35  36  37  38  39  40
41  42  43  44  45  46  47  48  49  50
51  52  53  54  55  56  57  58  59  60
61  62  63  64  65  66  67  68  69  70
71  72  73  74  75  76  77  78  79  80
81  82  83  84  85  86  87  88  89  90
91  92  93  94  95  96  97  98  99 100
```

The symbol indicates an exercise that can be used as a group discussion problem.

Mid-Chapter Quiz

Take this quiz as you would take a quiz in class. After you are done, check your work against the answers in the back of the book.

In Exercises 1–4, plot each real number as a point on the real line and place the correct inequality symbol (< or >) between the real numbers.
See Additional Answers.

1. $\frac{3}{16}$ < $\frac{3}{8}$

2. -2.5 > -4

3. -7 < 3

4. 2π > 6

In Exercises 5 and 6, evaluate the expression.

5. $-|-0.75|$ -0.75

6. $\left|-\frac{17}{19}\right|$ $\frac{17}{19}$

In Exercises 7 and 8, place the correct symbol (<, >, or =) between the real numbers.

7. $\left|\frac{7}{2}\right|$ = $|-3.5|$

8. $\left|\frac{3}{4}\right|$ > $-|0.75|$

9. Subtract -13 from -22. $-22 - (-13) = -9$

10. Find the absolute value of the sum of -54 and 26. $|-54 + 26| = 28$

In Exercises 11–22, evaluate the expression.

11. $34 + 65$ 99

12. $-24 + (-51)$ -75

13. $-15 - 12$ -27

14. $-35 - (-10)$ -25

15. $25 + (-75)$ -50

16. $72 - 134$ -62

17. $12 + (-6) - 8 + 10$ 8

18. $-9 - 17 + 36 + (-15)$ -5

19. $-6(10)$ -60

20. $(-7)(-13)$ 91

21. $\frac{-45}{-3}$ 15

22. $\frac{-24}{6}$ -4

23. Write the prime factorization of 144. $2 \cdot 2 \cdot 2 \cdot 2 \cdot 3 \cdot 3$

24. An electronics manufacturer's quarterly profits are shown in the bar graph at the left. What is the manufacturer's total profit for the year? $450,450

25. A cord of wood is a pile 8 feet long, 4 feet wide, and 4 feet high. The volume of a rectangular solid is its length times its width times its height. Find the number of cubic feet in a cord of wood.   128 cubic feet

26. It is necessary to cut a 90-foot rope into six pieces of equal length. What is the length of each piece?   15 feet

27. At the beginning of a month your account balance was $738. During the month, you withdrew $550, deposited $189, and payed a fee of $10. What was your balance at the end of the month? $367

28. When you left for class in the morning, the temperature was 60°F. By the time class ended, the temperature had increased by 15°. While you studied, the temperature increased by 2°. During your work study, the temperature decreased by 12°. What was the temperature after your work study?   65°

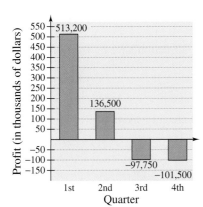

Figure for 24

1.4 Operations with Rational Numbers

Lon C. Diehl/PhotoEdit

What You Should Learn

① Rewrite fractions as equivalent fractions.

② Add and subtract fractions.

③ Multiply and divide fractions.

④ Add, subtract, multiply, and divide decimals.

Why You Should Learn It

Rational numbers are used to represent many real-life quantities. For instance, in Exercise 149 on page 46, you will use rational numbers to find the increase in the Dow Jones Industrial Average.

① Rewrite fractions as equivalent fractions.

Rewriting Fractions

A **fraction** is a number that is written as a quotient, with a *numerator* and a *denominator*. The terms *fraction* and *rational number* are related, but are not exactly the same. The term *fraction* refers to a number's form, whereas the term *rational number* refers to its classification. For instance, the number 2 is a fraction when it is written as $\frac{2}{1}$, but it is a rational number regardless of how it is written.

Rules of Signs for Fractions

1. If the numerator and denominator of a fraction have like signs, the value of the fraction is positive.

2. If the numerator and denominator of a fraction have unlike signs, the value of the fraction is negative.

All of the following fractions are positive and are equivalent to $\frac{2}{3}$.

$$\frac{2}{3}, \frac{-2}{-3}, -\frac{-2}{3}, -\frac{2}{-3}$$

All of the following fractions are negative and are equivalent to $-\frac{2}{3}$.

$$-\frac{2}{3}, \frac{-2}{3}, \frac{2}{-3}, -\frac{-2}{-3}$$

In both arithmetic and algebra, it is often beneficial to write a fraction in **simplest form** or reduced form, which means that the numerator and denominator have no common factors (other than 1). By finding the prime factors of the numerator and the denominator, you can determine what common factor(s) to divide out.

Writing a Fraction in Simplest Form

To write a fraction in simplest form, divide both the numerator and denominator by their greatest common factor (GCF).

Study Tip

To find the **greatest common factor** (or **GCF**) of two natural numbers, write the prime factorization of each number. The greatest common factor is the product of the common factors. For instance, from the prime factorizations

$$18 = 2 \cdot 3 \cdot 3$$

and

$$42 = 2 \cdot 3 \cdot 7$$

you can see that the common factors of 18 and 42 are 2 and 3. So, it follows that the greatest common factor is $2 \cdot 3$ or 6.

Example 1   Writing Fractions in Simplest Form

Write each fraction in simplest form.

a. $\dfrac{18}{24}$ **b.** $\dfrac{35}{21}$ **c.** $\dfrac{24}{72}$

Solution

a. $\dfrac{18}{24} = \dfrac{2 \cdot \overset{1}{\cancel{3}} \cdot 3}{2 \cdot 2 \cdot 2 \cdot \underset{1}{\cancel{3}}} = \dfrac{3}{4}$ Divide out GCF of 6.

b. $\dfrac{35}{21} = \dfrac{5 \cdot \overset{1}{\cancel{7}}}{3 \cdot \underset{1}{\cancel{7}}} = \dfrac{5}{3}$ Divide out GCF of 7.

c. $\dfrac{24}{72} = \dfrac{\overset{1}{\cancel{2}} \cdot \overset{1}{\cancel{2}} \cdot \overset{1}{\cancel{2}} \cdot \overset{1}{\cancel{3}}}{\underset{1}{\cancel{2}} \cdot \underset{1}{\cancel{2}} \cdot \underset{1}{\cancel{2}} \cdot \underset{1}{\cancel{3}} \cdot 3} = \dfrac{1}{3}$ Divide out GCF of 24.

You can obtain an **equivalent fraction** by multiplying the numerator and denominator by the same nonzero number or by dividing the numerator and denominator by the same nonzero number. Here are some examples.

Fraction	Equivalent Fraction	Operation
$\dfrac{9}{12} = \dfrac{\overset{1}{\cancel{3}} \cdot 3}{\underset{1}{\cancel{3}} \cdot 4}$	$\dfrac{3}{4}$	Divide numerator and denominator by 3. (See Figure 1.31.)
$\dfrac{6}{5} = \dfrac{6 \cdot 2}{5 \cdot 2}$	$\dfrac{12}{10}$	Multiply numerator and denominator by 2.
$\dfrac{-8}{12} = -\dfrac{\overset{1}{\cancel{2}} \cdot \overset{1}{\cancel{2}} \cdot 2}{\underset{1}{\cancel{2}} \cdot \underset{1}{\cancel{2}} \cdot 3}$	$-\dfrac{2}{3}$	Divide numerator and denominator by GCF of 4.

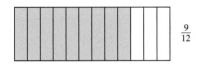

Figure 1.31   Equivalent Fractions

Example 2   Writing Equivalent Fractions

Write an equivalent fraction with the indicated denominator.

a. $\dfrac{2}{3} = \dfrac{}{15}$ **b.** $\dfrac{4}{7} = \dfrac{}{42}$ **c.** $\dfrac{9}{15} = \dfrac{}{35}$

Solution

a. $\dfrac{2}{3} = \dfrac{2 \cdot 5}{3 \cdot 5} = \dfrac{10}{15}$ Multiply numerator and denominator by 5.

b. $\dfrac{4}{7} = \dfrac{4 \cdot 6}{7 \cdot 6} = \dfrac{24}{42}$ Multiply numerator and denominator by 6.

c. $\dfrac{9}{15} = \dfrac{\cancel{3} \cdot 3}{\cancel{3} \cdot 5} = \dfrac{3 \cdot 7}{5 \cdot 7} = \dfrac{21}{35}$ Reduce first, then multiply numerator and denominator by 7.

2 Add and subtract fractions.

Adding and Subtracting Fractions

To add fractions with *like* denominators such as $\frac{3}{12}$ and $\frac{4}{12}$, add the numerators and write the sum over the like denominator.

$$\frac{3}{12} + \frac{4}{12} = \frac{3+4}{12}$$

$$= \frac{7}{12} \qquad \text{Add the numbers in the numerator.}$$

To add fractions with *unlike* denominators such as $\frac{1}{4}$ and $\frac{1}{3}$, rewrite the fractions as equivalent fractions with a common denominator.

$$\frac{1}{4} + \frac{1}{3} = \frac{1 \cdot 3}{4 \cdot 3} + \frac{1 \cdot 4}{3 \cdot 4} \qquad \text{Rewrite fractions in equivalent form.}$$

$$= \frac{3}{12} + \frac{4}{12} \qquad \text{Rewrite with like denominators.}$$

$$= \frac{7}{12} \qquad \text{Add numerators.}$$

To find a common denominator for two or more fractions, find the **least common multiple** (LCM) of their denominators. For instance, for the fractions $\frac{3}{8}$ and $-\frac{5}{12}$, the least common multiple of their denominators, 8 and 12, is 24. To see this, consider all multiples of 8 (8, 16, 24, 32, 40, 48, ...) and all multiples of 12 (12, 24, 36, 48, ...). The numbers 24 and 48 are common multiples, and the number 24 is the smallest of the common multiples.

$$\frac{3}{8} + \frac{-5}{12} = \frac{3(3)}{8(3)} + \frac{(-5)(2)}{12(2)} \qquad \text{LCM of 8 and 12 is 24.}$$

$$= \frac{9}{24} + \frac{-10}{24} \qquad \text{Rewrite with like denominators.}$$

$$= \frac{9-10}{24} \qquad \text{Add numerators.}$$

$$= \frac{-1}{24} \qquad \text{Simplify.}$$

$$= -\frac{1}{24}$$

Study Tip

Adding fractions with unlike denominators is an example of a basic problem-solving strategy that is used in mathematics—rewriting a given problem in a simpler or more familiar form.

Addition and Subtraction of Fractions

Let a, b, and c be integers with $c \neq 0$.

1. *With like denominators:*

$$\frac{a}{c} + \frac{b}{c} = \frac{a+b}{c} \quad \text{or} \quad \frac{a}{c} - \frac{b}{c} = \frac{a-b}{c}$$

2. *With unlike denominators:* rewrite both fractions so that they have like denominators. Then use the rule for adding and subtracting fractions with like denominators.

Study Tip

In Example 3, a common short-cut for writing $1\frac{4}{5}$ as $\frac{9}{5}$ is to multiply 1 by 5, add the result to 4, and then divide by 5, as follows.

$$1\frac{4}{5} = \frac{1(5) + 4}{5} = \frac{9}{5}$$

Additional Examples
Find the sum or difference.

a. $\frac{3}{5} + \frac{6}{7}$

b. $-2\frac{1}{3} - \frac{4}{9}$

Answers:

a. $\frac{51}{35}$

b. $-\frac{25}{9}$

Example 3   Adding Fractions

Add: $1\frac{4}{5} + \frac{11}{15}$.

Solution

To begin, rewrite the **mixed number** $1\frac{4}{5}$ as a fraction.

$$1\frac{4}{5} = 1 + \frac{4}{5} = \frac{5}{5} + \frac{4}{5} = \frac{9}{5}$$

Then add the two fractions as follows.

$$1\frac{4}{5} + \frac{11}{15} = \frac{9}{5} + \frac{11}{15} \qquad \text{Rewrite } 1\frac{4}{5} \text{ as } \frac{9}{5}.$$

$$= \frac{9(3)}{5(3)} + \frac{11}{15} \qquad \text{LCM of 5 and 15 is 15.}$$

$$= \frac{27}{15} + \frac{11}{15} \qquad \text{Rewrite with like denominators.}$$

$$= \frac{38}{15} \qquad \text{Add numerators.}$$

Example 4   Subtracting Fractions

Subtract: $\frac{7}{9} - \frac{11}{12}$.

Solution

$$\frac{7}{9} - \frac{11}{12} = \frac{7(4)}{9(4)} + \frac{-11(3)}{12(3)} \qquad \text{LCM of 9 and 12 is 36.}$$

$$= \frac{28}{36} + \frac{-33}{36} \qquad \text{Rewrite with like denominators.}$$

$$= \frac{-5}{36} \qquad \text{Add numerators.}$$

$$= -\frac{5}{36}$$

You can add or subtract *two* fractions, without first finding a common denominator, by using the following rule.

Alternative Rule for Adding and Subtracting Two Fractions

If a, b, c, and d are integers with $b \neq 0$ and $d \neq 0$, then

$$\frac{a}{b} + \frac{c}{d} = \frac{ad + bc}{bd} \quad \text{or} \quad \frac{a}{b} - \frac{c}{d} = \frac{ad - bc}{bd}.$$

On page 36, the sum of $\frac{3}{8}$ and $-\frac{5}{12}$ was found using the least common multiple of 8 and 12. Compare those solution steps with the following steps, which use the alternative rule for adding or subtracting two fractions.

$$\frac{3}{8} + \frac{-5}{12} = \frac{3(12) + 8(-5)}{8(12)}$$ Apply alternative rule.

$$= \frac{36 - 40}{96}$$ Simplify.

$$= \frac{-4}{96}$$ Simplify.

$$= -\frac{1}{24}$$ Write in simplest form.

Technology: Tip

When you use a scientific or graphing calculator to add or subtract fractions, your answer may appear in decimal form. An answer such as 0.583333333 is not as exact as $\frac{7}{12}$ and may introduce roundoff error. Refer to the user's manual for your calculator for instructions on adding and subtracting fractions and displaying answers in fraction form.

Example 5   Subtracting Fractions

$$\frac{5}{16} - \left(-\frac{7}{30}\right) = \frac{5}{16} + \frac{7}{30}$$ Add the opposite.

$$= \frac{5(30) + 16(7)}{16(30)}$$ Apply alternative rule.

$$= \frac{150 + 112}{480}$$ Simplfy.

$$= \frac{262}{480}$$ Simplify.

$$= \frac{131}{240}$$ Write in simplest form.

Example 6   Combining Three or More Fractions

Evaluate $\frac{5}{6} - \frac{7}{15} + \frac{3}{10} - 1$.

Solution

The least common denominator of 6, 15, and 10 is 30. So, you can rewrite the original expression as follows.

$$\frac{5}{6} - \frac{7}{15} + \frac{3}{10} - 1 = \frac{5(5)}{6(5)} + \frac{(-7)(2)}{15(2)} + \frac{3(3)}{10(3)} + \frac{(-1)(30)}{30}$$

$$= \frac{25}{30} + \frac{-14}{30} + \frac{9}{30} + \frac{-30}{30}$$ Rewrite with like denominators.

$$= \frac{25 - 14 + 9 - 30}{30}$$ Add numerators.

$$= \frac{-10}{30} = -\frac{1}{3}$$ Simplify.

③ Multiply and divide fractions.

Multiplying and Dividing Fractions

The procedure for multiplying fractions is simpler than those for adding and subtracting fractions. Regardless of whether the fractions have like or unlike denominators, you can find the product of two fractions by multiplying the numerators and multiplying the denominators.

Multiplication of Fractions

Let a, b, c, and d be integers with $b \neq 0$ and $d \neq 0$. Then the product of $\frac{a}{b}$ and $\frac{c}{d}$ is

$$\frac{a}{b} \cdot \frac{c}{d} = \frac{a \cdot c}{b \cdot d}.$$ Multiply numerators and denominators.

Emphasize that only common *factors* can be divided out as a fraction is simplified. For example, discuss

$$\frac{(3)(2)}{(3)(7)} \quad \text{and} \quad \frac{3 + 2}{(3)(7)}$$

and explain why the first fraction can be reduced but not the second fraction.

Example 7   Multiplying Fractions

a. $\dfrac{5}{8} \cdot \dfrac{3}{2} = \dfrac{5(3)}{8(2)}$ Multiply numerators and denominators.

$= \dfrac{15}{16}$ Simplify.

b. $\left(-\dfrac{7}{9}\right)\left(-\dfrac{5}{21}\right) = \dfrac{7}{9} \cdot \dfrac{5}{21}$ Product of two negatives is positive.

$= \dfrac{7(5)}{9(21)}$ Multiply numerators and denominators.

$= \dfrac{7(5)}{9(3)(7)}$ Divide out common factors.

$= \dfrac{5}{27}$ Write in simplest form.

Example 8   Multiplying Three Fractions

$\left(3\dfrac{1}{5}\right)\left(-\dfrac{7}{6}\right)\left(\dfrac{5}{3}\right) = \left(\dfrac{16}{5}\right)\left(-\dfrac{7}{6}\right)\left(\dfrac{5}{3}\right)$ Rewrite mixed number as a fraction.

$= \dfrac{16(-7)(5)}{5(6)(3)}$ Multiply numerators and denominators.

$= -\dfrac{(8)(2)(7)(5)}{(5)(3)(2)(3)}$ Divide out common factors.

$= -\dfrac{56}{9}$ Write in simplest form.

The **reciprocal** or **multiplicative inverse** of a number is the number by which it must be multiplied to obtain 1. For instance, the reciprocal of 3 is $\frac{1}{3}$ because $3\left(\frac{1}{3}\right) = 1$. Similarly, the reciprocal of $-\frac{2}{3}$ is $-\frac{3}{2}$ because

$$\left(-\frac{2}{3}\right)\left(-\frac{3}{2}\right) = 1.$$

To divide two fractions, multiply the first fraction by the *reciprocal* of the second fraction. Another way of saying this is "invert the divisor and multiply."

Division of Fractions

Let a, b, c, and d be integers with $b \neq 0$, $c \neq 0$, and $d \neq 0$. Then the quotient of $\frac{a}{b}$ and $\frac{c}{d}$ is

$$\frac{a}{b} \div \frac{c}{d} = \frac{a}{b} \cdot \frac{d}{c}. \qquad \text{Invert divisor and multiply.}$$

You might ask students to write some original exercises involving operations with fractions. Have the students do the operations with pencil and paper and then verify the results on their calculators. This exercise provides excellent practice. (Remind students that the calculator may introduce roundoff error.)

Example 9   Dividing Fractions

a. $\dfrac{5}{8} \div \dfrac{20}{12}$ **b.** $\dfrac{6}{13} \div \left(-\dfrac{9}{26}\right)$ **c.** $-\dfrac{1}{4} \div (-3)$

Solution

a. $\dfrac{5}{8} \div \dfrac{20}{12} = \dfrac{5}{8} \cdot \dfrac{12}{20}$ Invert divisor and multiply.

$\qquad\qquad = \dfrac{(5)(12)}{(8)(20)}$ Multiply numerators and denominators.

$\qquad\qquad = \dfrac{(5)(3)(4)}{(8)(4)(5)}$ Divide out common factors.

$\qquad\qquad = \dfrac{3}{8}$ Write in simplest form.

b. $\dfrac{6}{13} \div \left(-\dfrac{9}{26}\right) = \dfrac{6}{13} \cdot \left(-\dfrac{26}{9}\right)$ Invert divisor and multiply.

$\qquad\qquad = -\dfrac{(6)(26)}{(13)(9)}$ Multiply numerators and denominators.

$\qquad\qquad = -\dfrac{(2)(3)(2)(13)}{(13)(3)(3)}$ Divide out common factors.

$\qquad\qquad = -\dfrac{4}{3}$ Write in simplest form.

c. $-\dfrac{1}{4} \div (-3) = -\dfrac{1}{4} \cdot \left(-\dfrac{1}{3}\right)$ Invert divisor and multiply.

$\qquad\qquad = \dfrac{(-1)(-1)}{(4)(3)}$ Multiply numerators and denominators.

$\qquad\qquad = \dfrac{1}{12}$ Write in simplest form.

Additional Examples
Find the product or quotient.

a. $\left(\frac{6}{5}\right)\left(\frac{2}{8}\right)$

b. $\frac{3}{4} \div \frac{4}{9}$

Answers:

a. $\frac{3}{10}$

b. $\frac{27}{16}$

④ Add, subtract, multiply, and divide decimals.

Operations with Decimals

Rational numbers can be represented as **terminating** or **repeating decimals.** Here are some examples.

Terminating Decimals	Repeating Decimals
$\frac{1}{4} = 0.25$	$\frac{1}{6} = 0.1666\ldots$ or $0.1\overline{6}$
$\frac{3}{8} = 0.375$	$\frac{1}{3} = 0.3333\ldots$ or $0.\overline{3}$
$\frac{2}{10} = 0.2$	$\frac{1}{12} = 0.0833\ldots$ or $0.08\overline{3}$
$\frac{5}{16} = 0.3125$	$\frac{8}{33} = 0.2424\ldots$ or $0.\overline{24}$

Note that the bar notation is used to indicate the *repeated* digit (or digits) in the decimal notation. You can obtain the decimal representation of any fraction by long division. For instance, the decimal representation of $\frac{5}{12}$ is $0.41\overline{6}$, as can be seen from the following long division algorithm.

$$
\begin{array}{r}
0.4166\ \ldots = 0.41\overline{6} \\
12\overline{)5.0000} \\
\underline{4\,8} \\
20 \\
\underline{12} \\
80 \\
\underline{72} \\
80
\end{array}
$$

For calculations involving decimals such as $0.41666\ldots$, you must **round the decimal.** For instance, rounded to two decimal places, the number $0.41666\ldots$ is 0.42. Similarly, rounded to three decimal places, the number $0.41666\ldots$ is 0.417.

Rounding a Decimal

1. Determine the number of digits of accuracy you wish to keep. The digit in the last position you keep is called the **rounding digit,** and the digit in the first position you discard is called the **decision digit.**

2. If the decision digit is 5 or greater, round up by adding 1 to the rounding digit.

3. If the decision digit is 4 or less, round down by leaving the rounding digit unchanged.

Given Decimal	Rounded to Three Places
0.9763	0.976
0.9768	0.977
0.9765	0.977

Technology: Tip

You can use a calculator to round decimals. For instance, to round 0.9375 to two decimal places on a scientific calculator, enter

   [FIX] [2] .9375 [=]

On a graphing calculator, enter

   round (.9375, 2) [ENTER]

Without using a calculator, round -0.88247 to three decimal places. Verify your answer with a calculator. Name the rounding and decision digits.

See Technology Answers.

Example 10   Operations with Decimals

a. Add 0.583, 1.06, and 2.9104.

b. Multiply -3.57 and 0.032.

Solution

a. To add decimals, align the decimal points and proceed as in integer addition.

$$
\begin{array}{r}
1\ 1 \\
0.583 \\
1.06 \\
+\ \ 2.9104 \\
\hline
4.5534
\end{array}
$$

b. To multiply decimals, use integer multiplication and then place the decimal point (in the product) so that the number of decimal places equals the sum of the decimal places in the two factors.

$$
\begin{array}{r}
-3.57 \\
\times\quad 0.032 \\
\hline
714 \\
1071 \\
\hline
-0.11424
\end{array}
$$
    Two decimal places
    Three decimal places

    Five decimal places

Example 11   Dividing Decimal Fractions

Divide 1.483 by 0.56. Round the answer to two decimal places.

Solution

To divide 1.483 by 0.56, convert the divisor to an integer by moving its decimal point to the right. Move the decimal point in the dividend an equal number of places to the right. Place the decimal point in the quotient directly above the new decimal point in the dividend and then divide as with integers.

$$
\begin{array}{r}
2.648 \\
56\overline{)\ 148.300} \\
\underline{112} \\
36\ 3 \\
\underline{33\ 6} \\
2\ 70 \\
\underline{2\ 24} \\
460 \\
\underline{448}
\end{array}
$$

Rounded to two decimal places, the answer is 2.65. This answer can be written as

$$\frac{1.483}{0.56} \approx 2.65$$

where the symbol $\approx$ means **is approximately equal to.**

Example 12   Physical Fitness

To satisfy your health and fitness requirement, you decide to take a tennis class. You learn that you burn about 400 calories per hour playing tennis. In one week, you played tennis for $\frac{3}{4}$ hour on Tuesday, 2 hours on Wednesday, and $1\frac{1}{2}$ hours on Thursday. How many total calories did you burn playing tennis in one week? What was the average number of calories you burned playing tennis for the three days?

Solution

The total number of calories you burned playing tennis in one week was

$$400\left(\frac{3}{4}\right) + 400(2) + 400\left(1\frac{1}{2}\right) = 300 + 800 + 600 = 1700 \text{ calories.}$$

The average number of calories you burned playing tennis for the three days was

$$\frac{1700}{3} \approx 566.67 \text{ calories.}$$

Summary of Rules for Fractions

Let a, b, c, and d be real numbers.

Rule	Example
1. Rules of signs for fractions: $\dfrac{-a}{-b} = \dfrac{a}{b}$ $\dfrac{-a}{b} = \dfrac{a}{-b} = -\dfrac{a}{b}$	$\dfrac{-12}{-4} = \dfrac{12}{4}$ $\dfrac{-12}{4} = \dfrac{12}{-4} = -\dfrac{12}{4}$
2. Equivalent fractions: $\dfrac{a}{b} = \dfrac{a \cdot c}{b \cdot c},\ b \neq 0, c \neq 0$	$\dfrac{1}{4} = \dfrac{3}{12}$ because $\dfrac{1}{4} = \dfrac{1 \cdot 3}{4 \cdot 3} = \dfrac{3}{12}$
3. Addition of fractions: $\dfrac{a}{b} + \dfrac{c}{d} = \dfrac{ad + bc}{bd},\quad b \neq 0,\quad d \neq 0$	$\dfrac{1}{3} + \dfrac{2}{7} = \dfrac{1 \cdot 7 + 3 \cdot 2}{3 \cdot 7} = \dfrac{13}{21}$
4. Subtraction of fractions: $\dfrac{a}{b} - \dfrac{c}{d} = \dfrac{ad - bc}{bd},\quad b \neq 0,\quad d \neq 0$	$\dfrac{1}{3} - \dfrac{2}{7} = \dfrac{1 \cdot 7 - 3 \cdot 2}{3 \cdot 7} = \dfrac{1}{21}$
5. Multiplication of fractions: $\dfrac{a}{b} \cdot \dfrac{c}{d} = \dfrac{a \cdot c}{b \cdot d},\quad b \neq 0,\quad d \neq 0$	$\dfrac{1}{3} \cdot \dfrac{2}{7} = \dfrac{1(2)}{3(7)} = \dfrac{2}{21}$
6. Division of fractions: $\dfrac{a}{b} \div \dfrac{c}{d} = \dfrac{a}{b} \cdot \dfrac{d}{c},\ b \neq 0, c \neq 0, d \neq 0$	$\dfrac{1}{3} \div \dfrac{2}{7} = \dfrac{1}{3} \cdot \dfrac{7}{2} = \dfrac{7}{6}$

1.4 Exercises

Developing Skills

In Exercises 1–12, find the greatest common factor.

1. 6, 10   2

2. 6, 9   3

3. 20, 45   5

4. 48, 64   16

5. 45, 90   45

6. 27, 54   27

7. 18, 84, 90   6

8. 84, 98, 192   2

9. 240, 300, 360   60

10. 117, 195, 507   39

11. 134, 225, 315, 945   1

12. 80, 144, 214, 504   2

In Exercises 13–20, write the fraction in simplest form. See Example 1.

13. $\frac{2}{8}$ $\frac{1}{4}$

14. $\frac{3}{18}$ $\frac{1}{6}$

15. $\frac{12}{18}$ $\frac{2}{3}$

16. $\frac{16}{56}$ $\frac{2}{7}$

17. $\frac{60}{192}$ $\frac{5}{16}$

18. $\frac{45}{225}$ $\frac{1}{5}$

19. $\frac{28}{350}$ $\frac{2}{25}$

20. $\frac{88}{154}$ $\frac{4}{7}$

In Exercises 21–24, each figure is divided into regions of equal area. Write a fraction that represents the shaded portion of the figure. Then write the fraction in simplest form.

21.

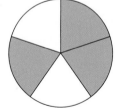

$\frac{3}{5}$

22.

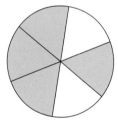

$\frac{4}{6} = \frac{2}{3}$

23.

$\frac{6}{10} = \frac{3}{5}$

24.

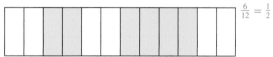

$\frac{6}{12} = \frac{1}{2}$

In Exercises 25–28, write an equivalent fraction with the indicated denominator. See Example 2.

25. $\frac{3}{8} = \frac{6}{16}$

26. $\frac{4}{5} = \frac{12}{15}$

27. $\frac{6}{15} = \frac{10}{25}$

28. $\frac{21}{49} = \frac{12}{28}$

In Exercises 29–42, find the sum or difference. Write the result in simplest form.

29. $\frac{7}{15} + \frac{2}{15}$ $\frac{3}{5}$

30. $\frac{13}{35} + \frac{5}{35}$ $\frac{18}{35}$

31. $\frac{9}{11} + \frac{5}{11}$ $\frac{14}{11}$

32. $\frac{5}{6} + \frac{13}{6}$ 3

33. $\frac{9}{16} - \frac{3}{16}$ $\frac{3}{8}$

34. $\frac{15}{32} - \frac{7}{32}$ $\frac{1}{4}$

35. $-\frac{23}{11} + \frac{12}{11}$ -1

36. $-\frac{39}{23} - \frac{11}{23}$ $-\frac{50}{23}$

37. $\frac{3}{4} - \frac{5}{4}$ $-\frac{1}{2}$

38. $\frac{3}{8} - \frac{5}{8}$ $-\frac{1}{4}$

39. $\frac{7}{10} + \left(-\frac{3}{10}\right)$ $\frac{2}{5}$

40. $\frac{11}{15} + \left(-\frac{2}{15}\right)$ $\frac{3}{5}$

41. $\frac{2}{5} + \frac{4}{5} + \frac{1}{5}$ $\frac{7}{5}$

42. $\frac{2}{9} + \frac{4}{9} + \frac{1}{9}$ $\frac{7}{9}$

In Exercises 43–66, evaluate the expression. Write the result in simplest form. See Examples 3, 4, and 5.

43. $\frac{1}{2} + \frac{1}{3}$ $\frac{5}{6}$

44. $\frac{3}{5} + \frac{1}{2}$ $\frac{11}{10}$

45. $\frac{1}{4} - \frac{1}{3}$ $-\frac{1}{12}$

46. $\frac{2}{3} - \frac{1}{6}$ $\frac{1}{2}$

47. $\frac{3}{16} + \frac{3}{8}$ $\frac{9}{16}$

48. $\frac{2}{3} + \frac{4}{9}$ $\frac{10}{9}$

49. $-\frac{1}{8} - \frac{1}{6}$ $-\frac{7}{24}$

50. $-\frac{13}{8} - \frac{3}{4}$ $-\frac{19}{8}$

51. $4 - \frac{8}{3}$ $\frac{4}{3}$

52. $2 - \frac{17}{25}$ $\frac{33}{25}$

53. $-\frac{7}{8} - \frac{5}{6}$ $-\frac{41}{24}$

54. $-\frac{5}{12} - \frac{1}{9}$ $-\frac{19}{36}$

55. $\frac{3}{4} - \frac{2}{5}$ $\frac{7}{20}$

56. $\frac{5}{8} - \frac{1}{6}$ $\frac{11}{24}$

57. $-\frac{5}{6} - \left(-\frac{3}{4}\right)$ $-\frac{1}{12}$

58. $-\frac{1}{9} - \left(-\frac{3}{5}\right)$ $\frac{22}{45}$

59. $3\frac{1}{2} + 5\frac{2}{3}$ $\frac{55}{6}$

60. $5\frac{3}{4} + 8\frac{1}{10}$ $\frac{277}{20}$

61. $1\frac{3}{16} - 2\frac{1}{4}$ $-\frac{17}{16}$

62. $5\frac{7}{8} - 2\frac{1}{2}$ $\frac{27}{8}$

63. $15\frac{5}{6} - 20\frac{1}{4}$ $-\frac{53}{12}$

64. $6 - 3\frac{5}{8}$ $\frac{19}{8}$

65. $-5\frac{2}{3} - 4\frac{5}{12}$ $-\frac{121}{12}$

66. $-2\frac{3}{4} - 3\frac{1}{5}$ $-\frac{119}{20}$

In Exercises 67–72, evaluate the expression. Write the result in simplest form. See Example 6.

67. $\frac{5}{12} - \frac{3}{8} + \frac{5}{16}$ $\frac{17}{48}$

68. $-\frac{3}{7} + \frac{5}{14} + \frac{3}{4}$ $\frac{19}{28}$

69. $3 + \frac{12}{3} + \frac{1}{9}$ $\frac{64}{9}$

70. $1 + \frac{2}{3} - \frac{5}{6}$ $\frac{5}{6}$

71. $2 - \frac{25}{6} - \frac{3}{4}$ $-\frac{35}{12}$

72. $2 - \frac{15}{16} - \frac{7}{8}$ $\frac{3}{16}$

In Exercises 73–76, determine the unknown fractional part of the circle graph.

73.

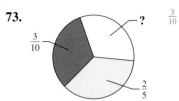

$\frac{3}{10}$

74.

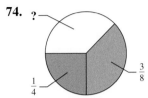

$\frac{3}{8}$

75.

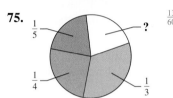

$\frac{13}{60}$

76.

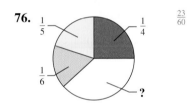

$\frac{23}{60}$

In Exercises 77–100, evaluate the expression. Write the result in simplest form. See Examples 7 and 8.

77. $\frac{1}{2} \cdot \frac{3}{4}$ $\frac{3}{8}$

78. $\frac{3}{5} \cdot \frac{1}{2}$ $\frac{3}{10}$

79. $-\frac{2}{3} \cdot \frac{5}{7}$ $-\frac{10}{21}$

80. $-\frac{5}{6} \cdot \frac{1}{2}$ $-\frac{5}{12}$

81. $\frac{2}{3}\left(-\frac{9}{16}\right)$ $-\frac{3}{8}$

82. $\left(\frac{5}{3}\right)\left(-\frac{3}{5}\right)$ -1

83. $\left(-\frac{3}{4}\right)\left(-\frac{4}{9}\right)$ $\frac{1}{3}$

84. $\left(-\frac{7}{16}\right)\left(-\frac{12}{5}\right)$ $\frac{21}{20}$

85. $\left(\frac{5}{18}\right)\left(\frac{3}{4}\right)$ $\frac{5}{24}$

86. $\left(\frac{3}{28}\right)\left(\frac{7}{8}\right)$ $\frac{3}{32}$

87. $\left(\frac{11}{12}\right)\left(-\frac{9}{44}\right)$ $-\frac{3}{16}$

88. $\left(\frac{5}{12}\right)\left(-\frac{6}{25}\right)$ $-\frac{1}{10}$

89. $\left(-\frac{3}{11}\right)\left(-\frac{11}{3}\right)$ 1

90. $\left(-\frac{7}{15}\right)\left(-\frac{15}{7}\right)$ 1

91. $9\left(\frac{4}{15}\right)$ $\frac{12}{5}$

92. $24\left(\frac{7}{18}\right)$ $\frac{28}{3}$

93. $\left(-\frac{3}{2}\right)\left(-\frac{15}{16}\right)\left(\frac{12}{25}\right)$ $\frac{27}{40}$

94. $\left(\frac{1}{2}\right)\left(-\frac{4}{15}\right)\left(-\frac{5}{24}\right)$ $\frac{1}{36}$

95. $6\left(\frac{3}{4}\right)\left(\frac{2}{9}\right)$ 1

96. $8\left(\frac{5}{12}\right)\left(\frac{3}{10}\right)$ 1

97. $2\frac{3}{4} \cdot 3\frac{2}{3}$ $\frac{121}{12}$

98. $2\frac{4}{5} \cdot 6\frac{2}{3}$ $\frac{56}{3}$

99. $-5\frac{2}{3} \cdot 4\frac{1}{2}$ $-\frac{51}{2}$

100. $-8\frac{1}{2} \cdot 3\frac{2}{5}$ $-\frac{289}{10}$

In Exercises 101–104, find the reciprocal of the number. Show that the product of the number and its reciprocal is 1.

101. 7 $\frac{1}{7}; 7 \cdot \frac{1}{7} = 1$

102. 14 $\frac{1}{14}; 14 \cdot \frac{1}{14} = 1$

103. $\frac{4}{7}$ $\frac{7}{4}; \frac{4}{7} \cdot \frac{7}{4} = 1$

104. $-\frac{5}{9}$ $-\frac{9}{5}; -\frac{5}{9} \cdot -\frac{9}{5} = 1$

In Exercises 105–122, evaluate the expression and write the result in simplest form. If it is not possible, explain why. See Example 9.

105. $\frac{3}{8} \div \frac{3}{4}$ $\frac{1}{2}$

106. $\frac{5}{16} \div \frac{25}{8}$ $\frac{1}{10}$

107. $-\frac{5}{12} \div \frac{45}{32}$ $-\frac{8}{27}$

108. $-\frac{16}{21} \div \frac{12}{27}$ $-\frac{12}{7}$

109. $\frac{3}{5} \div \frac{7}{5}$ $\frac{3}{7}$

110. $\frac{7}{8} \div \frac{3}{8}$ $\frac{7}{3}$

111. $\left(-\frac{5}{6}\right) \div \left(-\frac{8}{10}\right)$ $\frac{25}{24}$

112. $\left(-\frac{14}{15}\right) \div \left(-\frac{24}{25}\right)$ $\frac{35}{36}$

113. $-10 \div \frac{1}{9}$ -90

114. $-6 \div \frac{1}{3}$ -18

115. $0 \div (-21)$ 0

116. $0 \div (-33)$ 0

117. $\frac{3}{5} \div 0$

118. $\frac{11}{13} \div 0$

Division by zero is undefined.   Division by zero is undefined.

119. $3\frac{3}{4} \div 1\frac{1}{2}$ $\frac{5}{2}$

120. $2\frac{4}{9} \div 5\frac{1}{3}$ $\frac{11}{24}$

121. $3\frac{3}{4} \div 2\frac{5}{8}$ $\frac{10}{7}$

122. $1\frac{5}{6} \div 2\frac{1}{3}$ $\frac{11}{14}$

In Exercises 123–132, write the fraction in decimal form. (Use the bar notation for repeating digits.)

123. $\frac{3}{4}$ 0.75

124. $\frac{5}{8}$ 0.625

125. $\frac{9}{16}$ 0.5625

126. $\frac{7}{20}$ 0.35

127. $\frac{2}{3}$ $0.\overline{6}$

128. $\frac{5}{6}$ $0.8\overline{3}$

129. $\frac{7}{12}$ $0.58\overline{3}$

130. $\frac{8}{15}$ $0.5\overline{3}$

131. $\frac{5}{11}$ $0.\overline{45}$

132. $\frac{5}{21}$ $0.\overline{238095}$

In Exercises 133–146, evaluate the expression. Round your answer to two decimal places. See Examples 10 and 11.

133. $132.1 + (-25.45)$
   106.65

134. $408.9 + (-13.12)$
   395.78

135. $1.21 + 4.06 - 3.00$
   2.27

136. $3.4 + 1.062 - 5.13$
   -0.67

137. $-0.0005 - 2.01 + 0.111$ -1.90

138. $-1.0012 - 3.25 + 0.2$ -4.05

139. $(-6.3)(9.05)$ -57.02

140. $3.7(-14.8)$ -54.76

141. $(-0.05)(-85.95)$
   4.30

142. $(-0.09)(-0.45)$
   0.04

143. $4.69 \div 0.12$ 39.08

144. $7.14 \div 0.94$ 7.60

145. $1.062 \div (-2.1)$
   -0.51

146. $2.011 \div (-3.3)$
   -0.61

Estimation   In Exercises 147 and 148, estimate the sum to the nearest integer.

147. $\frac{3}{11} + \frac{7}{10}$ ≈ 1

148. $\frac{5}{8} + \frac{9}{7}$ ≈ 2

Solving Problems

149. *Stock Price*   On August 7, 2002, the Dow Jones Industrial Average closed at 8456.20 points. On August 8, 2002, it closed at 8712.00 points. Determine the increase in the Dow Jones Industrial Average.   255.80 points

150. *Sewing*   A pattern requires $3\frac{1}{6}$ yards of material to make a skirt and an additional $2\frac{3}{4}$ yards to make a matching jacket. Find the total amount of material required. $5\frac{11}{12}$ yards

151. *Agriculture*   During the months of January, February, and March, a farmer bought $8\frac{3}{4}$ tons, $7\frac{1}{5}$ tons, and $9\frac{3}{8}$ tons of feed, respectively. Find the total amount of feed purchased during the first quarter of the year. $\frac{1013}{40} = 25.325$ tons

152. *Cooking*   You are making a batch of cookies. You have placed 2 cups of flour, $\frac{1}{3}$ cup butter, $\frac{1}{2}$ cup brown sugar, and $\frac{1}{3}$ cup granulated sugar in a mixing bowl. How many cups of ingredients are in the mixing bowl? $\frac{19}{6}$ cups

153. *Construction Project*   The highway workers have a sign beside a construction project indicating what fraction of the work has been completed. At the beginnings of May and June the fractions of work completed were $\frac{5}{16}$ and $\frac{2}{3}$, respectively. What fraction of the work was completed during the month of May? $\frac{17}{48}$

154. *Fund Drive*   A charity is raising funds and has a display showing how close they are to reaching their goal. At the end of the first week of the fund drive, the display shows $\frac{1}{9}$ of the goal. At the end of the second week, the display shows $\frac{3}{5}$ of the goal. What fraction of the goal was gained during the second week? $\frac{22}{45}$

155. *Consumer Awareness*   At a convenience store you buy two gallons of milk at $2.59 per gallon and three loaves of bread at $1.68 per loaf. You give the clerk a 20-dollar bill. How much change will you receive? (Assume there is no sales tax.) $9.78

156. *Consumer Awareness*   A cellular phone company charges $1.16 for the first minute and $0.85 for each additional minute. Find the cost of a seven-minute cellular phone call. $6.26

157. *Cooking*   You make 60 ounces of dough for breadsticks. Each breadstick requires $\frac{5}{4}$ ounces of dough. How many breadsticks can you make?   48

158. *Unit Price*   A $2\frac{1}{2}$-pound can of food costs $4.95. What is the cost per pound? $1.98

159. *Consumer Awareness*   The sticker on a new car gives the fuel efficiency as 22.3 miles per gallon. The average cost of fuel is $1.259 per gallon. Estimate the annual fuel cost for a car that will be driven approximately 12,000 miles per year. $677.49

160. *Walking Time*   Your apartment is $\frac{3}{4}$ mile from the subway. You walk at the rate of $3\frac{1}{4}$ miles per hour. How long does it take you to walk to the subway? $\approx$ 14 minutes

161. *Stock Purchase*   You buy 200 shares of stock at $23.63 per share and 300 shares at $86.25 per share.
(a) Estimate the total cost of the stock.   Answers will vary.
(b) Use a calculator to find the total cost of the stock. $30,601

162. *Music*   Each day for a week, you practiced the saxophone for $\frac{2}{3}$ hour.
(a) Explain how to use mental math to estimate the number of hours of practice in a week.   Explanations will vary.
(b) Determine the actual number of hours you practiced during the week. Write the result in decimal form, rounding to one decimal place.   4.7 hours

163. *Consumer Awareness*   The prices per gallon of regular unleaded gasoline at three service stations are $1.259, $1.369, and $1.279, respectively. Find the average price per gallon. $1.302

164. *Consumer Awareness*   The prices of a 16-ounce bottle of soda at three different convenience stores are $1.09, $1.25, and $1.10, respectively. Find the average price for the bottle of soda. $1.15

Explaining Concepts

165. *(icon)* Answer parts (g)–(l) of Motivating the Chapter.

166. *Writing* Is it true that the sum of two fractions of like signs is positive? If not, give an example that shows the statement is false.   No. $-\frac{3}{4} + \left(-\frac{1}{8}\right) = -\frac{7}{8}$

167. *Writing* Does $\frac{2}{3} + \frac{3}{2} = (2 + 3)/(3 + 2) = 1$? Explain your answer.

No. Rewrite both fractions with like denominators. Then add their numerators and write the sum over the common denominator.

168. *Writing* In your own words, describe the rule for determining the sign of the product of two fractions.   If the fractions have the same sign, the product is positive. If the fractions have opposite signs, the product is negative.

169. *Writing* Is it true that $\frac{2}{3} = 0.67$? Explain your answer.   No. $\frac{2}{3} = 0.6$ (nonterminating)

170. *Writing* Use the figure to determine how many one-fourths are in 3. Explain how to obtain the same result by division.   12; Divide 3 by $\frac{1}{4}$.

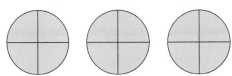

171. *Writing* Use the figure to determine how many one-sixths are in $\frac{2}{3}$. Explain how to obtain the same result by division.   4; Divide $\frac{2}{3}$ by $\frac{1}{6}$.

172. *Investigation* When using a calculator to perform operations with decimals, you should try to get in the habit of rounding your answers *only* after all the calculations are done. If you round the answer at a preliminary stage, you can introduce unnecessary roundoff error. The dimensions of a box are $l = 5.24$, $w = 3.03$, and $h = 2.749$. Find the volume, $l \cdot w \cdot h$, by multiplying the numbers and then rounding the answer to one decimal place. Now use a second method, first rounding each dimension to one decimal place and then multiplying the numbers. Compare your answers, and explain which of these techniques produces the more accurate answer.   Without rounding first: 43.6. Rounding first: 42.12. Rounding after calculations are done produces the more accurate answer.

True or False?   In Exercises 173–178, decide whether the statement is true or false. Justify your answer.

173. The reciprocal of every nonzero integer is an integer.   False. The reciprocal of 5 is $\frac{1}{5}$.

174. The reciprocal of every nonzero rational number is a rational number.   True. Fractions are rational numbers.

175. The product of two nonzero rational numbers is a rational number.   True. The product can always be written as a ratio of two integers.

176. The product of two positive rational numbers is greater than either factor.   False. $\frac{1}{2} \cdot \frac{1}{4} = \frac{1}{8}$

177. If $u > v$, then $u - v > 0$.   True. If you move v units to the left of u on the number line, the result will be to the right of zero.

178. If $u > 0$ and $v > 0$, then $u - v > 0$.   False. $6 > 0$ and $8 > 0$, but $6 - 8 < 0$.

179. *Estimation* Use mental math to determine whether $\left(5\frac{3}{4}\right) \times \left(4\frac{1}{8}\right)$ is less than 20. Explain your reasoning.   The product is greater than 20, because the factors are greater than factors that yield a product of 20.

180. Determine the placement of the digits 3, 4, 5, and 6 in the following addition problem so that you obtain the specified sum. Use each number only once.

$$\frac{3}{6} + \frac{4}{5} = \frac{13}{10}$$

181. If the fractions represented by the points P and R are multiplied, what point on the number line best represents their product: M, S, N, P, or T? (Source: National Council of Teachers of Mathematics)

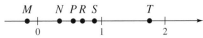

N. Since P and R are between 0 and 1, their product PR is less than the smaller of P and R but positive.

1.5 Exponents, Order of Operations, and Properties of Real Numbers

Michael Newman/PhotoEdit

What You Should Learn

① Rewrite repeated mutiplication in exponential form and evaluate exponential expressions.

② Evaluate expressions using order of operations.

③ Identify and use the properties of real numbers.

Why You Should Learn It

Properties of real numbers can be used to solve real-life problems. For instance, in Exercise 124 on page 57, you will use the Distributive Property to find the amount paid for a new truck.

① Rewrite repeated multiplication in exponential form and evaluate exponential expressions.

Exponents

In Section 1.3, you learned that multiplication by a positive integer can be described as repeated addition.

Repeated Addition	*Multiplication*
$7 + 7 + 7 + 7$	4×7

4 terms of 7

In a similar way, repeated multiplication can be described in **exponential form.**

Repeated Multiplication	*Exponential Form*
$7 \cdot 7 \cdot 7 \cdot 7$	7^4

4 factors of 7

In the exponential form 7^4, 7 is the **base** and it specifies the repeated factor. The number 4 is the **exponent** and it indicates how many times the base occurs as a factor.

When you write the exponential form 7^4, you can say that you are raising 7 to the fourth **power.** When a number is raised to the *first* power, you usually do not write the exponent 1. For instance, you would usually write 5 rather than 5^1. Here are some examples of how exponential expressions are read.

Exponential Expression	*Verbal Statement*
7^2	"seven to the second power" or "seven squared"
4^3	"four to the third power" or "four cubed"
$(-2)^4$	"negative two to the fourth power"
-2^4	"the opposite of two to the fourth power"

It is important to recognize how exponential forms such as $(-2)^4$ and -2^4 differ.

$$(-2)^4 = (-2)(-2)(-2)(-2) \qquad \text{The negative sign is part of the base.}$$
$$= 16 \qquad \text{The value of the expression is positive.}$$
$$-2^4 = -(2 \cdot 2 \cdot 2 \cdot 2) \qquad \text{The negative sign is not part of the base.}$$
$$= -16 \qquad \text{The value of the expression is negative.}$$

Technology: Discovery

When a negative number is raised to a power, the use of parentheses is very important. To discover why, use a calculator to evaluate $(-5)^4$ and -5^4. Write a statement explaining the results. Then use a calculator to evaluate $(-5)^3$ and -5^3. If necessary, write a new statement explaining your discoveries.

See Technology Answers.

Keep in mind that an exponent applies only to the factor (number) directly preceding it. Parentheses are needed to include a negative sign or other factors as part of the base.

Example 1   Evaluating Exponential Expressions

a. $2^5 = 2 \cdot 2 \cdot 2 \cdot 2 \cdot 2$ 　　　　　　　Rewrite expression as a product.

　　　$= 32$ 　　　　　　　　　　　　Simplify.

b. $\left(\dfrac{2}{3}\right)^4 = \dfrac{2}{3} \cdot \dfrac{2}{3} \cdot \dfrac{2}{3} \cdot \dfrac{2}{3}$ 　　　Rewrite expression as a product.

　　　$= \dfrac{2 \cdot 2 \cdot 2 \cdot 2}{3 \cdot 3 \cdot 3 \cdot 3}$ 　　　Multiply fractions.

　　　$= \dfrac{16}{81}$ 　　　　　　　　　Simplify.

Point out the distinction between $(-3)^4$ and -3^4. The failure to distinguish between such expressions is a common student error.

Example 2   Evaluating Exponential Expressions

a. $(-4)^3 = (-4)(-4)(-4)$ 　　　Rewrite expression as a product.

　　　$= -64$ 　　　　　　　　　　Simplify.

b. $(-3)^4 = (-3)(-3)(-3)(-3)$ 　　Rewrite expression as a product.

　　　$= 81$ 　　　　　　　　　　　Simplify.

c. $-3^4 = -(3 \cdot 3 \cdot 3 \cdot 3)$ 　　　Rewrite expression as a product.

　　　$= -81$ 　　　　　　　　　　Simplify.

In parts (a) and (b) of Example 2, note that when a negative number is raised to an odd power, the result is *negative*, and when a negative number is raised to an even power, the result is *positive*.

Example 3   Transporting Capacity

A truck can transport a load of motor oil that is 6 cases high, 6 cases wide, and 6 cases long. Each case contains 6 quarts of motor oil. How many quarts can the truck transport?

Solution

A sketch can help you solve this problem. From Figure 1.32, there are $6 \cdot 6 \cdot 6$ cases of motor oil and each case contains 6 quarts. You can see that 6 occurs as a factor four times, which implies that the total number of quarts is

$$(6 \cdot 6 \cdot 6) \cdot 6 = 6^4 = 1296.$$

So, the truck can transport 1296 quarts of oil.

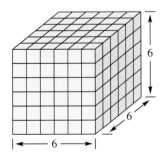

Figure 1.32

② Evaluate expressions using order of operations.

Order of Operations

Up to this point in the text, you have studied five operations of arithmetic—addition, subtraction, multiplication, division, and exponentiation (repeated multiplication). When you use more than one operation in a given problem, you face the question of which operation to do first. For example, without further guidelines, you could evaluate $4 + 3 \cdot 5$ in two ways.

Add First	Multiply First
$4 + 3 \cdot 5 \stackrel{?}{=} (4 + 3) \cdot 5$	$4 + 3 \cdot 5 \stackrel{?}{=} 4 + (3 \cdot 5)$
$= 7 \cdot 5$	$= 4 + 15$
$= 35$	$= 19$

According to the established **order of operations,** the second evaluation is correct. The reason for this is that multiplication has a higher priority than addition. The accepted priorities for order of operations are summarized below.

Order of Operations

1. Perform operations inside *symbols of grouping*—( ) or [ ]— or *absolute value symbols,* starting with the innermost symbols.

2. Evaluate all *exponential* expressions.

3. Perform all *multiplications* and *divisions* from left to right.

4. Perform all *additions* and *subtractions* from left to right.

In the priorities for order of operations, note that the highest priority is given to **symbols of grouping** such as parentheses or brackets. This means that when you want to be sure that you are communicating an expression correctly, you can insert symbols of grouping to specify which operations you intend to be performed first. For instance, if you want to make sure that $4 + 3 \cdot 5$ will be evaluated correctly, you can write it as $4 + (3 \cdot 5)$.

Technology: Discovery

To discover if your calculator performs the established order of operations, evaluate $7 + 5 \cdot 3 - 2^4 \div 4$ exactly as it appears. Does your calculator display 5 or 18? If your calculator performs the established order of operations, it will display 18.

Study Tip

When you use symbols of grouping in an expression, you should alternate between parentheses and brackets. For instance, the expression

$$10 - (3 - [4 - (5 + 7)])$$

is easier to understand than $10 - (3 - (4 - (5 + 7)))$.

Example 4   Order of Operations

a.
$$\begin{aligned} 7 - [(5 \cdot 3) + 2^3] &= 7 - [15 + 2^3] &&\text{Multiply inside the parentheses.}\\ &= 7 - [15 + 8] &&\text{Evaluate exponential expression.}\\ &= 7 - 23 &&\text{Add inside the brackets.}\\ &= -16 &&\text{Subtract.} \end{aligned}$$

b.
$$\begin{aligned} 36 \div (3^2 \cdot 2) - 6 &= 36 \div (9 \cdot 2) - 6 &&\text{Evaluate exponential expression.}\\ &= 36 \div 18 - 6 &&\text{Multiply inside the parentheses.}\\ &= 2 - 6 &&\text{Divide.}\\ &= -4 &&\text{Subtract.} \end{aligned}$$

Section 1.5 Exponents, Order of Operations, and Properties of Real Numbers   **51**

Example 5   Order of Operations

a. $\dfrac{3}{7} \div \dfrac{8}{7} + \left(-\dfrac{3}{5}\right)\left(\dfrac{1}{3}\right) = \dfrac{3}{7} \cdot \dfrac{7}{8} + \left(-\dfrac{3}{5}\right)\left(\dfrac{1}{3}\right)$ Invert divisor and multiply.

$= \dfrac{3}{8} + \left(-\dfrac{1}{5}\right)$ Multiply fractions.

$= \dfrac{15}{40} + \dfrac{-8}{40}$ Find common denominator.

$= \dfrac{7}{40}$ Add fractions.

b. $\dfrac{8}{3}\left(\dfrac{1}{6} + \dfrac{1}{4}\right) = \dfrac{8}{3}\left(\dfrac{2}{12} + \dfrac{3}{12}\right)$ Find common denominator.

$= \dfrac{8}{3}\left(\dfrac{5}{12}\right)$ Add inside the parentheses.

$= \dfrac{40}{36}$ Multiply fractions.

$= \dfrac{10}{9}$ Simplify.

You might tell students that there is no "best way" to solve a problem. For instance, the expression in Example 5(b) can be evaluated using the Distributive Property.

Additional Examples
Evaluate each expression.

a. $-4 + 2(-2 + 5)^2$

b. $\dfrac{2 \cdot 5^2 - 10}{3^2 - 4}$

Answers:

a. 14

b. 8

Example 6   Order of Operations

Evaluate the expression $6 + \dfrac{8 + 7}{3^2 - 4} - (-5)$.

Solution

Using the established order of operations, you can evaluate the expression as follows.

$6 + \dfrac{8 + 7}{3^2 - 4} - (-5) = 6 + \dfrac{8 + 7}{9 - 4} - (-5)$ Evaluate exponential expression.

$= 6 + \dfrac{15}{9 - 4} - (-5)$ Add in numerator.

$= 6 + \dfrac{15}{5} - (-5)$ Subtract in denominator.

$= 6 + 3 - (-5)$ Divide.

$= 9 + 5$ Add.

$= 14$ Add.

In Example 6, note that a fraction bar acts as a symbol of grouping. For instance,

You might ask students how they would enter these two expressions on their calculators.

$\dfrac{8 + 7}{3^2 - 4}$ means $(8 + 7) \div (3^2 - 4)$,   not $8 + 7 \div 3^2 - 4$.

3 Identify and use the properties of real numbers.

Properties of Real Numbers

You are now ready for the symbolic versions of the properties that are true about operations with real numbers. These properties are referred to as **properties of real numbers.** The table shows a verbal description and an illustrative example for each property. Keep in mind that the letters a, b, c, etc., represent real numbers, even though only rational numbers have been used to this point.

Properties of Real Numbers: Let a, b, and c be real numbers.

Property	Example
1. *Commutative Property of Addition:* Two real numbers can be added in either order. $$a + b = b + a$$	$3 + 5 = 5 + 3$
2. *Commutative Property of Multiplication:* Two real numbers can be multiplied in either order. $$ab = ba$$	$4 \cdot (-7) = -7 \cdot 4$
3. *Associative Property of Addition:* When three real numbers are added, it makes no difference which two are added first. $$(a + b) + c = a + (b + c)$$	$(2 + 6) + 5 = 2 + (6 + 5)$
4. *Associative Property of Multiplication:* When three real numbers are multiplied, it makes no difference which two are multiplied first. $$(ab)c = a(bc)$$	$(3 \cdot 5) \cdot 2 = 3 \cdot (5 \cdot 2)$
5. *Distributive Property:* Multiplication distributes over addition. $$a(b + c) = ab + ac$$ $$(a + b)c = ac + bc$$	$3(8 + 5) = 3 \cdot 8 + 3 \cdot 5$ $(3 + 8)5 = 3 \cdot 5 + 8 \cdot 5$
6. *Additive Identity Property:* The sum of zero and a real number equals the number itself. $$a + 0 = 0 + a = a$$	$3 + 0 = 0 + 3 = 3$
7. *Multiplicative Identity Property:* The product of 1 and a real number equals the number itself. $$a \cdot 1 = 1 \cdot a = a$$	$4 \cdot 1 = 1 \cdot 4 = 4$
8. *Additive Inverse Property:* The sum of a real number and its opposite is zero. $$a + (-a) = 0$$	$3 + (-3) = 0$
9. *Multiplicative Inverse Property:* The product of a nonzero real number and its reciprocal is 1. $$a \cdot \frac{1}{a} = 1, \ a \neq 0$$	$8 \cdot \frac{1}{8} = 1$

Example 7   Identifying Properties of Real Numbers

Identify the property of real numbers illustrated by each statement.

a. $3(a + 2) = 3 \cdot a + 3 \cdot 2$

b. $5 \cdot \dfrac{1}{5} = 1$

c. $7 + (5 + b) = (7 + 5) + b$

d. $(b + 3) + 0 = b + 3$

e. $4a = a(4)$

Solution

a. This statement illustrates the Distributive Property.

b. This statement illustrates the Multiplicative Inverse Property.

c. This statement illustrates the Associative Property of Addition.

d. This statement illustrates the Additive Identity Property.

e. This statement illustrates the Commutative Property of Multiplication.

Example 8   Using the Properties of Real Numbers

Complete each statement using the specified property of real numbers.

a. Commutative Property of Addition:

$5 + a = $

b. Associative Property of Multiplication:

$2(7c) = $

c. Distributive Property

$3 \cdot a + 3 \cdot 4 = $

Solution

a. By the Commutative Property of Addition, you can write

$5 + a = a + 5.$

b. By the Associative Property of Multiplication, you can write

$2(7c) = (2 \cdot 7)c.$

c. By the Distributive Property, you can write

$3 \cdot a + 3 \cdot 4 = 3(a + 4).$

One of the distinctive things about algebra is that its rules make sense. You don't have to accept them on "blind faith"—instead, you can learn the reasons that the rules work. For instance, the next example looks at some basic differences among the operations of addition, multiplication, subtraction, and division.

Example 9   Properties of Real Numbers

In the summary of properties of real numbers on page 52, why are all the properties listed in terms of addition and multiplication and not subtraction and division?

Solution

The reason for this is that subtraction and division lack many of the properties listed in the summary. For instance, subtraction and division are not commutative. To see this, consider the following.

$$7 - 5 \neq 5 - 7 \quad \text{and} \quad 12 \div 4 \neq 4 \div 12$$

Similarly, subtraction and division are not associative.

$$9 - (5 - 3) \neq (9 - 5) - 3 \quad \text{and} \quad 12 \div (4 \div 2) \neq (12 \div 4) \div 2$$

Example 10   Geometry: Area

You measure the width of a billboard and find that it is 60 feet. You are told that its height is 22 feet less than its width.

a. Write an expression for the area of the billboard.

b. Use the Distributive Property to rewrite the expression.

c. Find the area of the billboard.

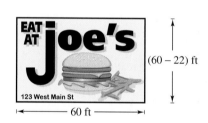

Figure 1.33

$(60 - 22)$ ft

60 ft

Solution

a. Begin by drawing and labeling a diagram, as shown in Figure 1.33. To find an expression for the area of the billboard, multiply the width by the height.

$$\text{Area} = \text{Width} \times \text{Height}$$
$$= 60(60 - 22)$$

b. To rewrite the expression $60(60 - 22)$ using the Distributive Property, distribute 60 over the subtraction.

$$60(60 - 22) = 60(60) - 60(22)$$

c. To find the area of the billboard, evaluate the expression from part (b) as follows.

$$60(60) - 60(22) = 3600 - 1320 \qquad \text{Multiply.}$$
$$= 2280 \qquad \text{Subtract.}$$

So, the area of the billboard is 2280 square feet.

From Example 10(b) you can see that the Distributive Property is also true for subtraction. For instance, the "subtraction form" of $a(b + c) = ab + ac$ is

$$a(b - c) = a[b + (-c)]$$
$$= ab + a(-c)$$
$$= ab - ac.$$

1.5 Exercises

Developing Skills

In Exercises 1–8, rewrite in exponential form.

1. $2 \cdot 2 \cdot 2 \cdot 2 \cdot 2$ 2^5

2. $4 \cdot 4 \cdot 4 \cdot 4 \cdot 4 \cdot 4$ 4^6

3. $(-5) \cdot (-5) \cdot (-5) \cdot (-5)$ $(-5)^4$

4. $(-3) \cdot (-3) \cdot (-3) \cdot (-3)$ $(-3)^4$

5. $\left(-\frac{1}{4}\right) \cdot \left(-\frac{1}{4}\right) \cdot \left(-\frac{1}{4}\right)$ $\left(-\frac{1}{4}\right)^3$

6. $\left(-\frac{3}{5}\right) \cdot \left(-\frac{3}{5}\right) \cdot \left(-\frac{3}{5}\right) \cdot \left(-\frac{3}{5}\right)$ $\left(-\frac{3}{5}\right)^4$

7. $(1.6) \cdot (1.6) \cdot (1.6) \cdot (1.6) \cdot (1.6)$ $(1.6)^5$

8. $(8.7) \cdot (8.7) \cdot (8.7)$ $(8.7)^3$

In Exercises 9–16, rewrite as a product.

9. $(-3)^6$
$(-3)(-3)(-3)(-3)(-3)(-3)$

10. $(-8)^4$
$(-8)(-8)(-8)(-8)$

11. $\left(\frac{3}{8}\right)^5$ $\left(\frac{3}{8}\right)\left(\frac{3}{8}\right)\left(\frac{3}{8}\right)\left(\frac{3}{8}\right)\left(\frac{3}{8}\right)$

12. $\left(\frac{3}{11}\right)^4$ $\left(\frac{3}{11}\right)\left(\frac{3}{11}\right)\left(\frac{3}{11}\right)\left(\frac{3}{11}\right)$

13. $\left(-\frac{1}{2}\right)^5$
$\left(-\frac{1}{2}\right)\left(-\frac{1}{2}\right)\left(-\frac{1}{2}\right)\left(-\frac{1}{2}\right)\left(-\frac{1}{2}\right)$

14. $\left(-\frac{4}{5}\right)^6$
$\left(-\frac{4}{5}\right)\left(-\frac{4}{5}\right)\left(-\frac{4}{5}\right)\left(-\frac{4}{5}\right)\left(-\frac{4}{5}\right)\left(-\frac{4}{5}\right)$

15. $(9.8)^3$ $(9.8)(9.8)(9.8)$

16. $(0.01)^8$ $(0.01)(0.01)(0.01)(0.01)(0.01)(0.01)(0.01)(0.01)$

In Exercises 17–28, evaluate the expression. See Examples 1 and 2.

17. 3^2 9

18. 4^3 64

19. 2^6 64

20. 5^3 125

21. $(-5)^3$ -125

22. $(-4)^2$ 16

23. -4^2 -16

24. $-(-6)^3$ 216

25. $\left(\frac{1}{4}\right)^3$ $\frac{1}{64}$

26. $\left(\frac{4}{5}\right)^3$ $\frac{64}{125}$

27. $(-1.2)^3$ -1.728

28. $(-1.5)^4$ 5.0625

In Exercises 29–70, evaluate the expression. If it is not possible, state the reason. Write fractional answers in simplest form. See Examples 4, 5, and 6.

29. $4 - 6 + 10$ 8

30. $8 + 9 - 12$ 5

31. $5 - (8 - 15)$ 12

32. $13 - (12 - 3)$ 4

33. $-|2 - (6 + 5)|$ -9

34. $125 - |10 - (25 - 3)|$ 113

35. $15 + 3 \cdot 4$ 27

36. $9 - 5 \cdot 2$ -1

37. $25 - 32 \div 4$ 17

38. $16 + 24 \div 8$ 19

39. $(16 - 5) \div (3 - 5)$ $-\frac{11}{2}$

40. $(19 - 4) \div (7 - 2)$ 3

41. $(10 - 16) \cdot (20 - 26)$ 36

42. $(14 - 17) \cdot (13 - 19)$ 18

43. $(45 \div 10) \cdot 2$ 9

44. $(38 \div 5) \cdot 3$ 22.8

45. $[360 - (8 + 12)] \div 10$ 34

46. $[127 - (13 + 4)] \div 11$ 10

47. $5 + (2^2 \cdot 3)$ 17

48. $181 - (13 \cdot 3^2)$ 64

49. $(-6)^2 - (5^2 \cdot 4)$ -64

50. $(-3)^3 + (12 \div 2^2)$ -24

51. $\left(3 \cdot \frac{5}{9}\right) + 1 - \frac{1}{3}$ $\frac{7}{3}$

52. $\frac{2}{3}\left(\frac{3}{4}\right) + 2 - \frac{1}{2}$ 2

53. $18\left(\frac{1}{2} + \frac{2}{3}\right)$ 21

54. $4\left(-\frac{2}{3} + \frac{4}{3}\right)$ $\frac{8}{3}$

55. $\frac{7}{25}\left(\frac{7}{16} - \frac{1}{8}\right)$ $\frac{7}{80}$

56. $\frac{3}{2}\left(\frac{2}{3} + \frac{1}{6}\right)$ $\frac{5}{4}$

57. $\dfrac{3 + [15 \div (-3)]}{16}$ $-\frac{1}{8}$

58. $\dfrac{5 + [(-12) \div 4]}{24}$ $\frac{1}{12}$

59. $\dfrac{3 \cdot 6 - 4 \cdot 6}{5 + 1}$ -1

60. $\dfrac{5 \cdot 3 + 5 \cdot 6}{7 - 2}$ 9

61. $\frac{7}{3}\left(\frac{2}{3}\right) \div \frac{28}{15}$ $\frac{5}{6}$

62. $\frac{3}{8}\left(\frac{1}{5}\right) \div \frac{25}{32}$ $\frac{12}{125}$

63. $\dfrac{1 - 3^2}{-2}$ 4

64. $\dfrac{3^2 + 4^2}{5}$ 5

65. $\dfrac{3^2 - 4^2}{0}$

Division by zero is undefined. 0

66. $\dfrac{0}{3^2 - 4^2}$ 0

67. $\dfrac{5^2 + 12^2}{13}$ 13

68. $\dfrac{4^2 - 2^3}{4}$ 2

69. $\dfrac{0}{5^2 + 1}$ 0

70. $\dfrac{3^2 + 1}{0}$

Division by zero is undefined.

In Exercises 71–74, use a calculator to evaluate the expression. Round your answer to two decimal places.

71. $300\left(1 + \dfrac{0.1}{12}\right)^{24}$ 366.12

72. $1000 \div \left(1 + \dfrac{0.09}{4}\right)^8$ 836.94

73. $\dfrac{1.32 + 4(3.68)}{1.5}$ 10.69

74. $\dfrac{4.19 - 7(2.27)}{14.8}$ -0.79

In Exercises 75–92, identify the property of real numbers illustrated by the statement. See Example 7.

75. $6(-3) = -3(6)$
Commutative Property of Multiplication

76. $16 + 10 = 10 + 16$
Commutative Property of Addition

77. $x + 10 = 10 + x$
Commutative Property of Addition

78. $8x = x(8)$
Commutative Property of Multiplication

79. $0 + 15 = 15$ Additive Identity Property

80. $1 \cdot 4 = 4$ Multiplicative Identity Property

81. $-16 + 16 = 0$ Additive Inverse Property

82. $(2 \cdot 3)4 = 2(3 \cdot 4)$
Associative Property of Multiplication

83. $(10 + 3) + 2 = 10 + (3 + 2)$
Associative Property of Addition

84. $25 + (-25) = 0$ Additive Inverse Property

85. $4(3 \cdot 10) = (4 \cdot 3)10$
Associative Property of Multiplication

86. $(32 + 8) + 5 = 32 + (8 + 5)$
Associative Property of Addition

87. $7\left(\frac{1}{7}\right) = 1$ Multiplicative Inverse Property

88. $14 + (-14) = 0$ Additive Inverse Property

89. $6(3 + x) = 6 \cdot 3 + 6x$ Distributive Property

90. $(14 + 2)3 = 14 \cdot 3 + 2 \cdot 3$
Distributive Property

91. $\frac{1}{a}(3 + y) = \frac{1}{a}(3) + \frac{1}{a}(y)$ Distributive Property

92. $[(x + y)u]v = (x + y)(uv)$
Associative Property of Multiplication

In Exercises 93–104, complete the statement using the specified property of real numbers. See Example 8.

93. Commutative Property of Addition:
$y + 5 = \quad 5 + y$

94. Commutative Property of Addition:
$3 + x = \quad x + 3$

95. Commutative Property of Multiplication:
$10(-3) = \quad -3(10)$

96. Commutative Property of Multiplication:
$5(u + v) = \quad (u + v)5$

97. Distributive Property:
$6(x + 2) = \quad 6x + 12$

98. Distributive Property:
$5(u + v) = \quad 5u + 5v$

99. Distributive Property:
$(4 + y)25 = \quad 100 + 25y$

100. Distributive Property:
$(4 - y)12 = \quad 48 - 12y$

101. Associative Property of Addition:
$3x + (2y + 5) = (3x + 2y) + 5$

102. Associative Property of Addition:
$10 + (x + 2y) = (10 + x) + 2y$

103. Associative Property of Multiplication:
$12(3 \cdot 4) = \quad (12 \cdot 3)4$

104. Associative Property of Multiplication:
$(6x)y = \quad 6(xy)$

In Exercises 105–112, find (a) the additive inverse and (b) the multiplicative inverse of the quantity.

105. 50 (a) -50 (b) $\frac{1}{50}$

106. 12 (a) -12 (b) $\frac{1}{12}$

107. -1 (a) 1 (b) -1

108. $-\frac{1}{2}$ (a) $\frac{1}{2}$ (b) -2

109. $2x$ (a) $-2x$ (b) $\frac{1}{2x}$

110. $5y$ (a) $-5y$ (b) $\frac{1}{5y}$

111. ab (a) $-ab$ (b) $\frac{1}{ab}$

112. uv (a) $-uv$ (b) $\frac{1}{uv}$

In Exercises 113–116, simplify the expression using (a) the Distributive Property and (b) order of operations.

113. $3(6 + 10)$ (a) 48 (b) 48

114. $4(8 - 3)$ (a) 20 (b) 20

115. $\frac{2}{3}(9 + 24)$ (a) 22 (b) 22

116. $\frac{1}{2}(4 - 2)$ (a) 1 (b) 1

In Exercises 117–120, identify the property of real numbers used to justify each step.

117. $7x + 9 + 2x$

$= 7x + 2x + 9$ Commutative Property of Addition

$= (7x + 2x) + 9$ Associative Property of Addition

$= (7 + 2)x + 9$ Distributive Property

$= 9x + 9$ Addition of Real Numbers

$= 9(x + 1)$ Distributive Property

118. $19 + 5x + 24$

$= 19 + 24 + 5x$ Commutative Property of Addition

$= (19 + 24) + 5x$ Associative Property of Addition

$= 43 + 5x$ Addition of Real Numbers

119. $3 + 10(x + 1)$

$= 3 + 10x + 10$ Distributive Property

$= 3 + 10 + 10x$ Commutative Property of Addition

$= (3 + 10) + 10x$ Associative Property of Addition

$= 13 + 10x$ Addition of Real Numbers

120. $2(x + 3) + x$

$= 2x + 2 \cdot 3 + x$ Distributive Property of Addition

$= 2x + x + 6$ Commutative Property of Addition

$= (2 + 1)x + 6$ Distributive Property

$= 3x + 6$ Addition of Real Numbers

$= 3(x + 2)$ Distributive Property

Solving Problems

▲ *Geometry*   In Exercises 121 and 122, find the area of the region.

121.   36 square units

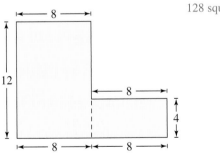

122.   128 square units

123. *Sales Tax*   You purchase a sweater for x dollars. There is a 6% sales tax, which implies that the total amount you must pay is $x + 0.06x$.

(a) Use the Distributive Property to rewrite the expression. $x(1 + 0.06) = 1.06x$

(b) The sweater costs $25.95. How much must you pay for the sweater including sales tax?
$27.51

124. *Cost of a Truck*   A new truck can be paid for by 48 monthly payments of x dollars each plus a down payment of 2.5 times the amount of the monthly payment. This implies that the total amount paid for the truck is $2.5x + 48x$.

(a) Use the Distributive Property to rewrite the expression. $x(2.5 + 48)$

(b) What is the total amount paid for a truck that has a monthly payment of $435? $21,967.50

125. ▲ *Geometry*   The width of a movie screen is 30 feet and its height is 8 feet less than the width. Write an expression for the area of the movie screen. Use the Distributive Property to rewrite the expression.

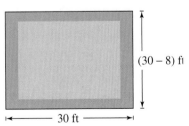

$30(30 - 8) = 30(30) - 30(8) = 660$ square units

126. ▲ *Geometry*   A picture frame is 36 inches wide and its height is 9 inches less than its width. Write an expression for the area of the picture frame. Use the Distributive Property to rewrite the expression.

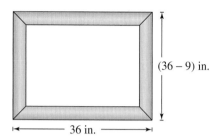

$36(36 - 9) = 36(36) - 36(9) = 972$ square inches

▲ *Geometry*   In Exercises 127 and 128, write an expression for the perimeter of the triangle shown in the figure. Use the properties of real numbers to simplify the expression.

127.

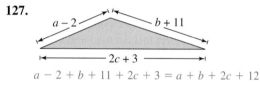

$a - 2 + b + 11 + 2c + 3 = a + b + 2c + 12$

128.

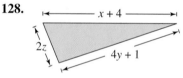

$x + 4 + 2z + 4y + 1 = x + 4y + 2z + 5$

▲ *Geometry*   In Exercises 129 and 130, find the area of the shaded rectangle in two ways. Explain how the results are related to the Distributive Property.

129.

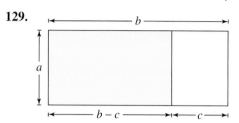

$a(b - c) = ab - ac$; Explanations will vary.

130.

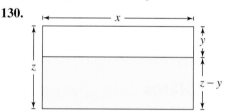

$x(z - y) = xz - xy$; Explanations will vary.

Think About It   In Exercises 131 and 132, determine whether the order in which the two activities are performed is "commutative." That is, do you obtain the same result regardless of which activity is performed first?

131. (a) "Drain the used oil from the engine."
   (b) "Fill the crankcase with 5 quarts of new oil." No

132. (a) "Weed the flower beds."
   (b) "Mow the lawn."   Yes

Explaining Concepts

133. Consider the expression 3^5.
   (a) What is the number 3 called?   Base
   (b) What is the number 5 called?   Exponent

134. *Writing*   Are -6^2 and $(-6)^2$ equal? Explain.
   No. $-6^2 = -36, (-6)^2 = 36$

135. *Writing*   Are $2 \cdot 5^2$ and 10^2 equal? Explain.
   No. $2 \cdot 5^2 = 2 \cdot 25 = 50, \ 10^2 = 100$

136. *Writing*   In your own words, describe the priorities for the established order of operations.
   (a) Perform operations inside symbols of grouping, starting with the innermost symbols.
   (b) Evaluate all exponential expressions.

   (c) Perform all multiplications and divisions from left to right.
   (d) Perform all additions and subtractions from left to right.

137. *Writing*   In your own words, state the Associative Properties of Addition and Multiplication. Give an example of each.
   Associative Property of Addition:
   $(a + b) + c = a + (b + c)$,
   $(x + 3) + 4 = x + (3 + 4)$

   Associative Property of Multiplication:
   $(ab)c = a(bc), (3 \cdot 4)x = 3(4x)$

138. *Writing* In your own words, state the Commutative Properties of Addition and Multiplication. Give an example of each.

Commutative Property of Addition:
$a + b = b + a$, $3 + x = x + 3$
Commutative Property of Multiplication:
$ab = ba$, $3x = x(3)$

Writing In Exercises 139–142, explain why the statement is true. (The symbol $\neq$ means "is not equal to.")

139. $4 \cdot 6^2 \neq 24^2$ $24^2 = (4 \cdot 6)^2 = 4^2 \cdot 6^2$

140. $4 - (6 - 2) \neq 4 - 6 - 2$
$4 - (6 - 2) = 4 - 6 + 2$

141. $-3^2 \neq (-3)(-3)$ **142.** $\dfrac{8-6}{2} \neq 4 - 6$

$-3^2 = -(3)(3) = -9$ $\dfrac{8-6}{2} = \dfrac{8}{2} - \dfrac{6}{2} = 1$

143. *Error Analysis* Describe the error.

$-9 + \dfrac{9+20}{3(5)} - (-3) = -9 + \dfrac{9}{3} + \dfrac{20}{5} - (-3)$

$= -9 + 3 + 4 - (-3)$

$= 1$

144. *Error Analysis* Describe the error.

$7 - 3(8 + 1) - 15 = 4(8 + 1) - 15$

$= 4(9) - 15$

$= 36 - 15$

$= 21$

$7 - 3(8 + 1) - 15 = 7 - 3(9) - 15$

$= 7 - 27 - 15$

$= -35$

Writing In Exercises 145–148, explain why the statement is true.

145. $5(x + 3) \neq 5x + 3$ **146.** $7(x - 2) \neq 7x - 2$
$5(x + 3) = 5x + 15$ $7(x - 2) = 7x - 14$

147. $\dfrac{8}{0} \neq 0$ **148.** $5\left(\dfrac{1}{5}\right) \neq 0$

Division by zero is undefined. $5\left(\dfrac{1}{5}\right) = 1$

149. Match each expression in the first column with its value in the second column.

Expression	Value	Expression	Value
$(6 + 2) \cdot (5 + 3)$	19	$(6 + 2) \cdot (5 + 3)$	$= 64$
$(6 + 2) \cdot 5 + 3$	22	$(6 + 2) \cdot 5 + 3$	$= 43$
$6 + 2 \cdot 5 + 3$	64	$6 + 2 \cdot 5 + 3$	$= 19$
$6 + 2 \cdot (5 + 3)$	43	$6 + 2 \cdot (5 + 3)$	$= 22$

150. Using the established order of operations, which of the following expressions has a value of 72? For those that don't, decide whether you can insert parentheses into the expression so that its value is 72.

(a) $4 + 2^3 - 7$
No

(b) $4 + 8 \cdot 6$
Yes; $(4 + 8) \cdot 6 = 72$

(c) $93 - 25 - 4$
Yes; $93 - (25 - 4) = 72$

(d) $70 + 10 \div 5$
$70 + 10 \div 5 = 72$

(e) $60 + 20 \div 2 + 32$
Yes; $(60 + 20) \div 2 + 32 = 72$

(f) $35 \cdot 2 + 2$
$35 \cdot 2 + 2 = 72$

151. Consider the rectangle shown in the figure.

(a) Find the area of the rectangle by adding the areas of regions I and II.
$2 \cdot 2 + 2 \cdot 3 = 4 + 6 = 10$

(b) Find the area of the rectangle by multiplying its length by its width. $2 \cdot 5 = 10$

(c) Explain how the results of parts (a) and (b) relate to the Distributive Property.
$2 \cdot 2 + 2 \cdot 3 = 2(2 + 3) = 2 \cdot 5 = 10$

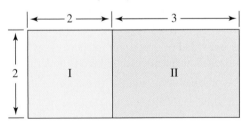

143. $-9 + \dfrac{9+20}{3(5)} - (-3) = -9 + \dfrac{29}{15} + 3$

$= -6 + \dfrac{29}{15}$

$= \dfrac{-90 + 29}{15}$

$= -\dfrac{61}{15}$

What Did You Learn?

Key Terms

real numbers, *p. 2*
natural numbers, *p. 2*
integers, *p. 2*
rational numbers, *p. 3*
irrational numbers, *p. 3*
real number line, *p. 4*

inequality symbol, *p. 5*
opposites, *p. 7*
absolute value, *p. 7*
expression, *p. 8*
evaluate, *p. 8*
additive inverse, *p. 13*

factor, *p. 24*
prime number, *p. 24*
greatest common factor,
   p. 35
reciprocal, *p. 40*
exponent, *p. 48*

Key Concepts

1.1 Ordering of real numbers
Use the real number line and an inequality symbol ($<$, $>$, $\leq$, or $\geq$) to order real numbers.

1.1 Absolute value
The absolute value of a number is its distance from zero on the real number line. The absolute value is either positive or zero.

1.2 Addition and subtraction of integers
To add two integers with like signs, add their absolute values and attach the common sign to the result.

To add two integers with different signs, subtract the smaller absolute value from the larger absolute value and attach the sign of the integer with the larger absolute value.

To subtract one integer from another, add the opposite of the integer being subtracted to the other integer.

1.3 Rules for multiplying and dividing integers
The product of an integer and zero is 0.

Zero divided by a nonzero integer is 0, whereas a nonzero integer divided by zero is undefined.

The product or quotient of two nonzero integers with like signs is positive.

The product or quotient of two nonzero integers with different signs is negative.

1.4 Addition and subtraction of fractions
1. Add or subtract two fractions with like denominators:

$$\frac{a}{c} + \frac{b}{c} = \frac{a+b}{c} \text{ or } \frac{a}{c} - \frac{b}{c} = \frac{a-b}{c}, c \neq 0$$

2. To add two fractions with unlike denominators, rewrite both fractions so that they have like denominators. Then use the rule for adding and subtracting fractions with like denominators.

1.4 Multiplication of fractions
$$\frac{a}{b} \cdot \frac{c}{d} = \frac{a \cdot c}{b \cdot d}, \ b \neq 0, d \neq 0$$

1.4 Division of fractions
$$\frac{a}{b} \div \frac{c}{d} = \frac{a}{b} \cdot \frac{d}{c}, \ b \neq 0, c \neq 0, d \neq 0$$

1.5 Order of operations
1. Perform operations inside symbols of grouping— ( ) or [ ]—or absolute value symbols, starting with the innermost symbols.

2. Evaluate all exponential expressions.

3. Perform all multiplications and divisions from left to right.

4. Perform all additions and subtractions from left to right.

1.5 Properties of real numbers
Commutative Property of Addition $a + b = b + a$
Commutative Property of Multiplication $ab = ba$
Associative Property of Addition
   $(a + b) + c = a + (b + c)$
Associative Property of Multiplication $(ab)c = a(bc)$
Distributive Property
   $a(b + c) = ab + ac$ $a(b - c) = ab - ac$
   $(a + b)c = ac + bc$ $(a - b)c = ac - bc$
Additive Identity Property $a + 0 = a$
Multiplicative Identity Property $a \cdot 1 = a$
Additive Inverse Property $a + (-a) = 0$

Multiplicative Inverse Property $a \cdot \dfrac{1}{a} = 1, \quad a \neq 0$

Review Exercises

1.1 Real Numbers: Order and Absolute Value

① Define sets and use them to classify numbers as natural, integer, rational, or irrational.

In Exercises 1 and 2, determine which of the numbers in the set are (a) natural numbers, (b) integers, (c) rational numbers, and (d) irrational numbers.

1. $\left\{-1, 4.5, \frac{2}{5}, -\frac{1}{7}, \sqrt{4}, \sqrt{5}\right\}$

   (a) none   (b) $-1, \sqrt{4}$

   (c) $-1, 4.5, \frac{2}{5}, -\frac{1}{7}, \sqrt{4}$ (d) $\sqrt{5}$

2. $\left\{10, -3, \frac{4}{5}, \pi, -3.\overline{16}, -\frac{19}{11}\right\}$

   (a) 10   (b) $10, -3$

   (c) $10, -3, \frac{4}{5}, -3.\overline{16}, -\frac{19}{11}$ (d) π

② Plot numbers on the real number line.

In Exercises 3–8, plot the numbers on the real number line.   See Additional Answers.

3. $-3, 5$ **4.** $-8, 11$

5. $-6, \frac{5}{4}$ **6.** $-\frac{7}{2}, 9$

7. $-1, 0, \frac{1}{2}$ **8.** $-2, -\frac{1}{3}, 5$

③ Use the real number line and inequality symbols to order real numbers.

In Exercises 9–12, plot each real number as a point on the real number line and place the correct inequality symbol (< or >) between the pair of real numbers. See Additional Answers.

9. $-\frac{1}{10}$ < 4

10. $\frac{25}{3}$ > $\frac{5}{3}$

11. -3 > -7

12. 10.6 > -3.5

13. Which is smaller: $\frac{2}{3}$ or 0.6? 0.6

14. Which is smaller: $-\frac{1}{3}$ or -0.3? $-\frac{1}{3}$

④ Find the absolute value of a number.

In Exercises 15–18, find the opposite of the number, and determine the distance of the number and its opposite from 0.

15. 152 $-152, 152$ **16.** -10.4 $10.4, 10.4$

17. $-\frac{7}{3}$ $\frac{7}{3}; \frac{7}{3}$ **18.** $\frac{2}{3}$ $-\frac{2}{3}; \frac{2}{3}$

In Exercises 19–22, evaluate the expression.

19. $|-8.5|$ 8.5 **20.** $|3.4|$ 3.4

21. $-|-8.5|$ -8.5 **22.** $|-9.6|$ 9.6

In Exercises 23–26, place the correct symbol (<, >, or =) between the pair of real numbers.

23. $|-84|$ = $|84|$

24. $|-10|$ > $|4|$

25. $\left|\frac{3}{10}\right|$ > $-\left|\frac{4}{5}\right|$

26. $|2.3|$ > $-|2.3|$

1.2 Adding and Subtracting Integers

① Add integers using a number line.

In Exercises 27–30, find the sum and demonstrate the addition on the real number line.   See Additional Answers.

27. $4 + 3$ 7 **28.** $15 + (-6)$ 9

29. $-1 + (-4)$ -5 **30.** $-6 + (-2)$ -8

② Add integers with like signs and with unlike signs.

In Exercises 31–40, find the sum.

31. $16 + (-5)$ 11 **32.** $25 + (-10)$ 15

33. $-125 + 30$ -95 **34.** $-54 + 12$ -42

35. $-13 + (-76)$ -89 **36.** $-24 + (-25)$ -49

37. $-10 + 21 + (-6)$ 5

38. $-23 + 4 + (-11)$ -30

39. $-17 + (-3) + (-9)$ -29

40. $-16 + (-2) + (-8)$ -26

41. *Profit*   A small software company had a profit of \$95,000 in January, a loss of \$64,400 in February, and a profit of \$51,800 in March. What was the company's overall profit (or loss) for the three months?   \$82,400

42. *Account Balance*   At the beginning of a month, your account balance was \$3090. During the month, you withdrew \$870 and \$465, deposited \$109, and earned interest of \$10.05. What was your balance at the end of the month?   \$1874.05

43. *Writing* Is the sum of two integers, one negative and one positive, negative? Explain.

The sum can be positive or negative. The sign is determined by the integer with the greater absolute value.

44. *Writing* Is the sum of two negative integers negative? Explain.

Yes, because to add two integers with like signs, you add their absolute values and attach the common sign to the result.

3 Subtract integers with like signs and with unlike signs.

In Exercises 45–54, find the difference.

45. $28 - 7$ 21   **46.** $43 - 12$ 31
47. $8 - 15$ -7 **48.** $17 - 26$ -9
49. $14 - (-19)$ 33   **50.** $28 - (-4)$ 32
51. $-18 - 4$ -22 **52.** $-37 - 14$ -51
53. $-12 - (-7)$ -5 **54.** $-26 - (-8)$ -18

55. Subtract -549 from 613. 1162

56. What number must be subtracted from -83 to obtain 43? -126

1.3 Multiplying and Dividing Integers

1 Multiply integers with like signs and with unlike signs.

In Exercises 57–68, find the product.

57. $15 \cdot 3$ 45   **58.** $21 \cdot 4$ 84
59. $-3 \cdot 24$ -72 **60.** $-2 \cdot 44$ -88
61. $6(-8)$ -48 **62.** $12(-5)$ -60
63. $(-5)(-9)$ 45   **64.** $(-10)(-81)$ 810
65. $3(-6)(3)$ -54 **66.** $15(-2)(7)$ -210
67. $(-4)(-5)(-2)$ -40 **68.** $(-12)(-2)(-6)$ -144

2 Divide integers with like signs and with unlike signs.

In Exercises 69–78, perform the division, if possible. If not possible, state the reason.

69. $72 \div 8$ 9   **70.** $63 \div 9$ 7
71. $\dfrac{-72}{6}$ -12 **72.** $\dfrac{-162}{9}$ -18
73. $75 \div (-5)$ -15 **74.** $48 \div (-4)$ -12
75. $\dfrac{-52}{-4}$ 13   **76.** $\dfrac{-64}{-4}$ 16
77. $815 \div 0$ Division by zero is undefined.
78. $135 \div 0$ Division by zero is undefined.

79. *Automobile Maintenance* You rotate the tires on your truck, including the spare, so that all five tires are used equally. After 40,000 miles, how many miles has each tire been driven? 32,000 miles

80. *Unit Price* At a garage sale, you buy a box of six glass canisters for a total of $78. All the canisters are of equal value. How much is each one worth? $13

3 Find factors and prime factors of an integer.

In Exercises 81–84, decide whether the number is prime or composite.

81. 839 Prime   **82.** 909 Composite
83. 1764 Composite   **84.** 1847 Prime

In Exercises 85–88, write the prime factorization of the number.

85. 378 $2 \cdot 3 \cdot 3 \cdot 3 \cdot 7$ **86.** 858 $2 \cdot 3 \cdot 11 \cdot 13$
87. 1612 $2 \cdot 2 \cdot 13 \cdot 31$ **88.** 1787 1787

4 Represent the definitions and rules of arithmetic symbolically.

In Exercises 89–92, complete the statement using the indicated definition or rule.

89. Rule for multiplying integers with unlike signs:
$12 \times (-3) = -36$

90. Definition of multiplication:
$(-4) + (-4) + (-4) = -12$

91. Definition of absolute value: $|-7| = 7$

92. Rule for adding integers with unlike signs:
$-9 + 5 = -4$

1.4 Operations with Rational Numbers

1 Rewrite fractions as equivalent fractions.

In Exercises 93–96, find the greatest common factor.

93. 54, 90 18   **94.** 154, 220 22
95. 63, 84, 441 21   **96.** 99, 132, 253 11

In Exercises 97–100, write an equivalent fraction with the indicated denominator.

97. $\dfrac{2}{3} = \dfrac{10}{15}$ **98.** $\dfrac{3}{7} = \dfrac{12}{28}$
99. $\dfrac{6}{10} = \dfrac{15}{25}$ **100.** $\dfrac{9}{12} = \dfrac{12}{16}$

② Add and subtract fractions.

In Exercises 101–112, evaluate the expression. Write the result in simplest form.

101. $\dfrac{3}{25} + \dfrac{7}{25}$ $\frac{2}{5}$ **102.** $\dfrac{9}{64} + \dfrac{7}{64}$ $\frac{1}{4}$

103. $\dfrac{27}{16} - \dfrac{15}{16}$ $\frac{3}{4}$ **104.** $-\dfrac{5}{12} + \dfrac{1}{12}$ $-\frac{1}{3}$

105. $-\dfrac{5}{9} + \dfrac{2}{3}$ $\frac{1}{9}$ **106.** $\dfrac{7}{15} - \dfrac{2}{25}$ $\frac{29}{75}$

107. $-\dfrac{25}{32} + \left(-\dfrac{7}{24}\right)$ $-\frac{103}{96}$ **108.** $-\dfrac{7}{8} - \dfrac{11}{12}$ $-\frac{43}{24}$

109. $5 - \dfrac{15}{4}$ $\frac{5}{4}$ **110.** $\dfrac{12}{5} - 3$ $-\frac{3}{5}$

111. $5\dfrac{3}{4} - 3\dfrac{5}{8}$ $\frac{17}{8}$ **112.** $-3\dfrac{7}{10} + 1\dfrac{1}{20}$ $-\frac{53}{20}$

113. *Meteorology* The table shows the amount of rainfall (in inches) during a five-day period. What was the total amount of rainfall for the five days? $2\frac{3}{4}$ inches

Day	Mon	Tue	Wed	Thu	Fri
Rainfall (in inches)	$\frac{3}{8}$	$\frac{1}{2}$	$\frac{1}{8}$	$1\frac{1}{4}$	$\frac{1}{2}$

114. *Fuel Consumption* The morning and evening readings of the fuel gauge on a car were $\frac{7}{8}$ and $\frac{1}{3}$, respectively. What fraction of the tank of fuel was used that day? $\frac{13}{24}$

③ Multiply and divide fractions.

In Exercises 115–126, evaluate the expression and write the result in simplest form. If it is not possible, explain why.

115. $\dfrac{5}{8} \cdot \dfrac{-2}{15}$ $-\frac{1}{12}$ **116.** $\dfrac{3}{32} \cdot \dfrac{32}{3}$ 1

117. $35\left(\dfrac{1}{35}\right)$ 1 **118.** $(-6)\left(\dfrac{5}{36}\right)$ $-\frac{5}{6}$

119. $\dfrac{3}{8}\left(-\dfrac{2}{27}\right)$ $-\frac{1}{36}$ **120.** $-\dfrac{5}{12}\left(-\dfrac{4}{25}\right)$ $\frac{1}{15}$

121. $\dfrac{5}{14} \div \dfrac{15}{28}$ $\frac{2}{3}$ **122.** $-\dfrac{7}{10} \div \dfrac{4}{15}$ $-\frac{21}{8}$

123. $\left(-\dfrac{3}{4}\right) \div \left(-\dfrac{7}{8}\right)$ $\frac{6}{7}$ **124.** $\dfrac{15}{32} \div \left(-\dfrac{5}{4}\right)$ $-\frac{3}{8}$

125. $-\dfrac{5}{9} \div 0$ **126.** $0 \div \dfrac{1}{12}$ 0

Division by zero is undefined.

127. *Meteorology* During an eight-hour period, $6\frac{3}{4}$ inches of snow fell. What was the average rate of snowfall per hour? $\frac{27}{32}$ inches per hour

128. *Sports* In three strokes on the golf course, you hit your ball a total distance of $64\frac{7}{8}$ meters. What is your average distance per stroke? $21\frac{5}{8}$ meters

④ Add, subtract, multiply, and divide decimals.

In Exercises 129–136, evaluate the expression. Round your answer to two decimal places.

129. $4.89 + 0.76$ 5.65 **130.** $1.29 + 0.44$ 1.73

131. $3.815 - 5.19$ -1.38 **132.** $7.234 - 8.16$ -0.93

133. $(1.49)(-0.5)$ -0.75 **134.** $(2.34)(-1.2)$ -2.81

135. $5.25 \div 0.25$ 21 **136.** $10.18 \div 1.6$ 6.36

137. *Consumer Awareness* A telephone company charges $0.64 for the first minute and $0.72 for each additional minute. Find the cost of a five-minute call. $3.52

138. *Consumer Awareness* A television costs $120.75 plus $27.56 each month for 18 months. Find the total cost of the television. $616.83

1.5 Exponents, Order of Operations, and Properties of Real Numbers

① Rewrite repeated multiplication in exponential form and evaluate exponential expressions.

In Exercises 139 and 140, rewrite in exponential form.

139. $6 \cdot 6 \cdot 6 \cdot 6 \cdot 6$ **140.** $(-3) \cdot (-3) \cdot (-3)$
6^5 $(-3)^3$

In Exercises 141 and 142, rewrite as a product.

141. $(-7)^4$ **142.** $\left(\dfrac{1}{2}\right)^5$
$(-7)(-7)(-7)(-7)$ $\left(\frac{1}{2}\right)\left(\frac{1}{2}\right)\left(\frac{1}{2}\right)\left(\frac{1}{2}\right)\left(\frac{1}{2}\right)$

In Exercises 143–146, evaluate the expression.

143. 2^4 16 **144.** $(-6)^2$ 36

145. $\left(-\dfrac{3}{4}\right)^3$ $-\frac{27}{64}$ **146.** $\left(\dfrac{2}{3}\right)^2$ $\frac{4}{9}$

② Evaluate expressions using order of operations.

In Exercises 147–166, evaluate the expression. Write fractional answers in simplest form.

147. $12 - 2 \cdot 3$ 6 **148.** $1 + 7 \cdot 3 - 10$ 12

149. $18 \div 6 \cdot 7$ 21 **150.** $3^2 \cdot 4 \div 2$ 18

151. $20 + (8^2 \div 2)$ 52 **152.** $(8 - 3) \div 15$ $\frac{1}{3}$

153. $240 - (4^2 \cdot 5)$ **154.** $5^2 - (625 \cdot 5^2)$
       160                              $-15,600$

155. $3^2(5-2)^2$ 81   **156.** $-5(10-7)^3$ -135

157. $\left(\frac{3}{4}\right)\left(\frac{5}{6}\right) + 4$ $\frac{37}{8}$ **158.** $75 - 24 \div 2^3$ 72

159. $122 - [45 - (32 + 8) - 23]$ 140

160. $-58 - (48 - 12) - (-30 - 4)$ -60

161. $\dfrac{6 \cdot 4 - 36}{4}$ -3 **162.** $\dfrac{144}{2 \cdot 3 \cdot 3}$ 8

163. $\dfrac{54 - 4 \cdot 3}{6}$ 7   **164.** $\dfrac{3 \cdot 5 + 125}{10}$ 14

165. $\dfrac{78 - |-78|}{5}$ **166.** $\dfrac{300}{15 - |-15|}$
          0   Division by zero is undefined.

In Exercises 167–170, use a calculator to evaluate the expression. Round your answer to two decimal places.

167. $(5.8)^4 - (3.2)^5$ 796.11 **168.** $\dfrac{(15.8)^3}{(2.3)^8}$ 5.04

169. $\dfrac{3000}{(1.05)^{10}}$ **170.** $500\left(1 + \dfrac{0.07}{4}\right)^{40}$
          1841.74   1000.80

171. *Depreciation*   After 3 years, the value of a $16,000 car is given by $16,000\left(\frac{3}{4}\right)^3$.

(a) What is the value of the car after 3 years?
$6750

(b) How much has the car depreciated during the 3 years? $9250

172. ▲ *Geometry*   The volume of water in a hot tub is given by $V = 6^2 \cdot 3$ (see figure). How many cubic feet of water will the hot tub hold? Find the total weight of the water in the tub. (Use the fact that 1 cubic foot of water weighs 62.4 pounds.)

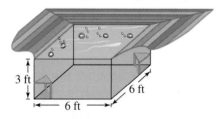

3 ft   6 ft
        6 ft

108 cubic feet, 6739.2 pounds

③ Identify and use the properties of real numbers.

In Exercises 173–180, identify the property of real numbers illustrated by the statement.

173. $123 - 123 = 0$ Additive Inverse Property

174. $9 \cdot \frac{1}{9} = 1$ Multiplicative Inverse Property

175. $14(3) = 3(14)$
Commutative Property of Multiplication

176. $5(3x) = (5 \cdot 3)x$
Associative Property of Multiplication

177. $17 \cdot 1 = 17$ Multiplicative Identity Property

178. $10 + 6 = 6 + 10$
Commutative Property of Addition

179. $-2(7 + x) = -2 \cdot 7 + (-2)x$
Distributive Property

180. $2 + (3 + x) = (2 + 3) + x$
Associative Property of Addition

In Exercises 181–184, complete the statement using the specified property of real numbers.

181. Additive Identity Property:
$(z + 1) + 0 =$ $0 + (z + 1) = z + 1$

182. Distributive Property:
$8(x + 2) =$ $8x + 16$

183. Commutative Property of Addition:
$2y + 1 =$ $1 + 2y$

184. Associative Property of Multiplication:
$9(4x) =$ $(9 \cdot 4)x$

185. ▲ *Geometry*   Find the area of the shaded rectangle in two ways. Explain how the results are related to the Distributive Property.

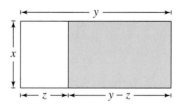

$x(y - z) = xy - xz$; Explanations will vary.

Take this test as you would take a test in class. After you are done, check your work against the answers in the back of the book.

1. Which of the following are (a) natural numbers, (b) integers, and (c) rational numbers?

   $\left\{4, -6, \frac{1}{2}, 0, \pi, \frac{7}{9}\right\}$ (a) 4 (b) 4, −6, 0 (c) 4, −6, $\frac{1}{2}$, 0, $\frac{7}{9}$

2. Place the correct inequality symbol (< or >) between the real numbers.

   $-\dfrac{3}{5}$ > $-|-2|$

In Exercises 3–18, evaluate the expression. Write fractional answers in simplest form.

3. $16 + (-20)$ -4

4. $-50 - (-60)$ 10

5. $7 + |-3|$ 10

6. $64 - (25 - 8)$ 47

7. $-5(32)$ -160

8. $\dfrac{-72}{-9}$ 8

9. $\dfrac{-15 + 6}{3}$ -3

10. $-\dfrac{(-2)(5)}{10}$ 1

11. $\frac{5}{6} - \frac{1}{8}$ $\frac{17}{24}$

12. $\left(-\frac{9}{50}\right)\left(-\frac{20}{27}\right)$ $\frac{2}{15}$

13. $\dfrac{7}{16} \div \dfrac{21}{28}$ $\frac{7}{12}$

14. $\dfrac{-8.1}{0.3}$ -27

15. $-(0.8)^2$ -0.64

16. $35 - (50 \div 5^2)$ 33

17. $5(3 + 4)^2 - 10$ 235

18. $18 - 7 \cdot 4 + 2^3$ -2

In Exercises 19–22, identify the property of real numbers illustrated by the statement.

19. $3(4 + 6) = 3 \cdot 4 + 3 \cdot 6$ Distributive Property

20. $5 \cdot \frac{1}{5} = 1$ Multiplicative Inverse Property

21. $3 + (4 + 8) = (3 + 4) + 8$ Associative Property of Addition

22. $3(x + 2) = (x + 2)3$ Commutative Property of Multiplication

23. Write the fraction $\frac{36}{162}$ in simplest form. $\frac{2}{9}$

24. Write the prime factorization of 324. $2 \cdot 2 \cdot 3 \cdot 3 \cdot 3 \cdot 3$

25. A jogger runs a race that is 8 miles long in 58 minutes. What is the average speed of the jogger in minutes per mile?   7.25 minutes per mile

26. At the grocery store, you buy two cartons of eggs at $1.59 a carton and three bottles of soda at $1.50 a bottle. You give the clerk a 20-dollar bill. How much change will you receive? (Assume there is no sales tax.) $12.32

Motivating the Chapter

 ## Beachwood Rental

Beachwood Rental is a rental company specializing in equipment for parties and special events. A wedding ceremony is to be held under a canopy that contains 15 rows of 12 chairs.

See Section 2.1, Exercise 91.

a. Let *c* represent the rental cost of a chair. Write an expression that represents the cost of renting all of the chairs under the canopy. The table at the right lists the rental prices for two types of chairs. Use the expression you wrote to find the cost of renting the plastic chairs and the cost of renting the wood chairs. $(15 \cdot 12)c = 180c$; Plastic chairs: $351; Wood chairs: $531

Chair rental	
Plastic	$1.95
Wood	$2.95

b. The table at the right lists the available canopy sizes. The rental rate for a canopy is $115 + 0.25t$ dollars, where *t* represents the size of the canopy in square feet. Find the cost of each canopy. (*Hint:* The total area under a 20 by 20 foot canopy is $20 \cdot 20 = 400$ square feet.)   Canopy 1: $215; Canopy 2: $265; Canopy 3: $415; Canopy 4: $565; Canopy 5: $715

Canopy sizes	
Canopy 1	20 by 20 feet
Canopy 2	20 by 30 feet
Canopy 3	30 by 40 feet
Canopy 4	30 by 60 feet
Canopy 5	40 by 60 feet

The figure at the right shows the arrangement of the chairs under the canopy. Beachwood Rental recommends the following.

> Width of center aisle—Three times the space between rows
> Width of side aisle—Two times the space between rows
> Depth of rear aisle—Two times the space between rows
> Depth of front region—Seven feet more than three times the space between rows

See Section 2.3, Exercise 85.

c. Let *x* represent the space (in feet) between rows of chairs. Write an expression for the width of the center aisle. Write an expression for the width of a side aisle. $3x$; $2x$

d. Each chair is 14 inches wide. Convert the width of a chair to feet. Write an expression for the width of the canopy.   14 in. $= \frac{7}{6}$ ft; $7x + 14$

e. Write an expression for the depth of the rear aisle. Write an expression for the depth of the front region. $2x$; $3x + 7$

f. Each chair is 12 inches deep. Convert the depth of a chair to feet. Write an expression for the depth of the canopy.   12 in. $= 1$ ft; $19x + 22$

g. When $x = 2$ feet, what is the width of the center aisle? What are the width and depth of the canopy? What size canopy do you need? What is the total rental cost of the canopy and chairs if the wood chairs are used?   6 feet; 28 feet; 60 feet; 30 by 60 feet; $1096.00

h. What could be done to save on the rental cost?   Rent plastic chairs.

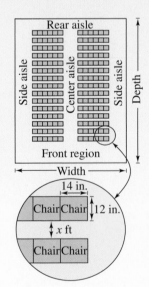

Mark Gibson/Unicorn Stock Photos

2

Fundamentals of Algebra

2.1 Writing and Evaluating Algebraic Expressions

Rubberball Production/Getty Images

Why You Should Learn It

Algebraic expressions can be used to represent real-life quantities, such as weekly income from a part-time job. See Example 1.

What You Should Learn

1. Define and identify terms, variables, and coefficients of algebraic expressions.
2. Define exponential form and interpret exponential expressions.
3. Evaluate algebraic expressions using real numbers.

Variables and Algebraic Expressions

One of the distinguishing characteristics of algebra is its use of symbols to represent quantities whose numerical values are unknown. Here is a simple example.

1. **Define and identify terms, variables, and coefficients of algebraic expressions.**

Example 1   Writing an Algebraic Expression

You accept a part-time job for \$7 per hour. The job offer states that you will be expected to work between 15 and 30 hours a week. Because you don't know how many hours you will work during a week, your total income for a week is unknown. Moreover, your income will probably *vary* from week to week. By representing the variable quantity (the number of hours worked) by the letter x, you can represent the weekly income by the following algebraic expression.

$$\underset{7x}{\overset{\substack{\$7\ per\qquad Number\ of\\ hour\qquad hours\ worked}}{\diagdown\quad\diagup}}$$

In the product $7x$, the number 7 is a *constant* and the letter x is a *variable*.

Definition of Algebraic Expression

A collection of letters (**variables**) and real numbers (**constants**) combined by using addition, subtraction, multiplication, or division is an **algebraic expression.**

Some examples of algebraic expressions are

$$3x + y, \quad -5a^3, \quad 2W - 7, \quad \frac{x}{y + 3}, \quad \text{and} \quad x^2 - 4x + 5.$$

The **terms** of an algebraic expression are those parts that are separated by *addition*. For example, the expression $x^2 - 4x + 5$ has three terms: x^2, $-4x$, and 5. Note that $-4x$, rather than $4x$, is a term of $x^2 - 4x + 5$ because

$$x^2 - 4x + 5 = x^2 + (-4x) + 5. \qquad \text{To subtract, add the opposite.}$$

For variable terms such as x^2 and $-4x$, the numerical factor is the **coefficient** of the term. Here, the coefficient of x^2 is 1 and the coefficient of $-4x$ is -4.

Point out to students that equivalent expressions have different terms. For instance, if Example 2(d) is rewritten as $5x - 15 + 3x - 4$, the terms are $5x$, -15, $3x$, and -4.

Example 2 Identifying the Terms of an Algebraic Expression

Identify the terms of each algebraic expression.

a. $x + 2$ **b.** $3x + \dfrac{1}{2}$

c. $2y - 5x - 7$ **d.** $5(x - 3) + 3x - 4$

e. $4 - 6x + \dfrac{x + 9}{3}$

Solution

Algebraic Expression	Terms
a. $x + 2$	$x, 2$
b. $3x + \dfrac{1}{2}$	$3x, \dfrac{1}{2}$
c. $2y - 5x - 7$	$2y, -5x, -7$
d. $5(x - 3) + 3x - 4$	$5(x - 3), 3x, -4$
e. $4 - 6x + \dfrac{x + 9}{3}$	$4, -6x, \dfrac{x + 9}{3}$

The terms of an algebraic expression depend on the way the expression is written. Rewriting the expression can (and, in fact, usually does) change its terms. For instance, the expression $2 + 4 - x$ has three terms, but the equivalent expression $6 - x$ has only two terms.

Example 3 Identifying Coefficients

Identify the coefficient of each term.

a. $-5x^2$ **b.** x^3

c. $\dfrac{2x}{3}$ **d.** $-\dfrac{x}{4}$

e. $-x^3$

Solution

Term	Coefficient	Comment
a. $-5x^2$	-5	Note that $-5x^2 = (-5)x^2$.
b. x^3	1	Note that $x^3 = 1 \cdot x^3$.
c. $\dfrac{2x}{3}$	$\dfrac{2}{3}$	Note that $\dfrac{2x}{3} = \dfrac{2}{3}(x)$.
d. $-\dfrac{x}{4}$	$-\dfrac{1}{4}$	Note that $-\dfrac{x}{4} = -\dfrac{1}{4}(x)$.
e. $-x^3$	-1	Note that $-x^3 = (-1)x^3$.

Additional Examples
Identify the terms and coefficients of each expression.
a. $-3y^2 - 5x + 7$
b. $-\dfrac{5}{x} + 4x - 1$
c. $3.4a^2 + 6b^2 + 2.1$
Terms
a. $-3y^2, -5x, 7$
b. $-\dfrac{5}{x}, 4x, -1$
c. $3.4a^2, 6b^2, 2.1$
Coefficients
a. $-3, -5, 7$
b. $-5, 4, -1$
c. $3.4, 6, 2.1$

② Define exponential form and interpret exponential expressions.

Exponential Form

You know from Section 1.5 that a number raised to a power can be evaluated by repeated multiplication. For example, 7^4 represents the product obtained by multiplying 7 by itself four times.

$$7^4 = \underbrace{7 \cdot 7 \cdot 7 \cdot 7}_{4 \text{ factors}}$$

Exponent ↗   Base ↓

In general, for any positive integer n and any real number a, you have

$$a^n = \underbrace{a \cdot a \cdot a \cdots a}_{n \text{ factors}}.$$

This rule applies to factors that are variables as well as to factors that are *algebraic expressions*.

Definition of Exponential Form

Let n be a positive integer and let a be a real number, a variable, or an algebraic expression.

$$a^n = \underbrace{a \cdot a \cdot a \cdots a}_{n \text{ factors}}$$

In this definition, remember that the letter a can be a number, a variable, or an algebraic expression. It may be helpful to think of a as a box into which you can place any algebraic expression.

$$\boxed{}^{\,n} = \boxed{} \cdot \boxed{} \cdots$$

The box may contain a number, a variable, or an algebraic expression.

Study Tip

Be sure you understand the difference between repeated addition

$$\underbrace{x + x + x + x}_{4 \text{ terms}} = 4x$$

and repeated multiplication

$$\underbrace{x \cdot x \cdot x \cdot x}_{4 \text{ factors}} = x^4.$$

Example 4   Interpreting Exponential Expressions

a. $3^4 = 3 \cdot 3 \cdot 3 \cdot 3$ **b.** $3x^4 = 3 \cdot x \cdot x \cdot x \cdot x$

c. $(-3x)^4 = (-3x)(-3x)(-3x)(-3x) = (-3)(-3)(-3)(-3) \cdot x \cdot x \cdot x \cdot x$

d. $(y + 2)^3 = (y + 2)(y + 2)(y + 2)$

e. $(5x)^2 y^3 = (5x)(5x)y \cdot y \cdot y = 5 \cdot 5 \cdot x \cdot x \cdot y \cdot y \cdot y$

Be sure you understand the priorities for order of operations involving exponents. Here are some examples that tend to cause problems.

Expression	Correct Evaluation	Incorrect Evaluation
-3^2	$-(3 \cdot 3) = -9$	$\cancel{(-3)(-3) = 9}$
$(-3)^2$	$(-3)(-3) = 9$	$\cancel{-(3 \cdot 3) = -9}$
$3x^2$	$3 \cdot x \cdot x$	$\cancel{(3x)(3x)}$
$-3x^2$	$-3 \cdot x \cdot x$	$\cancel{-(3x)(3x)}$
$(-3x)^2$	$(-3x)(-3x)$	$\cancel{-(3x)(3x)}$

3 Evaluate algebraic expressions using real numbers.

Evaluating Algebraic Expressions

In applications of algebra, you are often required to **evaluate** an algebraic expression. This means you are to find the value of an expression when its variables are substituted by real numbers. For instance, when $x = 2$, the value of the expression $2x + 3$ is as follows.

Expression	Substitute 2 for x.	Value of Expression
$2x + 3$	$2(2) + 3$	7

When finding the value of an algebraic expression, be sure to replace every occurrence of the specified variable with the appropriate real number. For instance, when $x = -2$, the value of $x^2 - x + 3$ is

$$(-2)^2 - (-2) + 3 = 4 + 2 + 3 = 9.$$

Example 5   Evaluating Algebraic Expressions

Evaluate each expression when $x = -3$ and $y = 5$.

a. $-x$

b. $x - y$

c. $3x + 2y$

d. $y - 2(x + y)$

e. $y^2 - 3y$

Solution

Encourage students to use parentheses when replacing a variable with a negative number or a fraction.

a. When $x = -3$, the value of $-x$ is

$$-x = -(-3) \qquad \text{Substitute } -3 \text{ for } x.$$
$$= 3. \qquad \text{Simplify.}$$

b. When $x = -3$ and $y = 5$, the value of $x - y$ is

$$x - y = -3 - 5 \qquad \text{Substitute } -3 \text{ for } x \text{ and } 5 \text{ for } y.$$
$$= -8. \qquad \text{Simplify.}$$

c. When $x = -3$ and $y = 5$, the value of $3x + 2y$ is

$$3x + 2y = 3(-3) + 2(5) \qquad \text{Substitute } -3 \text{ for } x \text{ and } 5 \text{ for } y.$$
$$= -9 + 10 \qquad \text{Simplify.}$$
$$= 1. \qquad \text{Simplify.}$$

d. When $x = -3$ and $y = 5$, the value of $y - 2(x + y)$ is

$$y - 2(x + y) = 5 - 2[(-3) + 5] \qquad \text{Substitute } -3 \text{ for } x \text{ and } 5 \text{ for } y.$$
$$= 5 - 2(2) \qquad \text{Simplify.}$$
$$= 1. \qquad \text{Simplify.}$$

e. When $y = 5$, the value of $y^2 - 3y$ is

$$y^2 - 3y = (5)^2 - 3(5) \qquad \text{Substitute } 5 \text{ for } y.$$
$$= 25 - 15 \qquad \text{Simplify.}$$
$$= 10. \qquad \text{Simplify.}$$

Study Tip

As shown in parts (a), (c), and (d) of Example 5, it is a good idea to use parentheses when substituting a negative number for a variable.

Example 6   Evaluating Algebraic Expressions

Evaluate each expression when $x = 4$ and $y = -6$.

a. y^2 **b.** $-y^2$ **c.** $y - x$ **d.** $|y - x|$ **e.** $|x - y|$

Solution

a. When $y = -6$, the value of the expression y^2 is

$$y^2 = (-6)^2 = 36.$$

b. When $y = -6$, the value of the expression $-y^2$ is

$$-y^2 = -(y^2) = -(-6)^2 = -36.$$

c. When $x = 4$ and $y = -6$, the value of the expression $y - x$ is

$$y - x = (-6) - 4 = -6 - 4 = -10.$$

d. When $x = 4$ and $y = -6$, the value of the expression $|y - x|$ is

$$|y - x| = |-6 - 4| = |-10| = 10.$$

e. When $x = 4$ and $y = -6$, the value of the expression $|x - y|$ is

$$|x - y| = |4 - (-6)| = |4 + 6| = |10| = 10.$$

Example 7   Evaluating Algebraic Expressions

Evaluate each expression when $x = -5$, $y = -2$, and $z = 3$.

a. $\dfrac{y + 2z}{5y - xz}$

b. $(y + 2z)(z - 3y)$

Solution

a. When $x = -5$, $y = -2$, and $z = 3$, the value of the expression is

$$\frac{y + 2z}{5y - xz} = \frac{-2 + 2(3)}{5(-2) - (-5)(3)} \qquad \text{Substitute for } x, y, \text{ and } z.$$

$$= \frac{-2 + 6}{-10 + 15} \qquad \text{Simplify.}$$

$$= \frac{4}{5}. \qquad \text{Simplify.}$$

b. When $y = -2$ and $z = 3$, the value of the expression is

$$(y + 2z)(z - 3y) = [(-2) + 2(3)][3 - 3(-2)] \qquad \text{Substitute for } y \text{ and } z.$$

$$= (-2 + 6)(3 + 6) \qquad \text{Simplify.}$$

$$= (4)(9) \qquad \text{Simplify.}$$

$$= 36. \qquad \text{Simplify.}$$

Remind students to follow the order of operations when evaluating expressions.

Additional Examples

Evaluate each expression.

a. $3x - 7y$ when $x = -2$ and $y = 3$

b. $\dfrac{5ab}{2a - 3b}$ when $a = 5$ and $b = 1$

c. $-x^3$ when $x = -1$

Answers:

a. -27

b. $\frac{25}{7}$

c. 1

Technology: Tip

If you have a graphing calcula-tor, try using it to store and evaluate the expression from Example 8. You can use the following steps to evaluate $-9x + 6$ when $x = 2$.

- Store the expression as Y_1.
- Store 2 in X.

  2 [STO▶] [X,T,Θ,n] [ENTER]

- Display Y_1.

  [VARS] [Y-VARS] [ENTER]

  [ENTER]

  and then press [ENTER] again.

On occasion you may need to evaluate an algebraic expression for *several* values of x. In such cases, a table format is a useful way to organize the values of the expression.

Example 8   Repeated Evaluation of an Expression

Complete the table by evaluating the expression $5x + 2$ for each value of x shown in the table.

x	-1	0	1	2
$5x + 2$				

Solution

Begin by substituting each value of x into the expression.

When $x = -1$: $5x + 2 = 5(-1) + 2 = -5 + 2 = -3$

When $x = 0$: $5x + 2 = 5(0) + 2 = 0 + 2 = 2$

When $x = 1$: $5x + 2 = 5(1) + 2 = 5 + 2 = 7$

When $x = 2$: $5x + 2 = 5(2) + 2 = 10 + 2 = 12$

Once you have evaluated the expression for each value of x, fill in the table with the values.

x	-1	0	1	2
$5x + 2$	-3	2	7	12

Example 9   Geometry: Area

Write an expression for the area of the rectangle shown in Figure 2.1. Then eval-uate the expression to find the area of the rectangle when $x = 7$.

Solution

Area of a rectangle = Length · Width

$$= (x + 5) \cdot x \qquad \text{Substitute.}$$

To evaluate the expression when $x = 7$, substitute 7 for x in the expression for the area of the rectangle.

$$(x + 5) \cdot x = (7 + 5) \cdot 7 \qquad \text{Substitute 7 for } x.$$

$$= 12 \cdot 7 \qquad \text{Simplify.}$$

$$= 84 \qquad \text{Simplify.}$$

So, the area of the rectangle is 84 square units.

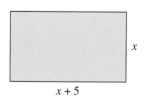

x

$x + 5$

Figure 2.1

2.1 Exercises

Review   Concepts, Skills, and Problem Solving

Keep mathematically in shape by doing these exercises *before* the problems of this section.

Properties and Definitions

In Exercises 1–4, identify the property of real numbers illustrated by the statement.

1. $x(5) = 5x$
   Commutative Property of Multiplication

2. $10 - 10 = 0$
   Additive Inverse Property

3. $3(t + 2) = 3 \cdot t + 3 \cdot 2$
   Distributive Property

4. $7 + (8 + z) = (7 + 8) + z$
   Associative Property of Addition

Simplifying Expressions

In Exercises 5–10, evaluate the expression.

5. $10 - |-7|$ 3

6. $6 - (10 - 12)$ 8

7. $\dfrac{3 - (5 - 20)}{4}$ $\frac{9}{2}$

8. $\dfrac{6}{7} - \dfrac{4}{7}$ $\frac{2}{7}$

9. $-\dfrac{3}{4}\left(\dfrac{28}{33}\right)$ $-\frac{7}{11}$

10. $\dfrac{5}{8} \div \dfrac{3}{16}$ $\frac{10}{3}$

Problem Solving

11. *Savings* You plan to save $50 per month for 10 years. How much money will you set aside during the 10 years? $6000

12. ▲ *Geometry* It is necessary to cut a 120-foot rope into eight pieces of equal length. What is the length of each piece?   15 feet

Developing Skills

In Exercises 1–4, write an algebraic expression for the statement. See Example 1.

1. The distance traveled in t hours if the average speed is 60 miles per hour $60t$

2. The cost of an amusement park ride for a family of n people if the cost per person is $1.25 $1.25n$

3. The cost of m pounds of meat if the cost per pound is $2.19 $2.19m$

4. The total weight of x bags of fertilizer if each bag weighs 50 pounds $50x$

In Exercises 5–10, identify the variables and constants in the expression.

5. $x + 3$
   Variable: x; Constant: 3

6. $y + 1$
   Variable: y; Constant: 1

7. $x + z$ Variables: x, z;
   Constants: none

8. $a + b$ Variables: a, b;
   Constants: none

9. $2^3 + x$
   Variable: x; Constant: 2^3

10. $3^2 + z$
   Variable: z; Constant: 3^2

In Exercises 11–24, identify the terms of the expression. See Example 2.

11. $4x + 3$ $4x, 3$

12. $3x^2 + 5$ $3x^2, 5$

13. $6x - 1$ $6x, -1$

14. $5 - 3t^2$ $5, -3t^2$

15. $\frac{5}{3} - 3y^3$ $\frac{5}{3}, -3y^3$

16. $6x - \frac{2}{3}$ $6x, -\frac{2}{3}$

17. $a^2 + 4ab + b^2$
   $a^2, 4ab, b^2$

18. $x^2 + 18xy + y^2$
   $x^2, 18xy, y^2$

19. $3(x + 5) + 10$
   $3(x + 5), 10$

20. $16 - (x + 1)$
   $16, -(x + 1)$

21. $15 + \dfrac{5}{x}$ $15, \frac{5}{x}$

22. $\dfrac{6}{t} + 22$ $\frac{6}{t}, 22$

23. $\dfrac{3}{x + 2} - 3x + 4$
   $\dfrac{3}{x + 2}, -3x, 4$

24. $\dfrac{5}{x - 5} - 7x^2 + 18$
   $\dfrac{5}{x - 5}, -7x^2, 18$

In Exercises 25–34, identify the coefficient of the term. See Example 3.

25. $14x$ 14

26. $25y$ 25

27. $-\frac{1}{3}y$ $-\frac{1}{3}$

28. $-\frac{2}{3}n$ $-\frac{2}{3}$

29. $\dfrac{2x}{5}$ $\frac{2}{5}$

30. $\dfrac{3x}{4}$ $\frac{3}{4}$

31. $2\pi x^2$ 2π

32. πt^4 π

33. $-3.06u$ -3.06

34. $-5.32b$ -5.32

In Exercises 35–52, expand the expression as a product of factors. See Example 4.

35. y^5 $y \cdot y \cdot y \cdot y \cdot y$ **36.** x^6 $x \cdot x \cdot x \cdot x \cdot x \cdot x$

37. $2^2 x^4$ $2 \cdot 2 \cdot x \cdot x \cdot x \cdot x$ **38.** $5^3 x^2$ $5 \cdot 5 \cdot 5 \cdot x \cdot x$

39. $4y^2 z^3$ $4 \cdot y \cdot y \cdot z \cdot z \cdot z$ **40.** $3uv^4$ $3 \cdot u \cdot v \cdot v \cdot v \cdot v$

41. $(a^2)^3$
$a^2 \cdot a^2 \cdot a^2 = a \cdot a \cdot a \cdot a \cdot a \cdot a$

42. $(z^3)^3$
$z^3 \cdot z^3 \cdot z^3 = z \cdot z \cdot z \cdot z \cdot z \cdot z \cdot z \cdot z \cdot z$

43. $4x^3 \cdot x^4$
$4 \cdot x \cdot x \cdot x \cdot x \cdot x \cdot x \cdot x$

44. $a^2 y^2 \cdot y^3$
$a \cdot a \cdot y \cdot y \cdot y \cdot y \cdot y$

45. $9(ab)^3$
$9 \cdot a \cdot a \cdot a \cdot b \cdot b \cdot b$

46. $2(xz)^4$
$2 \cdot x \cdot x \cdot x \cdot x \cdot z \cdot z \cdot z \cdot z$

47. $(x + y)^2$ $(x + y)(x + y)$

48. $(s - t)^5$ $(s - t)(s - t)(s - t)(s - t)(s - t)$

49. $\left(\dfrac{a}{3s}\right)^4$ $\left(\dfrac{a}{3s}\right)\left(\dfrac{a}{3s}\right)\left(\dfrac{a}{3s}\right)\left(\dfrac{a}{3s}\right)$

50. $\left(\dfrac{2}{5x}\right)^3$ $\left(\dfrac{2}{5x}\right)\left(\dfrac{2}{5x}\right)\left(\dfrac{2}{5x}\right)$

51. $[2(a - b)^3][2(a - b)^2]$
$2 \cdot 2 \cdot (a - b)(a - b)(a - b)(a - b)(a - b)$

52. $[3(r + s)^2][3(r + s)]^2$
$3 \cdot 3 \cdot 3 \cdot (r + s)(r + s)(r + s)(r + s)$

In Exercises 53–62, rewrite the product in exponential form.

53. $2 \cdot u \cdot u \cdot u \cdot u$
$2u^4$

54. $\frac{1}{3} \cdot x \cdot x \cdot x \cdot x \cdot x$
$\frac{1}{3}x^5$

55. $(2u) \cdot (2u) \cdot (2u) \cdot (2u)$
$(2u)^4$

56. $\frac{1}{3}x \cdot \frac{1}{3}x \cdot \frac{1}{3}x \cdot \frac{1}{3}x \cdot \frac{1}{3}x$
$\left(\frac{1}{3}x\right)^5$

57. $a \cdot a \cdot a \cdot b \cdot b$
$a^3 b^2$

58. $y \cdot y \cdot z \cdot z \cdot z \cdot z$
$y^2 z^4$

59. $3 \cdot (x - y) \cdot (x - y) \cdot 3 \cdot 3$ $3^3(x - y)^2$

60. $(u - v) \cdot (u - v) \cdot 8 \cdot 8 \cdot 8 \cdot (u - v)$ $8^3(u - v)^3$

61. $\dfrac{x + y}{4} \cdot \dfrac{x + y}{4} \cdot \dfrac{x + y}{4}$ $\left(\dfrac{x + y}{4}\right)^3$

62. $\dfrac{r - s}{5} \cdot \dfrac{r - s}{5} \cdot \dfrac{r - s}{5} \cdot \dfrac{r - s}{5}$ $\left(\dfrac{r - s}{5}\right)^4$

In Exercises 63–80, evaluate the algebraic expression for the given values of the variable(s). If it is not possible, state the reason. See Examples 5, 6, and 7.

Expression	Values
63. $2x - 1$	(a) $x = \frac{1}{2}$ 0 (b) $x = -4$ -9
64. $3x - 2$	(a) $x = \frac{4}{3}$ 2 (b) $x = -1$ -5
65. $2x^2 - 5$	(a) $x = -2$ 3
	(b) $x = 3$ 13
66. $64 - 16t^2$	(a) $t = 2$ 0 (b) $t = -3$ -80
67. $3x - 2y$	(a) $x = 4, y = 3$ 6
	(b) $x = \frac{2}{3}, y = -1$ 4
68. $10u - 3v$	(a) $u = 3, v = 10$ 0
	(b) $u = -2, v = \frac{4}{7}$ $-\frac{152}{7}$
69. $x - 3(x - y)$	(a) $x = 3, y = 3$ 3
	(b) $x = 4, y = -4$ -20
70. $-3x + 2(x + y)$	(a) $x = -2, y = 2$ 6
	(b) $x = 0, y = 5$ 10
71. $b^2 - 4ab$	(a) $a = 2, b = -3$ 33
	(b) $a = 6, b = -4$ 112
72. $a^2 + 2ab$	(a) $a = -2, b = 3$ -8
	(b) $a = 4, b = -2$ 0
73. $\dfrac{x - 2y}{x + 2y}$	(a) $x = 4, y = 2$ 0
	(b) $x = 4, y = -2$ Division by zero is undefined.
74. $\dfrac{5x}{y - 3}$	(a) $x = 2, y = 4$ 10
	(b) $x = 2, y = 3$ Division by zero is undefined.
75. $\dfrac{-y}{x^2 + y^2}$	(a) $x = 0, y = 5$ $-\frac{1}{5}$
	(b) $x = 1, y = -3$ $\frac{3}{10}$
76. $\dfrac{2x - y}{y^2 + 1}$	(a) $x = 1, y = 2$ 0
	(b) $x = 1, y = 3$ $-\frac{1}{10}$
77. *Area of a Triangle* $\frac{1}{2}bh$	(a) $b = 3, h = 5$ $\frac{15}{2}$
	(b) $b = 2, h = 10$ 10
78. *Distance Traveled* rt	(a) $r = 50, t = 3.5$ 175
	(b) $r = 35, t = 4$ 140
79. *Volume of a Rectangular Prism* lwh	(a) $l = 4, w = 2, h = 9$ 72
	(b) $l = 100, w = 0.8, h = 4$ 320

Expression   *Values*

80. *Simple Interest*

Prt (a) $P = 1000$, $r = 0.08$, $t = 3$

240

(b) $P = 500$, $r = 0.07$, $t = 5$

175

81. *Finding a Pattern*

(a) Complete the table by evaluating the expression $3x - 2$. See Example 8.

x	-1	0	1	2	3	4
$3x - 2$	-5	-2	1	4	7	10

(b) Use the table to find the increase in the value of the expression for each one-unit increase in x. 3

(c) From the pattern of parts (a) and (b), predict the increase in the algebraic expression $\frac{2}{3}x + 4$ for each one-unit increase in x. Then verify your prediction. $\frac{2}{3}$

82. *Finding a Pattern*

(a) Complete the table by evaluating the expression $3 - 2x$. See Example 8.

x	-1	0	1	2	3	4
$3 - 2x$	5	3	1	-1	-3	-5

(b) Use the table to find the change in the value of the expression for each one-unit increase in x. -2

(c) From the pattern of parts (a) and (b), predict the change in the algebraic expression $4 - \frac{3}{2}x$ for each one-unit increase in x. Then verify your prediction. $-\frac{3}{2}$

Solving Problems

Geometry   In Exercises 83–86, find an expression for the area of the figure. Then evaluate the expression for the given value(s) of the variable(s).

83. $n = 8$

$(n - 5)^2$, 9 square units

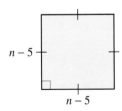

84. $x = 10$, $y = 3$

$(x + y)^2$, 169 square units

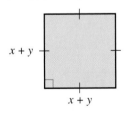

85. $a = 5$, $b = 4$ $a(a + b)$, 45 square units

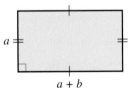

86. $x = 9$ $x(x + 3)$, 108 square units

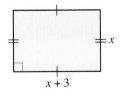

87. *Exploration*   For any natural number n, the sum of the numbers $1, 2, 3, \ldots, n$ is equal to

$$\frac{n(n + 1)}{2}, \quad n \geq 1$$

Verify the formula for (a) $n = 3$, (b) $n = 6$, and (c) $n = 10$.

(a) $\frac{3(4)}{2} = 6 = 1 + 2 + 3$

(b) $\frac{6(7)}{2} = 21 = 1 + 2 + 3 + 4 + 5 + 6$

(c) $\frac{10(11)}{2} = 55 = 1 + 2 + 3 + 4 + 5 + 6 + 7 + 8 + 9 + 10$

88. *Exploration*   A convex polygon with n sides has

$$\frac{n(n - 3)}{2}, \quad n \geq 4$$

diagonals. Verify the formula for (a) a square (two diagonals), (b) a pentagon (five diagonals), and (c) a hexagon (nine diagonals).

(a) Square: $\frac{4(1)}{2} = 2$ diagonals

(b) Pentagon: $\frac{5(2)}{2} = 5$ diagonals

(c) Hexagon: $\frac{6(3)}{2} = 9$ diagonals

89. ⊞ *Iteration and Exploration*   Once an expression has been evaluated for a specified value, the expression can be repeatedly evaluated by using the result of the preceding evaluation as the input for the next evaluation.

(a) The procedure for repeated evaluation of the algebraic expression $\frac{1}{2}x + 3$ can be accomplished on a graphing calculator, as follows.

- Clear the display.

- Enter 2 in the display and press ENTER.

- Enter $\frac{1}{2} *$ ANS $+ 3$ and press ENTER.

- Each time ENTER is pressed, the calculator will evaluate the expression at the value of x obtained in the preceding computation. Continue the process six more times. What value does the expression appear to be approaching? If necessary, round your answers to three decimal places.   4, 5, 5.5, 5.75, 5.875, 5.938, 5.969; Approaches 6.

(b) Repeat part (a) starting with $x = 12$.
9, 7.5, 6.75, 6.375, 6.188, 6.094, 6.047; Approaches 6.

90. ⊞ *Exploration*   Repeat Exercise 89 using the expression $\frac{3}{4}x + 2$. If necessary, round your answers to three decimal places.

(a) 3.5, 4.625, 5.469, 6.102, 6.576, 6.932, 7.199;
Approaches 8.

(b) 11, 10.25, 9.688, 9.266, 8.949, 8.712, 8.534;
Approaches 8.

Explaining Concepts

91. ⚙ Answer parts (a) and (b) of Motivating the Chapter on page 66.

92. *Writing*🖉 Discuss the difference between terms and factors.   Addition separates terms. Multiplication separates factors.

93. *Writing*🖉 Is $3x$ a term of $4 - 3x$? Explain.
No. The term includes the minus sign and is $-3x$.

94. In the expression $(10x)^3$, what is $10x$ called? What is 3 called? $10x$ is the base and 3 is the exponent.

95. *Writing*🖉 Is it possible to evaluate the expression

$$\frac{x + 2}{y - 3}$$

when $x = 5$ and $y = 3$? Explain.
No. When $y = 3$, the expression is undefined.

96. *Writing*🖉 Explain why the formulas in Exercises 87 and 88 will always yield natural numbers.

Either n or $n + 1$ is even. Therefore every product $n(n + 1)$ is divisible by 2. Either n or $n - 3$ is even. Therefore every product $n(n - 3)$ is divisible by 2.

97. *Writing*🖉 You are teaching an algebra class and one of your students hands in the following problem. Evaluate $y - 2(x - y)$ when $x = 2$ and $y = -4$.

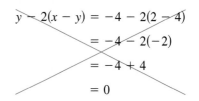

$$y - 2(x - y) = -4 - 2(2 - 4)$$
$$= -4 - 2(-2)$$
$$= -4 + 4$$
$$= 0$$

What is the error in this work? What are some possible related errors? Discuss ways of helping students avoid these types of errors.

$$y - 2(x - y) = -4 - 2[2 - (-4)]$$
$$= -4 - 2(2 + 4)$$
$$= -4 - 2(6)$$
$$= -4 - 12$$
$$= -16$$

Discussions will vary.

The symbol ⊞ indicates an exercise in which you are instructed to use a graphing calculator.

2.2 Simplifying Algebraic Expressions

Bill Pogue/Getty Images

What You Should Learn

1️⃣ Use the properties of algebra.

2️⃣ Combine like terms of an algebraic expression.

3️⃣ Simplify an algebraic expression by rewriting the terms.

4️⃣ Use the Distributive Property to remove symbols of grouping.

Why You Should Learn It

You can use an algebraic expression to find the area of a house lot, as shown in Exercise 157 on page 89.

1️⃣ **Use the properties of algebra.**

Study Tip

You'll discover as you review the table of properties at the right that they are the same as the properties of real numbers on page 52. The only difference is that the input for algebra rules can be real numbers, variables, or algebraic expressions.

Properties of Algebra

You are now ready to combine algebraic expressions using the properties below.

Properties of Algebra

Let a, b, and c represent real numbers, variables, or algebraic expressions.

Property	*Example*
Commutative Property of Addition:	
$a + b = b + a$	$3x + x^2 = x^2 + 3x$
Commutative Property of Multiplication:	
$ab = ba$	$(5 + x)x = x(5 + x)$
Associative Property of Addition:	
$(a + b) + c = a + (b + c)$	$(2x + 7) + x^2 = 2x + (7 + x^2)$
Associative Property of Multiplication:	
$(ab)c = a(bc)$	$(2x \cdot 5y) \cdot 7 = 2x \cdot (5y \cdot 7)$
Distributive Property:	
$a(b + c) = ab + ac$	$4x(7 + 3x) = 4x \cdot 7 + 4x \cdot 3x$
$(a + b)c = ac + bc$	$(2y + 5)y = 2y \cdot y + 5 \cdot y$
Additive Identity Property:	
$a + 0 = 0 + a = a$	$3y^2 + 0 = 0 + 3y^2 = 3y^2$
Multiplicative Identity Property:	
$a \cdot 1 = 1 \cdot a = a$	$(-2x^3) \cdot 1 = 1 \cdot (-2x^3) = -2x^3$
Additive Inverse Property:	
$a + (-a) = 0$	$3y^2 + (-3y^2) = 0$
Multiplicative Inverse Property:	
$a \cdot \dfrac{1}{a} = 1, \quad a \neq 0$	$(x^2 + 2) \cdot \dfrac{1}{x^2 + 2} = 1$

Example 1   Applying the Basic Rules of Algebra

Use the indicated rule to complete each statement.

a. Additive Identity Property: $(x - 2) + \quad = x - 2$

b. Commutative Property of Multiplication: $5(y + 6) =$

c. Commutative Property of Addition: $5(y + 6) =$

d. Distributive Property: $5(y + 6) =$

e. Associative Property of Addition: $(x^2 + 3) + 7 =$

f. Additive Inverse Property: $\quad + 4x = 0$

Solution

a. $(x - 2) + 0 = x - 2$

b. $5(y + 6) = (y + 6)5$

c. $5(y + 6) = 5(6 + y)$

d. $5(y + 6) = 5y + 5(6)$

e. $(x^2 + 3) + 7 = x^2 + (3 + 7)$

f. $-4x + 4x = 0$

Example 2 illustrates some common uses of the Distributive Property. Study this example carefully. Such uses of the Distributive Property are very important in algebra. Applying the Distributive Property as illustrated in Example 2 is called **expanding** an algebraic expression.

Example 2   Using the Distributive Property

Use the Distributive Property to expand each expression.

a. $2(7 - x)$ **b.** $(10 - 2y)3$ **c.** $2x(x + 4y)$ **d.** $-(1 - 2y + x)$

Solution

a. $2(7 - x) = 2 \cdot 7 - 2 \cdot x$

$\quad\quad\quad\quad = 14 - 2x$

b. $(10 - 2y)3 = 10(3) - 2y(3)$

$\quad\quad\quad\quad\quad = 30 - 6y$

c. $2x(x + 4y) = 2x(x) + 2x(4y)$

$\quad\quad\quad\quad\quad = 2x^2 + 8xy$

d. $-(1 - 2y + x) = (-1)(1 - 2y + x)$

$\quad\quad\quad\quad\quad\quad = (-1)(1) - (-1)(2y) + (-1)(x)$

$\quad\quad\quad\quad\quad\quad = -1 + 2y - x$

Study Tip

In Example 2(d), the negative sign is distributed over each term in the parentheses by multiplying each term by -1.

In the next example, note how area can be used to demonstrate the Distributive Property.

Example 3   The Distributive Property and Area

Write the area of each component of the figure. Then demonstrate the Distributive Property by writing the total area of each figure in two ways.

a. 2   4
3 | ⊢ 2 + 4 ⊣

b. a   b
a | ⊢ a + b ⊣

c. d   3a   c
2b | ⊢ d + 3a + c ⊣

Solution

a.   2   4
3 | 6 | 12

The total area is $3(2 + 4) = 3 \cdot 2 + 3 \cdot 4 = 6 + 12 = 18$.

b.   a   b
a | a^2 | ab

The total area is $a(a + b) = a \cdot a + a \cdot b = a^2 + ab$.

c.   d   3a   c
2b | $2bd$ | $6ab$ | $2bc$

The total area is $2b(d + 3a + c) = 2bd + 6ab + 2bc$.

② Combine like terms of an algebraic expression.

Combining Like Terms

Two or more terms of an algebraic expression can be combined only if they are **like terms**.

Definition of Like Terms

In an algebraic expression, two terms are said to be **like terms** if they are both constant terms or if they have the same variable factor(s). Factors such as x in $5x$ and ab in $6ab$ are called **variable factors.**

The terms $5x$ and $-3x$ are like terms because they have the same variable factor, x. Similarly, $3x^2y$, $-x^2y$, and $\frac{1}{3}(x^2y)$ are like terms because they have the same variable factors, x^2 and y.

Study Tip

Notice in Example 4(b) that x^2 and $3x$ are *not* like terms because the variable x is not raised to the same power in both terms.

Example 4   Identifying Like Terms in Expressions

Expression	Like Terms
a. $5xy + 1 - xy$	$5xy$ and $-xy$
b. $12 - x^2 + 3x - 5$	12 and -5
c. $7x - 3 - 2x + 5$	$7x$ and $-2x$, -3 and 5

Additional Examples

Simplify each expression by combining like terms.

a. $3a - 2b + 5b - 7a$

b. $-7 + 8 - 2x + 6x$

c. $3y - 7x + 6y - 8$

Answers:

a. $-4a + 3b$

b. $4x + 1$

c. $9y - 7x - 8$

To combine like terms in an algebraic expression, you can simply add their respective coefficients and attach the common variable factor. This is actually an application of the Distributive Property, as shown in Example 5.

Example 5   Combining Like Terms

Simplify each expression by combining like terms.

a. $5x + 2x - 4$ **b.** $-5 + 8 + 7y - 5y$ **c.** $2y - 3x - 4x$

Solution

a. $5x + 2x - 4 = (5 + 2)x - 4$		Distributive Property
$= 7x - 4$		Simplest form
b. $-5 + 8 + 7y - 5y = (-5 + 8) + (7 - 5)y$		Distributive Property
$= 3 + 2y$		Simplest form
c. $2y - 3x - 4x = 2y - x(3 + 4)$		Distributive Property
$= 2y - x(7)$		Simplify.
$= 2y - 7x$		Simplest form

Often, you need to use other rules of algebra before you can apply the Distributive Property to combine like terms. This is illustrated in the next example.

Example 6   Using Rules of Algebra to Combine Like Terms

Simplify each expression by combining like terms.

a. $7x + 3y - 4x$ **b.** $12a - 5 - 3a + 7$ **c.** $y - 4x - 7y + 9y$

Solution

a. $7x + 3y - 4x = 3y + 7x - 4x$	Commutative Property
$= 3y + (7x - 4x)$	Associative Property
$= 3y + (7 - 4)x$	Distributive Property
$= 3y + 3x$	Simplest form
b. $12a - 5 - 3a + 7 = 12a - 3a - 5 + 7$	Commutative Property
$= (12a - 3a) + (-5 + 7)$	Associative Property
$= (12 - 3)a + (-5 + 7)$	Distributive Property
$= 9a + 2$	Simplest form
c. $y - 4x - 7y + 9y = -4x + (y - 7y + 9y)$	Group like terms.
$= -4x + (1 - 7 + 9)y$	Distributive Property
$= -4x + 3y$	Simplest form

Study Tip

As you gain experience with the rules of algebra, you may want to combine some of the steps in your work. For instance, you might feel comfortable listing only the following steps to solve part (b) of Example 6.

$12a - 5 - 3a + 7$

$= (12a - 3a) + (-5 + 7)$

$= 9a + 2$

③ Simplify an algebraic expression by rewriting the terms.

Simplifying Algebraic Expressions

Simplifying an algebraic expression by rewriting it in a more usable form is one of the three most frequently used skills in algebra. You will study the other two—solving an equation and sketching the graph of an equation—later in this text.

To **simplify an algebraic expression** generally means to remove symbols of grouping and combine like terms. For instance, the expression $x + (3 + x)$ can be simplified as $2x + 3$.

Example 7   Simplifying Algebraic Expressions

Simplify each expression.

a. $-3(-5x)$ **b.** $7(-x)$

Solution

a. $-3(-5x) = (-3)(-5)x$ Associative Property

            $= 15x$ Simplest form

b. $7(-x) = 7(-1)(x)$ Coefficient of $-x$ is -1.

          $= -7x$ Simplest form

Example 8   Simplifying Algebraic Expressions

Simplify each expression.

a. $\dfrac{5x}{3} \cdot \dfrac{3}{5}$ **b.** $x^2(-2x^3)$ **c.** $(-2x)(4x)$ **d.** $(2rs)(r^2s)$

Solution

a. $\dfrac{5x}{3} \cdot \dfrac{3}{5} = \left(\dfrac{5}{3} \cdot x\right) \cdot \dfrac{3}{5}$ Coefficient of $\dfrac{5x}{3}$ is $\dfrac{5}{3}$.

            $= \left(\dfrac{5}{3} \cdot \dfrac{3}{5}\right) \cdot x$ Commutative and Associative Properties

            $= 1 \cdot x$ Multiplicative Inverse

            $= x$ Multiplicative Identity

b. $x^2(-2x^3) = (-2)(x^2 \cdot x^3)$ Commutative and Associative Properties

            $= -2 \cdot x \cdot x \cdot x \cdot x \cdot x$ Repeated multiplication

            $= -2x^5$ Exponential form

c. $(-2x)(4x) = (-2 \cdot 4)(x \cdot x)$ Commutative and Associative Properties

            $= -8x^2$ Exponential form

d. $(2rs)(r^2s) = 2(r \cdot r^2)(s \cdot s)$ Commutative and Associative Properties

            $= 2 \cdot r \cdot r \cdot r \cdot s \cdot s$ Repeated multiplication

            $= 2r^3s^2$ Exponential form

④ Use the Distributive Property to remove symbols of grouping.

Symbols of Grouping

The main tool for removing symbols of grouping is the Distributive Property, as illustrated in Example 9. You may want to review order of operations in Section 1.5.

Study Tip

When a parenthetical expression is preceded by a *plus* sign, you can remove the parentheses without changing the signs of the terms inside.

$$3y + (-2y + 7)$$
$$= 3y - 2y + 7$$

When a parenthetical expression is preceded by a *minus* sign, however, you must change the sign of each term to remove the parentheses.

$$3y - (2y - 7)$$
$$= 3y - 2y + 7$$

Remember that $-(2y - 7)$ is equal to $(-1)(2y - 7)$, and the Distributive Property can be used to "distribute the minus sign" to obtain $-2y + 7$.

Example 9   Removing Symbols of Grouping

Simplify each expression.

a. $-(2y - 7)$ **b.** $5x + (x - 7)2$

c. $-2(4x - 1) + 3x$ **d.** $3(y - 5) - (2y - 7)$

Solution

a. $-(2y - 7) = -2y + 7$ Distributive Property

b. $5x + (x - 7)2 = 5x + 2x - 14$ Distributive Property

$\qquad\qquad\qquad = 7x - 14$ Combine like terms.

c. $-2(4x - 1) + 3x = -8x + 2 + 3x$ Distributive Property

$\qquad\qquad\qquad = -8x + 3x + 2$ Commutative Property

$\qquad\qquad\qquad = -5x + 2$ Combine like terms.

d. $3(y - 5) - (2y - 7) = 3y - 15 - 2y + 7$ Distributive Property

$\qquad\qquad\qquad = (3y - 2y) + (-15 + 7)$ Group like terms.

$\qquad\qquad\qquad = y - 8$ Combine like terms.

Example 10   Removing Nested Symbols of Grouping

Simplify each expression.

a. $5x - 2[4x + 3(x - 1)]$

b. $-7y + 3[2y - (3 - 2y)] - 5y + 4$

Solution

a. $5x - 2[4x + 3(x - 1)]$

$\qquad = 5x - 2[4x + 3x - 3]$ Distributive Property

$\qquad = 5x - 2[7x - 3]$ Combine like terms.

$\qquad = 5x - 14x + 6$ Distributive Property

$\qquad = -9x + 6$ Combine like terms.

b. $-7y + 3[2y - (3 - 2y)] - 5y + 4$

$\qquad = -7y + 3[2y - 3 + 2y] - 5y + 4$ Distributive Property

$\qquad = -7y + 3[4y - 3] - 5y + 4$ Combine like terms.

$\qquad = -7y + 12y - 9 - 5y + 4$ Distributive Property

$\qquad = (-7y + 12y - 5y) + (-9 + 4)$ Group like terms.

$\qquad = -5$ Combine like terms.

Example 11   Simplifying an Algebraic Expression

Simplify $2x(x + 3y) + 4(5 - xy)$.

Solution

$$2x(x + 3y) + 4(5 - xy) = 2x \cdot x + 6xy + 20 - 4xy \qquad \text{Distributive Property}$$

$$= 2x^2 + 6xy - 4xy + 20 \qquad \text{Commutative Property}$$

$$= 2x^2 + 2xy + 20 \qquad \text{Combine like terms.}$$

The next example illustrates the use of the Distributive Property with a fractional expression.

Example 12   Simplifying a Fractional Expression

Simplify $\dfrac{x}{4} + \dfrac{2x}{7}$.

Solution

$$\frac{x}{4} + \frac{2x}{7} = \frac{1}{4}x + \frac{2}{7}x \qquad \text{Write with fractional coefficients.}$$

$$= \left(\frac{1}{4} + \frac{2}{7}\right)x \qquad \text{Distributive Property}$$

$$= \left[\frac{1(7)}{4(7)} + \frac{2(4)}{7(4)}\right]x \qquad \text{Common denominator}$$

$$= \frac{15}{28}x \qquad \text{Simplest form}$$

Example 13   Geometry: Perimeter and Area

Using Figure 2.2, write and simplify an expression for (a) the perimeter and (b) the area of the triangle.

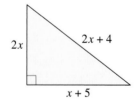

Figure 2.2

Solution

a. Perimeter of a Triangle $=$ Sum of the Three Sides

$$= 2x + (2x + 4) + (x + 5) \qquad \text{Substitute.}$$

$$= (2x + 2x + x) + (4 + 5) \qquad \text{Group like terms.}$$

$$= 5x + 9 \qquad \text{Combine like terms.}$$

b. Area of a Triangle $= \frac{1}{2} \cdot$ Base $\cdot$ Height

$$= \frac{1}{2}(x + 5)(2x) \qquad \text{Substitute.}$$

$$= \frac{1}{2}(2x)(x + 5) \qquad \text{Commutative Property}$$

$$= x(x + 5) \qquad \text{Multiply.}$$

$$= x^2 + 5x \qquad \text{Distributive Property}$$

2.2 Exercises

Review   Concepts, Skills, and Problem Solving

Keep in mathematical shape by doing these exercises *before* the problems of this section.

Properties and Definitions

1. *Writing*   Explain what it means to find the prime factorization of a number.   To find the prime factorization of a number is to write the number as a product of prime factors.

2. Identify the property of real numbers illustrated by the statement: $\frac{1}{2}(4x + 10) = 2x + 5$.
   Distributive Property

Simplifying Expressions

In Exercises 3–10, perform the operation.

3. $0 - (-12)$ 12

4. $60 - (-60)$ 120

5. $-12 - 2 + |-3|$ -11

6. $-730 + 1820 + 3150 + (-10,000)$ -5760

7. Find the sum of 72 and -37.   35

8. Subtract 600 from 250. -350

9. $\frac{5}{16} - \frac{3}{10}$ $\frac{1}{80}$

10. $\frac{9}{16} + 2\frac{3}{12}$ $\frac{45}{16}$

Problem Solving

11. *Profit*   An athletic shoe company showed a loss of $1,530,000 during the first 6 months of 2003. The company ended the year with an overall profit of $832,000. What was the profit during the last two quarters of the year?   2,362,000

12. *Average Speed*   A family on vacation traveled 676 miles in 13 hours. Determine their average speed in miles per hour.   52 miles per hour

Developing Skills

In Exercises 1–22, identify the property (or properties) of algebra illustrated by the statement. See Example 1.

1. $3a + 5b = 5b + 3a$ Commutative Property of Addition

2. $x + 2y = 2y + x$ Commutative Property of Addition

3. $-10(xy^2) = (-10x)y^2$
   Associative Property of Multiplication

4. $(9x)y = 9(xy)$ Associative Property of Multiplication

5. $rt + 0 = rt$ Additive Identity Property

6. $-8x + 0 = -8x$ Additive Identity Property

7. $(x^2 + y^2) \cdot 1 = x^2 + y^2$ Multiplicative Identity Property

8. $1 \cdot (5z + 12) = 5z + 12$
   Multiplicative Identity Property

9. $(3x + 2y) + z = 3x + (2y + z)$
   Associative Property of Addition

10. $-4a + (b^2 + 2c) = (-4a + b^2) + 2c$
    Associative Property of Addition

11. $2zy = 2yz$ Commutative Property of Multiplication

12. $-7a^2c = -7ca^2$
    Commutative Property of Multiplication

13. $-5x(y + z) = -5xy - 5xz$ Distributive Property

14. $x(y + z) = xy + xz$ Distributive Property

15. $(5m + 3) - (5m + 3) = 0$ Additive Inverse Property

16. $(2x - 10) - (2x - 10) = 0$ Additive Inverse Property

17. $16xy \cdot \dfrac{1}{16xy} = 1, \quad xy \neq 0$
    Multiplicative Inverse Property

18. $(x + y) \cdot \dfrac{1}{(x + y)} = 1, \quad x + y \neq 0$
    Multiplicative Inverse Property

19. $(x + 2)(x + y) = x(x + y) + 2(x + y)$
    Distributive Property

20. $(a + 6)(b + 2c) = (a + 6)b + (a + 6)2c$
    Distributive Property

21. $x^2 + (y^2 - y^2) = x^2$
    Additive Inverse Property, Additive Identity Property

22. $3y + (z^3 - z^3) = 3y$
    Additive Inverse Property, Additive Identity Property

In Exercises 23–34, complete the statement. Then state the property of algebra that you used. See Example 1.

23. $(-5r)s = -5(\;\;\; rs \;\;\;)$
    Associative Property of Multiplication

24. $(7x)y^2 = 7(\;\;\; xy^2 \;\;\;)$
    Associative Property of Multiplication

25. $v(2) = \;\;\; 2v$
    Commutative Property of Multiplication

26. $(2x - y)(-3) = -3 \;(2x - y)$
    Commutative Property of Multiplication

27. $5(t - 2) = 5(\quad 5t \quad) + 5(\quad -2 \quad)$
Distributive Property

28. $x(y + 4) = x(\quad y \quad) + x(\quad 4 \quad)$
Distributive Property

29. $(2z - 3) + \boxed{[-(2z - 3)]} = 0$
Additive Inverse Property

30. $(x + 10) + \boxed{[-(x + 10)]} = 0$
Additive Inverse Property

31. $5x(\quad 1/(5x) \quad) = 1, \quad x \neq 0$
Multiplicative Inverse Property

32. $4z^2(\quad 1/(4z^2) \quad) = 1, z \neq 0$
Multiplicative Inverse Property

33. $12 + (8 - x) = \quad (12 + 8) \quad - x$
Associative Property of Addition

34. $-11 + (5 + 2y) = \quad (-11 + 5) \quad + 2y$
Associative Property of Addition

In Exercises 35–62, use the Distributive Property to expand the expression. See Example 2.

35. $2(16 + 8z)$ $32 + 16z$

36. $5(7 + 3x)$ $35 + 15x$

37. $8(-3 + 5m)$ $-24 + 40m$

38. $12(-2 + y)$ $-24 + 12y$

39. $10(9 - 6x)$ $90 - 60x$

40. $3(7 - 4a)$ $21 - 12a$

41. $-8(2 + 5t)$ $-16 - 40t$

42. $-9(4 + 2b)$ $-36 - 18b$

43. $-5(2x - y)$ $-10x + 5y$

44. $-3(11y - 4)$ $-33y + 12$

45. $(x + 2)(3)$ $3x + 6$

46. $(r + 12)(2)$ $2r + 24$

47. $(4 - t)(-6)$ $-24 + 6t$

48. $(3 - x)(-5)$ $-15 + 5x$

49. $4(x + xy + y^2)$ $4x + 4xy + 4y^2$

50. $6(r - t + s)$ $6r - 6t + 6s$

51. $3(x^2 + x)$ $3x^2 + 3x$

52. $9(a^2 + a)$ $9a^2 + 9a$

53. $4(2y^2 - y)$ $8y^2 - 4y$

54. $5(3x^2 - x)$ $15x^2 - 5x$

55. $-z(5 - 2z)$ $-5z + 2z^2$

56. $-t(12 - 4t)$ $-12t + 4t^2$

57. $-4y(3y - 4)$ $-12y^2 + 16y$

58. $-6s(6s - 1)$ $-36s^2 + 6s$

59. $-(u - v)$ $-u + v$

60. $-(x + y)$ $-x - y$

61. $x(3x - 4y)$ $3x^2 - 4xy$

62. $r(2r^2 - t)$ $2r^3 - rt$

In Exercises 63–66, write the area of each component of the figure. Then demonstrate the Distributive Property by writing the total area of each figure in two ways. See Example 3.

63.

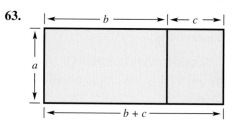

$ab;\ ac;\ a(b + c) = ab + ac$

64.

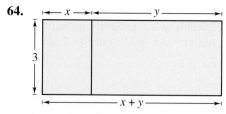

$3x;\ 3y;\ 3(x + y) = 3x + 3y$

65.

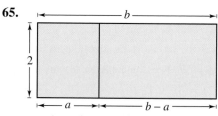

$2a;\ 2(b - a);\ 2a + 2(b - a) = 2b$

66.

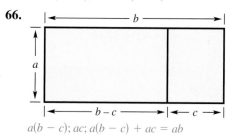

$a(b - c);\ ac;\ a(b - c) + ac = ab$

In Exercises 67–70, identify the terms of the expression and the coefficient of each term.

67. $6x^2 - 3xy + y^2$ $6x^2, -3xy, y^2; 6, -3, 1$

68. $4a^2 - 9ab + b^2$ $4a^2, -9ab, b^2; 4, -9, 1$

69. $-ab + 5ac - 7bc$ $-ab, 5ac, -7bc; -1, 5, -7$

70. $-4xy + 2xz - yz$ $-4xy, 2xz, -yz; -4, 2, -1$

In Exercises 71–76, identify the like terms. See Example 4.

71. $16t^3 + 4t - 5t + 3t^3$ $16t^3, 3t^3; 4t, -5t$

72. $-\frac{1}{4}x^2 - 3x + \frac{3}{4}x^2 + x$ $-\frac{1}{4}x^2, \frac{3}{4}x^2; -3x, x$

73. $4rs^2 - 2r^2s + 12rs^2$ $4rs^2, 12rs^2$

74. $6x^2y + 2xy - 4x^2y$ $6x^2y, -4x^2y$

75. $x^3 + 4x^2y - 2y^2 + 5xy^2 + 10x^2y + 3x^3$
$4x^2y, 10x^2y; x^3, 3x^3$

76. $a^2 + 5ab^2 - 3b^2 + 7a^2b - ab^2 + a^2$
$a^2, a^2; 5ab^2, -ab^2$

In Exercises 77–96, simplify the expression by combining like terms. See Examples 5 and 6.

77. $3y - 5y$ $-2y$

78. $-16x + 25x$ $9x$

79. $x + 5 - 3x$ $-2x + 5$

80. $7s + 3 - 3s$ $4s + 3$

81. $2x + 9x + 4$ $11x + 4$

82. $10x - 4 - 5x$ $5x - 4$

83. $5r + 6 - 2r + 1$ $3r + 7$

84. $2t - 4 + 8t + 9$ $10t + 5$

85. $x^2 - 2xy + 4 + xy$ $x^2 - xy + 4$

86. $r^2 + 3rs - 6 - rs$ $r^2 + 2rs - 6$

87. $5z - 5 + 10z + 2z + 16$ $17z + 11$

88. $7x - 4x + 8 + 3x - 6$ $6x + 2$

89. $z^3 + 2z^2 + z + z^2 + 2z + 1$ $z^3 + 3z^2 + 3z + 1$

90. $3x^2 - x^2 + 4x + 3x^2 - x + x^2$ $6x^2 + 3x$

91. $2x^2y + 5xy^2 - 3x^2y + 4xy + 7xy^2$
$-x^2y + 4xy + 12xy^2$

92. $6rt - 3r^2t + 2rt^2 - 4rt - 2r^2t$
$2rt - 5r^2t + 2rt^2$

93. $3\left(\frac{1}{x}\right) - \frac{1}{x} + 8$ $2\left(\frac{1}{x}\right) + 8$

94. $1.2\left(\frac{1}{x}\right) + 3.8\left(\frac{1}{x}\right) - 4x$ $5\left(\frac{1}{x}\right) - 4x$

95. $5\left(\frac{1}{t}\right) + 6\left(\frac{1}{t}\right) - 2t$ $11\left(\frac{1}{t}\right) - 2t$

96. $16\left(\frac{a}{b}\right) - 6\left(\frac{a}{b}\right) + \frac{3}{2} - \frac{1}{2}$ $10\left(\frac{a}{b}\right) + 1$

True or False? In Exercises 97–100, determine whether the statement is true or false. Justify your answer.

97. $3(x - 4) \stackrel{?}{=} 3x - 4$ False. $3(x - 4) = 3x - 12$

98. $-3(x - 4) \stackrel{?}{=} -3x - 12$
False. $-3(x - 4) = -3x + 12$

99. $6x - 4x \stackrel{?}{=} 2x$ True. $6x - 4x = 2x$

100. $12y^2 + 3y^2 \stackrel{?}{=} 36y^2$ False. $12y^2 + 3y^2 = 15y^2$

Mental Math In Exercises 101–108, use the Distributive Property to perform the required arithmetic *mentally*. For example, you work as a mechanic where the wage is $14 per hour and time-and-one-half for overtime. So, your hourly wage for overtime is

$$14(1.5) = 14\left(1 + \tfrac{1}{2}\right) = 14 + 7 = \$21.$$

101. $8(52) = 8(50 + 2)$ 416

102. $7(33) = 7(30 + 3)$ 231

103. $9(48) = 9(50 - 2)$ 432

104. $6(29) = 6(30 - 1)$ 174

105. $-4(59) = -4(60 - 1)$ -236

106. $-6(28) = -6(30 - 2)$ -168

107. $5(7.98) = 5(8 - 0.02)$ 39.9

108. $12(11.95) = 12(12 - 0.05)$ 143.4

In Exercises 109–122, simplify the expression. See Examples 7 and 8.

109. $2(6x)$ $12x$

110. $7(5a)$ $35a$

111. $-(4x)$ $-4x$

112. $-(5t)$ $-5t$

113. $(-2x)(-3x)$ $6x^2$

114. $-4(-3y)$ $12y$

115. $(-5z)(2z^2)$ $-10z^3$

116. $(10t)(-4t^2)$ $-40t^3$

117. $\frac{18a}{5} \cdot \frac{15}{6}$ $9a$

118. $\frac{5x}{8} \cdot \frac{16}{5}$ $2x$

119. $\left(-\frac{3x^2}{2}\right)\left(\frac{4x}{2}\right)$ $-3x^3$

120. $\left(\frac{4x}{3}\right)\left(\frac{3x}{2}\right)$ $2x^2$

121. $(12xy^2)(-2x^3y^2)$ $-24x^4y^4$

122. $(7r^2s^3)(3rs)$ $21r^3s^4$

In Exercises 123–142, simplify the expression by removing symbols of grouping and combining like terms. See Examples 9, 10, and 11.

123. $2(x - 2) + 4$ $2x$

124. $3(x - 5) - 2$ $3x - 17$

125. $6(2s - 1) + s + 4$ $13s - 2$

126. $(2x - 1)(2) + x$ $5x - 2$

127. $m - 3(m - 5)$ $-2m + 15$

128. $5l - 6(3l - 5)$ $-13l + 30$

129. $-6(1 - 2x) + 10(5 - x)$ $44 + 2x$

130. $3(r - 2s) - 5(3r - 5s)$ $-12r + 19s$

131. $\frac{2}{3}(12x + 15) + 16$ $8x + 26$

132. $\frac{3}{8}(4 - y) - \frac{5}{2} + 10$ $-\frac{3}{8}y + 9$

133. $3 - 2[6 + (4 - x)]$ $2x - 17$

134. $10x + 5[6 - (2x + 3)]$ 15

135. $7x(2 - x) - 4x$ $10x - 7x^2$

136. $-6x(x - 1) + x^2$ $-5x^2 + 6x$

137. $4x^2 + x(5 - x)$ $3x^2 + 5x$

138. $-z(z - 2) + 3z^2 + 5$ $2z^2 + 2z + 5$

139. $-3t(4 - t) + t(t + 1)$ $4t^2 - 11t$

140. $-2x(x - 1) + x(3x - 2)$ x^2

141. $3t[4 - (t - 3)] + t(t + 5)$ $26t - 2t^2$

142. $4y[5 - (y + 1)] + 3y(y + 1)$ $-y^2 + 19y$

In Exercises 143–150, use the Distributive Property to simplify the expression. See Example 12.

143. $\frac{2x}{3} - \frac{x}{3}$ $\frac{x}{3}$

144. $\frac{7y}{8} - \frac{3y}{8}$ $\frac{y}{2}$

145. $\frac{4z}{5} + \frac{3z}{5}$ $\frac{7z}{5}$

146. $\frac{5t}{12} + \frac{7t}{12}$ t

147. $\frac{x}{3} - \frac{5x}{4}$ $-\frac{11x}{12}$

148. $\frac{5x}{7} + \frac{2x}{3}$ $\frac{29x}{21}$

149. $\frac{3x}{10} - \frac{x}{10} + \frac{4x}{5}$ x

150. $\frac{3z}{4} - \frac{z}{2} - \frac{z}{3}$ $-\frac{z}{12}$

Solving Problems

 Geometry   In Exercises 151 and 152, write an expression for the perimeter of the triangle shown in the figure. Use the properties of algebra to simplify the expression.

151.

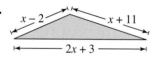

$5x + 9$

152.

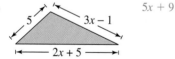

$4x + 12$

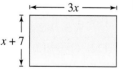 *Geometry*   In Exercises 153 and 154, write and simplify an expression for (a) the perimeter and (b) the area of the rectangle.

153.   (a) $8x + 14$ (b) $3x^2 + 21x$

154.   (a) $6x - 6$ (b) $2x^2 - 6x$

155. ▲ *Geometry*   The area of a trapezoid with parallel bases of lengths b_1 and b_2 and height h is $\frac{1}{2}h(b_1 + b_2)$ (see figure).

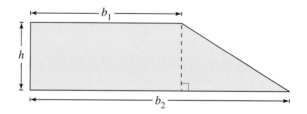

(a) Show that the area can also be expressed as $b_1 h + \frac{1}{2}(b_2 - b_1)h$, and give a geometric explanation for the area represented by each term in this expression.   Answers will vary.

(b) Find the area of a trapezoid with $b_1 = 7$, $b_2 = 12$, and $h = 3$. $\frac{57}{2}$

156. ⚡ ▲ *Geometry*   The remaining area of a square with side length x after a smaller square with side length y has been removed (see figure) is $(x + y)(x - y)$.

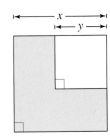

(a) Show that the remaining area can also be expressed as $x(x - y) + y(x - y)$, and give a geometric explanation for the area represented by each term in this expression.

$(x + y)(x - y) = x(x - y) + y(x - y)$ Distributive Property where x is the side length of the larger square, y is the side length of the smaller square, and $(x - y)$ is the difference of the lengths.

(b) Find the remaining area of a square with side length 9 after a square with side length 5 has been removed.   56 square units

▲ *Geometry*   In Exercises 157 and 158, use the formula for the area of a trapezoid, $\frac{1}{2}h(b_1 + b_2)$, to find the area of the trapezoidal house lot and tile.

157.

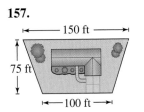

9375 square feet

158.

6 in.

5.2 in.

12 in.

46.8 square inches

Explaining Concepts

159. *Writing* ✎   Discuss the difference between $(6x)^4$ and $6x^4$. $(6x)^4 = (6x)(6x)(6x)(6x)$; $6x^4 = 6 \cdot x \cdot x \cdot x \cdot x$

160. ⚡ The expressions $4x$ and x^4 each represent repeated operations. What are the operations? Write the expressions showing the repeated operations.
Addition and multiplication
$4x = x + x + x + x$; $x^4 = x \cdot x \cdot x \cdot x$

161. *Writing* ✎   In your own words, state the definition of like terms. Give an example of like terms and an example of unlike terms.   Two terms are like terms if they are both constant or if they have the same variable factor(s). Like terms: $3x^2, -5x^2$; unlike terms: $3x^2, 5x$

162. *Writing* ✎   Describe how to combine like terms. What operations are used? Give an example of an expression that can be simplified by combining like terms.   To combine like terms, add the respective coefficients and attach the common variable factor(s). $3x^2 - 5x^2 = -2x^2$

Writing ✎   In Exercises 163 and 164, explain why the two expressions are not like terms.

163. $\frac{1}{2}x^2y, \frac{5}{2}xy^2$ The corresponding exponents of x and y are not raised to the same power.

164. $-16x^2y^3, 7x^2y$ The y exponents are not the same.

165. *Error Analysis*   Describe the error.

$$\frac{x}{3} + \frac{4x}{3} = \frac{5x}{6} \qquad \frac{x}{3} + \frac{4x}{3} = \frac{5x}{3}$$

166. *Writing* ✎   In your own words, describe the procedure for removing nested symbols of grouping.   Remove the innermost symbols first and combine like terms. A symbol of grouping preceded by a *minus* sign can be removed by changing the sign of each term within the symbols.

167. *Writing* ✎   Does the expression $[x - (3 \cdot 4)] \div 5$ change if the parentheses are removed? Does it change if the brackets are removed? Explain.
It does not change if the parentheses are removed because multiplication is a higher-order operation than subtraction. It does change if the brackets are removed because the division would be performed before the subtraction.

168. *Writing* ✎   In your own words, describe the priorities for order of operations.
(a) Perform operations inside symbols of grouping, starting with the innermost symbols.
(b) Evaluate all exponential expressions.
(c) Perform all multiplications and divisions from left to right.
(d) Perform all additions and subtractions from left to right.

Mid-Chapter Quiz

Take this quiz as you would take a quiz in class. After you are done, check your work against the answers in the back of the book.

In Exercises 1 and 2, evaluate the algebraic expression for the specified values of the variable(s). If it is not possible, state the reason.

1. $x^2 - 3x$ (a) $x = 3$ 0   (b) $x = -2$ 10

(c) $x = 0$ 0

2. $\dfrac{x}{y - 3}$ (a) $x = 2, y = 4$ 2   (b) $x = 0, y = -1$ 0

(c) $x = 5, y = 3$ Division by zero is undefined.

In Exercises 3 and 4, identify the terms and coefficients of the expression.

3. $4x^2 - 2x$
$4x^2, -2x; 4, -2$

4. $5x + 3y - 12z$
$5x, 3y, -12z; 5, 3, -12$

5. Rewrite each expression in exponential form.

(a) $3y \cdot 3y \cdot 3y \cdot 3y$ $(3y)^4$ (b) $2 \cdot (x - 3) \cdot (x - 3) \cdot 2 \cdot 2$ $2^3(x - 3)^2$

In Exercises 6–9, simplify the expression.

6. $-4(-5y^2)$ $20y^2$ **7.** $\dfrac{6}{7} \cdot \dfrac{7x}{6}$ x **8.** $(-3y)^2 y^3$ $9y^5$ **9.** $\dfrac{2z^2}{3y} \cdot \dfrac{5z}{7}$ $\dfrac{10z^3}{21y}$

In Exercises 10–13, identify the property of algebra illustrated by the statement.

10. $-3(2y) = (-3 \cdot 2)y$
Associative Property of Multiplication

11. $(x + 2)y = xy + 2y$
Distributive Property

12. $3y \cdot \dfrac{1}{3y} = 1, \quad y \neq 0$
Multiplicative Inverse Property

13. $x - x^2 + 2 = -x^2 + x + 2$
Commutative Property of Addition

In Exercises 14 and 15, use the Distributive Property to expand the expression.

14. $2x(3x - 1)$ $6x^2 - 2x$

15. $-4(2y - 3)$ $-8y + 12$

In Exercises 16 and 17, simplify the expression by combining like terms.

16. $y^2 - 3xy + y + 7xy$ $y^2 + 4xy + y$ **17.** $10\left(\dfrac{1}{u}\right) - 7\left(\dfrac{1}{u}\right) + 3u$ $3\left(\dfrac{1}{u}\right) + 3u$

In Exercises 18 and 19, simplify the expression by removing symbols of grouping and combining like terms.

18. $5(a - 2b) + 3(a + b)$ $8a - 7b$ **19.** $4x + 3[2 - 4(x + 6)]$ $-8x - 66$

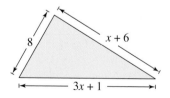

Figure for 20

20. Write and simplify an expression for the perimeter of the triangle (see figure).
$8 + (x + 6) + (3x + 1) = 4x + 15$

21. Evaluate the expression $4 \cdot 10^4 + 5 \cdot 10^3 + 7 \cdot 10^2$.   45,700

2.3 Algebra and Problem Solving

James Marshall/The Image Works

What You Should Learn

1. Define algebra as a problem-solving language.
2. Construct verbal mathematical models from written statements.
3. Translate verbal phrases into algebraic expressions.
4. Identify hidden operations when constructing algebraic expressions.
5. Use problem-solving strategies to solve application problems.

Why You Should Learn It

Translating verbal sentences and phrases into algebraic expressions enables you to solve real-life problems. For instance, in Exercise 58 on page 102, you will find an expression for the total distance traveled by an airplane.

1. Define algebra as a problem-solving language.

Study Tip

As you study this text, it is helpful to view algebra from the "big picture" as shown in Figure 2.3. The ability to write algebraic expressions and equations is needed in the major components of algebra — *simplifying* expressions, *solving* equations, and *graphing* functions.

What Is Algebra?

Algebra is a problem-solving language that is used to solve real-life problems. It has four basic components, which tend to nest within each other, as indicated in Figure 2.3.

1. Symbolic representations and applications of the rules of arithmetic
2. Rewriting (reducing, simplifying, factoring) algebraic expressions into equivalent forms
3. Creating and solving equations
4. Studying relationships among variables by the use of functions and graphs

1. Rules of arithmetic

2. Algebraic expressions: rewriting into equivalent forms

3. Algebraic equations: creating and solving

4. Functions and graphs: relationships among variables

Figure 2.3

Notice that one of the components deals with expressions and another deals with equations. As you study algebra, it is important to understand the difference between simplifying or rewriting an algebraic *expression,* and solving an algebraic *equation.* In general, remember that a mathematical expression *has no equal sign,* whereas a mathematical equation *must have an equal sign.*

When you use an equal sign to *rewrite* an expression, you are merely indicating the *equivalence* of the new expression and the previous one.

Original Expression	*equals*	*Equivalent Expression*
$(a + b)c$	$=$	$ac + bc$

2 Construct verbal mathematical models from written statements.

Constructing Verbal Models

In the first two sections of this chapter, you studied techniques for rewriting and simplifying algebraic expressions. In this section you will study ways to construct algebraic expressions from written statements by first constructing a **verbal mathematical model.**

Take another look at Example 1 in Section 2.1 (page 68). In that example, you are paid $7 per hour and your weekly pay can be represented by the verbal model

$$\boxed{\text{Pay per hour}} \cdot \boxed{\text{Number of hours}} = 7 \text{ dollars} \cdot x \text{ hours} = 7x.$$

Note the hidden operation of multiplication in this expression. Nowhere in the verbal problem does it say you are to multiply 7 times x. It is *implied* in the problem. This is often the case when algebra is used to solve real-life problems.

Constructing a verbal model is a helpful strategy when solving application problems. In class, encourage students to develop verbal models for several exercises before solving them.

Example 1   Constructing an Algebraic Expression

You are paid 5¢ for each aluminum soda can and 3¢ for each glass soda bottle you collect. Write an algebraic expression that represents the total weekly income for this recycling activity.

Solution

Before writing an algebraic expression for the weekly income, it is helpful to construct an informal verbal model. For instance, the following verbal model could be used.

$$\boxed{\text{Pay per can}} \cdot \boxed{\text{Number of cans}} + \boxed{\text{Pay per bottle}} \cdot \boxed{\text{Number of bottles}}$$

Note that the word *and* in the problem indicates addition. Because both the number of cans and the number of bottles can vary from week to week, you can use the two variables c and b, respectively, to write the following algebraic expression.

$$\boxed{5 \text{ cents}} \cdot \boxed{c \text{ cans}} + \boxed{3 \text{ cents}} \cdot \boxed{b \text{ bottles}} = 5c + 3b$$

In 2000, about 1 million tons of aluminum containers were recycled. This accounted for about 45% of all aluminum containers produced.   (Source: Franklin Associates, Ltd.)

In Example 1, notice that c is used to represent the number of *cans* and b is used to represent the number of *bottles*. When writing algebraic expressions, choose variables that can be identified with the unknown quantities.

The number of one kind of item can be expressed in terms of the number of another kind of item. Suppose the number of cans in Example 1 was said to be "three times the number of bottles." In this case, only one variable would be needed and the model could be written as

$$\boxed{5 \text{ cents}} \cdot \boxed{3 \cdot b \text{ cans}} + \boxed{3 \text{ cents}} \cdot \boxed{b \text{ bottles}} = 5(3b) + 3b$$
$$= 15b + 3b$$
$$= 18b.$$

③ Translate verbal phrases into algebraic expressions.

Translating verbal phrases into algebraic expressions is a helpful first step toward translating application problems into equations.

Translating Phrases

When translating verbal sentences and phrases into algebraic expressions, it is helpful to watch for key words and phrases that indicate the four different operations of arithmetic. The following list shows several examples.

Translating Phrases into Algebra Expressions

Key Words and Phrases	Verbal Description	Expression
Addition: Sum, plus, greater than, increased by, more than, exceeds, total of	The sum of 6 and x Eight more than y	$6 + x$ $y + 8$
Subtraction: Difference, minus, less than, decreased by, subtracted from, reduced by, the remainder	Five decreased by a Four less than z	$5 - a$ $z - 4$
Multiplication: Product, multiplied by, twice, times, percent of	Five times x	$5x$
Division: Quotient, divided by, ratio, per	The ratio of x and 3	$\dfrac{x}{3}$

Example 2 Translating Phrases Having Specified Variables

Translate each phrase into an algebraic expression.

a. Three less than m **b.** y decreased by 10

c. The product of 5 and x **d.** The quotient of n and 7

Solution

a. Three less than m

$m - 3$ Think: 3 subtracted from what?

b. y decreased by 10

$y - 10$ Think: What is subtracted from y?

c. The product of 5 and x

$5x$ Think: 5 times what?

d. The quotient of n and 7

$\dfrac{n}{7}$ Think: n is divided by what?

Example 3   Translating Phrases Having Specified Variables

Translate each phrase into an algebraic expression.

a. Six times the sum of x and 7

b. The product of 4 and x, divided by 3

c. k decreased by the product of 8 and m

Solution

a. Six times the sum of x and 7

$$6(x + 7)$$ Think: 6 multiplied by what?

b. The product of 4 and x, divided by 3

$$\frac{4x}{3}$$ Think: What is divided by 3?

c. k decreased by the product of 8 and m

$$k - 8m$$ Think: What is subtracted from k?

In most applications of algebra, the variables are not specified and it is your task to assign variables to the *appropriate* quantities. Although similar to the translations in Examples 2 and 3, the translations in the next example may seem more difficult because variables have not been assigned to the unknown quantities.

Example 4   Translating Phrases Having No Specified Variables

Translate each phrase into a variable expression.

a. The sum of 3 and a number

b. Five decreased by the product of 3 and a number

c. The difference of a number and 3, all divided by 12

Solution

In each case, let x be the unspecified number.

a. The sum of 3 and a number

$$3 + x$$ Think: 3 added to what?

b. Five decreased by the product of 3 and a number

$$5 - 3x$$ Think: What is subtracted from 5?

c. The difference of a number and 3, all divided by 12

$$\frac{x - 3}{12}$$ Think: What is divided by 12?

Study Tip

Any variable, such as b, k, n, r, or x, can be chosen to represent an unspecified number. The choice is a matter of preference. In Example 4, x was chosen as the variable.

A good way to learn algebra is to do it *forward* and *backward*. In the next example, algebraic expressions are translated into verbal form. Keep in mind that other key words could be used to describe the operations in each expression. Your goal is to use key words or phrases that keep the verbal expressions clear and concise.

Example 5 Translating Algebraic Expressions into Verbal Form

Without using a variable, write a verbal description for each expression.

a. $7x - 12$ **b.** $7(x - 12)$ **c.** $5 + \dfrac{x}{2}$ **d.** $\dfrac{5 + x}{2}$ **e.** $(3x)^2$

Solution

a. *Algebraic expression:* $7x - 12$

   Primary operation: Subtraction

   Terms: $7x$ and 12

   Verbal description: Twelve less than the product of 7 and a number

b. *Algebraic expression:* $7(x - 12)$

   Primary operation: Multiplication

   Factors: 7 and $(x - 12)$

   Verbal description: Seven times the difference of a number and 12

c. *Algebraic expression:* $5 + \dfrac{x}{2}$

   Primary operation: Addition

   Terms: 5 and $\dfrac{x}{2}$

   Verbal description: Five added to the quotient of a number and 2

d. *Algebraic expression:* $\dfrac{5 + x}{2}$

   Primary operation: Division

   Numerator, denominator: Numerator is $5 + x$; denominator is 2

   Verbal description: The sum of 5 and a number, all divided by 2

e. *Algebraic expression:* $(3x)^2$

   Primary operation: Raise to a power

   Base, power: $3x$ is the base, 2 is the power

   Verbal description: The square of the product of 3 and x

Translating algebraic expressions into verbal phrases is more difficult than it may appear. It is easy to write a phrase that is ambiguous. For instance, what does the phrase "the sum of 5 and a number times 2" mean? Without further information, this phrase could mean

$$5 + 2x \quad \text{or} \quad 2(5 + x).$$

④ Identify hidden operations when constructing algebraic expressions.

Verbal Models with Hidden Operations

Most real-life problems do not contain verbal expressions that clearly identify all the arithmetic operations involved. You need to rely on past experience and the physical nature of the problem in order to identify the operations hidden in the problem statement. Multiplication is the operation most commonly hidden in real life applications. Watch for *hidden operations* in the next two examples.

Example 6   Discovering Hidden Operations

a. A cash register contains n nickels and d dimes. Write an expression for this amount of money in cents.

b. A person riding a bicycle travels at a constant rate of 12 miles per hour. Write an expression showing how far the person can ride in t hours.

c. A person paid x dollars plus 6% sales tax for an automobile. Write an expression for the total cost of the automobile.

Solution

a. The amount of money is a sum of products.

Verbal Model: | Value of nickel | · | Number of nickels | + | Value of dime | · | Number of dimes |

Labels:
Value of nickel = 5   (cents)
Number of nickels = n (nickels)
Value of dime = 10   (cents)
Number of dimes = d (dimes)

Expression: $5n + 10d$ (cents)

b. The distance traveled is a product.

Verbal Model: | Rate of travel · Time traveled |

Labels:
Rate of travel = 12   (miles per hour)
Time traveled = t (hours)

Expression: $12t$ (miles)

c. The total cost is a sum.

Verbal Model: | Cost of automobile | + | Percent of sales tax | · | Cost of automobile |

Labels:
Percent of sales tax = 0.06   (decimal form)
Cost of automobile = x (dollars)

Expression: $x + 0.06x = (1 + 0.06)x$
                           $= 1.06x$

Study Tip

In Example 6(b), the final answer is listed in terms of miles. This makes sense as described below.

$$12 \frac{\text{miles}}{\text{hours}} \cdot t \text{ hours}$$

Note that the hours "divide out," leaving miles as the unit of measure. This technique is called *unit analysis* and can be very helpful in determining the final unit of measure.

Notice in part (c) of Example 6 that the equal sign is used to denote the equivalence of the three expressions. It is not an equation to be solved.

⑤ Use problem-solving strategies to solve application problems.

Additional Problem-Solving Strategies

In addition to constructing verbal models, there are other problem-solving strategies that can help you succeed in this course.

Summary of Additional Problem-Solving Strategies

1. **Guess, Check, and Revise** Guess a reasonable solution based on the given data. Check the guess, and revise it, if necessary. Continue guessing, checking, and revising until a correct solution is found.

2. **Make a Table/Look for a Pattern** Make a table using the data in the problem. Look for a number pattern. Then use the pattern to complete the table or find a solution.

3. **Draw a Diagram** Draw a diagram that shows the facts from the problem. Use the diagram to visualize the action of the problem. Use algebra to find a solution. Then check the solution against the facts.

4. **Solve a Simpler Problem** Construct a simpler problem that is similar to the original problem. Solve the simpler problem. Then use the same procedure to solve the original problem.

Encourage students to experiment with each of these four problem-solving strategies. Students should begin to realize that there are *many* correct ways to approach questions in mathematics.

Study Tip

The most common errors made when solving algebraic problems are arithmetic errors. Be sure to check your arithmetic when solving algebraic problems.

Example 7   Guess, Check, and Revise

You deposit $500 in an account that earns 6% interest compounded annually. The balance in the account after t years is $A = 500(1 + 0.06)^t$. How long will it take for your investment to double?

Solution

You can solve this problem using a guess, check, and revise strategy. For instance, you might guess that it takes 10 years for your investment to double. The balance in 10 years is

$$A = 500(1 + 0.06)^{10} \approx \$895.42.$$

Because the amount has not yet doubled, you increase your guess to 15 years.

$$A = 500(1 + 0.06)^{15} \approx \$1198.28$$

Because this amount is more than double the investment, your next guess should be a number between 10 and 15. After trying several more numbers, you can determine that your balance doubles in about 11.9 years.

Another strategy that works well for a problem such as Example 7 is to make a table of data values. You can use a calculator to create the following table.

t	2	4	6	8	10	12
A	561.80	631.24	709.26	796.92	895.42	1006.10

Example 8   Make a Table/Look for a Pattern

Find each product. Then describe the pattern and use your description to find the product of 14 and 16.

$$1 \cdot 3, \ 2 \cdot 4, \ 3 \cdot 5, \ 4 \cdot 6, \ 5 \cdot 7, \ 6 \cdot 8, \ 7 \cdot 9$$

Solution

One way to help find a pattern is to organize the results in a table.

Numbers	$1 \cdot 3$	$2 \cdot 4$	$3 \cdot 5$	$4 \cdot 6$	$5 \cdot 7$	$6 \cdot 8$	$7 \cdot 9$
Product	3	8	15	24	35	48	63

From the table, you can see that each of the products is 1 less than a perfect square. For instance, 3 is 1 less than 2^2 or 4, 8 is 1 less than 3^2 or 9, 15 is 1 less than 4^2 or 16, and so on.

If this pattern continues for other numbers, you can hypothesize that the product of 14 and 16 is 1 less than 15^2 or 225. That is,

$$14 \cdot 16 = 15^2 - 1$$
$$= 224.$$

You can confirm this result by actually multiplying 14 and 16.

Example 9   Draw a Diagram

The outer dimensions of a rectangular apartment are 25 feet by 40 feet. The combination living room, dining room, and kitchen areas occupy two-fifths of the apartment's area. Find the area of the remaining rooms.

Solution

For this problem, it helps to draw a diagram, as shown in Figure 2.4. From the figure, you can see that the total area of the apartment is

$$\text{Area} = (\text{Length})(\text{Width})$$
$$= (40)(25)$$
$$= 1000 \text{ square feet.}$$

The area occupied by the living room, dining room, and kitchen is

$$\frac{2}{5}(1000) = 400 \text{ square feet.}$$

This implies that the remaining rooms must have a total area of

$$1000 - 400 = 600 \text{ square feet.}$$

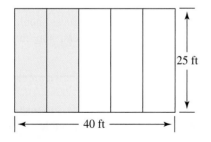

25 ft

40 ft

Figure 2.4

Example 10   Solve a Simpler Problem

You are driving on an interstate highway and are traveling at an average speed of 60 miles per hour. How far will you travel in $12\frac{1}{2}$ hours?

Solution

Distance and other related formulas can be found on the inside front cover of the text.

One way to solve this problem is to use the formula that relates distance, rate, and time. Suppose, however, that you have forgotten the formula. To help you remember, you could solve some simpler problems.

- If you travel 60 miles per hour for 1 hour, you will travel 60 miles.
- If you travel 60 miles per hour for 2 hours, you will travel 120 miles.
- If you travel 60 miles per hour for 3 hours, you will travel 180 miles.

From these examples, it appears that you can find the total miles traveled by multiplying the rate times the time. So, if you travel 60 miles per hour for $12\frac{1}{2}$ hours, you will travel a distance of

$$(60)(12.5) = 750 \text{ miles}.$$

Hidden operations are often involved when variable names (labels) are assigned to unknown quantities. A good strategy is to use a *specific* case to help you write a model for the *general* case. For instance, a specific case of finding three consecutive integers

$$3, 3 + 1, \text{ and } 3 + 2$$

may help you write a general case for finding three consecutive integers n, $n + 1$, and $n + 2$. This strategy is illustrated in Examples 11 and 12.

Example 11   Using a Specific Case to Find a General Case

In each of the following, use the variable to label the unknown quantity.

a. A person's weekly salary is d dollars. What is the annual salary?

b. A person's annual salary is y dollars. What is the monthly salary?

Solution

a. There are 52 weeks in a year.

Specific case: If the weekly salary is $200, then the annual salary (in dollars) is 52 · 200.

General case: If the weekly salary is d dollars, then the annual salary (in dollars) is 52 · d or 52d.

b. There are 12 months in a year.

Specific case: If the annual salary is $24,000, then the monthly salary (in dollars) is 24,000 ÷ 12.

General case: If the annual salary is y dollars, then the monthly salary (in dollars) is y ÷ 12 or $y/12$.

Example 12   Using a Specific Case to Find a General Case

In each of the following, use the variable to label the unknown quantity.

a. You are k inches shorter than a friend. You are 60 inches tall. How tall is your friend?

b. A consumer buys g gallons of gasoline for a total of d dollars. What is the price per gallon?

c. A person drives on the highway at an average speed of 60 miles per hour for t hours. How far has the person traveled?

Solution

a. You are k inches shorter than a friend.

Specific case: If you are 10 inches shorter than your friend, then your friend is $60 + 10$ inches tall.

General case: If you are k inches shorter than your friend, then your friend is $60 + k$ inches tall.

b. To obtain the price per gallon, divide the price by the number of gallons.

Specific case: If the total price is \$11.50 and the total number of gallons is 10, then the price per gallon is $11.50 \div 10$ dollars per gallon.

General case: If the total price is d dollars and the total number of gallons is g, then the price per gallon is $d \div g$ or d/g dollars per gallon.

c. To obtain the distance driven, multiply the speed by the number of hours.

Specific case: If the person has driven for 2 hours at a speed of 60 miles per hour, then the person has traveled $60 \cdot 2$ miles.

General case: If the person has driven for t hours at a speed of 60 miles per hour, then the person has traveled $60t$ miles.

Most of the verbal problems you encounter in a mathematics text have precisely the right amount of information necessary to solve the problem. In real life, however, you may need to collect additional information, as shown in Example 13.

Example 13   Enough Information?

Decide what additional information is needed to solve the following problem.

During a given week, a person worked 48 hours for the same employer. The hourly rate for overtime is \$14. Write an expression for the person's gross pay for the week, including any pay received for overtime.

Solution

To solve this problem, you would need to know how much the person is normally paid per hour. You would also need to be sure that the person normally works 40 hours per week and that overtime is paid on time worked beyond 40 hours.

2.3 Exercises

Review   *Concepts, Skills, and Problem Solving*

Keep mathematically in shape by doing these exercises *before* the problems of this section.

Properties and Definitions

1. The product of two real numbers is -35 and one of the factors is 5. What is the sign of the other factor?
   Negative

2. Determine the sum of the digits of 744. Since this sum is divisible by 3, the number 744 is divisible by what numbers?   15, 3

3. *True or False?* -4^2 is positive.
   False. $-4^2 = -1 \cdot 4 \cdot 4 = -16$

4. *True or False?* $(-4)^2$ is positive.
   True. $(-4)^2 = (-4)(-4) = 16$

Simplifying Expressions

In Exercises 5–10, evaluate the expression.

5. $(-6)(-13)$ 78
6. $|4(-6)(5)|$ 120
7. $\left(-\frac{4}{3}\right)\left(-\frac{9}{16}\right)$ $\frac{3}{4}$
8. $\frac{7}{8} \div \frac{3}{16}$ $\frac{14}{3}$
9. $\left|-\frac{5}{9}\right| + 2$ $\frac{23}{9}$
10. $-7\frac{3}{5} - 3\frac{1}{2}$ $-\frac{111}{10}$

Problem Solving

11. *Consumerism*   A coat costs $133.50, including tax. You save $30 a week. How many weeks must you save in order to buy the coat? How much money will you have left?   5 weeks, $16.50

12. ▲ *Geometry*   The length of a rectangle is $1\frac{1}{2}$ times its width. Its width is 8 meters. Find its perimeter.
    40 meters

Developing Skills

In Exercises 1–6, match the verbal phrase with the correct algebraic expression.

(a) $11 + \frac{1}{3}x$ (b) $3x - 12$
(c) $3(x - 12)$ (d) $12 - 3x$
(e) $11x + \frac{1}{3}$ (f) $12x + 3$

1. Twelve decreased by 3 times a number   (d)
2. Eleven more than $\frac{1}{3}$ of a number   (a)
3. Eleven times a number plus $\frac{1}{3}$ (e)
4. Three increased by 12 times a number   (f)
5. The difference between 3 times a number and 12   (b)
6. Three times the difference of a number and 12   (c)

In Exercises 7–30, translate the phrase into an algebraic expression. Let x represent the real number. See Examples 1, 2, 3, and 4.

7. A number increased by 5 $x + 5$
8. 17 more than a number $x + 17$
9. A number decreased by 25 $x - 25$
10. A number decreased by 7 $x - 7$
11. Six less than a number $x - 6$
12. Ten more than a number $x + 10$
13. Twice a number $2x$

14. The product of 30 and a number $30x$
15. A number divided by 3 $\frac{x}{3}$
16. A number divided by 100 $\frac{x}{100}$
17. The ratio of a number to 50 $\frac{x}{50}$
18. One-half of a number $\frac{1}{2}x$
19. Three-tenths of a number $\frac{3}{10}x$
20. Twenty-five hundredths of a number $0.25x$
21. A number is tripled and the product is increased by 5. $3x + 5$
22. A number is increased by 5 and the sum is tripled. $3(x + 5)$
23. Eight more than 5 times a number $5x + 8$
24. The quotient of a number and 5 is decreased by 15. $\frac{x}{5} - 15$
25. Ten times the sum of a number and 4 $10(x + 4)$
26. Seventeen less than 4 times a number $4x - 17$
27. The absolute value of the sum of a number and 4 $|x + 4|$
28. The absolute value of 4 less than twice a number $|2x - 4|$

29. The square of a number, increased by 1 $x^2 + 1$

30. Twice the square of a number, increased by 4
$2x^2 + 4$

In Exercises 31–44, write a verbal description of the algebraic expression, without using a variable. (There is more than one correct answer.) See Example 5.

31. $x - 10$ A number decreased by 10

32. $x + 9$ A number increased by 9

33. $3x + 2$ The product of 3 and a number, increased by 2

34. $4 - 7x$ Four decreased by 7 times a number

35. $\frac{1}{2}x - 6$ One-half a number decreased by 6

36. $9 - \frac{1}{4}x$ Nine decreased by $\frac{1}{4}$ of a number

37. $3(2 - x)$ Three times the difference of 2 and a number

38. $-10(t - 6)$ Negative 10 times the difference of a number and 6

39. $\dfrac{t + 1}{2}$ The sum of a number and 1, divided by 2

40. $\dfrac{y - 3}{4}$ One-fourth the difference of a number and 3

41. $\dfrac{1}{2} - \dfrac{t}{5}$ One-half decreased by a number divided by 5

42. $\dfrac{1}{4} + \dfrac{x}{8}$ One-fourth increased by $\frac{1}{8}$ of a number

43. $x^2 + 5$ The square of a number, increased by 5

44. $x^3 - 1$ The cube of a number, decreased by 1

In Exercises 45–52, translate the phrase into a mathematical expression. Simplify the expression.

45. The sum of x and 3 is multiplied by x.
$(x + 3)x = x^2 + 3x$

46. The sum of 6 and n is multiplied by 5.
$(6 + n)(5) = 30 + 5n$

47. The sum of 25 and x is added to x.
$(25 + x) + x = 25 + 2x$

48. The sum of 4 and x is added to the sum of x and -8.
$(4 + x) + [x + (-8)] = 2x - 4$

49. Nine is subtracted from x and the result is multiplied by 3. $(x - 9)(3) = 3x - 27$

50. The square of x is added to the product of x and $x + 1$. $x^2 + x(x + 1) = 2x^2 + x$

51. The product of 8 times the sum of x and 24 is divided by 2. $\dfrac{8(x + 24)}{2} = 4x + 96$

52. Fifteen is subtracted from x and the difference is multiplied by 4. $4(x - 15) = 4x - 60$

Solving Problems

53. *Money*   A cash register contains d dimes. Write an algebraic expression that represents the total amount of money (in dollars). See Example 6. $0.10d$

54. *Money*   A cash register contains d dimes and q quarters. Write an algebraic expression that represents the total amount of money (in dollars).
$0.10d + 0.25q$

55. *Sales Tax*   The sales tax on a purchase of L dollars is 6%. Write an algebraic expression that represents the total amount of sales tax. (*Hint:* Use the decimal form of 6%.) $0.06L$

56. *Income Tax*   The state income tax on a gross income of I dollars in Pennsylvania is 2.8%. Write an algebraic expression that represents the total amount of income tax. (*Hint:* Use the decimal form of 2.8%.) $0.028I$

57. *Travel Time*   A truck travels 100 miles at an average speed of r miles per hour. Write an algebraic expression that represents the total travel time. $\dfrac{100}{r}$

58. *Distance*   An airplane travels at the rate of r miles per hour for 3 hours. Write an algebraic expression that represents the total distance traveled by the airplane. $3r$

59. *Consumerism*   A campground charges $15 for adults and $2 for children. Write an algebraic expression that represents the total camping fee for m adults and n children. $15m + 2n$

60. *Hourly Wage*   The hourly wage for an employee is $12.50 per hour plus 75 cents for each of the q units produced during the hour. Write an algebraic expression that represents the total hourly earnings for the employee. $12.50 + 0.75q$

Guess, Check, and Revise   In Exercises 61–64, an expression for the balance in an account is given. Guess, check, and revise to determine the time (in years) necessary for the investment of $1000 to double. See Example 7.

61. Interest rate: 7%

$1000(1 + 0.07)^t$ $t = 10.2$ years

62. Interest rate: 5%

$1000(1 + 0.05)^t$ $t = 14.2$ years

63. Interest rate: 6%

$1000(1 + 0.06)^t$ $t = 11.9$ years

64. Interest rate: 8%

$1000(1 + 0.08)^t$ $t = 9.0$ years

Finding a Pattern   In Exercises 65 and 66, complete the table. The third row in the table is the difference between consecutive entries of the second row. Describe the pattern of the third row. See Example 8.

65.

n	0	1	2	3	4	5
$2n - 1$	-1	1	3	5	7	9
Differences		2	2	2	2	2

66.

n	0	1	2	3	4	5
$7n + 5$	5	12	19	26	33	40
Differences		7	7	7	7	7

Exploration   In Exercises 67 and 68, find values for a and b such that the expression $an + b$ yields the table values.

67.

n	0	1	2	3	4	5
$an + b$	4	9	14	19	24	29

$a = 5, b = 4$

68.

n	0	1	2	3	4	5
$an + b$	1	5	9	13	17	21

$a = 4, b = 1$

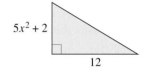 *Geometry*   In Exercises 69–74, write an algebraic expression that represents the area of the region. Use the rules of algebra to simplify the expression.

69.

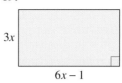

$3x$

$6x - 1$

$3x(6x - 1) = 18x^2 - 3x$

70.

$-4x$

$-4x$

$(-4x)(-4x) = 16x^2$

71.

$2x$

$14x + 3$

$\left(\frac{1}{2}\right)(14x + 3)(2x) = 14x^2 + 3x$

72.

$5x^2 + 2$

12

$\frac{1}{2}(12)(5x^2 + 2) = 30x^2 + 12$

73.

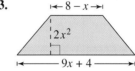

$8 - x$

$2x^2$

$9x + 4$

$\frac{1}{2}(2x^2)[(9x + 4) + (8 - x)] = 8x^3 + 12x^2$

74.

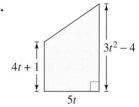

$3t^2 - 4$

$4t + 1$

$5t$

$\frac{1}{2}(5t)[(4t + 1) + (3t^2 - 4)] = \frac{15}{2}t^3 + 10t^2 - \frac{15}{2}t$

Drawing a Diagram   In Exercises 75 and 76, draw figures satisfying the specified conditions. See Example 9. See Additional Answers.

75. The sides of a square have length a centimeters. Draw the square. Draw the rectangle obtained by extending two parallel sides of the square 6 centimeters. Find expressions for the perimeter and area of each figure.

Perimeter of the square: $4a$ centimeters; Area of the square: a^2 square centimeters; Perimeter of the rectangle: $4a + 12$ centimeters; Area of the rectangle: $a(a + 6)$ square centimeters

76. The dimensions of a rectangular lawn are 150 feet by 250 feet. The property owner has the option of buying a rectangular strip x feet wide along one 250-foot side of the lawn. Draw diagrams representing the lawn before and after the purchase. Write an expression for the area of each.

Area of the original lawn: 37,500 square feet

Area of the expanded lawn: $250(150 + x)$ square feet

77. ▲ *Geometry*   A rectangle has sides of length $3w$ and w. Write an algebraic expression that represents the area of the rectangle. $3w^2$

78. ▲ *Geometry*   A square has sides of length s. Write an algebraic expression that represents the perimeter of the square. $4s$

79. ▲ *Geometry*   Write an algebraic expression that represents the perimeter of the picture frame in the figure. $5w$

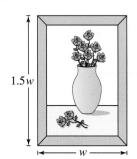

80. ▲ *Geometry*   A computer screen has sides of length s inches (see figure). Write an algebraic expression that represents the area of the screen. Write the area using the correct unit of measure.

s^2 square inches

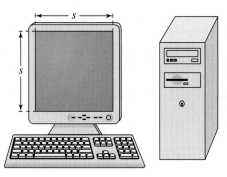

Figure for 80

In Exercises 81–84, decide what additional information is needed to solve the problem. (Do not solve the problem.) See Example 13.

81. *Distance*   A family taking a Sunday drive through the country travels at an average speed of 45 miles per hour. How far have they traveled by 3:00 P.M.?   The start time is missing.

82. *Consumer Awareness*   You purchase an MP3 player during a sale at an electronics store. The MP3 player is discounted by 15%. What is the sale price of the player?   The retail price is missing.

83. *Consumerism*   You decide to budget your money so that you can afford a new computer. The cost of the computer is $975. You put half of your weekly paycheck into your savings account to pay for the computer. How many hours will you have to work at your job in order to be able to afford the computer?   The amount of the paycheck and the number of hours worked on the paycheck are missing.

84. *Painting*   A painter is going to paint a rectangular room that is twice as long as it is wide. One gallon of paint covers 100 square feet. How much money will he have to spend on paint?   The cost of paint and the specific height and length (or width) of the room is missing.

Explaining Concepts

85. ⟳   Answer parts (c)–(h) of Motivating the Chapter on page 66.

86. The word *difference* indicates what operation?
Subtraction

87. The word *quotient* indicates what operation?
Division

88. Determine which phrase(s) is (are) equivalent to the expression $n + 4$.   (a), (b), (e)

(a) 4 more than n (b) the sum of n and 4

(c) n less than 4   (d) the ratio of n to 4

(e) the total of 4 and n

89. *Writing*   Determine whether order is important when translating each phrase into an algebraic expression. Explain.

(a) x increased by 10   No. Addition is commutative.

(b) 10 decreased by x Yes. Subtraction is not commutative.

(c) the product of x and 10   No. Multiplication is commutative.

(d) the quotient of x and 10   Yes. Division is not commutative.

90. Give two interpretations of "the quotient of 5 and a number times 3." $\left(\dfrac{5}{n}\right)3, \dfrac{5}{3n}$

2.4 Introduction to Equations

Byron Aughenbaugh/Getty Images

What You Should Learn

1. Distinguish between an algebraic expression and an algebraic equation.
2. Check whether a given value is a solution of an equation.
3. Use properties of equality to solve equations.
4. Use a verbal model to construct an algebraic equation.

Why You Should Learn It

You can use verbal models to write algebraic equations that model real-life situations. For instance, in Exercise 64 on page 114, you will write an equation to determine how far away a lightning strike is after hearing the thunder.

1. **Distinguish between an algebraic expression and an algebraic equation.**

Equations

An **equation** is a statement that two algebraic expressions are equal. For example,

$$x = 3, \quad 5x - 2 = 8, \quad \frac{x}{4} = 7, \quad \text{and} \quad x^2 - 9 = 0$$

are equations. To **solve** an equation involving the variable x means to find all values of x that make the equation true. Such values are called **solutions.** For instance, $x = 2$ is a solution of the equation

$$5x - 2 = 8$$

because

$$5(2) - 2 = 8$$

is a true statement. The solutions of an equation are said to **satisfy** the equation.

Be sure that you understand the distinction between an algebraic expression and an algebraic equation. The differences are summarized in the following table.

Algebraic Expression	Algebraic Equation
• Example: $4(x - 1)$ • Contains *no* equal sign • Can sometimes be *simplified* to an equivalent form: $4(x - 1)$ simplifies to $4x - 4$ • Can be evaluated for any real number for which the expression is defined	• Example: $4(x - 1) = 12$ • Contains an equal sign and is true for only certain values of the variable • Solution is found by forming equivalent equations using the properties of equality: $4(x - 1) = 12$ $4x - 4 = 12$ $4x = 16$ $x = 4$

Checking Solutions of Equations

To **check** whether a given solution is a solution to an equation, substitute the given value into the original equation. If the substitution results in a true statement, then the value is a solution of the equation. If the substitution results in a false statement, then the value is not a solution of the equation. This process is illustrated in Examples 1 and 2.

Example 1   Checking a Solution of an Equation

Determine whether $x = -2$ is a solution of $x^2 - 5 = 4x + 7$.

Solution

$$x^2 - 5 = 4x + 7 \qquad \text{Write original equation.}$$
$$(-2)^2 - 5 \stackrel{?}{=} 4(-2) + 7 \qquad \text{Substitute } -2 \text{ for } x.$$
$$4 - 5 \stackrel{?}{=} -8 + 7 \qquad \text{Simplify.}$$
$$-1 = -1 \qquad \text{Solution checks. } \checkmark$$

Because the substitution results in a true statement, you can conclude that $x = -2$ is a solution of the original equation.

> **Study Tip**
>
> When checking a solution, you should write a question mark over the equal sign to indicate that you are not sure of the validity of the equation.

Just because you have found one solution of an equation, you should not conclude that you have found all of the solutions. For instance, you can check that $x = 6$ is also a solution of the equation in Example 1 as follows.

$$x^2 - 5 = 4x + 7 \qquad \text{Write original equation.}$$
$$(6)^2 - 5 \stackrel{?}{=} 4(6) + 7 \qquad \text{Substitute 6 for } x.$$
$$36 - 5 \stackrel{?}{=} 24 + 7 \qquad \text{Simplify.}$$
$$31 = 31 \qquad \text{Solution checks. } \checkmark$$

Example 2   A Trial Solution That Does Not Check

Determine whether $x = 2$ is a solution of $x^2 - 5 = 4x + 7$.

Solution

$$x^2 - 5 = 4x + 7 \qquad \text{Write original equation.}$$
$$(2)^2 - 5 \stackrel{?}{=} 4(2) + 7 \qquad \text{Substitute 2 for } x.$$
$$4 - 5 \stackrel{?}{=} 8 + 7 \qquad \text{Simplify.}$$
$$-1 \neq 15 \qquad \text{Solution does not check. } ✗$$

Because the substitution results in a false statement, you can conclude that $x = 2$ is not a solution of the original equation.

③ Use properties of equality to solve equations.

Forming Equivalent Equations

It is helpful to think of an equation as having two sides that are in balance. Consequently, when you try to solve an equation, you must be careful to maintain that balance by performing the same operation on each side.

Two equations that have the same set of solutions are called **equivalent.** For instance, the equations

$$x = 3 \quad \text{and} \quad x - 3 = 0$$

are equivalent because both have only one solution—the number 3. When any one of the operations in the following list is applied to an equation, the resulting equation is equivalent to the original equation.

Forming Equivalent Equations: Properties of Equality

An equation can be transformed into an *equivalent equation* using one or more of the following procedures.

	Original Equation	*Equivalent Equation(s)*
1. *Simplify either side:* Remove symbols of grouping, combine like terms, or simplify fractions on one or both sides of the equation.	$3x - x = 8$	$2x = 8$
2. *Apply the Addition Property of Equality:* Add (or subtract) the same quantity to (from) *each* side of the equation.	$x - 2 = 5$	$x - 2 + 2 = 5 + 2$ $x = 7$
3. *Apply the Multiplication Property of Equality:* Multiply (or divide) each side of the equation by the same *nonzero* quantity.	$3x = 9$	$\dfrac{3x}{3} = \dfrac{9}{3}$ $x = 3$
4. *Interchange the two sides of the equation.*	$7 = x$	$x = 7$

The second and third operations in this list can be used to eliminate terms or factors in an equation. For example, to solve the equation $x - 5 = 1$, you need to eliminate the term -5 on the left side. This is accomplished by adding its opposite, 5, to each side.

$$x - 5 = 1 \qquad \text{Write original equation.}$$
$$x - 5 + 5 = 1 + 5 \qquad \text{Add 5 to each side.}$$
$$x + 0 = 6 \qquad \text{Combine like terms.}$$
$$x = 6 \qquad \text{Solution}$$

These four equations are equivalent, and they are called the **steps** of the solution.

The next example shows how the properties of equality can be used to solve equations. You will get many more opportunities to practice these skills in the next chapter. For now, your goal should be to understand why each step in the solution is valid. For instance, the second step in part (a) of Example 3 is valid because the Addition Property of Equality states that you can add the same quantity to each side of an equation.

Example 3   Operations Used to Solve Equations

Identify the property of equality used to solve each equation.

a. $x - 5 = 0$ Original equation

$x - 5 + 5 = 0 + 5$ Add 5 to each side.

$x = 5$ Solution

b. $\dfrac{x}{5} = -2$ Original equation

$\dfrac{x}{5}(5) = -2(5)$ Multiply each side by 5.

$x = -10$ Solution

c. $4x = 9$ Original equation

$\dfrac{4x}{4} = \dfrac{9}{4}$ Divide each side by 4.

$x = \dfrac{9}{4}$ Solution

d. $\dfrac{5}{3}x = 7$ Original equation

$\dfrac{3}{5} \cdot \dfrac{5}{3}x = \dfrac{3}{5} \cdot 7$ Multiply each side by $\frac{3}{5}$.

$x = \dfrac{21}{5}$ Solution

Study Tip

In Example 3(c), each side of the equation is divided by 4 to eliminate the coefficient 4 on the left side. You could just as easily *multiply* each side by $\frac{1}{4}$. Both techniques are legitimate—which one you decide to use is a matter of personal preference.

Solution

a. The Addition Property of Equality is used to add 5 to each side of the equation in the second step. Adding 5 eliminates the term -5 from the left side of the equation.

b. The Multiplication Property of Equality is used to multiply each side of the equation by 5 in the second step. Multiplying by 5 eliminates the denominator from the left side of the equation.

c. The Multiplication Property of Equality is used to divide each side of the equation by 4 $\left(\text{or multiply each side by } \frac{1}{4}\right)$ in the second step. Dividing by 4 eliminates the coefficient from the left side of the equation.

d. The Multiplication Property of Equality is used to multiply each side of the equation by $\frac{3}{5}$ in the second step. Multiplying by the *reciprocal* of the fraction $\frac{5}{3}$ eliminates the fraction from the left side of the equation.

④ Use a verbal model to construct an algebraic equation.

Constructing Equations

It is helpful to use two phases in constructing equations that model real-life situations, as shown below.

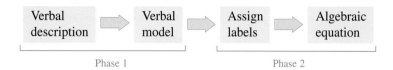

| Verbal description | ⇒ | Verbal model | ⇒ | Assign labels | ⇒ | Algebraic equation |

Phase 1   Phase 2

In the first phase, you translate the verbal description into a *verbal model.* In the second phase, you assign labels and translate the verbal model into a *mathematical model* or *algebraic equation.* Here are two examples of verbal models.

1. The sale price of a basketball is $28. The sale price is $7 less than the original price. What is the original price?

Verbal Model: Sale price = Original price − Discount

$28 = Original price − $7

2. The original price of a basketball is $35. The original price is discounted by $7. What is the sale price?

Verbal Model: Sale price = Original price − Discount

Sale price = $35 − $7

Verbal models help students organize and picture relationships, which can then be translated into equations.

Example 4 Using a Verbal Model to Construct an Equation

Write an algebraic equation for the following problem.

The total income that an employee received in 2003 was $31,550. How much was the employee paid each week? Assume that each weekly paycheck contained the same amount, and that the year consisted of 52 weeks.

Solution

Verbal Model: Income for year = 52 · Weekly pay

Labels: Income for year = 31,550   (dollars)
Weekly pay = x (dollars)

Algebraic Model: $31,550 = 52x$

When you construct an equation, be sure to check that both sides of the equation represent the *same* unit of measure. For instance, in Example 4, both sides of the equation $31,550 = 52x$ represent dollar amounts.

Example 5   Using a Verbal Model to Construct an Equation

Write an algebraic equation for the following problem.

Returning to college after spring break, you travel 3 hours and stop for lunch. You know that it takes 45 minutes to complete the last 36 miles of the 180-mile trip. What is the average speed during the first 3 hours of the trip?

Solution

Verbal Model:

$$\boxed{\text{Distance}} = \boxed{\text{Rate}} \cdot \boxed{\text{Time}}$$

Labels:

Distance = 180 − 36 = 144   (miles)
Rate = r (miles per hour)
Time = 3   (hours)

Algebraic Model: 144 = 3r

Example 6   Using a Verbal Model to Construct an Equation

Write an algebraic equation for the following problem.

Tickets for a concert cost $45 for each floor seat and $30 for each stadium seat. There were 800 seats on the main floor, and these were sold out. The total revenue from ticket sales was $54,000. How many stadium seats were sold?

Solution

Verbal Model:

$$\boxed{\begin{array}{c}\text{Total} \\ \text{revenue}\end{array}} = \boxed{\begin{array}{c}\text{Revenue from} \\ \text{floor seats}\end{array}} + \boxed{\begin{array}{c}\text{Revenue from} \\ \text{stadium seats}\end{array}}$$

Labels:

Total revenue = 54,000   (dollars)
Price per floor seat = 45   (dollars per seat)
Number of floor seats = 800   (seats)
Price per stadium seat = 30   (dollars per seat)
Number of stadium seats = x (seats)

Algebraic Model: 54,000 = 45(800) + 30x

In Example 6, you can use the following *unit analysis* to check that both sides of the equation are measured in dollars.

$$54{,}000 \text{ dollars} = \left(\frac{45 \text{ dollars}}{\text{seat}}\right)(800 \text{ seats}) + \left(\frac{30 \text{ dollars}}{\text{seat}}\right)(x \text{ seats})$$

In Section 3.1, you will study techniques for solving the equations constructed in Examples 4, 5, and 6.

2.4 Exercises

In Exercises 27–34, have your students review the examples in Sections 2.1 and 2.2.

Review   *Concepts, Skills, and Problem Solving*

Keep mathematically in shape by doing these exercises *before* the problems of this section.

Properties and Definitions

1. If the numerator and denominator of a fraction have unlike signs, the sign of the fraction is Negative .

2. *Writing* If a negative number is used as a factor eight times, what is the sign of the product? Explain.
   Positive. The product of an even number of negative factors is positive.

3. Complete the Commutative Property:

   $6 + 10 =$ 10 + 6 .

4. Identify the property of real numbers illustrated by $6\left(\frac{1}{6}\right) = 1$.   Multiplicative Inverse Property

Simplifying Expressions

In Exercises 5–10, simplify the expression.

5. $t^2 \cdot t^5$ t^7

6. $(-3y^3)y^2$ $-3y^5$

7. $6x + 9x$ $15x$

8. $4 - 3t + t$ $4 - 2t$

9. $-(-8b)$ $8b$

10. $7(-10x)$ $-70x$

Graphs and Models

△ *Geometry*   In Exercises 11 and 12, write and simplify expressions for the perimeter and area of the figure.

11.
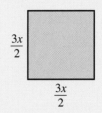

Perimeter: $6x$

Area: $\dfrac{9x^2}{4}$

12.
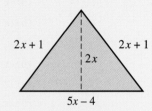

Perimeter: $9x - 2$

Area: $5x^2 - 4x$

Developing Skills

In Exercises 1–16, determine whether each value of *x* is a solution of the equation. See Examples 1 and 2.

	Equation		*Values*	
1.	$4x - 12 = 0$		(a) $x = 3$	(b) $x = 5$
	(a) Solution	(b) Not a solution		
2.	$3x - 3 = 0$		(a) $x = 4$	(b) $x = 1$
	(a) Not a solution	(b) Solution		
3.	$6x + 1 = -11$		(a) $x = 2$	(b) $x = -2$
	(a) Not a solution	(b) Solution		
4.	$2x + 5 = -15$		(a) $x = -10$	(b) $x = 5$
	(a) Solution	(b) Not a solution		
5.	$x + 5 = 2x$		(a) $x = -1$	(b) $x = 5$
	(a) Not a solution	(b) Solution		
6.	$2x - 3 = 5x$		(a) $x = 0$	(b) $x = -1$
	(a) Not a solution	(b) Solution		
7.	$x + 3 = 2(x - 4)$		(a) $x = 11$	(b) $x = -5$
	(a) Solution	(b) Not a solution		
8.	$5x - 1 = 3(x + 5)$		(a) $x = 8$	(b) $x = -2$
	(a) Solution	(b) Not a solution		

	Equation		*Values*	
9.	$2x + 10 = 7(x + 1)$		(a) $x = \frac{3}{5}$	(b) $x = -\frac{2}{3}$
	(a) Solution	(b) Not a solution		
10.	$3(3x + 2) = 9 - x$		(a) $x = -\frac{3}{4}$	(b) $x = \frac{3}{10}$
	(a) Not a solution	(b) Solution		
11.	$x^2 - 4 = x + 2$		(a) $x = 3$	(b) $x = -2$
	(a) Solution	(b) Solution		
12.	$x^2 = 8 - 2x$		(a) $x = 2$	(b) $x = -4$
	(a) Solution	(b) Solution		
13.	$\dfrac{2}{x} - \dfrac{1}{x} = 1$		(a) $x = 0$	(b) $x = \dfrac{1}{3}$
	(a) Not a solution	(b) Not a solution		
14.	$\dfrac{4}{x} + \dfrac{2}{x} = 1$		(a) $x = 0$	(b) $x = 6$
	(a) Not a solution	(b) Solution		
15.	$\dfrac{5}{x - 1} + \dfrac{1}{x} = 5$		(a) $x = 3$	(b) $x = \dfrac{1}{6}$
	(a) Not a solution	(b) Not a solution		
16.	$\dfrac{3}{x - 2} = x$		(a) $x = -1$	(b) $x = 3$
	(a) Solution	(b) Solution		

In Exercises 17–26, use a calculator to determine whether the value of x is a solution of the equation.

	Equation		Values	
17.	$x + 1.7 = 6.5$	(a)	$x = -3.1$	Not a solution
		(b)	$x = 4.8$	Solution
18.	$7.9 - x = 14.6$	(a)	$x = -6.7$	Solution
		(b)	$x = 5.4$	Not a solution
19.	$40x - 490 = 0$	(a)	$x = 12.25$	Solution
		(b)	$x = -12.25$	Not a solution
20.	$20x - 560 = 0$	(a)	$x = 27.5$	Not a solution
		(b)	$x = -27.5$	Not a solution
21.	$2x^2 - x - 10 = 0$	(a)	$x = \frac{5}{2}$	Solution
		(b)	$x = -1.09$	Not a solution
22.	$22x - 5x^2 = 17$	(a)	$x = 1$	Solution
		(b)	$x = 3.4$	Solution
23.	$\dfrac{1}{x} - \dfrac{9}{x-4} = 1$	(a)	$x = 0$	Not a solution
		(b)	$x = -2$	Solution
24.	$x = \dfrac{3}{4x+1}$	(a)	$x = -0.25$	Not a solution
		(b)	$x = 0.75$	Solution
25.	$x^3 - 1.728 = 0$	(a)	$x = \frac{6}{5}$	Solution
		(b)	$x = -\frac{6}{5}$	Not a solution
26.	$4x^2 - 10.24 = 0$	(a)	$x = \frac{8}{5}$	Solution
		(b)	$x = -\frac{8}{5}$	Solution

In Exercises 27–34, justify each step of the solution. See Example 3.

27.

$5x + 12 = 22$	Original equation
$5x + 12 - 12 = 22 - 12$	Subtract 12 from each side.
$5x = 10$	Combine like terms.
$\dfrac{5x}{5} = \dfrac{10}{5}$	Divide each side by 5.
$x = 2$	Solution

28.

$14 - 3x = 5$	Original equation
$14 - 3x - 14 = 5 - 14$	Subtract 14 from each side.
$14 - 14 - 3x = -9$	Commutative Property
$-3x = -9$	Additive Inverse Property
$\dfrac{-3x}{-3} = \dfrac{-9}{-3}$	Divide each side by -3.
$x = 3$	Solution

29.

$\dfrac{2}{3}x = 12$	Original equation
$\dfrac{3}{2}\left(\dfrac{2}{3}x\right) = \dfrac{3}{2}(12)$	Multiply each side by $\frac{3}{2}$.
$x = 18$	Solution

30.

$\dfrac{4}{5}x = -28$	Original equation
$\dfrac{5}{4}\left(\dfrac{4}{5}x\right) = \dfrac{5}{4}(-28)$	Multiply each side by $\frac{5}{4}$.
$x = -35$	Solution

31.

$2(x - 1) = x + 3$	Original equation
$2x - 2 = x + 3$	Distributive Property
$2x - x - 2 = x - x + 3$	Subtract x from each side.
$x - 2 = 3$	Combine like terms.
$x - 2 + 2 = 3 + 2$	Add 2 to each side.
$x = 5$	Solution

32.

$x + 6 = -6(4 - x)$	Original equation
$x + 6 = -24 + 6x$	Distributive Property
$x - x + 6 = -24 + 6x - x$	Subtract x from each side.
$6 = 5x - 24$	Combine like terms.
$6 + 24 = 5x - 24 + 24$	Add 24 to each side.
$30 = 5x$	Combine like terms.
$\dfrac{30}{5} = \dfrac{5x}{5}$	Divide each side by 5.
$6 = x$	Solution

33.

$x = -2(x + 3)$	Original equation
$x = -2x - 6$	Distributive Property
$x + 2x = -2x + 2x - 6$	Add $2x$ to each side.
$3x = 0 - 6$	Additive Inverse Property
$3x = -6$	Combine like terms.
$\dfrac{3x}{3} = \dfrac{-6}{3}$	Divide each side by 3.
$x = -2$	Solution

34.

$$\frac{x}{3} = x + 1 \qquad \text{Original equation}$$

$$3\left(\frac{x}{3}\right) = 3(x + 1) \qquad \text{Multiply each side by 3.}$$

$$x = 3x + 3 \qquad \text{Multiplicative Inverse and Distributive Properties}$$

$$x - 3x = 3x - 3x + 3 \qquad \text{Subtract } 3x \text{ from each side.}$$

$$-2x = 0 + 3 \qquad \text{Additive Inverse Property}$$

$$-2x = 3 \qquad \text{Additive Identity Property}$$

$$\frac{-2x}{-2} = \frac{3}{-2} \qquad \text{Divide each side by } -2.$$

$$x = -\frac{3}{2} \qquad \text{Solution}$$

In Exercises 35–38, use a property of equality to solve the equation. Check your solution. See Examples 1, 2, and 3.

35. $x - 8 = 5$ 13

36. $x + 3 = 19$ 16

37. $3x = 30$ 10

38. $\dfrac{x}{4} = 12$ 48

Solving Problems

In Exercises 39–44, write a verbal description of the algebraic equation without using a variable. (There is more than one correct answer.)

39. $2x + 5 = 21$ Twice a number increased by 5 is 21.

40. $3x - 2 = 7$ Three times a number decreased by 2 is 7.

41. $10(x - 3) = 8x$ Ten times the difference of a number and 3 is 8 times the number.

42. $2(x - 5) = 12$ Two times the difference of a number and 5 is 12.

43. $\dfrac{x + 1}{3} = 8$ The sum of a number and 1 divided by 3 is 8.

44. $\dfrac{x - 2}{10} = 6$ The difference of a number and 2 divided by 10 is 6.

In Exercises 45–68, construct an equation for the word problem. Do *not* solve the equation. See Examples 4, 5, and 6.

45. The sum of a number and 12 is 45. What is the number? $x + 12 = 45$

46. The sum of 3 times a number and 4 is 16. What is the number? $3x + 4 = 16$

47. Four times the sum of a number and 6 is 100. What is the number? $4(x + 6) = 100$

48. Find a number such that 6 times the number subtracted from 120 is 96. $120 - 6x = 96$

49. Find a number such that 2 times the number decreased by 14 equals the number divided by 3.
$2x - 14 = \dfrac{x}{3}$

50. The sum of a number and 8, divided by 4, is 32. What is the number? $\dfrac{x + 8}{4} = 32$

51. *Test Score*   After your instructor added 6 points to each student's test score, your score is 94. What was your original score? $x + 6 = 94$

52. *Meteorology*   With the 1.2-inch rainfall today, the total for the month is 4.5 inches. How much had been recorded for the month before today's rainfall?
$4.5 = x + 1.2$

53. *Consumerism*   You have $1044 saved for the purchase of a new computer that will cost $1926. How much more must you save? $1044 + x = 1926$

54. *List Price*   The sale price of a coat is $225.98. The discount is $64. What is the list (original) price?
$225.98 = x - 64$

55. *Travel Costs*   A company pays its sales representatives 35 cents per mile if they use their personal cars. A sales representative submitted a bill to be reimbursed for $148.05 for driving. How many miles did the sales representative drive? $0.35x = 148.05$

56. *Money*   A student has n quarters and seven $1 bills totaling $8.75. How many quarters does the student have? $0.25n + 7 = 8.75$

57. ▲ *Geometry*   The base of a rectangular box is 4 feet by 6 feet and its volume is 72 cubic feet (see figure). What is the height of the box? $24h = 72$

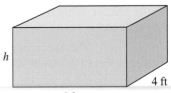

58. ▲ *Geometry*   The width of a rectangular mirror is one-third its length, as shown in the figure. The perimeter of the mirror is 96 inches. What are the dimensions of the mirror? $2l + 2\left(\frac{1}{3}l\right) = 96$

59. *Average Speed*   After traveling for 3 hours, your family is still 25 miles from completing a 160-mile trip (see figure). What was the average speed during the first 3 hours of the trip? $3r + 25 = 160$

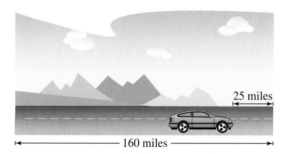

25 miles

160 miles

60. *Average Speed*   After traveling for 4 hours, you are still 24 miles from completing a 200-mile trip. It requires one-half hour to travel the last 24 miles. What was the average speed during the first 4 hours of the trip? $4r + 24 = 200$

61. *Average Speed*   A group of students plans to take two cars to a soccer tournament. The first car leaves on time, travels at an average speed of 45 miles per hour, and arrives at the destination in 3 hours. The second car leaves one-half hour after the first car and arrives at the tournament at the same time as the students in the first car. What is the average speed of the second car? $135 = 2.5x$

62. *Dow Jones Average*   The Dow Jones average fell 58 points during a week and was 8695 at the close of the market on Friday. What was the average at the close of the market on the previous Friday? $x - 58 = 8695$

63. *Consumer Awareness*   The price of a gold ring has increased by $45 over the past year. It is now selling for $375. What was the price one year ago? $p + 45 = 375$

64. *Meteorology*   You hear thunder 3 seconds after seeing the lightning. The speed of sound is 1100 feet per second. How far away is the lightning? $\dfrac{d}{1100} = 3$

65. *Depreciation*   A textile corporation buys equipment with an initial purchase price of $750,000. It is estimated that its useful life will be 3 years and at that time its value will be $75,000. The total depreciation is divided equally among the three years. (Depreciation is the difference between the initial price of an item and its current value.) What is the total amount of depreciation declared each year? $750,000 - 3D = 75,000$

66. *Car Payments*   You make 48 monthly payments of $158 each to buy a used car. The total amount financed is $6000. What is the total amount of interest that you paid? $48(158) - 6000 = I$

67. *Fund Raising*   A student group is selling boxes of greeting cards at a profit of $1.75 each. The group needs $2000 more to have enough money for a trip to Washington, D.C. How many boxes does the group need to sell to earn $2000? $1.75n = 2000$

68. *Consumer Awareness*   The price of a compact car increased $1432 over the past year. The price of the car was $9850 two years ago and $10,120 one year ago. What is its current price? $10,120 + 1432 = x$

Unit Analysis   In Exercises 69–76, simplify the expression. State the units of the simplified value.

69. $\dfrac{3 \text{ dollars}}{\text{unit}} \cdot (5 \text{ units})$ 15 dollars

70. $\dfrac{25 \text{ miles}}{\text{gallon}} \cdot (15 \text{ gallons})$ 375 miles

71. $\dfrac{50 \text{ pounds}}{\text{foot}} \cdot (3 \text{ feet})$ 150 pounds

72. $\dfrac{3 \text{ dollars}}{\text{pound}} \cdot (5 \text{ pounds})$ 15 dollars

73. $\dfrac{5 \text{ feet}}{\text{second}} \cdot \dfrac{60 \text{ seconds}}{\text{minute}} \cdot (20 \text{ minutes})$ 6000 feet

74. $\dfrac{12 \text{ dollars}}{\text{hour}} \cdot \dfrac{1 \text{ hour}}{60 \text{ minutes}} \cdot (45 \text{ minutes})$ 9 dollars

75. $\dfrac{100 \text{ centimeters}}{\text{meter}} \cdot (2.4 \text{ meters})$ 240 centimeters

76. $\dfrac{1000 \text{ milliliters}}{\text{liter}} \cdot (5.6 \text{ liters})$ 5600 milliliters

Explaining Concepts

77. *Writing*   Explain how to decide whether a real number is a solution of an equation. Give an example of an equation with a solution that checks and one that does not check.   Substitute the real number into the equation. If the equation is true, the real number is a solution. Given the equation $2x - 3 = 5$, $x = 4$ is a solution and $x = -2$ is not a solution.

78. *Writing*   In your own words, explain what is meant by the term *equivalent equations*.   Equivalent equations have the same solution set.

79. *Writing*   Explain the difference between simplifying an expression and solving an equation. Give an example of each.   Simplifying an expression means removing all symbols of grouping and combining like terms. Solving an equation means finding all values of the variable for which the equation is true.
Simplify: $3(x - 2) - 4(x + 1) = 3x - 6 - 4x - 4$
$$= -x - 10$$
Solve: $3(x - 2) = 6$
$$3x - 6 = 6$$
$$3x = 12 \rightarrow x = 4$$

80. *Writing*   Describe a real-life problem that uses the following verbal model.

Revenue of $840	=	$35 per case	·	Number of cases

The total cost of a shipment of bulbs is $840. Find the number of cases of bulbs if each case costs $35.

81. *Writing*   Describe, from memory, the steps that can be used to transform an equation into an equivalent equation.

(a) Simplify each side by removing symbols of grouping, combining like terms, and reducing fractions on one or both sides.

(b) Add (or subtract) the same quantity to (from) each side of the equation.

(c) Multiply (or divide) each side of the equation by the same nonzero real number.

(d) Interchange the two sides of the equation.

What Did You Learn?

Key Terms

variables, *p. 68*
constants, *p. 68*
algebraic expression, *p. 68*
terms, *p. 68*
coefficient, *p. 68*

evaluate an algebraic
  expression, *p. 71*
expanding an algebraic
  expression, *p. 79*
like terms, *p. 80*
simplify an algebraic
  expression, *p. 82*

verbal mathematical
  model, *p. 92*
equation, *p. 105*
solutions, *p. 105*
satisfy, *p. 105*
equivalent equations,
  p. 107

Key Concepts

2.1 ⬤ Exponential form

Repeated multiplication can be expressed in exponential form using a base a and an exponent n, where a is a real number, variable, or algebraic expression and n is a positive integer.

$$a^n = a \cdot a \cdots a$$

2.1 ⬤ Evaluating algebraic expressions

To evaluate an algebraic expression, substitute every occurrence of the variable in the expression with the appropriate real number and perform the operation(s).

2.2 ⬤ Properties of algebra

Commutative Property:
Addition $a + b = b + a$
Multiplication $ab = ba$

Associative Property:
Addition $(a + b) + c = a + (b + c)$
Multiplication $(ab)c = a(bc)$

Distributive Property:
$a(b + c) = ab + ac$ $a(b - c) = ab - ac$
$(a + b)c = ac + bc$ $(a - b)c = ac - bc$

Identities:
Additive $a + 0 = 0 + a = a$
Multiplicative $a \cdot 1 = 1 \cdot a = a$

Inverses:
Additive $a + (-a) = 0$

Multiplicative $a \cdot \dfrac{1}{a} = 1, \ a \neq 0$

2.2 ⬤ Combining like terms

To combine like terms in an algebraic expression, add their respective coefficients and attach the common variable factor.

2.2 ⬤ Simplifying an algebraic expression

To simplify an algebraic expression, remove symbols of grouping and combine like terms.

2.3 ⬤ Additional problem-solving strategies

Additional problem-solving strategies are listed below.

1. Guess, check, and revise
2. Make a table/look for a pattern
3. Draw a diagram
4. Solve a simpler problem

2.4 ⬤ Checking solutions of equations

To check a solution, substitute the given solution for each occurrence of the variable in the original equation. Evaluate each side of the equation. If both sides are equivalent, the solution checks.

2.4 ⬤ Properties of equality

Addition: Add (or subtract) the same quantity to (from) each side of the equation.

Multiplication: Multiply (or divide) each side of the equation by the same nonzero quantity.

2.4 ⬤ Constructing equations

From the verbal description, write a verbal mathematical model. Assign labels to the known and unknown quantities, and write an algebraic model.

Review Exercises

2.1 Writing and Evaluating Algebraic Expressions

① Define and identify terms, variables, and coefficients of algebraic expressions.

In Exercises 1 and 2, identify the variable and the constant in the expression.

1. $15 - x$ $-x, 15$

2. $t - 5^2$ $t, -5^2$

In Exercises 3–8, identify the terms and the coefficients of the expression.

3. $12y + y^2$
   $12y, y^2; 12, 1$

4. $4x - \frac{1}{2}x^3$
   $4x, -\frac{1}{2}x^3; 4, -\frac{1}{2}$

5. $5x^2 - 3xy + 10y^2$ $5x^2, -3xy, 10y^2; 5, -3, 10$

6. $y^2 - 10yz + \frac{2}{3}z^2$ $y^2, -10yz, \frac{2}{3}z^2; 1, -10, \frac{2}{3}$

7. $\frac{2y}{3} - \frac{4x}{y}$
   $\frac{2y}{3}, -\frac{4x}{y}; \frac{2}{3}, -4$

8. $-\frac{4b}{9} + \frac{11a}{b}$
   $-\frac{4b}{9}, \frac{11a}{b}; -\frac{4}{9}, 11$

② Define exponential form and interpret exponential expressions.

In Exercises 9–12, rewrite the product in exponential form.

9. $5z \cdot 5z \cdot 5z$ $(5z)^3$

10. $\frac{3}{8}y \cdot \frac{3}{8}y \cdot \frac{3}{8}y \cdot \frac{3}{8}y$ $\left(\frac{3}{8}y\right)^4$

11. $(b - c) \cdot (b - c) \cdot 6 \cdot 6$ $6^2(b - c)^2$

12. $2 \cdot (a + b) \cdot 2 \cdot (a + b) \cdot 2$ $2^3(a + b)^2$

③ Evaluate algebraic expressions using real numbers.

In Exercises 13–18, evaluate the algebraic expression for the given values of the variable(s).

Expression	*Values*
13. $x^2 - 2x + 5$	(a) $x = 0$ (b) $x = 2$
(a) 5 (b) 5	
14. $x^3 - 8$	(a) $x = 2$ (b) $x = 4$
(a) 0 (b) 56	
15. $x^2 - x(y + 1)$	(a) $x = 2, y = -1$
(a) 4 (b) -2	(b) $x = 1, y = 2$
16. $2r + r(t^2 - 3)$	(a) $r = 3, t = -2$
(a) 9 (b) -16	(b) $r = -2, t = 3$
17. $\dfrac{x + 5}{y}$	(a) $x = -5, y = 3$
(a) 0 (b) -7	(b) $x = 2, y = -1$

Expression	*Values*
18. $\dfrac{a - 9}{2b}$	(a) $a = 7, b = -3$
(a) $\frac{1}{3}$ (b) $-\frac{13}{10}$	(b) $a = -4, b = 5$

2.2 Simplifying Algebraic Expressions

① Use the properties of algebra.

In Exercises 19–24, identify the property of algebra illustrated by the statement.

19. $xy \cdot \dfrac{1}{xy} = 1$ Multiplicative Inverse Property

20. $u(vw) = (uv)w$ Associative Property of Multiplication

21. $(x - y)(2) = 2(x - y)$
   Commutative Property of Multiplication

22. $(a + b) + 0 = a + b$ Additive Identity Property

23. $2x + (3y - z) = (2x + 3y) - z$
   Associative Property of Addition

24. $x(y + z) = xy + xz$ Distributive Property

In Exercises 25–32, use the Distributive Property to expand the expression.

25. $4(x + 3y)$
   $4x + 12y$

26. $3(8s - 12t)$
   $24s - 36t$

27. $-5(2u - 3v)$
   $-10u + 15v$

28. $-3(-2x - 8y)$
   $6x + 24y$

29. $x(8x + 5y)$
   $8x^2 + 5xy$

30. $-u(3u - 10v)$
   $-3u^2 + 10uv$

31. $-(-a + 3b)$
   $a - 3b$

32. $(7 - 2j)(-6)$
   $-42 + 12j$

② Combine like terms of an algebraic expression.

In Exercises 33–44, simplify the expression by combining like terms.

33. $3a - 5a$ $-2a$

34. $6c - 2c$ $4c$

35. $3p - 4q + q + 8p$ $11p - 3q$

36. $10x - 4y - 25x + 6y$ $-15x + 2y$

37. $\frac{1}{4}s - 6t + \frac{7}{2}s + t$ $\frac{15}{4}s - 5t$

38. $\frac{2}{3}a + \frac{3}{5}a - \frac{1}{2}b + \frac{2}{3}b$ $\frac{19}{15}a + \frac{1}{6}b$

39. $x^2 + 3xy - xy + 4$ $x^2 + 2xy + 4$

40. $uv^2 + 10 - 2uv^2 + 2$ $-uv^2 + 12$

41. $5x - 5y + 3xy - 2x + 2y$ $3x - 3y + 3xy$

42. $y^3 + 2y^2 + 2y^3 - 3y^2 + 1$ $3y^3 - y^2 + 1$

43. $5\left(1 + \dfrac{r}{n}\right)^2 - 2\left(1 + \dfrac{r}{n}\right)^2$ $3\left(1 + \dfrac{r}{n}\right)^2$

44. $-7\left(\dfrac{1}{u}\right) + 4\left(\dfrac{1}{u^2}\right) + 3\left(\dfrac{1}{u}\right)$ $-4\left(\dfrac{1}{u}\right) + 4\left(\dfrac{1}{u^2}\right)$

③ Simplify an algebraic expression by rewriting the terms.

In Exercises 45–52, simplify the expression.

45. $12(4t)$ $48t$

46. $8(7x)$ $56x$

47. $-5(-9x^2)$ $45x^2$

48. $-10(-3b^3)$ $30b^3$

49. $(-6x)(2x^2)$ $-12x^3$

50. $(-3y^2)(15y)$ $-45y^3$

51. $\dfrac{12x}{5} \cdot \dfrac{10}{3}$ $8x$

52. $\dfrac{4z}{15} \cdot \dfrac{9}{2}$ $\dfrac{6z}{5}$

④ Use the Distributive Property to remove symbols of grouping.

In Exercises 53–64, simplify the expression by removing symbols of grouping and combining like terms.

53. $5(u - 4) + 10$ $5u - 10$

54. $16 - 3(v + 2)$ $10 - 3v$

55. $3s - (r - 2s)$ $5s - r$

56. $50x - (30x + 100)$ $20x - 100$

57. $-3(1 - 10z) + 2(1 - 10z)$ $10z - 1$

58. $8(15 - 3y) - 5(15 - 3y)$ $45 - 9y$

59. $\frac{1}{3}(42 - 18z) - 2(8 - 4z)$ $2z - 2$

60. $\frac{1}{4}(100 + 36s) - (15 - 4s)$ $13s + 10$

61. $10 - [8(5 - x) + 2]$ $8x - 32$

62. $3[2(4x - 5) + 4] - 3$ $24x - 21$

63. $2[x + 2(y - x)]$ $-2x + 4y$

64. $2t[4 - (3 - t)] + 5t$ $2t^2 + 7t$

65. *Depreciation*   You pay P dollars for new equipment. Its value after 5 years is given by

$$P\left(\dfrac{9}{10}\right)\left(\dfrac{9}{10}\right)\left(\dfrac{9}{10}\right)\left(\dfrac{9}{10}\right)\left(\dfrac{9}{10}\right).$$

Simplify the expression. $P\left(\frac{9}{10}\right)^5$

66. ▲ *Geometry*   The height of a triangle is $1\frac{1}{2}$ times its base. Its area is given by $\frac{1}{2}b\left(\frac{3}{2}b\right)$. Simplify the expression. $\frac{3}{4}b^2$

67. Simplify the algebraic expression that represents the sum of three consecutive odd integers, $2n - 1$, $2n + 1$, and $2n + 3$.
$(2n - 1) + (2n + 1) + (2n + 3) = 6n + 3$

68. Simplify the algebraic expression that represents the sum of three consecutive even integers, $2n$, $2n + 2$, $2n + 4$. $2n + (2n + 2) + (2n + 4) = 6n + 6$

69. ▲ *Geometry*   The face of a DVD player has the dimensions shown in the figure. Write an algebraic expression that represents the area of the face of the DVD player excluding the compartment holding the disc. $58x^2$

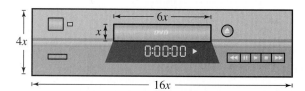

70. ▲ *Geometry*   Write an expression for the perimeter of the figure. Use the rules of algebra to simplify the expression. $6x - 2$

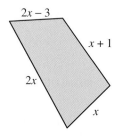

2.3 Algebra and Problem Solving

② Construct verbal mathematical models from written statements.

In Exercises 71 and 72, construct a verbal model and then write an algebraic expression that represents the specified quantity.

71. The total hourly wage for an employee when the base pay is $8.25 per hour and an additional $0.60 is paid for each unit produced per hour
Verbal model:

Base pay per hour $+$ Additional pay per unit $\cdot$ Number of units produced per hour

Algebraic expression: $8.25 + 0.60x$

72. The total cost for a family to stay one night at a campground if the charge is $18 for the parents plus $3 for each of the children
Verbal model:

Cost of parents $+$ Cost per child $\cdot$ Number of children

Algebraic expression: $18 + 3x$

3 Translate verbal phrases into algebraic expressions.

In Exercises 73–82, translate the phrase into an algebraic expression. Let x represent the real number.

73. Two-thirds of a number, plus 5 $\frac{2}{3}x + 5$

74. One hundred, decreased by 5 times a number
$100 - 5x$

75. Ten less than twice a number $2x - 10$

76. The ratio of a number to 10 $\frac{x}{10}$

77. Fifty increased by the product of 7 and a number
$50 + 7x$

78. Ten decreased by the quotient of a number and 2
$10 - \frac{x}{2}$

79. The sum of a number and 10 all divided by 8
$\frac{x + 10}{8}$

80. The product of 15 and a number, decreased by 2
$15x - 2$

81. The sum of the square of a real number and 64
$x^2 + 64$

82. The absolute value of the sum of a number and -10
$|x + (-10)|$

In Exercises 83–86, write a verbal description of the expression without using a variable. (There is more than one correct answer.)

83. $x + 3$ A number plus 3

84. $3x - 2$ Three times a number decreased by 2

85. $\frac{y - 2}{3}$ A number decreased by 2, divided by 3

86. $4(x + 5)$ Four times the sum of a number and 5

4 Identify hidden operations when constructing algebraic expressions.

87. *Commission*   A salesperson earns 5% commission on his total weekly sales, x. Write an algebraic expression that represents the amount in commissions that the salesperson earns in a week. $0.05x$

88. *Sale Price*   A cordless phone is advertised for 20% off the list price of L dollars. Write an algebraic expression that represents the sale price of the phone.
$0.8L$

89. *Rent*   The monthly rent for your apartment is $625 for n months. Write an algebraic expression that represents the total rent. $625n$

90. *Distance*   A car travels for 10 hours at an average speed of s miles per hour. Write an algebraic expression that represents the total distance traveled by the car. $10s$

6 Use problem-solving strategies to solve application problems.

91. *Finding a Pattern*

(a) Complete the table. The third row in the table is the difference between consecutive entries of the second row. The fourth row is the difference between consecutive entries of the third row.

n	0	1	2	3	4	5
$n^2 + 3n + 2$	2	6	12	20	30	42
Differences		4	6	8	10	12
Differences			2	2	2	2

(b) Describe the patterns of the third and fourth rows.   Third row: entries increase by 2; Fourth row: constant 2

92. *Finding a Pattern*   Find values for a and b such that the expression $an + b$ yields the table values.

n	0	1	2	3	4	5
$an + b$	4	9	14	19	24	29

$a = 5, b = 4$

2.4 Introduction to Equations

2 Check whether a given value is a solution of an equation.

In Exercises 93–102, determine whether each value of x is a solution of the equation.

Equation	Values
93. $5x + 6 = 36$	(a) $x = 3$ (b) $x = 6$
(a) Not a solution	(b) Solution
94. $17 - 3x = 8$	(a) $x = 3$ (b) $x = -3$
(a) Solution	(b) Not a solution
95. $3x - 12 = x$	(a) $x = -1$ (b) $x = 6$
(a) Not a solution	(b) Solution
96. $8x + 24 = 2x$	(a) $x = 0$ (b) $x = -4$
(a) Not a solution	(b) Solution

Equation   Values

97. $4(2 - x) = 3(2 + x)$ (a) $x = \dfrac{2}{7}$ (b) $x = -\dfrac{2}{3}$

    (a) Solution   (b) Not a solution

98. $5x + 2 = 3(x + 10)$ (a) $x = 14$ (b) $x = -10$

    (a) Solution   (b) Not a solution

99. $\dfrac{4}{x} - \dfrac{2}{x} = 5$ (a) $x = -1$ (b) $x = \dfrac{2}{5}$

    (a) Not a solution   (b) Solution

100. $\dfrac{x}{3} + \dfrac{x}{6} = 1$ (a) $x = \dfrac{2}{9}$ (b) $x = -\dfrac{2}{9}$

    (a) Not a solution   (b) Not a solution

101. $x(x - 7) = -12$ (a) $x = 3$ (b) $x = 4$

    (a) Solution   (b) Solution

102. $x(x + 1) = 2$ (a) $x = 1$ (b) $x = -2$

    (a) Solution   (b) Solution

❸ Use properties of equality to solve equations.

In Exercises 103 and 104, justify each step of the solution.

103.

$3(x - 2) = x + 2$	Original equation
$3x - 6 = x + 2$	Distributive Property
$3x - x - 6 = x - x + 2$	Subtract x from each side.
$2x - 6 = 2$	Combine like terms.
$2x - 6 + 6 = 2 + 6$	Add 6 to each side.
$2x = 8$	Combine like terms.
$\dfrac{2x}{2} = \dfrac{8}{2}$	Divide each side by 2.
$x = 4$	Solution

104.

$x = -(x - 14)$	Original equation
$x = -x + 14$	Distributive Property
$x + x = -x + x + 14$	Add x to each side.
$2x = 14$	Combine like terms.
$\dfrac{2x}{2} = \dfrac{14}{2}$	Divide each side by 2.
$x = 7$	Solution

❹ Use a verbal model to construct an algebraic equation.

In Exercises 105–108, construct an equation for the word problem. Do *not* solve the equation.

105. The sum of a number and its reciprocal is $\frac{37}{6}$. What is the number? $x + \dfrac{1}{x} = \dfrac{37}{6}$

106. *Distance*   A car travels 135 miles in t hours with an average speed of 45 miles per hour (see figure). How many hours did the car travel? $135 = 45t$

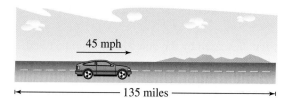

45 mph

135 miles

107. ▲ *Geometry*   The area of the shaded region in the figure is 24 square inches. What is the length of the rectangle? $6x - \dfrac{1}{2}(6x) = \dfrac{1}{2}(6x) = 24$

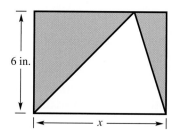

6 in.

x

108. ▲ *Geometry*   The perimeter of the face of a rectangular traffic light is 72 inches (see figure). What are the dimensions of the traffic light?

$2L + 2(0.35L) = 2.7L = 72$

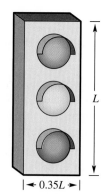

L

$0.35L$

Chapter Test

Take this test as you would take a test in class. After you are done, check your work against the answers in the back of the book.

1. Identify the terms and coefficients of the expression.

   $2x^2 - 7xy + 3y^3$ $2x^2, 2; -7xy, -7; 3y^3, 3$

2. Rewrite the product in exponential form.

   $x \cdot (x + y) \cdot x \cdot (x + y) \cdot x$ $x^3(x + y)^2$

In Exercises 3–6, identify the property of algebra illustrated by the statement.

3. $(5x)y = 5(xy)$ Associative Property of Multiplication

4. $2 + (x - y) = (x - y) + 2$ Commutative Property of Addition

5. $7xy + 0 = 7xy$ Additive Identity Property

6. $(x + 5) \cdot \dfrac{1}{(x + 5)} = 1$ Multiplicative Inverse Property

In Exercises 7–10, use the Distributive Property to expand the expression.

7. $3(x + 8)$ $3x + 24$

8. $5(4r - s)$ $20r - 5s$

9. $-y(3 - 2y)$ $-3y + 2y^2$

10. $-9(4 - 2x + x^2)$ $-36 + 18x - 9x^2$

In Exercises 11–14, simplify the expression.

11. $3b - 2a + a - 10b$ $-a - 7b$

12. $15(u - v) - 7(u - v)$ $8u - 8v$

13. $3z - (4 - z)$ $4z - 4$

14. $2[10 - (t + 1)]$ $18 - 2t$

15. Evaluate the expression when $x = 3$ and $y = -12$.

    (a) $x^3 - 2$ (b) $x^2 + 4(y + 2)$ (a) 25   (b) -31

16. Explain why it is not possible to evaluate $\dfrac{a + 2b}{3a - b}$ when $a = 2$ and $b = 6$.
    Division by zero is undefined.

17. Translate the phrase, "one-fifth of a number, increased by two," into an algebraic expression. Let n represent the number. $\frac{1}{5}n + 2$

18. (a) Write expressions for the perimeter and area of the rectangle at the left.

    Perimeter: $2w + 2(2w - 4)$; Area: $w(2w - 4)$

    (b) Simplify the expressions.   Perimeter: $6w - 8$; Area: $2w^2 - 4w$

    (c) Identify the unit of measure for each expression.   Perimeter: unit of length; Area: square units

    (d) Evaluate each expression when $w = 12$ feet.   Perimeter: 64 feet; Area: 240 square feet

w

$2w - 4$

Figure for 18

19. The prices of concert tickets for adults and children are $15 and $10, respectively. Write an algebraic expression that represents the total income from the concert for m adults and n children. $15m + 10n$

20. Determine whether the values of x are solutions of $6(3 - x) - 5(2x - 1) = 7$.

    (a) $x = -2$ (b) $x = 1$ (a) Not a solution   (b) Solution

121

Motivating the Chapter

⚡ Talk Is Cheap?

You plan to purchase a cellular phone with a service contract. For a price of $99, one package includes the phone and 3 months of service. You will be billed a *per minute usage rate* each time you make or receive a call. After 3 months you will be billed a monthly service charge of $19.50 and the per minute usage rate.

A second cellular phone package costs $80, which includes the phone and 1 month of service. You will be billed a per minute usage rate each time you make or receive a call. After the first month you will be billed a monthly service charge of $24 and the per minute usage rate.

See Section 3.3, Exercise 105.

a. Write an equation to find the cost of the phone in the first package. Solve the equation to find the cost of the phone. $3(19.50) + x = 99$, $40.50

b. Write an equation to find the cost of the phone in the second package. Solve the equation to find the cost of the phone. Which phone costs more, the one in the first package or the one in the second package?
   $24 + x = 80$, $56.00, Second package

c. What percent of the purchase price of $99 goes toward the price of the cellular phone in the first package? Use an equation to answer the question. $40.50 = p(99)$, 40.9%

d. What percent of the purchase price of $80 goes toward the price of the cellular phone in the second package? Use an equation to answer the question. $56 = p(80)$, 70%

e. The sales tax on your purchase is 5%. What is the total cost of purchasing the first cellular phone package? Use an equation to answer the question.
   $x = 99 + 0.05(99)$, $103.95

f. You decide to buy the first cellular phone package. Your total cellular phone bill for the fourth month of use is $92.46 for 3.2 hours of use. What is the per minute usage rate? Use an equation to answer the question. $19.50 + 60(3.2)x = 92.46$, $0.38

See Section 3.4, Exercise 87.

g. For the fifth month you were billed the monthly service charge and $47.50 for 125 minutes of use. You estimate that during the next month you spent 150 minutes on calls. Use a proportion to find the charge for 150 minutes of use. (Use the first package.) $57.00

See Section 3.6, Exercise 85.

h. You determine that the most you can spend each month on phone calls is $75. Write an inequality that describes the maximum number of minutes you can spend talking on the cellular phone each month if the per minute usage rate is $0.35. Solve the inequality. (Use the first package.) $0.35x + 19.50 \le 75.00$; $x \le 158.57$ minutes

3

Steven Poe/Alamy

Linear Equations and Problem Solving

3.1 Solving Linear Equations

Amy Etra/PhotoEdit

Why You Should Learn It

Linear equations are used in many real-life applications. For instance, in Exercise 65 on page 133, you will use a linear equation to determine the number of hours spent repairing your car.

What You Should Learn

1. Solve linear equations in standard form.
2. Solve linear equations in nonstandard form.
3. Use linear equations to solve application problems.

1 Solve linear equations in standard form.

Linear Equations in the Standard Form $ax + b = 0$

This is an important step in your study of algebra. In the first two chapters, you were introduced to the rules of algebra, and you learned to use these rules to rewrite and simplify algebraic expressions. In Sections 2.3 and 2.4, you gained experience in translating verbal expressions and problems into algebraic forms. You are now ready to use these skills and experiences to *solve equations*.

In this section, you will learn how the rules of algebra and the properties of equality can be used to solve the most common type of equation—a linear equation in one variable.

Definition of Linear Equation

A **linear equation** in one variable x is an equation that can be written in the standard form

$$ax + b = 0$$

where a and b are real numbers with $a \neq 0$.

A linear equation in one variable is also called a **first-degree equation** because its variable has an (implied) exponent of 1. Some examples of linear equations in standard form are

$$2x = 0, \quad x - 7 = 0, \quad 4x + 6 = 0, \quad \text{and} \quad \frac{x}{2} - 1 = 0.$$

Remember that to *solve* an equation involving x means to find all values of x that satisfy the equation. For the linear equation $ax + b = 0$, the goal is to *isolate* x by rewriting the equation in the form

$$x = \boxed{\text{a number}}.\qquad \text{Isolate the variable } x.$$

To obtain this form, you use the techniques discussed in Section 2.4. That is, beginning with the original equation, you write a sequence of equivalent equations, each having the same solution as the original equation. For instance, to solve the linear equation $x - 2 = 0$, you can add 2 to each side of the equation to obtain $x = 2$. As mentioned in Section 2.4, each equivalent equation is called a **step** of the solution.

Example 1   Solving a Linear Equation in Standard Form

Solve $3x - 15 = 0$. Then check the solution.

Solution

$$3x - 15 = 0 \qquad \text{Write original equation.}$$
$$3x - 15 + 15 = 0 + 15 \qquad \text{Add 15 to each side.}$$
$$3x = 15 \qquad \text{Combine like terms.}$$
$$\frac{3x}{3} = \frac{15}{3} \qquad \text{Divide each side by 3.}$$
$$x = 5 \qquad \text{Simplify.}$$

It appears that the solution is $x = 5$. You can check this as follows:

Check

$$3x - 15 = 0 \qquad \text{Write original equation.}$$
$$3(5) - 15 \overset{?}{=} 0 \qquad \text{Substitute 5 for } x.$$
$$15 - 15 \overset{?}{=} 0 \qquad \text{Simplify.}$$
$$0 = 0 \qquad \text{Solution checks. } \checkmark$$

In Example 1, be sure you see that solving an equation has two basic stages. The first stage is to *find* the solution (or solutions). The second stage is to *check* that each solution you find actually satisfies the original equation. You can improve your accuracy in algebra by developing the habit of checking each solution.

A common question in algebra is

"How do I know which step to do *first* to isolate x?"

The answer is that you need practice. By solving many linear equations, you will find that your skill will improve. The key thing to remember is that you can "get rid of" terms and factors by using *inverse* operations. Here are some guidelines and examples.

Guideline	*Equation*	*Inverse Operation*
1. Subtract to remove a sum.	$x + 3 = 0$	Subtract 3 from each side.
2. Add to remove a difference.	$x - 5 = 0$	Add 5 to each side.
3. Divide to remove a product.	$4x = 20$	Divide each side by 4.
4. Multiply to remove a quotient.	$\frac{x}{8} = 2$	Multiply each side by 8.

For additional examples, review Example 3 on page 108. In each case of that example, note how inverse operations are used to isolate the variable.

Example 2   Solving a Linear Equation in Standard Form

Solve $2x + 3 = 0$. Then check the solution.

Solution

$$2x + 3 = 0 \qquad \text{Write original equation.}$$

$$2x + 3 - 3 = 0 - 3 \qquad \text{Subtract 3 from each side.}$$

$$2x = -3 \qquad \text{Combine like terms.}$$

$$\frac{2x}{2} = -\frac{3}{2} \qquad \text{Divide each side by 2.}$$

$$x = -\frac{3}{2} \qquad \text{Simplify.}$$

Check

$$2x + 3 = 0 \qquad \text{Write original equation.}$$

$$2\left(-\frac{3}{2}\right) + 3 \stackrel{?}{=} 0 \qquad \text{Substitute } -\tfrac{3}{2} \text{ for } x.$$

$$-3 + 3 \stackrel{?}{=} 0 \qquad \text{Simplify.}$$

$$0 = 0 \qquad \text{Solution checks. } \checkmark$$

So, the solution is $x = -\frac{3}{2}$.

Example 3   Solving a Linear Equation in Standard Form

Solve $5x - 12 = 0$. Then check the solution.

Solution

$$5x - 12 = 0 \qquad \text{Write original equation.}$$

$$5x - 12 + 12 = 0 + 12 \qquad \text{Add 12 to each side.}$$

$$5x = 12 \qquad \text{Combine like terms.}$$

$$\frac{5x}{5} = \frac{12}{5} \qquad \text{Divide each side by 5.}$$

$$x = \frac{12}{5} \qquad \text{Simplify.}$$

Check

$$5x - 12 = 0 \qquad \text{Write original equation.}$$

$$5\left(\frac{12}{5}\right) - 12 \stackrel{?}{=} 0 \qquad \text{Substitute } \tfrac{12}{5} \text{ for } x.$$

$$12 - 12 \stackrel{?}{=} 0 \qquad \text{Simplify.}$$

$$0 = 0 \qquad \text{Solution checks. } \checkmark$$

So, the solution is $x = \frac{12}{5}$.

Study Tip

To eliminate a fractional coefficient, it may be easier to multiply each side by the *reciprocal* of the fraction than to divide by the fraction itself. Here is an example.

$$-\frac{2}{3}x = 4$$

$$\left(-\frac{3}{2}\right)\left(-\frac{2}{3}\right)x = \left(-\frac{3}{2}\right)4$$

$$x = -\frac{12}{2}$$

$$x = -6$$

Additional Example

Solve $\frac{2x}{16} + 4 = 0$.

Answer: $x = -32$

Technology: Tip

Remember to check your solution in the original equation. This can be done efficiently with a graphing calculator.

② Solve linear equations in nonstandard form.

Example 4   Solving a Linear Equation in Standard Form

Solve $\frac{x}{3} + 3 = 0$. Then check the solution.

Solution

$$\frac{x}{3} + 3 = 0 \qquad \text{Write original equation.}$$

$$\frac{x}{3} + 3 - 3 = 0 - 3 \qquad \text{Subtract 3 from each side.}$$

$$\frac{x}{3} = -3 \qquad \text{Combine like terms.}$$

$$3\left(\frac{x}{3}\right) = 3(-3) \qquad \text{Multiply each side by 3.}$$

$$x = -9 \qquad \text{Simplify.}$$

Check

$$\frac{x}{3} + 3 = 0 \qquad \text{Write original equation.}$$

$$\frac{-9}{3} + 3 \stackrel{?}{=} 0 \qquad \text{Substitute } -9 \text{ for } x.$$

$$-3 + 3 \stackrel{?}{=} 0 \qquad \text{Simplify.}$$

$$0 = 0 \qquad \text{Solution checks. } \checkmark$$

So, the solution is $x = -9$.

As you gain experience in solving linear equations, you will probably find that you can perform some of the solution steps in your head. For instance, you might solve the equation given in Example 4 by writing only the following steps.

$$\frac{x}{3} + 3 = 0 \qquad \text{Write original equation.}$$

$$\frac{x}{3} = -3 \qquad \text{Subtract 3 from each side.}$$

$$x = -9 \qquad \text{Multiply each side by 3.}$$

Solving a Linear Equation in Nonstandard Form

The definition of linear equation contains the phrase "that can be written in the standard form $ax + b = 0$." This suggests that some linear equations may come in nonstandard or disguised form.

A common form of linear equations is one in which the variable terms are not combined into one term. In such cases, you can begin the solution by *combining like terms*. Note how this is done in the next two examples.

Example 5 **Solving a Linear Equation in Nonstandard Form**

Solve $3y + 8 - 5y = 4$. Then check your solution.

Solution

$3y + 8 - 5y = 4$	Write original equation.
$3y - 5y + 8 = 4$	Group like terms.
$-2y + 8 = 4$	Combine like terms.
$-2y + 8 - 8 = 4 - 8$	Subtract 8 from each side.
$-2y = -4$	Combine like terms.
$\dfrac{-2y}{-2} = \dfrac{-4}{-2}$	Divide each side by -2.
$y = 2$	Simplify.

Check

$3y + 8 - 5y = 4$	Write original equation.
$3(2) + 8 - 5(2) \stackrel{?}{=} 4$	Substitute 2 for y.
$6 + 8 - 10 \stackrel{?}{=} 4$	Simplify.
$4 = 4$	Solution checks. ✓

So, the solution is $y = 2$.

Additional Examples
Solve each equation.

a. $5x - 7 = 3x + 2$

b. $7(x + 1) = 14x - 8$

Answers:

a. $x = \frac{9}{2}$

b. $x = \frac{15}{7}$

The solution for Example 5 began by collecting like terms. You can use any of the properties of algebra to attain your goal of "isolating the variable." The next example shows how to solve a linear equation using the Distributive Property.

Example 6 **Using the Distributive Property**

Solve $x + 6 = 2(x - 3)$.

Solution

$x + 6 = 2(x - 3)$	Write original equation.
$x + 6 = 2x - 6$	Distributive Property
$x - 2x + 6 = 2x - 2x - 6$	Subtract $2x$ from each side.
$-x + 6 = -6$	Combine like terms.
$-x + 6 - 6 = -6 - 6$	Subtract 6 from each side.
$-x = -12$	Combine like terms.
$(-1)(-x) = (-1)(-12)$	Multiply each side by -1.
$x = 12$	Simplify.

The solution is $x = 12$. Check this in the original equation.

Study Tip

You can isolate the variable term on either side of the equal sign. For instance, Example 6 could have been solved in the following way.

$$x + 6 = 2(x - 3)$$
$$x + 6 = 2x - 6$$
$$x - x + 6 = 2x - x - 6$$
$$6 = x - 6$$
$$6 + 6 = x - 6 + 6$$
$$12 = x$$

There are three different situations that can be encountered when solving linear equations in one variable. The first situation occurs when the linear equation has *exactly one* solution. You can show this with the steps below.

$$ax + b = 0 \qquad \text{Write original equation, with } a \neq 0.$$

$$ax = 0 - b \qquad \text{Subtract } b \text{ from each side.}$$

$$x = \frac{-b}{a} \qquad \text{Divide each side by } a.$$

So, the *linear* equation has exactly one solution: $x = -b/a$. The other two situations are the possibilities for the equation to have either *no solution* or *infinitely many solutions*. These two special cases are demonstrated below.

No Solution	Infinitely Many Solutions
$2x + 3 \overset{?}{=} 2(x + 4)$	$2(x + 3) = 2x + 6$
$2x + 3 \overset{?}{=} 2x + 8$	$2x + 6 = 2x + 6$ Identity equation
$2x - 2x + 3 \overset{?}{=} 2x - 2x + 8$	$2x - 2x + 6 - 6 = 2x - 2x + 6 - 6$
$3 \neq 8$	$0 = 0$

Watch out for these types of equations in the exercise set.

Applications

Example 7   Geometry: Dimensions of a Dog Pen

You have 96 feet of fencing to enclose a rectangular pen for your dog. To provide sufficient running space for the dog to exercise, the pen is to be three times as long as it is wide. Find the dimensions of the pen.

Solution

Begin by drawing and labeling a diagram, as shown in Figure 3.1. The perimeter of a rectangle is the sum of twice its length and twice its width.

Verbal Model: Perimeter $= 2 \cdot$ Length $+ 2 \cdot$ Width

Algebraic Model: $96 = 2(3x) + 2x$

You can solve this equation as follows.

$$96 = 6x + 2x \qquad \text{Multiply.}$$

$$96 = 8x \qquad \text{Combine like terms.}$$

$$\frac{96}{8} = \frac{8x}{8} \qquad \text{Divide each side by 8.}$$

$$12 = x \qquad \text{Simplify.}$$

So, the width of the pen is 12 feet and its length is 36 feet.

Study Tip

In the *No Solution* equation, the result is not true because $3 \neq 8$. This means that there is no value of x that will make the equation true.

In the *Infinitely Many Solutions* equation, the result is true. This means that *any* real number is a solution to the equation. This type of equation is called an **identity.**

---

3 Use linear equations to solve application problems.

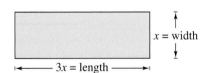

$x = $ width

$3x = $ length

Figure 3.1

Example 8   Ticket Sales

Tickets for a concert are $40 for each floor seat and $20 for each stadium seat. There are 800 seats on the main floor, and these are sold out. The total revenue from ticket sales is $92,000. How many stadium seats were sold?

Solution

*Verbal
Model:*

$$\boxed{\text{Total revenue}} = \boxed{\text{Revenue from floor seats}} + \boxed{\text{Revenue from stadium seats}}$$

Labels: Total revenue = 92,000   (dollars)
              Price per floor seat = 40   (dollars per seat)
              Number of floor seats = 800   (seats)
              Price per stadium seat = 20   (dollars per seat)
              Number of stadium seats = x (seats)

*Algebraic
Model:* $92,000 = 40(800) + 20x$

Now that you have written an algebraic equation to represent the problem, you can solve the equation as follows.

$92,000 = 40(800) + 20x$	Write original equation.
$92,000 = 32,000 + 20x$	Simplify.
$92,000 - 32,000 = 32,000 - 32,000 + 20x$	Subtract 32,000 from each side.
$60,000 = 20x$	Combine like terms.
$\dfrac{60,000}{20} = \dfrac{20x}{20}$	Divide each side by 20.
$3000 = x$	Simplify.

There were 3000 stadium seats sold. To check this solution, you should go back to the original statement of the problem and substitute 3000 stadium seats into the equation. You will find that the total revenue is $92,000.

Two integers are called **consecutive integers** if they differ by 1. So, for any integer n, its next two larger consecutive integers are $n + 1$ and $(n + 1) + 1$ or $n + 2$. You can denote three consecutive integers by $n, n + 1,$ and $n + 2$.

Expressions for Special Types of Integers

Let n be an integer. Then the following expressions can be used to denote even integers, odd integers, and consecutive integers, respectively.

1. $2n$ denotes an *even* integer.

2. $2n - 1$ and $2n + 1$ denote *odd* integers.

3. The set $\{n, n + 1, n + 2\}$ denotes three *consecutive* integers.

Example 9   Consecutive Integers

Find three consecutive integers whose sum is 48.

Solution

Verbal Model: First integer + Second integer + Third integer = 48

Labels: First integer $= n$
Second integer $= n + 1$
Third integer $= n + 2$

Equation:

$n + (n + 1) + (n + 2) = 48$	Original equation
$3n + 3 = 48$	Combine like terms.
$3n + 3 - 3 = 48 - 3$	Subtract 3 from each side.
$3n = 45$	Combine like terms.
$\dfrac{3n}{3} = \dfrac{45}{3}$	Divide each side by 3.
$n = 15$	Simplify.

So, the first integer is 15, the second integer is $15 + 1 = 16$, and the third integer is $15 + 2 = 17$. Check this in the original statement of the problem.

Example 10   Consecutive Even Integers

Find two consecutive even integers such that the sum of the first even integer and three times the second is 78.

Solution

Verbal Model: First even integer $+ 3 \cdot$ Second even integer = 78

Labels: First even integer $= 2n$
Second even integer $= 2n + 2$

Equation:

$2n + 3(2n + 2) = 78$	Original equation
$2n + 6n + 6 = 78$	Distributive Property
$8n + 6 = 78$	Combine like terms.
$8n + 6 - 6 = 78 - 6$	Subtract 6 from each side.
$8n = 72$	Combine like terms.
$\dfrac{8n}{8} = \dfrac{72}{8}$	Divide each side by 8.
$n = 9$	Simplify.

So, the first even integer is $2 \cdot 9 = 18$, and the second even integer is $2 \cdot 9 + 2 = 20$. Check this in the original statement of the problem.

Study Tip

When solving a word problem, be sure to ask yourself whether your solution makes sense. For example, a problem asks you to find the height of the ceiling of a room. The answer you obtain is 3 square meters. This answer does not make sense because height is measured in meters, not square meters.

3.1 Exercises

Review   Concepts, Skills, and Problem Solving

Keep mathematically in shape by doing these exercises *before* the problems of this section.

Properties and Definitions

1. Identify the property of real numbers illustrated by $(3y - 9) \cdot 1 = 3y - 9$.   Multiplicative Identity Property

2. Identify the property of real numbers illustrated by $(2x + 5) + 8 = 2x + (5 + 8)$.   Associative Property of Addition

Simplifying Expressions

In Exercises 3–10, simplify the expression.

3. $3 - 2x + 14 + 7x$ $5x + 17$

4. $4a + 2ab - b^2 + 5ab - b^2$ $-2b^2 + 7ab + 4a$

5. $-3(x - 5)^2(x - 5)^3$
   $-3(x - 5)^5$

6. $(4rs)(-5r^2)(2s^3)$
   $-40r^3s^4$

7. $\dfrac{2m^2}{3n} \cdot \dfrac{3m}{5n^3}$ $\dfrac{2m^3}{5n^4}$

8. $\dfrac{5(x + 3)^2}{10(x + 8)}$ $\dfrac{(x + 3)^2}{2(x + 8)}$

9. $-3(3x - 2y) + 5y$
   $-9x + 11y$

10. $3v - (4 - 5v)$
    $8v - 4$

Problem Solving

11. *Distance*   The length of a relay race is $\frac{3}{4}$ mile. The last change of runners occurs at the $\frac{2}{3}$ mile marker. How far does the last person run? $\frac{1}{12}$ mile

12. *Agriculture*   During the months of January, February, and March, a farmer bought $10\frac{1}{3}$ tons, $7\frac{3}{5}$ tons, and $12\frac{5}{6}$ tons of soybeans, respectively. Find the total amount of soybeans purchased during the first quarter of the year. $30\frac{23}{30}$ tons

Developing Skills

Mental Math   In Exercises 1–8, solve the equation mentally.

1. $x + 6 = 0$ -6

2. $a + 5 = 0$ -5

3. $x - 9 = 4$ 13

4. $u - 3 = 8$ 11

5. $7y = 28$ 4

6. $4s = 12$ 3

7. $4z = -36$ -9

8. $9z = -63$ -7

In Exercises 9–12, justify each step of the solution. See Examples 1–6.

9.

$5x + 15 = 0$	Original equation
$5x + 15 - 15 = 0 - 15$	Subtract 15 from each side.
$5x = -15$	Combine like terms.
$\dfrac{5x}{5} = \dfrac{-15}{5}$	Divide each side by 5.
$x = -3$	Simplify.

10.

$7x - 14 = 0$	Original equation
$7x - 14 + 14 = 0 + 14$	Add 14 to each side.
$7x = 14$	Combine like terms.
$\dfrac{7x}{7} = \dfrac{14}{7}$	Divide each side by 7.
$x = 2$	Simplify.

11.

$-2x + 5 = 13$	Original equation
$-2x + 5 - 5 = 13 - 5$	Subtract 5 from each side.
$-2x = 8$	Combine like terms.
$\dfrac{-2x}{-2} = \dfrac{8}{-2}$	Divide each side by -2.
$x = -4$	Simplify.

12.

$22 - 3x = 10$	Original equation
$22 - 3x + 3x = 10 + 3x$	Add $3x$ to each side.
$22 = 10 + 3x$	Combine like terms.
$22 - 10 = 10 + 3x - 10$	Subtract 10 from each side.
$12 = 3x$	Combine like terms.
$\dfrac{12}{3} = \dfrac{3x}{3}$	Divide each side by 3.
$4 = x$	Simplify.

In Exercises 13–60, solve the equation and check your solution. (Some equations have no solution.) See Examples 1–6.

13. $8x - 16 = 0$ 2

14. $4x - 24 = 0$ 6

15. $3x + 21 = 0$ -7

16. $2x + 52 = 0$ -26

17. $5x = 30$ 6

18. $12x = 18$ $\frac{3}{2}$

19. $9x = -21$ $-\frac{7}{3}$ **20.** $-14x = 42$ -3

21. $8x - 4 = 20$ 3 **22.** $-7x + 24 = 3$ 3

23. $25x - 4 = 46$ 2 **24.** $15x - 18 = 12$ 2

25. $10 - 4x = -6$ 4 **26.** $15 - 3x = -15$ 10

27. $6x - 4 = 0$ $\frac{2}{3}$ **28.** $8z - 2 = 0$ $\frac{1}{4}$

29. $3y - 2 = 2y$ 2 **30.** $2s - 13 = 28s$ $-\frac{1}{2}$

31. $4 - 7x = 5x$ $\frac{1}{3}$ **32.** $24 - 5x = x$ 4

33. $4 - 5t = 16 + t$ -2 **34.** $3x + 4 = x + 10$ 3

35. $-3t + 5 = -3t$ **36.** $4z + 2 = 4z$

No solution   No solution

37. $15x - 3 = 15 - 3x$ **38.** $2x - 5 = 7x + 10$

1 -3

39. $7a - 18 = 3a - 2$ 4 **40.** $4x - 2 = 3x + 1$ 3

41. $7x + 9 = 3x + 1$ **42.** $6t - 3 = 8t + 1$

-2 -2

43. $4x - 6 = 4x - 6$ **44.** $5 - 3x = 5 - 3x$

Identity   Identity

45. $2x + 4 = -3(x - 2)$ **46.** $4(y + 1) = -y + 5$

$\frac{2}{5}$ $\frac{1}{5}$

47. $2x = -3x$ 0 **48.** $6t = 9t$ 0

49. $2x - 5 + 10x = 3$ **50.** $-4x + 10 + 10x = 4$

$\frac{2}{3}$ -1

51. $\frac{x}{3} = 10$ 30 **52.** $-\frac{x}{2} = 3$ -6

53. $x - \frac{1}{3} = \frac{4}{3}$ $\frac{5}{3}$ **54.** $x + \frac{5}{2} = \frac{9}{2}$ 2

55. $t - \frac{1}{3} = \frac{1}{2}$ $\frac{5}{6}$ **56.** $z + \frac{2}{5} = -\frac{3}{10}$ $-\frac{7}{10}$

57. $5t - 4 + 3t = 4(2t - 1)$ Identity

58. $7z - 5z - 8 = 2(z - 4)$ Identity

59. $2(y - 9) = -5y - 4$ 2

60. $6 - 21x = 3(4 - 7x)$ No solution

Solving Problems

61. ▲ *Geometry*   The perimeter of a rectangle is 240 inches. The length is twice its width. Find the dimensions of the rectangle.   80 inches × 40 inches

62. ▲ *Geometry*   The length of a tennis court is 6 feet more than twice the width (see figure). Find the width of the court if the length is 78 feet.   36 feet

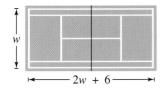

Figure for 62

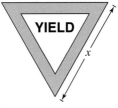

Figure for 63

63. ▲ *Geometry*   The sign in the figure has the shape of an equilateral triangle (sides have the same length). The perimeter of the sign is 225 centimeters. Find the length of its sides.   75 centimeters

64. ▲ *Geometry*   You are asked to cut a 12-foot board into three pieces. Two pieces are to have the same length and the third is to be twice as long as the others. How long are the pieces?   3 feet, 3 feet, 6 feet

65. *Car Repair*   The bill (including parts and labor) for the repair of your car is shown. Some of the bill is unreadable. From what is given, can you determine how many hours were spent on labor? Explain.

Yes. Subtract the cost of parts from the total to find the cost of labor. Then divide by 32 to find the number of hours spent on labor. $2\frac{1}{4}$ hours

| Parts . . . .$285.00 |
| Labor ($32 per hour) . . . . $ |
| **Total** . . . . **$357.00** |

Bill for 65

66. *Car Repair*   The bill for the repair of your car was $439. The cost for parts was $265. The cost for labor was $29 per hour. How many hours did the repair work take?   6 hours

67. *Ticket Sales*   Tickets for a community theater are $10 for main floor seats and $8 for balcony seats. There are 400 seats on the main floor, and these were sold out for the evening performance. The total revenue from ticket sales was $5200. How many balcony seats were sold?   150 seats

68. *Ticket Sales*   Tickets for a marching band competition are $5 for 50-yard-line seats and $3 for bleacher seats. Eight hundred 50-yard-line seats were sold. The total revenue from ticket sales was $5500. How many bleacher seats were sold?

500 seats

69. *Summer Jobs*   You have two summer jobs. In the first job, you work 40 hours a week and earn $9.25 an hour at a coffee shop. In the second job, you tutor for $7.50 an hour and can work as many hours as you want. You want to earn a combined total of $425 a week. How many hours must you tutor?

7 hours 20 minutes

70. *Summer Jobs*   You have two summer jobs. In the first job, you work 30 hours a week and earn $8.75 an hour at a gas station. In the second job, you work as a landscaper for $11.00 an hour and can work as many hours as you want. You want to earn a combined total of $400 a week. How many hours must you work at the second job?   12 hours 30 minutes

71. *Number Problem*   Five times the sum of a number and 16 is 100. Find the number.   4

72. *Number Problem*   Find a number such that the sum of twice that number and 31 is 69.   19

73. *Number Problem*   The sum of two consecutive odd integers is 72. Find the two integers.   35, 37

74. *Number Problem*   The sum of two consecutive even integers is 154. Find the two integers.   76, 78

75. *Number Problem*   The sum of three consecutive odd integers is 159. Find the three integers.   51, 53, 55

76. *Number Problem*   The sum of three consecutive even integers is 192. Find the three integers.   62, 64, 66

Explaining Concepts

77. The scale below is balanced. Each blue box weighs 1 ounce. How much does the red box weigh? If you removed three blue boxes from each side, would the scale still balance? What property of equality does this illustrate?   The red box weighs 6 ounces. If you removed three blue boxes from each side, the scale would still balance. The Addition (or Subtraction) Property of Equality

78. *Writing*   In your own words, describe the steps that can be used to transform an equation into an equivalent equation.

79. *Writing*   Explain how to solve the equation $x + 5 = 32$. What property of equality are you using?   Subtract 5 from each side of the equation. Addition Property of Equality

80. *Writing*   Explain how to solve the equation $3x = 5$. What property of equality are you using?   Divide each side of the equation by 3. Multiplication Property of Equality

81. *True or False?*   Subtracting 0 from each side of an equation yields an equivalent equation. Justify your answer.   True. Subtracting 0 from each side does not change any values. The equation remains the same.

82. *True or False?*   Multiplying each side of an equation by 0 yields an equivalent equation. Justify your answer.   False. Multiplying each side by 0 yields $0 = 0$.

83. *Finding a Pattern*   The length of a rectangle is t times its width (see figure). The rectangle has a perimeter of 1200 meters, which implies that $2w + 2(tw) = 1200$, where w is the width of the rectangle.

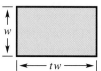

(a) Complete the table.

t	1	1.5	2
Width	300	240	200
Length	300	360	400
Area	90,000	86,400	80,000

t	3	4	5
Width	150	120	100
Length	450	480	500
Area	67,500	57,600	50,000

(b) Use the completed table to draw a conclusion concerning the area of a rectangle of given perimeter as the length increases relative to its width.   The area decreases.

78.
(a) Simplify each side by removing symbols of grouping, combining like terms, and reducing fractions on one or both sides.
(b) Add (or subtract) the same quantity to (from) each side of the equation.
(c) Multiply (or divide) each side of the equation by the same nonzero real number.
(d) Interchange the two sides of the equation.

3.2 Equations That Reduce to Linear Form

What You Should Learn

① Solve linear equations containing symbols of grouping.

② Solve linear equations involving fractions.

③ Solve linear equations involving decimals.

Why You Should Learn It

Many real-life applications can be modeled with linear equations involving decimals. For instance, Exercise 81 on page 144 shows how a linear equation can model the projected number of persons 65 years and older in the United States.

① Solve linear equations containing symbols of grouping.

Equations Containing Symbols of Grouping

In this section you will continue your study of linear equations by looking at more complicated types of linear equations. To solve a linear equation that contains symbols of grouping, use the following guidelines.

1. Remove symbols of grouping from each side by using the Distributive Property.
2. Combine like terms.
3. Isolate the variable in the usual way using properties of equality.
4. Check your solution in the original equation.

Example 1   Solving a Linear Equation Involving Parentheses

Solve $4(x - 3) = 8$. Then check your solution.

Solution

$4(x - 3) = 8$	Write original equation.
$4 \cdot x - 4 \cdot 3 = 8$	Distributive Property
$4x - 12 = 8$	Simplify.
$4x - 12 + 12 = 8 + 12$	Add 12 to each side.
$4x = 20$	Combine like terms.
$\dfrac{4x}{4} = \dfrac{20}{4}$	Divide each side by 4.
$x = 5$	Simplify.

Check

$4(5 - 3) \overset{?}{=} 8$	Substitute 5 for x in original equation.
$4(2) \overset{?}{=} 8$	Simplify.
$8 = 8$	Solution checks. ✓

The solution is $x = 5$.

Study Tip

Notice in the check of Example 1 that you do not need to use the Distributive Property to remove the parentheses. Simply evaluate the expression within the parentheses and then multiply.

Example 2   Solving a Linear Equation Involving Parentheses

Solve $3(2x - 1) + x = 11$. Then check your solution.

Solution

$3(2x - 1) + x = 11$	Write original equation.
$3 \cdot 2x - 3 \cdot 1 + x = 11$	Distributive Property
$6x - 3 + x = 11$	Simplify.
$6x + x - 3 = 11$	Group like terms.
$7x - 3 = 11$	Combine like terms.
$7x - 3 + 3 = 11 + 3$	Add 3 to each side.
$7x = 14$	Combine like terms.
$\dfrac{7x}{7} = \dfrac{14}{7}$	Divide each side by 7.
$x = 2$	Simplify.

Additional Example

Solve $6(y - 1) + 4y = 3(7y + 1)$.

Answer: $y = -\dfrac{9}{11}$

Check

$3(2x - 1) + x = 11$	Write original equation.
$3[2(2) - 1] + 2 \stackrel{?}{=} 11$	Substitute 2 for x.
$3(4 - 1) + 2 \stackrel{?}{=} 11$	Simplify.
$3(3) + 2 \stackrel{?}{=} 11$	Simplify.
$9 + 2 \stackrel{?}{=} 11$	Simplify.
$11 = 11$	Solution checks. ✓

The solution is $x = 2$.

Example 3   Solving a Linear Equation Involving Parentheses

Solve $5(x + 2) = 2(x - 1)$.

Solution

$5(x + 2) = 2(x - 1)$	Write original equation.
$5x + 10 = 2x - 2$	Distributive Property
$5x - 2x + 10 = 2x - 2x - 2$	Subtract $2x$ from each side.
$3x + 10 = -2$	Combine like terms.
$3x + 10 - 10 = -2 - 10$	Subtract 10 from each side.
$3x = -12$	Combine like terms.
$x = -4$	Divide each side by 3.

The solution is $x = -4$. Check this in the original equation.

Example 4   Solving a Linear Equation Involving Parentheses

Solve $2(x - 7) - 3(x + 4) = 4 - (5x - 2)$.

Solution

$$2(x - 7) - 3(x + 4) = 4 - (5x - 2) \qquad \text{Write original equation.}$$
$$2x - 14 - 3x - 12 = 4 - 5x + 2 \qquad \text{Distributive Property}$$
$$-x - 26 = -5x + 6 \qquad \text{Combine like terms.}$$
$$-x + 5x - 26 = -5x + 5x + 6 \qquad \text{Add } 5x \text{ to each side.}$$
$$4x - 26 = 6 \qquad \text{Combine like terms.}$$
$$4x - 26 + 26 = 6 + 26 \qquad \text{Add 26 to each side.}$$
$$4x = 32 \qquad \text{Combine like terms.}$$
$$x = 8 \qquad \text{Divide each side by 4.}$$

The solution is $x = 8$. Check this in the original equation.

The linear equation in the next example involves both brackets and parentheses. Watch out for nested symbols of grouping such as these. The *innermost symbols of grouping* should be removed first.

Example 5   An Equation Involving Nested Symbols of Grouping

Solve $5x - 2[4x + 3(x - 1)] = 8 - 3x$.

Solution

$$5x - 2[4x + 3(x - 1)] = 8 - 3x \qquad \text{Write original equation.}$$
$$5x - 2[4x + 3x - 3] = 8 - 3x \qquad \text{Distributive Property}$$
$$5x - 2[7x - 3] = 8 - 3x \qquad \text{Combine like terms inside brackets.}$$
$$5x - 14x + 6 = 8 - 3x \qquad \text{Distributive Property}$$
$$-9x + 6 = 8 - 3x \qquad \text{Combine like terms.}$$
$$-9x + 3x + 6 = 8 - 3x + 3x \qquad \text{Add } 3x \text{ to each side.}$$
$$-6x + 6 = 8 \qquad \text{Combine like terms.}$$
$$-6x + 6 - 6 = 8 - 6 \qquad \text{Subtract 6 from each side.}$$
$$-6x = 2 \qquad \text{Combine like terms.}$$
$$\frac{-6x}{-6} = \frac{2}{-6} \qquad \text{Divide each side by } -6.$$
$$x = -\frac{1}{3} \qquad \text{Simplify.}$$

The solution is $x = -\frac{1}{3}$. Check this in the original equation.

Technology: Tip

Try using your graphing calculator to check the solution found in Example 5. You will need to nest some parentheses inside other parentheses. This will give you practice working with nested parentheses on a graphing calculator.

Left side of equation

$$5\left(-\frac{1}{3}\right) - 2\left(4\left(-\frac{1}{3}\right)\right. \\ \left. + 3\left(\left(-\frac{1}{3}\right) - 1\right)\right)$$

Right side of equation

$$8 - 3\left(-\frac{1}{3}\right)$$

② Solve linear equations involving fractions

Equations Involving Fractions or Decimals

To solve a linear equation that contains one or more fractions, it is usually best to first *clear the equation of fractions.*

Clearing an Equation of Fractions

An equation such as

$$\frac{x}{a} + \frac{b}{c} = d$$

that contains one or more fractions can be cleared of fractions by multiplying each side by the least common multiple (LCM) of a and c.

For example, the equation

$$\frac{3x}{2} - \frac{1}{3} = 2$$

can be cleared of fractions by multiplying each side by 6, the LCM of 2 and 3. Notice how this is done in the next example.

Example 6   Solving a Linear Equation Involving Fractions

Solve $\dfrac{3x}{2} - \dfrac{1}{3} = 2$.

Solution

$$6\left(\frac{3x}{2} - \frac{1}{3}\right) = 6 \cdot 2 \qquad \text{Multiply each side by LCM 6.}$$

$$6 \cdot \frac{3x}{2} - 6 \cdot \frac{1}{3} = 12 \qquad \text{Distributive Property}$$

$$9x - 2 = 12 \qquad \text{Clear fractions.}$$

$$9x = 14 \qquad \text{Add 2 to each side.}$$

$$x = \frac{14}{9} \qquad \text{Divide each side by 9.}$$

The solution is $x = \frac{14}{9}$. Check this in the original equation.

To check a fractional solution such as $\frac{14}{9}$ in Example 6, it is helpful to rewrite the variable term as a product.

$$\frac{3}{2} \cdot x - \frac{1}{3} = 2 \qquad \text{Write fraction as a product.}$$

In this form the substitution of $\frac{14}{9}$ for x is easier to calculate.

Example 7   Solving a Linear Equation Involving Fractions

Solve $\dfrac{x}{5} + \dfrac{3x}{4} = 19$. Then check your solution.

Solution

$$\dfrac{x}{5} + \dfrac{3x}{4} = 19 \qquad \text{Write original equation.}$$

$$20\left(\dfrac{x}{5}\right) + 20\left(\dfrac{3x}{4}\right) = 20(19) \qquad \text{Multiply each side by LCM 20.}$$

$$4x + 15x = 380 \qquad \text{Simplify.}$$

$$19x = 380 \qquad \text{Combine like terms.}$$

$$x = 20 \qquad \text{Divide each side by 19.}$$

Check

$$\dfrac{20}{5} + \dfrac{3(20)}{4} \stackrel{?}{=} 19 \qquad \text{Substitute 20 for } x \text{ in original equation.}$$

$$4 + 15 \stackrel{?}{=} 19 \qquad \text{Simplify.}$$

$$19 = 19 \qquad \text{Solution checks.} \checkmark$$

The solution is $x = 20$.

Study Tip

Notice in Example 8 that to clear all fractions in the equation, you multiply by 12, which is the LCM of 3, 4, and 2.

Example 8   Solving a Linear Equation Involving Fractions

Solve $\dfrac{2}{3}\left(x + \dfrac{1}{4}\right) = \dfrac{1}{2}$.

Solution

$$\dfrac{2}{3}\left(x + \dfrac{1}{4}\right) = \dfrac{1}{2} \qquad \text{Write original equation.}$$

$$\dfrac{2}{3}x + \dfrac{2}{12} = \dfrac{1}{2} \qquad \text{Distributive Property}$$

$$12 \cdot \dfrac{2}{3}x + 12 \cdot \dfrac{2}{12} = 12 \cdot \dfrac{1}{2} \qquad \text{Multiply each side by LCM 12.}$$

$$8x + 2 = 6 \qquad \text{Simplify.}$$

$$8x = 4 \qquad \text{Subtract 2 from each side.}$$

$$x = \dfrac{4}{8} \qquad \text{Divide each side by 8.}$$

$$x = \dfrac{1}{2} \qquad \text{Simplify.}$$

The solution is $x = \frac{1}{2}$. Check this in the original equation.

A common type of linear equation is one that equates two fractions. To solve such an equation, consider the fractions to be **equivalent** and use **cross-multiplication.** That is, if

$$\frac{a}{b} = \frac{c}{d}, \quad \text{then} \quad a \cdot d = b \cdot c.$$

Note how cross-multiplication is used in the next example.

You might point out that cross-multiplication would *not* be an appropriate first step in equations such as $\frac{x+2}{3} + 4 = \frac{8}{5}$ and $\frac{x+2}{3} = \frac{8}{5} - 2$.

Example 9   Using Cross-Multiplication

Use cross-multiplication to solve $\dfrac{x+2}{3} = \dfrac{8}{5}$. Then check your solution.

Solution

$$\frac{x+2}{3} = \frac{8}{5} \qquad \text{Write original equation.}$$

$$5(x+2) = 3(8) \qquad \text{Cross multiply.}$$

$$5x + 10 = 24 \qquad \text{Distributive Property}$$

$$5x = 14 \qquad \text{Subtract 10 from each side.}$$

$$x = \frac{14}{5} \qquad \text{Divide each side by 5.}$$

Checking solutions may sometimes be challenging for students, but the checking can improve students' accuracy and reinforce their computational skills.

Check

$$\frac{x+2}{3} = \frac{8}{5} \qquad \text{Write original equation.}$$

$$\frac{\left(\frac{14}{5} + 2\right)}{3} \stackrel{?}{=} \frac{8}{5} \qquad \text{Substitute } \tfrac{14}{5} \text{ for } x.$$

$$\frac{\left(\frac{14}{5} + \frac{10}{5}\right)}{3} \stackrel{?}{=} \frac{8}{5} \qquad \text{Write 2 as } \tfrac{10}{5}.$$

$$\frac{\frac{24}{5}}{3} \stackrel{?}{=} \frac{8}{5} \qquad \text{Simplify.}$$

$$\frac{24}{5}\left(\frac{1}{3}\right) \stackrel{?}{=} \frac{8}{5} \qquad \text{Invert and multiply.}$$

$$\frac{8}{5} = \frac{8}{5} \qquad \text{Solution checks. } \checkmark$$

The solution is $x = \frac{14}{5}$.

Additional Example
Use cross-multiplication to solve $\dfrac{x+5}{2} = \dfrac{x-6}{3}$.

Answer: $x = -27$

Bear in mind that cross-multiplication can be used only with equations written in a form that equates two fractions. Try rewriting the equation in Example 6 in this form and then use cross-multiplication to solve for x.

More extensive applications of cross-multiplication will be discussed when you study ratios and proportions later in this chapter.

③ Solve linear equations involving decimals.

Many real-life applications of linear equations involve decimal coefficients. To solve such an equation, you can clear it of decimals in much the same way you clear an equation of fractions. Multiply each side by a power of 10 that converts all decimal coefficients to integers, as shown in the next example.

Example 10   Solving a Linear Equation Involving Decimals

Solve $0.3x + 0.2(10 - x) = 0.15(30)$. Then check your solution.

Solution

$0.3x + 0.2(10 - x) = 0.15(30)$	Write original equation.
$0.3x + 2 - 0.2x = 4.5$	Distributive Property
$0.1x + 2 = 4.5$	Combine like terms.
$10(0.1x + 2) = 10(4.5)$	Multiply each side by 10.
$x + 20 = 45$	Clear decimals.
$x = 25$	Subtract 20 from each side.

Check

$0.3(25) + 0.2(10 - 25) \overset{?}{=} 0.15(30)$	Substitute 25 for x in original equation.
$0.3(25) + 0.2(-15) \overset{?}{=} 0.15(30)$	Perform subtraction within parentheses.
$7.5 - 3.0 \overset{?}{=} 4.5$	Multiply.
$4.5 = 4.5$	Solution checks. ✓

The solution is $x = 25$.

Example 11   ACT Participants

The number y (in thousands) of students who took the ACT from 1996 to 2002 can be approximated by the linear model $y = 30.5t + 746$, where t represents the year, with $t = 6$ corresponding to 1996. Assuming that this linear pattern continues, find the year in which there will be 1234 thousand students taking the ACT. (Source: The ACT, Inc.)

Solution

To find the year in which there will be 1234 thousand students taking the ACT, substitute 1234 for y in the original equation and solve the equation for t.

$1234 = 30.5t + 746$	Substitute 1234 for y in original equation.
$488 = 30.5t$	Subtract 746 from each side.
$16 = t$	Divide each side by 30.5.

Because $t = 6$ corresponds to 1996, $t = 16$ must represent 2006. So, from this model, there will be 1234 thousand students taking the ACT in 2006. Check this in the original statement of the problem.

3.2 Exercises

Review   Concepts, Skills, and Problem Solving

Keep mathematically in shape by doing these exercises *before* the problems of this section.

Properties and Definitions

1. *Writing✐*   In your own words, describe how you add the following fractions.

(a) $\frac{1}{5} + \frac{7}{5}$ Add the numerators and write the sum over the like denominator. The result is $\frac{8}{5}$.

(b) $\frac{1}{5} + \frac{7}{3}$ Find equivalent fractions with a common denominator. Add the numerators and write the sum over the like denominator. The result is $\frac{38}{15}$.

2. Create two examples of algebraic expressions.
Answers will vary. Examples are given.

$$3x^2 + 2\sqrt{x};\ \frac{4}{x^2 + 1}$$

Simplifying Expressions

In Exercises 3–10, simplify the expression.

3. $(-2x)^2 x^4$ $4x^6$

4. $-y^2(-2y)^3$ $8y^5$

5. $5z^3(z^2)$
$5z^5$

6. $a^2 + 3a + 4 - 2a - 6$
$a^2 + a - 2$

7. $\dfrac{5x}{3} - \dfrac{2x}{3} - 4$
$x - 4$

8. $2x^2 - 4 + 5 - 3x^2$
$-x^2 + 1$

9. $-y^2(y^2 + 4) + 6y^2$
$-y^4 + 2y^2$

10. $5t(2 - t) + t^2$
$10t - 4t^2$

Problem Solving

11. *Fuel Usage*   At the beginning of the day, a gasoline tank was full. The tank holds 20 gallons. At the end of the day, the fuel gauge indicates that the tank is $\frac{5}{8}$ full. How many gallons of gasoline were used?   7.5 gallons

12. *Consumerism*   You buy a pickup truck for $1800 down and 36 monthly payments of $625 each.

(a) What is the total amount you will pay? $24,300

(b) The final cost of the pickup is $19,999. How much extra did you pay in finance charges and other fees? $4301

Developing Skills

In Exercises 1–52, solve the equation and check your solution. (Some of the equations have no solution.) See Examples 1–8.

1. $2(y - 4) = 0$ 4

2. $9(y - 7) = 0$ 7

3. $-5(t + 3) = 10$ -5

4. $-3(x + 1) = 18$ -7

5. $25(z - 2) = 60$ $\frac{22}{5}$

6. $2(x - 3) = 4$ 5

7. $7(x + 5) = 49$ 2

8. $4(x + 1) = 24$ 5

9. $4 - (z + 6) = 8$
-10

10. $25 - (y + 3) = 15$
7

11. $3 - (2x - 4) = 3$
2

12. $16 - (3x - 10) = 5$
7

13. $12(x - 3) = 0$ 3

14. $4(z - 2) = 0$ 2

15. $3(2x - 1) = 3(2x + 5)$ No solution

16. $4(z - 2) = 2(2z - 4)$ Identity

17. $-3(x + 4) = 4(x + 4)$ -4

18. $-8(x - 6) = 3(x - 6)$ 6

19. $7 = 3(x + 2) - 3(x - 5)$ No solution

20. $24 = 12(z + 1) - 3(4z - 2)$ No solution

21. $7x - 2(x - 2) = 12$ $\frac{8}{5}$

22. $15(x + 1) - 8x = 29$ 2

23. $6 = 3(y + 1) - 4(1 - y)$ 1

24. $100 = 4(y - 6) - (y - 1)$ 41

25. $-6(3 + x) + 2(3x + 5) = 0$ No solution

26. $-3(5x + 2) + 5(1 + 3x) = 0$ No solution

27. $2[(3x + 5) - 7] = 3(5x - 2)$ $\frac{2}{9}$

28. $3[(5x + 1) - 4] = 4(2x - 3)$ $-\frac{3}{7}$

29. $4x + 3[x - 2(2x - 1)] = 4 - 3x$ 1

30. $16 + 4[5x - 4(x + 2)] = 7 - 2x$ $\frac{23}{6}$

31. $\dfrac{y}{5} = \dfrac{3}{5}$ 3

32. $\dfrac{z}{3} = \dfrac{10}{3}$ 10

33. $\dfrac{y}{5} = -\dfrac{3}{10}$ $-\frac{3}{2}$

34. $\dfrac{v}{4} = -\dfrac{7}{8}$ $-\frac{7}{2}$

35. $\dfrac{6x}{25} = \dfrac{3}{5}$ $\frac{5}{2}$

36. $\dfrac{8x}{9} = \dfrac{2}{3}$ $\frac{3}{4}$

37. $\dfrac{5x}{4} + \dfrac{1}{2} = 0$ $-\frac{2}{5}$

38. $\dfrac{3z}{7} + \dfrac{6}{11} = 0$ $-\frac{14}{11}$

39. $\frac{x}{5} - \frac{1}{2} = 3$ $\frac{35}{2}$

40. $\frac{y}{4} - \frac{5}{8} = 2$ $\frac{21}{2}$

41. $\frac{x}{5} - \frac{x}{2} = 1$ $-\frac{10}{3}$

42. $\frac{x}{3} + \frac{x}{4} = 1$ $\frac{12}{7}$

43. $2s + \frac{3}{2} = 2s + 2$ No solution

44. $\frac{3}{4} + 5s = -2 + 5s$ No solution

45. $3x + \frac{1}{4} = \frac{3}{4}$ $\frac{1}{6}$

46. $2x - \frac{3}{8} = \frac{5}{8}$ $\frac{1}{2}$

47. $\frac{1}{5}x + 1 = \frac{3}{10}x - 4$ 50

48. $\frac{1}{8}x + 3 = \frac{1}{4}x + 5$ -16

49. $\frac{2}{3}(z + 5) - \frac{1}{4}(z + 24) = 0$ $\frac{32}{5}$

50. $\frac{3x}{2} + \frac{1}{4}(x - 2) = 10$ 6

51. $\frac{100 - 4u}{3} = \frac{5u + 6}{4} + 6$ 10

52. $\frac{8 - 3x}{2} - 4 = \frac{x}{6}$ 0

In Exercises 53–62, use cross-multiplication to solve the equation. See Example 9.

53. $\frac{t + 4}{6} = \frac{2}{3}$ 0

54. $\frac{x - 6}{10} = \frac{3}{5}$ 12

55. $\frac{x - 2}{5} = \frac{2}{3}$ $\frac{16}{3}$

56. $\frac{2x + 1}{3} = \frac{5}{2}$ $\frac{13}{4}$

57. $\frac{5x - 4}{4} = \frac{2}{3}$ $\frac{4}{3}$

58. $\frac{10x + 3}{6} = \frac{1}{2}$ 0

59. $\frac{x}{4} = \frac{1 - 2x}{3}$ $\frac{4}{11}$

60. $\frac{x + 1}{6} = \frac{3x}{10}$ $\frac{5}{4}$

61. $\frac{10 - x}{2} = \frac{x + 4}{5}$ 6

62. $\frac{2x + 3}{5} = \frac{3 - 4x}{8}$ $-\frac{1}{4}$

In Exercises 63–72, solve the equation. Round your answer to two decimal places. See Example 10.

63. $0.2x + 5 = 6$
5.00

64. $4 - 0.3x = 1$
10.00

65. $0.234x + 1 = 2.805$
7.71

66. $275x - 3130 = 512$
13.24

67. $0.02x - 0.96 = 1.50$
123.00

68. $1.35x + 14.50 = 6.34$
-6.04

69. $\frac{x}{3.25} + 1 = 2.08$
3.51

70. $\frac{x}{4.08} + 7.2 = 5.14$
-8.40

71. $\frac{x}{3.155} = 2.850$ 8.99

72. $\frac{3x}{4.5} = \frac{1}{8}$ 0.19

Solving Problems

73. *Time to Complete a Task* Two people can complete 80% of a task in t hours, where t must satisfy the equation $t/10 + t/15 = 0.8$. How long will it take for the two people to complete 80% of the task?
4.8 hours

74. *Time to Complete a Task* Two machines can complete a task in t hours, where t must satisfy the equation $t/10 + t/15 = 1$. How long will it take for the two machines to complete the task?   6 hours

75. *Course Grade* To get an A in a course, you must have an average of at least 90 points for four tests of 100 points each. For the first three tests, your scores are 87, 92, and 84. What must you score on the fourth exam to earn a 90% average for the course?
97

76. *Course Grade* Repeat Exercise 75 if the fourth test is weighted so that it counts for twice as much as each of the first three tests.   93.5

In Exercises 77–80, use the equation and solve for x.

$$p_1 x + p_2(a - x) = p_3 a$$

77. *Mixture Problem* Determine the number of quarts of a 10% solution that must be mixed with a 30% solution to obtain 100 quarts of a 25% solution. ($p_1 = 0.1$, $p_2 = 0.3$, $p_3 = 0.25$, and $a = 100$.)
25 quarts

78. *Mixture Problem* Determine the number of gallons of a 25% solution that must be mixed with a 50% solution to obtain 5 gallons of a 30% solution. ($p_1 = 0.25$, $p_2 = 0.5$, $p_3 = 0.3$, and $a = 5$.)
4 gallons

79. *Mixture Problem* An eight-quart automobile cooling system is filled with coolant that is 40% antifreeze. Determine the amount that must be withdrawn and replaced with pure antifreeze so that the 8 quarts of coolant will be 50% antifreeze. ($p_1 = 1$, $p_2 = 0.4$, $p_3 = 0.5$, and $a = 8$.) $1\frac{1}{3}$ quarts

80. *Mixture Problem*   A grocer mixes two kinds of nuts costing $2.49 per pound and $3.89 per pound to make 100 pounds of a mixture costing $3.19 per pound. How many pounds of the nuts costing $2.49 per pound must be put into the mixture? ($p_1 = 2.49$, $p_2 = 3.89$, $p_3 = 3.19$, and $a = 100$.)   50 pounds

81. *Data Analysis*   The table shows the projected numbers N (in millions) of persons 65 years of age or older in the United States.   (Source: U.S. Census Bureau)

Year	2005	2015	2025	2035
N	36.4	46.0	62.6	74.8

A model for the data is

$N = 1.32t + 28.6$

where t represents time in years, with $t = 5$ corresponding to the year 2005. According to the model, in what year will the population of those 65 or older exceed 80 million?   2038

82. *Fireplace Construction*   A fireplace is 93 inches wide. Each brick in the fireplace has a length of 8 inches and there is $\frac{1}{2}$ inch of mortar between adjoining bricks (see figure). Let n be the number of bricks per row.

(a) Explain why the number of bricks per row is the solution of the equation $8n + \frac{1}{2}(n - 1) = 93$.

Each of the n bricks is 8 inches long. Each of the $(n - 1)$ mortar joints is $\frac{1}{2}$ inch wide. The total length is 93 inches.

(b) Find the number of bricks per row in the fireplace.   11

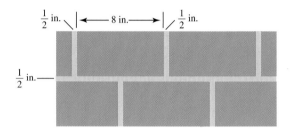

Explaining Concepts

83. *Writing*   In your own words, describe the procedure for removing symbols of grouping. Give some examples.

84. You could solve $3(x - 7) = 15$ by applying the Distributive Property as the first step. However, there is another way to begin. What is it?   Divide each side by 3.

85. *Error Analysis*   Describe the error.

$$-2(x - 5) = 8$$
$$-2x - 5 = 8 \qquad -2(x - 5) = -2x + 10$$

86. *Writing*   Explain what happens when you divide each side of an equation by a variable factor.

Dividing by a variable assumes that it does not equal zero, which may yield a false solution.

87. *Writing*   What is meant by the least common multiple of the denominators of two or more fractions? Discuss the method for finding the least common multiple of the denominators of fractions.

88. *Writing*   When solving an equation that contains fractions, explain what is accomplished by multiplying each side of the equation by the least common multiple of the denominators of the fractions.

It clears the equation of fractions.

89. *Writing*   When simplifying an algebraic *expression* involving fractions, why can't you simplify the expression by multiplying by the least common multiple of the denominators?   Because the expression is not an equation, there are not two sides to multiply by the least common multiple of the denominators.

83. Use the Distributive Property to remove symbols of grouping. Remove the innermost symbols first and combine like terms. Symbols of grouping preceded by a *minus* sign can be removed by changing the sign of each term within the symbols.

$$2x - [3 + (x - 1)] = 2x - [3 + x - 1]$$
$$= 2x - [2 + x]$$
$$= 2x - 2 - x = x - 2$$

87. The least common multiple of the denominators is the simplest expression that is a multiple of all the denominators. The least common multiple of the denominators contains each prime factor of the denominators repeated the maximum number of times it occurs in any one of the factorizations of the denominators.

3.3 Problem Solving with Percents

Bob Mahoney/The Image Works

What You Should Learn

① Convert percents to decimals and fractions and convert decimals and fractions to percents.

② Solve linear equations involving percents.

③ Solve application problems involving markups and discounts.

① Convert percents to decimals and fractions and convert decimals and fractions to percents.

Percents

In applications involving percents, you usually must convert the percents to decimal (or fractional) form before performing any arithmetic operations. Consequently, you need to be able to convert from percents to decimals (or fractions), and vice versa. The following verbal model can be used to perform the conversions.

$$\boxed{\text{Decimal or fraction}} \cdot \boxed{100\%} = \boxed{\text{Percent}}$$

For example, the decimal 0.38 corresponds to 38 percent. That is,

$$0.38(100\%) = 38\%.$$

Example 1   Converting Decimals and Fractions to Percents

Convert each number to a percent.

a. $\dfrac{3}{5}$ **b.** 1.20

Solution

a. *Verbal Model:* $\boxed{\text{Fraction}} \cdot \boxed{100\%} = \boxed{\text{Percent}}$

Equation: $\dfrac{3}{5}(100\%) = \dfrac{300}{5}\%$

$$= 60\%$$

So, the fraction $\frac{3}{5}$ corresponds to 60%.

b. *Verbal Model:* $\boxed{\text{Decimal}} \cdot \boxed{100\%} = \boxed{\text{Percent}}$

Equation: $(1.20)(100\%) = 120\%$

So, the decimal 1.20 corresponds to 120%.

Study Tip

Note in Example 1(b) that it is possible to have percents that are larger than 100%. It is also possible to have percents that are less than 1%, such as $\frac{1}{2}\%$ or 0.78%.

Study Tip

In Examples 1 and 2, there is a quick way to convert between percent form and decimal form.

• To convert from percent form to decimal form, move the decimal point two places to the *left*. For instance,

$$3.5\% = 0.035.$$

• To convert from decimal form to percent form, move the decimal point two places to the *right*. For instance,

$$1.20 = 120\%.$$

• Decimal-to-fraction or fraction-to-decimal conversions can be done on a calculator. Consult your user's guide.

Example 2 Converting Percents to Decimals and Fractions

a. Convert 3.5% to a decimal.

b. Convert 55% to a fraction.

Solution

a. *Verbal Model:*   Decimal · 100% = Percent

Label: x = decimal

Equation: $x(100\%) = 3.5\%$ Original equation

$$x = \frac{3.5\%}{100\%}$$ Divide each side by 100%.

$$x = 0.035$$ Simplify.

So, 3.5% corresponds to the decimal 0.035.

b. *Verbal Model:*   Fraction · 100% = Percent

Label: x = fraction

Equation: $x(100\%) = 55\%$ Original equation

$$x = \frac{55\%}{100\%}$$ Divide each side by 100%.

$$x = \frac{11}{20}$$ Simplify.

So, 55% corresponds to the fraction $\frac{11}{20}$.

Some percents occur so commonly that it is helpful to memorize their conversions. For instance, 100% corresponds to 1 and 200% corresponds to 2. The table below shows the decimal and fraction conversions for several percents.

Percent	10%	12½%	20%	25%	33⅓%	50%	66⅔%	75%
Decimal	0.1	0.125	0.2	0.25	$0.\overline{3}$	0.5	$0.\overline{6}$	0.75
Fraction	$\frac{1}{10}$	$\frac{1}{8}$	$\frac{1}{5}$	$\frac{1}{4}$	$\frac{1}{3}$	$\frac{1}{2}$	$\frac{2}{3}$	$\frac{3}{4}$

Percent means *per hundred* or *parts of 100*. (The Latin word for 100 is *centum*.) For example, 20% means 20 parts of 100, which is equivalent to the fraction 20/100 or $\frac{1}{5}$. In applications involving percent, many people like to state percent in terms of a portion. For instance, the statement "20% of the population lives in apartments" is often stated as "1 out of every 5 people lives in an apartment."

② Solve linear equations involving percents.

The Percent Equation

The primary use of percents is to compare two numbers. For example, 2 is 50% of 4, and 5 is 25% of 20. The following model is helpful.

Verbal Model: $a = p$ percent of b

Labels: $b =$ base number
$p =$ percent (in decimal form)
$a =$ number being compared to b

Equation: $a = p \cdot b$

Example 3   Solving Percent Equations

a. What number is 30% of 70?

b. Fourteen is 25% of what number?

c. One hundred thirty-five is what percent of 27?

Solution

a. *Verbal Model:*   What number $=$ 30% of 70

Label: $a =$ unknown number

Equation: $a = (0.3)(70) = 21$

So, 21 is 30% of 70.

b. *Verbal Model:*   14 $=$ 25% of what number

Label: $b =$ unknown number

Equation: $14 = 0.25b$

$$\frac{14}{0.25} = b$$

$$56 = b$$

So, 14 is 25% of 56.

c. *Verbal Model:*   135 $=$ What percent of 27

Label: $p =$ unknown percent (in decimal form)

Equation: $135 = p(27)$

$$\frac{135}{27} = p$$

$$5 = p$$

So, 135 is 500% of 27.

Additional Examples

a. 225 is what percent of 500?

b. What number is 25% of 104?

c. 36 is 12% of what number?

Answers:

a. 45%

b. 26

c. 300

From Example 3, you can see that there are three basic types of percent problems. Each can be solved by substituting the two given quantities into the percent equation and solving for the third quantity.

Question	*Given*	*Percent Equation*
a is what percent of b?	a and b	Solve for p.
What number is p percent of b?	p and b	Solve for a.
a is p percent of what number?	a and p	Solve for b.

For instance, part (b) of Example 3 fits the form "a is p percent of what number?"

In most real-life applications, the base number b and the number a are much more disguised than they are in Example 3. It sometimes helps to think of a as a "new" amount and b as the "original" amount.

Example 4   Real Estate Commission

A real estate agency receives a commission of $8092.50 for the sale of a $124,500 house. What percent commission is this?

Solution

Verbal Model:
$$\text{Commission} = \frac{\text{Percent}}{\text{(in decimal form)}} \cdot \text{Sale price}$$

Labels:

Commission = 8092.50	(dollars)
Percent = p	(in decimal form)
Sale price = 124,500	(dollars)

Equation:

$8092.50 = p \cdot (124,500)$ Original equation

$\dfrac{8092.50}{124,500} = p$ Divide each side by 124,500.

$0.065 = p$ Simplify.

So, the real estate agency receives a commission of 6.5%.

Example 5   Cost-of-Living Raise

A union negotiates for a cost-of-living raise of 7%. What is the raise for a union member whose salary is $23,240? What is this person's new salary?

Solution

Verbal Model:
$$\text{Raise} = \frac{\text{Percent}}{\text{(in decimal form)}} \cdot \text{Salary}$$

Labels:

Raise = a	(dollars)
Percent = 7% = 0.07	(in decimal form)
Salary = 23,240	(dollars)

Equation: $a = 0.07(23,240) = 1626.80$

So, the raise is $1626.80 and the new salary is

$23,240.00 + 1626.80 = \$24,866.80.$

Example 6   Course Grade

You missed an A in your chemistry course by only three points. Your point total for the course is 402. How many points were possible in the course? (Assume that you needed 90% of the course total for an A.)

Solution

Verbal Model:	$\boxed{\begin{array}{c}\text{Your}\\\text{points}\end{array}} + \boxed{\begin{array}{c}3\\\text{points}\end{array}} = \boxed{\begin{array}{c}\text{Percent}\\\text{(in decimal form)}\end{array}} \cdot \boxed{\begin{array}{c}\text{Total}\\\text{points}\end{array}}$

Labels: Your points = 402   (points)
              Percent = 90% = 0.9   (in decimal form)
              Total points for course = b (points)

Equation: $402 + 3 = 0.9b$ Original equation

              $405 = 0.9b$ Add.

              $\dfrac{405}{0.9} = b$ Divide each side by 0.9.

              $450 = b$ Simplify.

So, there were 450 total points for the course. You can check your solution as follows.

$402 + 3 = 0.9b$ Write original equation.

$402 + 3 \overset{?}{=} 0.9(450)$ Substitute 450 for b.

$405 = 405$ Solution checks. ✓

Example 7   Percent Increase

The monthly basic cable TV rate was $7.69 in 1980 and $30.08 in 2000. Find the percent increase in the monthly basic cable TV rate from 1980 to 2000. (Source: Paul Kagan Associates, Inc.)

Solution

Verbal Model:	$\boxed{\begin{array}{c}2000\\\text{price}\end{array}} = \boxed{\begin{array}{c}1980\\\text{price}\end{array}} \cdot \boxed{\begin{array}{c}\text{Percent increase}\\\text{(in decimal form)}\end{array}} + \boxed{\begin{array}{c}1980\\\text{price}\end{array}}$

Labels: 2000 price = 30.08   (dollars)
              Percent increase = p (in decimal form)
              1980 price = 7.69   (dollars)

Equation: $30.08 = 7.69p + 7.69$ Original equation

              $22.39 = 7.69p$ Subtract 7.69 from each side.

              $2.91 \approx p$ Divide each side by 7.69.

So, the percent increase in the monthly basic cable TV rate from 1980 to 2000 is approximately 291%. Check this in the original statement of the problem.

③ Solve application problems involving markups and discounts.

Markups and Discounts

You may have had the experience of buying an item at one store and later finding that you could have paid less for the same item at another store. The basic reason for this price difference is **markup,** which is the difference between the **cost** (the amount a retailer pays for the item) and the **price** (the amount at which the retailer sells the item to the consumer). A verbal model for this problem is as follows.

$$\boxed{\text{Selling price}} \; = \; \boxed{\text{Cost}} \; + \; \boxed{\text{Markup}}$$

In such a problem, the markup may be known or it may be expressed as a percent of the cost. This percent is called the **markup rate.**

$$\boxed{\text{Markup}} \; = \; \boxed{\text{Markup rate}} \; \cdot \; \boxed{\text{Cost}}$$

Markup is one of those "hidden operations" referred to in Section 2.3.

In business and economics, the terms *cost* and *price* do not mean the same thing. The cost of an item is the amount a business pays for the item. The price of an item is the amount for which the business sells the item.

Example 8   Finding the Selling Price

A sporting goods store uses a markup rate of 55% on all items. The cost of a golf bag is $45. What is the selling price of the bag?

Solution

Verbal Model: $\boxed{\text{Selling price}} = \boxed{\text{Cost}} + \boxed{\text{Markup}}$

Labels:
Selling price $= x$ (dollars)
Cost $= 45$ (dollars)
Markup rate $= 0.55$ (rate in decimal form)
Markup $= (0.55)(45)$ (dollars)

Equation:
$x = 45 + (0.55)(45)$ Original equation.

$ = 45 + 24.75$ Multiply.

$ = \$69.75$ Simplify.

The selling price is $69.75. You can check your solution as follows:

$x = 45 + (0.55)(45)$ Write original equation.

$69.75 \stackrel{?}{=} 45 + (0.55)(45)$ Substitute 69.75 for x.

$69.75 = 69.75$ Solution checks. ✓

In Example 8, you are given the cost and are asked to find the selling price. Example 9 illustrates the reverse problem. That is, in Example 9 you are given the selling price and are asked to find the cost.

Example 9   Finding the Cost of an Item

The selling price of a pair of ski boots is $98. The markup rate is 60%. What is the cost of the boots?

Solution

Verbal Model: Selling price = Cost + Markup

Labels:

Selling price = 98   (dollars)
Cost = x (dollars)
Markup rate = 0.60   (rate in decimal form)
Markup = 0.60x (dollars)

Equation:

$98 = x + 0.60x$	Original equation
$98 = 1.60x$	Combine like terms.
$61.25 = x$	Divide each side by 1.60.

The cost is $61.25. Check this in the original statement of the problem.

Example 10   Finding the Markup Rate

A pair of walking shoes sells for $60. The cost of the walking shoes is $24. What is the markup rate?

Solution

Verbal Model: Selling price = Cost + Markup

Labels:

Selling price = 60   (dollars)
Cost = 24   (dollars)
Markup rate = p (rate in decimal form)
Markup = $p(24)$ (dollars)

Equation:

$60 = 24 + p(24)$	Original equation
$36 = 24p$	Subtract 24 from each side.
$1.5 = p$	Divide each side by 24.

Because $p = 1.5$, it follows that the markup rate is 150%.

The mathematics of a discount is similar to that of a markup. The model for this situation is

Selling price = List price − Discount

where the **discount** is given in dollars, and the **discount rate** is given as a percent of the list price. Notice the "hidden operation" in the discount.

Discount = Discount rate · List price

Example 11   Finding the Discount Rate

During a midsummer sale, a lawn mower listed at $199.95 is on sale for $139.95. What is the discount rate?

Solution

Verbal Model:

$$\boxed{\text{Discount}} = \boxed{\begin{array}{c}\text{Discount}\\\text{rate}\end{array}} \cdot \boxed{\begin{array}{c}\text{List}\\\text{price}\end{array}}$$

Labels:

Discount = 199.95 − 139.95 = 60	(dollars)
List price = 199.95	(dollars)
Discount rate = p	(rate in decimal form)

Equation: $60 = p(199.95)$ Original equation

                    $0.30 \approx p$ Divide each side by 199.95.

Because $p \approx 0.30$, it follows that the discount rate is approximately 30%.

> **Study Tip**
>
> Recall from Section 1.1 that the symbol $\approx$ means "is approximately equal to."

Example 12   Finding the Sale Price

A drug store advertises 40% off the prices of all summer tanning products. A bottle of suntan oil lists for $3.49. What is the sale price?

Solution

Verbal Model:

$$\boxed{\begin{array}{c}\text{Sale}\\\text{price}\end{array}} = \boxed{\begin{array}{c}\text{List}\\\text{price}\end{array}} - \boxed{\text{Discount}}$$

Labels:

List price = 3.49	(dollars)
Discount rate = 0.4	(rate in decimal form)
Discount = 0.4(3.49)	(dollars)
Sale price = x	(dollars)

Equation: $x = 3.49 - (0.4)(3.49) \approx \$2.09$

The sale price is $2.09. Check this in the original statement of the problem.

The following guidelines summarize the problem-solving strategy that you should use when solving word problems.

> ### Guidelines for Solving Word Problems
>
> 1. Write a *verbal model* that describes the problem.
> 2. Assign *labels* to fixed quantities and variable quantities.
> 3. Rewrite the verbal model as an *algebraic equation* using the assigned labels.
> 4. *Solve* the resulting algebraic equation.
> 5. *Check* to see that your solution satisfies the original problem as stated.

3.3 Exercises

Review *Concepts, Skills, and Problem Solving*

Keep mathematically in shape by doing these exercises *before* the problems of this section.

Properties and Definitions

1. *Writing* Explain how to put the two numbers 63 and -28 in order.   Plot the numbers on a number line. -28 is less than 63 because -28 is to the left of 63.

2. For any real number, its distance from 0 on the real number line is its absolute value.

Simplifying Expressions

In Exercises 3–6, evaluate the expression.

3. $8 - |-7 + 11| + (-4)$ 0

4. $34 - [54 - (-16 + 4) + 6]$ -38

5. Subtract 230 from -300. -530

6. Find the absolute value of the difference of 17 and -12.   29

In Exercises 7 and 8, use the Distributive Property to expand the expression.

7. $4(2x - 5)$ $8x - 20$ 8. $-z(xz - 2y^2)$ $-xz^2 + 2y^2z$

In Exercises 9 and 10, evaluate the algebraic expression for the specified values of the variables. (If not possible, state the reason.)

9. $x^2 - y^2$
(a) $x = 4, y = 3$ 7
(b) $x = -5, y = 3$ 16

10. $\dfrac{z^2 + 2}{x^2 - 1}$
(a) $x = 1, z = 1$ Division by zero is undefined.
(b) $x = 2, z = 2$ 2

Problem Solving

11. *Consumer Awareness* A telephone company charges $1.37 for the first minute of a long-distance telephone call and $0.95 for each additional minute. Find the cost of a 15-minute telephone call. $14.67

12. *Distance* A train travels at the rate of r miles per hour for 5 hours. Write an algebraic expression that represents the total distance traveled by the train. $5r$

Developing Skills

In Exercises 1–12, complete the table showing the equivalent forms of a percent. See Examples 1 and 2.

	Percent	Parts out of 100	Decimal	Fraction
1.	40%	40	0.40	$\frac{2}{5}$
2.	16%	16	0.16	$\frac{4}{25}$
3.	7.5%	7.5	0.075	$\frac{3}{40}$
4.	75%	75	0.75	$\frac{3}{4}$
5.	63%	63	0.63	$\frac{63}{100}$
6.	10.5%	10.5	0.105	$\frac{21}{200}$
7.	15.5%	15.5	0.155	$\frac{31}{200}$
8.	80%	80	0.80	$\frac{4}{5}$
9.	60%	60	0.60	$\frac{3}{5}$
10.	15%	15	0.15	$\frac{3}{20}$
11.	150%	150	1.50	$\frac{3}{2}$
12.	125%	125	1.25	$\frac{5}{4}$

In Exercises 13–20, convert the decimal to a percent. See Example 1.

13. 0.62   62%
14. 0.57   57%
15. 0.20   20%
16. 0.38   38%
17. 0.075   7.5%
18. 0.005   0.5%
19. 2.38   238%
20. 1.75   175%

In Exercises 21–28, convert the percent to a decimal. See Example 2.

21. 12.5%   0.125   22. 95%   0.95
23. 125%   1.25   24. 250%   2.50
25. 8.5%   0.085   26. 0.3%   0.003
27. $\frac{3}{4}$%   0.0075   28. $4\frac{4}{5}$%   0.048

In Exercises 29–36, convert the fraction to a percent. See Example 1.

29. $\frac{4}{5}$ 80%

30. $\frac{1}{4}$ 25%

31. $\frac{5}{4}$ 125%

32. $\frac{6}{5}$ 120%

33. $\frac{5}{6}$ $83\frac{1}{3}$%

34. $\frac{2}{3}$ $66\frac{2}{3}$%

35. $\frac{21}{20}$ 105%

36. $\frac{5}{2}$ 250%

In Exercises 37–40, what percent of the figure is shaded? (There are a total of 360° in a circle.)

37. $37\frac{1}{2}$%

38. $66\frac{2}{3}$%

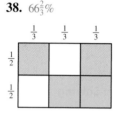

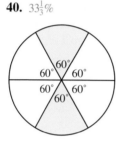

39. $41\frac{2}{3}$%

40. $33\frac{1}{3}$%

In Exercises 41–64, solve the percent equation. See Example 3.

41. What number is 30% of 150?   45

42. What number is 62% of 1200?   744

43. What number is $66\frac{2}{3}$% of 816?   544

44. What number is $33\frac{1}{3}$% of 516?   172

45. What number is 0.75% of 56?   0.42

46. What number is 0.2% of 100,000?   200

47. What number is 200% of 88?   176

48. What number is 325% of 450?   1462.5

49. 903 is 43% of what number?   2100

50. 425 is 85% of what number?   500

51. 275 is $12\frac{1}{2}$% of what number?   2200

52. 814 is $66\frac{2}{3}$% of what number?   1221

53. 594 is 450% of what number?   132

54. 210 is 250% of what number?   84

55. 2.16 is 0.6% of what number?   360

56. 51.2 is 0.08% of what number?   64,000

57. 576 is what percent of 800?   72%

58. 1950 is what percent of 5000?   39%

59. 45 is what percent of 360?   12.5%

60. 38 is what percent of 5700? $\frac{2}{3}$%

61. 22 is what percent of 800?   2.75%

62. 110 is what percent of 110?   100%

63. 1000 is what percent of 200?   500%

64. 148.8 is what percent of 960?   15.5%

In Exercises 65–74, find the missing quantities. See Examples 8, 9, and 10.

	Cost	Selling Price	Markup	Markup Rate
65.	$26.97	$49.95	$22.98	85.2%
66.	$71.97	$119.95	$47.98	$66\frac{2}{3}$%
67.	$40.98	$74.38	$33.40	81.5%
68.	$45.01	$69.99	$24.98	55.5%
69.	$69.29	$125.98	$56.69	81.8%
70.	$269.23	$350.00	$80.77	30%
71.	$13,250.00	$15,900.00	$2650.00	20%
72.	$149.79	$224.87	$75.08	50.1%
73.	$107.97	$199.96	$91.99	85.2%
74.	$680.00	$906.67	$226.67	$33\frac{1}{3}$%

In Exercises 75–84, find the missing quantities. See Examples 11 and 12.

	List Price	Sale Price	Discount	Discount Rate
75.	$39.95	$29.95	$10.00	25%
76.	$50.99	$45.99	$5.00	9.8%
77.	$23.69	$18.95	$4.74	20%
78.	$315.00	$189.00	$126.00	40%
79.	$189.99	$159.99	$30.00	15.8%
80.	$18.95	$10.95	$8.00	42.2%
81.	$119.96	$59.98	$59.98	50%
82.	$84.95	$29.73	$55.22	65%
83.	$995.00	$695.00	$300.00	30.2%
84.	$394.97	$259.97	$135.00	34.2%

Solving Problems

85. *Rent*   You spend 17% of your monthly income of $3200 for rent. What is your monthly payment? $544

86. *Cost of Housing*   You budget 30% of your annual after-tax income for housing. Your after-tax income is $38,500. What amount can you spend on housing? $11,550

87. *Retirement Plan*   You budget $7\frac{1}{2}$% of your gross income for an individual retirement plan. Your annual gross income is $45,800. How much will you put in your retirement plan each year? $3435

88. *Enrollment*   In the fall of 2001, 41% of the students enrolled at Alabama State University were freshmen. The enrollment of the college was 5590. Find the number of freshmen enrolled in the fall of 2001. (Source: Alabama State University)   2292 students

89. *Meteorology*   During the winter of 2000–2001, 33.6 inches of snow fell in Detroit, Michigan. Of that amount, 25.1 inches fell in December. What percent of the total snowfall amount fell in December? (Source: National Weather Service)   74.7%

90. *Inflation Rate*   You purchase a lawn tractor for $3750 and 1 year later you note that the cost has increased to $3900. Determine the inflation rate (as a percent) for the tractor.   4%

91. *Unemployment Rate*   During a recession, 72 out of 1000 workers in the population were unemployed. Find the unemployment rate (as a percent).   7.2%

92. *Layoff*   Because of slumping sales, a small company laid off 30 of its 153 employees.

  (a) What percent of the work force was laid off?   19.6%

  (b) Complete the statement: "About 1 out of every   5   workers was laid off."

93. *Original Price*   A coat sells for $250 during a 20% off storewide clearance sale. What was the original price of the coat? $312.50

94. *Course Grade*   You were six points shy of a B in your mathematics course. Your point total for the course was 394. How many points were possible in the course? (Assume that you needed 80% of the course total for a B.)   500 points

95. *Consumer Awareness*   The price of a new van is approximately 110% of what it was 3 years ago. The current price is $26,850. What was the approximate price 3 years ago? $24,409

96. *Membership Drive*   Because of a membership drive for a public television station, the current membership is 125% of what it was a year ago. The current number of members is 7815. How many members did the station have last year?   6252 members

97. *Eligible Voters*   The news media reported that 6432 votes were cast in the last election and that this represented 63% of the eligible voters of a district. How many eligible voters are in the district?   10,210 eligible votes

98. *Quality Control*   A quality control engineer tested several parts and found two to be defective. The engineer reported that 2.5% were defective. How many were tested?   80 parts

99. ▲ *Geometry*   A rectangular plot of land measures 650 feet by 825 feet (see figure). A square garage with sides of length 24 feet is built on the plot of land. What percentage of the plot of land is occupied by the garage?   0.107%

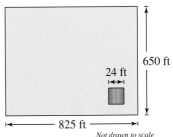

Not drawn to scale

100. ▲ *Geometry*   A circular target is attached to a rectangular board, as shown in the figure. The radius of the circle is $4\frac{1}{2}$ inches, and the measurements of the board are 12 inches by 15 inches. What percentage of the board is covered by the target? (The area of a circle is $A = \pi r^2$, where r is the radius of the circle.)   35.3%

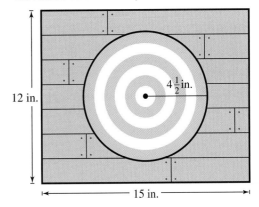

101. *Data Analysis*   In 1999 there were 841.3 million visits to office-based physicians. The circle graph classifies the age groups of those making the visits. Approximate the number of Americans in each of the classifications. (Source: U.S. National Center for Health Statistics)

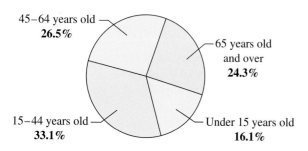

45–64 years old **26.5%**

65 years old and over **24.3%**

15–44 years old **33.1%**

Under 15 years old **16.1%**

   < 15   135.45 million;   15–44   278.47 million
   45–64   222.94 million;   > 64   204.44 million

102. *Graphical Estimation*   The bar graph shows the numbers (in thousands) of criminal cases commenced in the United States District Courts from 1997 through 2001. (Source: Administrative Office of the U.S. Courts)

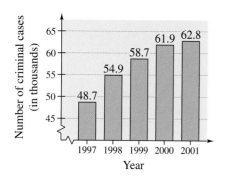

(a) Determine the percent increase in cases from 1997 to 1998.   12.7%

(b) Determine the percent increase in cases from 1998 to 2001.   14.4%

103. *Interpreting a Table*   The table shows the numbers of women scientists and the percents of women scientists in the United States in three fields for the years 1983 and 2000. (Source: U.S. Bureau of Labor Statistics)

Field	1983		2000	
	Number	%	Number	%
Math/Computer	137,048	29.6%	651,236	31.4%
Chemistry	22,834	23.3%	46,359	30.3%
Biology	22,440	40.8%	51,756	45.4%

(a) Find the total number of mathematicians and computer scientists (men and women) in 2000. 2,074,000

(b) Find the total number of chemists (men and women) in 1983.   98,000

(c) Find the total number of biologists (men and women) in 2000.   114,000

104. *Data Analysis*   The table shows the approximate population (in millions) of Bangladesh for each decade from 1960 through 2000. Approximate the percent growth rate for each decade. If the growth rate of the 1990s continued until the year 2020, approximate the population in 2020. (Source: U.S. Bureau of the Census, International Data Base)   1960s 23.4%; 1970s 30.7%; 1980s 24.7%   1990s 17.6%   178.7 million

Year	1960	1970	1980	1990	2000
Population	54.6	67.4	88.1	109.9	129.2

Explaining Concepts

105.   Answer parts (a)–(f) of Motivating the Chapter on page 122.

106. *Writing*   Explain the meaning of the word "percent."   Percent means part of 100.

107. *Writing*   Explain the concept of "rate."   A rate is a fixed ratio.

108. *Writing*   Can any positive decimal be written as a percent? Explain.   Yes. Multiply by 100 and affix the percent sign.

109. *Writing*   Is it true that $\frac{1}{2}\% = 50\%$? Explain.   No. $\frac{1}{2}\% = 0.5\% = 0.005$

Mid-Chapter Quiz

Take this quiz as you would take a quiz in class. After you are done, check your work against the answers in the back of the book.

In Exercises 1–10, solve the equation.

1. $74 - 12x = 2$ 6

2. $10(y - 8) = 0$ 8

3. $3x + 1 = x + 20$ $\frac{19}{2}$

4. $6x + 8 = 8 - 2x$ 0

5. $-10x + \frac{2}{3} = \frac{7}{3} - 5x$ $-\frac{1}{3}$

6. $\frac{x}{5} + \frac{x}{8} = 1$ $\frac{40}{13}$

7. $\frac{9 + x}{3} = 15$ 36

8. $7 - 2(5 - x) = -7$ -2

9. $\frac{x + 3}{6} = \frac{4}{3}$ 5

10. $\frac{x + 7}{5} = \frac{x + 9}{7}$ -2

In Exercises 11 and 12, solve the equation. Round your answer to two decimal places. In your own words, explain how to check the solution.

11. $32.86 - 10.5x = 11.25$ 2.06

Substitute 2.06 for x. After simplifying, the equation should be an identity.

12. $\frac{x}{5.45} + 3.2 = 12.6$ 51.23

Substitute 51.23 for x. After simplifying, the equation should be an identity.

13. What number is 62% of 25?   15.5

14. What number is $\frac{1}{2}$% of 8400?   42

15. 300 is what percent of 150?   200%

16. 145.6 is 32% of what number?   455

17. The perimeter of a rectangle is 60 meters. The length is $1\frac{1}{2}$ times its width. Find the dimensions of the rectangle.   12 meters × 18 meters

18. You have two jobs. In the first job, you work 40 hours a week at a candy store and earn $7.50 per hour. In the second job, you earn $6.00 per hour baby-sitting and can work as many hours as you want. You want to earn $360 a week. How many hours must you work at the second job?   10 hours

19. A region has an area of 42 square meters. It must be divided into three subregions so that the second has twice the area of the first, and the third has twice the area of the second. Find the area of each subregion.   6 square meters, 12 square meters, 24 square meters

20. To get an A in a psychology course, you must have an average of at least 90 points for three tests of 100 points each. For the first two tests, your scores are 84 and 93. What must you score on the third test to earn a 90% average for the course?   93

21. The price of a television set is approximately 108% of what it was 2 years ago. The current price is $535. What was the approximate price 2 years ago? $495.37

Endangered Wildlife and Plant Species

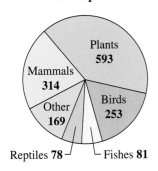

Reptiles 78 ⌐ Fishes 81

Figure for 22

22. The circle graph at the left shows the number of endangered wildlife and plant species for the year 2001. What percent of the total endangered wildlife and plant species were birds?   (Source: U.S. Fish and Wildlife Service)   17%

23. Two people can paint a room in t hours, where t must satisfy the equation $t/4 + t/12 = 1$. How long will it take for the two people to paint the room?   3 hours

3.4 Ratios and Proportions

Eunice Harris/Photo Researchers, Inc.

Why You Should Learn It

Ratios can be used to represent many real-life quantities. For instance, in Exercise 60 on page 167, you will find the gear ratios for a five-speed bicycle.

What You Should Learn

① Compare relative sizes using ratios.

② Find the unit price of a consumer item.

③ Solve proportions that equate two ratios.

④ Solve application problems using the Consumer Price Index.

① Compare relative sizes using ratios.

Setting Up Ratios

A **ratio** is a comparison of one number to another by division. For example, in a class of 29 students made up of 16 women and 13 men, the ratio of women to men is 16 to 13 or $\frac{16}{13}$. Some other ratios for this class are as follows.

Men to women: $\frac{13}{16}$ Men to students: $\frac{13}{29}$ Students to women: $\frac{29}{16}$

Note the order implied by a ratio. The ratio of a to b means a/b, whereas the ratio of b to a means b/a.

Definition of Ratio

The **ratio** of the real number a to the real number b is given by

$$\frac{a}{b}.$$

The ratio of a to b is sometimes written as $a : b$.

Example 1   Writing Ratios in Fractional Form

a. The ratio of 7 to 5 is given by $\frac{7}{5}$.

b. The ratio of 12 to 8 is given by $\frac{12}{8} = \frac{3}{2}$.

   Note that the fraction $\frac{12}{8}$ can be written in simplest form as $\frac{3}{2}$.

c. The ratio of $3\frac{1}{2}$ to $5\frac{1}{4}$ is given by

$$\frac{3\frac{1}{2}}{5\frac{1}{4}} = \frac{\frac{7}{2}}{\frac{21}{4}} \qquad \text{Rewrite mixed numbers as fractions.}$$

$$= \frac{7}{2} \cdot \frac{4}{21} \qquad \text{Invert divisor and multiply.}$$

$$= \frac{2}{3}. \qquad \text{Simplify.}$$

There are many real-life applications of ratios. For instance, ratios are used to describe opinion surveys (for/against), populations (male/female, unemployed/employed), and mixtures (oil/gasoline, water/alcohol).

When comparing two *measurements* by a ratio, you should use the same unit of measurement in both the numerator and the denominator. For example, to find the ratio of 4 feet to 8 inches, you could convert 4 feet to 48 inches (by multiplying by 12) to obtain

$$\frac{4 \text{ feet}}{8 \text{ inches}} = \frac{48 \text{ inches}}{8 \text{ inches}} = \frac{48}{8} = \frac{6}{1}$$

or you could convert 8 inches to $\frac{8}{12}$ feet (by dividing by 12) to obtain

$$\frac{4 \text{ feet}}{8 \text{ inches}} = \frac{4 \text{ feet}}{\frac{8}{12} \text{ feet}} = 4 \cdot \frac{12}{8} = \frac{6}{1}.$$

If you use different units of measurement in the numerator and denominator, then you *must* include the units. If you use the same units of measurement in the numerator and denominator, then it is not necessary to write the units. A list of common conversion factors is found on the inside back cover.

Example 2   Comparing Measurements

Find ratios to compare the relative sizes of the following.

a. 5 gallons to 7 gallons   **b.** 3 meters to 40 centimeters

c. 200 cents to 3 dollars   **d.** 30 months to $1\frac{1}{2}$ years

Solution

a. Because the units of measurement are the same, the ratio is $\frac{5}{7}$.

b. Because the units of measurement are different, begin by converting meters to centimeters *or* centimeters to meters. Here, it is easier to convert meters to centimeters by multiplying by 100.

$$\frac{3 \text{ meters}}{40 \text{ centimeters}} = \frac{3(100) \text{ centimeters}}{40 \text{ centimeters}}$$ Convert meters to centimeters.

$$= \frac{300}{40}$$ Multiply numerator.

$$= \frac{15}{2}$$ Simplify.

c. Because 200 cents is the same as 2 dollars, the ratio is

$$\frac{200 \text{ cents}}{3 \text{ dollars}} = \frac{2 \text{ dollars}}{3 \text{ dollars}} = \frac{2}{3}.$$

d. Because $1\frac{1}{2}$ years $= 18$ months, the ratio is

$$\frac{30 \text{ months}}{1\frac{1}{2} \text{ years}} = \frac{30 \text{ months}}{18 \text{ months}} = \frac{30}{18} = \frac{5}{3}.$$

② Find the unit price of a consumer item.

Unit Prices

As a consumer, you must be able to determine the unit prices of items you buy in order to make the best use of your money. The **unit price** of an item is given by the ratio of the total price to the total units.

$$\text{Unit price} = \frac{\text{Total price}}{\text{Total units}}$$

The word *per* is used to state unit prices. For instance, the unit price for a particular brand of coffee might be 4.69 dollars *per* pound, or $4.69 per pound.

Example 3   Finding a Unit Price

Find the unit price (in dollars per ounce) for a five-pound, four-ounce box of detergent that sells for $4.62.

Solution

Begin by writing the weight in ounces. That is,

$$5 \text{ pounds} + 4 \text{ ounces} = 5 \text{ pounds} \left(\frac{16 \text{ ounces}}{1 \text{ pound}} \right) + 4 \text{ ounces}$$

$$= 80 \text{ ounces} + 4 \text{ ounces}$$

$$= 84 \text{ ounces}.$$

Next, determine the unit price as follows.

Verbal Model: $\text{Unit price} = \dfrac{\text{Total price}}{\text{Total units}}$

Unit Price: $\dfrac{\$4.62}{84 \text{ ounces}} = \$0.055 \text{ per ounce}$

Example 4   Comparing Unit Prices

Which has the lower unit price: a 12-ounce box of breakfast cereal for $2.69 or a 16-ounce box of the same cereal for $3.49?

Solution

The unit price for the smaller box is

$$\text{Unit price} = \frac{\text{Total price}}{\text{Total units}} = \frac{\$2.69}{12 \text{ ounces}} \approx \$0.224 \text{ per ounce}.$$

The unit price for the larger box is

$$\text{Unit price} = \frac{\text{Total price}}{\text{Total units}} = \frac{\$3.49}{16 \text{ ounces}} \approx \$0.218 \text{ per ounce}.$$

So, the larger box has a slightly lower unit price.

③ Solve proportions that equate two ratios.

Solving Proportions

A **proportion** is a statement that equates two ratios. For example, if the ratio of a to b is the same as the ratio of c to d, you can write the proportion as

$$\frac{a}{b} = \frac{c}{d}.$$

In typical applications, you know the values of three of the letters (quantities) and are required to find the value of the fourth. To solve such a fractional equation, you can use the *cross-multiplication* procedure introduced in Section 3.2.

Solving a Proportion

If

$$\frac{a}{b} = \frac{c}{d}$$

then $ad = bc$. The quantities a and d are called the **extremes** of the proportion, whereas b and c are called the **means** of the proportion.

Example 5   Solving Proportions

Solve each proportion.

a. $\dfrac{50}{x} = \dfrac{2}{28}$ **b.** $\dfrac{x}{3} = \dfrac{10}{6}$

Solution

a.
$$\frac{50}{x} = \frac{2}{28} \qquad \text{Write original proportion.}$$
$$50(28) = 2x \qquad \text{Cross-multiply.}$$
$$\frac{1400}{2} = x \qquad \text{Divide each side by 2.}$$
$$700 = x \qquad \text{Simplify.}$$

So, the ratio of 50 to 700 is the same as the ratio of 2 to 28.

b.
$$\frac{x}{3} = \frac{10}{6} \qquad \text{Write original proportion.}$$
$$x = \frac{30}{6} \qquad \text{Multiply each side by 3.}$$
$$x = 5 \qquad \text{Simplify.}$$

So, the ratio of 5 to 3 is the same as the ratio of 10 to 6.

Additional Examples
Solve each proportion.

a. $\dfrac{8}{x} = \dfrac{5}{2}$

b. $\dfrac{20}{184} = \dfrac{45}{x}$

Answers:

a. $x = \dfrac{16}{5}$

b. $x = 414$

To solve an equation, you want to isolate the variable. In Example 5(b), this was done by multiplying each side by 3 instead of cross-multiplying. In this case, multiplying each side by 3 was the only step needed to isolate the x-variable. However, either method is valid for solving the equation.

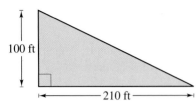

100 ft

210 ft

Triangular lot

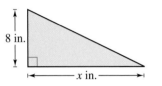

8 in.

x in.

Sketch

Figure 3.2

Example 6 Geometry: Similar Triangles

A triangular lot has perpendicular sides of lengths 100 feet and 210 feet. You are to make a proportional sketch of this lot using 8 inches as the length of the shorter side. How long should you make the other side?

Solution

This is a case of similar triangles in which the ratios of the corresponding sides are equal. The triangles are shown in Figure 3.2.

$$\frac{\text{Shorter side of lot}}{\text{Longer side of lot}} = \frac{\text{Shorter side of sketch}}{\text{Longer side of sketch}}$$ Proportion for similar triangles

$$\frac{100}{210} = \frac{8}{x}$$ Substitute.

$$x \cdot 100 = 210 \cdot 8$$ Cross-multiply.

$$x = \frac{1680}{100} = 16.8$$ Divide each side by 100.

So, the length of the longer side of the sketch should be 16.8 inches.

Example 7 Resizing a Picture

You have a 7-by-8-inch picture of a graph that you want to paste into a research paper, but you have only a 6-by-6-inch space in which to put it. You go to the copier that has five options for resizing your graph: 64%, 78%, 100%, 121%, and 129%.

a. Which option should you choose?

b. What are the measurements of the resized picture?

Solution

a. Because the longest side must be reduced from 8 inches to no more than 6 inches, consider the proportion

$$\frac{\text{New length}}{\text{Old length}} = \frac{\text{New percent}}{\text{Old percent}}$$ Original proportion

$$\frac{6}{8} = \frac{x}{100}$$ Substitute.

$$\frac{6}{8} \cdot 100 = x$$ Multiply each side by 100.

$$75 = x.$$ Simplify.

To guarantee a fit, you should choose the 64% option, because 78% is greater than the required 75%.

b. To find the measurements of the resized picture, multiply by 64% or 0.64.

Length $= 0.64(8) = 5.12$ inches   Width $= 0.64(7) = 4.48$ inches

The size of the reduced picture is 5.12 inches by 4.48 inches.

Example 8   Gasoline Cost

You are driving from New York to Phoenix, a trip of 2450 miles. You begin the trip with a full tank of gas and after traveling 424 miles, you refill the tank for $24.00. How much should you plan to spend on gasoline for the entire trip?

Solution

Verbal Model:

$$\frac{\text{Cost for trip}}{\text{Cost for tank}} = \frac{\text{Miles for trip}}{\text{Miles for tank}}$$

Labels:

Cost of gas for entire trip $= x$	(dollars)
Cost of gas for tank $= 24$	(dollars)
Miles for entire trip $= 2450$	(miles)
Miles for tank $= 424$	(miles)

Proportion: $\dfrac{x}{24} = \dfrac{2450}{424}$ Original proportion

$x = 24\left(\dfrac{2450}{424}\right)$ Multiply each side by 24.

$x \approx 138.68$ Simplify.

You should plan to spend approximately $138.68 for gasoline on the trip. Check this in the original statement of the problem.

In examples such as Example 8, you might point out that an "approximate" answer will not check "exactly" in the original statement of the problem. However, the process of checking solutions is still important.

④ Solve application problems using the Consumer Price Index.

The Consumer Price Index

The rate of inflation is important to all of us. Simply stated, *inflation* is an economic condition in which the price of a fixed amount of goods or services increases. So, a fixed amount of money buys less in a given year than in previous years.

The most widely used measurement of inflation in the United States is the *Consumer Price Index* (CPI), often called the *Cost-of-Living Index*. The table below shows the "All Items" or general index for the years 1970 to 2001. (Source: Bureau of Labor Statistics)

Year	CPI	Year	CPI	Year	CPI	Year	CPI
1970	38.8	1978	65.2	1986	109.6	1994	148.2
1971	40.5	1979	72.6	1987	113.6	1995	152.4
1972	41.8	1980	82.4	1988	118.6	1996	156.9
1973	44.4	1981	90.9	1989	124.0	1997	160.5
1974	49.3	1982	96.5	1990	130.7	1998	163.0
1975	53.8	1983	99.6	1991	136.2	1999	166.6
1976	56.9	1984	103.9	1992	140.3	2000	172.2
1977	60.6	1985	107.6	1993	144.5	2001	177.1

To determine (from the CPI) the change in the buying power of a dollar from one year to another, use the following proportion.

$$\frac{\text{Price in year } n}{\text{Price in year } m} = \frac{\text{Index in year } n}{\text{Index in year } m}$$

Example 9   Using the Consumer Price Index

You paid $35,000 for a house in 1971. What is the amount you would pay for the same house in 2000?

Solution

Verbal Model: $\dfrac{\text{Price in 2000}}{\text{Price in 1971}} = \dfrac{\text{Index in 2000}}{\text{Index in 1971}}$

Labels: Price in 2000 = x (dollars)
Price in 1971 = 35,000   (dollars)
Index in 2000 = 172.2
Index in 1971 = 40.5

Proportion: $\dfrac{x}{35,000} = \dfrac{172.2}{40.5}$ Original proportion

$x = \dfrac{172.2}{40.5} \cdot 35,000$ Multiply each side by 35,000.

$x \approx \$148,815$ Simplify.

So, you would pay approximately $148,815 for the house in 2000. Check this in the original statement of the problem.

Example 10   Using the Consumer Price Index

You inherited a diamond pendant from your grandmother in 1999. The pendant was appraised at $1300. What was the value of the pendant when your grandmother bought it in 1973?

Solution

Verbal Model: $\dfrac{\text{Price in 1999}}{\text{Price in 1973}} = \dfrac{\text{Index in 1999}}{\text{Index in 1973}}$

Labels: Price in 1999 = 1300   (dollars)
Price in 1973 = x (dollars)
Index in 1999 = 166.6
Index in 1973 = 44.4

Proportion: $\dfrac{1300}{x} = \dfrac{166.6}{44.4}$ Original proportion

$57,720 = 166.6x$ Cross-multiply.

$346 \approx x$ Divide each side by 166.6.

So, the value of the pendant in 1973 was approximately $346. Check this in the original statement of the problem.

3.4 Exercises

Review   Concepts, Skills, and Problem Solving

Keep mathematically in shape by doing these exercises *before* the problems of this section.

Properties and Definitions

1. *Writing* ✎   Explain how to write $\frac{15}{12}$ in simplest form.

   Divide both the numerator and denominator by 3.

2. *Writing* ✎   Explain how to divide $\frac{3}{5}$ by $\frac{x}{2}$.

   Multiply $\frac{3}{5}$ by $\frac{2}{x}$.

3. Complete the Associative Property: $(3x)y = $ $3(xy)$.

4. Identify the property of real numbers illustrated by $x^2 + 0 = x^2$.   Additive Identity Property

Simplifying Expressions

In Exercises 5–10, evaluate the expression.

5. $3^2 - (-4)$ 13

6. $(-5)^3 + 3$ -122

7. 9.3×10^6 9,300,000

8. $\dfrac{-|7 + 3^2|}{4}$ -4

9. $(-4)^2 - (30 \div 50)$ $\frac{77}{5}$ or 15.4

10. $(8 \cdot 9) + (-4)^3$ 8

Writing Models

In Exercises 11 and 12, translate the phrase into an algebraic expression.

11. Twice the difference of a number and 10

    $2(n - 10)$

12. The area of a triangle with base b and height $\frac{1}{2}(b + 6)$ $\frac{1}{4}b(b + 6)$

Developing Skills

In Exercises 1–8, write the ratio as a fraction in simplest form. See Example 1.

1. 36 to 9 $\frac{4}{1}$

2. 45 to 15 $\frac{3}{1}$

3. 27 to 54 $\frac{1}{2}$

4. 27 to 63 $\frac{3}{7}$

5. 14 : 21 $\frac{2}{3}$

6. 12 : 30 $\frac{2}{5}$

7. 144 : 16 $\frac{9}{1}$

8. 60 : 45 $\frac{4}{3}$

In Exercises 9–26, find a ratio that compares the relative sizes of the quantities. (Use the same units of measurement for both quantities.) See Example 2.

9. Forty-two inches to 21 inches $\frac{2}{1}$

10. Eighty-one feet to 27 feet $\frac{3}{1}$

11. Forty dollars to $60 $\frac{2}{3}$

12. Twenty-four pounds to 30 pounds $\frac{4}{5}$

13. One quart to 1 gallon $\frac{1}{4}$

14. Three inches to 2 feet $\frac{1}{8}$

15. Seven nickels to 3 quarters $\frac{7}{15}$

16. Twenty-four ounces to 3 pounds $\frac{1}{2}$

17. Three hours to 90 minutes $\frac{2}{1}$

18. Twenty-one feet to 35 yards $\frac{1}{5}$

19. Seventy-five centimeters to 2 meters $\frac{3}{8}$

20. Three meters to 128 centimeters $\frac{75}{32}$

21. Sixty milliliters to 1 liter $\frac{3}{50}$

22. Fifty cubic centimeters to 1 liter $\frac{1}{20}$

23. Ninety minutes to 2 hours $\frac{3}{4}$

24. Five and one-half pints to 2 quarts $\frac{11}{8}$

25. Three thousand pounds to 5 tons $\frac{3}{10}$

26. Twelve thousand pounds to 2 tons $\frac{3}{1}$

In Exercises 27–30, find the unit price (in dollars per ounce). See Example 3.

27. A 20-ounce can of pineapple for 98¢ $0.049

28. An 18-ounce box of cereal for $4.29 $0.2383

29. A one-pound, four-ounce loaf of bread for $1.46 $0.073

30. A one-pound package of cheese for $3.08 $0.1925

In Exercises 31–36, which product has the lower unit price? See Example 4.

31. (a) A $27\frac{3}{4}$-ounce can of spaghetti sauce for $1.68

    (b) A 32-ounce jar of spaghetti sauce for $1.87

    32-ounce jar

32. (a) A 16-ounce package of margarine quarters for $1.54

(b) A three-pound tub of margarine for $3.62
3-pound tub

33. (a) A 10-ounce package of frozen green beans for 72¢

(b) A 16-ounce package of frozen green beans for 93¢   16-ounce package

34. (a) An 18-ounce jar of peanut butter for $1.92

(b) A 28-ounce jar of peanut butter for $3.18
18-ounce jar

35. (a) A two-liter bottle (67.6 ounces) of soft drink for $1.09

(b) Six 12-ounce cans of soft drink for $1.69
2-liter bottle

36. (a) A one-quart container of oil for $2.12

(b) A 2.5-gallon container of oil for $19.99
2.5-gallon container

In Exercises 37–52, solve the proportion. See Example 5.

37. $\dfrac{5}{3} = \dfrac{20}{y}$ 12

38. $\dfrac{9}{x} = \dfrac{18}{5}$ $\frac{5}{2}$

39. $\dfrac{4}{t} = \dfrac{2}{25}$ 50

40. $\dfrac{5}{x} = \dfrac{3}{2}$ $\frac{10}{3}$

41. $\dfrac{y}{25} = \dfrac{12}{10}$ 30

42. $\dfrac{z}{35} = \dfrac{5}{14}$ $\frac{25}{2}$

43. $\dfrac{8}{3} = \dfrac{t}{6}$ 16

44. $\dfrac{x}{6} = \dfrac{7}{12}$ $\frac{7}{2}$

45. $\dfrac{0.5}{0.8} = \dfrac{n}{0.3}$ $\frac{3}{16}$

46. $\dfrac{2}{4.5} = \dfrac{t}{0.5}$ $\frac{2}{9}$

47. $\dfrac{x+1}{5} = \dfrac{3}{10}$ $\frac{1}{2}$

48. $\dfrac{z-3}{8} = \dfrac{3}{16}$ $\frac{9}{2}$

49. $\dfrac{x+6}{3} = \dfrac{x-5}{2}$ 27

50. $\dfrac{x-2}{4} = \dfrac{x+10}{10}$ 10

51. $\dfrac{x+2}{8} = \dfrac{x-1}{3}$ $\frac{14}{5}$

52. $\dfrac{x-4}{5} = \dfrac{x}{6}$ 24

Solving Problems

In Exercises 53–62, express the statement as a ratio in simplest form. (Use the same units of measurement for both quantities.)

53. *Study Hours*   You study 4 hours per day and are in class 6 hours per day. Find the ratio of the number of study hours to class hours. $\frac{2}{3}$

54. *Income Tax*   You have $16.50 of state tax withheld from your paycheck per week when your gross pay is $750. Find the ratio of tax to gross pay. $\frac{11}{500}$

55. *Consumer Awareness*   Last month, you used your cellular phone for 36 long-distance minutes and 184 local minutes. Find the ratio of local minutes to long-distance minutes. $\frac{46}{9}$

56. *Education*   There are 2921 students and 127 faculty members at your school. Find the ratio of the number of students to the number of faculty members. $\frac{23}{1}$

57. *Compression Ratio*   The *compression ratio* of an engine is the ratio of the expanded volume of gas in one of its cylinders to the compressed volume of gas in the cylinder (see figure). A cylinder in a diesel engine has an expanded volume of 345 cubic centimeters and a compressed volume of 17.25 cubic centimeters. What is the compression ratio of this engine? $\frac{20}{1}$

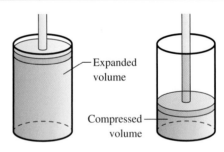

Figure for 57

58. *Turn Ratio*   The *turn ratio* of a transformer is the ratio of the number of turns on the secondary winding to the number of turns on the primary winding (see figure). A transformer has a primary winding with 250 turns and a secondary winding with 750 turns. What is its turn ratio? $\frac{3}{1}$

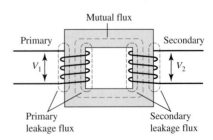

59. *Gear Ratio*   The gear ratio of two gears is the ratio of the number of teeth on one gear to the number of teeth on the other gear. Find the gear ratio of the larger gear to the smaller gear for the gears shown in the figure. $\frac{3}{2}$

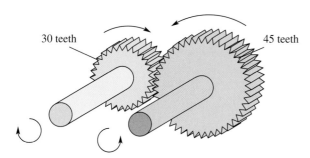

30 teeth   45 teeth

60. *Gear Ratio*   On a five-speed bicycle, the ratio of the pedal gear to the axle gear depends on which axle gear is engaged. Use the table to find the gear ratios for the five different gears. For which gear is it easiest to pedal? Why? $\frac{13}{7}, \frac{13}{6}, \frac{13}{5}, \frac{52}{17}, \frac{26}{7}$; 1st gear; 1st gear has the smallest gear ratio.

Gear	1st	2nd	3rd	4th	5th
Teeth on pedal gear	52	52	52	52	52
Teeth on axle gear	28	24	20	17	14

61. ▲ *Geometry*   A large pizza has a radius of 10 inches and a small pizza has a radius of 7 inches. Find the ratio of the area of the large pizza to the area of the small pizza. (*Note:* The area of a circle is $A = \pi r^2$.)
$\frac{100}{49}$

62. *Specific Gravity*   The *specific gravity* of a substance is the ratio of its weight to the weight of an equal volume of water. Kerosene weighs 0.82 gram per cubic centimeter and water weighs 1 gram per cubic centimeter. What is the specific gravity of kerosene?
0.82

63. *Gasoline Cost*   A car uses 20 gallons of gasoline for a trip of 500 miles. How many gallons would be used on a trip of 400 miles?
16 gallons

64. *Amount of Fuel*   A tractor requires 4 gallons of diesel fuel to plow for 90 minutes. How many gallons of fuel would be required to plow for 8 hours?
$21\frac{1}{3}$ gallons

65. *Building Material*   One hundred cement blocks are required to build a 16-foot wall. How many blocks are needed to build a 40-foot wall?   250 blocks

66. *Force on a Spring*   A force of 50 pounds stretches a spring 4 inches. How much force is required to stretch the spring 6 inches?   75 pounds

67. *Real Estate Taxes*   The tax on a property with an assessed value of $65,000 is $825. Find the tax on a property with an assessed value of $90,000. $1142

68. *Real Estate Taxes*   The tax on a property with an assessed value of $65,000 is $1100. Find the tax on a property with an assessed value of $90,000.
$1523

69. *Polling Results*   In a poll, 624 people from a sample of 1100 indicated they would vote for the republican candidate. How many votes can the candidate expect to receive from 40,000 votes cast?
22,691

70. *Quality Control*   A quality control engineer found two defective units in a sample of 50. At this rate, what is the expected number of defective units in a shipment of 10,000 units?   400

71. *Pumping Time*   A pump can fill a 750-gallon tank in 35 minutes. How long will it take to fill a 1000-gallon tank with this pump? $46\frac{2}{3}$ minutes

72. *Recipe*   Two cups of flour are required to make one batch of cookies. How many cups are required for $2\frac{1}{2}$ batches?   5 cups

73. *Amount of Gasoline*   The gasoline-to-oil ratio for a two-cycle engine is 40 to 1. How much gasoline is required to produce a mixture that contains one-half pint of oil?
20 pints

74. *Building Material*   The ratio of cement to sand in an 80-pound bag of dry mix is 1 to 4. Find the number of pounds of sand in the bag. (*Note:* Dry mix is composed of only cement and sand.)
64 pounds

75. *Map Scale*   On a map, $1\frac{1}{4}$ inch represents 80 miles. Estimate the distance between two cities that are 6 inches apart on the map.
384 miles

76. *Map Scale*   On a map, $1\frac{1}{2}$ inches represents 40 miles. Estimate the distance between two cities that are 4 inches apart on the map.
$106\frac{2}{3}$ miles

▲ *Geometry*   In Exercises 77 and 78, find the length *x* of the side of the larger triangle. (Assume that the two triangles are similar, and use the fact that corresponding sides of similar triangles are proportional.)

77. $\frac{5}{2}$

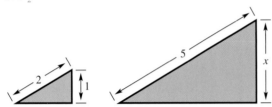

78. 10

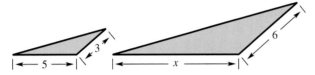

79. ▲ *Geometry*   Find the length of the shadow of the man shown in the figure. (*Hint:* Use similar triangles to create a proportion.) $6\frac{2}{3}$ feet

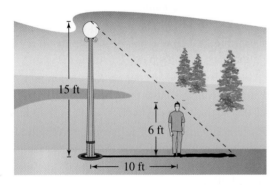

80. ▲ *Geometry*   Find the height of the tree shown in the figure. (*Hint:* Use similar triangles to create a proportion.)   81 feet

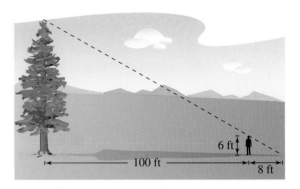

Figure for 80

81. *Resizing a Picture*   You have an 8-by-10-inch photo of a soccer player that must be reduced to a size of 1.6 by 2 inches for the school yearbook. What percent does the photo need to be reduced by in order to fit the allotted space?   80%

82. *Resizing a Picture*   You have a 7-by-5-inch photo of the math club that must be reduced to a size of 5.6 by 4 inches for the school yearbook. What percent does the photo need to be reduced by in order to fit the allotted space?   20%

In Exercises 83–86, use the Consumer Price Index table on page 163 to estimate the price of the item in the indicated year.

83. The 1999 price of a lawn tractor that cost $2875 in 1978 $7346

84. The 2000 price of a watch that cost $158 in 1988 $229

85. The 1970 price of a gallon of milk that cost $2.75 in 1996 $0.68

86. The 1980 price of a coat that cost $225 in 2001 $105

Explaining Concepts

87. ⊘ Answer part (g) of Motivating the Chapter on page 122.

88. *Writing*✐   In your own words, describe the term *ratio.*   A ratio is a comparison of one number to another by division.

89. *Writing*✐   You are told that the ratio of men to women in a class is 2 to 1. Does this information tell you the total number of people in the class? Explain.   No. It is necessary to know one of the following: the number of men in the class or the number of women in the class.

90. *Writing*✐   Explain the following statement. "When setting up a ratio, be sure you are comparing apples to apples and not apples to oranges."   The units must be the same.

91. *Writing*✐   In your own words, describe the term *proportion.*   A proportion is a statement that equates two ratios.

92. Create a proportion problem. Exchange problems with another student and solve the problem you receive.   Answers will vary.

3.5 Geometric and Scientific Applications

Esbin-Anderson/The Image Works

What You Should Learn

1. Use common formulas to solve application problems.
2. Solve mixture problems involving hidden products.
3. Solve work-rate problems.

Why You Should Learn It

The formula for distance can be used whenever you decide to take a road trip. For instance, in Exercise 52 on page 179, you will use the formula for distance to find the travel time for an automobile trip.

① Use common formulas to solve application problems.

Using Formulas

Some formulas occur so frequently in problem solving that it is to your benefit to memorize them. For instance, the following formulas for area, perimeter, and volume are often used to create verbal models for word problems. In the geometric formulas below, A represents area, P represents perimeter, C represents circumference, and V represents volume.

Common Formulas for Area, Perimeter, and Volume

Square	Rectangle	Circle	Triangle
$A = s^2$	$A = lw$	$A = \pi r^2$	$A = \frac{1}{2}bh$
$P = 4s$	$P = 2l + 2w$	$C = 2\pi r$	$P = a + b + c$

Cube	Rectangular Solid	Circular Cylinder	Sphere
$V = s^3$	$V = lwh$	$V = \pi r^2 h$	$V = \frac{4}{3}\pi r^3$

Study Tip

When solving problems involving perimeter, area, or volume, be sure you list the units of measurement for your answers.

- *Perimeter* is always measured in linear units, such as inches, feet, miles, centimeters, meters, and kilometers.
- *Area* is always measured in square units, such as square inches, square feet, square centimeters, and square meters.
- *Volume* is always measured in cubic units, such as cubic inches, cubic feet, cubic centimeters, and cubic meters.

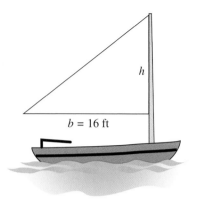

Figure 3.3

Example 1   Using a Geometric Formula

A sailboat has a triangular sail with an area of 96 square feet and a base that is 16 feet long, as shown in Figure 3.3. What is the height of the sail?

Solution

Because the sail is triangular, and you are given its area, you should begin with the formula for the area of a triangle.

$$A = \frac{1}{2}bh \qquad \text{Area of a triangle}$$

$$96 = \frac{1}{2}(16)h \qquad \text{Substitute 96 for } A \text{ and 16 for } b.$$

$$96 = 8h \qquad \text{Simplify.}$$

$$12 = h \qquad \text{Divide each side by 8.}$$

The height of the sail is 12 feet.

In Example 1, notice that b and h are measured in feet. When they are multiplied in the formula $\frac{1}{2}bh$, the resulting area is measured in *square* feet.

$$A = \frac{1}{2}(16 \text{ feet})(12 \text{ feet}) = 96 \text{ feet}^2$$

Note that square feet can be written as feet^2.

Example 2   Using a Geometric Formula

The local municipality is planning to develop the street along which you own a rectangular lot that is 500 feet deep and has an area of 100,000 square feet. To help pay for the new sewer system, each lot owner will be assessed $5.50 per foot of lot frontage.

a. Find the width of the frontage of your lot.

b. How much will you be assessed for the new sewer system?

Solution

a. To solve this problem, it helps to begin by drawing a diagram such as the one shown in Figure 3.4. In the diagram, label the depth of the property as $l = 500$ feet and the unknown frontage as w.

$$A = lw \qquad \text{Area of a rectangle}$$

$$100{,}000 = 500(w) \qquad \text{Substitute 100,000 for } A \text{ and 500 for } l.$$

$$200 = w \qquad \text{Divide each side by 500 and simplify.}$$

The frontage of the rectangular lot is 200 feet.

b. If each foot of frontage costs $5.50, then your total assessment will be 200(5.50) = $1100.

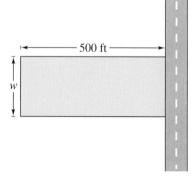

Figure 3.4

Miscellaneous Common Formulas

Temperature: F = degrees Fahrenheit, C = degrees Celsius

$$F = \frac{9}{5}C + 32$$

Simple Interest: I = interest, P = principal, r = interest rate, t = time

$$I = Prt$$

Distance: d = distance traveled, r = rate, t = time

$$d = rt$$

In some applications, it helps to rewrite a common formula by solving for a different variable. For instance, using the common formula for temperature you can obtain a formula for C (degrees Celsius) in terms of F (degrees Fahrenheit) as follows.

$$F = \frac{9}{5}C + 32 \qquad \text{Temperature formula}$$

$$F - 32 = \frac{9}{5}C \qquad \text{Subtract 32 from each side.}$$

$$\frac{5}{9}(F - 32) = C \qquad \text{Multiply each side by } \tfrac{5}{9}.$$

$$C = \frac{5}{9}(F - 32) \qquad \text{Formula}$$

> Remind students that because Example 3 asks for the annual interest rate, time must be expressed in *years*. Six months should be interpreted as $\frac{1}{2}$ year.

Example 3   Simple Interest

An amount of $5000 is deposited in an account paying simple interest. After 6 months, the account has earned $162.50 in interest. What is the annual interest rate for this account?

Solution

$$I = Prt \qquad \text{Simple interest formula}$$

$$162.50 = 5000(r)\left(\frac{1}{2}\right) \qquad \text{Substitute for } I, P, \text{ and } t.$$

$$162.50 = 2500r \qquad \text{Simplify.}$$

$$\frac{162.50}{2500} = r \qquad \text{Divide each side by 2500.}$$

$$0.065 = r \qquad \text{Simplify.}$$

The annual interest rate is $r = 0.065$ (or 6.5%). Check this solution in the original statement of the problem.

Technology: Tip

You can use a graphing calculator to solve simple interest problems by using the program found at our website *math.college.hmco.com/students.* Use the program to check the results of Example 3. Then use the program and the guess, check, and revise method to find P when $I = \$5269$, $r = 11\%$, and $t = 5$ years. See Technology Answers.

One of the most familiar rate problems and most often used formulas in real life is the one that relates distance, rate (or speed), and time: $d = rt$. For instance, if you travel at a constant (or average) rate of 50 miles per hour for 45 minutes, the total distance traveled is given by

$$\left(50\,\frac{\text{miles}}{\text{hour}}\right)\left(\frac{45}{60}\,\text{hour}\right) = 37.5 \text{ miles.}$$

As with all problems involving applications, be sure to check that the units in the model make sense. For instance, in this problem the rate is given in *miles per hour*. So, in order for the solution to be given in *miles,* you must convert the time (from minutes) to *hours*. In the model, you can think of dividing out the 2 "hours," as follows:

$$\left(50\,\frac{\text{miles}}{\text{hour}}\right)\left(\frac{45}{60}\,\text{hour}\right) = 37.5 \text{ miles}$$

Example 4   A Distance-Rate-Time Problem

You jog at an average rate of 8 kilometers per hour. How long will it take you to jog 14 kilometers?

Solution

Verbal Model:

| Distance | = | Rate | · | Time |

Labels: Distance = 14   (kilometers)
               Rate = 8   (kilometers per hour)
               Time = t (hours)

Equation: $14 = 8(t)$

                   $\dfrac{14}{8} = t$

                   $1.75 = t$

It will take you 1.75 hours (or 1 hour and 45 minutes). Check this in the original statement of the problem.

If you are having trouble solving a distance-rate-time problem, consider making a table such as that shown below for Example 4.

Distance = Rate · Time

Rate (km/hr)	8	8	8	8	8	8	8	8
Time (hours)	0.25	0.50	0.75	1.00	1.25	1.50	1.75	2.00
Distance (kilometers)	2	4	6	8	10	12	14	16

In 2001, about 24.5 million Americans ran or jogged on a regular basis. Almost three times that many walked regularly for exercise.

Additional Examples

a. How long will it take you to drive 325 miles at 65 miles per hour?

b. A train can travel a distance of 252 kilometers in 2 hours and 15 minutes without making any stops. What is the average speed of the train?

Answers:

a. 5 hours

b. 112 kilometers per hour

② Solve mixture problems involving hidden products.

Solving Mixture Problems

Many real-world problems involve combinations of two or more quantities that make up a new or different quantity. Such problems are called **mixture problems.** They are usually composed of the sum of two or more "hidden products" that involve *rate factors.* Here is the generic form of the verbal model for mixture problems.

$$\underbrace{\boxed{\text{First rate}} \cdot \boxed{\text{Amount}}}_{\text{First component}} + \underbrace{\boxed{\text{Second rate}} \cdot \boxed{\text{Amount}}}_{\text{Second component}} = \underbrace{\boxed{\text{Final rate}} \cdot \boxed{\text{Final amount}}}_{\text{Final mixture}}$$

The rate factors are usually expressed as *percents* or *percents of measure* such as dollars per pound, jobs per hour, or gallons per minute.

Example 5   A Nut Mixture Problem

A grocer wants to mix cashew nuts worth $7 per pound with 15 pounds of peanuts worth $2.50 per pound. To obtain a nut mixture worth $4 per pound, how many pounds of cashews are needed? How many pounds of mixed nuts will be produced for the grocer to sell?

Solution

In this problem, the rates are the *unit prices* of the nuts.

Verbal Model: $\boxed{\text{Total cost of cashews}} + \boxed{\text{Total cost of peanuts}} = \boxed{\text{Total cost of mixed nuts}}$

Labels:
Unit price of cashews = 7   (dollars per pound)
Unit price of peanuts = 2.5   (dollars per pound)
Unit price of mixed nuts = 4   (dollars per pound)
Amount of cashews = x (pounds)
Amount of peanuts = 15   (pounds)
Amount of mixed nuts = $x + 15$ (pounds)

Equation: $7(x) + 2.5(15) = 4(x + 15)$

$$7x + 37.5 = 4x + 60$$

$$3x = 22.5$$

$$x = \frac{22.5}{3} = 7.5$$

The grocer needs 7.5 pounds of cashews. This will result in $x + 15 = 7.5 + 15 = 22.5$ pounds of mixed nuts. You can check these results as follows.

$$\underbrace{(\$7.00/\text{lb})(7.5 \text{ lb})}_{\text{Cashews}} + \underbrace{(\$2.50/\text{lb})(15 \text{ lb})}_{\text{Peanuts}} \overset{?}{=} \underbrace{(\$4.00/\text{lb})(22.5 \text{ lb})}_{\text{Mixed Nuts}}$$

$$\$52.50 + \$37.50 \overset{?}{=} \$90.00$$

$$\$90.00 = \$90.00 \qquad \text{Solution checks. } ✔$$

In Chapter 8, similar mixture problems will be solved using a system of linear equations.

Example 6   A Solution Mixture Problem

A pharmacist needs to strengthen a 15% alcohol solution with a pure alcohol solution to obtain a 32% solution. How much pure alcohol should be added to 100 milliliters of the 15% solution (see Figure 3.5)?

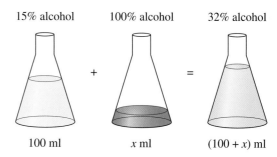

15% alcohol   100% alcohol   32% alcohol

100 ml x ml $(100 + x)$ ml

Figure 3.5

Solution

In this problem, the rates are the alcohol *percents* of the solutions.

Verbal Model: Amount of 15% alcohol solution + Amount of 100% alcohol solution = Amount of final alcohol solution

Labels: 15% solution: Percent alcohol = 0.15   (decimal form)
                            Amount of alcohol solution = 100   (milliliters)
            100% solution: Percent alcohol = 1.00   (decimal form)
                            Amount of alcohol solution = x (milliliters)
            32% solution: Percent alcohol = 0.32   (decimal form)
                            Amount of alcohol solution = $100 + x$ (milliliters)

Equation: $0.15(100) + 1.00(x) = 0.32(100 + x)$
$$15 + x = 32 + 0.32x$$
$$0.68x = 17$$
$$x = \frac{17}{0.68}$$
$$= 25 \text{ ml}$$

So, the pharmacist should add 25 milliliters of pure alcohol to the 15% solution. You can check this in the original statement of the problem as follows.

$$0.15(100) + 1.00(25) \overset{?}{=} 0.32(125)$$
$$15 + 25 \overset{?}{=} 40$$
$$40 = 40 \quad \text{Solution checks. } \checkmark$$

Remember that mixture problems are sums of two or more hidden products that involve different rates. Watch for such problems in the exercises.

Additional Example
Five quarts of a 25% alcohol solution must be mixed with a 50% alcohol solution to obtain a 40% alcohol solution. How much of the 50% solution should be added to the 25% solution?

Answer: 7.5 quarts

Mixture problems can also involve a "mix" of investments, as shown in the next example.

Example 7   Investment Mixture

You invested a total of $10,000 at $4\frac{1}{2}\%$ and $5\frac{1}{2}\%$ simple interest. During 1 year the two accounts earned $508.75. How much did you invest in each account?

Solution

Verbal Model:

$$\boxed{\text{Interest earned from } 4\tfrac{1}{2}\%} + \boxed{\text{Interest earned from } 5\tfrac{1}{2}\%} = \boxed{\text{Total interest earned}}$$

Labels:

Amount invested at $4\frac{1}{2}\% = x$	(dollars)
Amount invested at $5\frac{1}{2}\% = 10{,}000 - x$	(dollars)
Interest earned from $4\frac{1}{2}\% = (x)(0.045)(1)$	(dollars)
Interest earned from $5\frac{1}{2}\% = (10{,}000 - x)(0.055)(1)$	(dollars)
Total interest earned $= 508.75$	(dollars)

Equation:

$$0.045x + 0.055(10{,}000 - x) = 508.75$$
$$0.045x + 550 - 0.055x = 508.75$$
$$550 - 0.01x = 508.75$$
$$-0.01x = -41.25$$
$$x = 4125$$

So, you invested $4125 at $4\frac{1}{2}\%$ and $10{,}000 - x = 10{,}000 - 4125 = \$5875$ at $5\frac{1}{2}\%$. Check this in the original statement of the problem.

---

③ Solve work-rate problems.

Solving Work-Rate Problems

Although not generally referred to as such, most **work-rate problems** are actually *mixture* problems because they involve two or more rates. Here is the generic form of the verbal model for work-rate problems.

$$\boxed{\text{First rate}} \cdot \boxed{\text{Time}} + \boxed{\text{Second rate}} \cdot \boxed{\text{Time}} = \boxed{\begin{array}{c}1 \\ \text{(one whole job completed)}\end{array}}$$

In work-rate problems, the work rate is the *reciprocal* of the time needed to do the entire job. For instance, if it takes 7 hours to complete a job, the per-hour work rate is

$$\frac{1}{7} \text{ job per hour.}$$

Similarly, if it takes $4\frac{1}{2}$ minutes to complete a job, the per-minute rate is

$$\frac{1}{4\frac{1}{2}} = \frac{1}{\frac{9}{2}} = \frac{2}{9} \text{ job per minute.}$$

Remind students that the work rate is the reciprocal of the time required to do the entire job. Machine 1 in Example 8, which requires 3 hours, has a work rate of $\frac{1}{3}$ job per hour. Machine 2, which requires $2\frac{1}{2}$ hours, has a work rate that is the reciprocal of $2\frac{1}{2}$; this work rate is

$$\frac{1}{2\frac{1}{2}} = \frac{1}{\frac{5}{2}} = \frac{2}{5} \text{ job per hour.}$$

Example 8   A Work-Rate Problem

Consider two machines in a paper manufacturing plant. Machine 1 can complete one job (2000 pounds of paper) in 3 hours. Machine 2 is newer and can complete one job in $2\frac{1}{2}$ hours. How long will it take the two machines working together to complete one job?

Solution

Verbal Model:

$$\boxed{\text{Portion done by machine 1}} + \boxed{\text{Portion done by machine 2}} = \boxed{\begin{array}{c}1\\ \text{(one whole job completed)}\end{array}}$$

Labels:
One whole job completed $= 1$	(job)
Rate (machine 1) $= \frac{1}{3}$	(job per hour)
Time (machine 1) $= t$	(hours)
Rate (machine 2) $= \frac{2}{5}$	(job per hour)
Time (machine 2) $= t$	(hours)

Equation:
$$\left(\tfrac{1}{3}\right)(t) + \left(\tfrac{2}{5}\right)(t) = 1$$
$$\left(\tfrac{1}{3} + \tfrac{2}{5}\right)(t) = 1$$
$$\left(\tfrac{11}{15}\right)(t) = 1$$
$$t = \tfrac{15}{11}$$

It will take $\frac{15}{11}$ hours (or about 1.36 hours) for the machines to complete the job working together. Check this solution in the original statement of the problem.

Study Tip

Note in Example 8 that the "2000 pounds" of paper is unnecessary information. The 2000 pounds is represented as "one complete job." This unnecessary information is called a *red herring*.

Example 9   A Fluid-Rate Problem

An above-ground swimming pool has a capacity of 15,600 gallons, as shown in Figure 3.6. A drain pipe can empty the pool in $6\frac{1}{2}$ hours. At what rate (in gallons per minute) does the water flow through the drain pipe?

Solution

To begin, change the time from hours to minutes by multiplying by 60. That is, $6\frac{1}{2}$ hours is equal to $(6.5)(60)$ or 390 minutes.

Verbal Model:
$$\boxed{\text{Volume of pool}} = \boxed{\text{Rate}} \cdot \boxed{\text{Time}}$$

Labels:
Volume $= 15,600$	(gallons)
Rate $= r$	(gallons per minute)
Time $= 390$	(minutes)

Equation:
$$15,600 = r(390)$$
$$\frac{15,600}{390} = r$$
$$40 = r$$

The water is flowing through the drain pipe at a rate of 40 gallons per minute. Check this solution in the original statement of the problem.

Figure 3.6

3.5 Exercises

Review   Concepts, Skills, and Problem Solving

Keep mathematically in shape by doing these exercises *before* the problems of this section.

Properties and Definitions

1. If n is an integer, distinguish between $2n$ and $2n + 1$. $2n$ is an even integer and $2n + 1$ is an odd integer.

2. Demonstrate the Addition Property of Equality for the equation $2x - 3 = 10$.
$$2x - 3 = 10$$
$$2x - 3 + 3 = 10 + 3$$
$$2x = 13$$

Simplifying Expressions

In Exercises 3–10, simplify the expression.

3. $(-3.5y^2)(8y)$ $-28y^3$

4. $(-3x^2)(x^4)$ $-3x^6$

5. $\left(\dfrac{24u}{15}\right)\left(\dfrac{25u^2}{6}\right)$ $\dfrac{20u^3}{3}$

6. $12\left(\dfrac{3y}{18}\right)$ $2y$

7. $5x(2 - x) + 3x$ $13x - 5x^2$

8. $3t - 4(2t - 8)$ $-5t + 32$

9. $3(v - 4) + 7(v - 4)$ $10v - 40$

10. $5[6 - 2(x - 3)]$ $60 - 10x$

Problem Solving

11. *Sales Tax*   You buy a computer for $1150 and your total bill is $1219. Find the sales tax rate.   6%

12. *Consumer Awareness*   A mail-order catalog lists an area rug for $109.95, plus a shipping charge of $14.25. A local store has a sale on the same rug with 20% off a list price of $139.99. Which is the better bargain?   20% off

Developing Skills

In Exercises 1–14, solve for the specified variable.

1. Solve for h: $A = \frac{1}{2}bh$ $\dfrac{2A}{b}$

2. Solve for R: $E = IR$ $\dfrac{E}{I}$

3. Solve for r: $A = P + Prt$ $\dfrac{A - P}{Pt}$

4. Solve for L: $P = 2L + 2W$ $\dfrac{P - 2W}{2}$

5. Solve for l: $V = lwh$ $\dfrac{V}{wh}$

6. Solve for r: $C = 2\pi r$ $\dfrac{C}{2\pi}$

7. Solve for C: $S = C + RC$ $\dfrac{S}{1 + R}$

8. Solve for L: $S = L - RL$ $\dfrac{S}{1 - R}$

9. Solve for m_2: $F = \alpha\dfrac{m_1 m_2}{r^2}$ $\dfrac{Fr^2}{\alpha m_1}$

10. Solve for b: $V = \frac{4}{3}\pi a^2 b$ $\dfrac{3V}{4\pi a^2}$

11. Solve for b: $A = \frac{1}{2}(a + b)h$ $\dfrac{2A - ah}{h}$

12. Solve for r: $V = \frac{1}{3}\pi h^2(3r - h)$ $\dfrac{3V + \pi h^3}{3\pi h^2}$

13. Solve for a: $h = v_0 t + \frac{1}{2}at^2$ $\dfrac{2(h - v_0 t)}{t^2}$

14. Solve for a: $S = \dfrac{n}{2}[2a + (n - 1)d]$ $\dfrac{2S - n^2 d + nd}{2n}$

In Exercises 15–18, evaluate the formula for the specified values of the variables. (List the *units* of the answer.)

15. *Volume of a Right Circular Cylinder:* $V = \pi r^2 h$
$r = 5$ meters, $h = 4$ meters 100π cubic meters

16. *Body Mass Index:* $B = \dfrac{703w}{h^2}$
$w = 127$ pounds, $h = 61$ inches   24 pounds per square inch

17. *Electric Power:* $I = \dfrac{P}{V}$
$P = 1500$ watts, $V = 110$ volts $\dfrac{150}{11}$ watts per volt

18. *Statistical z-score:* $z = \dfrac{x - m}{s}$
$x = 100$ points, $m = 80$ points, $s = 10$ points   2

In Exercises 19–24, find the missing distance, rate, or time. See Example 4.

	Distance, d	Rate, r	Time, t
19.	48 meters	4 m/min	12 min
20.	155 miles	62 mi/hr	$2\frac{1}{2}$ hr
21.	128 km	8 km/hr	16 hours
22.	210 mi	50 mi/hr	4.2 hours
23.	2054 m	$114.\overline{1}$ m/sec	18 sec
24.	482 ft	12.05 ft/min	40 min

Solving Problems

In Exercises 25–32, use a common geometric formula to solve the problem. See Examples 1 and 2.

25. ▲ *Geometry* Each room in the floor plan of a house is square (see figure). The perimeter of the bathroom is 32 feet. The perimeter of the kitchen is 80 feet. Find the area of the living room.   784 square feet

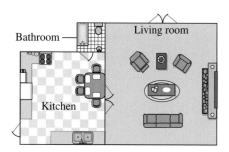

26. ▲ *Geometry* A rectangle has a perimeter of 10 feet and a width of 2 feet. Find the length of the rectangle.
3 feet

27. ▲ *Geometry* A triangle has an area of 48 square meters and a height of 12 meters. Find the length of the base.
8 meters

28. ▲ *Geometry* The perimeter of a square is 48 feet. Find its area.
144 square feet

29. ▲ *Geometry* The circumference of a wheel is 30π inches. Find the diameter of the wheel.
30 inches

30. ▲ *Geometry* A circle has a circumference of 15 meters. What is the radius of the circle? Round your answer to two decimal places.
2.39 meters

31. ▲ *Geometry* A circle has a circumference of 25 meters. Find the radius and area of the circle. Round your answers to two decimal places.
Radius: 3.98 inches; Area: 49.74 square inches

32. ▲ *Geometry* The volume of a right circular cylinder is $V = \pi r^2 h$. Find the volume of a right circular cylinder that has a radius of 2 meters and a height of 3 meters. List the units of measurement for your result. 12π cubic meters

▲ *Geometry* In Exercises 33–36, use the closed rectangular box shown in the figure to solve the problem.

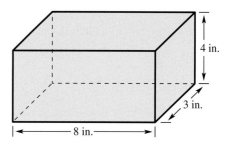

33. Find the area of the base.   24 square inches
34. Find the perimeter of the base.   22 inches
35. Find the volume of the box.   96 cubic inches
36. Find the surface area of the box. (*Note:* This is the combined area of the six surfaces.)   136 square inches

Simple Interest   In Exercises 37–44, use the formula for simple interest. See Example 3.

37. Find the interest on a $1000 bond paying an annual rate of 9% for 6 years.
$540

38. A $1000 corporate bond pays an annual rate of $7\frac{1}{2}$%. The bond matures in $3\frac{1}{2}$ years. Find the interest on the bond.
$262.50

39. You borrow $15,000 for $\frac{1}{2}$ year. You promise to pay back the principal and the interest in one lump sum. The annual interest rate is 13%. What is your payment?
$15,975

40. You have a balance of $650 on your credit card that you cannot pay this month. The annual interest rate on an unpaid balance is 19%. Find the lump sum of principal and interest due in 1 month. $660.29

41. Find the annual rate on a savings account that earns $110 interest in 1 year on a principal of $1000.   11%

42. Find the annual interest rate on a certificate of deposit that earned $128.98 interest in 1 year on a principal of $1500.   8.6%

43. How long must $700 be invested at an annual interest rate of 6.25% to earn $460 interest?   10.51 years

44. How long must $1000 be invested at an annual interest rate of $7\frac{1}{2}$% to earn $225 interest?   3 years

In Exercises 45–54, use the formula for distance to solve the problem. See Example 4.

45. *Space Shuttle*   The speed of the space shuttle (see figure) is 17,500 miles per hour. How long will it take the shuttle to travel a distance of 3000 miles?   0.17 hour

46. *Speed of Light*   The speed of light is 670,616,629.4 miles per hour, and the distance between Earth and the sun is 93,000,000 miles. How long does it take light from the sun to reach Earth?   0.139 hour or ≈ 8.3 minutes

47. *Average Speed*   Determine the average speed of an experimental plane that can travel 3000 miles in 2.6 hours.   1154 miles per hour

48. *Average Speed*   Determine the average speed of an Olympic runner who completes the 10,000-meter race in 27 minutes and 45 seconds.   360 meters per minute

49. *Distance*   Two cars start at a given point and travel in the same direction at average speeds of 45 miles per hour and 52 miles per hour (see figure). How far apart will they be in 4 hours?   28 miles

Figure for 49

50. *Distance*   Two planes leave Orlando International Airport at approximately the same time and fly in opposite directions (see figure). Their speeds are 510 miles per hour and 600 miles per hour. How far apart will the planes be after $1\frac{1}{2}$ hours?   1665 miles

51. *Travel Time*   Two cars start at the same point and travel in the same direction at average speeds of 40 miles per hour and 55 miles per hour. How much time must elapse before the two cars are 5 miles apart? $\frac{1}{3}$ hour

52. *Travel Time*   On the first part of a 225-mile automobile trip you averaged 55 miles per hour. On the last part of the trip you averaged 48 miles per hour because of increased traffic congestion. The total trip took 4 hours and 15 minutes. Find the travel time for each part of the trip.   55 miles per hour for 3 hours; 48 miles per hour for 1.25 hours

53. *Think About It*   A truck traveled at an average speed of 60 miles per hour on a 200-mile trip to pick up a load of freight. On the return trip, with the truck fully loaded, the average speed was 40 miles per hour.

(a) Guess the average speed for the round trip.   Answers will vary.

(b) Calculate the average speed for the round trip. Is the result the same as in part (a)? Explain.   48 miles per hour; Answers will vary.

54. *Time*   A jogger leaves a point on a fitness trail running at a rate of 4 miles per hour. Ten minutes later a second jogger leaves from the same location running at 5 miles per hour. How long will it take the second jogger to overtake the first? How far will each have run at that point?   40 minutes after the second jogger leaves; $3\frac{1}{3}$ miles

Mixture Problem In Exercises 55–58, determine the numbers of units of solutions 1 and 2 required to obtain the desired amount and percent alcohol concentration of the final solution. See Example 6.

	Concentration Solution 1	Concentration Solution 2	Concentration Final Solution	Amount of Final Solution
55.	10%	30%	25%	100 gal

Solution 1: 25 gallons; Solution 2: 75 gallons

56.	25%	50%	30%	5 L

Solution 1: 4 liters; Solution 2: 1 liter

57.	15%	45%	30%	10 qt

Solution 1: 5 quarts; Solution 2: 5 quarts

58.	70%	90%	75%	25 gal

Solution 1: 18.75 gallons; Solution 2: 6.25 gallons

59. *Number of Stamps* You have 100 stamps that have a total value of $31.02. Some of the stamps are worth 24¢ each and the others are worth 37¢ each. How many stamps of each type do you have? 46 stamps at 24¢, 54 stamps at 37¢

60. *Number of Stamps* You have 20 stamps that have a total value of $6.62. Some of the stamps are worth 24¢ each and others are worth 37¢ each. How many stamps of each type do you have? 6 stamps at 24¢, 14 stamps at 37¢

61. *Number of Coins* A person has 20 coins in nickels and dimes with a combined value of $1.60. Determine the number of coins of each type. 8 nickels, 12 dimes

62. *Number of Coins* A person has 50 coins in dimes and quarters with a combined value of $7.70. Determine the number of coins of each type. 32 dimes, 18 quarters

63. *Nut Mixture* A grocer mixes two kinds of nuts that cost $2.49 and $3.89 per pound to make 100 pounds of a mixture that costs $3.47 per pound. How many pounds of each kind of nut are put into the mixture? See Example 5. 30 pounds at $2.49 per pound, 70 pounds at $3.89 per pound

64. *Flower Order* A floral shop receives a $384 order for roses and carnations. The prices per dozen for the roses and carnations are $18 and $12, respectively. The order contains twice as many roses as carnations. How many of each type of flower are in the order? 16 dozen roses, 8 dozen carnations

65. *Antifreeze* The cooling system in a truck contains 4 gallons of coolant that is 30% antifreeze. How much must be withdrawn and replaced with 100% antifreeze to bring the coolant in the system to 50% antifreeze? $\frac{8}{7}$ gallons

66. *Ticket Sales* Ticket sales for a play total $1700. The number of tickets sold to adults is three times the number sold to children. The prices of the tickets for adults and children are $5 and $2, respectively. How many of each type were sold? 100 children, 300 adults

67. *Investment Mixture* You divided $6000 between two investments earning 7% and 9% simple interest. During 1 year the two accounts earned $500. How much did you invest in each account? See Example 7. $2000 at 7%, $4000 at 9%

68. *Investment Mixture* You divided an inheritance of $30,000 into two investments earning 8.5% and 10% simple interest. During 1 year, the two accounts earned $2700. How much did you invest in each account? $20,000 at 8.5%, $10,000 at 10%

69. *Interpreting a Table* An agricultural corporation must purchase 100 tons of cattle feed. The feed is to be a mixture of soybeans, which cost $200 per ton, and corn, which costs $125 per ton.

(a) Complete the table, where x is the number of tons of corn in the mixture.

Corn, x	Soybeans, $100 - x$	Price per ton of the mixture
0	100	$200
20	80	$185
40	60	$170
60	40	$155
80	20	$140
100	0	$125

(b) How does an increase in the number of tons of corn affect the number of tons of soybeans in the mixture? Decreases

(c) How does an increase in the number of tons of corn affect the price per ton of the mixture? Decreases

(d) If there were equal weights of corn and soybeans in the mixture, how would the price of the mixture relate to the price of each component? Average of the two prices

70. *Interpreting a Table*   A metallurgist is making 5 ounces of an alloy of metal A, which costs $52 per ounce, and metal B, which costs $16 per ounce.

(a) Complete the table, where x is the number of ounces of metal A in the alloy.

Metal A, x	Metal B, $5 - x$	Price per ounce of the alloy
0	5	$16.00
1	4	$23.20
2	3	$30.40
3	2	$37.60
4	1	$44.80
5	0	$52.00

(b) How does an increase in the number of ounces of metal A in the alloy affect the number of ounces of metal B in the alloy?   Decreases

(c) How does an increase in the number of ounces of metal A in the alloy affect the price of the alloy?   Increases

(d) If there were equal amounts of metal A and metal B in the alloy, how would the price of the alloy relate to the price of each of the components?   Average of the two prices

71. *Work Rate*   You can mow a lawn in 2 hours using a riding mower, and in 3 hours using a push mower. Using both machines together, how long will it take you and a friend to mow the lawn? See Example 8. $1\frac{1}{5}$ hours

72. *Work Rate*   One person can complete a typing project in 6 hours, and another can complete the same project in 8 hours. If they both work on the project, in how many hours can it be completed? $3\frac{3}{7}$ hours

73. *Work Rate*   One worker can complete a task in m minutes while a second can complete the task in $9m$ minutes. Show that by working together they can complete the task in $t = \frac{9}{10}m$ minutes.   Answers will vary.

74. *Work Rate*   One worker can complete a task in h hours while a second can complete the task in $3h$ hours. Show that by working together they can complete the task in $t = \frac{3}{4}h$ hours.   Answers will vary.

75. *Age Problem*   A mother was 30 years old when her son was born. How old will the son be when his age is $\frac{1}{3}$ his mother's age?   15 years

76. *Age Problem*   The difference in age between a father and daughter is 32 years. Determine the age of the father when his age is twice that of his daughter.   64 years

77. *Poll Results*   One thousand people were surveyed in an opinion poll. Candidates A and B received approximately the same number of votes. Candidate C received twice as many votes as either of the other two candidates. How many votes did each candidate receive?   Candidate A: 250 votes, Candidate B: 250 votes, Candidate C: 500 votes

78. *Poll Results*   One thousand people were surveyed in an opinion poll. The numbers of votes for candidates A, B, and C had ratios 5 to 3 to 2, respectively. How many people voted for each candidate?   Candidate A: 500 votes, Candidate B: 300 votes, Candidate C: 200 votes

Explaining Concepts

79. *Writing*   In your own words, describe the units of measure used for perimeter, area, and volume. Give examples of each.
Perimeter: linear units—inches, feet, meters; Area: square units—square inches, square meters; Volume: cubic units—cubic inches, cubic centimeters

80. *Writing*   If the height of a triangle is doubled, does the area of the triangle double? Explain.   Yes. $A = \frac{1}{2}bh$. If h is doubled, you have $A = \frac{1}{2}b(2h) = 2\left(\frac{1}{2}bh\right)$.

81. *Writing*   If the radius of a circle is doubled, does its circumference double? Does its area double? Explain.

82. *Writing*   It takes you 4 hours to drive 180 miles. Explain how to use mental math to find your average speed. Then explain how your method is related to the formula $d = rt$.   Divide by 2 to obtain 90 miles per 2 hours and divide by 2 again to obtain 45 miles per hour.

$$r = \frac{d}{t} = \frac{180}{4} = 45 \text{ miles per hour}$$

83. It takes you 5 hours to complete a job. What portion do you complete each hour? $\frac{1}{5}$

81. The circumference would double; the area would quadruple.
Circumference: $C = 2\pi r$, Area: $A = \pi r^2$
If r is doubled, you have $C = 2\pi(2r) = 2(2\pi r)$
and $A = \pi(2r)^2 = 4\pi r^2$.

3.6 Linear Inequalities

Superstock

Why You Should Learn It

Linear inequalities can be used to model and solve real-life problems. For instance, in Exercise 79 on page 191, you will use an inequality to determine a budget for a class trip to an amusement park.

What You Should Learn

1 Graph the solution set of an inequality on the real number line and determine whether a given value is a solution of an inequality.

2 Become familiar with the properties of inequalities.

3 Solve a linear inequality and graph its solution set.

4 Translate verbal statements into linear inequalities and solve application problems.

Inequalities and Their Graphs

1 Graph the solution set of an inequality on the real number line and determine whether a given value is a solution of an inequality.

In this section you will study **algebraic inequalities,** which are inequalities that contain one or more variable terms. Here are some examples.

$$x \leq 3, \quad x \geq -2, \quad x - 5 < 2, \quad \text{and} \quad 5x - 7 < 3x + 9$$

Each of these inequalities is a **linear inequality** in the variable x because the (implied) exponent of x is 1.

As with an equation, you can **solve an inequality** in the variable x by finding all values of x for which the inequality is true. Such values are **solutions** and are said to **satisfy** the inequality. The **solution set** of an inequality is the set of all real numbers that are solutions of the inequality.

Often, the solution set of an inequality will consist of infinitely many real numbers. To get a visual image of the solution set, it is helpful to sketch its **graph** on the real number line. For instance, the graph of the solution set of $x < 2$ consists of all points on the real number line that are to the left of 2. A parenthesis is used to *exclude* an endpoint from the solution interval. A square bracket is used to *include* an endpoint in the solution interval. This is illustrated in the following example.

Example 1   Graphs of Inequalities

Inequality	Graph of Solution Set	Verbal Description
a. $x < 2$	*(graph: open parenthesis at 2, shaded to the left; number line marked −1, 0, 1, 2, 3, 4)*	x is less than 2.
b. $x > 2$	*(graph: open parenthesis at 2, shaded to the right; number line marked −1, 0, 1, 2, 3, 4)*	x is greater than 2.
c. $x \geq -2$	*(graph: square bracket at −2, shaded to the right; number line marked −3, −2, −1, 0, 1, 2)*	x is greater than or equal to −2.
d. $x \leq 0$	*(graph: square bracket at 0, shaded to the left; number line marked −2, −1, 0, 1, 2, 3)*	x is less than or equal to 0.

The inequalities in Example 1 are considered simple inequalities because they contain only one inequality symbol. Inequalities that contain two inequality symbols are **compound inequalities.** A compound inequality is two inequalities in one. For instance, the compound inequality $0 < x < 4$ means $0 < x$ *and* $x < 4$ and is read as "x is greater than zero *and* less than 4."

Remind students that the following pairs of inequalities are equivalent.

$x < 7$ and $7 > x$
$b > -4$ and $-4 < b$
$c \geq 9$ and $9 \leq c$
$-3 \leq y \leq 5$ and $5 \geq y \geq -3$

Example 2 Graphs of Compound Inequalities

Inequality	Graph of Solution Set	Verbal Description
a. $-1 \leq x \leq 2$		x is greater than or equal to -1 *and* less than or equal to 2.
b. $2 \leq x < 5$		x is greater than or equal to 2 *and* less than 5.
c. $-3 < x \leq -1$		x is greater than -3 *and* less than or equal to -1.

The procedure for checking solutions of inequalities is similar to checking solutions of equations, as shown in Example 3.

Example 3 Checking a Solution of an Inequality

Determine whether (a) $x = 9$ or (b) $x = -5$ is a solution of $-3x + 8 \leq 11$.

Solution

a.
$-3x + 8 \leq 11$	Write original inequality.
$-3(9) + 8 \leq 11$	Substitute 9 for x.
$-27 + 8 \leq 11$	Multiply.
$-19 \leq 11$	Solution checks. ✓

Because -19 is less than 11, you can conclude that $x = 9$ *is* a solution of the original inequality.

b.
$-3x + 8 \leq 11$	Write original inequality.
$-3(-5) + 8 \leq 11$	Substitute -5 for x.
$15 + 8 \leq 11$	Multiply.
$23 \nleq 11$	Solution does not check. ✗

Because 23 is not less than or equal to 11, you can conclude that $x = -5$ *is not* a solution of the original inequality.

Technology: Tip

Linear inequalities can be graphed using a graphing calculator. The inequality $x > -2$ is shown in the graph below. Notice that the graph appears above the x-axis. Consult the user's manual of your graphing calculator for instructions.

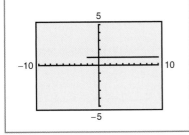

② Become familiar with the properties of inequalities.

Properties of Inequalities

The procedures for solving linear inequalities in one variable are much like those for solving linear equations. To isolate the variable, you can use the **properties of inequalities.** These properties are similar to the properties of equality, but there are two important exceptions. *When both sides of an inequality are multiplied or divided by a negative number, the direction of the inequality symbol must be reversed.* Here is an example.

$$-2 < 5 \qquad \text{Original inequality}$$

$$(-3)(-2) > (-3)(5) \qquad \text{Multiply each side by } -3 \text{ and reverse inequality.}$$

$$6 > -15 \qquad \text{Simplify.}$$

Two inequalities that have the same solution set are called **equivalent.** The list below describes operations that can be used to create equivalent inequalities.

Properties of Inequalities

Let a, b, and c be real numbers, variables, or algebraic expressions.

Property	Verbal Description	Algebraic Description
Addition	Add the same quantity to each side.	If $a < b$, then $a + c < b + c$.
Subtraction	Subtract the same quantity from each side.	If $a < b$, then $a - c < b - c$.
Multiplication	Multiply each side by a *positive* quantity.	If $a < b$ and c is positive, then $ac < bc$.
	Multiply each side by a *negative* quantity and reverse the inequality symbol.	If $a < b$ and c is negative, then $ac > bc$.
Division	Divide each side by a *positive* quantity.	If $a < b$ and c is positive, then $\dfrac{a}{c} < \dfrac{b}{c}$.
	Divide each side by a *negative* quantity and reverse the inequality symbol.	If $a < b$ and c is negative, then $\dfrac{a}{c} > \dfrac{b}{c}$.
Transitive	Consider three quantities of which the first is less that the second, and the second is less than the third. It follows that the first quantity must be less than the third quantity.	If $a < b$ and $b < c$, then $a < c$.

Each of the properties above is true if the symbol $<$ is replaced by $\leq$ and/or the symbol $>$ is replaced by $\geq$. Note that you cannot multiply or divide each side of an inequality by zero.

3 Solve a linear inequality and graph its solution set.

Solving Inequalities

The solution set of a linear inequality can be written in set notation. For the solution $x > 1$, the set notation is $\{x \mid x > 1\}$ and is read "the set of all x such that x is greater than 1."

In Examples 6 and 7, pay special attention to the steps in which the inequality symbol is reversed. Remember that when you multiply or divide an inequality by a negative number, you must reverse the inequality symbol.

Example 4   Solving a Linear Inequality

Solve and graph the inequality $x + 5 < 8$.

Solution

$x + 5 < 8$	Write original inequality.
$x + 5 - 5 < 8 - 5$	Subtract 5 from each side.
$x < 3$	Solution set

The solution set is $x < 3$ or, in set notation, $\{x \mid x < 3\}$. The graph of the solution set is shown in Figure 3.7.

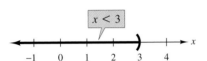

Figure 3.7   All real numbers that are less than 3

Checking the solution set of an inequality is not as simple as checking the solution set of an equation. (There are usually too many x-values to substitute back into the original inequality.) You can, however, get an indication of the validity of a solution set by substituting a few convenient values of x. For instance, in Example 4, the solution of $x + 5 < 8$ was found to be $x < 3$. Try checking that $x = 0$ satisfies the original inequality, whereas $x = 4$ does not.

Example 5   Solving a Linear Inequality

Ask students to compare the solving of this inequality with the solving of the equation $3y - 1 = -13$.

Solve and graph the inequality $3y - 1 \le -13$.

Solution

$3y - 1 \le -13$	Write original inequality.
$3y - 1 + 1 \le -13 + 1$	Add 1 to each side.
$3y \le -12$	Combine like terms.
$\dfrac{3y}{3} \le \dfrac{-12}{3}$	Divide each side by (positive) 3.
$y \le -4$	Solution set

The solution set is $y \le -4$ or, in set notation, $\{y \mid y \le -4\}$. The graph of the solution set is shown in Figure 3.8.

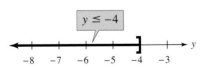

Figure 3.8   All real numbers that are less than or equal to -4

Forgetting to reverse the inequality symbol is a common student error in solving inequalities.

Example 6 Solving a Linear Inequality

Solve and graph the inequality $3(2x - 6) < 10x + 2$.

Solution

$3(2x - 6) < 10x + 2$	Write original inequality.
$6x - 18 < 10x + 2$	Distributive Property
$6x - 10x - 18 < 2$	Subtract $10x$ from each side.
$-4x - 18 < 2$	Combine like terms.
$-4x < 2 + 18$	Add 18 to each side.
$-4x < 20$	Combine like terms.
$x > -5$	Divide each side by -4 and reverse inequality.

The solution set is $x > -5$ or, in set notation, $\{x \mid x > -5\}$. The graph of the solution set is shown in Figure 3.9.

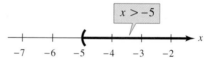

Figure 3.9   All real numbers that are greater than -5

Example 7 Solving a Linear Inequality

Solve and graph the inequality $1 - \dfrac{3x}{2} \geq x - 4$.

Solution

$1 - \dfrac{3x}{2} \geq x - 4$	Write original inequality.
$2 - 3x \geq 2x - 8$	Multiply each side by 2.
$-3x \geq 2x - 10$	Subtract 2 from each side.
$-5x \geq -10$	Subtract $2x$ from each side.
$x \leq 2$	Divide each side by -5 and reverse inequality.

The solution set is $x \leq 2$ or, in set notation, $\{x \mid x \leq 2\}$. The graph of the solution set is shown in Figure 3.10.

Figure 3.10   All real numbers that are less than or equal to 2

④ Translate verbal statements into linear inequalities and solve application problems.

Students frequently need to translate such phrases into inequalities when they are solving application problems.

Applications

Linear inequalities in real-life problems arise from statements that involve key phrases such as "at least," "no more than," "minimum value," and so on. Study the meanings of the key phrases in the next example.

Example 8   Translating Verbal Statements

Verbal Statement	Inequality
a. x is at most 2.	$x \le 2$
b. x is no more than 2.	$x \le 2$
c. x is at least 2.	$x \ge 2$
d. x is more than 2.	$x > 2$
e. x is less than 2.	$x < 2$
f. x is a minimum of 2.	$x \ge 2$

When translating inequalities, remember that "at most" means "less than or equal to," and "at least" means "greater than or equal to." Also, be sure to distinguish between the *sum* "2 more than a number" $(x + 2)$ and the *inequality* "2 is more than a number" $(2 > x)$. It is generally preferable to read an inequality from left to right.

To solve real-life problems involving linear inequalities, you can use the same "verbal-model approach" you use with linear equations.

Example 9   Course Grade

You are taking a history course in which your grade is based on six 100-point exams. To earn an A in the course, you must have a total of at least 90% of the points. On the first five exams, your scores were 85, 92, 88, 96, and 87. How many points do you have to obtain on the sixth test in order to earn an A in the course?

Solution

Verbal Model: Total points $\ge$ 90% of 600

Labels: Score for sixth exam $= x$ (points)
Total points $= (85 + 92 + 88 + 96 + 87) + x$ (points)

Inequality: $(85 + 92 + 88 + 96 + 87) + x \ge 0.90(600)$

$$448 + x \ge 540$$

$$x \ge 92$$

You must get at least 92 points on the sixth exam to earn an A in the course. Check this solution in the original statement of the problem.

Example 10   Finding the Maximum Width of a Package

An overnight delivery service will not accept any package whose combined length and girth (perimeter of a cross section) exceeds 132 inches. You are sending a rectangular package that has square cross sections. The length of the package is 68 inches. What is the maximum width of the sides of its square cross section?

Solution

Begin by making a sketch. In Figure 3.11, notice that the length of the package is 68 inches, and each side is x inches wide.

Figure 3.11

Verbal Model: | Length | + | Girth | ≤ | 132 inches |

Labels: Width of a side $= x$ (inches)
             Length $= 68$ (inches)
             Girth $= 4x$ (inches)

Inequality: $68 + 4x \leq 132$

$$4x \leq 64$$

$$x \leq 16$$

The width of each side of the package must be less than or equal to 16 inches. Check this in the original statement of the problem.

Example 11   Comparing Costs

A subcompact car can be rented from Company A for $240 per week with no extra charge for mileage. A similar car can be rented from Company B for $100 per week, plus an additional 25 cents for each mile driven. How many miles must you drive in a week so that the rental fee for Company B is more than that for Company A?

Solution

Miles driven	Company A	Company B
557	$240.00	$239.25
558	$240.00	$239.50
559	$240.00	$239.75
560	$240.00	$240.00
561	$240.00	$240.25
562	$240.00	$240.50
563	$240.00	$240.75

Verbal Model: | Weekly cost for B | > | Weekly cost for A |

Labels: Number of miles driven in 1 week $= m$ (miles)
             Weekly cost for A $= 240$ (dollars)
             Weekly cost for B $= 100 + 0.25m$ (dollars)

Inequality: $100 + 0.25m > 240$

$$0.25m > 140$$

$$m > 560$$

So, the car from Company B is more expensive if you drive more than 560 miles in a week. The table at the left confirms this conclusion.

3.6 Exercises

Review   Concepts, Skills, and Problem Solving

Keep mathematically in shape by doing these exercises *before* the problems of this section.

Properties and Definitions

1. Identify the property of real numbers illustrated by $3x(x + 1) = 3x^2 + 3x$.   Distributive Property

2. Complete the Associative Property:
$(x + 2) - 4 = \boxed{x + (2 - 4)}$.

3. If $a < 0$, then $|a| = \boxed{-a}$.

4. If $a < 0$ and $b > 0$, then $a \cdot b \boxed{<} 0$.

In Exercises 5–8, place the correct inequality symbol between the two real numbers.

5. $-\frac{1}{2} \boxed{>} -7$

6. $-\frac{1}{3} \boxed{<} -\frac{1}{6}$

7. $-\pi \boxed{<} -3$

8. $-6 \boxed{>} -\frac{13}{2}$

Graphs and Models

▲ *Geometry*   In Exercises 9 and 10, write expressions for the perimeter and area of the triangle. Then simplify the expressions.

9.

$x^2 + 2x$ x^2

$5x - 3$

Perimeter: $2x^2 + 7x - 3$
Area: $\frac{5}{2}x^3 - \frac{3}{2}x^2$

10.

$2y - 1$ y $2y - 1$

$2y^2 - 4y + 2$

Perimeter: $2y^2$
Area: $y^3 - 2y^2 + y$

Problem Solving

11. *Profit*   A small software company had a first-quarter loss of \$312,500, a second-quarter profit of \$275,500, a third-quarter profit of \$297,750, and a fourth-quarter profit of \$71,300. What was the profit for the year?   \$332,050

12. *Average Speed*   A family on vacation traveled 371 miles in 7 hours. Determine their average speed.
53 miles per hour

Developing Skills

In Exercises 1–10, write a verbal description of the inequality and sketch its graph. See Examples 1 and 2.
See Additional Answers.

1. $x \geq 3$ x is greater than or equal to 3.

2. $z > 8$ z is greater than 8.

3. $x \leq 10$ x is less than or equal to 10.

4. $t < 0$ t is less than 0.

5. $y < -9$ y is less than -9.

6. $x \leq -2$ x is less than or equal to -2.

7. $5 \leq z \leq 10$ z is greater than or equal to 5 *and* less than or equal to 10.

8. $-3 < x < 4$ x is greater than -3 *and* less than 4.

9. $-\frac{3}{2} < y \leq 5$ y is greater than $-\frac{3}{2}$ *and* less than or equal to 5.

10. $-3 \geq t > -3.8$ t is greater than -3.8 *and* less than or equal to -3.

In Exercises 11–20, determine whether the value of x is a solution of the inequality. See Example 3.

Inequality	*Values*

11. $5x - 12 > 0$ (a) $x = 3$ (b) $x = -3$
   Yes   No
   (c) $x = \frac{5}{2}$ Yes   (d) $x = \frac{3}{2}$ No

12. $2x + 1 < 3$ (a) $x = 0$ Yes   (b) $x = 4$ No
   (c) $x = -\frac{2}{5}$ Yes   (d) $x = \frac{1}{2}$ Yes

13. $3 - \frac{1}{2}x > 0$ (a) $x = 10$ No   (b) $x = 6$ No
   (c) $x = -\frac{3}{4}$ Yes   (d) $x = 0$ Yes

14. $\frac{2}{3}x + 4 < 6$ (a) $x = 7$ No   (b) $x = 0$ Yes
   (c) $x = -\frac{1}{2}$ Yes   (d) $x = 3$ No

15. $5(x - 2) + 1 < 12$ (a) $x = 4$ Yes   (b) $x = 1$ Yes
   (c) $x = 5$ (d) $x = -3$
   No   Yes

Inequality   *Values*

16. $3(x + 5) - 4 > 2$ (a) $x = 3$ Yes   (b) $x = 0$ Yes

(c) $x = -4$ (d) $x = -10$
         No   No

17. $15 - (x + 8) \geq 13$ (a) $x = 0$ (b) $x = -6$
                                      No   Yes

(c) $x = 2$ (d) $x = -10$
         No   Yes

18. $9 - (x + 3) \leq 10$ (a) $x = -4$ Yes (b) $x = 4$ Yes

(c) $x = 0$ (d) $x = -6$
         Yes   No

19. $5x + 3 \leq x - 5$ (a) $x = 1$ (b) $x = -2$
                                      No   Yes

(c) $x = -1$ No (d) $x = 2$ No

20. $4 - 3x \geq x + 12$ (a) $x = 0$ No   (b) $x = 3$ No

(c) $x = -2$ (d) $x = -4$
         Yes   Yes

In Exercises 21–26, match the inequality with its graph.
[The graphs are labeled (a), (b), (c), (d), (e), and (f).]

(a)

(b)

(c)

(d)

(e)

(f)

21. $x < 4$ b   **22.** $x \geq 6$ f

23. $-3 \leq x < 2$ c   **24.** $-\frac{1}{2} < x \leq \frac{5}{2}$ a

25. $8 > x > \frac{3}{2}$ d **26.** $5 \geq x \geq -5$ e

In Exercises 27–68, solve and graph the inequality. See
Examples 4–7.   See Additional Answers.

27. $t - 3 \geq 2$ $t \geq 5$ **28.** $t + 1 < 6$ $t < 5$

29. $x + 4 \leq 6$ $x \leq 2$ **30.** $z - 2 > 0$ $z > 2$

31. $4x < 12$ $x < 3$ **32.** $2x > 3$ $x > \frac{3}{2}$

33. $-10x < 40$ $x > -4$ **34.** $-6x > 18$ $x < -3$

35. $\frac{2}{3}x \leq 12$ $x \leq 18$ **36.** $\frac{1}{5}x > 3$ $x > 15$

37. $-\frac{5}{8}x \geq 10$ $x \leq -16$ **38.** $-\frac{3}{4}t \leq 9$ $t \geq -12$

39. $3x \geq -\frac{4}{5}$ $x \geq -\frac{4}{15}$ **40.** $7x < -\frac{1}{2}$ $x < -\frac{1}{14}$

41. $2x - 5 > 7$ $x > 6$ **42.** $4z - 1 \leq 5$ $z \leq \frac{3}{2}$

43. $3x + 2 \leq 14$ $x \leq 4$ **44.** $8y + 4 > 0$ $y > -\frac{1}{2}$

45. $4 - 2x < 3$ $x > \frac{1}{2}$ **46.** $14 - 3x > 5$ $x < 3$

47. $12 - x > 4$ $x < 8$ **48.** $18 - y \leq 5$ $y \geq 13$

49. $3x + 9 < 2x$
$x < -9$

50. $12 + 5x \geq 3x$
$x \geq -6$

51. $7t + 9 < 14 + 6t$
$t < 5$

52. $25x + 4 \leq 10x + 19$
$x \leq 1$

53. $2x - 5 > -x + 6$
$x > \frac{11}{3}$

54. $6 - x \leq 3x + 4$
$x \geq \frac{1}{2}$

55. $2(x + 7) > 12$
$x > -1$

56. $9(y - 4) \leq 36$
$y \leq 8$

57. $-3(x + 11) \leq 6$
$x \geq -13$

58. $-7(z + 4) > 14$
$z < -6$

59. $-2(z + 1) \geq 3(z + 1)$ $z \leq -1$

60. $8(t - 3) < 4(t - 3)$ $t < 3$

61. $3(x + 1) \geq 2(x + 5)$ $x \geq 7$

62. $8(z - 2) < 4(z + 1)$ $z < 5$

63. $10(1 - y) < -4(y - 2)$ $y > \frac{1}{3}$

64. $6(3 - z) \geq 5(3 + z)$ $z \leq \frac{3}{11}$

65. $\frac{x}{4} + \frac{1}{2} > 0$ $x > -2$ **66.** $\frac{y}{4} - \frac{5}{8} < 2$ $y < \frac{21}{2}$

67. $\frac{x}{5} - \frac{x}{2} \leq 1$ $x \geq -\frac{10}{3}$ **68.** $\frac{x}{3} + \frac{x}{4} \geq 1$ $x \geq \frac{12}{7}$

In Exercises 69–76, translate the verbal statement into a
linear inequality. See Example 8.

69. x is nonnegative.   **70.** P is no more than 2.
$x \geq 0$ $P \leq 2$

71. y is more than -6.   **72.** z is at least 3.
$y > -6$ $z \geq 3$

73. x is at least 4. $x \geq 4$ **74.** t is less than 8. $t < 8$

75. x is greater than 0 and less than or equal to 6.
$0 < x \leq 6$

76. x is greater than -4 and less than or equal to 3.
$-4 < x \leq 3$

Solving Problems

77. *Astronomy* Mars is farther from the sun than Venus, and Venus is farther from the sun than Mercury. What can be said about the relationship between the distances of Mars and Mercury from the sun? Identify the property of inequalities that is demonstrated. Mars is farther from the sun than Mercury. Transitive Property

78. *Budgets* The personnel department's budget is less than the finance department's budget, and the finance department's budget is less than the research and development department's budget. What can you say about the relationship between the budgets of the personnel department and the research and development department? Identify the property of inequalities that is demonstrated. The personnel department's budget is less than the research and development department's budget. Transitive Property

79. *Budget* You have $225 budgeted for your class trip to an amusement park. The ticket for admission to the park is $80. To stay within your budget, all other costs must total no more than what amount? $145

80. *Consumer Awareness* The cost of a cellular phone call is $0.46 for the first minute and $0.31 for each additional minute. The total cost of the call cannot exceed $4. Find the interval of time that is available for the call. $0 \le t \le 12.4$ minutes

81. *Cargo Weight* The weight of a truck is 4350 pounds. The legal gross weight of the loaded truck is 6000 pounds. Find an interval for the number of bushels of grain that the truck can haul if each bushel weighs 48 pounds. $0 \le x \le 34$ bushels

82. *Operating Costs* A utility company has a fleet of vans. The annual operating cost C (in dollars) per van is $C = 0.45m + 3200$, where m is the number of miles traveled by a van in a year. What is the maximum number of miles that will yield an annual operating cost per van that is less than $12,000? $m \le 19{,}555$ miles

83. *Comparing Costs* You can rent a minivan from Company A for $270 per week with unlimited mileage. A similar minivan can be rented from Company B for $180 a week plus an additional 25 cents for each mile driven. How many miles must you drive in a week so that the rental fee for Company B is more than that for Company A? $m > 360$ miles

84. ▲ *Geometry* The lengths of the sides of the triangle in the figure are a, b, and c. Find the inequality that relates $a + b$ and c. $a + b > c$

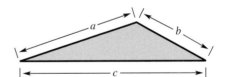

Explaining Concepts

85. ◐ Answer part (h) of Motivating the Chapter on page 122.

86. *Writing* Is dividing each side of an inequality by 5 the same as multiplying each side by $\frac{1}{5}$? Explain. Yes. Multiplication and Division Properties of Inequalities.

True or False? In Exercises 87–90, determine whether the statement is true or false. Justify your answer.

87. The inequality $x + 6 > 0$ is equivalent to $x > -6$. True. Subtract 6 from each side of the inequality.

88. The statement that z is nonnegative is equivalent to the inequality $z > 0$. False. $z \ge 0$

89. The statement that u is at least 10 is equivalent to the inequality $u \ge 10$. True. u is equal to 10 or greater than 10.

90. The inequality $-\frac{1}{2}x + 6 > 0$ is equivalent to $x > 12$. False.
$$-\tfrac{1}{2}x + 6 > 0$$
$$-\tfrac{1}{2}x > -6$$
$$x < 12$$

91. How many numbers are in the solution set of a linear inequality? Give an example. Infinitely many numbers. $x - 3 > 0 \to x > 3$

92. *Writing* Compare solving linear equations with solving linear inequalities. The process is similar except that the direction of the inequality must be reversed if each side of the inequality is multiplied or divided by a negative number.

What Did You Learn?

Key Terms

linear equation, *p. 124*

consecutive integers, *p. 130*

cross-multiplication, *p. 140*

markup, *p. 150*

discount, *p. 151*

ratio, *p. 158*

unit price, *p. 160*

proportion, *p. 161*

mixture problems, *p. 173*

work-rate problems, *p. 175*

linear inequality, *p. 182*

solution set, *p. 182*

Key Concepts

3.1 ◯ Solving a linear equation

Solve a linear equation using inverse operations to isolate the variable.

3.1 ◯ Expressions for special types of integers

Let *n* be an integer.

1. $2n$ denotes an *even* integer.

2. $2n - 1$ and $2n + 1$ denote *odd* integers.

3. The set $\{n, n + 1, n + 2\}$ denotes three *consecutive* integers.

3.2 ◯ Solving equations containing symbols of grouping

1. Remove symbols of grouping from each side by using the Distributive Property.

2. Combine like terms.

3. Isolate the variable in the usual way using properties of equality.

4. Check your solution in the original equation.

3.2 ◯ Equations involving fractions or decimals

1. Clear an equation of fractions by multiplying each side by the least common multiple (LCM) of the denominators.

2. Use cross-multiplication to solve a linear equation that equates two fractions. That is, if

$$\frac{a}{b} = \frac{c}{d}, \text{ then } a \cdot d = b \cdot c.$$

3. To solve a linear equation with decimal coefficients, multiply each side by a power of 10 that converts all decimal coefficients to integers.

3.3 ◯ The percent equation

The percent equation $a = p \cdot b$ compares two numbers.

b is the base number.

p is the percent in decimal form.

a is the number being compared to *b*.

3.3 ◯ Markups and discounts

1. A markup is the difference between the cost (the amount the retailer pays) and the price (the amount the consumer pays).

2. A discount is the amount off the list price (what the consumer pays).

3.3 ◯ Guidelines for solving word problems

1. Write a *verbal model* that describes the problem.

2. Assign *labels* to fixed quantities and variable quantities.

3. Rewrite the verbal model as an *algebraic equation* using the assigned labels.

4. *Solve* the resulting algebraic equation.

5. *Check* to see that your solution satisfies the original problem as stated.

3.4 ◯ Solving a proportion

A proportion equates two ratios.

If $\dfrac{a}{b} = \dfrac{c}{d}$, then $ad = bc$.

3.6 ◯ Properties of inequalities

Let *a*, *b*, and *c* be real numbers, variables, or algebraic expressions.

Addition: If $a < b$, then $a + c < b + c$.

Subtraction: If $a < b$, then $a - c < b - c$.

Multiplication: If $a < b$ and $c > 0$, then $ac < bc$.

If $a < b$ and $c < 0$, then $ac > bc$.

Division: If $a < b$ and $c > 0$, then $\dfrac{a}{c} < \dfrac{b}{c}$.

If $a < b$ and $c < 0$, then $\dfrac{a}{c} > \dfrac{b}{c}$.

Transitive: If $a < b$ and $b < c$, then $a < c$.

3.6 ◯ Solving a linear inequality or a compound inequality

Solve a linear inequality by performing inverse operations on all parts of the inequality.

Review Exercises

3.1 Solving Linear Equations

① Solve linear equations in standard form.

In Exercises 1–4, solve the equation and check your solution.

1. $2x - 10 = 0$ 5
2. $12y + 72 = 0$ −6
3. $-3y - 12 = 0$ −4
4. $-7x + 21 = 0$ 3

② Solve linear equations in nonstandard form.

In Exercises 5–18, solve the equation and check your solution.

5. $x + 10 = 13$ 3
6. $x - 3 = 8$ 11
7. $5 - x = 2$ 3
8. $3 = 8 - x$ 5
9. $10x = 50$ 5
10. $-3x = 21$ −7
11. $8x + 7 = 39$ 4
12. $12x - 5 = 43$ 4
13. $24 - 7x = 3$ 3
14. $13 + 6x = 61$ 8
15. $15x - 4 = 16$ $\frac{4}{3}$
16. $3x - 8 = 2$ $\frac{10}{3}$
17. $\dfrac{x}{5} = 4$ 20
18. $-\dfrac{x}{14} = \dfrac{1}{2}$ −7

③ Use linear equations to solve application problems.

19. *Hourly Wage*   Your hourly wage is $8.30 per hour plus 60 cents for each unit you produce. How many units must you produce in an hour so that your hourly wage is $15.50?   12 units

20. *Consumer Awareness*   A long-distance carrier's connection fee for a phone call is $1.25. There is also a charge of $0.10 per minute. How long was a phone call that cost $3.05?   18 minutes

21. ▲ *Geometry*   The perimeter of a rectangle is 260 meters. The length is 30 meters greater than its width. Find the dimensions of the rectangle.
80 × 50 meters

22. ▲ *Geometry*   A 10-foot board is cut so that one piece is 4 times as long as the other. Find the length of each piece.   2 feet, 8 feet

3.2 Equations That Reduce to Linear Form

① Solve linear equations containing symbols of grouping.

In Exercises 23–28, solve the equation and check your solution.

23. $3x - 2(x + 5) = 10$ 20

24. $4x + 2(7 - x) = 5$ $-\frac{9}{2}$
25. $2(x + 3) = 6(x - 3)$ 6
26. $8(x - 2) = 3(x + 2)$ $\frac{22}{5}$
27. $7 - [2(3x + 4) - 5] = x - 3$ 1
28. $14 + [3(6x - 15) + 4] = 5x - 1$ 2

② Solve linear equations involving fractions.

In Exercises 29–36, solve the equation and check your solution.

29. $\frac{2}{3}x - \frac{1}{6} = \frac{9}{2}$ 7
30. $\frac{1}{8}x + \frac{3}{4} = \frac{5}{2}$ 14
31. $\dfrac{x}{3} - \dfrac{1}{9} = 2$ $\frac{19}{3}$
32. $\dfrac{1}{2} - \dfrac{x}{8} = 7$ −52
33. $\dfrac{u}{10} + \dfrac{u}{5} = 6$ 20
34. $\dfrac{x}{3} + \dfrac{x}{5} = 1$ $\frac{15}{8}$
35. $\dfrac{2x}{9} = \dfrac{2}{3}$ 3
36. $\dfrac{5y}{13} = \dfrac{2}{5}$ $\frac{26}{25}$

③ Solve linear equations involving decimals.

In Exercises 37–40, solve the equation. Round your answer to two decimal places.

37. $5.16x - 87.5 = 32.5$ 23.26
38. $2.825x + 3.125 = 12.5$ 3.32
39. $\dfrac{x}{4.625} = 48.5$ 224.31
40. $5x + \dfrac{1}{4.5} = 18.125$ 3.58

3.3 Problem Solving with Percents

① Convert percents to decimals and fractions and convert decimals and fractions to percents.

In Exercises 41 and 42, complete the table showing the equivalent forms of a percent.

Percent	Parts out of 100	Decimal	Fraction
41. 35%	35	0.35	$\frac{7}{20}$
42. 80%	80	0.80	$\frac{4}{5}$

② Solve linear equations involving percents.

In Exercises 43–48, solve the percent equation.

43. What number is 125% of 16?   20
44. What number is 0.8% of 3250?   26

45. 150 is $37\frac{1}{2}\%$ of what number?   400

46. 323 is 95% of what number?   340

47. 150 is what percent of 250?   60%

48. 130.6 is what percent of 3265?   4%

③ Solve application problems involving markups and discounts.

49. *Selling Price*   An electronics store uses a markup rate of 62% on all items. The cost of a CD player is $48. What is the selling price of the CD player?
$77.76

50. *Sale Price*   A clothing store advertises 30% off the list price of all sweaters. A turtleneck sweater has a list price of $120. What is the sale price? $84

51. *Sales*   The sales (in millions) for the Yankee Candle Company in the years 2000 and 2001 were $338.8 and $379.8, respectively. Determine the percent increase in sales from 2000 to 2001.   (Source: The Yankee Candle Company)
12.1%

52. *Price Increase*   The manufacturer's suggested retail price for a car is $18,459. Estimate the price of a comparably equipped car for the next model year if the price will increase by $4\frac{1}{2}\%$.
$19,290

3.4 Ratios and Proportions

① Compare relative sizes using ratios.

In Exercises 53–56, find a ratio that compares the relative sizes of the quantities. (Use the same units of measurement for both quantities.)

53. Eighteen inches to 4 yards $\frac{1}{8}$

54. One pint to 2 gallons $\frac{1}{16}$

55. Two hours to 90 minutes $\frac{4}{3}$

56. Four meters to 150 centimeters $\frac{8}{3}$

② Find the unit price of a consumer item.

In Exercises 57 and 58, which product has the lower unit price?

57. (a) An 18 ounce container of cooking oil for $0.89

(b) A 24-ounce container of cooking oil for $1.12
24-ounce container

58. (a) A 17.4-ounce box of pasta noodles for $1.32

(b) A 32-ounce box of pasta noodles for $2.62
17.4-ounce box

③ Solve proportions that equate two ratios.

In Exercises 59–64, solve the proportion.

59. $\dfrac{7}{16} = \dfrac{z}{8}$ $\frac{7}{2}$

60. $\dfrac{x}{12} = \dfrac{5}{4}$ 15

61. $\dfrac{x+2}{4} = -\dfrac{1}{3}$ $-\frac{10}{3}$

62. $\dfrac{x-4}{1} = \dfrac{9}{4}$ $\frac{25}{4}$

63. $\dfrac{x-3}{2} = \dfrac{x+6}{5}$ 9

64. $\dfrac{x+1}{3} = \dfrac{x+2}{4}$ 2

④ Solve application problems using the Consumer Price Index.

In Exercises 65 and 66, use the Consumer Price Index table on page 163 to estimate the price of the item in the indicated year.

65. The 2001 price of a recliner chair that cost $78 in 1984 $133

66. The 1986 price of a microwave oven that cost $120 in 1999 $79

3.5 Geometric and Scientific Applications

① Use common formulas to solve application problems.

In Exercises 67 and 68, solve for the specified variable.

67. Solve for w: $P = 2l + 2w$ $w = \dfrac{P - 2l}{2}$

68. Solve for t: $I = Prt$ $t = \dfrac{I}{Pr}$

In Exercises 69–72, find the missing distance, rate, or time.

Distance, d	Rate, r	Time, t
69. 520 mi	65 mi/hr	8 hr
70. 855 m	5 m/min	171 min
71. 3000 mi	60 m/hr	50 hr
72. 1000 km	40 km/hr	25 hr

73. *Distance*   An airplane has an average speed of 475 miles per hour. How far will it travel in $2\frac{1}{3}$ hours?
≈ 1108.3 miles

74. *Average Speed*   You can walk 20 kilometers in 3 hours and 47 minutes. What is your average speed?
≈ 5.3 kilometers per hour

75. ▲ *Geometry*   The width of a rectangular swimming pool is 4 feet less than its length. The perimeter of the pool is 112 feet. Find the dimensions of the pool. 30×26 feet

76. ▲ *Geometry*   The perimeter of an isosceles triangle is 65 centimeters. Find the length of the two equal sides if each is 10 centimeters longer than the third side. (An isosceles triangle has two sides of equal length.)   25 centimeters

Simple Interest   **In Exercises 77 and 78, use the simple interest formula.**

77. Find the total interest you will earn on a $1000 corporate bond that matures in 5 years and has an annual interest rate of 9.5%. $475

78. Find the annual interest rate on a certificate of deposit that pays $60 per year in interest on a principal of $750.   8%

② Solve mixture problems involving hidden products.

79. *Number of Coins*   You have 30 coins in dimes and quarters with a combined value of $5.55. Determine the number of coins of each type.

13 dimes, 17 quarters

80. *Bird Seed Mixture*   A pet store owner mixes two types of bird seed that cost $1.25 per pound and $2.20 per pound to make 20 pounds of a mixture that costs $1.65 per pound. How many pounds of each kind of birdseed are in the mixture?   12 pounds at $1.25 per pound, 8 pounds at $2.20 per pound

③ Solve work-rate problems.

81. *Work Rate*   One person can complete a task in 5 hours, and another can complete the same task in 6 hours. How long will it take both people working together to complete the task? $\frac{30}{11} \approx 2.7$ hours

82. *Work Rate*   The person in Exercise 79 who can complete the task in 5 hours has already worked 1 hour when the second person starts. How long will they work together to complete the task? $\frac{24}{11} \approx 2.2$ hours

3.6 Linear Inequalities

① Graph the solution set of an inequality on the real number line and determine whether a given value is a solution of an inequality.

In Exercises 83–86, write a verbal description of the inequality and sketch its graph.
See Additional Answers.

83. $x < 3$ *x* is less than 3.

84. $x \geq 20$ *x* is greater than or equal to 20.

85. $1 \leq x < 4$
x is greater than or equal to 1 *and* less than 4.

86. $-3 < x < 0$ *x* is greater than -3 *and* less than 0.

In Exercises 87 and 88, determine whether each value of *x* is a solution of the inequality.

Inequality	Values
87. $7x - 10 > 0$	(a) $x = 3$ Yes (b) $x = \frac{1}{2}$ No
88. $3x + 2 < 1$	(a) $x = 0$ No (b) $x = -4$ Yes

③ Solve a linear inequality and graph its solution set.

In Exercises 89–100, solve and graph the inequality.
See Additional Answers.

89. $x + 5 \geq 7$ $x \geq 2$ **90.** $x - 2 \leq 1$ $x \leq 3$

91. $3x - 8 < 1$ $x < 3$ **92.** $4x + 3 > 15$ $x > 3$

93. $-11x \leq -22$ $x \geq 2$ **94.** $-7x \geq 21$ $x \leq -3$

95. $\frac{4}{5}x > 8$ $x > 10$ **96.** $\frac{2}{3}n < -4$ $n < -6$

97. $14 - \frac{1}{2}t < 12$ **98.** $32 + \frac{7}{8}k > 11$
$t > 4$ $k > -24$

99. $3 - 3y \geq 2(4 + y)$ **100.** $4 - 3y \leq 8(10 - y)$
$y \leq -1$ $y \leq \frac{76}{5}$

④ Translate verbal statements into linear inequalities and solve application problems.

In Exercises 101–104, translate the verbal statement into a linear inequality.

101. *z* is at least 10.   **102.** *x* is nonnegative.
$z \geq 10$ $x \geq 0$

103. The area *A* is no more than 100 square feet.
$A \leq 100$

104. The volume *V* is less than 12 cubic feet. $V < 12$

105. *Budget*   You have budgeted $85 for a car rental on your vacation. The rental agency charges $50 a week plus an additional $0.75 per mile. To stay within your budget, how many miles can you drive the rental car?   46 miles

106. *Fund Raiser*   A neighbor sponsors you in a bowl-a-thon for a school fund raiser. Your neighbor is going to give you $1.60 plus $0.18 per pin that you hit. How many pins will you need to hit to earn at least $25.00 from your neighbor?   130 pins

Chapter Test

Take this test as you would take a test in class. After you are done, check your work against the answers in the back of the book.

In Exercises 1–6, solve the equation and check your solution.

1. $8x + 104 = 0$ -13

2. $4x - 3 = 18$ $\frac{21}{4}$

3. $5 - 3x = -2x - 2$ 7

4. $10 - (2 - x) = 2x + 1$ 7

5. $\dfrac{3x}{4} = \dfrac{5}{2} + x$ -10

6. $\dfrac{t + 2}{3} = \dfrac{2t}{5}$ 10

7. Solve $4.08(x + 10) = 9.50(x - 2)$. Round your answer to two decimal places.
   11.03

8. The bill (including parts and labor) for the repair of a home appliance is $142. The cost of parts is $62 and the cost of labor is $32 per hour. How many hours were spent repairing the appliance? $2\frac{1}{2}$ hours

9. Write the fraction $\frac{3}{8}$ as a percent and as a decimal. $37\frac{1}{2}\%, 0.375$

10. 324 is 27% of what number? 1200

11. 90 is what percent of 250? 36%

12. Write the ratio of 40 inches to 2 yards as a fraction in simplest form. Use the same units for both quantities, and explain how you made this conversion.
    $\frac{5}{9}$; 2 yards = 6 feet = 72 inches

13. Solve the proportion $\dfrac{2x}{3} = \dfrac{x + 4}{5}$. $\frac{12}{7}$

14. Find the length x of the side of the larger triangle shown in the figure at the left. (Assume that the two triangles are similar, and use the fact that corresponding sides of similar triangles are proportional.) 5

15. You traveled 264 miles in $5\frac{1}{2}$ hours. What was your average speed?
    48 miles per hour

16. You can paint a building in 9 hours. Your friend can paint the same building in 12 hours. Working together, how long will it take the two of you to paint the building? $\frac{36}{7} \approx 5.1$ hours

17. Solve for R in the formula $S = C + RC$. $\dfrac{S - C}{C}$

18. How much must you deposit in an account to earn $500 per year at 8% simple interest? $\$6250$

In Exercises 19–24, solve and graph the inequality.   See Additional Answers.

19. $x + 3 \leq 7$ $x \leq 4$

20. $-\dfrac{2x}{3} > 4$ $x < -6$

21. $21 - 3x \leq 6$ $x \geq 5$

22. $3(6 + 2x) > 8$ $x > -\frac{5}{3}$

23. $-4(9 + 2x) \geq -40$ $x \leq \frac{1}{2}$

24. $-(3 + x) < 2(3x - 5)$ $x > 1$

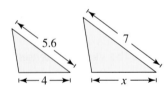

5.6   7

4   x

Figure for 14

Cumulative Test: Chapters 1–3

Take this test as you would take a test in class. After you are done, check your
work against the answers in the back of the book.

1. Place the correct symbol (< or >) between the numbers: $-\frac{3}{4}$ < $\left|-\frac{7}{8}\right|$.

Cumulative Tests provide a useful
progress check that students can use
to assess how well they are retaining
various algebraic skills and concepts.

In Exercises 2–7, evaluate the expression.

2. $(-200)(2)(-3)$ 1200   **3.** $\frac{3}{8} - \frac{5}{6}$ $-\frac{11}{24}$ **4.** $-\frac{2}{9} \div \frac{8}{75}$ $-\frac{25}{12}$

5. $-(-2)^3$ 8   **6.** $3 + 2(6) - 1$ 14   **7.** $24 + 12 \div 3$ 28

In Exercises 8 and 9, evaluate the expression when $x = -2$ and $y = 3$.

8. $-3x - (2y)^2$ -30 **9.** $4y - x^3$ 20

10. Use exponential form to write the product $3 \cdot (x + y) \cdot (x + y) \cdot 3 \cdot 3$.
$3^3(x + y)^2$

11. Use the Distributive Property to expand $-2x(x - 3)$. $-2x^2 + 6x$

12. Identify the property of real numbers illustrated by

$$2 + (3 + x) = (2 + 3) + x.$$ Associative Property of Addition

In Exercises 13–15, simplify the expression.

13. $(3x^3)(5x^4)$ $15x^7$

14. $2x^2 - 3x + 5x^2 - (2 + 3x)$ $7x^2 - 6x - 2$

15. $3(x^2 + x) - 2(2x - x^2)$ $5x^2 - x$

In Exercises 16–18, solve the equation and check your solution.

16. $12x - 3 = 7x + 27$ 6   **17.** $2x - \dfrac{5x}{4} = 13$ $\frac{52}{3}$

18. $2(x - 3) + 3 = 12 - x$ 5

19. Solve and graph the inequality.

$$12 - 3x \leq -15 \quad x \geq 9$$

20. The sticker on a new car gives the fuel efficiency as 28.3 miles per gallon. In
your own words, explain how to estimate the annual fuel cost for the buyer if
the car will be driven approximately 15,000 miles per year and the fuel cost
is $1.179 per gallon.

$$\frac{15{,}000 \text{ miles}}{1 \text{ year}} \cdot \frac{1 \text{ gallon}}{28.3 \text{ miles}} \cdot \frac{\$1.179}{1 \text{ gallon}} \approx \$624.91 \text{ per year}$$

21. Write the ratio "24 ounces to 2 pounds" as a fraction in simplest form. $\frac{3}{4}$

22. The suggested retail price of a digital camcorder is $1150. The camcorder is
on sale for "20% off" the list price. Find the sale price. $920

23. The figure at the left shows two pieces of property. The assessed values of the
properties are proportional to their areas. The value of the larger piece is
$95,000. What is the value of the smaller piece? $57,000

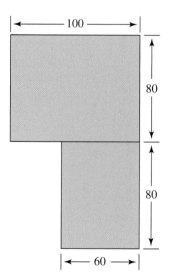

Figure for 23

Motivating the Chapter

⬢ *Salary Plus Commission*

You work as a sales representative for an advertising agency. You are paid a weekly salary, plus a commission on all ads placed by your accounts. The table shows your sales and your total weekly earnings.

	Week 1	Week 2	Week 3	Week 4
Weekly sales	$24,000	$7,000	$0	$36,000
Weekly earnings	$980	$640	$500	$1,220

See Section 4.1, Exercise 77.

a. Rewrite the data as a set of ordered pairs.
   (24,000, 980), (7000, 640), (0, 500), (36,000, 1220)

b. Plot the ordered pairs on a rectangular coordinate system.
   See Additional Answers.

See Section 4.4, Exercise 109.

c. Explain how to determine whether the data in the table follows a linear pattern.   The function is linear if the slopes are the same between the points (x, y), where x is the weekly sales and y is weekly earnings.

d. Determine the slope of the line passing through the ordered pairs for week 1 and week 2. (Let x represent the weekly sales and let y represent the weekly earnings.) What is the *rate* at which the weekly pay increases for each unit increase in ad sales? What is the rate called in the context of the problem? $m = 0.02$; 2%; Commission rate

e. Write an equation that describes the linear relationship between weekly sales and weekly earnings. $y = 500 + 0.02x$

f. Sketch a graph of the equation. Identify the y-intercept and explain its meaning in the context of the problem. Identify the x-intercept. Does the x-intercept have any meaning in the context of the problem? If so, what is it?   See Additional Answers.   (0, 500) The y-intercept is the weekly earnings when no ads are sold. $(-25,000, 0)$ The x-intercept does not have meaning.

See Section 4.5, Exercise 71.

g. What amount of ad sales is needed to guarantee a weekly pay of at least $840?   At least $17,000

Michael Newman/PhotoEdit

Equations and Inequalities in Two Variables

4.1 Ordered Pairs and Graphs

Why You Should Learn It

The Cartesian plane can be used to represent relationships between two variables. For instance, Exercises 67–70 on page 210 show how to represent graphically the number of new privately owned housing starts in the United States.

1 Plot and find the coordinates of a point on a rectangular coordinate system.

What You Should Learn

1 Plot and find the coordinates of a point on a rectangular coordinate system.

2 Construct a table of values for equations and determine whether ordered pairs are solutions of equations.

3 Use the verbal problem-solving method to plot points on a rectangular coordinate system.

The Rectangular Coordinate System

Just as you can represent real numbers by points on the real number line, you can represent **ordered pairs** of real numbers by points in a plane. This plane is called a **rectangular coordinate system** or the **Cartesian plane,** after the French mathematician René Descartes (1596–1650).

A rectangular coordinate system is formed by two real lines intersecting at right angles, as shown in Figure 4.1. The horizontal number line is usually called the **x-axis** and the vertical number line is usually called the **y-axis.** (The plural of axis is *axes.*) The point of intersection of the two axes is called the **origin,** and the axes separate the plane into four regions called **quadrants.**

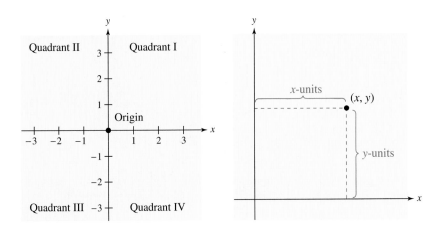

Figure 4.1   **Figure 4.2**

Each point in the plane corresponds to an **ordered pair** (x, y) of real numbers x and y, called the **coordinates** of the point. The first number (or **x-coordinate**) tells how far to the left or right the point is from the vertical axis, and the second number (or **y-coordinate**) tells how far up or down the point is from the horizontal axis, as shown in Figure 4.2.

A positive x-coordinate implies that the point lies to the *right* of the vertical axis; a negative x-coordinate implies that the point lies to the *left* of the vertical axis; and an x-coordinate of zero implies that the point lies *on* the vertical axis. Similarly, a positive y-coordinate implies that the point lies *above* the horizontal axis, and a negative y-coordinate implies that the point lies *below* the horizontal axis.

Locating a point in a plane is called **plotting** the point. This procedure is demonstrated in Example 1.

Example 1   Plotting Points on a Rectangular Coordinate System

Plot the points $(-1, 2)$, $(3, 0)$, $(2, -1)$, $(3, 4)$, $(0, 0)$, and $(-2, -3)$ on a rectangular coordinate system.

Solution

The point $(-1, 2)$ is one unit to the *left* of the vertical axis and two units *above* the horizontal axis.

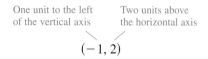

Similarly, the point $(3, 0)$ is three units to the *right* of the vertical axis and *on* the horizontal axis. (It is on the horizontal axis because the y-coordinate is zero.) The other four points can be plotted in a similar way, as shown in Figure 4.3.

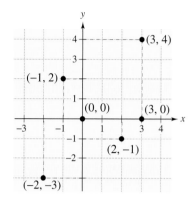

Figure 4.3

In Example 1 you were given the coordinates of several points and were asked to plot the points on a rectangular coordinate system. Example 2 looks at the reverse problem—that is, you are given points on a rectangular coordinate system and are asked to determine their coordinates.

Example 2   Finding Coordinates of Points

Determine the coordinates of each of the points shown in Figure 4.4.

Solution

Point A lies three units to the *left* of the vertical axis and two units *above* the horizontal axis. So, point A must be given by the ordered pair $(-3, 2)$. The coordinates of the other four points can be determined in a similar way, and the results are summarized as follows.

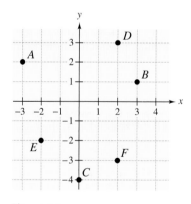

Figure 4.4

Point	Position	Coordinates
A	Three units *left*, two units *up*	$(-3, 2)$
B	Three units *right*, one unit *up*	$(3, 1)$
C	Zero units *left* (or right), four units *down*	$(0, -4)$
D	Two units *right*, three units *up*	$(2, 3)$
E	Two units *left*, two units *down*	$(-2, -2)$
F	Two units *right*, three units *down*	$(2, -3)$

In Example 2, note that point A $(-3, 2)$ and point F $(2, -3)$ are different points. The order in which the numbers appear in an ordered pair is important. Notice that because point C lies on the y-axis, it has an x-coordinate of 0.

Each year since 1967, the winners of the American Football Conference and the National Football Conference have played in the Super Bowl. The first Super Bowl was played between the Green Bay Packers and the Kansas City Chiefs.

Example 3   Super Bowl Scores

The scores of the winning and losing football teams in the Super Bowl games from 1983 through 2003 are shown in the table. Plot these points on a rectangular coordinate system.   (Source: National Football League)

Year	1983	1984	1985	1986	1987	1988	1989
Winning score	27	38	38	46	39	42	20
Losing score	17	9	16	10	20	10	16

Year	1990	1991	1992	1993	1994	1995	1996
Winning score	55	20	37	52	30	49	27
Losing score	10	19	24	17	13	26	17

Year	1997	1998	1999	2000	2001	2002	2003
Winning score	35	31	34	23	34	20	48
Losing score	21	24	19	16	7	17	21

Solution

The x-coordinates of the points represent the year of the game, and the y-coordinates represent either the winning score or the losing score. In Figure 4.5, the winning scores are shown as black dots, and the losing scores are shown as blue dots. Note that the break in the x-axis indicates that the numbers between 0 and 1983 have been omitted.

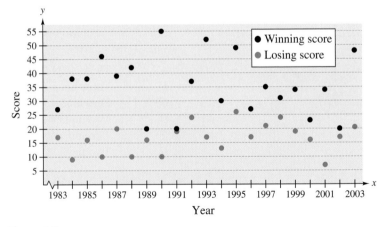

Figure 4.5

② Construct a table of values for equations and determine whether ordered pairs are solutions of equations.

Ordered Pairs as Solutions of Equations

In Example 3, the relationship between the year and the Super Bowl scores was given by a **table of values.** In mathematics, the relationship between the variables x and y is often given by an equation. From the equation, you can construct your own table of values. For instance, consider the equation

$$y = 2x + 1.$$

To construct a table of values for this equation, choose several x-values and then calculate the corresponding y-values. For example, if you choose $x = 1$, the corresponding y-value is

$$y = 2(1) + 1 \qquad \text{Substitute 1 for } x.$$

$$y = 3. \qquad \text{Simplify.}$$

The corresponding ordered pair $(x, y) = (1, 3)$ is a **solution point** (or **solution**) of the equation. The table below is a table of values (and the corresponding solution points) using x-values of $-3, -2, -1, 0, 1, 2,$ and 3. These x-values are arbitrary. You should try to use x-values that are convenient and simple to use.

Choose x	Calculate y from $y = 2x + 1$	Solution point
$x = -3$	$y = 2(-3) + 1 = -5$	$(-3, -5)$
$x = -2$	$y = 2(-2) + 1 = -3$	$(-2, -3)$
$x = -1$	$y = 2(-1) + 1 = -1$	$(-1, -1)$
$x = 0$	$y = 2(0) + 1 = 1$	$(0, 1)$
$x = 1$	$y = 2(1) + 1 = 3$	$(1, 3)$
$x = 2$	$y = 2(2) + 1 = 5$	$(2, 5)$
$x = 3$	$y = 2(3) + 1 = 7$	$(3, 7)$

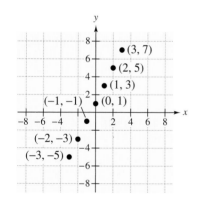

Figure 4.6

Once you have constructed a table of values, you can get a visual idea of the relationship between the variables x and y by plotting the solution points on a rectangular coordinate system. For instance, the solution points shown in the table are plotted in Figure 4.6.

In many places throughout this course, you will see that approaching a problem in different ways can help you understand the problem better. For instance, the discussion above looks at solutions of an equation in three ways.

Three Approaches to Problem Solving

1. **Algebraic Approach** Use algebra to find several solutions.

2. **Numerical Approach** Construct a table that shows several solutions.

3. **Graphical Approach** Draw a graph that shows several solutions.

When constructing a table of values for an equation, it is helpful first to solve the equation for y. For instance, the equation $4x + 2y = -8$ can be solved for y as follows.

$4x + 2y = -8$	Write original equation.
$4x - 4x + 2y = -8 - 4x$	Subtract $4x$ from each side.
$2y = -8 - 4x$	Combine like terms.
$\dfrac{2y}{2} = \dfrac{-8 - 4x}{2}$	Divide each side by 2.
$y = -4 - 2x$	Simplify.

This procedure is further demonstrated in Example 4.

Example 4   Constructing a Table of Values

Construct a table of values showing five solution points for the equation

$$6x - 2y = 4.$$

Then plot the solution points on a rectangular coordinate system. Choose x-values of $-2, -1, 0, 1,$ and 2.

Solution

$6x - 2y = 4$	Write original equation.
$6x - 6x - 2y = 4 - 6x$	Subtract $6x$ from each side.
$-2y = -6x + 4$	Combine like terms.
$\dfrac{-2y}{-2} = \dfrac{-6x + 4}{-2}$	Divide each side by -2.
$y = 3x - 2$	Simplify.

Now, using the equation $y = 3x - 2$, you can construct a table of values, as shown below.

x	-2	-1	0	1	2
$y = 3x - 2$	-8	-5	-2	1	4
Solution point	$(-2, -8)$	$(-1, -5)$	$(0, -2)$	$(1, 1)$	$(2, 4)$

Finally, from the table you can plot the five solution points on a rectangular coordinate system, as shown in Figure 4.7.

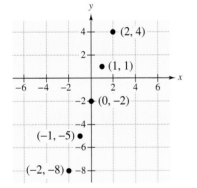

Figure 4.7

In the next example, you are given several ordered pairs and are asked to determine whether they are solutions of the original equation. To do this, you need to substitute the values of x and y into the equation. If the substitution produces a true equation, the ordered pair (x, y) is a solution and is said to **satisfy** the equation.

Guidelines for Verifying Solutions

To verify that an ordered pair (x, y) is a solution to an equation with variables x and y, use the following steps.

1. Substitute the values of x and y into the equation.

2. Simplify each side of the equation.

3. If each side simplifies to the same number, the ordered pair is a solution. If the two sides yield different numbers, the ordered pair is not a solution.

Example 5 Verifying Solutions of an Equation

Determine whether each of the ordered pairs is a solution of $x + 3y = 6$.

a. $(1, 2)$　　　**b.** $\left(-2, \frac{8}{3}\right)$　　　**c.** $(0, 2)$

Solution

a. For the ordered pair $(1, 2)$, substitute $x = 1$ and $y = 2$ into the original equation.

$$x + 3y = 6 \qquad \text{Write original equation.}$$
$$1 + 3(2) \overset{?}{=} 6 \qquad \text{Substitute 1 for } x \text{ and 2 for } y.$$
$$7 \neq 6 \qquad \text{Is not a solution. } \times$$

Because the substitution does not satisfy the original equation, you can conclude that the ordered pair $(1, 2)$ is *not* a solution of the original equation.

b. For the ordered pair $\left(-2, \frac{8}{3}\right)$, substitute $x = -2$ and $y = \frac{8}{3}$ into the original equation.

$$x + 3y = 6 \qquad \text{Write original equation.}$$
$$(-2) + 3\left(\tfrac{8}{3}\right) \overset{?}{=} 6 \qquad \text{Substitute } -2 \text{ for } x \text{ and } \tfrac{8}{3} \text{ for } y.$$
$$-2 + 8 \overset{?}{=} 6 \qquad \text{Simplify.}$$
$$6 = 6 \qquad \text{Is a solution. } \checkmark$$

Because the substitution satisfies the original equation, you can conclude that the ordered pair $\left(-2, \frac{8}{3}\right)$ *is* a solution of the original equation.

c. For the ordered pair $(0, 2)$, substitute $x = 0$ and $y = 2$ into the original equation.

$$x + 3y = 6 \qquad \text{Write original equation.}$$
$$0 + 3(2) \overset{?}{=} 6 \qquad \text{Substitute 0 for } x \text{ and 2 for } y.$$
$$6 = 6 \qquad \text{Is a solution. } \checkmark$$

Because the substitution satisfies the original equation, you can conclude that the ordered pair $(0, 2)$ *is* a solution of the original equation.

3 Use the verbal problem-solving method to plot points on a rectangular coordinate system.

Application

Example 6   Total Cost

You set up a small business to assemble computer keyboards. Your initial cost is $120,000, and your unit cost to assemble each keyboard is $40. Write an equation that relates your total cost to the number of keyboards produced. Then plot the total costs of producing 1000, 2000, 3000, 4000, and 5000 keyboards.

Solution

The total cost equation must represent both the unit cost and the initial cost. A verbal model for this problem is as follows.

Verbal Model: $\boxed{\text{Total cost}} = \boxed{\text{Unit cost}} \cdot \boxed{\text{Number of keyboards}} + \boxed{\text{Initial cost}}$

Labels: Total cost $= C$ (dollars)
             Unit cost $= 40$ (dollars per keyboard)
             Number of keyboards $= x$ (keyboards)
             Initial cost $= 120{,}000$ (dollars)

Algebraic Model: $C = 40x + 120{,}000$

Using this equation, you can construct the following table of values.

x	1,000	2,000	3,000	4,000	5,000
$C = 40x + 120{,}000$	160,000	200,000	240,000	280,000	320,000

From the table, you can plot the ordered pairs, as shown in Figure 4.8.

Figure 4.8

Although graphs can help you visualize relationships between two variables, they can also be misleading. The graphs shown in Figure 4.9 and Figure 4.10 represent the yearly profits for a truck rental company. The graph in Figure 4.9 is misleading. The scale on the vertical axis makes it appear that the change in profits from 1998 to 2002 is dramatic, but the total change is only $3000, which is small in comparison with $3,000,000.

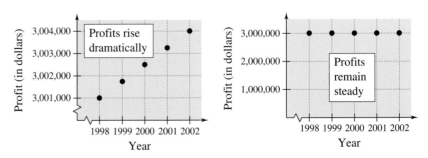

Figure 4.9   **Figure 4.10**

4.1 Exercises

Review   Concepts, Skills, and Problem Solving

Keep mathematically in shape by doing these exercises *before* the problems of this section.

Properties and Definitions

1. *Writing* Is $3x = 7$ a linear equation? Explain. Is $x^2 + 3x = 2$ a linear equation? Explain. $3x = 7$ is a linear equation since it has the form $ax + b = c$. $x^2 + 3x = 2$ is not of that form, and therefore is not linear.

2. *Writing* Explain how to check whether $x = 3$ is a solution to the equation $5x - 4 = 11$.   Substitute 3 for x in the equation to verify that it satisfies the equation.

Solving Equations

In Exercises 3–10, solve the equation.

3. $-y = 10$ -10 4. $10 - t = 6$ 4

5. $3x - 42 = 0$ 14   6. $64 - 16x = 0$ 4

7. $125(r - 1) = 625$ 6   8. $2(3 - y) = 7y + 5$ $\frac{1}{9}$

9. $20 - \frac{1}{9}x = 4$ 144   10. $0.35x = 70$ 200

Problem Solving

11. *Cost*   The total cost of a lot and house is $154,000. The cost of constructing the house is 7 times the cost of the lot. What is the cost of the lot? $19,250

12. *Summer Jobs*   You have two summer jobs. In the first job, you work 40 hours a week and earn $9.50 an hour. In the second job, you work as many hours as you want and earn $8 an hour. You want to earn $450 a week. How many hours a week should you work at the second job?   8 hours 45 minutes

Developing Skills

In Exercises 1–10, plot the points on a rectangular coordinate system. See Example 1.   See Additional Answers.

1. $(3, 2), (-4, 2), (2, -4)$

2. $(-1, 6), (-1, -6), (4, 6)$

3. $(-10, -4), (4, -4), (0, 0)$

4. $(-6, 4), (0, 0), (3, -2)$

5. $(-3, 4), (0, -1), (2, -2), (5, 0)$

6. $(-1, 3), (0, 2), (-4, -4), (-1, 0)$

7. $\left(\frac{3}{2}, -1\right), \left(-3, \frac{3}{4}\right), \left(\frac{1}{2}, -\frac{1}{2}\right)$

8. $\left(-\frac{2}{3}, 4\right), \left(\frac{1}{2}, -\frac{5}{2}\right), \left(-4, -\frac{5}{4}\right)$

9. $(3, -4), \left(\frac{5}{2}, 0\right), (0, 3)$ 10. $\left(\frac{5}{2}, 2\right), \left(-3, \frac{4}{3}\right), \left(\frac{3}{4}, \frac{9}{4}\right)$

In Exercises 11–14, determine the coordinates of the points. See Example 2.

11.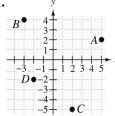

A: $(5, 2)$, B: $(-3, 4)$,
C: $(2, -5)$, D: $(-2, -2)$

12.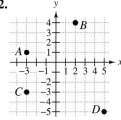

A: $(-3, 1)$, B: $(2, 4)$,
C: $(-3, -3)$, D: $(5, -5)$

13.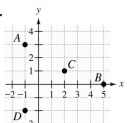

A: $(-1, 3)$, B: $(5, 0)$,
C: $(2, 1)$, D: $(-1, -2)$

14.

A: $(0, 3)$, B: $(4, 0)$,
C: $(-2, -2)$, D: $(3, -1)$

In Exercises 15–20, determine the quadrant in which the point is located without plotting it.

15. $(-3, 1)$ Quadrant II   16. $(4, -3)$ Quadrant IV

17. $\left(-\frac{1}{8}, -\frac{2}{7}\right)$ Quadrant III   18. $\left(\frac{3}{11}, \frac{7}{8}\right)$ Quadrant I

19. $(-100, -365.6)$ 20. $(-157.4, 305.6)$
    Quadrant III   Quadrant II

In Exercises 21–26, determine the quadrant(s) in which the point is located without plotting it. Assume $x \neq 0$ and $y \neq 0$.

21. $(-5, y)$, y is a real number.   Quadrant II or III

22. $(6, y)$, y is a real number.   Quadrant I or IV

23. $(x, -2)$, x is a real number.   Quadrant III or IV

24. $(x, 3)$, x is a real number.   Quadrant I or II

25. (x, y), $xy < 0$ Quadrant II or IV

26. (x, y), $xy > 0$ Quadrant I or III

In Exercises 27–34, plot the points and connect them with line segments to form the figure.   See Additional Answers.

27. Triangle: $(-1, 1)$, $(2, -1)$, $(3, 4)$

28. Triangle: $(0, 3)$, $(-1, -2)$, $(4, 8)$

29. Square: $(2, 4)$, $(5, 1)$, $(2, -2)$, $(-1, 1)$

30. Rectangle: $(2, 1)$, $(4, 2)$, $(-1, 7)$, $(1, 8)$

31. Parallelogram: $(5, 2)$, $(7, 0)$, $(1, -2)$, $(-1, 0)$

32. Parallelogram: $(-1, 1)$, $(0, 4)$, $(4, -2)$, $(5, 1)$

33. Rhombus: $(0, 0)$, $(3, 2)$, $(2, 3)$, $(5, 5)$

34. Rhombus: $(0, 0)$, $(1, 2)$, $(2, 1)$, $(3, 3)$

In Exercises 35–40, complete the table of values. Then plot the solution points on a rectangular coordinate system. See Example 4.   See Additional Answers.

35.

x	-2	0	2	4	6
$y = 3x - 4$	-10	-4	2	8	14

36.

x	-2	0	2	4	6
$y = 2x + 1$	-3	1	5	9	13

37.

x	-4	-2	4	6	8
$y = -\frac{3}{2}x + 5$	11	8	-1	-4	-7

38.

x	-4	-2	0	2	4
$y = -\frac{1}{2}x + 3$	5	4	3	2	1

39.

x	-2	-1	0	1	2
$y = 2x - 1$	-5	-3	-1	1	3

40.

x	-2	0	$\frac{1}{2}$	2	4
$y = -\frac{7}{2}x + 3$	10	3	$\frac{5}{4}$	-4	-11

In Exercises 41–50, solve the equation for y.

41. $7x + y = 8$

$y = -7x + 8$

42. $2x + y = 1$

$y = -2x + 1$

43. $10x - y = 2$

$y = 10x - 2$

44. $12x - y = 7$

$y = 12x - 7$

45. $6x - 3y = 3$

$y = 2x - 1$

46. $15x - 5y = 25$

$y = 3x - 5$

47. $x + 4y = 8$

$y = -\frac{1}{4}x + 2$

48. $x - 2y = -6$

$y = \frac{1}{2}x + 3$

49. $4x - 5y = 3$

$y = \frac{4}{5}x - \frac{3}{5}$

50. $4y - 3x = 7$

$y = \frac{3}{4}x + \frac{7}{4}$

In Exercises 51–58, determine whether the ordered pairs are solutions of the equation. See Example 5.

51. $y = 2x + 4$ (a) $(3, 10)$ (b) $(-1, 3)$
                 (c) $(0, 0)$ (d) $(-2, 0)$

(a) Solution   (b) Not a solution
(c) Not a solution   (d) Solution

52. $y = 5x - 2$ (a) $(2, 0)$ (b) $(-2, -12)$
                 (c) $(6, 28)$ (d) $(1, 1)$

(a) Not a solution   (b) Solution
(c) Solution   (d) Not a solution

53. $2y - 3x + 1 = 0$ (a) $(1, 1)$ (b) $(5, 7)$
                    (c) $(-3, -1)$ (d) $(-3, -5)$

(a) Solution   (b) Solution
(c) Not a solution   (d) Solution

54. $x - 8y + 10 = 0$ (a) $(-2, 1)$ (b) $(6, 2)$
                    (c) $(0, -1)$ (d) $(2, -4)$

(a) Solution   (b) Solution
(c) Not a solution   (d) Not a solution

55. $y = \frac{2}{3}x$ (a) $(6, 6)$ (b) $(-9, -6)$
           (c) $(0, 0)$ (d) $\left(-1, \frac{2}{3}\right)$

(a) Not a solution   (b) Solution
(c) Solution   (d) Not a solution

56. $y = -\frac{7}{8}x$ (a) $(-5, -2)$ (b) $(0, 0)$
           (c) $(8, 8)$ (d) $\left(\frac{3}{5}, 1\right)$

(a) Not a solution   (b) Solution
(c) Not a solution   (d) Not a solution

57. $y = 3 - 4x$ (a) $\left(-\frac{1}{2}, 5\right)$ (b) $(1, 7)$
           (c) $(0, 0)$ (d) $\left(-\frac{3}{4}, 0\right)$

(a) Solution   (b) Not a solution
(c) Not a solution   (d) Not a solution

58. $y = \frac{3}{2}x + 1$ (a) $\left(0, \frac{3}{2}\right)$ (b) $(4, 7)$
           (c) $\left(\frac{2}{3}, 2\right)$ (d) $(-2, -2)$

(a) Not a solution   (b) Solution
(c) Solution   (d) Solution

Solving Problems

59. *Organizing Data*   The distance *y* (in centimeters) a spring is compressed by a force *x* (in kilograms) is given by $y = 0.066x$. Complete a table of values for $x = 20, 40, 60, 80,$ and 100 to determine the distance the spring is compressed for each of the specified forces. Plot the results on a rectangular coordinate system.   See Additional Answers.

60. *Organizing Data*   A company buys a new copier for $9500. Its value *y* after *x* years is given by $y = -800x + 9500$. Complete a table of values for $x = 0, 2, 4, 6,$ and 8 to determine the value of the copier at the specified times. Plot the results on a rectangular coordinate system.   See Additional Answers.

61. *Organizing Data*   With an initial cost of $5000, a company will produce *x* units of a video game at $25 per unit. Write an equation that relates the total cost of producing *x* units to the number of units produced. Plot the cost for producing 100, 150, 200, 250, and 300 units. $y = 25x + 5000$ See Additional Answers.

62. *Organizing Data*   An employee earns $10 plus $0.50 for every *x* units produced per hour. Write an equation that relates the employee's total hourly wage to the number of units produced. Plot the hourly wage for producing 2, 5, 8, 10, and 20 units per hour.   See Additional Answers.

63. *Organizing Data*   The table shows the normal temperatures *y* (in degrees Fahrenheit) for Anchorage, Alaska for each month *x* of the year, with $x = 1$ corresponding to January. (Source: National Climatic Data Center)

x	1	2	3	4	5	6
y	15	19	26	36	47	54

x	7	8	9	10	11	12
y	58	56	48	35	21	16

(a) Plot the data shown in the table. Did you use the same scale on both axes? Explain.   See Additional Answers.   No, because there are only 12 months, but the temperature ranges from 15°F to 58°F.

(b) Using the graph, find the three consecutive months when the normal temperature changes the least.   June, July, August

64. *Organizing Data*   The table shows the speed of a car *x* (in miles per hour) and the approximate stopping distance *y* (in feet).

x	20	30	40	50	60
y	63	109	164	229	303

(a) Plot the data in the table.   See Additional Answers.

(b) The *x*-coordinates increase at equal increments of 10 miles per hour. Describe the pattern for the *y*-coordinates. What are the implications for the driver?   Increasing at an increasing rate; Answers will vary.

65. *Graphical Interpretation*   The table shows the numbers of hours *x* that a student studied for five different algebra exams and the resulting scores *y*.

x	3.5	1	8	4.5	0.5
y	72	67	95	81	53

(a) Plot the data in the table.   See Additional Answers.

(b) Use the graph to describe the relationship between the number of hours studied and the resulting exam score.   Scores increase with increased study time.

66. *Graphical Interpretation*   The table shows the net income per share of common stock *y* (in dollars) for the Dow Chemical Company for the years 1991 through 2000, where *x* represents the year. (Source: Dow Chemical Company 2000 Annual Report)

x	1991	1992	1993	1994	1995
y	1.15	0.33	0.78	1.12	2.57

x	1996	1997	1998	1999	2000
y	2.57	2.60	1.94	2.01	2.24

(a) Plot the data in the table.   See Additional Answers.

(b) Use the graph to determine the year that had the greatest increase and the year that had the greatest decrease in the net income per share of common stock.   Greatest increase: 1995, Greatest decrease: 1992

Graphical Estimation In Exercises 67–70, use the scatter plot, which shows new privately owned housing unit starts (in thousands) in the United States from 1988 through 2000. (Source: U.S. Census Bureau)

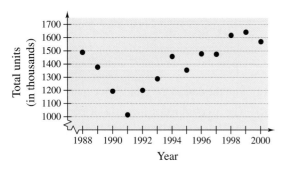

Year

67. Estimate the number of new housing unit starts in 1989. 1,380,000

68. Estimate the number of new housing unit starts in 1994. 1,450,000

69. Estimate the increase and the percent increase in housing unit starts from 1997 to 1998. 150,000; 10%

70. Estimate the decrease and the percent decrease in housing unit starts from 1999 to 2000. 70,000; 4%

Graphical Estimation In Exercises 71–74, use the scatter plot, which shows the per capita personal income in the United States from 1993 through 2000. (Source: U.S. Bureau of Economic Analysis)

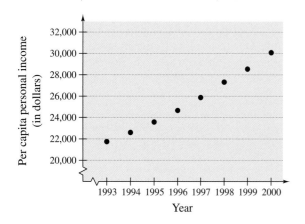

Year

71. Estimate the per capita personal income in 1994. $22,500

72. Estimate the per capita personal income in 1995. $23,500

73. Estimate the percent increase in per capita personal income from 1999 to 2000. 5%

74. The per capita personal income in 1980 was $10,205. Estimate the percent increase in per capita personal income from 1980 to 1993. 114%

Graphical Estimation In Exercises 75 and 76, use the bar graph, which shows the percents of gross domestic product spent on health care in several countries in 2000. (Source: Organization for Economic Cooperation and Development)

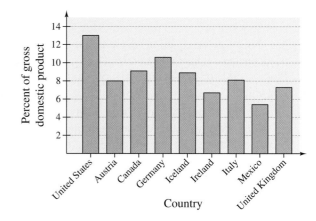

Country

75. Estimate the percent of gross domestic product spent on health care in Mexico. 5%

76. Estimate the percent of gross domestic product spent on health care in the United States. 13%

Explaining Concepts

77. 🌀 Answer parts (a) and (b) of Motivating the Chapter on page 198.

78. What is the *x*-coordinate of any point on the *y*-axis? What is the *y*-coordinate of any point on the *x*-axis?
The *x*-coordinate of any point on the *y*-axis is 0. The *y*-coordinate of any point on the *x*-axis is 0.

79. *Writing* 📝 Describe the signs of the *x*- and *y*-coordinates of points that lie in the first and second quadrants. First quadrant: $(+, +)$, Second quadrant: $(-, +)$

80. *Writing* Describe the signs of the *x*- and *y*-coordinates of points that lie in the third and fourth quadrants. Third quadrant: $(-, -)$, Fourth quadrant: $(+, -)$

81. (a) Plot the points $(3, 2)$, $(-5, 4)$, and $(6, -4)$ on a rectangular coordinate system. See Additional Answers.

(b) Change the sign of the *y*-coordinate of each point plotted in part (a). Plot the three new points on the same rectangular coordinate system used in part (a). See Additional Answers.

(c) What can you infer about the location of a point when the sign of its *y*-coordinate is changed? Reflection in the *x*-axis

82. (a) Plot the points $(3, 2)$, $(-5, 4)$, and $(6, -4)$ on a rectangular coordinate system. See Additional Answers.

(b) Change the sign of the *x*-coordinate of each point plotted in part (a). Plot the three new points on the same rectangular coordinate system used in part (a). See Additional Answers.

(c) What can you infer about the location of a point when the sign of its *x*-coordinate is changed? Reflection in the *y*-axis

83. *Writing* Discuss the significance of the word "ordered" when referring to an ordered pair (x, y). Order is significant because each number in the pair has a particular interpretation. The first measures horizontal distance and the second measures vertical distance.

84. *Writing* When the point (x, y) is plotted, what does the *x*-coordinate measure? What does the *y*-coordinate measure? The *x*-coordinate measures the distance from the *y*-axis to the point. The *y*-coordinate measures the distance from the *x*-axis to the point.

85. In a rectangular coordinate system, must the scales on the *x*-axis and *y*-axis be the same? If not, give an example in which the scales differ. No. The scales are determined by the magnitudes of the quantities being measured by *x* and *y*. If *y* is measuring revenue for a product and *x* is measuring time in years, the scale on the *y*-axis may be in units of $100,000 and the scale on the *x*-axis may be in units of 1 year.

86. *Writing* Review the tables in Exercises 35–40 and observe that in some cases the *y*-coordinates of the solution points increase and in others the *y*-coordinates decrease. What factor in the equation causes this? Explain. The *y*-coordinates increase if the coefficient of *x* is positive and decrease if the coefficient is negative.

4.2 Graphs of Equations in Two Variables

Tom & Dee Ann McCarthy/Corbis

What You Should Learn

① Sketch graphs of equations using the point-plotting method.

② Find and use *x*- and *y*-intercepts as aids to sketching graphs.

③ Use the verbal problem-solving method to write an equation and sketch its graph.

Why You Should Learn It

The graph of an equation can help you see relationships between real-life quantities. For instance, in Exercise 83 on page 221, a graph can be used to illustrate the change over time in the life expectancy for a child at birth.

① Sketch graphs of equations using the point-plotting method.

The Graph of an Equation in Two Variables

You have already seen that the solutions of an equation involving two variables can be represented by points on a rectangular coordinate system. The set of *all* such points is called the **graph** of the equation.

To see how to sketch a graph, consider the equation

$$y = 2x - 1.$$

To begin sketching the graph of this equation, construct a table of values, as shown at the left. Next, plot the solution points on a rectangular coordinate system, as shown in Figure 4.11. Finally, find a pattern for the plotted points and use the pattern to connect the points with a smooth curve or line, as shown in Figure 4.12.

x	$y = 2x - 1$	Solution point
-3	-7	$(-3, -7)$
-2	-5	$(-2, -5)$
-1	-3	$(-1, -3)$
0	-1	$(0, -1)$
1	1	$(1, 1)$
2	3	$(2, 3)$
3	5	$(3, 5)$

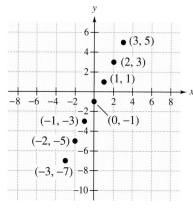

Figure 4.11

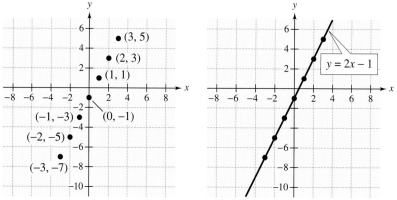

Figure 4.12

The Point-Plotting Method of Sketching a Graph

1. If possible, rewrite the equation by isolating one of the variables.

2. Make a table of values showing several solution points.

3. Plot these points on a rectangular coordinate system.

4. Connect the points with a smooth curve or line.

Example 1   Sketching the Graph of an Equation

Sketch the graph of $3x + y = 5$.

Solution

Begin by solving the equation for y, so that y is isolated on the left.

$$3x + y = 5 \qquad \text{Write original equation.}$$

$$3x - 3x + y = -3x + 5 \qquad \text{Subtract } 3x \text{ from each side.}$$

$$y = -3x + 5 \qquad \text{Simplify.}$$

Next, create a table of values, as shown below.

x	-2	-1	0	1	2	3
$y = -3x + 5$	11	8	5	2	-1	-4
Solution point	$(-2, 11)$	$(-1, 8)$	$(0, 5)$	$(1, 2)$	$(2, -1)$	$(3, -4)$

Now, plot the solution points, as shown in Figure 4.13. It appears that all six points lie on a line, so complete the sketch by drawing a line through the points, as shown in Figure 4.14.

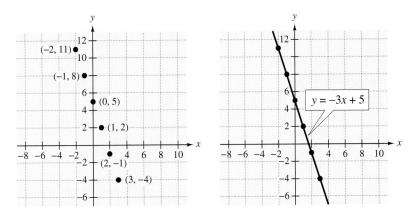

Figure 4.13   Figure 4.14

When creating a table of values, you are generally free to choose any x-values. When doing this, however, remember that the more x-values you choose, the easier it will be to recognize a pattern.

The equation in Example 1 is an example of a **linear equation in two variables**—the variables are raised to the first power and the graph of the equation is a line. As shown in the next two examples, graphs of nonlinear equations are not lines.

Example 2   Sketching the Graph of a Nonlinear Equation

Sketch the graph of $x^2 + y = 4$.

Solution

Begin by solving the equation for y, so that y is isolated on the left.

$$x^2 + y = 4 \qquad \text{Write original equation.}$$
$$x^2 - x^2 + y = -x^2 + 4 \qquad \text{Subtract } x^2 \text{ from each side.}$$
$$y = -x^2 + 4 \qquad \text{Simplify.}$$

Next, create a table of values, as shown below. Be careful with the signs of the numbers when creating a table. For instance, when $x = -3$, the value of y is

$$y = -(-3)^2 + 4$$
$$= -9 + 4$$
$$= -5.$$

x	-3	-2	-1	0	1	2	3
$y = -x^2 + 4$	-5	0	3	4	3	0	-5
Solution point	$(-3, -5)$	$(-2, 0)$	$(-1, 3)$	$(0, 4)$	$(1, 3)$	$(2, 0)$	$(3, -5)$

Now, plot the solution points, as shown in Figure 4.15. Finally, connect the points with a smooth curve, as shown in Figure 4.16.

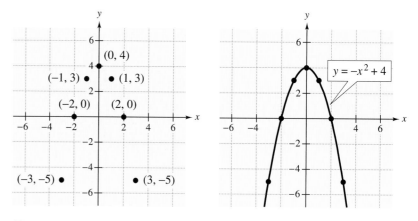

Figure 4.15   Figure 4.16

The graph of the equation in Example 2 is called a **parabola.** You will study this type of graph in a later chapter.

Example 3 examines the graph of an equation that involves an absolute value. Remember that the absolute value of a number is its distance from zero on the real number line. For instance, $|-5| = 5$, $|2| = 2$, and $|0| = 0$.

Example 3   The Graph of an Absolute Value Equation

Sketch the graph of $y = |x - 1|$.

Solution

This equation is already written in a form with y isolated on the left. You can begin by creating a table of values, as shown below. Be sure to check the values in this table to make sure that you understand how the absolute value is working. For instance, when $x = -2$, the value of y is

$$y = |-2 - 1|$$
$$= |-3|$$
$$= 3.$$

Similarly, when $x = 2$, the value of y is $|2 - 1| = 1$.

x	-2	-1	0	1	2	3	4
$y = \|x - 1\|$	3	2	1	0	1	2	3
Solution point	$(-2, 3)$	$(-1, 2)$	$(0, 1)$	$(1, 0)$	$(2, 1)$	$(3, 2)$	$(4, 3)$

Plot the solution points, as shown in Figure 4.17. It appears that the points lie in a "V-shaped" pattern, with the point (1, 0) lying at the bottom of the "V." Following this pattern, connect the points to form the graph shown in Figure 4.18.

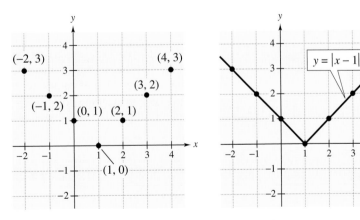

Figure 4.17   Figure 4.18

2 Find and use *x*- and *y*-intercepts as aids to sketching graphs.

Intercepts: Aids to Sketching Graphs

Two types of solution points that are especially useful are those having zero as either the *x*- or *y*-coordinate. Such points are called **intercepts** because they are the points at which the graph intersects the *x*- or *y*-axis.

Definition of Intercepts

The point $(a, 0)$ is called an **x-intercept** of the graph of an equation if it is a solution point of the equation. To find the *x*-intercept(s), let $y = 0$ and solve the equation for *x*.

The point $(0, b)$ is called a **y-intercept** of the graph of an equation if it is a solution point of the equation. To find the *y*-intercept(s), let $x = 0$ and solve the equation for *y*.

Example 4   Finding the Intercepts of a Graph

Find the intercepts and sketch the graph of $y = 2x - 5$.

Solution

To find any *x*-intercepts, let $y = 0$ and solve the resulting equation for *x*.

$$y = 2x - 5 \qquad \text{Write original equation.}$$

$$0 = 2x - 5 \qquad \text{Let } y = 0.$$

$$\frac{5}{2} = x \qquad \text{Solve equation for } x.$$

To find any *y*-intercepts, let $x = 0$ and solve the resulting equation for *y*.

$$y = 2x - 5 \qquad \text{Write original equation.}$$

$$y = 2(0) - 5 \qquad \text{Let } x = 0.$$

$$y = -5 \qquad \text{Solve equation for } y.$$

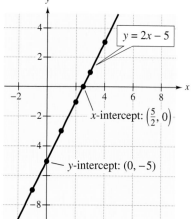

Figure 4.19

So, the graph has one *x*-intercept, which occurs at the point $\left(\frac{5}{2}, 0\right)$, and one *y*-intercept, which occurs at the point $(0, -5)$. To sketch the graph of the equation, create a table of values. (Include the intercepts in the table.) Then plot the points and connect the points with a line, as shown in Figure 4.19.

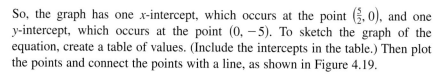

x	-1	0	1	2	$\frac{5}{2}$	3	4
$y = 2x - 5$	-7	-5	-3	-1	0	1	3
Solution point	$(-1, -7)$	$(0, -5)$	$(1, -3)$	$(2, -1)$	$\left(\frac{5}{2}, 0\right)$	$(3, 1)$	$(4, 3)$

When you create a table of values, include any intercepts you have found. You should also include points to the left and to the right of the intercepts. This helps to give a more complete view of the graph.

③ Use the verbal problem-solving method to write an equation and sketch its graph.

Application

Example 5 Depreciation

The value of a $35,500 sport utility vehicle (SUV) depreciates over 10 years (the depreciation is the same each year). At the end of the 10 years, the salvage value is expected to be $5500.

a. Find an equation that relates the value of the SUV to the number of years.

b. Sketch the graph of the equation.

c. What is the y-intercept of the graph and what does it represent in the context of the problem?

Solution

a. The total depreciation over the 10 years is $35,500 - 5500 = \$30,000$. Because the same amount is depreciated each year, it follows that the annual depreciation is $30,000/10 = \$3000$.

Verbal Model:	Value after t years	=	Original value	−	Annual depreciation	·	Number of years

Labels: Value after t years $= y$ (dollars)
            Original value $= 35,500$ (dollars)
            Annual depreciation $= 3000$ (dollars per year)
            Number of years $= t$ (years)

Algebraic Model: $y = 35,500 - 3000t$

b. A sketch of the graph of the depreciation equation is shown in Figure 4.20.

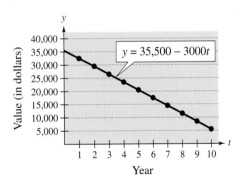

$$y = 35,500 - 3000t$$

Figure 4.20

c. To find the y-intercept of the graph, let $t = 0$ and solve the equation for y.

$y = 35,500 - 3000t$ Write original equation.

$y = 35,000 - 3000(0)$ Substitute 0 for t.

$y = 35,500$ Simplify.

So, the y-intercept is $(0, 35,500)$, and it corresponds to the original value of the SUV.

4.2 Exercises

Review   *Concepts, Skills, and Problem Solving*

Keep mathematically in shape by doing these exercises *before* the problems of this section.

Properties and Definitions

1. *Writing* ✏ If $x - 2 > 5$ and c is an algebraic expression, then what is the relationship between $x - 2 + c$ and $5 + c$? $x - 2 + c > 5 + c$

2. *Writing* ✏ If $x - 2 < 5$ and $c < 0$, then what is the relationship between $(x - 2)c$ and $5c$?
   $(x - 2)c > 5c$

3. Complete the Multiplicative Inverse Property:
   $x(1/x) = \boxed{1}$.

4. Identify the property of real numbers illustrated by $x + y = y + x$.   Commutative Property of Addition

Simplifying Expressions

In Exercises 5–10, simplify the expression.

5. $-3(3x - 2y) + 5y$
   $-9x + 11y$

6. $3z - (4 - 5z)$
   $8z - 4$

7. $-y^2(y^2 + 4) + 6y^2$
   $-y^4 + 2y^2$

8. $5t(2 - t) + t^2$
   $10t - 4t^2$

9. $3[6x - 5(x - 2)]$
   $3x + 30$

10. $5(t - 2) - 5(t - 2)$
    0

Problem Solving

11. *Company Reimbursement*   A company reimburses its sales representatives \$30 per day plus 35 cents per mile for the use of their personal cars. A sales representative submits a bill for \$52.75 for driving her own car.

    (a) How many miles did she drive?   65 miles

    (b) How many days did she drive? Explain.
        1 day, since $2(30) > 52.75$

12. 🔺   *Geometry*   The width of a rectangular mirror is $\frac{3}{5}$ its length. The perimeter of the mirror is 80 inches. What are the measurements of the mirror?
    25×15 inches

Developing Skills

In Exercises 1–8, match the equation with its graph. [The graphs are labeled (a), (b), (c), (d), (e), (f), (g), and (h).]

(a)

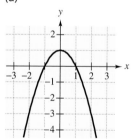

(b)

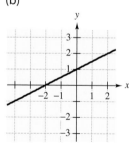

(c)

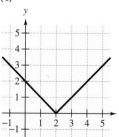

(d)

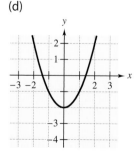

(e)

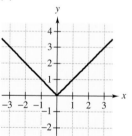

(f)

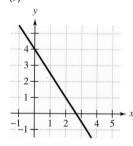

(g)

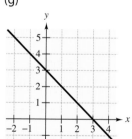

(h)

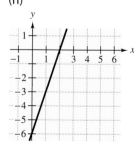

1. $y = 3 - x$ g

2. $y = \frac{1}{2}x + 1$ b

3. $y = -x^2 + 1$ a

4. $y = |x|$ e

5. $y = 3x - 6$ h

6. $y = |x - 2|$ c

7. $y = x^2 - 2$ d

8. $y = 4 - \frac{3}{2}x$ f

In Exercises 9–16, complete the table and use the results to sketch the graph of the equation. See Examples 1–3. See Additional Answers.

9. $y = 9 - x$

x	-2	-1	0	1	2
y	11	10	9	8	7

10. $y = x - 1$

x	-2	-1	0	1	2
y	-3	-2	-1	0	1

11. $x + 2y = 4$

x	-2	0	2	4	6
y	3	2	1	0	-1

12. $3x - 2y = 6$

x	-2	0	2	4	6
y	-6	-3	0	3	6

13. $y = (x - 1)^2$

x	-1	0	1	2	3
y	4	1	0	1	4

14. $y = x^2 + 3$

x	-2	-1	0	1	2
y	7	4	3	4	7

15. $y = |x + 1|$

x	-3	-2	-1	0	1
y	2	1	0	1	2

16. $y = |x| - 2$

x	-2	-1	0	1	2
y	0	-1	-2	-1	0

In Exercises 17–24, graphically estimate the *x*- and *y*-intercepts of the graph. Then check your results algebraically.

17. $4x - 2y = -8$

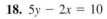
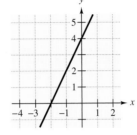

$(-2, 0), (0, 4)$

18. $5y - 2x = 10$

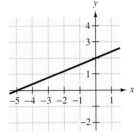

$(-5, 0), (0, 2)$

19. $x + 3y = 6$

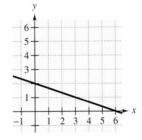

$(6, 0), (0, 2)$

20. $4x + 3y = 12$

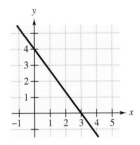

$(3, 0), (0, 4)$

21. $y = |x| - 3$

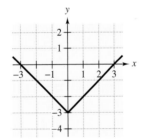

$(-3, 0), (3, 0), (0, -3)$

22. $y = 4 - |x|$

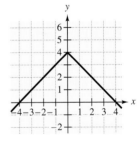

$(-4, 0), (4, 0), (0, 4)$

23. $y = 16 - x^2$

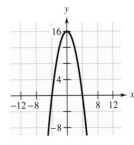

$(-4, 0), (4, 0), (0, 16)$

24. $y = x^2 - 4$

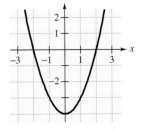

$(-2, 0), (2, 0), (0, -4)$

In Exercises 25–36, find the *x*- and *y*-intercepts (if any) of the graph of the equation. See Example 4.

25. $y = -2x + 7$
$\left(\frac{7}{2}, 0\right), (0, 7)$

26. $y = 5x - 3$
$\left(\frac{3}{5}, 0\right), (0, -3)$

27. $y = \frac{1}{2}x - 1$
$(2, 0), (0, -1)$

28. $y = -\frac{1}{2}x + 3$
$(6, 0), (0, 3)$

29. $x - y = 1$
$(1, 0), (0, -1)$

30. $x + y = 10$
$(10, 0), (0, 10)$

31. $2x + y = 4$
$(2, 0), (0, 4)$

32. $3x - 2y = 1$
$\left(\frac{1}{3}, 0\right), \left(0, -\frac{1}{2}\right)$

33. $2x + 6y - 9 = 0$
$\left(\frac{9}{2}, 0\right), \left(0, \frac{3}{2}\right)$

34. $2x - 5y + 50 = 0$
$(-25, 0), (0, 10)$

35. $\frac{3}{4}x - \frac{1}{2}y = 3$
$(4, 0), (0, -6)$

36. $\frac{1}{2}x + \frac{2}{3}y = 1$
$(2, 0), \left(0, \frac{3}{2}\right)$

In Exercises 37–62, sketch the graph of the equation and label the coordinates of at least three solution points.   See Additional Answers.

37. $y = 2 - x$

38. $y = 12 - x$

39. $y = x - 1$

40. $y = x + 8$

41. $y = 3x$

42. $y = -2x$

43. $4x + y = 6$

44. $2x + y = -2$

45. $10x + 5y = 20$

46. $7x + 7y = 14$

47. $4x - y = 2$

48. $2x - y = 5$

49. $y = \frac{3}{8}x + 15$

50. $y = 14 - \frac{2}{3}x$

51. $y = \frac{2}{3}x - 5$

52. $y = \frac{3}{2}x + 3$

53. $y = x^2$

54. $y = -x^2$

55. $y = -x^2 + 9$

56. $y = x^2 - 1$

57. $y = (x - 3)^2$

58. $y = -(x + 2)^2$

59. $y = |x - 5|$

60. $y = |x + 3|$

61. $y = 5 - |x|$

62. $y = |x| - 3$

In Exercises 63–66, use a graphing calculator to graph both equations in the same viewing window. Are the graphs identical? If so, what property of real numbers is being illustrated?   See Additional Answers.

63. $y_1 = \frac{1}{3}x - 1$
$y_2 = -1 + \frac{1}{3}x$
Yes; Commutative Property of Addition

64. $y_1 = 3\left(\frac{1}{4}x\right)$
$y_2 = \left(3 \cdot \frac{1}{4}\right)x$
Yes; Associative Property of Multiplication

65. $y_1 = 2(x - 2)$
$y_2 = 2x - 4$
Yes; Distributive Property

66. $y_1 = 2 + (x + 4)$
$y_2 = (2 + x) + 4$
Yes; Associative Property of Addition

In Exercises 67–74, use a graphing calculator to graph the equation. (Use a standard viewing window.)
See Additional Answers.

67. $y = 4x$

68. $y = -2x$

69. $y = -\frac{1}{3}x$

70. $y = \frac{1}{2}x$

71. $y = -2x^2 + 5$

72. $y = x^2 - 7$

73. $y = |x + 1| - 2$

74. $y = 4 - |x - 2|$

In Exercises 75 and 76, use a graphing calculator and the given viewing window to graph the equation.
See Additional Answers.

75. $y = 25 - 5x$

Xmin = -5
Xmax = 7
Xscl = 1
Ymin = -5
Ymax = 30
Yscl = 5

76. $y = 1.7 - 0.1x$

Xmin = -10
Xmax = 25
Xscl = 5
Ymin = -5
Ymax = 5
Yscl = .5

In Exercises 77–80, use a graphing calculator to graph the equation and find a viewing window that yields a graph that matches the one shown.
See Additional Answers.

77. $y = \frac{1}{2}x + 2$

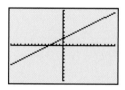

78. $y = 2x - 1$

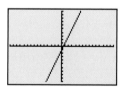

79. $y = \frac{1}{4}x^2 - 4x + 12$

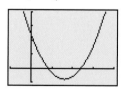

80. $y = 16 - 4x - x^2$

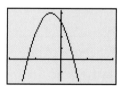

Solving Problems

81. *Creating a Model*   Let *y* represent the distance traveled by a car that is moving at a constant speed of 35 miles per hour. Let *t* represent the number of hours the car has traveled. Write an equation that relates *y* to *t* and sketch its graph. $y = 35t$ See Additional Answers.

82. *Creating a Model*   The cost of printing a book is $500, plus $5 per book. Let *C* represent the total cost and let *x* represent the number of books. Write an equation that relates *C* and *x* and sketch its graph. $C = 500 + 5x$ See Additional Answers.

83. *Life Expectancy*   The table shows the life expectancy *y* (in years) in the United States for a child at birth for the years 1994 through 1999.

Year	1995	1996	1997	1998	1999	2000
y	75.8	76.1	76.5	76.7	76.7	76.9

A model for this data is $y = 0.21t + 74.8$, where *t* is the time in years, with $t = 5$ corresponding to 1995. (Source: U.S. National Center for Health Statistics and U.S. Census Bureau)

(a) Plot the data and graph the model on the same set of coordinate axes.   See Additional Answers.

(b) Use the model to predict the life expectancy for a child born in 2010.   79.0 years

84. *Graphical Comparisons*   The graphs of two types of depreciation are shown. In one type, called *straight-line depreciation*, the value depreciates by the same amount each year. In the other type, called *declining balances*, the value depreciates by the same percent each year. Which is which?   Left: Declining balances; Right: Straight-line depreciation

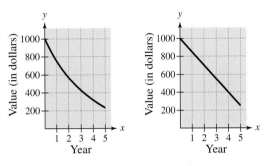

85. *Graphical Interpretation*   In Exercise 84, what is the original cost of the equipment that is being depreciated? $1000

86. *Writing*   Compare the benefits and disadvantages of the two types of depreciation shown in Exercise 84.   Straight-line depreciation is easier to compute. The declining balances method yields a more realistic approximation of the higher rate of depreciation early in the useful lifetime of the equipment.

Explaining Concepts

87. *Writing*   In your own words, define what is meant by the *graph* of an equation.   The set of all solutions of an equation plotted on a rectangular coordinate system is called its graph.

88. *Writing*   In your own words, describe the point-plotting method of sketching the graph of an equation.   Make up a table of values showing several solution points. Plot these points on a rectangular coordinate system and connect them with a smooth curve or line.

89. *Writing*   In your own words, describe how you can check that an ordered pair (x, y) is a solution of an equation.   Substitute the coordinates for the respective variables in the equation and determine if the equation is true.

90. *Writing*   Explain how to find the *x*- and *y*-intercepts of a graph.   To find the *x*-intercept(s), let $y = 0$ and solve the equation for *x*. To find the *y*-intercept(s), let $x = 0$ and solve the equation for *y*.

91. You are walking toward a tree. Let *x* represent the time (in seconds) and let *y* represent the distance (in feet) between you and the tree. Sketch a possible graph that shows how *x* and *y* are related.   See Additional Answers.

92. How many solution points can an equation in two variables have? How many points do you need to determine the general shape of the graph?   An equation in two variables has an infinite number of solutions. The number of points you need to graph an equation depends on the complexity of the graph. A line requires only two points.

4.3 Slope and Graphs of Linear Equations

Kent Meireis/The Image Works

What You Should Learn

(1) Determine the slope of a line through two points.

(2) Write linear equations in slope-intercept form and graph the equations.

(3) Use slopes to determine whether lines are parallel, perpendicular, or neither.

Why You Should Learn It

Slopes of lines can be used in many business applications. For instance, in Exercise 92 on page 234, you will interpret the meaning of the slopes of linear equations that model the predicted profit for an outerwear manufacturer.

(1) **Determine the slope of a line through two points.**

The Slope of a Line

The **slope** of a nonvertical line is the number of units the line rises or falls vertically for each unit of horizontal change from left to right. For example, the line in Figure 4.21 rises two units for each unit of horizontal change from left to right, and so this line has a slope of $m = 2$.

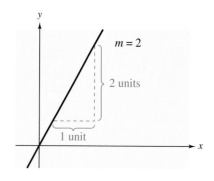

Figure 4.21

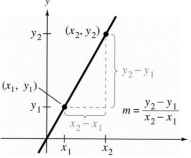

Figure 4.22

Study Tip

In the definition at the right, the *rise* is the vertical change between the points and the *run* is the horizontal change between the points.

Definition of the Slope of a Line

The **slope** m of a nonvertical line passing through the points (x_1, y_1) and (x_2, y_2) is

$$m = \frac{y_2 - y_1}{x_2 - x_1} = \frac{\text{Change in } y}{\text{Change in } x} = \frac{\text{Rise}}{\text{Run}}$$

where $x_1 \neq x_2$ (see Figure 4.22).

When the formula for slope is used, the *order of subtraction* is important. Given two points on a line, you are free to label either of them (x_1, y_1) and the other (x_2, y_2). However, once this has been done, you must form the numerator and denominator using the same order of subtraction.

$$m = \frac{y_2 - y_1}{x_2 - x_1} \qquad m = \frac{y_1 - y_2}{x_1 - x_2} \qquad m = \frac{y_2 - y_1}{x_1 - x_2}$$

  Correct   Correct   Incorrect

Example 1   Finding the Slope of a Line Through Two Points

Find the slope of the line passing through each pair of points.

a. $(-2, 0)$ and $(3, 1)$ **b.** $(0, 0)$ and $(1, -1)$

Solution

You might point out that the subtraction could be done in the opposite order for both x-coordinates and y-coordinates, and the result would be the same.

$$m = \frac{0 - 1}{-2 - 3} = \frac{-1}{-5} = \frac{1}{5}$$

a. Let $(x_1, y_1) = (-2, 0)$ and $(x_2, y_2) = (3, 1)$. The slope of the line through these points is

$$m = \frac{y_2 - y_1}{x_2 - x_1}$$

$$= \frac{1 - 0}{3 - (-2)} \qquad \text{Difference in } y\text{-values}$$
$$\qquad\qquad\qquad \text{Difference in } x\text{-values}$$

$$= \frac{1}{5}. \qquad \text{Simplify.}$$

The graph of the line is shown in Figure 4.23.

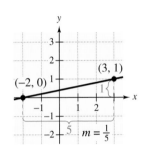

Figure 4.23

The slope of a nonvertical line can be described as the *ratio* of vertical change to horizontal change between any two points on the line.

b. The slope of the line through $(0, 0)$ and $(1, -1)$ is

$$m = \frac{-1 - 0}{1 - 0} \qquad \text{Difference in } y\text{-values}$$
$$\qquad\qquad\qquad \text{Difference in } x\text{-values}$$

$$= \frac{-1}{1} \qquad \text{Simplify.}$$

$$= -1. \qquad \text{Simplify.}$$

The graph of the line is shown in Figure 4.24.

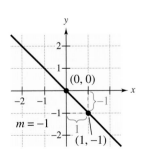

Figure 4.24

Example 2   Horizontal and Vertical Lines and Slope

Find the slope of the line passing through each pair of points.

a. $(-1, 2)$ and $(2, 2)$ **b.** $(2, 4)$ and $(2, 1)$

Solution

a. The line through $(-1, 2)$ and $(2, 2)$ is horizontal because its y-coordinates are the same. The slope of this horizontal line is

$$m = \frac{2 - 2}{2 - (-1)} \qquad \text{Difference in } y\text{-values}$$
$$\qquad\qquad\qquad \text{Difference in } x\text{-values}$$

$$= \frac{0}{3} \qquad\qquad \text{Simplify.}$$

$$= 0. \qquad\qquad \text{Simplify.}$$

The graph of the line is shown in Figure 4.25.

b. The line through $(2, 4)$ and $(2, 1)$ is vertical because its x-coordinates are the same. Applying the formula for slope, you have

$$\frac{4 - 1}{2 - 2} = \frac{3}{0}. \qquad \text{Division by 0 is undefined.}$$

Because division by zero is not defined, the slope of a vertical line is not defined. The graph of the line is shown in Figure 4.26.

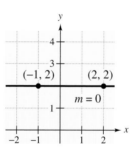

Figure 4.25

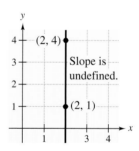

Figure 4.26

From the slopes of the lines shown in Figures 4.23–4.26, you can make several generalizations about the slope of a line.

Slope of a Line

1. A line with positive slope ($m > 0$) *rises* from left to right.

2. A line with negative slope ($m < 0$) *falls* from left to right.

3. A line with zero slope ($m = 0$) is *horizontal.*

4. A line with undefined slope is *vertical.*

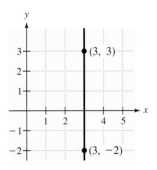

Vertical line: undefined slope
Figure 4.27

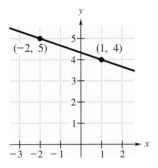

Line falls: negative slope
Figure 4.28

Example 3   Using Slope to Describe Lines

Describe the line through each pair of points.

a. $(3, -2)$ and $(3, 3)$ **b.** $(-2, 5)$ and $(1, 4)$

Solution

a. Let $(x_1, y_1) = (3, -2)$ and $(x_2, y_2) = (3, 3)$.

$$m = \frac{3 - (-2)}{3 - 3} = \frac{5}{0} \qquad \text{Undefined slope (See Figure 4.27.)}$$

Because the slope is undefined, the line is vertical.

b. Let $(x_1, y_1) = (-2, 5)$ and $(x_2, y_2) = (1, 4)$.

$$m = \frac{4 - 5}{1 - (-2)} = -\frac{1}{3} < 0 \qquad \text{Negative slope (See Figure 4.28.)}$$

Because the slope is negative, the line falls from left to right.

Example 4   Using Slope to Describe Lines

Describe the line through each pair of points.

a. $(-4, -3)$ and $(0, -3)$ **b.** $(1, 0)$ and $(4, 6)$

Solution

a. Let $(x_1, y_1) = (-4, -3)$ and $(x_2, y_2) = (0, -3)$.

$$m = \frac{-3 - (-3)}{0 - (-4)} = \frac{0}{4} = 0 \qquad \text{Zero slope (See Figure 4.29.)}$$

Because the slope is zero, the line is horizontal.

b. Let $(x_1, y_1) = (1, 0)$ and $(x_2, y_2) = (4, 6)$.

$$m = \frac{6 - 0}{4 - 1} = \frac{6}{3} = 2 > 0 \qquad \text{Positive slope (See Figure 4.30.)}$$

Because the slope is positive, the line rises from left to right.

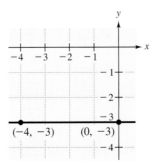

Horizontal line: zero slope
Figure 4.29

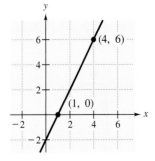

Line rises: positive slope
Figure 4.30

Any two points on a nonvertical line can be used to calculate its slope. This is demonstrated in the next two examples.

Example 5 Finding the Slope of a Ladder

Find the slope of the ladder leading up to the tree house in Figure 4.31.

Solution

Consider the tree trunk as the y-axis and the level ground as the x-axis. The endpoints of the ladder are $(0, 12)$ and $(5, 0)$. So, the slope of the ladder is

$$m = \frac{y_2 - y_1}{x_2 - x_1} = \frac{0 - 12}{5 - 0} = -\frac{12}{5}.$$

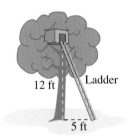

Figure 4.31

Example 6 Finding the Slope of a Line

Sketch the graph of the line $3x - 2y = 4$. Then find the slope of the line. (Choose two different pairs of points on the line and show that the same slope is obtained from either pair.)

Solution

Begin by solving the equation for y.

$$y = \frac{3}{2}x - 2$$

Then, construct a table of values, as shown below.

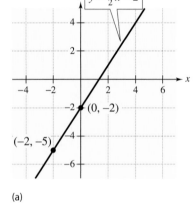

(a)

x	-2	0	2	4
$y = \frac{3}{2}x - 2$	-5	-2	1	4
Solution point	$(-2, -5)$	$(0, -2)$	$(2, 1)$	$(4, 4)$

From the solution points shown in the table, sketch the line, as shown in Figure 4.32. To calculate the slope of the line using two different sets of points, first use the points $(-2, -5)$ and $(0, -2)$, as shown in Figure 4.32(a), and obtain a slope of

$$m = \frac{-2 - (-5)}{0 - (-2)} = \frac{3}{2}.$$

Next, use the points $(2, 1)$ and $(4, 4)$, as shown in Figure 4.32(b), and obtain a slope of

$$m = \frac{4 - 1}{4 - 2} = \frac{3}{2}.$$

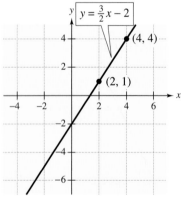

(b)

Figure 4.32

Try some other pairs of points on the line to see that you obtain a slope of $m = \frac{3}{2}$ regardless of which two points you use.

2 Write linear equations in slope-intercept form and graph the equations.

Slope as a Graphing Aid

You saw in Section 4.1 that before creating a table of values for an equation, it is helpful first to solve the equation for y. When you do this for a linear equation, you obtain some very useful information. Consider the results of Example 6.

$3x - 2y = 4$	Write original equation.
$3x - 3x - 2y = -3x + 4$	Subtract $3x$ from each side.
$-2y = -3x + 4$	Simplify.
$\dfrac{-2y}{-2} = \dfrac{-3x + 4}{-2}$	Divide each side by -2.
$y = \dfrac{3}{2}x - 2$	Simplify.

Observe that the coefficient of x is the slope of the graph of this equation (see Example 6). Moreover, the constant term, -2, gives the y-intercept of the graph.

$$y = \underset{\text{slope}}{\boxed{\dfrac{3}{2}}}\, x + \underset{y\text{-intercept } (0,\,-2)}{\boxed{-2}}$$

This form is called the **slope-intercept form** of the equation of the line.

Technology: Tip

Setting the viewing window on a graphing calculator affects the appearance of a line's slope. When you are using a graphing calculator, remember that you cannot judge whether a slope is steep or shallow unless you use a *square* setting—a setting that shows equal spacing of the units on both axes. For many graphing calculators, a square setting is obtained by using the ratio of 10 vertical units to 15 horizontal units.

Slope-Intercept Form of the Equation of a Line

The graph of the equation

$$y = mx + b$$

is a line whose slope is m and whose y-intercept is $(0, b)$. (See Figure 4.33.)

Study Tip

Remember that slope is a *rate of change*. In the slope-intercept equation

$$y = mx + b$$

the slope m is the rate of change of y with respect to x.

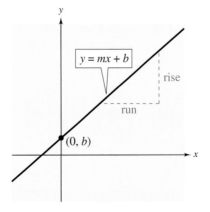

Figure 4.33

So far, you have been plotting several points to sketch the equation of a line. However, now that you can recognize equations of lines, you don't have to plot as many points—two points are enough. (You might remember from geometry that *two points are all that are necessary to determine a line.*) The next example shows how to use the slope to help sketch a line.

Example 7   Using the Slope and *y*-Intercept to Sketch a Line

Use the slope and *y*-intercept to sketch the graph of

$$x - 3y = -6.$$

Solution

First, write the equation in slope-intercept form.

$x - 3y = -6$	Write original equation.
$-3y = -x - 6$	Subtract x from each side.
$y = \dfrac{-x - 6}{-3}$	Divide each side by -3.
$y = \dfrac{1}{3}x + 2$	Simplify to slope-intercept form.

So, the slope of the line is $m = \frac{1}{3}$ and the *y*-intercept is $(0, b) = (0, 2)$. Now you can sketch the graph of the equation. First, plot the *y*-intercept, as shown in Figure 4.34. Then, using a slope of $\frac{1}{3}$,

$$m = \frac{1}{3} = \frac{\text{Change in } y}{\text{Change in } x}$$

locate a second point on the line by moving three units to the right and one unit up (or one unit up and three units to the right), also shown in Figure 4.34. Finally, obtain the graph by drawing a line through the two points (see Figure 4.35).

Point out that the *larger* the *positive* slope of a line, the more steeply the line rises from left to right.

Additional Examples
Use the slope and *y*-intercept to sketch the graph of each equation.

a. $y = \frac{1}{3}x + 1$

b. $2x + y - 3 = 0$

Answers:

a. 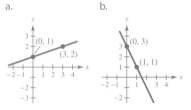 b.

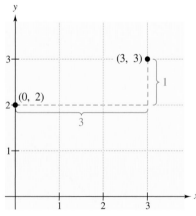

Figure 4.34

Figure 4.35

③ Use slopes to determine whether lines are parallel, perpendicular, or neither.

Parallel and Perpendicular Lines

You know from geometry that two lines in a plane are **parallel** if they do not intersect. What this means in terms of their slopes is shown in Example 8.

Example 8   Lines That Have the Same Slope

On the same set of coordinate axes, sketch the lines $y = 3x$ and $y = 3x - 4$.

Solution

For the line

$$y = 3x$$

the slope is $m = 3$ and the y-intercept is $(0, 0)$. For the line

$$y = 3x - 4$$

the slope is also $m = 3$ and the y-intercept is $(0, -4)$. The graphs of these two lines are shown in Figure 4.36.

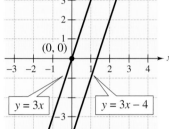

Figure 4.36

In Example 8, notice that the two lines have the same slope and that the two lines appear to be parallel. The following rule states that this is always the case.

Parallel Lines

Two distinct nonvertical lines are parallel if and only if they have the same slope.

The phrase "if and only if" in this rule is used in mathematics as a way to write two statements in one. The first statement says that *if two distinct nonvertical lines have the same slope, they must be parallel*. The second (or reverse) statement says that *if two distinct nonvertical lines are parallel, they must have the same slope*.

Example 9   Lines That Have Negative Reciprocal Slopes

On the same set of coordinate axes, sketch the lines $y = 5x + 2$ and $y = -\frac{1}{5}x - 4$.

Solution

For the line

$$y = 5x + 2$$

the slope is $m = 5$ and the y-intercept is $(0, 2)$. For the line

$$y = -\frac{1}{5}x - 4$$

the slope is $m = -\frac{1}{5}$ and the y-intercept is $(0, -4)$. The graphs of these two lines are shown in Figure 4.37.

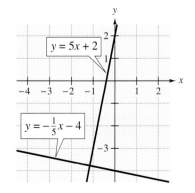

Figure 4.37

Point out that a negative slope indicates that the line falls from left to right.

In Example 9, notice that the two lines have slopes that are negative reciprocals of each other and that the two lines appear to be perpendicular. Another rule from geometry is that two lines in a plane are **perpendicular** if they intersect at right angles. In terms of their slopes, this means that two nonvertical lines are perpendicular if their slopes are negative reciprocals of each other.

Perpendicular Lines

Consider two nonvertical lines whose slopes are m_1 and m_2. The two lines are perpendicular if and only if their slopes are *negative reciprocals* of each other. That is,

$$m_1 = -\frac{1}{m_2}$$

or, equivalently,

$$m_1 \cdot m_2 = -1.$$

Example 10   Parallel or Perpendicular?

Determine whether the pairs of lines are parallel, perpendicular, or neither.

a. $y = -3x - 2$, $y = \frac{1}{3}x + 1$
b. $y = \frac{1}{2}x + 1$, $y = \frac{1}{2}x - 1$

Solution

Consider this additional pair of equations: Line 1: $y = 3x + 1$ and Line 2: $y = -3x + 2$. Because the slopes are not the same and are not negative reciprocals, the lines are neither parallel nor perpendicular. Because they are not parallel, the lines must intersect, but they do not intersect at right angles.

a. The first line has a slope of $m_1 = -3$ and the second line has a slope of $m_2 = \frac{1}{3}$. Because these slopes are negative reciprocals of each other, the two lines must be perpendicular, as shown in Figure 4.38.

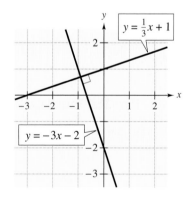

Figure 4.38

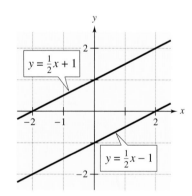

Figure 4.39

b. Both lines have a slope of $m = \frac{1}{2}$. So, the two lines must be parallel, as shown in Figure 4.39.

4.3 Exercises

Review   *Concepts, Skills, and Problem Solving*

Keep mathematically in shape by doing these exercises *before* the problems of this section.

Properties and Definitions

1. Two equations that have the same set of solutions are called   equivalent equations   .

2. Use the Addition Property of Equality to fill in the blank.

$$5x - 2 = 6$$
$$5x = 6 + \boxed{2}$$

Simplifying Expressions

In Exercises 3–10, simplify the expression.

3. $x^2 \cdot x^3$ x^5

4. $z^2 \cdot z^2$ z^4

5. $(-y^2)y$ $-y^3$

6. $5x^2(x^5)$ $5x^7$

7. $(25x^3)(2x^2)$ $50x^5$

8. $(3yz)(6yz^3)$ $18y^2z^4$

9. $x^2 - 2x - x^2 + 3x + 2$ $x + 2$

10. $x^2 - 5x - 2 + x$ $x^2 - 4x - 2$

Problem Solving

11. *Carpentry*   A carpenter must cut a 10-foot board into three pieces. Two are to have the same length and the third is to be three times as long as the two of equal length. Find the lengths of the three pieces.   2 feet, 2 feet, 6 feet

12. *Repair Bill*   The bill for the repair of your dishwasher was $113. The cost for parts was $65. The cost for labor was $32 per hour. How many hours did the repair work take?   1.5 hours

Developing Skills

In Exercises 1–10, estimate the slope (if it exists) of the line from its graph.

1.
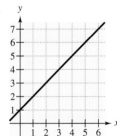
1

2.
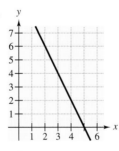
−2

3.
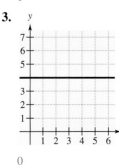
0

4.
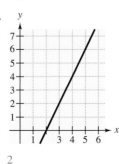
2

5.
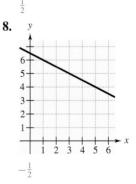
$-\frac{1}{3}$

6.

$\frac{1}{2}$

7.
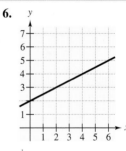
Undefined

8.
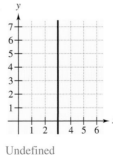
$-\frac{1}{2}$

9.

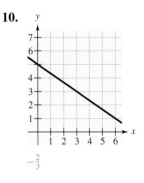

$\frac{5}{4}$

10.

$-\frac{2}{3}$

In Exercises 11 and 12, identify the line in the figure that has each slope.

11. (a) $m = \frac{3}{2}$

(b) $m = 0$

(c) $m = -\frac{2}{3}$

(d) $m = -2$

(a) L_2 (b) L_3

(c) L_4 (d) L_1

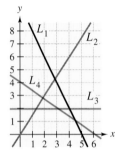

12. (a) $m = -\frac{3}{4}$

(b) $m = \frac{1}{2}$

(c) m is undefined.

(d) $m = 3$

(a) L_2 (b) L_4

(c) L_3 (d) L_1

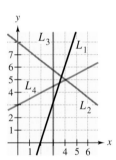

In Exercises 13–32, plot the points and find the slope (if possible) of the line passing through the points. State whether the line rises, falls, is horizontal, or is vertical. See Examples 1–4.   See Additional Answers.

13. $(0, 0), (4, 5)$
$m = \frac{5}{4}$; The line rises.

14. $(0, 0), (3, 6)$
$m = 2$; The line rises.

15. $(0, 0), (8, -4)$
$m = -\frac{1}{2}$; The line falls.

16. $(0, 0), (-1, 3)$
$m = -3$; The line falls.

17. $(0, 6), (8, 0)$
$m = -\frac{3}{4}$; The line falls.

18. $(5, 0), (0, 7)$
$m = -\frac{7}{5}$; The line falls.

19. $(-3, -2), (1, 6)$
$m = 2$; The line rises.

20. $(2, 4), (-4, -4)$
$m = \frac{4}{3}$; The line rises.

21. $(-6, -1), (-6, 4)$
m is undefined.
The line is vertical.

22. $(-4, -10), (-4, 0)$
m is undefined.
The line is vertical.

23. $(3, -4), (8, -4)$
$m = 0$
The line is horizontal.

24. $(1, -2), (-2, -2)$
$m = 0$
The line is horizontal.

25. $\left(\frac{1}{4}, \frac{3}{2}\right), \left(\frac{9}{2}, -3\right)$
$m = -\frac{18}{17}$; The line falls.

26. $\left(\frac{5}{4}, \frac{1}{4}\right), \left(\frac{7}{8}, 2\right)$
$m = -\frac{14}{3}$; The line falls.

27. $(3.2, -1), (-3.2, 4)$
$m = -\frac{25}{32}$; The line falls.

28. $(1.4, 3), (-1.4, 5)$
$m = -\frac{5}{7}$; The line falls.

29. $(3.5, -1), (5.75, 4.25)$
$m = \frac{7}{3}$; The line rises.

30. $(6, 6.4), (-3.1, 5.2)$
$m = \frac{12}{91}$; The line rises.

31. $(a, 3), (4, 3),\ a \neq 4$
$m = 0$
The line is horizontal.

32. $(4, a), (4, 2),\ a \neq 2$
m is undefined.
The line is vertical.

In Exercises 33 and 34, complete the table. Use two different pairs of solution points to show that the same slope is obtained using either pair. See Example 6.
See Additional Answers.

x	-2	0	2	4
y				
Solution point				

33. $y = -2x - 2$ $m = -2$ **34.** $y = 3x + 4$ $m = 3$

In Exercises 35–38, use the formula for slope to find the value of y such that the line through the points has the given slope.

35. Points: $(3, -2), (0, y)$
Slope: $m = -8$
$y = 22$

36. Points: $(-3, y), (8, 2)$
Slope: $m = 2$
$y = -20$

37. Points: $(-4, y), (7, 6)$
Slope: $m = \frac{5}{2}$
$y = -\frac{43}{2}$

38. Points: $(0, 10), (6, y)$
Slope: $m = -\frac{1}{3}$
$y = 8$

In Exercises 39–50, a point on a line and the slope of the line are given. Plot the point and use the slope to find two additional points on the line. (There are many correct answers.)   See Additional Answers.

39. $(2, 1),\ m = 0$
$(0, 1), (1, 1)$

40. $(5, 10),\ m = 0$
$(-2, 10), (8, 10)$

41. $(1, -6),\ m = 2$
$(2, -4), (3, -2)$

42. $(-2, 4),\ m = 1$
$(-3, 3), (-1, 5)$

43. $(0, 1),\ m = -2$
$(1, -1), (2, -3)$

44. $(5, 6),\ m = -3$
$(4, 9), (6, 3)$

45. $(-4, 0)$, $m = \frac{2}{3}$
$(-1, 2), (2, 4)$

46. $(-1, -1)$, $m = \frac{1}{4}$
$(3, 0), (7, 1)$

47. $(3, 5)$, $m = -\frac{1}{2}$
$(5, 4), (7, 3)$

48. $(1, 3)$, $m = -\frac{4}{3}$
$(-2, 7), (4, -1)$

49. $(-8, 1)$
m is undefined.
$(-8, 0), (-8, -1)$

50. $(6, -4)$
m is undefined.
$(6, -1), (6, 2)$

69. $x - 3y + 6 = 0$
$y = \frac{1}{3}x + 2$

70. $3x - 2y - 2 = 0$
$y = \frac{3}{2}x - 1$

71. $x + 2y - 2 = 0$
$y = -\frac{1}{2}x + 1$

72. $10x + 6y - 3 = 0$
$y = -\frac{5}{3}x + \frac{1}{2}$

73. $3x - 4y + 2 = 0$
$y = \frac{3}{4}x + \frac{1}{2}$

74. $2x - 3y + 1 = 0$
$y = \frac{2}{3}x + \frac{1}{3}$

75. $y + 5 = 0$ $y = -5$

76. $y - 3 = 0$ $y = 3$

In Exercises 51–56, sketch the graph of a line through the point $(0, 2)$ having the given slope.   See Additional Answers.

51. $m = 0$

52. m is undefined.

53. $m = 3$

54. $m = -1$

55. $m = -\frac{2}{3}$

56. $m = \frac{3}{4}$

In Exercises 77–80, determine whether the lines L_1 and L_2 passing through the pairs of points are parallel, perpendicular, or neither.

77. L_1: $(0, -1), (5, 9)$
L_2: $(0, 3), (4, 1)$
Perpendicular

78. L_1: $(-2, -1), (1, 5)$
L_2: $(1, 3), (5, 5)$
Neither

79. L_1: $(3, 6), (-6, 0)$
L_2: $(0, -1), \left(5, \frac{7}{3}\right)$
Parallel

80. L_1: $(4, 8), (-4, 2)$
L_2: $(3, -5), \left(-1, \frac{1}{3}\right)$
Perpendicular

In Exercises 57–62, plot the x- and y-intercepts and sketch the graph of the line.   See Additional Answers.

57. $2x + 3y + 6 = 0$

58. $3x + 4y + 12 = 0$

59. $5x - 2y - 10 = 0$

60. $3x - 7y - 21 = 0$

61. $6x - 4y + 12 = 0$

62. $2x - 5y - 20 = 0$

In Exercises 63–76, write the equation in slope-intercept form. Use the slope and y-intercept to sketch the line. See Example 7.   See Additional Answers.

63. $x - y = 0$
$y = x$

64. $x + y = 0$
$y = -x$

65. $\frac{1}{2}x + y = 0$
$y = -\frac{1}{2}x$

66. $\frac{3}{4}x - y = 0$
$y = \frac{3}{4}x$

67. $2x - y - 3 = 0$
$y = 2x - 3$

68. $x - y + 2 = 0$
$y = x + 2$

In Exercises 81–84, sketch the graphs of the two lines on the same rectangular coordinate system. Determine whether the lines are parallel, perpendicular, or neither. Use a graphing calculator to verify your result. (Use a square setting.) See Examples 8–10.   See Additional Answers.

81. $y_1 = 2x - 3$
$y_2 = 2x + 1$
Parallel

82. $y_1 = -\frac{1}{3}x - 3$
$y_2 = \frac{1}{3}x + 1$
Neither

83. $y_1 = 2x - 3$
$y_2 = -\frac{1}{2}x + 1$
Perpendicular

84. $y_1 = -\frac{1}{3}x - 3$
$y_2 = 3x + 1$
Perpendicular

Solving Problems

85. *Roof Pitch*   Determine the slope, or pitch, of the roof of the house shown in the figure. $\frac{2}{5}$

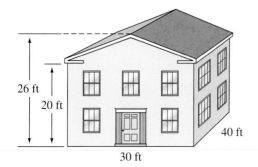

26 ft
20 ft
40 ft
30 ft

86. *Ladder*   Find the slope of the ladder shown in the figure. $-\frac{40}{9}$

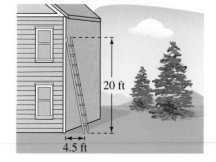

20 ft
4.5 ft

87. *Subway Track*   A subway track rises 3 feet over a 200-foot horizontal distance.

(a) Draw a diagram of the track and label the rise and run.   See Additional Answers.

(b) Find the slope of the track. $\frac{3}{200}$

(c) Would the slope be steeper if the track rose 3 feet over a distance of 100 feet? Explain.

Yes; $\left|\frac{3}{100}\right| > \left|\frac{3}{200}\right|$

88. *Water-Ski Ramp*   In tournament water-ski jumping, the ramp rises to a height of 6 feet on a raft that is 21 feet long.

(a) Draw a diagram of the ramp and label the rise and run.   See Additional Answers.

(b) Find the slope of the ramp. $\frac{2}{7}$

(c) Would the slope be steeper if the ramp rose 6 feet over a distance of 24 feet? Explain.   No; $\left|\frac{2}{8}\right| < \left|\frac{2}{7}\right|$

89. *Flight Path*   An airplane leaves an airport. As it flies over a town, its altitude is 4 miles. The town is about 20 miles from the airport. Approximate the slope of the linear path followed by the airplane during takeoff. $\frac{1}{5}$

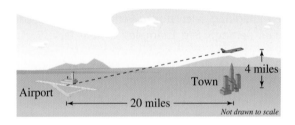

90. *Slide*   The ladder of a straight slide in a playground is 8 feet high. The distance along the ground from the ladder to the foot of the slide is 12 feet. Approximate the slope of the slide. $-\frac{2}{3}$

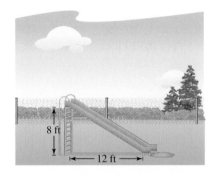

91. *Net Sales*   The graph shows the net sales (in billions of dollars) for Wal-Mart for the years 1996 through 2000.   (Source: 2000 Wal-Mart Annual Report)

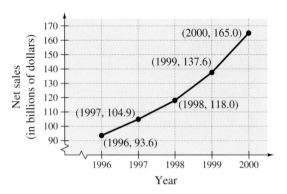

Figure for 91

(a) Find the slopes of the four line segments.
11.3, 13.1, 19.6, 27.4

(b) Find the slope of the line segment connecting the years 1996 and 2000. Interpret the meaning of this slope in the context of the problem.   17.85 is the average annual increase in net sales from 1996 to 2000.

92. *Profit*   Based on different assumptions, the marketing department of an outerwear manufacturer develops two linear models to predict the annual profit of the company over the next 10 years. The models are $P_1 = 0.2t + 2.4$ and $P_2 = 0.3t + 2.4$ where P_1 and P_2 represent profit in millions of dollars and t is time in years ($0 \le t \le 10$).

(a) Interpret the slopes of the two linear models in the context of the problem.
Estimated yearly increase in profits

(b) Which model predicts a faster increase in profits? P_2

(c) Use each model to predict profits when $t = 10$.
$P_1(10) = 4.4$ million, $P_2(10) = 5.4$ million

(d) 🖩 Use a graphing calculator to graph the models in the same viewing window.   See Additional Answers.

Rate of Change   In Exercises 93–98, the slopes of lines representing annual sales y in terms of time t in years are given. Use the slopes to determine any change in annual sales for a 1-year increase in time t.

93. $m = 76$ Sales increase by 76 units.

94. $m = 0$ Sales do not change.

95. $m = 18$ Sales increase by 18 units.

96. $m = 0.5$ Sales increase by 0.5 unit.

97. $m = -14$ Sales decrease by 14 units.

98. $m = -4$ Sales decrease by 4 units.

Explaining Concepts

99. *Writing* Is the slope of a line a ratio? Explain. Yes. The slope is the ratio of the change in y to the change in x.

100. *Writing* Explain how you can visually determine the sign of the slope of a line by observing the graph of the line. The slope is positive if the line rises to the right and negative if it falls to the right.

101. *True or False?* If both the x- and y-intercepts of a line are positive, then the slope of the line is positive. Justify your answer. False. Both the x- and y-intercepts of the line $y = -x + 5$ are positive, but the slope is negative.

102. *Writing* Which slope is steeper: -5 or 2? Explain. -5; The steeper line is the one whose slope has the greater absolute value.

103. *Writing* Is it possible to have two perpendicular lines with positive slopes? Explain. No. The slopes of nonvertical perpendicular lines have opposite signs. The slopes are the negative reciprocals of each other.

104. *Writing* The slope of a line is $\frac{3}{2}$. x is increased by eight units. How much will y change? Explain. For each 2-unit increase in x, y will increase by 3 units. Because there are four 2-unit increases in x, y will increase by 12 units.

105. When a quantity y is increasing or decreasing at a constant rate over time t, the graph of y versus t is a line. What is another name for the rate of change? The slope

106. *Writing* Explain how to use slopes to determine if the points $(-2, -3)$, $(1, 1)$, and $(3, 4)$ lie on the same line. If the points lie on the same line, the slopes of the lines between any two pairs of points will be the same.

107. *Writing* When determining the slope of the line through two points, does the order of subtracting coordinates of the points matter? Explain. Yes. You are free to label either one of the points as (x_1, y_1) and the other as (x_2, y_2). However, once this is done, you must form the numerator and denominator using the same order of subtraction.

108. *Misleading Graphs*

(a) Use a graphing calculator to graph the line $y = 0.75x - 2$ for each viewing window. See Additional Answers.

Xmin = -10	Xmin = 0
Xmax = 10	Xmax = 1
Xscl = 2	Xscl = 0.5
Ymin = -100	Ymin = -2
Ymax = 100	Ymax = -1.5
Yscl = 10	Yscl = 0.1

(b) Do the lines appear to have the same slope? No

(c) Does either of the lines appear to have a slope of 0.75? If not, find a viewing window that will make the line appear to have a slope of 0.75. No. Use the square feature.

(d) Describe real-life situations in which it would be to your advantage to use the two given settings. Answers will vary.

Mid-Chapter Quiz

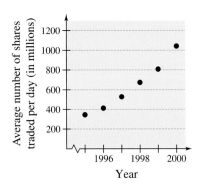

Figure for 4

4. 1995:   340 million
     1996:   410 million
     1997:   530 million
     1998:   670 million
     1999:   810 million
     2000: 1040 million

13. $m = -3$; The line falls.

14. $m = 0$; The line is horizontal.

15. $m = \frac{3}{5}$; The line rises.

18. Substitute the coordinates for the respective variables in the equation and determine if the equation is true.

19. (a) $y = 3000 - 500t$

   (b) See Additional Answers.

   (c) $(0, 3000)$; The value of the computer system when it is first introduced into the market

Take this quiz as you would take a quiz in class. After you are done, check your work against the answers in the back of the book.

1. Plot the points $(4, -2)$ and $\left(-1, -\frac{5}{2}\right)$ on a rectangular coordinate system.
See Additional Answers.

2. Determine the quadrant(s) in which the point $(x, 5)$ is located without plotting it. (x is a real number.)   Quadrants I and II

3. Determine whether each ordered pair is a solution of the equation $y = 9 - |x|$.

   (a) $(2, 7)$ (b) $(-3, 12)$ (c) $(-9, 0)$ (d) $(0, -9)$

   (a) Solution   (b) Not a solution   (c) Solution   (d) Not a solution

4. The scatter plot at the left shows the average number of shares traded per day (in millions) on the New York Stock Exchange for the years 1995 through 2000. Estimate the average number of shares traded per day for each year from 1995 to 2000.   (Source: The New York Stock Exchange, Inc.)

In Exercises 5 and 6, find the x- and y-intercepts of the graph of the equation.

5. $x - 3y = 12$ $(12, 0), (0, -4)$ **6.** $y = -7x + 2$ $\left(\frac{2}{7}, 0\right), (0, 2)$

In Exercises 7–12, sketch the graph of the equation.   See Additional Answers.

7. $y = x - 1$ **8.** $y = 5 - 2x$ **9.** $y = 4 - x^2$

10. $y = (x + 2)^2$ **11.** $y = |x + 3|$ **12.** $y = 1 - |x|$

In Exercises 13–15, plot the points and find the slope (if possible) of the line passing through the points. State whether the line rises, falls, is horizontal, or is vertical.   See Additional Answers.

13. $(0, 0), (-3, 9)$ **14.** $(3, -5), (7, -5)$ **15.** $\left(4, \frac{1}{2}\right), \left(-1, -\frac{5}{2}\right)$

In Exercises 16 and 17, write the equation in slope-intercept form. Use the slope and y-intercept to sketch the line.   See Additional Answers.

16. $3x - 3y + 9 = 0$ $y = x + 3$ **17.** $-2x + 3y - 6 = 0$ $y = \frac{2}{3}x + 2$

18. 📟 Use a graphing calculator to graph $y = 3.6x - 2.4$. Graphically estimate the intercepts of the graph. Explain how to verify your estimates algebraically.   See Additional Answers.

19. A new computer system sells for approximately $3000 and depreciates at the rate of $500 per year for 4 years.

   (a) Find an equation that relates the value of the computer system to the number of years.

   (b) Sketch the graph of the equation.

   (c) What is the y-intercept of the graph and what does it represent in the context of the problem?

4.4 Equations of Lines

Davis Barber/PhotoEdit

Why You Should Learn It

Real-life problems can be modeled and solved using linear equations. For instance, in Example 8 on page 243, a linear equation is used to model the relationship between the time and the height of a mountain climber.

What You Should Learn

① Write equations of lines using the point-slope form.

② Write the equations of horizontal and vertical lines.

③ Use linear models to solve application problems.

① Write equations of lines using the point-slope form.

The Point-Slope Form of the Equation of a Line

In Sections 4.1 through 4.3, you studied analytic (or coordinate) geometry. Analytic geometry uses a coordinate plane to give visual representations of algebraic concepts, such as equations or functions.

There are two basic types of problems in analytic geometry.

1. Given an equation, sketch its graph.

Algebra ⟹ Geometry

2. Given a graph, write its equation.

Geometry ⟹ Algebra

In Section 4.3, you worked primarily with the first type of problem. In this section, you will study the second type. Specifically, you will learn how to write the equation of a line when you are given its slope and a point on the line. Before a general formula for doing this is given, consider the following example.

Example 1   Writing an Equation of a Line

A line has a slope of $\frac{5}{3}$ and passes through the point $(2, 1)$. Find its equation.

Solution

Begin by sketching the line, as shown in Figure 4.40. The slope of a line is the same through any two points on the line. So, using *any* representative point (x, y) and the given point $(2, 1)$, it follows that the slope of the line is

$$m = \frac{y - 1}{x - 2}.$$

Difference in y-coordinates
Difference in x-coordinates

By substituting $\frac{5}{3}$ for m, you obtain the equation of the line.

$$\frac{5}{3} = \frac{y - 1}{x - 2}$$ Slope formula

$$5(x - 2) = 3(y - 1)$$ Cross-multiply.

$$5x - 10 = 3y - 3$$ Distributive Property

$$5x - 3y = 7$$ Equation of line

Figure 4.40

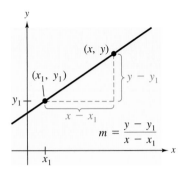

Figure 4.41

Point out the relationship between the point-slope form of the equation of a line and the definition of slope.

The procedure in Example 1 can be used to derive a *formula* for the equation of a line given its slope and a point on the line. In Figure 4.41, let (x_1, y_1) be a given point on a line whose slope is m. If (x, y) is any *other* point on the line, it follows that

$$\frac{y - y_1}{x - x_1} = m.$$

This equation in variables x and y can be rewritten in the form

$$y - y_1 = m(x - x_1)$$

which is called the **point-slope form** of the equation of a line.

Point-Slope Form of the Equation of a Line

The **point-slope form** of the equation of a line with slope m and passing through the point (x_1, y_1) is

$$y - y_1 = m(x - x_1).$$

Example 2   The Point-Slope Form of the Equation of a Line

Write an equation of the line that passes through the point $(1, -2)$ and has slope $m = 3$.

Solution

Use the point-slope form with $(x_1, y_1) = (1, -2)$ and $m = 3$.

$$y - y_1 = m(x - x_1) \qquad \text{Point-slope form}$$
$$y - (-2) = 3(x - 1) \qquad \text{Substitute } -2 \text{ for } y_1, 1 \text{ for } x_1, \text{ and } 3 \text{ for } m.$$
$$y + 2 = 3x - 3 \qquad \text{Simplify.}$$
$$y = 3x - 5 \qquad \text{Equation of line}$$

So, an equation of the line is $y = 3x - 5$. Note that this is the slope-intercept form of the equation. The graph of this line is shown in Figure 4.42.

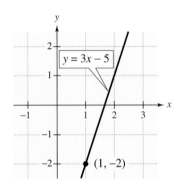

Figure 4.42

In Example 2, note that it was concluded that $y = 3x - 5$ is "an" equation of the line rather than "the" equation of the line. The reason for this is that every equation can be written in many equivalent forms. For instance,

$$y = 3x - 5, \quad 3x - y = 5, \quad \text{and} \quad 3x - y - 5 = 0$$

are all equations of the line in Example 2. The first of these equations $(y = 3x - 5)$ is in the slope-intercept form

$$y = mx + b \qquad \text{Slope-intercept form}$$

and it provides the most information about the line. The last of these equations $(3x - y - 5 = 0)$ is in the general form of the equation of a line.

$$ax + by + c = 0 \qquad \text{General form}$$

The point-slope form can be used to find an equation of a line passing through any two points (x_1, y_1) and (x_2, y_2). First, use the formula for the slope of a line passing through these two points.

$$m = \frac{y_2 - y_1}{x_2 - x_1}$$

Then, knowing the slope, use the point-slope form to obtain the equation

$$y - y_1 = \frac{y_2 - y_1}{x_2 - x_1}(x - x_1). \qquad \text{Two-point form}$$

This is sometimes called the **two-point form** of the equation of a line.

Example 3   An Equation of a Line Passing Through Two Points

Write an equation of the line that passes through the points $(3, 1)$ and $(-3, 4)$.

Solution

Let $(x_1, y_1) = (3, 1)$ and $(x_2, y_2) = (-3, 4)$. The slope of a line passing through these points is

$$m = \frac{y_2 - y_1}{x_2 - x_1} \qquad \text{Formula for slope}$$

$$= \frac{4 - 1}{-3 - 3} \qquad \text{Substitute for } x_1, y_1, x_2, \text{ and } y_2.$$

$$= \frac{3}{-6} \qquad \text{Simplify.}$$

$$= -\frac{1}{2}. \qquad \text{Simplify.}$$

Now, use the point-slope form to find an equation of the line.

$$y - y_1 = m(x - x_1) \qquad \text{Point-slope form}$$

$$y - 1 = -\frac{1}{2}(x - 3) \qquad \text{Substitute 1 for } y_1, \text{ 3 for } x_1, \text{ and } -\frac{1}{2} \text{ for } m.$$

$$y - 1 = -\frac{1}{2}x + \frac{3}{2} \qquad \text{Simplify.}$$

$$y = -\frac{1}{2}x + \frac{5}{2} \qquad \text{Equation of line}$$

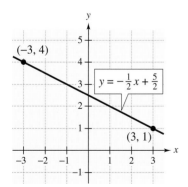

Figure 4.43

The graph of this line is shown in Figure 4.43.

In Example 3, it does not matter which of the two points is labeled (x_1, y_1) and which is labeled (x_2, y_2). Try switching these labels to $(x_1, y_1) = (-3, 4)$ and $(x_2, y_2) = (3, 1)$ and reworking the problem to see that you obtain the same equation.

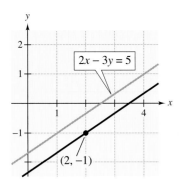

Figure 4.44

Example 4   Equations of Parallel Lines

Write an equation of the line that passes through the point $(2, -1)$ and is parallel to the line

$$2x - 3y = 5$$

as shown in Figure 4.44.

Solution

To begin, write the original equation in slope-intercept form.

$2x - 3y = 5$	Write original equation.
$-3y = -2x + 5$	Subtract $2x$ from each side.
$y = \dfrac{2}{3}x - \dfrac{5}{3}$	Divide each side by -3.

Because the line has a slope of $m = \frac{2}{3}$, it follows that any parallel line must have the same slope. So, an equation of the line through $(2, -1)$, parallel to the original line, is

$y - y_1 = m(x - x_1)$	Point-slope form
$y - (-1) = \dfrac{2}{3}(x - 2)$	Substitute -1 for y_1, 2 for x_1, and $\frac{2}{3}$ for m.
$y + 1 = \dfrac{2}{3}x - \dfrac{4}{3}$	Distributive Property
$y = \dfrac{2}{3}x - \dfrac{7}{3}.$	Equation of parallel line

---

Technology: Tip

With a graphing calculator, parallel lines appear to be parallel in both *square* and *nonsquare* window settings. Verify this by graphing $y = 2x - 3$ and $y = 2x + 1$ in both a square and a nonsquare window.

Such is not the case with perpendicular lines, as you can see by graphing $y = 2x - 3$ and $y = -\frac{1}{2}x + 1$ in a square and a nonsquare window.

Example 5   Equations of Perpendicular Lines

Write an equation of the line that passes through the point $(2, -1)$ and is perpendicular to the line $2x - 3y = 5$, as shown in Figure 4.45.

Solution

From Example 4, the original line has a slope of $\frac{2}{3}$. So, any line perpendicular to this line must have a slope of $-\frac{3}{2}$. So, an equation of the line through $(2, -1)$, perpendicular to the original line, is

$y - y_1 = m(x - x_1)$	Point-slope form
$y - (-1) = -\dfrac{3}{2}(x - 2)$	Substitute -1 for y_1, 2 for x_1, and $-\frac{3}{2}$ for m.
$y + 1 = -\dfrac{3}{2}x + 3$	Distributive Property
$y = -\dfrac{3}{2}x + 2.$	Equation of perpendicular line

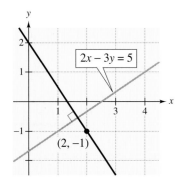

Figure 4.45

2 Write the equations of horizontal and vertical lines.

Equations of Horizontal and Vertical Lines

Recall from Section 4.3 that a horizontal line has a slope of zero. From the slope-intercept form of the equation of a line, you can see that a horizontal line has an equation of the form

$$y = (0)x + b \quad \text{or} \quad y = b. \qquad \text{Horizontal line}$$

Students may have difficulty recognizing equations of horizontal and vertical lines. Here are some examples.

$x = 8$	Vertical
$y = -3$	Horizontal
$x + 7 = 0$	Vertical
$y - 0 = 1$	Horizontal

This is consistent with the fact that each point on a horizontal line through $(0, b)$ has a y-coordinate of b. Similarly, each point on a vertical line through $(a, 0)$ has an x-coordinate of a. Because you know that a vertical line has an undefined slope, you know that it has an equation of the form

$$x = a. \qquad \text{Vertical line}$$

Every line has an equation that can be written in the **general form**

$$ax + by + c = 0 \qquad \text{General form}$$

where a and b are not *both* zero.

Example 6   Writing Equations of Horizontal and Vertical Lines

Write an equation for each line.

a. Vertical line through $(-3, 2)$
b. Line passing through $(-1, 2)$ and $(4, 2)$
c. Line passing through $(0, 2)$ and $(0, -2)$
d. Horizontal line through $(0, -4)$

Solution

a. Because the line is vertical and passes through the point $(-3, 2)$, every point on the line has an x-coordinate of -3. So, the equation of the line is

$$x = -3. \qquad \text{Vertical line}$$

b. Because both points have the same y-coordinate, the line through $(-1, 2)$ and $(4, 2)$ is horizontal. So, its equation is

$$y = 2. \qquad \text{Horizontal line}$$

c. Because both points have the same x-coordinate, the line through $(0, 2)$ and $(0, -2)$ is vertical. So, its equation is

$$x = 0. \qquad \text{Vertical line (y-axis)}$$

d. Because the line is horizontal and passes through the point $(0, -4)$, every point on the line has a y-coordinate of -4. So, the equation of the line is

$$y = -4. \qquad \text{Horizontal line}$$

The graphs of the lines are shown in Figure 4.46.

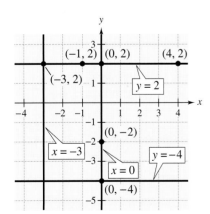

Figure 4.46

In Example 6(c), note that the equation $x = 0$ represents the y-axis. In a similar way, you can show that the equation $y = 0$ represents the x-axis.

3 Use linear models to solve application problems.

Applications

Example 7   Predicting Sales

Harley-Davidson, Inc. had total sales of $2452.9 million in 1999 and $2906.4 million in 2000. Using only this information, write a linear equation that models the sales in terms of the year. Then predict the sales for 2001.   (Source: Harley-Davidson, Inc.)

Solution

Let $t = 9$ represent 1999. Then the two given values are represented by the data points $(9, 2452.9)$ and $(10, 2906.4)$. The slope of the line through these points is

$$m = \frac{2906.4 - 2452.9}{10 - 9}$$

$$= 453.5.$$

Using the point-slope form, you can find the equation that relates the sales y and the year t to be

$y - y_1 = m(t - t_1)$	Point-slope form
$y - 2452.9 = 453.5(t - 9)$	Substitute for y_1, m, and t_1.
$y - 2452.9 = 453.5t - 4081.5$	Distributive Property
$y = 453.5t - 1628.6.$	Write in slope-intercept form.

Using this equation, a prediction of the sales in 2001 ($t = 11$) is

$$y = 453.5(11) - 1628.6 = \$3359.9 \text{ million.}$$

In this case, the prediction is quite good—the actual sales in 2001 were $3363.4 million. The graph of this equation is shown in Figure 4.47.

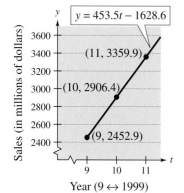

Figure 4.47

The prediction method illustrated in Example 7 is called **linear extrapolation.** Note in Figure 4.48 that for linear extrapolation, the estimated point lies *to the right* of the given points. When the estimated point lies *between* two given points, the method is called **linear interpolation,** as shown in Figure 4.49.

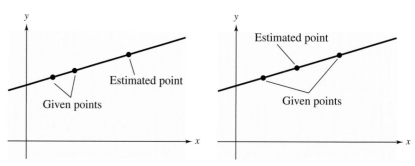

Linear Extrapolation
Figure 4.48

Linear Interpolation
Figure 4.49

In the linear equation $y = mx + b$, you know that m represents the slope of the line. In applications, the slope of a line can often be interpreted as the *rate of change of y with respect to x*. Rates of change should always be described with appropriate units of measure.

Example 8   Using Slope as a Rate of Change

A mountain climber is climbing up a 500-foot cliff. By 1 P.M., the mountain climber has climbed 115 feet up the cliff. By 4 P.M., the climber has reached a height of 280 feet, as shown in Figure 4.50. Find the average rate of change of the climber and use this rate of change to find a linear model that relates the height of the climber to the time.

Solution

Let y represent the height of the climber and let t represent the time. Then the two points that represent the climber's two positions are $(t_1, y_1) = (1, 115)$ and $(t_2, y_2) = (4, 280)$. So, the average rate of change of the climber is

$$\text{Average rate of change} = \frac{y_2 - y_1}{t_2 - t_1} = \frac{280 - 115}{4 - 1} = 55 \text{ feet per hour.}$$

So, an equation that relates the height of the climber to the time is

$$y - y_1 = m(t - t_1) \qquad \text{Point-slope form}$$

$$y - 115 = 55(t - 1) \qquad \text{Substitute } y_1 = 115, t_1 = 1, \text{ and } m = 55.$$

$$y = 55t + 60. \qquad \text{Linear model}$$

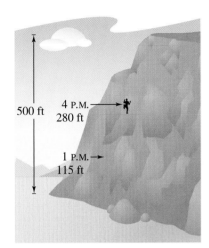

Figure 4.50

You have now studied several formulas that relate to equations of lines. In the summary below, remember that the formulas that deal with slope cannot be applied to vertical lines. For instance, the lines $x = 2$ and $y = 3$ are perpendicular, but they do not follow the "negative reciprocal property" of perpendicular lines because the line $x = 2$ is vertical (and has no slope).

Summary of Equations of Lines

1. Slope of the line through (x_1, y_1) and (x_2, y_2): $m = \dfrac{y_2 - y_1}{x_2 - x_1}$

2. General form of equation of line: $ax + by + c = 0$

3. Equation of vertical line: $x = a$

4. Equation of horizontal line: $y = b$

5. Slope-intercept form of equation of line: $y = mx + b$

6. Point-slope form of equation of line: $y - y_1 = m(x - x_1)$

7. Parallel lines have *equal* slopes: $m_1 = m_2$

8. Perpendicular lines have *negative reciprocal* slopes: $m_1 = -\dfrac{1}{m_2}$

Study Tip

The slope-intercept form of the equation of a line is better suited for *sketching a line*. On the other hand, the point-slope form of the equation of a line is better suited for *creating the equation of a line,* given its slope and a point on the line.

4.4 Exercises

Review   Concepts, Skills, and Problem Solving

Keep mathematically in shape by doing these exercises *before* the problems of this section.

Properties and Definitions

1. *Writing*   Find the greatest common factor of 180 and 300 and explain how you arrived at your answer.
60; The greatest common factor is the product of the common prime factors.

2. *Writing*   Find the least common multiple of 180 and 300 and explain how you arrived at your answer.
900; The least common multiple is the product of the highest powers of the prime factors of the numbers.

Simplifying Expressions

In Exercises 3–6, simplify the expression.

3. $4(3 - 2x)$ $12 - 8x$ **4.** $x^2(xy^3)$ x^3y^3

5. $3x - 2(x - 5)$ $x + 10$ **6.** $u - [3 + (u - 4)]$ 1

Solving Equations

In Exercises 7–10, solve for y in terms of x.

7. $3x + y = 4$ **8.** $4 - y + x = 0$
$y = -3x + 4$ $y = x + 4$

9. $4x - 5y = -2$ **10.** $3x + 4y - 5 = 0$
$y = \frac{4}{5}x + \frac{2}{5}$ $y = -\frac{3}{4}x + \frac{5}{4}$

Developing Skills

In Exercises 1–14, write an equation of the line that passes through the point and has the specified slope. Sketch the line. See Example 1.   See Additional Answers.

1. $(0, 0), m = -2$
$2x + y = 0$

2. $(0, -2), m = 3$
$3x - y = 2$

3. $(6, 0), m = \frac{1}{2}$
$x - 2y = 6$

4. $(0, 10), m = -\frac{1}{4}$
$x + 4y = 40$

5. $(-2, 1), m = 2$
$2x - y = -5$

6. $(3, -5), m = -1$
$x + y = -2$

7. $(-8, -1), m = -\frac{1}{4}$
$x + 4y = -12$

8. $(12, 4), m = -\frac{2}{3}$
$2x + 3y = 36$

9. $\left(\frac{1}{2}, -3\right), m = 0$
$y = -3$

10. $\left(-\frac{5}{4}, 6\right), m = 0$
$y = 6$

11. $\left(0, \frac{3}{2}\right), m = \frac{2}{3}$
$4x - 6y = -9$

12. $\left(0, -\frac{5}{2}\right), m = \frac{3}{4}$
$3x - 4y = 10$

13. $(2, 4), m = -0.8$
$4x + 5y = 28$

14. $(6, -3), m = 0.67$
$67x - 100y = 702$

In Exercises 15–26, use the point-slope form to write an equation of the line that passes through the point and has the specified slope. Write the equation in slope-intercept form. See Example 2.

15. $(0, -4), m = 3$
$y = 3x - 4$

16. $(-7, 0), m = 2$
$y = 2x + 14$

17. $(-3, 6), m = -2$
$y = -2x$

18. $(-4, 1), m = -4$
$y = -4x - 15$

19. $(9, 0), m = -\frac{1}{3}$
$y = -\frac{1}{3}x + 3$

20. $(0, 2), m = \frac{3}{5}$
$y = \frac{3}{5}x + 2$

21. $(-10, 4), m = 0$
$y = 4$

22. $(-2, -5), m = 0$
$y = -5$

23. $(8, 1), m = -\frac{3}{4}$
$y = -\frac{3}{4}x + 7$

24. $(1, 10), m = -\frac{1}{3}$
$y = -\frac{1}{3}x + \frac{31}{3}$

25. $(-2, 1), m = \frac{2}{3}$
$y = \frac{2}{3}x + \frac{7}{3}$

26. $(1, -3), m = \frac{1}{2}$
$y = \frac{1}{2}x - \frac{7}{2}$

In Exercises 27–38, determine the slope of the line. If it is not possible, explain why.

27. $y = \frac{3}{8}x - 4$ $\frac{3}{8}$

28. $y = -\frac{3}{5}x - 2$ $-\frac{3}{5}$

29. $y - 2 = 5(x + 3)$ 5

30. $y + 3 = -2(x - 6)$ -2

31. $y + \frac{5}{6} = \frac{2}{3}(x + 4)$ $\frac{2}{3}$

32. $y - \frac{1}{4} = \frac{5}{8}\left(x - \frac{13}{5}\right)$ $\frac{5}{8}$

33. $y + 9 = 0$ 0

34. $y - 6 = 0$ 0

35. $x - 12 = 0$ Undefined

36. $x + 5 = 0$ Undefined

37. $3x - 2y + 10 = 0$ $\frac{3}{2}$

38. $5x + 4y - 8 = 0$ $-\frac{5}{4}$

In Exercises 39–42, write the slope-intercept form of the line that has the specified y-intercept and slope.

39.

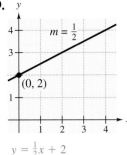

$y = \frac{1}{2}x + 2$

40.

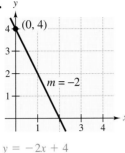

$y = -2x + 4$

41.

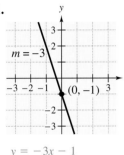

$y = -3x - 1$

42.

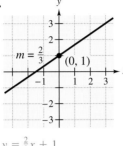

$y = \frac{2}{3}x + 1$

In Exercises 43–46, write the point-slope form of the equation of the line.

43.

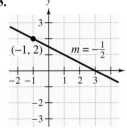

$y - 2 = -\frac{1}{2}(x + 1)$

44.

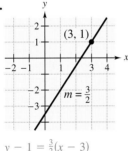

$y - 1 = \frac{3}{2}(x - 3)$

45.

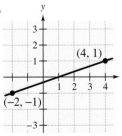

$y + 1 = \frac{1}{3}(x + 2)$ or
$y - 1 = \frac{1}{3}(x - 4)$

46.

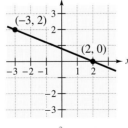

$y - 2 = -\frac{2}{5}(x + 3)$ or
$y - 0 = -\frac{2}{5}(x - 2)$

In Exercises 47–58, write an equation of the line that passes through the points. When possible, write the equation in slope-intercept form. Sketch the line. See Example 3. See Additional Answers.

47. $(0, 0), (4, -4)$
$y = -x$

48. $(0, 0), (-2, 4)$
$y = -2x$

49. $(0, 0), (2, -4)$
$y = -2x$

50. $(6, -1), (3, 3)$
$y = -\frac{4}{3}x + 7$

51. $(-2, 3), (6, -5)$
$y = -x + 1$

52. $(-4, 6), (-2, 3)$
$y = -\frac{3}{2}x$

53. $(-6, 2), (3, 5)$
$y = \frac{1}{3}x + 4$

54. $(-9, 7), (-4, 4)$
$y = -\frac{3}{5}x + \frac{8}{5}$

55. $(5, -1), (3, 2)$
$y = -\frac{3}{2}x + \frac{13}{2}$

56. $(0, 3), (5, 3)$
$y = 3$

57. $\left(\frac{5}{2}, -1\right), \left(\frac{9}{2}, 7\right)$
$y = 4x - 11$

58. $\left(4, \frac{5}{3}\right), \left(-1, \frac{2}{3}\right)$
$y = \frac{1}{5}x + \frac{13}{15}$

In Exercises 59–72, write an equation of the line passing through the points. Write the equation in general form.

59. $(0, 3), (3, 0)$
$x + y - 3 = 0$

60. $(0, -2), (-2, 0)$
$x + y + 2 = 0$

61. $(5, -1), (-5, 5)$
$3x + 5y - 10 = 0$

62. $(4, 3), (-4, 5)$
$x + 4y - 16 = 0$

63. $(5, 4), (1, -4)$
$2x - y - 6 = 0$

64. $(-5, 7), (-2, 1)$
$2x + y + 3 = 0$

65. $(5, -1), (7, -4)$
$3x + 2y - 13 = 0$

66. $(3, 5), (1, 6)$
$x + 2y - 13 = 0$

67. $(-3, 8), (2, 5)$
$3x + 5y - 31 = 0$

68. $(9, -9), (7, -5)$
$2x + y - 9 = 0$

69. $\left(2, \frac{1}{2}\right), \left(\frac{1}{2}, \frac{5}{2}\right)$
$8x + 6y - 19 = 0$

70. $\left(\frac{1}{4}, 1\right), \left(-\frac{3}{4}, -\frac{2}{3}\right)$
$20x - 12y + 7 = 0$

71. $(1, 0.6), (2, -0.6)$
$6x + 5y - 9 = 0$

72. $(-8, 0.6), (2, -2.4)$
$3x + 10y + 18 = 0$

In Exercises 73–82, write equations of the lines through the point (a) parallel and (b) perpendicular to the given line. See Examples 4 and 5.

73. $(2, 1)$
$x - y = 3$
(a) $x - y - 1 = 0$
(b) $x + y - 3 = 0$

74. $(-3, 2)$
$x + y = 7$
(a) $x + y + 1 = 0$
(b) $x - y + 5 = 0$

75. $(-12, 4)$
$3x + 4y = 7$
(a) $3x + 4y + 20 = 0$
(b) $4x - 3y + 60 = 0$

76. $(15, -2)$
$5x + 3y = 0$
(a) $5x + 3y - 69 = 0$
(b) $3x - 5y - 55 = 0$

77. $(1, 3)$
  $2x + y = 0$
  (a) $2x + y - 5 = 0$
  (b) $x - 2y + 5 = 0$

78. $(5, -2)$
  $x + 5y = 3$
  (a) $x + 5y + 5 = 0$
  (b) $5x - y - 27 = 0$

79. $(-1, 0)$
  $y + 3 = 0$
  (a) $y = 0$
  (b) $x + 1 = 0$

80. $(2, 5)$
  $x - 4 = 0$
  (a) $x - 2 = 0$
  (b) $y - 5 = 0$

81. $(4, -1)$
  $3y - 2x = 7$
  (a) $2x - 3y - 11 = 0$
  (b) $3x + 2y - 10 = 0$

82. $(-6, 5)$
  $4x - 5y = 2$
  (a) $4x - 5y + 49 = 0$
  (b) $5x + 4y + 10 = 0$

In Exercises 83–90, write an equation of the line. See Example 6.

83. Vertical line through $(-2, 4)$ $x = -2$
84. Horizontal line through $(7, 3)$ $y = 3$
85. Horizontal line through $\left(\frac{1}{2}, \frac{2}{3}\right)$ $y = \frac{2}{3}$
86. Vertical line through $\left(\frac{1}{4}, 0\right)$ $x = \frac{1}{4}$
87. Line passing through $(4, 1)$ and $(4, 8)$ $x = 4$
88. Line passing through $(-1, 5)$ and $(6, 5)$ $y = 5$
89. Line passing through $(1, -8)$ and $(7, -8)$ $y = -8$
90. Line passing through $(3, 0)$ and $(3, 5)$ $x = 3$

▦ *Graphical Exploration*   In Exercises 91–94, use a graphing calculator to graph the lines in the same viewing window. Use the square setting. Are the lines parallel, perpendicular, or neither? See Additional Answers.

91. $y_1 = -0.4x + 3$

    $y_2 = \frac{5}{2}x - 1$
    Perpendicular

92. $y_1 = \dfrac{2x - 3}{3}$

    $y_2 = \dfrac{4x + 3}{6}$
    Parallel

93. $y_1 = 0.4x + 1$

    $y_2 = x + 2.5$
    Neither

94. $y_1 = \frac{3}{4}x - 5$

    $y_2 = -\frac{3}{4}x + 2$
    Neither

▦ *Graphical Exploration*   In Exercises 95 and 96, use a graphing calculator to graph the equations in the same viewing window. Use the square setting. What can you conclude? See Additional Answers.

95. $y_1 = \frac{1}{3}x + 2$

    $y_2 = -3x + 2$ y_1 and y_2 are perpendicular.

96. $y_1 = 4x + 2$

    $y_2 = -\frac{1}{4}x + 2$ y_1 and y_2 are perpendicular.

Solving Problems

97. *Wages*   A sales representative receives a monthly salary of $2000 plus a commission of 2% of the total monthly sales. Write a linear model that relates total monthly wages W to sales S.
$W = 2000 + 0.02S$

98. *Wages*   A sales representative receives a salary of $2300 per month plus a commission of 3% of the total monthly sales. Write a linear model that relates wages W to sales S. $W = 2300 + 0.03S$

99. *Reimbursed Expenses*   A sales representative is reimbursed $225 per day for lodging and meals plus $0.35 per mile driven. Write a linear model that relates the daily cost C to the number of miles driven x. $C = 225 + 0.35x$

100. *Reimbursed Expenses*   A sales representative is reimbursed $250 per day for lodging and meals plus $0.30 per mile driven. Write a linear model that relates the daily cost C to the number of miles driven x. $C = 250 + 0.30x$

101. *Average Speed*   A car travels for t hours at an average speed of 50 miles per hour. Write a linear model that relates distance d to time t. Graph the model for $0 \le t \le 5$.
$d = 50t$ See Additional Answers.

102. *Discount*   A department store is offering a 20% discount on all items in its inventory.

  (a) Write a linear model that relates the sale price S to the list price L. $S = L - 0.2L = 0.8L$

  (b) ▦ Use a graphing calculator to graph the model. See Additional Answers.

  (c) ▦ Use the graph to estimate the sale price of a coffee maker whose list price is $49.98. Verify your estimate algebraically. $39.98

103. *Depreciation*   A school district purchases a high-volume printer, copier, and scanner for $25,000. After 1 year, its depreciated value is $22,700. The depreciation is linear. See Example 7.

(a) Write a linear model that relates the value V of the equipment to the time t in years.

$V = -2300t + 25{,}000$

(b) Use the model to estimate the value of the equipment after 3 years. $18{,}100

104. *Depreciation*   A sub shop purchases a used pizza oven for $875. After 1 year, its depreciated value is $790. The depreciation is linear.

(a) Write a linear model that relates the value V of the oven to the time t in years.

$V = -85t + 875$

(b) Use the model to estimate the value of the oven after 5 years. $450

105. *Rental Demand*   A real estate office handles an apartment complex with 50 units. When the rent per unit is $580 per month, all 50 units are occupied. However, when the rent is $625 per month, the average number of occupied units drops to 47. Assume that the relationship between the monthly rent p and the demand x is linear.

(a) Represent the given information as two ordered pairs of the form (x, p). Plot these ordered pairs.

$(50, 580), (47, 625)$ See Additional Answers.

(b) Write a linear model that relates the monthly rent p to the demand x. Graph the model and describe the relationship between the rent and the demand. $p = -15x + 1330$; As the rent increases, the demand decreases.   See Additional Answers.

(c) *Linear Extrapolation*   Use the model in part (b) to predict the number of units occupied if the rent is raised to $655.   45 units

(d) *Linear Interpolation*   Use the model in part (b) to estimate the number of units occupied if the rent is $595.   49 units

106. *Soft Drink Demand*   When soft drinks sold for $0.80 per can at football games, approximately 6000 cans were sold. When the price was raised to $1.00 per can, the demand dropped to 4000. Assume that the relationship between the price p and the demand x is linear.

(a) Represent the given information as two ordered pairs of the form (x, p). Plot these ordered pairs.

$(6000, 0.8), (4000, 1)$ See Additional Answers.

(b) Write a linear model that relates the price p to the demand x. Graph the model and describe the relationship between the price and the demand. $p = -0.0001x + 1.4$; As the price increases, the demand decreases.   See Additional Answers.

(c) *Linear Extrapolation*   Use the model in part (b) to predict the number of soft drinks sold if the price is raised to $1.10.   3000 cans

(d) *Linear Interpolation*   Use the model in part (b) to estimate the number of soft drinks sold if the price is $0.90.   5000 cans

107. *Graphical Interpretation*   Match each situation labeled (a), (b), (c), and (d) with one of the graphs labeled (e), (f), (g), and (h). Then determine the slope of each line and interpret the slope in the context of the real-life situation.

(a) A friend is paying you $10 per week to repay a $100 loan.   (f): $m = -10$; Loan decreases by $10 per week.

(b) An employee is paid $12.50 per hour plus $1.50 for each unit produced per hour.   (e): $m = 1.50$; Pay increases by $1.50 per unit.

(c) A sales representative receives $40 per day for food plus $0.32 for each mile traveled.   (g): $m = 0.32$; Amount increases by $0.32 per mile.

(d) A television purchased for $600 depreciates $100 per year.   (h): $m = -100$; Annual depreciation is $100.

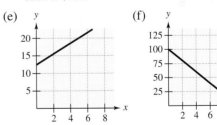

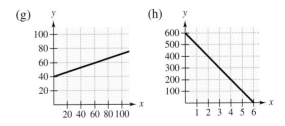

108. *Rate of Change*   You are given the dollar value of a product in 2005 and the rate at which the value is expected to change during the next 5 years. Use this information to write a linear equation that gives the dollar value V of the product in terms of the year t. (Let $t = 5$ represent 2005.)

2005 Value	Rate
(a) $2540	$125 increase per year

$V = 125t + 1915$

(b) $156 $4.50 increase per year

$V = 4.5t + 133.5$

2005 Value	Rate
(c) $20,400	$2000 decrease per year

$V = -2000t + 30,400$

(d) $45,000 $2300 decrease per year

$V = -2300t + 56,500$

(e) $31 $0.75 increase per year

$V = 0.75t + 27.25$

(f) $4500 $800 decrease per year

$V = -800t + 8500$

Explaining Concepts

109.   Answer parts (c)–(f) of Motivating the Chapter on page 198.

110. *Writing*   Can any pair of points on a line be used to calculate the slope of the line? Explain.   Yes. When different pairs of points are selected, the change in y and the change in x are the lengths of the sides of similar triangles. Corresponding sides of similar triangles are proportional.

111. *Writing*   Can the equation of a vertical line be written in slope-intercept form? Explain.   No. The slope is undefined.

112. In the equation $y = mx + b$, what do m and b represent? m is the slope of the line and $(0, b)$ is the y-intercept.

113. In the equation $y - y_1 = m(x - x_1)$, what do x_1 and y_1 represent?   The coordinates of a point on the line

114. *Writing*   Explain how to find analytically the x-intercept of the line given by $y = mx + b$.   Set $y = 0$ and solve the resulting equation for x.   The x-intercept is $\left(-\dfrac{b}{m}, 0\right)$.

115. *Think About It*   Find the slope of the line for the equation $5x + 7y - 21 = 0$. Use the same process to find a formula for the slope of the line $ax + by + c = 0$ where $b \neq 0$. $-\dfrac{5}{7}, -\dfrac{a}{b}$

116. What is implied about the graphs of the lines $a_1x + b_1y + c_1 = 0$ and $a_2x + b_2y + c_2 = 0$ if $\dfrac{a_1}{b_1} = \dfrac{a_2}{b_2}$?   The lines are parallel.

117. *Research Project*   Use a newspaper or weekly news magazine to find an example of data that is *increasing* linearly with time. Find a linear model that relates the data to time. Repeat the project for data that is decreasing.   Answers will vary.

4.5 Graphs of Linear Inequalities

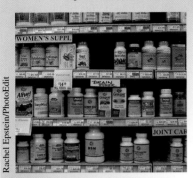

Rachel Epstein/PhotoEdit

Why You Should Learn It

Linear inequalities can be used to model and solve real-life problems. For instance, in Exercise 70 on page 257, you will use a linear inequality to analyze the components of dietary supplements.

What You Should Learn

1. Determine whether an ordered pair is a solution of a linear inequality in two variables.
2. Sketch graphs of linear inequalities in two variables.
3. Use linear inequalities to model and solve real-life problems.

Linear Inequalities in Two Variables

A **linear inequality in two variables,** x and y, is an inequality that can be written in one of the forms below (where a and b are not both zero).

$$ax + by < c, \quad ax + by > c, \quad ax + by \le c, \quad ax + by \ge c$$

Some examples include: $x - y > 2$, $3x - 2y \le 6$, $x \ge 5$, and $y < -1$. An ordered pair (x_1, y_1) is a **solution** of a linear inequality in x and y if the inequality is true when x_1 and y_1 are substituted for x and y, respectively. For instance, the ordered pair $(3, 2)$ is a solution of the inequality $x - y > 0$ because $3 - 2 > 0$ is a true statement.

1. Determine whether an ordered pair is a solution of a linear inequality in two variables.

Example 1   Verifying Solutions of Linear Inequalities

Determine whether each point is a solution of $3x - y \ge -1$.

a. $(0, 0)$ **b.** $(1, 4)$ **c.** $(-1, 2)$

Solution

a. $3x - y \ge -1$ Write original inequality.

$3(0) - 0 \overset{?}{\ge} -1$ Substitute 0 for x and 0 for y.

$0 \ge -1$ Inequality is satisfied. ✓

Because the inequality is satisfied, the point $(0, 0)$ *is* a solution.

b. $3x - y \ge -1$ Write original inequality.

$3(1) - 4 \overset{?}{\ge} -1$ Substitute 1 for x and 4 for y.

$-1 \ge -1$ Inequality is satisfied. ✓

Because the inequality is satisfied, the point $(1, 4)$ *is* a solution.

c. $3x - y \ge -1$ Write original inequality.

$3(-1) - 2 \overset{?}{\ge} -1$ Substitute -1 for x and 2 for y.

$-5 \not\ge -1$ Inequality is not satisfied. ✗

Because the inequality is not satisfied, the point $(-1, 2)$ *is not* a solution.

② Sketch graphs of linear inequalities in two variables.

The Graph of a Linear Inequality in Two Variables

The **graph** of an inequality is the collection of all solution points of the inequality. To sketch the graph of a linear inequality such as

$$3x - 2y < 6 \qquad \text{Original linear inequality}$$

begin by sketching the graph of the *corresponding linear equation*

$$3x - 2y = 6. \qquad \text{Corresponding linear equation}$$

Use *dashed* lines for the inequalities $<$ and $>$ and *solid* lines for the inequalities $\leq$ and $\geq$. The graph of the equation separates the plane into two regions, called **half-planes.** In each half-plane, one of the following *must* be true.

1. All points in the half-plane are solutions of the inequality.

2. No point in the half-plane is a solution of the inequality.

So, you can determine whether the points in an entire half-plane satisfy the inequality by simply testing *one* point in the region. This graphing procedure is summarized as follows.

Sketching the Graph of a Linear Inequality in Two Variables

1. Replace the inequality sign by an equal sign and sketch the graph of the resulting equation. (Use a dashed line for $<$ or $>$ and a solid line for $\leq$ or $\geq$.)

2. Test one point in each of the half-planes formed by the graph in Step 1. If the point satisfies the inequality, then shade the entire half-plane to denote that every point in the region satisfies the inequality.

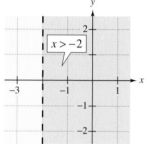

Figure 4.51

Example 2   Sketching the Graph of a Linear Inequality

Sketch the graph of each linear inequality.

a. $x > -2$

b. $y \leq 3$

Solution

a. The graph of the corresponding equation $x = -2$ is a vertical line. The points (x, y) that satisfy the inequality $x > -2$ are those lying to the right of this line, as shown in Figure 4.51.

b. The graph of the corresponding equation $y = 3$ is a horizontal line. The points (x, y) that satisfy the inequality $y \leq 3$ are those lying below (or on) this line, as shown in Figure 4.52.

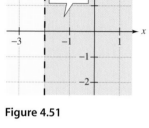

Figure 4.52

Notice that a dashed line is used for the graph of $x > -2$ and a solid line is used for the graph of $y \leq 3$.

Study Tip

You can use any point that is not on the line as a test point. However, the origin is often the most convenient test point because it is easy to evaluate expressions in which 0 is substituted for each variable.

Example 3   Sketching the Graph of a Linear Inequality

Sketch the graph of the linear inequality

$$x - y < 2.$$

Solution

The graph of the corresponding equation

$$x - y = 2 \qquad \text{Write corresponding linear equation.}$$

is a line, as shown in Figure 4.53. Because the origin $(0, 0)$ does not lie on the line, use it as the test point.

$$x - y < 2 \qquad \text{Write original inequality.}$$
$$0 - 0 \overset{?}{<} 2 \qquad \text{Substitute 0 for } x \text{ and 0 for } y.$$
$$0 < 2 \qquad \text{Inequality is satisfied. } \checkmark$$

Because $(0, 0)$ satisfies the inequality, the graph consists of the half-plane lying above the line. Try checking a point below the line. Regardless of the point you choose, you will see that it does not satisfy the inequality.

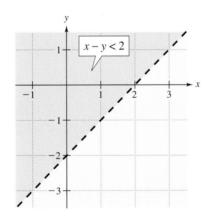

Figure 4.53

Technology: Tip

Many graphing calculators are capable of graphing linear inequalities. Consult the user's guide of your graphing calculator for specific instructions.

The graph of $y \le -x + 2$ is shown at the right.

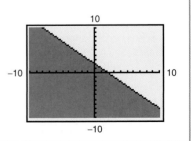

Study Tip

The solution of the inequality

$$y < \frac{3}{2}x - \frac{5}{2}$$

is a half-plane with the line

$$y = \frac{3}{2}x - \frac{5}{2}$$

as its boundary. The y-values that are less than those yielded by this equation make the inequality true. So, you want to shade the half-plane with the smaller y-values, as shown in Figure 4.54.

Additional Example

Sketch the graph of the linear inequality.

$$x - 3y > 6$$

Answer:

For a linear inequality in two variables, you can sometimes simplify the graphing procedure by writing the inequality in *slope-intercept* form. For instance, by writing $x - y < 2$ in the form $y > x - 2$, you can see that the solution points lie *above* the line $y = x - 2$, as shown in Figure 4.53. Similarly, by writing the inequality $3x - 2y > 5$ in the form

$$y < \frac{3}{2}x - \frac{5}{2}$$

you can see that the solutions lie *below* the line $y = \frac{3}{2}x - \frac{5}{2}$, as shown in Figure 4.54.

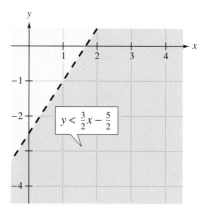

$$y < \frac{3}{2}x - \frac{5}{2}$$

Figure 4.54

Example 4   Sketching the Graph of a Linear Inequality

Use the slope-intercept form of a linear equation as an aid in sketching the graph of the inequality $5x + 4y \le 12$.

Solution

To begin, rewrite the inequality in slope-intercept form.

$5x + 4y \le 12$	Write original inequality.
$4y \le -5x + 12$	Subtract $5x$ from each side.
$y \le -\dfrac{5}{4}x + 3$	Write in slope-intercept form.

From this form, you can conclude that the solution is the half-plane lying *on* or *below* the line $y = -\frac{5}{4}x + 3$. The graph is shown in Figure 4.55. You can verify this by testing the solution point $(0, 0)$.

$5x + 4y \le 12$	Write original inequality.
$5(0) + 4(0) \overset{?}{\le} 12$	Substitute 0 for x and 0 for y.
$0 \le 12$	Inequality is satisfied. ✓

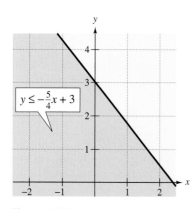

$$y \le -\frac{5}{4}x + 3$$

Figure 4.55

③ Use linear inequalities to model and solve real-life problems.

Application

Example 5   Working to Meet a Budget

Your budget requires you to earn *at least* $160 per week. You work two part-time jobs. One is at a fast-food restaurant, which pays $6 per hour, and the other is tutoring for $8 per hour. Let x represent the number of hours worked at the restaurant and let y represent the number of hours tutoring.

a. Write an inequality that represents the number of hours worked at each job in order to meet your budget requirements.

b. Graph the inequality and identify at least two ordered pairs (x, y) that identify the number of hours you must work at each job in order to meet your budget requirements.

Solution

a. To write the inequality, use the problem-solving method.

Verbal Model:	Hourly pay at fast-food restaurant	·	Number of hours at fast-food restaurant	+	Hourly pay tutoring	·	Number of hours tutoring	≥	Earnings in a week

Labels:	Hourly pay at fast-food restaurant $= 6$	(dollars per hour)
	Number of hours at fast-food restaurant $= x$	(hours)
	Hourly pay tutoring $= 8$	(dollars per hour)
	Number of hours tutoring $= y$	(hours)
	Earnings in a week $= 160$	(dollars)

Algebraic Inequality: $6x + 8y \geq 160$

b. To sketch the graph, rewrite the inequality in slope-intercept form.

$$6x + 8y \geq 160 \qquad \text{Write original inequality.}$$

$$8y \geq -6x + 160 \qquad \text{Subtract } 6x \text{ from each side.}$$

$$y \geq -\frac{3}{4}x + 20 \qquad \text{Divide each side by 8.}$$

Graph the corresponding equation

$$y = -\frac{3}{4}x + 20$$

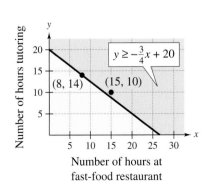

Number of hours at fast-food restaurant

Figure 4.56

and shade the half-plane lying above the line, as shown in Figure 4.56. From the graph, you can see that two solutions that will yield the desired weekly earnings of at least $160 are (8, 14) and (15, 10). In other words, you could work 8 hours at the restaurant and 14 hours as a tutor, or 15 hours at the restaurant and 10 hours as a tutor, to meet your budget requirements. There are many other solutions.

4.5 Exercises

Review *Concepts, Skills, and Problem Solving*

Keep mathematically in shape by doing these exercises *before* the problems of this section.

Properties and Definitions

In Exercises 1–4, complete the property of inequalities by inserting the correct inequality symbol. (Let a, b, and c be real numbers, variables, or algebraic expressions.)

1. If $a < b$, then $a + 5$ $<$ $b + 5$.

2. If $a < b$, then $2a$ $<$ $2b$.

3. If $a < b$, then $-3a$ $>$ $-3b$.

4. If $a < b$ and $b < c$, then a $<$ c.

Solving Inequalities

In Exercises 5–10, solve the inequality and sketch the solution on the real number line.

See Additional Answers.

5. $x + 3 > 0$ $x > -3$ 6. $2 - x \geq 0$ $x \leq 2$

7. $2t - 11 \leq 5$ $t \leq 8$

8. $\frac{3}{2}y + 8 < 20$ $y < 8$

9. $2(x - 5) > 13$ $x > 11.5$

10. $-4(x + 7) \geq -36$ $x \leq 2$

Problem Solving

11. *Sales Commission* A sales representative receives a commission of 4.5% of the total monthly sales. Determine the sales of a representative who earned $544.50 as a sales commission. $12,100.00

12. *Work Rate* One person can complete a typing project in 3 hours, and another can complete the same project in 4 hours. If they both work on the project, in how many hours can it be completed? $\frac{12}{7}$ hours

Developing Skills

In Exercises 1–8, determine whether the points are solutions of the inequality. See Example 1.

Inequality	Points	
1. $x + 4y > 10$	(a) $(0, 0)$	Not a solution
	(b) $(3, 2)$	Solution
	(c) $(1, 2)$	Not a solution
	(d) $(-2, 4)$	Solution
2. $2x + 3y > 9$	(a) $(0, 0)$	Not a solution
	(b) $(1, 1)$	Not a solution
	(c) $(2, 2)$	Solution
	(d) $(-2, 5)$	Solution
3. $-3x + 5y \leq 12$	(a) $(1, 2)$	Solution
	(b) $(2, -3)$	Solution
	(c) $(1, 3)$	Solution
	(d) $(2, 8)$	Not a solution
4. $5x + 3y < 100$	(a) $(25, 10)$	Not a solution
	(b) $(6, 10)$	Solution
	(c) $(0, -12)$	Solution
	(d) $(4, 5)$	Solution

Inequality	Points	
5. $3x - 2y < 2$	(a) $(1, 3)$	Solution
	(b) $(2, 0)$	Not a solution
	(c) $(0, 0)$	Solution
	(d) $(3, -5)$	Not a solution
6. $y - 2x > 5$	(a) $(4, 13)$	Not a solution
	(b) $(8, 1)$	Not a solution
	(c) $(0, 7)$	Solution
	(d) $(1, -3)$	Not a solution
7. $5x + 4y \geq 6$	(a) $(-2, 4)$	Solution
	(b) $(5, 5)$	Solution
	(c) $(7, 0)$	Solution
	(d) $(-2, 5)$	Solution
8. $5y + 8x \leq 14$	(a) $(-3, 8)$	Not a solution
	(b) $(7, -6)$	Not a solution
	(c) $(1, 1)$	Solution
	(d) $(3, 0)$	Not a solution

In Exercises 9–12, state whether the boundary of the graph of the inequality should be dashed or solid.

9. $2x + 3y < 6$ Dashed

10. $2x + 3y \leq 6$ Solid

11. $2x + 3y \geq 6$ Solid

12. $2x + 3y > 6$ Dashed

In Exercises 13–16, match the inequality with its graph. [The graphs are labeled (a), (b), (c), and (d).]

(a)

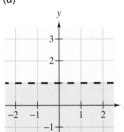

(b)

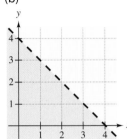

(c)

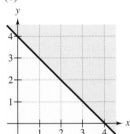

(d)

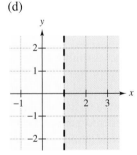

13. $x + y < 4$ b

14. $x + y \geq 4$ c

15. $x > 1$ d

16. $y < 1$ a

In Exercises 17–20, match the inequality with its graph. [The graphs are labeled (a), (b), (c), and (d).]

(a)

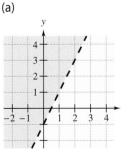

(b)

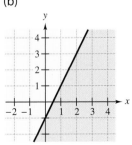

(c)

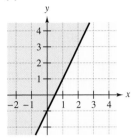

(d)

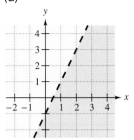

17. $2x - y \leq 1$ c

18. $2x - y < 1$ a

19. $2x - y \geq 1$ b

20. $2x - y > 1$ d

In Exercises 21–50, sketch the graph of the linear inequality. See Examples 2–4.   See Additional Answers.

21. $y \geq 3$

22. $x \leq 0$

23. $x > -4$

24. $y < -2$

25. $y < 3x$

26. $y > 5x$

27. $x - y < 0$

28. $x + y > 0$

29. $y \leq 2x - 1$

30. $y \geq -x + 3$

31. $y \leq 2 - x$

32. $y \geq 2x + 1$

33. $y > 2 - x$

34. $y < -x + 3$

35. $y > -2x + 10$

36. $y < 3x + 1$

37. $y \geq \frac{2}{3}x + \frac{1}{3}$

38. $y \leq -\frac{3}{4}x + 2$

39. $-3x + 2y - 6 < 0$

40. $x - 2y + 6 \leq 0$

41. $2x + y - 3 \geq 3$

42. $x + 4y + 2 \geq 2$

43. $5x + 2y < 5$

44. $5x + 2y > 5$

45. $x \geq 3y - 5$

46. $x > -2y + 10$

47. $y - 3 < \frac{1}{2}(x - 4)$

48. $y + 1 < -2(x - 3)$

49. $\frac{x}{3} + \frac{y}{4} < 1$

50. $\frac{x}{-2} + \frac{y}{2} > 1$

In Exercises 51–58, use a graphing calculator to graph the linear inequality.   See Additional Answers.

51. $y \geq 2x - 1$

52. $y \leq 4 - 0.5x$

53. $y \leq -2x + 4$

54. $y \geq x - 3$

55. $y \geq \frac{1}{2}x + 2$

56. $y \leq -\frac{2}{3}x + 6$

57. $6x + 10y - 15 \leq 0$

58. $3x - 2y + 4 \geq 0$

In Exercises 59–64, write an inequality for the shaded region shown in the figure.

59.

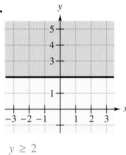

$y \geq 2$

60.

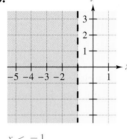

$x < -1$

61.

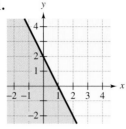

$2x + y \leq 2$

62.

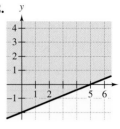

$2x - 5y \leq 10$

63.

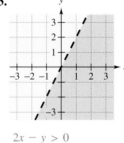

$2x - y > 0$

64.

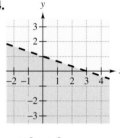

$x + 3y < 3$

Solving Problems

65. *Part-Time Jobs*   You work two part-time jobs. One is at a grocery store, which pays $9 per hour, and the other is mowing lawns, which pays $6 per hour. Between the two jobs, you want to earn at least $150 a week. Write a linear inequality that shows the different numbers of hours you can work at each job, and sketch the graph of the inequality. From the graph, find several ordered pairs with positive integer coordinates that are solutions of the inequality.
$9x + 6y \geq 150$; (x, y): $(20, 0)$, $(10, 15)$, $(5, 30)$
See Additional Answers.

66. *Money*   A cash register must have at least $25 in change consisting of d dimes and q quarters. Write a linear inequality that shows the different numbers of coins that can be in the cash register, and sketch the graph of the inequality. From the graph, find several ordered pairs with positive integer coordinates that are solutions of the inequality.
$0.10d + 0.25q \geq 25$; (d, q): $(250, 0)$, $(0, 100)$, $(300, 50)$
See Additional Answers.

67. *Manufacturing*   Each table produced by a furniture company requires 1 hour in the assembly center. The matching chair requires $1\frac{1}{2}$ hours in the assembly center. A total of 12 hours per day is available in the assembly center. Write a linear inequality that shows the different numbers of hours that can be spent assembling tables and chairs, and sketch the graph of the inequality. From the graph, find several ordered pairs with positive integer coordinates that are solutions of the inequality.
$T + \frac{3}{2}C \leq 12$; (T, C): $(5, 4)$, $(2, 6)$, $(0, 8)$
See Additional Answers.

68. *Inventory*   A store sells two models of computers. The costs to the store of the two models are $2000 and $3000, and the owner of the store does not want more than $30,000 invested in the inventory for these two models. Write a linear inequality that represents the different numbers of each model that can be held in inventory, and sketch the graph of the inequality. From the graph, find several ordered pairs with positive integer coordinates that are solutions of the inequality.
$2x + 3y \leq 30$; (x, y): $(0, 10)$, $(15, 0)$, $(10, 3)$
See Additional Answers.

69. *Sports*   Your hockey team needs at least 60 points for the season in order to advance to the playoffs. Your team finishes with w wins, each worth 2 points, and t ties, each worth 1 point. Write a linear inequality that shows the different numbers of points your team can score to advance to the playoffs, and sketch the graph of the inequality. From the graph, find several ordered pairs with positive integer coordinates that are solutions of the inequality.

$2w + t \geq 60$; (w, t): $(30, 0)$, $(20, 25)$, $(0, 60)$

See Additional Answers.

70. *Nutrition*   A dietitian is asked to design a special dietary supplement using two different foods. Each ounce of food X contains 20 units of calcium and each ounce of food Y contains 10 units of calcium. The minimum daily requirement in the diet is 300 units of calcium. Write a linear inequality that shows the different numbers of units of food X and food Y required, and sketch the graph of the inequality. From the graph, find several ordered pairs with positive integer coordinates that are solutions of the inequality.

$20x + 10y \geq 300$; (x, y): $(10, 10)$, $(5, 20)$, $(0, 30)$

See Additional Answers.

Explaining Concepts

71. ⊘   Answer part (g) of Motivating the Chapter on page 198.

72. List the four forms of a linear inequality in variables x and y.

$ax + by < c$, $ax + by > c$, $ax + by \leq c$, $ax + by \geq c$

73. What is meant by saying that (x_1, y_1) is a solution of a linear inequality in x and y?   The inequality is true when x_1 and y_1 are substituted for x and y, respectively.

74. *Writing*   Explain the difference between graphs that have dashed lines and those that have solid lines.

Use dashed lines for the inequalities $<$ and $>$ and solid lines for the inequalities $\leq$ and $\geq$.

75. *Writing*   After graphing the boundary, explain how you determine which half-plane is the graph of a linear inequality.

Test a point in one of the half-planes.

76. *Writing*   Explain the difference between graphing the solution to the inequality $x \geq 1$ (a) on the real number line and (b) on a rectangular coordinate system.

(a) The solution is an unbounded interval on the x-axis.

(b) The solution is a half-plane.

77. Write the inequality whose graph consists of all points above the x-axis. $y > 0$

78. *Writing*   Does $2x < 2y$ have the same graph as $y > x$? Explain.   Yes; $2x < 2y = 2y > 2x$. Divide each side by 2 to obtain $y > x$.

79. Write an inequality whose graph has no points in the first quadrant. $x + y < 0$

What Did You Learn?

Key Terms

rectangular coordinate
 system, *p. 200*
ordered pair, *p. 200*
x-coordinate, *p. 200*
y-coordinate, *p. 200*
solution point, *p. 203*
graph, *p. 212*

linear equation in two
 variables, *p. 213*
x-intercept, *p. 216*
y-intercept, *p. 216*
slope, *p. 222*
slope-intercept form,
 p. 227

parallel lines, *p. 229*
perpendicular lines, *p. 230*
point-slope form, *p. 238*
general form, *p. 241*
linear inequality in two
 variables, *p. 249*
half-plane, *p. 250*

Key Concepts

4.1 ◯ Rectangular coordinate system

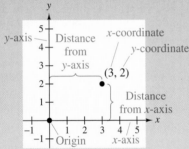

4.2 ◯ Point-plotting method of sketching a graph

1. If possible, rewrite the equation by isolating one of the variables.

2. Make a table of values showing several solution points.

3. Plot these points on a rectangular coordinate system.

4. Connect the points with a smooth curve or line.

4.2 ◯ Finding *x*- and *y*-intercepts

1. To find the *x*-intercept(s), let $y = 0$ and solve the equation for *x*.

2. To find the *y*-intercept(s), let $x = 0$ and solve the equation for *y*.

4.3 ◯ Slope of a line
The slope *m* of a nonvertical line passing through points (x_1, y_1) and (x_2, y_2) is

$$m = \frac{y_2 - y_1}{x_2 - x_1} = \frac{\text{Change in } y}{\text{Change in } x} = \frac{\text{Rise}}{\text{Run}}$$

where $x_1 \neq x_2$.

1. If $m > 0$, the line rises from left to right.

2. If $m < 0$, the line falls from left to right.

3. If $m = 0$, the line is horizontal.

4. If *m* is undefined $(x_1 = x_2)$, the line is vertical.

4.4 ◯ Summary of equations of lines
1. Slope of the line through (x_1, y_1) and (x_2, y_2):

$$m = \frac{y_2 - y_1}{x_2 - x_1}$$

2. General form of equation of line: $ax + by + c = 0$

3. Equation of vertical line: $x = a$

4. Equation of horizontal line: $y = b$

5. Slope-intercept form of equation of line:

$$y = mx + b$$

6. Point-slope form of equation of line:

$$y - y_1 = m(x - x_1)$$

7. Parallel lines have equal slopes: $m_1 = m_2$

8. Perpendicular lines have negative reciprocal slopes:

$$m_1 = -\frac{1}{m_2}$$

4.5 ◯ Sketching the graph of a linear inequality in two variables
1. Replace the inequality sign by an equal sign and sketch the graph of the resulting equation. (Use a dashed line for < or > and a solid line for ≤ or ≥.)

2. Test one point in each of the half-planes formed by the graph in Step 1. If the point satisfies the inequality, then shade the entire half-plane to denote that every point in the region satisfies the inequality.

Review Exercises

4.1 Ordered Pairs and Graphs

1 Plot and find the coordinates of a point on a rectangular coordinate system.

In Exercises 1–4, plot the points on a rectangular coordinate system.   See Additional Answers.

1. $(-1, 6)$, $(4, -3)$, $(-2, 2)$, $(3, 5)$
2. $(0, -1)$, $(-4, 2)$, $(5, 1)$, $(3, -4)$
3. $(-2, 0)$, $\left(\frac{3}{2}, 4\right)$, $(-1, -3)$
4. $\left(3, -\frac{5}{2}\right)$, $\left(-5, 2\frac{3}{4}\right)$, $(4, 6)$

In Exercises 5 and 6, determine the coordinates of the points.

5.
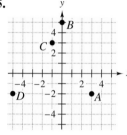

6.
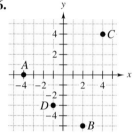

A: $(3, -2)$; B: $(0, 5)$;
C: $(-1, 3)$; D: $(-5, -2)$

A: $(-4, 0)$; B: $(2, -5)$;
C: $(4, 4)$; D: $(-1, -3)$

In Exercises 7–14, determine the quadrant(s) in which the point is located or the axis on which the point is located without plotting it.

7. $(-5, 3)$ Quadrant II
8. $(4, -6)$ Quadrant IV
9. $(4, 0)$ x-axis
10. $(0, -3)$ y-axis
11. $(x, 5)$, $x < 0$
    Quadrant II
12. $(-3, y)$, $y > 0$
    Quadrant II
13. $(-6, y)$, y is a real number.
    Quadrant II or III, or on x-axis
14. $(x, -1)$, x is a real number.
    Quadrant III or IV, or on y-axis

2 Construct a table of values for equations and determine whether ordered pairs are solutions of equations.

In Exercises 15 and 16, complete the table of values. Then plot the solution points on a rectangular coordinate system.   See Additional Answers.

15.

x	-1	0	1	2
$y = 4x - 1$	-5	-1	3	7

16.

x	-1	0	1	2
$y = -\frac{1}{2}x - 1$	$-\frac{1}{2}$	-1	$-\frac{3}{2}$	-2

In Exercises 17–20, solve the equation for y.

17. $3x + 4y = 12$
    $y = -\frac{3}{4}x + 3$
18. $2x + 3y = 6$
    $y = -\frac{2}{3}x + 2$
19. $x - 2y = 8$
    $y = \frac{1}{2}x - 4$
20. $-x - 3y = 9$
    $y = -\frac{1}{3}x - 3$

In Exercises 21–24, determine whether the ordered pairs are solutions of the equation.

21. $x - 3y = 4$
    (a) $(1, -1)$ Solution
    (b) $(0, 0)$ Not a solution
    (c) $(2, 1)$ Not a solution
    (d) $(5, -2)$ Not a solution
22. $y - 2x = -1$
    (a) $(3, 7)$ Not a solution
    (b) $(0, -1)$ Solution
    (c) $(-2, -5)$ Solution
    (d) $(-1, 0)$ Not a solution
23. $y = \frac{2}{3}x + 3$
    (a) $(3, 5)$ Solution
    (b) $(-3, 1)$ Solution
    (c) $(-6, 0)$ Not a solution
    (d) $(0, 3)$ Solution
24. $y = \frac{1}{4}x + 2$
    (a) $(-4, 1)$ Solution
    (b) $(-8, 0)$ Solution
    (c) $(12, 5)$ Solution
    (d) $(0, 2)$ Solution

3 Use the verbal problem-solving method to plot points on a rectangular coordinate system.

25. *Organizing Data*   The data from a study measuring the relationship between the wattage x of a standard 120-volt light bulb and the energy rate y (in lumens) is shown in the table.

x	25	40	60	100	150	200
y	235	495	840	1675	2650	3675

(a) Plot the data shown in the table.
    See Additional Answers.
(b) Use the graph to describe the relationship between the wattage and energy rate.
    Approximately linear

26. *Organizing Data*   The table shows the average salaries (in thousands of dollars) for professional baseball players in the United States for the years 1997 through 2002, where x represents the year. (Source: Major League Baseball and the Associated Press)

x	1997	1998	1999	2000	2001	2002
y	1314	1385	1572	1834	2089	2341

(a) Plot the data shown in the table. See Additional Answers.

(b) Use the graph to describe the relationship between the year and the average salary.
Approximately linear

(c) Find the percent increase in average salaries for baseball players from 1997 to 2002.   78%

4.2 Graphs of Equations in Two Variables

① Sketch graphs of equations using the point-plotting method.

In Exercises 27–38, sketch the graph of the equation using the point-plotting method.
See Additional Answers.

27. $y = 7$

28. $x = -2$

29. $y = 3x$

30. $y = -2x$

31. $y = 4 - \frac{1}{2}x$

32. $y = \frac{3}{2}x - 3$

33. $y - 2x - 4 = 0$

34. $3x + 2y + 6 = 0$

35. $y = 2x - 1$

36. $y = 5 - 4x$

37. $y = \frac{1}{4}x + 2$

38. $y = -\frac{2}{3}x - 2$

② Find and use x- and y-intercepts as aids to sketching graphs.

In Exercises 39–46, find the x- and y-intercepts (if any) of the graph of the equation. Then sketch the graph of the equation and label the x- and y-intercepts.
See Additional Answers.

39. $y = 6x + 2$

40. $y = -3x + 5$

41. $y = \frac{2}{5}x - 2$

42. $y = \frac{1}{3}x + 1$

43. $2x - y = 4$

44. $3x - y = 10$

45. $4x + 2y = 8$

46. $9x + 3y = 6$

③ Use the verbal problem-solving method to write an equation and sketch its graph.

47. *Creating a Model*   The cost of producing a DVD is $125, plus $3 per DVD. Let C represent the total cost and let x represent the number of DVDs. Write an equation that relates C and x and sketch its graph.
$C = 3x + 125$ See Additional Answers.

48. *Creating a Model*   Let y represent the distance traveled by a train that is moving at a constant speed of 80 miles per hour. Let t represent the number of hours the train has traveled. Write an equation that relates y to t and sketch its graph. $y = 80t$ See Additional Answers.

4.3 Slope and Graphs of Linear Equations

① Determine the slope of a line through two points.

In Exercises 49 and 50, estimate the slope of the line from its graph.

49.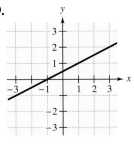

$\frac{1}{2}$

50.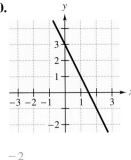

-2

In Exercises 51–62, plot the points and find the slope (if possible) of the line passing through the points. State whether the line rises, falls, is horizontal, or is vertical. See Additional Answers.

51. $(2, 1), (14, 6)$ $m = \frac{5}{12}$; The line rises.

52. $(-2, 2), (3, -10)$ $m = -\frac{12}{5}$; The line falls.

53. $(-1, 0), (6, 2)$ $m = \frac{2}{7}$; The line rises.

54. $(1, 6), (4, 2)$ $m = -\frac{4}{3}$; The line falls.

55. $(4, 0), (4, 6)$ m is undefined; The line is vertical.

56. $(1, 3), (4, 3)$ $m = 0$; The line is horizontal.

57. $(-2, 5), (1, 1)$ $m = -\frac{4}{3}$; The line falls.

58. $(-6, 1), (10, 5)$ $m = \frac{1}{4}$; The line rises.

59. $(1, -4), (5, 10)$ $m = \frac{7}{2}$; The line rises.

60. $(-3, 3), (8, 6)$ $m = \frac{3}{11}$; The line rises.

61. $\left(0, \frac{5}{2}\right), \left(\frac{5}{6}, 0\right)$ $m = -3$; The line falls.

62. $(0, 0), \left(3, \frac{4}{5}\right)$ $m = \frac{4}{15}$; The line rises.

63. *Truck* The floor of a truck is 4 feet above ground level. The end of the ramp used in loading the truck rests on the ground 6 feet behind the truck. Determine the slope of the ramp. $\frac{2}{3}$

64. *Flight Path* An aircraft is on its approach to an airport. Radar shows its altitude to be 15,000 feet when it is 10 miles from touchdown. Approximate the slope of the linear path followed by the aircraft during landing. $-\frac{25}{88}$

② Write linear equations in slope-intercept form and graph the equations.

In Exercises 65–72, write the equation in slope-intercept form. Use the slope and *y*-intercept to sketch the line. See Additional Answers.

65. $2x - y = -1$
$y = 2x + 1$

66. $-4x + y = -2$
$y = 4x - 2$

67. $12x + 4y = 8$
$y = -3x + 2$

68. $2x - 2y = 12$
$y = x - 6$

69. $3x + 6y = 12$
$y = -\frac{1}{2}x + 2$

70. $7x + 21y = -14$
$y = -\frac{1}{3}x - \frac{2}{3}$

71. $5y - 2x = 5$
$y = \frac{2}{5}x + 1$

72. $3y - x = 6$
$y = \frac{1}{3}x + 2$

③ Use slopes to determine whether lines are parallel, perpendicular, or neither.

In Exercises 73–76, determine whether lines L_1 and L_2 passing through the pairs of points are parallel, perpendicular, or neither.

73. L_1: $(0, 3), (-2, 1)$
L_2: $(-8, -3), (4, 9)$
Parallel

74. L_1: $(-3, -1), (2, 5)$
L_2: $(2, 11), (8, 6)$
Perpendicular

75. L_1: $(3, 6), (-1, -5)$
L_2: $(-2, 3), (4, 7)$
Neither

76. L_1: $(-1, 2), (-1, 4)$
L_2: $(7, 3), (4, 7)$
Neither

4.4 Equations of Lines

① Write equations of lines using the point-slope form.

In Exercises 77–86, use the point-slope form to write an equation of the line that passes through the point and has the specified slope. Write the equation in slope-intercept form.

77. $(4, -1), m = 2$
$y = 2x - 9$

78. $(-5, 2), m = 3$
$y = 3x + 17$

79. $(1, 2), m = -4$
$y = -4x + 6$

80. $(7, -3), m = -1$
$y = -x + 4$

81. $(-5, -2), m = \frac{4}{5}$
$y = \frac{4}{5}x + 2$

82. $(12, -4), m = -\frac{1}{6}$
$y = -\frac{1}{6}x - 2$

83. $(-1, 3), m = -\frac{8}{3}$
$y = -\frac{8}{3}x + \frac{1}{3}$

84. $(4, -2), m = \frac{8}{5}$
$y = \frac{8}{5}x - \frac{42}{5}$

85. $(3, 8), m$ is undefined.
$x = 3$

86. $(-4, 6), m = 0$
$y = 6$

In Exercises 87–94, write an equation of the line passing through the points. Write the equation in general form.

87. $(-4, 0), (0, -2)$
$x + 2y + 4 = 0$

88. $(-4, -2), (4, 6)$
$x - y + 2 = 0$

89. $(0, 8), (6, 8)$
$y - 8 = 0$

90. $(2, -6), (2, 5)$
$x - 2 = 0$

91. $(-1, 2), (4, 7)$
$x - y + 3 = 0$

92. $\left(0, \frac{4}{3}\right), (3, 0)$
$4x + 9y - 12 = 0$

93. $(2.4, 3.3), (6, 7.8)$
$25x - 20y + 6 = 0$

94. $(-1.4, 0), (3.2, 9.2)$
$10x - 5y + 14 = 0$

In Exercises 95–98, write equations of the lines through the point (a) parallel and (b) perpendicular to the given line.

95. $(-6, 3)$
$2x + 3y = 1$
(a) $2x + 3y + 3 = 0$
(b) $3x - 2y + 24 = 0$

96. $\left(\frac{1}{5}, -\frac{4}{5}\right)$
$5x + y = 2$
(a) $25x + 5y - 1 = 0$
(b) $5x - 25y - 21 = 0$

97. $\left(\frac{3}{8}, 4\right)$
$4x + 3y = 16$
(a) $8x + 6y - 27 = 0$
(b) $24x - 32y + 119 = 0$

98. $(-2, 1)$
$5x = 2$
(a) $x + 2 = 0$
(b) $y - 1 = 0$

② Write the equations of horizontal and vertical lines.

In Exercises 99–102, write an equation of the line.

99. Horizontal line through $(-4, 5)$ $y = 5$

100. Horizontal line through $(3, -7)$ $y = -7$

101. Vertical line through $(5, -1)$ $x = 5$

102. Vertical line through $(-10, 4)$ $x = -10$

③ Use linear models to solve application problems.

103. *Wages* A pharmaceutical salesperson receives a monthly salary of $2500 plus a commission of 7% of the total monthly sales. Write a linear model that relates total monthly wages W to sales S.
$W = 2500 + 0.07S$

104. *Rental Demand*   A real estate office handles an apartment complex with 50 units. When the rent per unit is $380 per month, all 50 units are occupied. However, when the rent is $425 per month, the average number of occupied units drops to 47. Assume that the relationship between the monthly rent p and the demand x is linear.

(a) Represent the given information as two ordered pairs of the form (x, p). Plot these ordered pairs.   (50, 380), (47, 425)   See Additional Answers.

(b) Write a linear model that relates the monthly rent p to the demand x. Graph the model and describe the relationship between the rent and the demand. $p = -15x + 1130$; As the rent increases, the demand decreases.   See Additional Answers.

(c) *Linear Extrapolation*   Use the model in part (b) to predict the number of units occupied if the rent is raised to $485.   43 units

(d) *Linear Interpolation*   Use the model in part (b) to estimate the number of units occupied if the rent is $410.   48 units

4.5 Graphs of Linear Inequalities

① **Determine whether an ordered pair is a solution of a linear inequality in two variables.**

In Exercises 105 and 106, determine whether the points are solutions of the inequality.

105. $x - y > 4$

(a) $(-1, -5)$ Not a solution

(b) $(0, 0)$ Not a solution

(c) $(3, -2)$ Solution

(d) $(8, 1)$ Solution

106. $y - 2x \le -1$

(a) $(0, 0)$ Not a solution

(b) $(-2, 1)$ Not a solution

(c) $(-3, 4)$ Not a solution

(d) $(-1, -6)$ Solution

② **Sketch graphs of linear inequalities in two variables.**

In Exercises 107–112, sketch the graph of the linear inequality.   See Additional Answers.

107. $x - 2 \ge 0$ **108.** $y + 3 < 0$

109. $2x + y < 1$ **110.** $3x - 4y > 2$

111. $x \le 4y - 2$ **112.** $x \ge 3 - 2y$

In Exercises 113–116, write an inequality for the shaded region shown in the figure.

113. $y < 2$

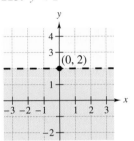

114. $x \ge -1$

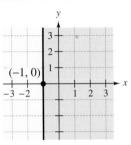

115. $y \le x + 1$

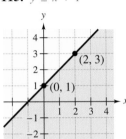

116. $y > -\frac{1}{3}x$

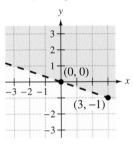

③ **Use linear inequalities to model and solve real-life problems.**

117. *Manufacturing*   Each VCR produced by an electronics manufacturer requires 2 hours in the assembly center. Each camcorder produced by the same manufacturer requires 3 hours in the assembly center. A total of 120 hours per week is available in the assembly center. Write a linear inequality that shows the different numbers of hours that can be spent assembling VCRs and camcorders, and sketch the graph of the inequality. From the graph, find several ordered pairs with positive integer coordinates that are solutions of the inequality. $2x + 3y \le 120$; (x, y): (10, 15), (20, 20), (30, 20)   See Additional Answers.

118. *Manufacturing*   A company produces two types of wood chippers, Economy and Deluxe. The Deluxe model requires 3 hours in the assembly center and the Economy model requires $1\frac{1}{2}$ hours in the assembly center. A total of 24 hours per day is available in the assembly center. Write a linear inequality that shows the different numbers of hours that can be spent assembling the two models, and sketch the graph of the inequality. From the graph, find several ordered pairs with positive integer coordinates that are solutions of the inequality. $3x + \frac{3}{2}y \le 24$; (x, y): (8, 0), (0, 16), (4, 4)   See Additional Answers.

Chapter Test

Take this test as you would take a test in class. After you are done, check your work against the answers in the back of the book.

1. Plot the points $(-1, 2)$, $(1, 4)$, and $(2, -1)$ on a rectangular coordinate system. Connect the points with line segments to form a right triangle.
   See Additional Answers.

2. Determine whether the ordered pairs are solutions of $y = |x| + |x - 2|$.
   (a) $(0, -2)$ (b) $(0, 2)$ (c) $(-4, 10)$ (d) $(-2, -2)$
   (a) Not a solution   (b) Solution   (c) Solution   (d) Not a solution

3. What is the y-coordinate of any point on the x-axis?   0

4. Find the x- and y-intercepts of the graph of $3x - 4y = -12$.
   $(-4, 0), (0, 3)$

5. Complete the table and use the results to sketch the graph of the equation.
   $x - 2y = 6$ See Additional Answers.

x	-2	-1	0	1	2
y	-4	$-\frac{7}{2}$	-3	$-\frac{5}{2}$	-2

In Exercises 6–9, sketch the graph of the equation.   See Additional Answers.

6. $x + 2y = 6$

7. $y = \frac{1}{4}x - 1$

8. $y = |x + 2|$

9. $y = (x - 3)^2$

10. Find the slope of the line passing through the points $(-5, 0)$ and $\left(2, \frac{3}{2}\right)$. Then write an equation of the line in slope-intercept form. $\frac{3}{14}$; $y = \frac{3}{14}x + \frac{15}{14}$

11. A line with slope $m = -2$ passes through the point $(-3, 4)$. Plot the point and use the slope to find two additional points on the line. (There are many correct answers.)   See Additional Answers. $(-2, 2), (-1, 0)$

12. Find the slope of a line *perpendicular* to the line $3x - 5y + 2 = 0$. $-\frac{5}{3}$

13. Find an equation of the line that passes through the point $(0, 6)$ with slope $m = -\frac{3}{8}$. $3x + 8y - 48 = 0$

14. Write an equation of the vertical line that passes through the point $(3, -7)$.
    $x = 3$

15. (a) Solution
    (b) Solution
    (c) Solution
    (d) Solution

15. Determine whether the points are solutions of $3x + 5y \le 16$.
    (a) $(2, 2)$ (b) $(6, -1)$ (c) $(-2, 4)$ (d) $(7, -1)$

In Exercises 16–19, sketch the graph of the linear inequality.
See Additional Answers.

16. $y \ge -2$

17. $y < 5 - 2x$

18. $x \ge 2$

19. $y \le 5$

20. The sales y of a product are modeled by $y = 230x + 5000$, where x is time in years. Interpret the meaning of the slope in this model.
    Sales are increasing at a rate of 230 units per year.

Motivating the Chapter

⬤ Packaging Restrictions

A shipping company has the following restrictions on the dimensions and weight of packages.

1. The maximum weight is 150 pounds.
2. The maximum length is 108 inches.
3. The sum of the length and girth can be at most 130 inches.

The girth of a package is the minimum distance around the package, as shown in the figure.

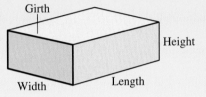

$$\text{Girth} = 2(\text{Height} + \text{Width})$$

You are shipping a package that has a height of x inches. The length of the package is twice the square of the height, and the width is 5 inches more than 3 times the height.

See Section 5.2, Exercise 103.

a. Write an expression for the length of the package in terms of the height x. Write an expression for the width of the package in terms of the height x. Length: $(2x^2)$ inches; Width: $(3x + 5)$ inches

b. Write an expression for the *perimeter* of the base of the package. Simplify the expression. $(4x^2 + 6x + 10)$ inches

c. Write an expression for the *girth* of the package. Simplify the expression. Write an expression for the sum of the length and the girth. If the height of the package is 5 inches, does the package meet the second and third restrictions? Explain. Girth: $(8x + 10)$ inches; Length and girth: $(2x^2 + 8x + 10)$ inches; Yes. Substituting 5 for x in $2x^2$, you find that the length is 50 inches. Substituting 5 for x in $2x^2 + 8x + 10$, you find that the sum of the length and girth is 100 inches.

See Section 5.3, Exercise 133.

d. Write an expression for the *surface area* of the package. Simplify the expression. (The surface area is the sum of the areas of the six sides of the package.) $(16x^3 + 26x^2 + 10x)$ square inches

e. The length of the package is changed to match its width (5 inches more than 3 times its height). Write an expression for the area of the base. Simplify the expression. $(3x + 5)^2 = (9x^2 + 30x + 25)$ square inches

f. Write an expression for the *volume* of the package in part (e). Simplify the expression. $(9x^3 + 30x^2 + 25x)$ cubic inches

See Section 5.4, Exercise 75.

g. The width of the package is the same as in part (a). The area of the base is $6x^2 + 7x - 5$. What is the length of the package? $(2x - 1)$ inches

Najlah Feanny/Corbis SABA

5

Exponents and Polynomials

5.1 Negative Exponents and Scientific Notation

Spencer Grant/PhotoEdit

What You Should Learn

1. Apply the rules of exponents to rewrite exponential expressions.
2. Use the negative exponent rule to rewrite exponential expressions.
3. Write very large and very small numbers in scientific notation.

Why You Should Learn It

Scientific notation can be used to represent very large numbers. For instance, in Exercise 190 on page 278, you will use scientific notation to represent the storage capacity of a computer network.

1. Apply the rules of exponents to rewrite exponential expressions.

Rules of Exponents

Recall from Section 1.5 that repeated multiplication can be represented in *exponential form*. When simplifying algebraic expressions, you often need to use some rules for operating with exponential forms. Consider the following.

1. Multiplying exponential forms with like bases: *Rule*

$$a^3 \cdot a^2 = \underbrace{(a \cdot a \cdot a)}_{\text{3 factors}} \cdot \underbrace{(a \cdot a)}_{\text{2 factors}}$$

Add exponents.

$$= \underbrace{a \cdot a \cdot a \cdot a \cdot a}_{\text{5 factors}} = a^5 = a^{3+2}$$

2. Raising an exponential form to a power: *Rule*

$$(a^3)^2 = \underbrace{a^3 \cdot a^3}_{\text{2 factors of } a^3}$$

Multiply exponents.

$$= \underbrace{(a \cdot a \cdot a)}_{\text{3 factors}} \cdot \underbrace{(a \cdot a \cdot a)}_{\text{3 factors}} = a^6 = a^{3 \cdot 2}$$

3. Raising a product to a power: *Rule*

$$(a \cdot b)^3 = \underbrace{(a \cdot b) \cdot (a \cdot b) \cdot (a \cdot b)}_{\text{3 factors of } (a \cdot b)}$$

Apply exponent to each factor.

$$= \underbrace{(a \cdot a \cdot a)}_{\text{3 factors}} \cdot \underbrace{(b \cdot b \cdot b)}_{\text{3 factors}} = a^3 \cdot b^3$$

Study Tip

Rules 1 and 3 can be extended to three or more factors, such as $a^m \cdot a^n \cdot a^k = a^{m+n+k}$ and $(abc)^m = a^m \cdot b^m \cdot c^m$.

These illustrations suggest the following rules for exponential forms.

Product and Power Rules of Exponents

Let m and n be positive integers, and let a and b be real numbers, variables, or algebraic expressions.

1. Product Rule: $a^m \cdot a^n = a^{m+n}$

2. Power-to-Power Rule: $(a^m)^n = a^{m \cdot n}$

3. Product-to-Power Rule: $(ab)^m = a^m \cdot b^m$

Example 1   Applying the Product and Power Rules of Exponents

Use the product and power rules of exponents to simplify each expression.

a. $5^2 \cdot 5^6 \cdot 5$ **b.** $b^4 b^2 b$ **c.** $3^2 x^3 \cdot x$ **d.** $(-x)^4$

Solution

a. $5^2 \cdot 5^6 \cdot 5 = 5^{2+6+1} = 5^9$

b. $b^4 b^2 b = b^{4+2+1} = b^7$

c. $3^2 x^3 \cdot x = (3^2)(x^{3+1}) = 9x^4$

d. $(-x)^4 = (-1)^4 x^4 = x^4$

Be sure you see the difference between the expressions

$$x^3 \cdot x^4 \text{ and } x^3 + x^4.$$

The first is a *product* of exponential forms, whereas the second is a *sum* of exponential forms. The rule for multiplying exponential forms having the same base can be applied to the first expression, but *not* to the second expression.

Example 2   Applying the Product and Power Rules of Exponents

Use the product and power rules of exponents to simplify each expression.

a. $(2^3)^4$ **b.** $(y^2)^3$ **c.** $(2x^2)^3$ **d.** $(3x)^3$

Solution

a. $(2^3)^4 = 2^{3 \cdot 4} = 2^{12} = 4096$

b. $(y^2)^3 = y^{2 \cdot 3} = y^6$

c. $(2x^2)^3 = 2^3(x^2)^3 = 2^3 x^{2 \cdot 3} = 8x^6$

d. $(3x)^3 = 3^3 \cdot x^3 = 27x^3$

Example 3   Applying the Product and Power Rules of Exponents

Use the product and power rules of exponents to simplify each expression.

a. $(2x^2y)(-xy^4)$ **b.** $x(x^3y^2)^3$ **c.** $(-9x^2)^2(-3x^5)$

Solution

a. $(2x^2y)(-xy^4) = (2)(-1)(x^2 \cdot x)(y \cdot y^4) = -2x^{2+1}y^{1+4} = -2x^3y^5$

b. $x(x^3y^2)^3 = x(x^{3 \cdot 3}y^{2 \cdot 3}) = x(x^9y^6) = x^{1+9}y^6 = x^{10}y^6$

c. $(-9x^2)^2(-3x^5) = (-9)^2(x^2)^2(-3x^5)$

$$= 81 \cdot (-3) \cdot x^{2 \cdot 2} \cdot x^5$$

$$= -243 \cdot x^{4+5}$$

$$= -243x^9$$

It is important to recognize that the product and power rules of exponents apply to products and not to sums or differences. Note the following illustrations.

Product	*Example*
$x^5 \cdot x^4 = x^{5+4}$	$2^5 \cdot 2^4 \overset{?}{=} 2^{5+4}$
	$512 = 512$
Sum	
$x^5 + x^4 \neq x^{5+4}$	$2^5 + 2^4 \neq 2^{5+4}$
	$48 \neq 512$

To simplify the quotient of two exponential forms with like bases, you make use of the *quotient* rules of exponents illustrated in the following examples. (In each of the following, assume that the variable is *not zero*.)

By Reducing

$$\frac{x^4}{x^2} = \frac{x \cdot x \cdot \cancel{x} \cdot \cancel{x}}{\cancel{x} \cdot \cancel{x}} = x^2$$

$$\frac{y^3}{y^3} = \frac{\cancel{y} \cdot \cancel{y} \cdot \cancel{y}}{\cancel{y} \cdot \cancel{y} \cdot \cancel{y}} = 1$$

$$\frac{5y^7}{2y^5} = \frac{5 \cdot y \cdot y \cdot \cancel{y} \cdot \cancel{y} \cdot \cancel{y} \cdot \cancel{y} \cdot \cancel{y}}{2 \cdot \cancel{y} \cdot \cancel{y} \cdot \cancel{y} \cdot \cancel{y} \cdot \cancel{y}} = \frac{5y^2}{2}$$

$$\frac{2x^3}{x} = \frac{2 \cdot x \cdot x \cdot \cancel{x}}{\cancel{x}} = 2x^2$$

By Subtracting Exponents

$$\frac{x^4}{x^2} = x^{4-2} = x^2$$

$$\frac{y^3}{y^3} = y^{3-3} = y^0 = 1$$

$$\frac{5y^7}{2y^5} = \frac{5y^{7-5}}{2} = \frac{5y^2}{2}$$

$$\frac{2x^3}{x} = 2x^{3-1} = 2x^2$$

The examples above show that you can simplify the quotient of two exponential forms with like bases by subtracting exponents.

Quotient Rules of Exponents

Let m and n be positive integers and let a represent a real number, a variable, or an algebraic expression.

1. $\dfrac{a^m}{a^n} = a^{m-n}$, $m > n$, $a \neq 0$

2. $\dfrac{a^n}{a^n} = 1 = a^0$, $a \neq 0$

Note the special definition for raising a *nonzero* quantity to the zero power. That is, if $a \neq 0$, then $a^0 = 1$.

If you remember the first rule above, you can use it to derive the second rule. That is,

$$1 = \frac{a^n}{a^n}$$

$$= a^{n-n}$$

$$= a^0, \qquad a \neq 0.$$

Additional Examples
Use the quotient rules of exponents to simplify each expression.

a. $\dfrac{14x^5}{7x}$ b. $\dfrac{2a^4b^5}{a^3b^2}$

Answers:

a. $2x^4$ b. $2ab^3$

Example 4   Applying the Quotient Rules of Exponents

Use the quotient rules of exponents to simplify each expression. (In each case, assume that $x \neq 0$ and $y \neq 0$.)

a. $\dfrac{2y^8}{y^5}$ b. $\dfrac{16x^4}{4x^2}$ c. $\dfrac{16x^4}{3x^4}$ d. $\dfrac{32y^6}{8y^5}$

Solution

a. $\dfrac{2y^8}{y^5} = 2 \cdot y^{8-5} = 2y^3$

b. $\dfrac{16x^4}{4x^2} = \dfrac{16}{4} \cdot \dfrac{x^4}{x^2} = \dfrac{16}{4} \cdot x^{4-2} = 4x^2$

c. $\dfrac{16x^4}{3x^4} = \dfrac{16}{3} \cdot \dfrac{x^4}{x^4} = \dfrac{16}{3} \cdot x^{4-4} = \dfrac{16}{3}(1) = \dfrac{16}{3}$

d. $\dfrac{32y^6}{8y^5} = \dfrac{32}{8} \cdot \dfrac{y^6}{y^5} = \dfrac{32}{8} \cdot y^{6-5} = 4y$

Example 5   Applying the Quotient Rules of Exponents

Use the quotient rules of exponents to simplify each expression. (In each case, assume that $x \neq 0$ and $y \neq 0$.)

a. $\dfrac{12x^8y^2}{4x^4y}$ b. $\dfrac{3x^3y^6}{4xy^3}$

Solution

a. $\dfrac{12x^8y^2}{4x^4y} = \dfrac{12}{4} \cdot \dfrac{x^8}{x^4} \cdot \dfrac{y^2}{y} = \dfrac{12}{4} \cdot x^{8-4} \cdot y^{2-1} = 3x^4y$

b. $\dfrac{3x^3y^6}{4xy^3} = \dfrac{3}{4} \cdot \dfrac{x^3}{x} \cdot \dfrac{y^6}{y^3} = \dfrac{3}{4} \cdot x^{3-1} \cdot y^{6-3} = \dfrac{3}{4}x^2y^3$

Study Tip

The quotient rules of exponents work only for expressions with the *same variable* for a base. For instance, the quotient rule does not apply to x^5/y^3 because you cannot divide out any variables. So, for the expression

$$\dfrac{x^5}{y^3} = \dfrac{x \cdot x \cdot x \cdot x \cdot x}{y \cdot y \cdot y}$$

no simplifying is possible.

Although Examples 4 and 5 are straightforward, you should study them carefully. Be sure you can justify each step. Also remember that there are often several ways to solve a given problem in algebra. As you gain practice and confidence, you will discover that you like some techniques better than others. For instance, which one of the following techniques seems best to you?

1. $\dfrac{6x^3}{2x} = \dfrac{3 \cdot 2 \cdot x \cdot x \cdot \cancel{x}}{2 \cdot \cancel{x}} = 3x^2, \quad x \neq 0$

2. $\dfrac{6x^3}{2x} = \left(\dfrac{6}{2}\right)\left(\dfrac{x^3}{x}\right) = (3)(x^{3-1}) = 3x^2, \quad x \neq 0$

3. $\dfrac{6x^3}{2x} = \dfrac{3x^3}{x} = 3x^{3-1} = 3x^2, \quad x \neq 0$

Point out that there can be *many* correct ways to solve an algebraic problem.

When two different people (even math instructors) are writing out the steps of a solution, rarely will the steps be the same, so don't worry if your steps don't look exactly like someone else's. Just be sure that each step can be justified by the rules of algebra.

② Use the negative exponent rule to rewrite exponential expressions.

Negative Exponents

The rules of exponents can be extended to include **negative exponents.** Consider the product rule

$$a^m \cdot a^n = a^{m+n}, \quad a \neq 0.$$

If this rule is to hold for negative exponents, then the statement

$$a^2 \cdot a^{-2} = a^{2+(-2)} = a^0 = 1$$

implies that a^{-2} is the *reciprocal* of a^2. In other words, it must be true that

$$a^{-2} = \frac{1}{a^2}.$$

Informally, you can think of this property as allowing you to "move" powers from the numerator to the denominator (or vice versa) by changing the sign of the exponent.

Negative Exponent Rule

Let n be an integer and let a be a real number, variable, or algebraic expression such that $a \neq 0$.

$$a^{-n} = \frac{1}{a^n} \quad \text{and} \quad \frac{1}{a^{-n}} = a^n$$

Technology: Tip

The keystrokes used to evaluate expressions with negative exponents vary. For instance, to evaluate 13^{-2} on a calculator, you can try one of the following keystroke sequences.

Keystrokes

13 y^x 2 $+/-$ $=$ Scientific

13 $\wedge$ $(-)$ 2 $ENTER$ Graphing

With either of these sequences, your calculator should display .00591716. If it doesn't, consult the user's guide for your calculator to find the correct keystrokes.

Example 6   Negative Exponents

a. $6^{-2} = \frac{1}{6^2} = \frac{1}{36}$ — Move 6^{-2} to denominator and change the sign of the exponent.

b. $x^{-7} = \frac{1}{x^7}$ — Move x^{-7} to denominator and change the sign of the exponent.

c. $5x^{-4} = \frac{5}{x^4}$ — Move x^{-4} to denominator and change the sign of the exponent.

d. $\frac{1}{2x^{-3}} = \frac{x^3}{2}$ — Move x^{-3} to numerator and change the sign of the exponent.

e. $x^{-2}y^3 = \frac{y^3}{x^2}$ — Move x^{-2} to denominator and change the sign of the exponent.

Be sure you see that the negative exponent rule allows you to move only *factors* in a numerator (or denominator), *not terms*. For example,

$$\frac{x^{-2} \cdot y}{4} = \frac{y}{4x^2}$$ — Can move *factor* to denominator

whereas the following is not true.

$$\frac{x^{-2} + y}{4} = \frac{y}{4x^2}$$ — Cannot move *term* to denominator

All of the rules of exponents apply to negative exponents. For convenience, these rules are summarized below. Remember that these rules apply to real numbers, variables, or algebraic expressions.

Technology: Discovery

Use a calculator to evaluate the expressions below.

$$\frac{3.4^{5.8}}{3.4^{1.6}} \quad \text{and} \quad 3.4^{4.2}$$

How are these two expressions related? Use your calculator to verify other rules of exponents.

See Technology Answers.

Summary of Rules of Exponents

Let m and n be integers, and let a and b be real numbers, variables, or algebraic expressions, such that $a \neq 0$ and $b \neq 0$.

Rule	Example
1. $a^m a^n = a^{m+n}$	$y^2 \cdot y^4 = y^{2+4} = y^6$
2. $\dfrac{a^m}{a^n} = a^{m-n}$	$\dfrac{x^7}{x^4} = x^{7-4} = x^3$
3. $(ab)^m = a^m b^m$	$(5x)^4 = 5^4 x^4$
4. $\left(\dfrac{a}{b}\right)^m = \dfrac{a^m}{b^m}$	$\left(\dfrac{2}{x}\right)^3 = \dfrac{2^3}{x^3}$
5. $(a^m)^n = a^{mn}$	$(y^3)^{-4} = y^{3(-4)} = y^{-12}$
6. $a^{-n} = \dfrac{1}{a^n}$	$y^{-4} = \dfrac{1}{y^4}$
7. $a^0 = 1$	$(x^2 + 1)^0 = 1$

Example 7   Using Rules of Exponents

a. $x^3(2x^{-4}) = 2(x^3)(x^{-4})$ Regroup factors.

$\qquad\qquad\quad = 2x^{3+(-4)}$ Apply rules of exponents.

$\qquad\qquad\quad = 2x^{-1}$ Simplify.

$\qquad\qquad\quad = \dfrac{2}{x}$ Simplify.

b. $(-3ab^4)(4ab^{-3}) = (-3)(4)(a)(a)(b^4)(b^{-3})$ Regroup factors.

$\qquad\qquad\qquad\quad = (-12)(a^{1+1})(b^{4+(-3)})$ Apply rules of exponents.

$\qquad\qquad\qquad\quad = -12a^2 b$ Simplify.

Study Tip

There is more than one way to solve problems such as the one in Example 8. For instance, you might prefer to write Example 8 as

$$\frac{y^{-2}}{3y^{-5}} = \frac{y^5}{3y^2} = \frac{y^{5-2}}{3} = \frac{y^3}{3}.$$

Example 8   Using Rules of Exponents

$$\frac{y^{-2}}{3y^{-5}} = \frac{1}{3}y^{-2-(-5)}$$ Apply rules of exponents.

$$\qquad\quad = \frac{1}{3}y^3$$ Simplify.

$$\qquad\quad = \frac{y^3}{3}$$ Simplify.

You could ask students to compare these examples.

$$5^0 = 1 \qquad 5x^0 = 5$$
$$-5^0 = -1 \qquad -(5x)^0 = -1$$
$$(-5)^0 = 1 \qquad (-5x^3)^0 = 1$$

Assure students that there is more than one way to solve these problems. Encourage students to justify each step mentally by the rules of exponents.

Example 9   Using Rules of Exponents

Use rules of exponents to simplify each expression using only positive exponents. (Assume that no variable is equal to zero.)

a. $3x^{-1}(-4x^2y)^0$

b. $\left(\dfrac{5x^3}{y^{-1}}\right)^2$

c. $\left(\dfrac{a^2}{3}\right)^{-2}$

d. $\left(\dfrac{x^{-2}y^3}{2}\right)^{-3}$

Solution

a. Because any nonzero number raised to the zero power is 1, you can write

$$3x^{-1}(-4x^2y)^0 = 3x^{-1}(1) \qquad \text{Apply rules of exponents.}$$
$$= \frac{3}{x}. \qquad \text{Simplify.}$$

b. The important thing to realize is that the *entire fraction* $(5x^3/y^{-1})$ is raised to the second power. This means that you must apply the exponent 2 to each factor of the numerator and denominator, as follows.

$$\left(\frac{5x^3}{y^{-1}}\right)^2 = \frac{(5x^3)^2}{(y^{-1})^2} \qquad \text{Apply rules of exponents.}$$
$$= \frac{5^2(x^3)^2}{y^{-2}} \qquad \text{Apply rules of exponents.}$$
$$= \frac{25x^6}{y^{-2}} \qquad \text{Apply rules of exponents.}$$
$$= 25x^6y^2 \qquad \text{Simplify.}$$

c. $\left(\dfrac{a^2}{3}\right)^{-2} = \dfrac{a^{-4}}{3^{-2}} \qquad \text{Apply rules of exponents.}$
$$= \frac{3^2}{a^4} \qquad \text{Apply rules of exponents.}$$
$$= \frac{9}{a^4} \qquad \text{Simplify.}$$

d. $\left(\dfrac{x^{-2}y^3}{2}\right)^{-3} = \dfrac{x^6y^{-9}}{2^{-3}} \qquad \text{Apply rules of exponents.}$
$$= \frac{2^3x^6}{y^9} \qquad \text{Apply rules of exponents.}$$
$$= \frac{8x^6}{y^9} \qquad \text{Simplify.}$$

③ Write very large and very small numbers in scientific notation.

Scientific Notation

Exponents provide an efficient way of writing and computing with the very large (or very small) numbers used in science. For instance, a drop of water contains more than 33 billion billion molecules. That is 33 followed by 18 zeros. It is convenient to write such numbers in **scientific notation.** This notation has the form $c \times 10^n$, where $1 \le c < 10$ and n is an integer. So, the number of molecules in a drop of water can be written in scientific notation as follows.

$$33,000,000,000,000,000,000 = 3.3 \times 10^{19}$$

19 places

The *positive* exponent 19 indicates that the number is large (10 or more) and that the decimal point has been moved 19 places.

A *negative* exponent in scientific notation indicates that the number is *small* (less than 1). For instance, the mass (in grams) of one electron is approximately as follows.

$$9.0 \times 10^{-28} = 0.00000000000000000000000000009$$

28 places

Additional Examples
Write each number in scientific notation.

a. 3400

b. 0.00034

Answers:

a. 3.4×10^3

b. 3.4×10^{-4}

Example 10   Converting from Decimal to Scientific Notation

a. The speed of light is

$$299,792,458 = 2.99792458 \times 10^8 \text{ meters per second.}$$

Eight places

Large number yields positive exponent.

b. The relative density of hydrogen is

$$0.08988 = 8.988 \times 10^{-2} \text{ grams per cubic centimeter.}$$

Two places

Small number yields negative exponent.

Example 11   Converting from Scientific to Decimal Notation

a. The number of air sacs in human lungs is

$$3.5 \times 10^8 = 350,000,000.$$

Eight places

Positive exponent yields large number.

b. The width of a human hair is

$$9.0 \times 10^{-4} = 0.0009 \text{ meter.}$$

Four places

Negative exponent yields small number.

Technology: Tip

Most scientific calculators automatically switch to scientific notation when they are displaying large (or small) numbers that exceed the display range. Try multiplying $98,900,000 \times 5000$. If your calculator follows standard conventions, its display should show

> 4.945   11

or

> 4.945 E 11 .

This means that $c = 4.945$ and the exponent of 10 is $n = 11$, which implies that the number is 4.945×10^{11}.

For entering numbers in scientific notation, your calculator should have an exponential entry key labeled [EE] or [EXP].

Example 12   Using Scientific Notation

Use a calculator to evaluate $78,000 \times 2,400,000,000$.

Solution

Because $78,000 = 7.8 \times 10^4$ and $2,400,000,000 = 2.4 \times 10^9$, you can evaluate the product as follows.

7.8 [EE] 4 [×] 2.4 [EE] 9 [=]   Calculator with EE key

7.8 [EXP] 4[×] 2.4 [EXP] 9 [ENTER]   Calculator with EXP key

After these keystrokes have been entered, the calculator display should show $\boxed{1.872\ 14}$ or $\boxed{1.872\ \text{E}\ 14}$. So, the product of the two numbers is

$$(7.8 \times 10^4)(2.4 \times 10^9) = 1.872 \times 10^{14}$$
$$= 187,200,000,000,000.$$

Example 13   Federal Debt

In July 2000, the estimated population of the United States was 275 million people, and the estimated federal debt was \$5.629 trillion. Use these two numbers to determine the amount each person would have to pay to eliminate the debt. (Source: U.S. Census Bureau and U.S. Office of Management and Budget)

Solution

To determine the amount each person would have to pay to eliminate the federal debt, divide the federal debt by the population.

$$\text{Amount per person} = \frac{\text{Federal debt}}{\text{Population}} \qquad \text{Write verbal model.}$$

$$= \frac{5.629 \text{ trillion}}{275 \text{ million}} \qquad \text{Substitute.}$$

$$= \frac{5.629 \times 10^{12}}{2.75 \times 10^8} \qquad \text{Write in scientific notation.}$$

$$= \frac{5.629}{2.75} \times 10^4 \qquad \text{Quotient rule of exponents}$$

$$\approx 20,469.09 \qquad \text{Simplify.}$$

So, the amount each person would have to pay to eliminate the federal debt is approximately \$20,469.09.

5.1 Exercises

Review   Concepts, Skills, and Problem Solving

Keep mathematically in shape by doing these exercises *before* the problems of this section.

Properties and Definitions

1. *Writing* In your own words, describe how you would add two integers without like signs.   Subtract the smaller absolute value from the larger absolute value and attach the sign of the integer with the larger absolute value.

2. *Writing* In your own words, explain what it means if a is a divisor of b. a is a divisor of b if there is an integer c such that $a \cdot c = b$.

3. Complete the Distributive Property:
$(4x - 2y^2)(-3x) = -12x^2 + 6xy^2$.

4. Identify the property of real numbers illustrated by
$(7ab^2 - 11c) \cdot 1 = 7ab^2 - 11c$.
   Multiplicative Identity Property

Simplifying Expressions

In Exercises 5–8, simplify the expression. (Assume that no denominator is zero.)

5. $\dfrac{a^2 b^3}{c} \cdot \dfrac{2a}{3}$ $\dfrac{2a^3 b^3}{3c}$

6. $\dfrac{3xy^2}{4} \cdot \dfrac{8}{9}$ $\dfrac{2xy^2}{3}$

7. $(-3z)(-3z)(-3z)$ $-27z^3$

8. $-\left(\dfrac{2x}{y}\right)\left(\dfrac{2x}{y}\right)$ $-\dfrac{4x^2}{y^2}$

Graphing Equations

In Exercises 9–12, use a graphing calculator to graph the equation. Identify any intercepts.
See Additional Answers.

9. $3x + y = 4$
$(0, 4), \left(\frac{4}{3}, 0\right)$

10. $y = |2x + 1|$
$\left(-\frac{1}{2}, 0\right), (0, 1)$

11. $y = x^2 - 2x + 1$
$(1, 0), (0, 1)$

12. $y = \sqrt{x + 4}$
$(-4, 0), (0, 2)$

Developing Skills

In Exercises 1–26, simplify the expression. See Examples 1–3.

1. $u^2 \cdot u^4$ u^6
2. $z^3 \cdot z$ z^4
3. $3x^3 \cdot x^4$ $3x^7$
4. $4y^3 \cdot y$ $4y^4$
5. $5x(x^6)$ $5x^7$
6. $(-6x^2)x^4$ $-6x^6$
7. $(-5z^3)(3z^2)$ $-15z^5$
8. $(-2x^2)(-4x)$ $8x^3$
9. $(-xz)(-2y^2z)$ $2xy^2z^2$
10. $(6u^2v)(3uv^2)$ $18u^3v^3$
11. $2b^4(-ab)(3b^2)$ $-6ab^7$
12. $4xy(-3x^2)(-2y^3)$ $24x^3y^4$
13. $(t^2)^4$ t^8
14. $(v^3)^2$ v^6
15. $5(uv)^5$ $5u^5v^5$
16. $3(pq)^4$ $3p^4q^4$
17. $(-2s)^3$ $-8s^3$
18. $(-3z)^2$ $9z^2$
19. $(a^2b)^3(ab^2)^4$ $a^{10}b^{11}$
20. $(st)^5(s^2t)^4$ $s^{13}t^9$
21. $(u^2v^3)(-2uv^2)^4$ $16u^6v^{11}$
22. $(-3y^2z)^2(2yz^2)^3$ $72y^7z^8$
23. $[(x - 3)^4]^2$ $(x - 3)^8$
24. $[(t + 1)^2]^5$ $(t + 1)^{10}$
25. $(x - 2y)^3(x - 2y)^3$ $(x - 2y)^6$
26. $(x - 3)^2(x - 3)^5$ $(x - 3)^7$

In Exercises 27–40, simplify the expression by dividing out *and* by subtracting exponents. (Assume that no denominator is zero.)

27. $\dfrac{x^5}{x^2}$ x^3
28. $\dfrac{y^7}{y^3}$ y^4
29. $\dfrac{x^2}{x}$ x
30. $\dfrac{y^3}{y}$ y^2
31. $\dfrac{z^7}{z^4}$ z^3
32. $\dfrac{y^8}{y^3}$ y^5
33. $\dfrac{3u^4}{u^3}$ $3u$
34. $\dfrac{z^6}{5z^4}$ $\dfrac{z^2}{5}$
35. $\dfrac{2^3y^4}{2^2y^2}$ $2y^2$
36. $\dfrac{3^5x^7}{3^3x^2}$ $9x^5$
37. $\dfrac{4^5x^5}{4x^3}$ $256x^2$
38. $\dfrac{6z^5}{6z^5}$ 1
39. $\dfrac{3^4(ab)^3}{3(ab)^2}$ $27ab$
40. $\dfrac{8^2u^4v^5}{8^3u^4v^2}$ $\dfrac{v^3}{8}$

In Exercises 41–58, simplify the expression. (Assume that no denominator is zero.) See Examples 4 and 5.

41. $\dfrac{-3x^2}{x}$ $-3x$

42. $\dfrac{-4a^6}{-a^2}$ $4a^4$

43. $\dfrac{4x^6}{x^3}$ $4x^3$

44. $\dfrac{-16v^3}{v^2}$ $-16v$

45. $\dfrac{-12z^3}{-3z}$ $4z^2$

46. $\dfrac{16y^5}{8y^3}$ $2y^2$

47. $\dfrac{32b^4}{12b^3}$ $\dfrac{8b}{3}$

48. $\dfrac{-7c^5}{8c^2}$ $-\dfrac{7c^3}{8}$

49. $\dfrac{-22y^2}{4y}$ $-\dfrac{11y}{2}$

50. $\dfrac{54x^4}{-24x^2}$ $-\dfrac{9x^2}{4}$

51. $\dfrac{-18s^4}{-12r^2s}$ $\dfrac{3s^3}{2r^2}$

52. $\dfrac{-21v^3}{12u^2v}$ $-\dfrac{7v^2}{4u^2}$

53. $\dfrac{(-3z)^3}{18z^2}$ $-\dfrac{3z}{2}$

54. $\dfrac{4a^3}{(-8a)^2}$ $\dfrac{a}{16}$

55. $\dfrac{(2x^2y)^3}{(4y)^2x^4}$ $\dfrac{x^2y}{2}$

56. $\dfrac{15(uv^4)^3}{(-3u)^3v^5}$ $-\dfrac{5v^7}{9}$

57. $\dfrac{24u^2v^6}{18u^2v^4}$ $\dfrac{4v^2}{3}$

58. $\dfrac{15x^3y^8}{27x^3y^6}$ $\dfrac{5y^2}{9}$

In Exercises 59–66, determine the exponent that makes the statement true.

59. $5^{-2} = \dfrac{1}{25}$

60. $2^{-4} = \dfrac{1}{16}$

61. $\dfrac{1}{3^{-2}} = 9$

62. $\dfrac{1}{4^{-2}} = 16$

63. $(x^{-2}y^3)^{-2} = \dfrac{x^4}{y^6}$

64. $(x^{-3}y^{-2})^3 = \dfrac{1}{x^9y^6}$

65. $(x^4y^{-3})^{-3} = \dfrac{y^9}{x^{12}}$

66. $(x^{-2}y^5)^{-1} = \dfrac{x^2}{y^5}$

In Exercises 67–78, rewrite the expression using only positive exponents. See Example 6.

67. 3^{-3} $\dfrac{1}{3^3}$

68. 4^{-2} $\dfrac{1}{4^2}$

69. y^{-5} $\dfrac{1}{y^5}$

70. z^{-2} $\dfrac{1}{z^2}$

71. $8x^{-7}$ $\dfrac{8}{x^7}$

72. $6x^{-2}y^{-3}$ $\dfrac{6}{x^2y^3}$

73. $7x^{-4}y^{-1}$ $\dfrac{7}{x^4y}$

74. $9u^{-5}v^{-2}$ $\dfrac{9}{u^5v^2}$

75. $\dfrac{1}{2z^{-4}}$ $\dfrac{z^4}{2}$

76. $\dfrac{7x^2}{y^{-3}}$ $7x^2y^3$

77. $\dfrac{2x}{3y^{-2}}$ $\dfrac{2xy^2}{3}$

78. $\dfrac{5u^2}{6v^{-4}}$ $\dfrac{5u^2v^4}{6}$

In Exercises 79–88, rewrite the expression using only negative exponents.

79. $\dfrac{1}{4}$ 4^{-1}

80. $\dfrac{1}{3^2}$ 3^{-2}

81. $\dfrac{1}{x^2}$ x^{-2}

82. $\dfrac{7}{y^3}$ $7y^{-3}$

83. $\dfrac{10}{t^5}$ $10t^{-5}$

84. $\dfrac{3}{z^n}$ $3z^{-n}$

85. $\dfrac{5}{x^n}$ $5x^{-n}$

86. $\dfrac{9}{y^n}$ $9y^{-n}$

87. $\dfrac{2x^2}{y^4}$ $2y^{-4}x^{-2}$

88. $\dfrac{5x^3}{y^6}$ $5y^{-6}x^{-3}$

In Exercises 89–102, rewrite the expression using only positive exponents. Then evaluate the expression.

89. 3^{-2} $\dfrac{1}{3^2} = \dfrac{1}{9}$

90. 5^{-3} $\dfrac{1}{5^3} = \dfrac{1}{125}$

91. $(-4)^{-3}$
$\dfrac{1}{(-4)^3} = -\dfrac{1}{64}$

92. $(-6)^{-2}$
$\dfrac{1}{(-6)^2} = \dfrac{1}{36}$

93. $\dfrac{1}{4^{-2}}$ $4^2 = 16$

94. $\dfrac{1}{16^{-1}}$ 16

95. $\dfrac{2}{3^{-4}}$ $2(3^4) = 162$

96. $\dfrac{4}{3^{-2}}$ $4(3^2) = 36$

97. $\dfrac{2^{-4}}{3}$ $\dfrac{1}{3(2^4)} = \dfrac{1}{48}$

98. $\dfrac{4^{-3}}{2}$ $\dfrac{1}{2(4^3)} = \dfrac{1}{128}$

99. $\dfrac{4^{-2}}{3^{-4}}$ $\dfrac{3^4}{4^2} = \dfrac{81}{16}$

100. $\dfrac{5^{-3}}{3^{-1}}$ $\dfrac{3^1}{5^3} = \dfrac{3}{125}$

101. $\left(\dfrac{2}{3}\right)^{-2}$ $\left(\dfrac{3}{2}\right)^2 = \dfrac{9}{4}$

102. $\left(\dfrac{5}{4}\right)^{-3}$ $\left(\dfrac{4}{5}\right)^3 = \dfrac{64}{125}$

In Exercises 103–106, use a calculator to evaluate the expression. Round your answer to four decimal places.

103. 3.8^{-4} 0.0048

104. 6.2^{-3} 0.0042

105. $100(1.06)^{-15}$
41.7265

106. $500(1.08)^{-20}$
107.2741

In Exercises 107–148, use rules of exponents to simplify the expression using only positive exponents. (Assume that no variable is zero.) See Examples 7–9.

107. $4^{-2} \cdot 4^3$ 4

108. $5^{-3} \cdot 5^2$ $\dfrac{1}{5}$

109. $x^{-4} \cdot x^6$ x^2

110. $a^{-5} \cdot a^2$ $\dfrac{1}{a^3}$

111. $u^{-6} \cdot u^3$ $\dfrac{1}{u^3}$

112. $t^{-2} \cdot t^2$ 1

113. $xy^{-3} \cdot y^2$ $\dfrac{x}{y}$

114. $u^{-2}v \cdot u^2$ v

115. $\dfrac{x^2}{x^{-3}}$ x^5

116. $\dfrac{z^4}{z^{-2}}$ z^6

117. $\dfrac{y^{-5}}{y}$ $\dfrac{1}{y^6}$

118. $\dfrac{x^{-3}}{x^2}$ $\dfrac{1}{x^5}$

119. $\dfrac{x^{-4}}{x^{-2}}$ $\dfrac{1}{x^2}$

120. $\dfrac{t^{-5}}{t^{-1}}$ $\dfrac{1}{t^4}$

121. $(y^{-3})^2$ $\dfrac{1}{y^6}$

122. $(z^{-2})^3$ $\dfrac{1}{z^6}$

123. $(s^2)^{-1}$ $\dfrac{1}{s^2}$

124. $(a^3)^{-3}$ $\dfrac{1}{a^9}$

125. $(2x^{-2})^0$ 1

126. $(2x^{-5})^0$ 1

127. $\dfrac{b^2 \cdot b^{-3}}{b^4}$ $\dfrac{1}{b^5}$

128. $\dfrac{c^{-3} \cdot c^4}{c^{-1}}$ c^2

129. $(3x^2y)^{-2}$ $\dfrac{1}{9x^4y^2}$

130. $(4x^{-3}y^2)^{-3}$ $\dfrac{x^9}{64y^6}$

131. $(4a^{-2}b^3)^{-3}$
$\dfrac{a^6}{64b^9}$

132. $(-2s^{-1}t^{-2})^{-1}$
$-\dfrac{st^2}{2}$

133. $(-2x^2)(4x^{-3})$ $-\dfrac{8}{x}$

134. $(4y^{-2})(3y^4)$ $12y^2$

135. $\left(\dfrac{x}{10}\right)^{-1}$ $\dfrac{10}{x}$

136. $\left(\dfrac{4}{z}\right)^{-2}$ $\dfrac{z^2}{16}$

137. $\left(\dfrac{3z^2}{x}\right)^{-2}$ $\dfrac{x^2}{9z^4}$

138. $\left(\dfrac{x^{-3}y^4}{5}\right)^{-3}$ $\dfrac{125x^9}{y^{12}}$

139. $\dfrac{(2y)^{-4}}{(2y)^{-4}}$ 1

140. $\dfrac{(3z)^{-2}}{(3z)^{-2}}$ 1

141. $\dfrac{3}{2} \cdot \left(\dfrac{-2}{3}\right)^{-3}$ $-\dfrac{81}{16}$

142. $\dfrac{3}{8} \cdot \left(\dfrac{-5}{2}\right)^{-3}$ $-\dfrac{3}{125}$

143. $\dfrac{(-2x)^{-3}}{-4x^{-2}}$ $\dfrac{1}{32x}$

144. $\dfrac{2x^{-3}}{(5x)^{-1}}$ $\dfrac{10}{x^2}$

145. $(5x^2y^4z^6)^3(5x^2y^4z^6)^{-3}$
1

146. $(8x^3y^2z^5)^6(8x^3y^2z^5)^{-6}$
1

147. $(x+y)^{-8}(x+y)^8$
1

148. $(u^2-v)^4(u^2-v)^{-4}$
1

In Exercises 149–158, write the number in scientific notation. See Example 10.

149. 93,000,000
9.3×10^7

150. 900,000,000
9×10^8

151. 1,637,000,000
1.637×10^9

152. 16,000,000
1.6×10^7

153. 0.000435
4.35×10^{-4}

154. 0.008367
8.367×10^{-3}

155. 0.004392
4.392×10^{-3}

156. 0.00000045
4.5×10^{-7}

157. 0.0678 6.78×10^{-2}

158. 0.00082 8.2×10^{-4}

In Exercises 159–168, write the number in decimal notation. See Example 11.

159. 1.09×10^6
1,090,000

160. 2.345×10^8
234,500,000

161. 8.67×10^2
867

162. 9.4675×10^4
94,675

163. 8.52×10^{-3}
0.00852

164. 7.021×10^{-5}
0.00007021

165. 6.21×10^{-6}
0.00000621

166. 4.73×10^{-8}
0.0000000473

167. $(8 \times 10^3) + (3 \times 10^0) + (5 \times 10^{-2})$ 8003.05

168. $(6 \times 10^4) + (9 \times 10^0) + (4 \times 10^{-1})$ 60,009.4

In Exercises 169–182, use a calculator to evaluate the expression. Write your answer in scientific notation. Round your answer to four decimal places. See Example 12.

169. $8,000,000 \times 623,000$ 4.9840×10^{12}

170. $93,200,000 \times 1,657,000$ 1.5443×10^{14}

171. $0.000345 \times 8,980,000,000$ 3.0981×10^6

172. $345,000 \times 0.000086$ 2.9670×10^1

173. $3,200,000^5$
3.3554×10^{32}

174. $75,000,000^6$
1.7798×10^{47}

175. $(3.28 \times 10^{-6})^4$
1.1574×10^{-22}

176. $(4.5 \times 10^{-5})^3$
9.1125×10^{-14}

177. $\dfrac{848,000,000}{1,620,000}$
5.2346×10^2

178. $\dfrac{67,000,000}{0.0052}$
1.2885×10^{10}

179. $(4.85 \times 10^5)(2.04 \times 10^8)$ 9.8940×10^{13}

180. $(3.29 \times 10^3)(5.08 \times 10^6)$ 1.6713×10^{10}

181. $\dfrac{8.6 \times 10^4}{3.9 \times 10^7}$
2.2051×10^{-3}

182. $\dfrac{4.1 \times 10^5}{9.6 \times 10^9}$
4.2708×10^{-5}

Solving Problems

183. *Astronomy* The star Beta Andromeda is approximately 75 light-years from Earth (see figure). (A light-year is the distance light can travel in 1 year.) A light-year is approximately 5.8657×10^{12} miles. Estimate the distance from Earth to the star.

4.3993×10^{14} miles

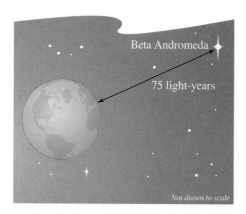

Beta Andromeda

75 light-years

Not drawn to scale

184. *Astronomy* One astronomical unit AU is the average distance between the sun and Earth (approximately 149,605,000 kilometers). The table shows the average distances between selected planets and the sun in astronomical units. Approximate each distance in kilometers and write the answer in scientific notation.

Planet	Mercury	Saturn	Neptune	Pluto
AU	0.39	9.54	30.06	39.53
	5.83×10^{7}	1.43×10^{9}	4.50×10^{9}	5.91×10^{9}

185. *Time for Light to Travel* Light travels from the sun to Earth in approximately

$$\frac{9.3 \times 10^{7}}{1.1 \times 10^{7}}$$

minutes. Write this time in decimal notation.

8.45 minutes

186. *Boltzmann's Constant* The study of the kinetic energy of an ideal gas uses Boltzmann's constant. This constant k is given by

$$k = \frac{8.31}{6.02 \times 10^{23}}.$$

Evaluate the expression and write the result in scientific notation. 1.38×10^{-23}

187. *Depreciation* A new car is purchased for $24,000. Its value V after t years is

$$V = 24,000(1.2)^{-t}.$$

(a) Use the model to complete the table.

t	0	2	4
$24,000(1.2)^{-t}$	$24,000	$16,667	$11,574

t	6	8
$24,000(1.2)^{-t}$	$8,038	$5,582

(b) Plot the data in the table.
    See Additional Answers.

(c) *Guess, Check, and Revise* When will the car be valued at less than $1000?
    When the car is 18 years old

188. *Numerical and Graphical Analysis*

(a) Complete the table by evaluating the indicated powers of 2.

x	-1	-2	-3	-4	-5
$y = 2^{x}$	$\frac{1}{2}$	$\frac{1}{4}$	$\frac{1}{8}$	$\frac{1}{16}$	$\frac{1}{32}$

(b) Plot the data in the table.
    See Additional Answers.

(c) Use the table or the graph to describe the value of 2^{-n} when n is very large. Will the value of 2^{-n} ever be negative?   Approaches 0; No.

189. *Hydraulic Compression* A hydraulic cylinder in a large press contains 2 gallons of oil. When the cylinder is under full pressure, the actual volume of oil is decreased by $2(150)(20 \times 10^{-6})$ gallons. Write the decreased volume in decimal notation.

1.994 gallons

190. *Computer Storage* A gigabyte is a measure of a computer's storage capacity. One gigabyte holds about one billion bytes of information. A firm's computer network contains 2500 gigabytes of memory. How many bytes are in the network? Write your result in scientific notation.

2.5×10^{12} bytes

Explaining Concepts

True or False? In Exercises 191–196, state whether the equation is true or false. If it is false, find values of x and y that show it to be false.

191. $x^3y^3 = xy^3$ False. Let $x = 5$ and $y = 2$.

192. $x^{-1}y^{-1} = \dfrac{1}{xy}$ True

193. $x^{-1} + y^{-1} = \dfrac{1}{x+y}$ False. Let $x = 5$ and $y = 2$.

194. $\dfrac{x^{-4}}{x^{-3}} = x$ False. Let $x = 2$.

195. $(x \times 10^3)^4 = x^4 \times 10^{12}$ True

196. $\dfrac{2x \times 10^{-5}}{x \times 10^{-3}} = 2 \times 10^{-2}$ True

197. *Writing* Does $a^{-1}b^{-1} = \dfrac{1}{ab}$? Explain your answer.

Yes. $a^{-1}b^{-1} = a^{-1} \cdot b^{-1} = \dfrac{1}{a} \cdot \dfrac{1}{b} = \dfrac{1}{ab}$

198. *Writing* Does $a^{-1} + b^{-1} = \dfrac{1}{a+b}$? Explain your answer.

No. $a^{-1} + b^{-1} = \dfrac{1}{a} + \dfrac{1}{b} \neq \dfrac{1}{a+b}$

199. Without looking back at page 271, state as many of the seven rules of exponents as you can.

$a^m a^n = a^{m+n}$

$\dfrac{a^m}{a^n} = a^{m-n}$

$(ab)^m = a^m b^m$

$\left(\dfrac{a}{b}\right)^m = \dfrac{a^m}{b^m}$

$(a^m)^n = a^{mn}$

$a^{-n} = \dfrac{1}{a^n}$

$a^0 = 1$

200. Give examples of large and small numbers written in scientific notation.

$3.4 \times 10^7 = 34,000,000$

$3.4 \times 10^{-6} = 0.0000034$

201. Find the reciprocal of 4×10^{-3}. $\dfrac{10^3}{4}$

202. (a) Find as many equivalent pairs as possible among the following exponential expressions.

$\dfrac{2}{x^{-3}}$, $\dfrac{1}{2x^3}$, $2x^{-3}$, $\dfrac{1}{(2x)^{-3}}$, $2x^3$, $\dfrac{x^3}{8}$, $\dfrac{1}{8x^3}$, $\dfrac{x^{-3}}{2}$,

$8x^3$, $(2x)^{-3}$

Equivalent pairs: $\dfrac{2}{x^{-3}}$ and $2x^3$, $\dfrac{1}{2x^3}$ and $\dfrac{x^{-3}}{2}$, $\dfrac{1}{(2x)^{-3}}$ and $8x^3$, and $\dfrac{1}{8x^3}$ and $(2x)^{-3}$

(b) Use a calculator, with $x = 3$, to illustrate the equivalence of each pair. Organize your work into a table with four columns–an expression, the equivalent expression, and each expression evaluated at $x = 3$. See Additional Answers.

203. Justify each step.
$$(3 \times 10^5)(4 \times 10^6) = (3 \times 10^5)(10^6 \times 4)$$
$$= 3(10^5 \times 10^6)(4)$$
$$= 3(10^{5+6})(4)$$
$$= (3 \cdot 4)10^{11}$$
$$= 12 \times 10^{11} = 1.2 \times 10^{12}$$

$(3 \times 10^5)(4 \times 10^6)$	
$= (3 \times 10^5)(10^6 \times 4)$	Commutative Property of Multiplication
$= 3(10^5 \times 10^6)(4)$	Associative Property of Multiplication
$= 3(10^{5+6})(4)$	Rule of exponents
$= (3 \cdot 4)10^{11}$	Commutative Property of Multiplication
$= 12 \times 10^{11}$	Multiplication
$= 1.2 \times 10^{12}$	Scientific notation

5.2 Adding and Subtracting Polynomials

David Lassman/The Image Works

Why You Should Learn It

Polynomials can be used to model and solve real-life problems. For instance, in Exercise 101 on page 288, polynomials are used to model the numbers of daily morning and evening newspapers in the United States.

1 Identify the degrees and leading coefficients of polynomials.

What You Should Learn

1 Identify the degrees and leading coefficients of polynomials.

2 Add polynomials using a horizontal or vertical format.

3 Subtract polynomials using a horizontal or vertical format.

Basic Definitions

Recall from Section 2.1 that the *terms* of an algebraic expression are those parts separated by addition. An algebraic expression whose terms are all of the form ax^k, where a is any real number and k is a nonnegative integer, is called a **polynomial in one variable,** or simply a **polynomial.** Here are some examples of polynomials in one variable.

$$2x + 5, \quad x^2 - 3x + 7, \quad 9x^5, \quad \text{and} \quad x^3 + 8$$

In the term ax^k, a is the **coefficient** of the term and k is the **degree** of the term. Note that the degree of the term ax is 1, and the degree of a constant term is 0. Because a polynomial is an algebraic *sum*, the coefficients take on the signs between the terms. For instance,

$$x^4 + 2x^3 - 5x^2 + 7 = (1)x^4 + 2x^3 + (-5)x^2 + (0)x + 7$$

has coefficients 1, 2, -5, 0, and 7. For this polynomial, the last term, 7, is the **constant term.** Polynomials are usually written in the order of descending powers of the variable. This is called **standard form.** Here are three examples.

Nonstandard Form	*Standard Form*
$4 + x$	$x + 4$
$3x^2 - 5 - x^3 + 2x$	$-x^3 + 3x^2 + 2x - 5$
$18 - x^2 + 3$	$-x^2 + 21$

The **degree of a polynomial** is the degree of the term with the highest power, and the coefficient of this term is the **leading coefficient** of the polynomial. For instance, the polynomial

Degree
/
$$-3x^4 + 4x^2 + x + 7$$
/
Leading coefficient

is of fourth degree, and its leading coefficient is -3. The reasons why the degree of a polynomial is important will become clear as you study factoring and problem solving in Chapter 6.

Definition of a Polynomial in x

Let $a_n, a_{n-1}, \ldots, a_2, a_1, a_0$ be real numbers and let n be a nonnegative integer. A **polynomial in x** is an expression of the form

$$a_n x^n + a_{n-1} x^{n-1} + \cdots + a_2 x^2 + a_1 x + a_0$$

where $a_n \neq 0$. The polynomial is of **degree n,** and the number a_n is called the **leading coefficient.** The number a_0 is called the **constant term.**

Example 1　Identifying Polynomials

Identify which of the following are polynomials, and for any that are not polynomials, state why.

a. $3x^4 - 8x + x^{-1}$　　**b.** $x^2 - 3x + 1$

c. $x^3 + 3x^{1/2}$　　　**d.** $-\dfrac{1}{3}x + \dfrac{x^3}{4}$

Solution

a. $3x^4 - 8x + x^{-1}$ is *not* a polynomial because the third term, x^{-1}, has a negative exponent.

b. $x^2 - 3x + 1$ is a polynomial of degree 2 with integer coefficients.

c. $x^3 + 3x^{1/2}$ is *not* a polynomial because the exponent in the second term, $3x^{1/2}$, is not an integer.

d. $-\dfrac{1}{3}x + \dfrac{x^3}{4}$ is a polynomial of degree 3 with rational coefficients.

Example 2　Determining Degrees and Leading Coefficients

Write each polynomial in standard form and identify the degree and leading coefficient.

	Polynomial	Standard Form	Degree	Leading Coefficient
a.	$4x^2 - 5x^7 - 2 + 3x$	$-5x^7 + 4x^2 + 3x - 2$	7	-5
b.	$4 - 9x^2$	$-9x^2 + 4$	2	-9
c.	8	8	0	8
d.	$2 + x^3 - 5x^2$	$x^3 - 5x^2 + 2$	3	1

In part (c), note that a polynomial with only a constant term has a degree of zero.

A polynomial with only one term is called a **monomial.** Polynomials with two *unlike* terms are called **binomials,** and those with three *unlike* terms are called **trinomials.** For example, $3x^2$ is a *monomial,* $-3x + 1$ is a *binomial,* and $4x^3 - 5x + 6$ is a *trinomial.*

② Add polynomials using a horizontal or vertical format.

Adding Polynomials

As with algebraic expressions, the key to adding two polynomials is to recognize *like* terms—those having the *same degree.* By the Distributive Property, you can then combine the like terms using either a horizontal or a vertical format of terms. For instance, the polynomials $2x^2 + 3x + 1$ and $x^2 - 2x + 2$ can be added horizontally to obtain

$$(2x^2 + 3x + 1) + (x^2 - 2x + 2) = (2x^2 + x^2) + (3x - 2x) + (1 + 2)$$
$$= 3x^2 + x + 3$$

or they can be added vertically to obtain the same result.

$$2x^2 + 3x + 1 \qquad \text{\small Vertical format}$$
$$\underline{x^2 - 2x + 2}$$
$$3x^2 + x + 3$$

Technology: Tip

You can use a graphing calculator to check the results of adding or subtracting polynomials. For instance, try graphing

$$y_1 = (2x + 1) + (-3x - 4)$$

and

$$y_2 = -x - 3$$

in the same viewing window, as shown below. Because both graphs are the same, you can conclude that

$$(2x + 1) + (-3x - 4) = -x - 3.$$

This graphing technique is called "graph the left side and graph the right side."

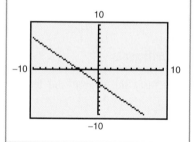

Study Tip

When you use a vertical format to add polynomials, be sure that you line up the *like terms.*

Example 3   Adding Polynomials Horizontally

Use a horizontal format to find each sum.

a. $(2x^2 + 4x - 1) + (x^2 - 3)$ Original polynomials

$\quad = (2x^2 + x^2) + (4x) + (-1 - 3)$ Group like terms.

$\quad = 3x^2 + 4x - 4$ Combine like terms.

b. $(x^3 + 2x^2 + 4) + (3x^2 - x + 5)$ Original polynomials

$\quad = (x^3) + (2x^2 + 3x^2) + (-x) + (4 + 5)$ Group like terms.

$\quad = x^3 + 5x^2 - x + 9$ Combine like terms.

c. $(2x^2 - x + 3) + (4x^2 - 7x + 2) + (-x^2 + x - 2)$ Original polynomials

$\quad = (2x^2 + 4x^2 - x^2) + (-x - 7x + x) + (3 + 2 - 2)$ Group like terms.

$\quad = 5x^2 - 7x + 3$ Combine like terms.

Example 4   Adding Polynomials Vertically

Use a vertical format to find each sum.

a. $(-4x^3 - 2x^2 + x - 5) + (2x^3 + 3x + 4)$

b. $(5x^3 + 2x^2 - x + 7) + (3x^2 - 4x + 7) + (-x^3 + 4x^2 - 2x - 8)$

Solution

a. $-4x^3 - 2x^2 + x - 5$

$\quad\underline{ 2x^3 + 3x + 4}$

$\quad-2x^3 - 2x^2 + 4x - 1$

b. $5x^3 + 2x^2 - x + 7$

$\quad 3x^2 - 4x + 7$

$\quad\underline{-x^3 + 4x^2 - 2x - 8}$

$\quad4x^3 + 9x^2 - 7x + 6$

③ Subtract polynomials using
a horizontal or vertical format.

Subtracting Polynomials

To subtract one polynomial from another, you *add the opposite* by changing the sign of each term of the polynomial that is being subtracted and then adding the resulting like terms. Note how $(x^2 - 1)$ is subtracted from $(2x^2 - 4)$.

$$(2x^2 - 4) - (x^2 - 1) = 2x^2 - 4 - x^2 + 1 \qquad \text{Distributive Property}$$
$$= (2x^2 - x^2) + (-4 + 1) \qquad \text{Group like terms.}$$
$$= x^2 - 3 \qquad \text{Combine like terms.}$$

Recall from the Distributive Property that

$$-(x^2 - 1) = (-1)(x^2 - 1) = -x^2 + 1.$$

Example 5   Subtracting Polynomials Horizontally

Use a horizontal format to find each difference.

a. $(2x^2 + 3) - (3x^2 - 4)$

b. $(3x^3 - 4x^2 + 3) - (x^3 + 3x^2 - x - 4)$

Solution

a. $(2x^2 + 3) - (3x^2 - 4) = 2x^2 + 3 - 3x^2 + 4 \qquad \text{Distributive Property}$
$$= (2x^2 - 3x^2) + (3 + 4) \qquad \text{Group like terms.}$$
$$= -x^2 + 7 \qquad \text{Combine like terms.}$$

b. $(3x^3 - 4x^2 + 3) - (x^3 + 3x^2 - x - 4) \qquad \text{Original polynomials}$
$$= 3x^3 - 4x^2 + 3 - x^3 - 3x^2 + x + 4 \qquad \text{Distributive Property}$$
$$= (3x^3 - x^3) + (-4x^2 - 3x^2) + (x) + (3 + 4) \qquad \text{Group like terms.}$$
$$= 2x^3 - 7x^2 + x + 7 \qquad \text{Combine like terms.}$$

Students may be able to omit some of these steps. However, point out that changing signs incorrectly is one of the most common algebraic errors.

Additional Examples
Use a horizontal format to perform the indicated operations.

a.
$(2y^4 - 3y^2 + y - 6) + (-y^3 + 6y^2 + 8)$

b. $(4x^4 - x^2 + 1) - (x^4 - 2x^3 - x^2)$

Answers:

a. $2y^4 - y^3 + 3y^2 + y + 2$

b. $3x^4 + 2x^3 + 1$

Example 6   Combining Polynomials Horizontally

Use a horizontal format to perform the indicated operations.

$$(x^2 - 2x + 1) - [(x^2 + x - 3) + (-2x^2 - 4x)]$$

Solution

$$(x^2 - 2x + 1) - [(x^2 + x - 3) + (-2x^2 - 4x)] \qquad \text{Original polynomials}$$
$$= (x^2 - 2x + 1) - [(x^2 - 2x^2) + (x - 4x) + (-3)] \qquad \text{Group like terms.}$$
$$= (x^2 - 2x + 1) - [-x^2 - 3x - 3] \qquad \text{Combine like terms.}$$
$$= x^2 - 2x + 1 + x^2 + 3x + 3 \qquad \text{Distributive Property}$$
$$= (x^2 + x^2) + (-2x + 3x) + (1 + 3) \qquad \text{Group like terms.}$$
$$= 2x^2 + x + 4 \qquad \text{Combine like terms.}$$

Be especially careful to use the correct signs when subtracting one polynomial from another. One of the most common mistakes in algebra is to forget to change signs correctly when subtracting one expression from another. Here is an example.

Wrong sign

$$(x^2 + 3) - (x^2 + 2x - 2) \neq x^2 + 3 - x^2 + 2x - 2 \qquad \text{Common error}$$

Wrong sign

Note that the error is forgetting to change *all* of the signs in the polynomial that is being subtracted. Here is the correct way to perform the subtraction.

Correct sign

$$(x^2 + 3) - (x^2 + 2x - 2) = x^2 + 3 - x^2 - 2x + 2 \qquad \text{Correct}$$

Correct sign

Just as you did for addition, you can use a vertical format to subtract one polynomial from another. (The vertical format does not work well with subtractions involving three or more polynomials.) When using a vertical format, write the polynomial being subtracted underneath the one from which it is being subtracted. Be sure to line up like terms in vertical columns.

Example 7 Subtracting Polynomials Vertically

Use a vertical format to find each difference.

a. $(3x^2 + 7x - 6) - (3x^2 + 7x)$
b. $(5x^3 - 2x^2 + x) - (4x^2 - 3x + 2)$
c. $(4x^4 - 2x^3 + 5x^2 - x + 8) - (3x^4 - 2x^3 + 3x - 4)$

Solution

a.
$$\begin{array}{l} (3x^2 + 7x - 6) \\ -(3x^2 + 7x \quad) \end{array} \implies \begin{array}{l} 3x^2 + 7x - 6 \\ -3x^2 - 7x \\ \hline \qquad\qquad -6 \end{array} \quad \text{Change signs and add.}$$

b.
$$\begin{array}{l} (5x^3 - 2x^2 + \ x \quad) \\ -(\quad 4x^2 - 3x + 2) \end{array} \implies \begin{array}{l} 5x^3 - 2x^2 + \ x \\ \quad - 4x^2 + 3x - 2 \\ \hline 5x^3 - 6x^2 + 4x - 2 \end{array} \quad \text{Change signs and add.}$$

c.
$$\begin{array}{l} (4x^4 - 2x^3 + 5x^2 - \ x + 8) \\ -(3x^4 - 2x^3 \qquad + 3x - 4) \end{array} \implies \begin{array}{l} 4x^4 - 2x^3 + 5x^2 - \ x + \ 8 \\ -3x^4 + 2x^3 \qquad - 3x + \ 4 \\ \hline x^4 \qquad + 5x^2 - 4x + 12 \end{array}$$

In Example 7, try using a horizontal arrangement to perform the subtractions.

Example 8 Combining Polynomials

Perform the indicated operations and simplify.

$$(3x^2 - 7x + 2) - (4x^2 + 6x - 1) + (-x^2 + 4x + 5)$$

Remind students that they can use graphing calculators to verify these results.

Solution

$$\begin{aligned}
&(3x^2 - 7x + 2) - (4x^2 + 6x - 1) + (-x^2 + 4x + 5)\\
&= 3x^2 - 7x + 2 - 4x^2 - 6x + 1 - x^2 + 4x + 5\\
&= (3x^2 - 4x^2 - x^2) + (-7x - 6x + 4x) + (2 + 1 + 5)\\
&= -2x^2 - 9x + 8
\end{aligned}$$

Additional Example
Perform the indicated operations.

$3(x^2 - 2x + 1) - 2(x^2 + x - 3)$

Answer:

$x^2 - 8x + 9$

Example 9 Combining Polynomials

Perform the indicated operations and simplify.

$$(-2x^2 + 4x - 3) - [(4x^2 - 5x + 8) - 2(-x^2 + x + 3)]$$

Solution

$$\begin{aligned}
&(-2x^2 + 4x - 3) - [(4x^2 - 5x + 8) - 2(-x^2 + x + 3)]\\
&= (-2x^2 + 4x - 3) - [4x^2 - 5x + 8 + 2x^2 - 2x - 6]\\
&= (-2x^2 + 4x - 3) - [(4x^2 + 2x^2) + (-5x - 2x) + (8 - 6)]\\
&= (-2x^2 + 4x - 3) - [6x^2 - 7x + 2]\\
&= -2x^2 + 4x - 3 - 6x^2 + 7x - 2\\
&= (-2x^2 - 6x^2) + (4x + 7x) + (-3 - 2)\\
&= -8x^2 + 11x - 5
\end{aligned}$$

Example 10 Geometry: Area of a Region

Find an expression for the area of the shaded region shown in Figure 5.1.

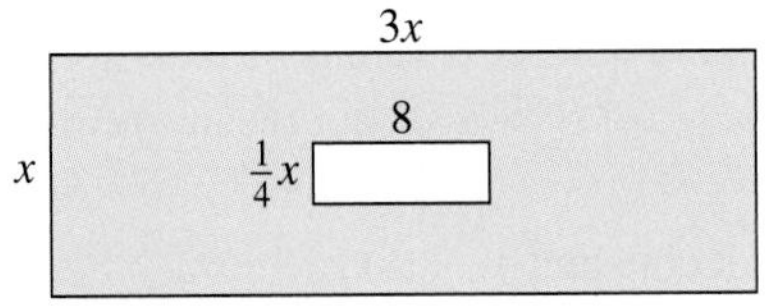

Figure 5.1

Solution

To find a polynomial that represents the area of the shaded region, subtract the area of the inner rectangle from the area of the outer rectangle, as follows.

$$\begin{aligned}
\text{Area of shaded region} &= \text{Area of outer rectangle} - \text{Area of inner rectangle}\\
&= 3x(x) - 8\left(\frac{1}{4}x\right)\\
&= 3x^2 - 2x
\end{aligned}$$

5.2 Exercises

Review   Concepts, Skills, and Problem Solving

Keep mathematically in shape by doing these exercises *before* the problems of this section.

Properties and Definitions

1. *Writing* 🖉   In your own words, state the definition of an algebraic expression.   An algebraic expression is a collection of letters (variables) and real numbers (constants) combined by using addition, subtraction, multiplication, or division.

2. *Writing* 🖉   State the definition of the terms of an algebraic expression.   The terms of an algebraic expression are those parts separated by addition.

Simplifying Expressions

In Exercises 3–6, use the Distributive Property to expand the expression.

3. $10(x - 1)$ $10x - 10$ 4. $4(3 - 2z)$ $12 - 8z$

5. $-\frac{1}{2}(4 - 6x)$ $-2 + 3x$

6. $-25(2x - 3)$ $-50x + 75$

In Exercises 7–10, simplify the expression.

7. $8y - 2x + 7x - 10y$ $5x - 2y$

8. $\frac{5}{6}x - \frac{2}{3}x + 8$ $\frac{1}{6}x + 8$

9. $10(x - 1) - 3(x + 2)$ $7x - 16$

10. $-3[x + (2 + 3x)]$ $-12x - 6$

Graphing Equations

In Exercises 11 and 12, graph the equation. Use a graphing calculator to verify your graph.
See Additional Answers.

11. $y = 2 + \frac{3}{2}x$

12. $y = |x - 1| + x$

Developing Skills

In Exercises 1–8, determine whether the expression is a polynomial. If it is not, explain why. See Example 1.

1. $9 - z$ Polynomial   2. $t^2 - 4$ Polynomial

3. $x^{2/3} + 8$ Not a polynomial because the exponent in the first term is not an integer.

4. $9 - z^{1/2}$ Not a polynomial because the exponent in the second term is not an integer.

5. $6x^{-1}$ Not a polynomial because the exponent is negative.

6. $1 - 4x^{-2}$ Not a polynomial because the exponent in the second term is negative.

7. $z^2 - 3z + \frac{1}{4}$ Polynomial

8. $t^3 - 3t + 4$ Polynomial

In Exercises 9–18, write the polynomial in standard form. Then identify its degree and leading coefficient. See Example 2.

9. $12x + 9$ Standard form: $12x + 9$; Degree: 1; Leading coefficient: 12

10. $4 - 7y$ Standard form: $-7y + 4$; Degree: 1; Leading coefficient: -7

11. $7x - 5x^2 + 10$ Standard form: $-5x^2 + 7x + 10$; Degree: 2; Leading coefficient: -5

12. $5 - x + 15x^2$ Standard form: $15x^2 - x + 5$; Degree: 2; Leading coefficient: 15

13. $8x + 2x^5 - x^2 - 1$ Standard form: $2x^5 - x^2 + 8x - 1$; Degree: 5; Leading coefficient: 2

14. $5x^3 - 3x^2 + 10$ Standard form: $5x^3 - 3x^2 + 10$; Degree: 3; Leading coefficient: 5

15. 10 Standard form: 10; Degree: 0; Leading coefficient: 10

16. -32 Standard form: -32; Degree: 0; Leading coefficient: -32;

17. $v_0t - 16t^2$ (v_0 is a constant.)   Standard form: $-16t^2 + v_0 t$; Degree: 2; Leading coefficient: -16

18. $64 - \frac{1}{2}at^2$ (a is a constant.)   Standard form: $-\frac{1}{2}at^2 + 64$; Degree: 2; Leading coefficient: $-\frac{1}{2}a$

In Exercises 19–24, determine whether the polynomial is a monomial, binomial, or trinomial.

19. $14y - 2$ Binomial   20. -16 Monomial

21. $93z^2$ Monomial   22. $a^2 + 2a - 9$ Trinomial

23. $4x + 18x^2 - 5$ Trinomial

24. $6x^2 + x$ Binomial

In Exercises 25–30, give an example of a polynomial in one variable satisfying the condition. (*Note:* There are many correct answers.)

25. A binomial of degree 3 $5x^3 - 10$

26. A trinomial of degree 4 $2z^4 + 7z - 2$

27. A monomial of degree 2 $3y^2$

28. A binomial of degree 5 $6 - 2v^5$

29. A trinomial of degree 6 $x^6 - 4x^3 - 2$

30. A monomial of degree 0 7

In Exercises 31–42, use a horizontal format to find the sum. See Example 3.

31. $(11x - 2) + (3x + 8)$ $14x + 6$

32. $(-2x + 4) + (x - 6)$ $-x - 2$

33. $(3z^2 - z + 2) + (z^2 - 4)$ $4z^2 - z - 2$

34. $(6x^4 + 8x) + (4x - 6)$ $6x^4 + 12x - 6$

35. $b^2 + (b^3 - 2b^2 + 3) + (b^3 - 3)$ $2b^3 - b^2$

36. $(3x^2 - x) + 5x^3 + (-4x^3 + x^2 - 8)$
$x^3 + 4x^2 - x - 8$

37. $(2ab - 3) + (a^2 - 2ab) + (4b^2 - a^2)$ $4b^2 - 3$

38. $(uv - 3) + (4uv + 1)$ $5uv - 2$

39. $\left(\frac{2}{3}y^2 - \frac{3}{4}\right) + \left(\frac{5}{6}y^2 + 2\right)$ $\frac{3}{2}y^2 + \frac{5}{4}$

40. $\left(\frac{3}{4}x^3 - \frac{1}{2}\right) + \left(\frac{1}{8}x^3 + 3\right)$ $\frac{7}{8}x^3 + \frac{5}{2}$

41. $(0.1t^3 - 3.4t^2) + (1.5t^3 - 7.3)$ $1.6t^3 - 3.4t^2 - 7.3$

42. $(0.7x^2 - 0.2x + 2.5) + (7.4x - 3.9)$
$0.7x^2 + 7.2x - 1.4$

In Exercises 43–56, use a vertical format to find the sum. See Example 4.

43. $2x + 5$
$\underline{3x + 8}$ $5x + 13$

44. $10x - 7$
$\underline{6x + 4}$ $16x - 3$

45. $-2x + 10$
$\underline{x - 38}$ $-x - 28$

46. $4x^2 + 13$
$\underline{3x^2 - 11}$ $7x^2 + 2$

47. $(-x^3 + 3) + (3x^3 + 2x^2 + 5)$
$2x^3 + 2x^2 + 8$

48. $(2z^3 + 3z - 2) + (z^2 - 2z)$
$2z^3 + z^2 + z - 2$

49. $(3x^4 - 2x^3 - 4x^2 + 2x - 5) + (x^2 - 7x + 5)$
$3x^4 - 2x^3 - 3x^2 - 5x$

50. $(x^5 - 4x^3 + x + 9) + (2x^4 + 3x^3 - 3)$
$x^5 + 2x^4 - x^3 + x + 6$

51. $(x^2 - 4) + (2x^2 + 6)$ $3x^2 + 2$

52. $(x^3 + 2x - 3) + (4x + 5)$ $x^3 + 6x + 2$

53. $(x^2 - 2x + 2) + (x^2 + 4x) + 2x^2$ $4x^2 + 2x + 2$

54. $(5y + 10) + (y^2 - 3y - 2) + (2y^2 + 4y - 3)$
$3y^2 + 6y + 5$

55. Add $8y^3 + 7$ to $5 - 3y^3$. $5y^3 + 12$

56. Add $2z - 8z^2 - 3$ to $z^2 + 5z$. $-7z^2 + 7z - 3$

In Exercises 57–66, use a horizontal format to find the difference. See Examples 5 and 6.

57. $(11x - 8) - (2x + 3)$ $9x - 11$

58. $(9x + 2) - (15x - 4)$ $-6x + 6$

59. $(x^2 - x) - (x - 2)$ $x^2 - 2x + 2$

60. $(x^2 - 4) - (x^2 - 4)$ 0

61. $(4 - 2x - x^3) - (3 - 2x + 2x^3)$ $-3x^3 + 1$

62. $(t^4 - 2t^2) - (3t^2 - t^4 - 5)$ $2t^4 - 5t^2 + 5$

63. $10 - (u^2 + 5)$ $-u^2 + 5$

64. $(z^3 + z^2 + 1) - z^2$ $z^3 + 1$

65. $(x^5 - 3x^4 + x^3 - 5x + 1) - (4x^5 - x^3 + x - 5)$
$-3x^5 - 3x^4 + 2x^3 - 6x + 6$

66. $(t^4 + 5t^3 - t^2 + 8t - 10) -$
$(t^4 + t^3 + 2t^2 + 4t - 7)$ $4t^3 - 3t^2 + 4t - 3$

In Exercises 67–80, use a vertical format to find the difference. See Example 7.

67. $2x - 2$
$\underline{-(x - 1)}$ $x - 1$

68. $9x + 7$
$\underline{-(3x + 9)}$ $6x - 2$

69. $2x^2 - x + 2$
$\underline{-(3x^2 + x - 1)}$ $-x^2 - 2x + 3$

70. $y^4 - 2$
$\underline{-(y^4 + 2)}$ -4

71. $(-3x^3 - 4x^2 + 2x - 5) - (2x^4 + 2x^3 - 4x + 5)$
$-2x^4 - 5x^3 - 4x^2 + 6x - 10$

72. $(12x^3 + 25x^2 - 15) - (-2x^3 + 18x^2 - 3x)$
$14x^3 + 7x^2 + 3x - 15$

73. $(2 - x^3) - (2 + x^3)$ $-2x^3$

74. $(4z^3 - 6) - (-z^3 + z - 2)$ $5z^3 - z - 4$

75. $(4t^3 - 3t + 5) - (3t^2 - 3t - 10)$ $4t^3 - 3t^2 + 15$

76. $(-s^2 - 3) - (2s^2 + 10s)$ $-3s^2 - 10s - 3$

77. $(6x^3 - 3x^2 + x) - [(x^3 + 3x^2 + 3) + (x - 3)]$
$5x^3 - 6x^2$

78. $(y^2 - y) - [(2y^2 + y) - (4y^2 - y + 2)]$
$3y^2 - 3y + 2$

79. Subtract $7x^3 - 4x + 5$ from $10x^3 + 15$.
$3x^3 + 4x + 10$

80. Subtract $y^5 - y^4$ from $y^2 + 3y^4$. $-y^5 + 4y^4 + y^2$

In Exercises 81–94, perform the indicated operations and simplify. See Examples 8 and 9.

81. $(6x - 5) - (8x + 15)$ $-2x - 20$

82. $(2x^2 + 1) + (x^2 - 2x + 1)$ $3x^2 - 2x + 2$

83. $-(x^3 - 2) + (4x^3 - 2x)$ $3x^3 - 2x + 2$

84. $-(5x^2 - 1) - (-3x^2 + 5)$ $-2x^2 - 4$

85. $2(x^4 + 2x) + (5x + 2)$ $2x^4 + 9x + 2$

86. $(z^4 - 2z^2) + 3(z^4 + 4)$ $4z^4 - 2z^2 + 12$

87. $(15x^2 - 6) - (-8x^3 - 14x^2 - 17)$
$8x^3 + 29x^2 + 11$

88. $(15x^4 - 18x - 19) - (-13x^4 - 5x + 15)$
$28x^4 - 13x - 34$

89. $5z - [3z - (10z + 8)]$ $12z + 8$

90. $(y^3 + 1) - [(y^2 + 1) + (3y - 7)]$
$y^3 - y^2 - 3y + 7$

91. $2(t^2 + 5) - 3(t^2 + 5) + 5(t^2 + 5)$ $4t^2 + 20$

92. $-10(u + 1) + 8(u - 1) - 3(u + 6)$ $-5u - 36$

93. $8v - 6(3v - v^2) + 10(10v + 3)$ $6v^2 + 90v + 30$

94. $3(x^2 - 2x + 3) - 4(4x + 1) - (3x^2 - 2x)$
$-20x + 5$

Solving Problems

▲ *Geometry*   In Exercises 95 and 96, find an expression for the perimeter of the figure.

95.

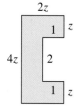

2z

1   z

4z   2

1   z

10z + 4

96.

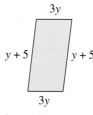

3y

y + 5   y + 5

3y

8y + 10

▲ *Geometry*   In Exercises 97–100, find an expression for the area of the shaded region of the figure. See Example 10.

97.

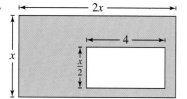

2x $2x^2 - 2x$

4

x

$\frac{x}{2}$

98.

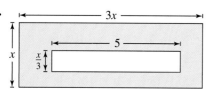

3x $3x^2 - \frac{5}{3}x$

5

x $\frac{x}{3}$

99.

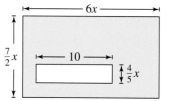

6x $21x^2 - 8x$

$\frac{7}{2}x$ 10

$\frac{4}{5}x$

100.

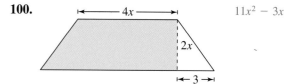

4x $11x^2 - 3x$

2x

3

7x

101. *Comparing Models*   The numbers of daily morning M and evening E newspapers for the years 1995 through 2000 can be modeled by

$$M = -0.29t^2 + 24.7t + 543, \quad 5 \le t \le 10$$

and

$$E = -31.8t + 1042, \quad 5 \le t \le 10$$

where t represents the year, with $t = 5$ corresponding to 1995.   (Source: Editor & Publisher Co.)

(a) Add the polynomials to find a model for the total number T of daily newspapers.
$T = -0.29t^2 - 7.1t + 1585, \quad 5 \le t \le 10$

(b) 🖩 Use a graphing calculator to graph all three models.   See Additional Answers.

(c) 🖩 Use the graphs from part (b) to determine whether the numbers of morning, evening, and total newspapers are increasing or decreasing.
Increasing, decreasing, decreasing

102. *Cost, Revenue, and Profit* The cost C of producing x units of a product is $C = 100 + 30x$. The revenue R for selling x units is $R = 90x - x^2$, where $0 \le x \le 40$. The profit P is the difference between revenue and cost.

(a) Perform the subtraction required to find the polynomial representing profit P.
$P = -x^2 + 60x - 100, \quad 0 \le x \le 40$

(b) 🖩 Use a graphing calculator to graph the polynomial representing profit. See Additional Answers.

(c) 🖩 Determine the profit when 30 units are produced and sold. Use the graph in part (b) to predict the change in profit if x is some value other than 30. 800; If x is some value other than 30, the profit is less than $800.

Explaining Concepts

103. Answer parts (a)–(c) of Motivating the Chapter on page 264.

104. *Writing* Explain the difference between the degree of a term of a polynomial and the degree of a polynomial. The degree of a term ax^k is k. The degree of a polynomial is the degree of its highest-degree term.

105. *Writing* Determine which of the two statements is always true. Is the statement not selected always false? Explain.

(a) "A polynomial is a trinomial." Sometimes true. $x^3 - 2x^2 + x + 1$ is a polynomial that is not a trinomial.

(b) "A trinomial is a polynomial." True

106. *Writing* In your own words, define "like terms." What is the only factor of like terms that can differ? Two terms are like terms if they are both constant or if they have the same variable factor(s). Numerical coefficients

107. *Writing* Describe how to combine like terms. What operations are used? Add (or subtract) their respective coefficients and attach the common variable factor.

108. *Writing* Is a polynomial an algebraic expression? Explain. Yes. A polynomial is an algebraic expression whose terms are all of the form ax^k, where a is any real number and k is a nonnegative integer.

109. *Writing* Is the sum of two binomials always a binomial? Explain. No. $(x^2 - 2) + (5 - x^2) = 3$

110. *Writing* Write a paragraph that explains how the adage "You can't add apples and oranges" might relate to adding two polynomials. Include several examples to illustrate the applicability of this statement. Answers will vary. The key point is that you can combine only like terms.

111. *Writing* In your own words, explain how to subtract polynomials. Give an example. To subtract one polynomial from another, add the opposite. You can do this by changing the sign of each of the terms of the polynomial that is being subtracted and then adding the resulting like terms. Examples will vary.

Mid-Chapter Quiz

Take this quiz as you would take a quiz in class. After you are done, check your work against the answers in the back of the book.

In Exercises 1–4, simplify the expression. (Assume that no denominator is zero.)

1. $(3a^2b)^2$ $9a^4b^2$

2. $(-3xy)^2(2x^2y)^3$ $72x^8y^5$

3. $\dfrac{-12x^3y}{9x^5y^2}$ $-\dfrac{4}{3x^2y}$

4. $\dfrac{3t^3}{(-6t)^2}$ $\dfrac{t}{12}$

In Exercises 5 and 6, rewrite the expression using only positive exponents.

5. $5x^{-2}y^{-3}$ $\dfrac{5}{x^2y^3}$

6. $\dfrac{3x^{-2}y}{5z^{-1}}$ $\dfrac{3yz}{5x^2}$

In Exercises 7 and 8, use rules of exponents to simplify the expression using only positive exponents. (Assume that no variable is zero.)

7. $(3a^{-3}b^2)^{-2}$ $\dfrac{a^6}{9b^4}$

8. $(4t^{-3})^0$ 1

9. Write the number 9,460,000,000 in scientific notation. 9.46×10^9

10. Write the number 5.021×10^{-8} in decimal notation. 0.00000005021

11. Explain why $x^2 + 2x - 3x^{-1}$ is not a polynomial.   Because the exponent of the third term is negative.

12. Determine the degree and the leading coefficient of the polynomial $-3x^4 + 2x^2 - x$.   Degree: 4; Leading coefficient: -3

13. Give an example of a trinomial in one variable of degree 5. $3x^5 - 3x + 1$

In Exercises 14–17, perform the indicated operations and simplify.

14. $(y^2 + 3y - 1) + (4 + 3y)$
$y^2 + 6y + 3$

15. $(3v^2 - 5) - (v^3 + 2v^2 - 6v)$
$-v^3 + v^2 + 6v - 5$

16. $9s - [6 - (s - 5) + 7s]$
$3s - 11$

17. $-3(4 - x) + 4(x^2 + 2) - (x^2 - 2x)$
$3x^2 + 5x - 4$

In Exercises 18 and 19, use a vertical format to find the sum.

18.
$$5x^4 \quad\ + 2x^2 + \ x - 3$$
$$\underline{\quad\quad 3x^3 - 2x^2 - 3x + 5}$$
$$5x^4 + 3x^3 \quad\quad - 2x + 2$$

19.
$$2x^3 + x^2 \quad\quad\ - 8$$
$$\underline{\quad\quad\quad 5x^2 - 3x - 9}$$
$$2x^3 + 6x^2 - 3x - 17$$

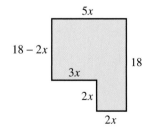

Figure for 20

20. Find an expression for the perimeter of the figure. $10x + 36$

5.3 Multiplying Polynomials: Special Products

Holly Harris/Getty Images

What You Should Learn

1. Find products with monomial multipliers.
2. Multiply binomials using the Distributive Property and the FOIL Method.
3. Multiply polynomials using a horizontal or vertical format.
4. Identify and use special binomial products.

Why You Should Learn It

Multiplying polynomials enables you to model and solve real-life problems. For instance, in Exercise 129 on page 303, you will multiply polynomials to find the total consumption of milk in the United States.

Monomial Multipliers

To multiply polynomials, you use many of the rules for simplifying algebraic expressions. You may want to review these rules from Section 2.2 and Section 5.1.

1. The Distributive Property
2. Combining like terms
3. Removing symbols of grouping
4. Rules of exponents

The simplest type of polynomial multiplication involves a monomial multiplier. The product is obtained by direct application of the Distributive Property. For instance, to multiply the monomial x by the polynomial $(2x + 5)$, multiply *each* term of the polynomial by x.

$$(x)(2x + 5) = (x)(2x) + (x)(5) = 2x^2 + 5x$$

1 **Find products with monomial multipliers.**

Example 1   Finding Products with Monomial Multipliers

Find each product.

a. $(3x - 7)(-2x)$ **b.** $3x^2(5x - x^3 + 2)$ **c.** $(-x)(2x^2 - 3x)$

Solution

a. $(3x - 7)(-2x) = 3x(-2x) - 7(-2x)$ Distributive Property

$$= -6x^2 + 14x \qquad\qquad \text{Write in standard form.}$$

b. $3x^2(5x - x^3 + 2)$

$$= (3x^2)(5x) - (3x^2)(x^3) + (3x^2)(2) \qquad \text{Distributive Property}$$

$$= 15x^3 - 3x^5 + 6x^2 \qquad\qquad \text{Rules of exponents}$$

$$= -3x^5 + 15x^3 + 6x^2 \qquad\qquad \text{Write in standard form.}$$

c. $(-x)(2x^2 - 3x) = (-x)(2x^2) - (-x)(3x)$ Distributive Property

$$= -2x^3 + 3x^2 \qquad\qquad \text{Write in standard form.}$$

Additional Examples
Find each product.

a. $(-x^2)(4x + 7)$

b. $(2x)(3x^2 - 4x + 1)$

Answers:

a. $-4x^3 - 7x^2$

b. $6x^3 - 8x^2 + 2x$

② Multiply binomials using the Distributive Property and the FOIL Method.

Multiplying Binomials

To multiply two binomials, you can use both (left and right) forms of the Distributive Property. For example, if you treat the binomial $(5x + 7)$ as a single quantity, you can multiply $(3x - 2)$ by $(5x + 7)$ as follows.

$$(3x - 2)(5x + 7) = 3x(5x + 7) - 2(5x + 7)$$

$$= (3x)(5x) + (3x)(7) - (2)(5x) - 2(7)$$

$$= 15x^2 + 21x - 10x - 14$$

| Product of **First terms** | Product of **Outer terms** | Product of **Inner terms** | Product of **Last terms** |

$$= 15x^2 + 11x - 14$$

With practice, you should be able to multiply two binomials without writing out all of the steps above. In fact, the four products in the boxes above suggest that you can write the product of two binomials in just one step. This is called the **FOIL Method.** Note that the words *first, outer, inner,* and *last* refer to the positions of the terms in the original product.

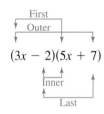

Technology: Tip

Remember that you can use a graphing calculator to check whether you have performed a polynomial operation correctly. For instance, to check if

$(x - 1)(x + 5) = x^2 + 4x - 5$

you can "graph the left side and graph the right side" in the same viewing window, as shown below. Because both graphs are the same, you can conclude that the multiplication was performed correctly.

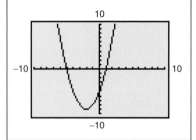

Example 2   Multiplying with the Distributive Property

Use the Distributive Property to find each product.

a. $(x - 1)(x + 5)$

b. $(2x + 3)(x - 2)$

Solution

a. $(x - 1)(x + 5) = x(x + 5) - 1(x + 5)$ Right Distributive Property

$$= x^2 + 5x - x - 5$$ Left Distributive Property

$$= x^2 + (5x - x) - 5$$ Group like terms.

$$= x^2 + 4x - 5$$ Combine like terms.

b. $(2x + 3)(x - 2) = 2x(x - 2) + 3(x - 2)$ Right Distributive Property

$$= 2x^2 - 4x + 3x - 6$$ Left Distributive Property

$$= 2x^2 + (-4x + 3x) - 6$$ Group like terms.

$$= 2x^2 - x - 6$$ Combine like terms.

Example 3 Multiplying Binomials Using the FOIL Method

Use the FOIL Method to find each product.

a. $(x + 4)(x - 4)$

b. $(3x + 5)(2x + 1)$

Solution

$$\text{a. } (x + 4)(x - 4) = \overset{F}{x^2} - \overset{O}{4x} + \overset{I}{4x} - \overset{L}{16}$$

$$= x^2 - 16 \qquad \text{Combine like terms.}$$

$$\text{b. } (3x + 5)(2x + 1) = \overset{F}{6x^2} + \overset{O}{3x} + \overset{I}{10x} + \overset{L}{5}$$

$$= 6x^2 + 13x + 5 \qquad \text{Combine like terms.}$$

In Example 3(a), note that the outer and inner products add up to zero.

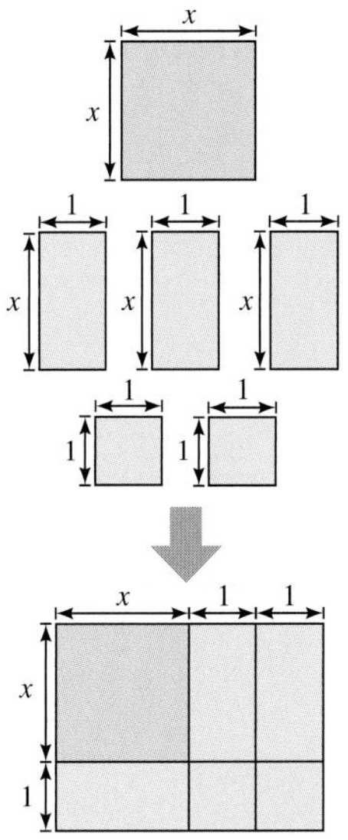

Figure 5.2

Example 4 A Geometric Model of a Polynomial Product

Use the geometric model shown in Figure 5.2 to show that

$$x^2 + 3x + 2 = (x + 1)(x + 2).$$

Solution

The top of the figure shows that the sum of the areas of the six rectangles is

$$x^2 + (x + x + x) + (1 + 1) = x^2 + 3x + 2.$$

The bottom of the figure shows that the area of the rectangle is

$$(x + 1)(x + 2) = x^2 + 2x + x + 2$$

$$= x^2 + 3x + 2.$$

So, $x^2 + 3x + 2 = (x + 1)(x + 2)$.

Example 5 Simplifying a Polynomial Expression

Simplify the expression and write the result in standard form.

$$(4x + 5)^2$$

Solution

$$(4x + 5)^2 = (4x + 5)(4x + 5) \qquad \text{Repeated multiplication}$$

$$= 16x^2 + 20x + 20x + 25 \qquad \text{Use FOIL Method.}$$

$$= 16x^2 + 40x + 25 \qquad \text{Combine like terms.}$$

Additional Examples
Use the FOIL Method to find each product.

a. $(2y + 3)(5y - 9)$

b. $(6a - 1)(2a - 3)$

Answers:

a. $10y^2 - 3y - 27$

b. $12a^2 - 20a + 3$

Example 6 **Simplifying a Polynomial Expression**

Simplify the expression and write the result in standard form.

$$(3x^2 - 2)(4x + 7) - (4x)^2$$

Solution

$$(3x^2 - 2)(4x + 7) - (4x)^2$$

$$= 12x^3 + 21x^2 - 8x - 14 - (4x)^2 \qquad \text{Use FOIL Method.}$$

$$= 12x^3 + 21x^2 - 8x - 14 - 16x^2 \qquad \text{Square monomial.}$$

$$= 12x^3 + 5x^2 - 8x - 14 \qquad \text{Combine like terms.}$$

③ Multiply polynomials using a horizontal or vertical format.

Multiplying Polynomials

The FOIL Method for multiplying two binomials is simply a device for guaranteeing that *each term of one binomial is multiplied by each term of the other binomial.*

$$(ax + b)(cx + d) = ax(cx) + ax(d) + b(cx) + b(d)$$

$$\text{F} \qquad \text{O} \qquad \text{I} \qquad \text{L}$$

This same rule applies to the product of any two polynomials: *each term of one polynomial must be multiplied by each term of the other polynomial.* This can be accomplished using either a horizontal or a vertical format.

Example 7 **Multiplying Polynomials (Horizontal Format)**

You could show students that Example 7(b) could also be written as

$2x^2(4x + 3) - 7x(4x + 3) + 1(4x + 3)$

$= 8x^3 + 6x^2 - 28x^2 - 21x + 4x + 3$

$= 8x^3 - 22x^2 - 17x + 3.$

Use a horizontal format to find each product.

a. $(x - 4)(x^2 - 4x + 2)$ **b.** $(2x^2 - 7x + 1)(4x + 3)$

Solution

a. $(x - 4)(x^2 - 4x + 2)$

$$= x(x^2 - 4x + 2) - 4(x^2 - 4x + 2) \qquad \text{Distributive Property}$$

$$= x^3 - 4x^2 + 2x - 4x^2 + 16x - 8 \qquad \text{Distributive Property}$$

$$= x^3 - 8x^2 + 18x - 8 \qquad \text{Combine like terms.}$$

b. $(2x^2 - 7x + 1)(4x + 3)$

$$= (2x^2 - 7x + 1)(4x) + (2x^2 - 7x + 1)(3) \qquad \text{Distributive Property}$$

$$= 8x^3 - 28x^2 + 4x + 6x^2 - 21x + 3 \qquad \text{Distributive Property}$$

$$= 8x^3 - 22x^2 - 17x + 3 \qquad \text{Combine like terms.}$$

Example 8   **Multiplying Polynomials (Vertical Format)**

Use a vertical format to find the product of $(3x^2 + x - 5)$ and $(2x - 1)$.

Solution

With a vertical format, line up like terms in the same vertical columns, just as you align digits in whole number multiplication.

$$
\begin{array}{r}
3x^2 + x - 5 \\
\times 2x - 1 \\
\hline
-3x^2 - x + 5 \\
6x^3 + 2x^2 - 10x \\
\hline
6x^3 - x^2 - 11x + 5
\end{array}
$$

Place polynomial with most terms on top.

Line up like terms.

$-1(3x^2 + x - 5)$

$2x(3x^2 + x - 5)$

Combine like terms in columns.

Example 9   **Multiplying Polynomials (Vertical Format)**

It's helpful for students to practice with both the horizontal and vertical formats on problems such as these.

Use a vertical format to find the product of $(4x^3 + 8x - 1)$ and $(2x^2 + 3)$.

Solution

$$
\begin{array}{r}
4x^3 + 8x - 1 \\
\times 2x^2 + 3 \\
\hline
12x^3 + 24x - 3 \\
8x^5 + 16x^3 - 2x^2 \\
\hline
8x^5 + 28x^3 - 2x^2 + 24x - 3
\end{array}
$$

Place polynomial with most terms on top.

Line up like terms.

$3(4x^3 + 8x - 1)$

$2x^2(4x^3 + 8x - 1)$

Combine like terms in columns.

When multiplying two polynomials, it is best to write each in standard form before using either the horizontal or vertical format. This is illustrated in the next example.

Example 10   **Multiplying Polynomials (Vertical Format)**

Write the polynomials in standard form and use a vertical format to find the product of $(x + 3x^2 - 4)$ and $(5 + 3x - x^2)$.

Solution

$$
\begin{array}{r}
3x^2 + x - 4 \\
\times -x^2 + 3x + 5 \\
\hline
15x^2 + 5x - 20 \\
9x^3 + 3x^2 - 12x \\
-3x^4 - x^3 + 4x^2 \\
\hline
-3x^4 + 8x^3 + 22x^2 - 7x - 20
\end{array}
$$

Write in standard form.

Write in standard form.

$5(3x^2 + x - 4)$

$3x(3x^2 + x - 4)$

$-x^2(3x^2 + x - 4)$

Combine like terms.

Example 11   Multiplying Polynomials

Multiply $(x - 3)^3$.

Solution

To raise $(x - 3)$ to the third power, you can use two steps. First, because $(x - 3)^3 = (x - 3)^2(x - 3)$, find the product $(x - 3)^2$.

$$(x - 3)^2 = (x - 3)(x - 3) \qquad \text{Repeated multiplication}$$

$$= x^2 - 3x - 3x + 9 \qquad \text{Use FOIL Method.}$$

$$= x^2 - 6x + 9 \qquad \text{Combine like terms.}$$

Now multiply $x^2 - 6x + 9$ by $x - 3$, as follows.

$$(x^2 - 6x + 9)(x - 3) = (x^2 - 6x + 9)(x) - (x^2 - 6x + 9)(3)$$

$$= x^3 - 6x^2 + 9x - 3x^2 + 18x - 27$$

$$= x^3 - 9x^2 + 27x - 27.$$

So, $(x - 3)^3 = x^3 - 9x^2 + 27x - 27$.

④ Identify and use special binomial products.

Special Products

Some binomial products, such as those in Examples 3(a) and 5, have special forms that occur frequently in algebra. The product

$$(x + 4)(x - 4)$$

is called a **product of the sum and difference of two terms.** With such products, the two middle terms cancel, as follows.

$$(x + 4)(x - 4) = x^2 - 4x + 4x - 16 \qquad \text{Sum and difference of two terms}$$

$$= x^2 - 16 \qquad \text{Product has no middle term.}$$

Another common type of product is the **square of a binomial.**

$$(4x + 5)^2 = (4x + 5)(4x + 5) \qquad \text{Square of a binomial}$$

$$= 16x^2 + 20x + 20x + 25 \qquad \text{Use FOIL Method.}$$

$$= 16x^2 + 40x + 25 \qquad \begin{array}{l}\text{Middle term is twice the product}\\ \text{of the terms of the binomial.}\end{array}$$

Study Tip

You should learn to recognize the patterns of the two special products at the right. The FOIL Method can be used to verify each rule.

In general, when a binomial is squared, the resulting middle term is always twice the product of the two terms.

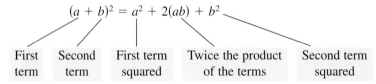

$$(a + b)^2 = a^2 + 2(ab) + b^2$$

First term | Second term | First term squared | Twice the product of the terms | Second term squared

Be sure to include the middle term. For instance, $(a + b)^2$ is not equal to $a^2 + b^2$.

Emphasize the importance of these special products.

Special Products

Let a and b be real numbers, variables, or algebraic expressions.

Special Product	*Example*

Sum and Difference of Two Terms:

$$(a + b)(a - b) = a^2 - b^2 \qquad\qquad (2x - 5)(2x + 5) = 4x^2 - 25$$

Square of a Binomial:

$$(a + b)^2 = a^2 + 2ab + b^2 \qquad\qquad (3x + 4)^2 = 9x^2 + 2(3x)(4) + 16$$
$$= 9x^2 + 24x + 16$$

$$(a - b)^2 = a^2 - 2ab + b^2 \qquad\qquad (x - 7)^2 = x^2 - 2(x)(7) + 49$$
$$= x^2 - 14x + 49$$

Additional Example

Multiply $(2 + 3x)(2 - 3x)$.

Answer:

$4 - 9x^2$

Example 12   Finding the Product of the Sum and Difference of Two Terms

Multiply $(x + 2)(x - 2)$.

Solution

Sum   Difference   (1st term)²   (2nd term)²

$$(x + 2)(x - 2) = (x)^2 - (2)^2$$
$$= x^2 - 4$$

Example 13   Finding the Product of the Sum and Difference of Two Terms

Multiply $(5x - 6)(5x + 6)$.

Solution

Difference   Sum   (1st term)²   (2nd term)²

$$(5x - 6)(5x + 6) = (5x)^2 - (6)^2$$
$$= 25x^2 - 36$$

Example 14   Squaring a Binomial

Multiply $(4x - 9)^2$.

Solution

2nd term   Twice the product of the terms
1st term   (1st term)²   (2nd term)²

$$(4x - 9)^2 = (4x)^2 - 2(4x)(9) + (9)^2$$
$$= 16x^2 - 72x + 81$$

Example 15   Squaring a Binomial

Multiply $(3x + 7)^2$.

Solution

$$
\begin{array}{c}
\text{1st term} \quad \text{2nd term} \quad \text{(1st term)}^2 \quad \text{Twice the product of the terms} \quad \text{(2nd term)}^2 \\
(3x + 7)^2 = (3x)^2 + 2(3x)(7) + (7)^2
\end{array}
$$

$$= 9x^2 + 42x + 49$$

Example 16   Squaring a Binomial

Multiply $(6 - 5x^2)^2$.

Solution

$$
\begin{array}{c}
\text{1st term} \quad \text{2nd term} \quad \text{(1st term)}^2 \quad \text{Twice the product of the terms} \quad \text{(2nd term)}^2 \\
(6 - 5x^2)^2 = (6)^2 - 2(6)(5x^2) + (5x^2)^2
\end{array}
$$

$$= 36 - 60x^2 + (5)^2(x^2)^2$$

$$= 36 - 60x^2 + 25x^4$$

Example 17   Finding the Dimensions of a Golf Tee

A landscaper wants to reshape a square tee area for the ninth hole of a golf course. The new tee area is to have one side 2 feet longer and the adjacent side 6 feet longer than the original tee. (See Figure 5.3.) The new tee has 204 square feet more area than the original tee. What are the dimensions of the original tee?

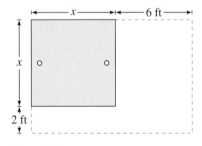

Figure 5.3

Solution

Verbal Model:
$$\boxed{\text{New area}} = \boxed{\text{Old area}} + 204$$

Labels: Original length = original width = x (feet)
        New length = $x + 6$ (feet)
        New width = $x + 2$ (feet)

Equation: $(x + 6)(x + 2) = x^2 + 204$ x^2 is original area.

$$x^2 + 8x + 12 = x^2 + 204 \quad \text{Multiply factors.}$$

$$8x + 12 = 204 \quad \text{Subtract } x^2 \text{ from each side.}$$

$$8x = 192 \quad \text{Subtract 12 from each side.}$$

$$x = 24 \quad \text{Simplify.}$$

The original tee measured 24 feet by 24 feet.

5.3 Exercises

Review   Concepts, Skills, and Problem Solving

Keep mathematically in shape by doing these exercises *before* the problems of this section.

Properties and Definitions

1. *Writing* ✎   Relative to the x- and y-axes, explain the meaning of each coordinate of the point $(3, -2)$.
The point represented by $(3, -2)$ is located three units to the right of the y-axis and two units below the x-axis.

2. A point lies four units from the x-axis and three units from the y-axis. Give the ordered pair for such a point in each quadrant.
$(3, 4), (-3, 4), (-3, -4), (3, -4)$

Simplifying Expressions

In Exercises 3–8, simplify the expression.

3. $\frac{3}{4}x - \frac{5}{2} + \frac{3}{2}x$ $\frac{9}{4}x - \frac{5}{2}$

4. $4 - 2(3 - x)$ $2x - 2$

5. $2(x - 4) + 5x$
$7x - 8$

6. $4(3 - y) + 2(y + 1)$
$-2y + 14$

7. $-3(z - 2) - (z - 6)$
$-4z + 12$

8. $(u - 2) - 3(2u + 1)$
$-5u - 5$

Problem Solving

9. *Sales Commission*   Your sales commission rate is 5.5%. Your commission is $1600. How much did you sell? $29,090.91

10. *Distance*   A jogger leaves a location on a fitness trail running at a rate of 4 miles per hour. Fifteen minutes later, a second jogger leaves from the same location running at 5 miles per hour. How long will it take the second jogger to overtake the first, and how far will each have run at that point? Use a diagram to help solve the problem.   1 hour; 5 miles

Graphing Equations

⊞ In Exercises 11 and 12, use a graphing calculator to graph the equation. Identify any intercepts.
See Additional Answers.

11. $y = 4 - \frac{1}{2}x$
$(0, 4), (8, 0)$

12. $y = x(x - 4)$
$(0, 0), (4, 0)$

Developing Skills

In Exercises 1–50, perform the multiplication and simplify. See Examples 1–3, 5, and 6.

1. $x(-2x)$ $-2x^2$

2. $y(-3y)$ $-3y^2$

3. $t^2(4t)$ $4t^3$

4. $3u(u^4)$ $3u^5$

5. $\left(\dfrac{x}{4}\right)(10x)$ $\frac{5}{2}x^2$

6. $9x\left(\dfrac{x}{12}\right)$ $\frac{3}{4}x^2$

7. $(-2b^2)(-3b)$ $6b^3$

8. $(-4x)(-5x)$ $20x^2$

9. $y(3 - y)$ $3y - y^2$

10. $z(z - 3)$ $z^2 - 3z$

11. $-x(x^2 - 4)$
$-x^3 + 4x$

12. $-t(10 - 9t^2)$
$-10t + 9t^3$

13. $-3x(2x^2 + 5)$
$-6x^3 - 15x$

14. $-5u(u^2 + 4)$
$-5u^3 - 20u$

15. $-4x(3 + 3x^2 - 6x^3)$
$-12x - 12x^3 + 24x^4$

16. $5v(5 - 4v + 5v^2)$
$25v - 20v^2 + 25v^3$

17. $3x(x^2 - 2x + 1)$
$3x^3 - 6x^2 + 3x$

18. $y(4y^2 + 2y - 3)$
$4y^3 + 2y^2 - 3y$

19. $2x(x^2 - 2x + 8)$
$2x^3 - 4x^2 + 16x$

20. $3y(y^2 - y + 5)$
$3y^3 - 3y^2 + 15y$

21. $4t^3(t - 3)$
$4t^4 - 12t^3$

22. $-2t^4(t + 6)$
$-2t^5 - 12t^4$

23. $x^2(4x^2 - 3x + 1)$
$4x^4 - 3x^3 + x^2$

24. $y^2(2y^2 + y - 5)$
$2y^4 + y^3 - 5y^2$

25. $-3x^3(4x^2 - 6x + 2)$
$-12x^5 + 18x^4 - 6x^3$

26. $5u^4(2u^3 - 3u + 3)$
$10u^7 - 15u^5 + 15u^4$

27. $-2x(-3x)(5x + 2)$
$30x^3 + 12x^2$

28. $4x(-2x)(x^2 - 1)$
$-8x^4 + 8x^2$

29. $-2x(-6x^4) - 3x^2(2x^2)$ $12x^5 - 6x^4$

30. $-8y(-5y^4) - 2y^2(5y^3)$ $30y^5$

31. $(x + 3)(x + 4)$
$x^2 + 7x + 12$

32. $(x - 5)(x + 10)$
$x^2 + 5x - 50$

33. $(3x - 5)(2x + 1)$
$6x^2 - 7x - 5$

34. $(7x - 2)(4x - 3)$
$28x^2 - 29x + 6$

35. $(2x - y)(x - 2y)$
$2x^2 - 5xy + 2y^2$

36. $(x + y)(x + 2y)$
$x^2 + 3xy + 2y^2$

37. $(2x + 4)(x + 1)$
$2x^2 + 6x + 4$

38. $(4x + 3)(2x - 1)$
$8x^2 + 2x - 3$

39. $(6 - 2x)(4x + 3)$
$-8x^2 + 18x + 18$

40. $(8x - 6)(5 - 4x)$
$-32x^2 + 64x - 30$

41. $(3x - 2y)(x - y)$
$3x^2 - 5xy + 2y^2$

42. $(7x + 5y)(x + y)$
$7x^2 + 12xy + 5y^2$

43. $(3x^2 - 4)(x + 2)$
$3x^3 + 6x^2 - 4x - 8$

44. $(5x^2 - 2)(x - 1)$
$5x^3 - 5x^2 - 2x + 2$

45. $(2x^2 + 4)(x^2 + 6)$
$2x^4 + 16x^2 + 24$

46. $(7x^2 - 3)(2x^2 - 4)$
$14x^4 - 34x^2 + 12$

47. $(3s + 1)(3s + 4) - (3s)^2$ $15s + 4$

48. $(2t + 5)(4t - 2) - (2t)^2$ $4t^2 + 16t - 10$

49. $(4x^2 - 1)(2x + 8) + (-x^2)^3$
$-x^6 + 8x^3 + 32x^2 - 2x - 8$

50. $(3 - 3x^2)(4 - 5x^2) - (-x^3)^2$
$-x^6 + 15x^4 - 27x^2 + 12$

In Exercises 51–64, use a horizontal format to find the product. See Example 7.

51. $(x + 10)(x + 2)$
$x^2 + 12x + 20$

52. $(x - 1)(x + 3)$
$x^2 + 2x - 3$

53. $(2x - 5)(x + 2)$
$2x^2 - x - 10$

54. $(3x - 2)(2x - 3)$
$6x^2 - 13x + 6$

55. $(x + 1)(x^2 + 2x - 1)$
$x^3 + 3x^2 + x - 1$

56. $(x - 3)(x^2 - 3x + 4)$
$x^3 - 6x^2 + 13x - 12$

57. $(x^3 - 2x + 1)(x - 5)$
$x^4 - 5x^3 - 2x^2 + 11x - 5$

58. $(x + 1)(x^2 - x + 1)$
$x^3 + 1$

59. $(x - 2)(x^2 + 2x + 4)$
$x^3 - 8$

60. $(x + 9)(x^2 - x - 4)$
$x^3 + 8x^2 - 13x - 36$

61. $(x^2 + 3)(x^2 - 6x + 2)$
$x^4 - 6x^3 + 5x^2 - 18x + 6$

62. $(x^2 + 3)(x^2 - 2x + 3)$
$x^4 - 2x^3 + 6x^2 - 6x + 9$

63. $(3x^2 + 1)(x^2 - 4x - 2)$
$3x^4 - 12x^3 - 5x^2 - 4x - 2$

64. $(x^2 + 2x + 5)(4x^3 - 2)$
$4x^5 + 8x^4 + 20x^3 - 2x^2 - 4x - 10$

In Exercises 65–80, use a vertical format to find the product. See Examples 8–10.

65. $x + 3$
$\underline{\times\ x - 2}$
$x^2 + x - 6$

66. $2x - 1$
$\underline{\times\ 5x + 1}$
$10x^2 - 3x - 1$

67. $4x^4 - 6x^2 + 9$
$\underline{\times\qquad 2x\ + 3}$
$8x^5 + 12x^4 - 12x^3 - 18x^2 + 18x + 27$

68. $x^2 - 3x + 9$
$\underline{\times\qquad x + 3}$ $x^3 + 27$

69. $(x^2 - x + 2)(x^2 + x - 2)$ $x^4 - x^2 + 4x - 4$

70. $(x^2 + 2x + 5)(2x^2 - x - 1)$
$2x^4 + 3x^3 + 7x^2 - 7x - 5$

71. $(x^3 + x + 3)(x^2 + 5x - 4)$
$x^5 + 5x^4 - 3x^3 + 8x^2 + 11x - 12$

72. $(x^3 - x - 1)(x^2 + x + 1)$
$x^5 + x^4 - 2x^2 - 2x - 1$

73. $(x - 2)^3$
$x^3 - 6x^2 + 12x - 8$

74. $(x + 3)^3$
$x^3 + 9x^2 + 27x + 27$

75. $(x - 1)^2(x - 1)^2$ $x^4 - 4x^3 + 6x^2 - 4x + 1$

76. $(x + 4)^2(x + 4)^2$
$x^4 + 16x^3 + 96x^2 + 256x + 256$

77. $(x + 2)^2(x - 4)$
$x^3 - 12x - 16$

78. $(x - 4)^2(x - 1)$
$x^3 - 9x^2 + 24x - 16$

79. $(u - 1)(2u + 3)(2u + 1)$ $4u^3 + 4u^2 - 5u - 3$

80. $(2x + 5)(x - 2)(5x - 3)$ $10x^3 - x^2 - 53x + 30$

In Exercises 81–110, use a special product pattern to find the product. See Examples 12–16.

81. $(x + 3)(x - 3)$
$x^2 - 9$

82. $(x - 5)(x + 5)$
$x^2 - 25$

83. $(x + 20)(x - 20)$
$x^2 - 400$

84. $(y + 9)(y - 9)$
$y^2 - 81$

85. $(2u + 3)(2u - 3)$
$4u^2 - 9$

86. $(3z + 4)(3z - 4)$
$9z^2 - 16$

87. $(4t - 6)(4t + 6)$
$16t^2 - 36$

88. $(3u + 7)(3u - 7)$
$9u^2 - 49$

89. $(2x + 3y)(2x - 3y)$ $4x^2 - 9y^2$

90. $(5u + 12v)(5u - 12v)$ $25u^2 - 144v^2$

91. $(4u - 3v)(4u + 3v)$ $16u^2 - 9v^2$

92. $(8a - 5b)(8a + 5b)$ $64a^2 - 25b^2$

93. $(2x^2 + 5)(2x^2 - 5)$ $4x^4 - 25$

94. $(4t^2 + 6)(4t^2 - 6)$ $16t^4 - 36$

95. $(x + 6)^2$ $x^2 + 12x + 36$

96. $(a - 2)^2$ $a^2 - 4a + 4$

97. $(t - 3)^2$ $t^2 - 6t + 9$

98. $(x + 10)^2$ $x^2 + 20x + 100$

99. $(3x + 2)^2$ $9x^2 + 12x + 4$

100. $(2x - 8)^2$ $4x^2 - 32x + 64$

101. $(8 - 3z)^2$ $64 - 48z + 9z^2$

102. $(1 - 5t)^2$ $1 - 10t + 25t^2$

103. $(2x - 5y)^2$ $4x^2 - 20xy + 25y^2$

104. $(4s + 3t)^2$ $16s^2 + 24st + 9t^2$

105. $(6t + 5s)^2$ $36t^2 + 60st + 25s^2$

106. $(3u - 8v)^2$ $9u^2 - 48uv + 64v^2$

107. $[(x + 1) + y]^2$
$x^2 + y^2 + 2xy + 2x + 2y + 1$

108. $[(x - 3) - y]^2$
$x^2 + y^2 - 2xy - 6x + 6y + 9$

109. $[u - (v - 3)]^2$
$u^2 + v^2 - 2uv + 6u - 6v + 9$

110. $[2u + (v + 1)]^2$
$4u^2 + v^2 + 4uv + 4u + 2v + 1$

In Exercises 111 and 112, perform the multiplication and simplify.

111. $(x + 2)^2 - (x - 2)^2$ **112.** $(u + 5)^2 + (u - 5)^2$
   $8x$ $2u^2 + 50$

Think About It   In Exercises 113 and 114, is the equation an identity? Explain.

113. $(x + y)^3 = x^3 + 3x^2y + 3xy^2 + y^3$ Yes

114. $(x - y)^3 = x^3 - 3x^2y + 3xy^2 - y^3$ Yes

In Exercises 115 and 116, use the results of Exercise 113 to find the product.

115. $(x + 2)^3$ **116.** $(x + 1)^3$
   $x^3 + 6x^2 + 12x + 8$ $x^3 + 3x^2 + 3x + 1$

Solving Problems

117. ▲ *Geometry*   The base of a triangular sail is $2x$ feet and its height is $(x + 10)$ feet (see figure). Find an expression for the area of the sail.
$(x^2 + 10x)$ square feet

118. ▲ *Geometry*   The height of a rectangular sign is twice its width w (see figure). Find an expression for (a) the perimeter and (b) the area of the sign.
(a) $6w$ (b) $2w^2$

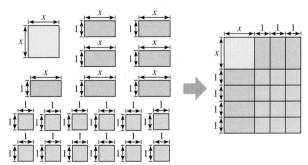

119. ▲ *Geometry*   In Exercises 119–122, what polynomial product is represented by the geometric model? Explain. See Example 4.

119. $x^2 + 5x + 4 = (x + 1)(x + 4)$

120. $x^2 + 7x + 12 = (x + 4)(x + 3)$

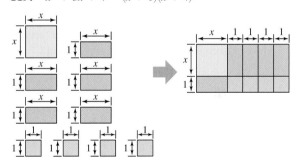

121. $4x^2 + 6x + 2 = (2x + 1)(2x + 2)$

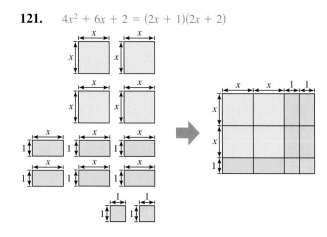

122. $2x^2 + 4x = 2x(x + 2)$

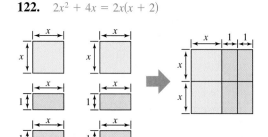

123. ▲ *Geometry* Add the areas of the four rectangular regions shown in the figure. What special product does the geometric model represent?

$(x + 2)^2 = x^2 + 4x + 4$; Square of a binomial

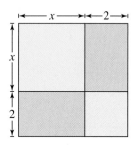

124. ▲ *Geometry* Add the areas of the four rectangular regions shown in the figure. What special method does the geometric model represent?

$x^2 + bx + ax + ab$; The FOIL Method

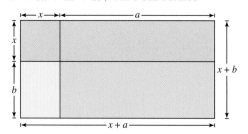

▲ *Geometry* In Exercises 125 and 126, find a polynomial product that represents the area of the region. Then simplify the product.

125.

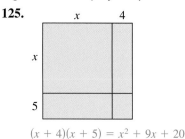

$(x + 4)(x + 5) = x^2 + 9x + 20$

126.

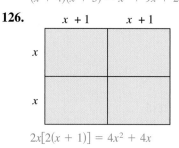

$2x[2(x + 1)] = 4x^2 + 4x$

▲ *Geometry* In Exercises 127 and 128, find two different expressions that represent the area of the shaded portion of the figure.

127.

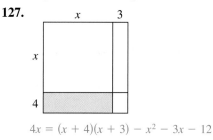

$4x = (x + 4)(x + 3) - x^2 - 3x - 12$

128.

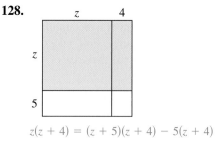

$z(z + 4) = (z + 5)(z + 4) - 5(z + 4)$

129. *Milk Consumption*   The per capita consumption (average consumption per person) of milk M (in gallons) in the United States for the years 1990 through 2000 is given by

$$M = -0.32t + 24.3, \quad 0 \le t \le 10.$$

The population P (in millions) of the United States during the same time period is given by

$$P = -0.016t^2 + 2.69t + 250.1, \quad 0 \le t \le 10.$$

In both models, t represents the year, with $t = 0$ corresponding to 1990.   (Source: USDA/Economic Research Service and U.S. Census Bureau)

(a) Multiply the polynomials to find a model for the total consumption of milk T in the United States.

$T = 0.00512t^3 - 1.2496t^2 - 14.665t + 6077.43,$
$0 \le t \le 10$

(b) 🖩 Use a graphing calculator to graph the model from part (a).   See Additional Answers.

(c) 🖩 Use the graph from part (b) to estimate the total consumption of milk in 1998.

Approximately 5883 million gallons

130. 🖩 *Interpreting Graphs*   When x units of a home theater system are sold, the revenue R is given by

$$R = x(900 - 0.5x).$$

(a) Use a graphing calculator to graph the equation. See Additional Answers.

(b) Multiply the factors in the expression for revenue and use a graphing calculator to graph the product in the same viewing window you used in part (a). Verify that the graph is the same as in part (a). $900x - 0.5x^2$

(c) Find the revenue when 500 units are sold. Use the graph to determine if revenue would increase or decrease if more units were sold.

$325,000; Increase

131. *Compound Interest*   After 2 years, an investment of $500 compounded annually at interest rate r will yield an amount $500(1 + r)^2$. Find this product.
$500r^2 + 1000r + 500$

132. *Compound Interest*   After 2 years, an investment of $1200 compounded annually at interest rate r will yield an amount $1200(1 + r)^2$. Find this product.
$1200r^2 + 2400r + 1200$

Explaining Concepts

133. 🟢   Answer parts (d)–(f) of Motivating the Chapter on page 264.

134. *Writing*✏️   Explain why an understanding of the Distributive Property is essential in multiplying polynomials. Illustrate your explanation with an example.   Multiplying a polynomial by a monomial is a direct application of the Distributive Property. Multiplying two polynomials requires repeated application of the Distributive Property. $3x(x + 4) = 3x^2 + 12x$

135. *Writing*✏️   Describe the rules of exponents that are used to multiply polynomials. Give examples.
Product Rule: $3^2 \cdot 3^4 = 3^{2+4}$
Product-to-Power Rule: $(5 \cdot 2)^8 = 5^8 \cdot 2^8$
Power-to-Power Rule: $(2^3)^2 = 2^{3 \cdot 2}$

136. *Writing*✏️   Discuss any differences between the expressions $(3x)^2$ and $3x^2$.
$(3x)^2 = 3^2 \cdot x^2 = 9x^2 \ne 3x^2$

137. *Writing*✏️   Explain the meaning of each letter of "FOIL" as it relates to multiplying two binomials.
First, Outer, Inner, Last

138. *Writing*✏️   What is the degree of the product of two polynomials of degrees m and n? Explain.

139. *Writing*✏️   A polynomial with m terms is multiplied by a polynomial with n terms. How many monomial-by-monomial products must be found? Explain. mn; Each term of the first factor must be multiplied by each term of the second factor.

140. *True or False?*   Because the product of two monomials is a monomial, it follows that the product of two binomials is a binomial. Justify your answer.   False. $(x - 2)(x + 3) = x^2 + x - 6$

141. *Finding a Pattern*   Perform each multiplication.

(a) $(x - 1)(x + 1)$ $x^2 - 1$

(b) $(x - 1)(x^2 + x + 1)$ $x^3 - 1$

(c) $(x - 1)(x^3 + x^2 + x + 1)$ $x^4 - 1$

(d) From the pattern formed in the first three products, can you predict the product of

$$(x - 1)(x^4 + x^3 + x^2 + x + 1)?$$

Verify your prediction by multiplying. $x^5 - 1$

138. The product of the terms of highest degree in each polynomial will be of the form $(ax^m)(bx^n) = abx^{m+n}$. This will be the term of highest degree in the product, and therefore the degree of the product is $m + n$.

5.4 Dividing Polynomials

Superstock

What You Should Learn

1 Divide polynomials by monomials.

2 Divide polynomials by binomials.

Why You Should Learn It

Dividing polynomials enables you to model and solve real-life problems. For instance, in Exercise 73 on page 311, you will divide polynomials to find the ratio of the ages of two children in a family.

1 **Divide polynomials by monomials.**

Dividing a Polynomial by a Monomial

The quotient rules of exponents in Section 5.1 show how to divide a *monomial* by a monomial. To divide a *polynomial* by a monomial, use the reverse form of the rule for adding two fractions with a common denominator. In Section 1.4, you added two fractions with like denominators using the rule

$$\frac{a}{c} + \frac{b}{c} = \frac{a+b}{c}. \qquad \text{Add fractions.}$$

Here you can use the rule in the *reverse* order and divide a polynomial by a monomial by dividing each term of the polynomial by the monomial. That is,

$$\frac{a+b}{c} = \frac{a}{c} + \frac{b}{c}. \qquad \text{Divide by a monomial.}$$

Here is an example.

$$\frac{x^3 - 5x^2}{x^2} = \frac{x^3}{x^2} - \frac{5x^2}{x^2} = x - 5, \quad x \neq 0$$

The essence of this problem is to separate the original division problem into two division problems, each involving the division of a monomial by a monomial.

Caution students to avoid these common errors.

$$\frac{x+6}{x} \neq 6 \text{ and } \frac{x+6}{3} \neq x+2$$

Instead, $\dfrac{x+6}{x} = \dfrac{x}{x} + \dfrac{6}{x} = 1 + \dfrac{6}{x}$,

and $\dfrac{x+6}{3} = \dfrac{x}{3} + \dfrac{6}{3} = \dfrac{x}{3} + 2$.

Dividing a Polynomial by a Monomial

Let a, b, and c be real numbers, variables, or algebraic expressions, such that $c \neq 0$.

1. $\dfrac{a+b}{c} = \dfrac{a}{c} + \dfrac{b}{c}$ 2. $\dfrac{a-b}{c} = \dfrac{a}{c} - \dfrac{b}{c}$

When dividing a polynomial by a monomial, remember to reduce the resulting expressions to simplest form, as illustrated in Example 1.

Example 1   Dividing a Polynomial by a Monomial

a. $\dfrac{6x+5}{3} = \dfrac{6x}{3} + \dfrac{5}{3} = 2x + \dfrac{5}{3}$ **b.** $\dfrac{4x^2-3x}{3x} = \dfrac{4x^2}{3x} - \dfrac{3x}{3x} = \dfrac{4x}{3} - 1, \quad x \neq 0$

Technology: Tip

As with other types of operations with polynomials, you can use a graphing calculator to help check division problems. For instance, graph

$$y_1 = \frac{6x - 5}{3} \quad \text{and} \quad y_2 = 2x - \frac{5}{3}$$

in the same viewing window, as shown below. Because both graphs are the same, you can conclude that

$$\frac{6x - 5}{3} = 2x - \frac{5}{3}.$$

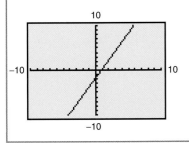

Example 2   Dividing a Polynomial by a Monomial

Perform the division and simplify.

$$\frac{8x^3 - 6x^2 + 10x}{2x}$$

Solution

$$\frac{8x^3 - 6x^2 + 10x}{2x} = \frac{8x^3}{2x} - \frac{6x^2}{2x} + \frac{10x}{2x} = 4x^2 - 3x + 5, \quad x \neq 0$$

Example 3   Dividing a Polynomial by a Monomial

Perform the division. (Assume $x \neq 0$.)

a. $(5x^3 - 4x^2 - x + 6) \div 2x$

b. $(8x^4 + 6x^3 + 3x^2 - 2x) \div 3x^2$

Solution

a. $\dfrac{5x^3 - 4x^2 - x + 6}{2x} = \dfrac{5x^3}{2x} - \dfrac{4x^2}{2x} - \dfrac{x}{2x} + \dfrac{6}{2x}$ Divide each term separately.

$$= \frac{5x^2}{2} - 2x - \frac{1}{2} + \frac{3}{x}$$ Use rules of exponents.

b. $\dfrac{8x^4 + 6x^3 + 3x^2 - 2x}{3x^2} = \dfrac{8x^4}{3x^2} + \dfrac{6x^3}{3x^2} + \dfrac{3x^2}{3x^2} - \dfrac{2x}{3x^2}$ Divide each term separately.

$$= \frac{8x^2}{3} + 2x + 1 - \frac{2}{3x}$$ Use rules of exponents.

2 Divide polynomials by binomials.

Dividing a Polynomial by a Binomial

To divide a polynomial by a *binomial,* follow the *long division algorithm* used for dividing positive integers. Recall that you divide 6982 by 27 as follows.

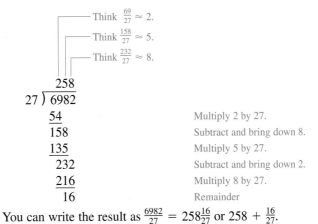

You can write the result as $\frac{6982}{27} = 258\frac{16}{27}$ or $258 + \frac{16}{27}$.

Study Tip

In the illustration at the right, the number 6982 is the **dividend,** 27 is the **divisor,** 258 is the **quotient,** and 16 is the **remainder.**

Along with the long division algorithm, follow these steps when performing long division of polynomials.

Long Division of Polynomials

1. Write the dividend and divisor in descending powers of the variable.

2. Insert placeholders with zero coefficients for missing powers of the variable. (See Examples 7 and 8).

3. Perform the long division of the polynomials as you would with integers.

4. Continue the process until the degree of the remainder is less than the degree of the divisor.

In the next several examples, you will see how the long division algorithm can be extended to cover division of a polynomial by a binomial.

Example 4 **Long Division Algorithm for Polynomials**

Use the long division algorithm to divide $x^2 + 3x + 5$ by $x + 1$.

Solution

Think $\dfrac{x^2}{x} = x.$

Think $\dfrac{2x}{x} = 2.$

$$
\begin{array}{r}
x + 2 \\
x + 1 \,\overline{)\, x^2 + 3x + 5} \\
\underline{x^2 + x} \\
2x + 5 \\
\underline{2x + 2} \\
3
\end{array}
$$

Multiply x by $(x + 1)$.
Subtract and bring down 5.
Multiply 2 by $(x + 1)$.
Remainder

Considering the remainder as a fractional part of the divisor, the result is

$$
\underbrace{\frac{x^2 + 3x + 5}{x + 1}}_{\substack{\text{Dividend} \\ \text{Divisor}}} = \overbrace{x + 2}^{\text{Quotient}} + \underbrace{\frac{\overset{\text{Remainder}}{3}}{x + 1}}_{\text{Divisor}}.
$$

Remember that a division problem can be checked by multiplying the *divisor* by the *quotient* plus the *remainder* to obtain the *dividend*. For instance, you can check the result of Example 4, as follows.

$$
\overbrace{(x + 1)}^{\text{Divisor}}\overbrace{(x + 2)}^{\text{Quotient}} + 3 \overset{?}{=} \overbrace{x^2 + 3x + 5}^{\text{Dividend}}
$$
$$
x^2 + 3x + 2 + 3 \overset{?}{=} x^2 + 3x + 5
$$
$$
x^2 + 3x + 5 = x^2 + 3x + 5 \quad \checkmark
$$

Example 5   Long Division Algorithm for Polynomials

Use the long division algorithm to divide $6x^3 - 19x^2 + 16x - 4$ by $x - 2$.

Solution

$$
\begin{array}{r}
6x^2 - 7x + 2 \\
x - 2 \overline{)\, 6x^3 - 19x^2 + 16x - 4} \\
\end{array}
$$

Think $\dfrac{6x^3}{x} = 6x^2$.

Think $-\dfrac{7x^2}{x} = -7x$.

Think $\dfrac{2x}{x} = 2$.

$6x^3 - 12x^2$ Multiply $6x^2$ by $(x - 2)$.

$-\; 7x^2 + 16x$ Subtract and bring down $16x$.

$-\; 7x^2 + 14x$ Multiply $-7x$ by $(x - 2)$.

$2x - 4$ Subtract and bring down -4.

$2x - 4$ Multiply 2 by $(x - 2)$.

0 Remainder

So, $\underbrace{6x^3 - 19x^2 + 16x - 4}_{\text{Dividend}} \div \underbrace{x - 2}_{\text{Divisor}} = \underbrace{6x^2 - 7x + 2}_{\text{Quotient}}, \quad x \neq 2.$

Check this result by multiplying the quotient $6x^2 - 7x + 2$ by the divisor $x - 2$.

In Example 5, the remainder is zero. In such cases, the denominator (or divisor) is said to **divide evenly** into the numerator (or dividend).

Example 6   Long Division Algorithm for Polynomials

Use long division to divide $-13x^3 + 10x^4 + 8x - 7x^2 + 4$ by $3 - 2x$.

Solution

$$
\begin{array}{r}
-5x^3 - x^2 + 2x - 1 \\
-2x + 3 \overline{)\, 10x^4 - 13x^3 - 7x^2 + 8x + 4} \\
\end{array}
$$

Write in standard form.

$10x^4 - 15x^3$ Multiply $-5x^3$ by $(-2x + 3)$.

$2x^3 - 7x^2$ Subtract and bring down $-7x^2$.

$2x^3 - 3x^2$ Multiply $-x^2$ by $(-2x + 3)$.

$-4x^2 + 8x$ Subtract and bring down $8x$.

$-4x^2 + 6x$ Multiply $2x$ by $(-2x + 3)$.

$2x + 4$ Subtract and bring down 4.

$2x - 3$ Multiply -1 by $(-2x + 3)$.

7 Remainder

> **Study Tip**
>
> When using long division to divide polynomials, remember to write the divisor and dividend in standard form before using the algorithm.

You might advise students that errors in long division problems frequently occur in the *subtraction* steps. In Example 6, notice that

$$-13x^3 - (-15x^3) = -13x^3 + 15x^3$$
$$= 2x^3.$$

This shows that

$$\underbrace{\dfrac{10x^4 - 13x^3 - 7x^2 + 8x + 4}{\underbrace{-2x + 3}_{\text{Divisor}}}}_{\text{Dividend}} = \underbrace{-5x^3 - x^2 + 2x - 1}_{\text{Quotient}} + \underbrace{\dfrac{7}{-2x + 3}}_{\text{Remainder}}.$$

When the dividend is missing one or more powers of x, the long division algorithm requires that you account for the missing powers (using zero coefficients), as shown in Examples 7 and 8.

Example 7   Accounting for Missing Powers of x

Use the long division algorithm to perform the division.

$$\frac{x^3 - 1}{x - 1}$$

Solution

Because there are no x^2- or x-terms in the dividend, you can line up the subtractions by using *zero* coefficients for the missing terms.

$$
\require{enclose}
\begin{array}{r}
x^2 + x + 1 \\[-2pt]
x - 1 \enclose{longdiv}{x^3 + 0x^2 + 0x - 1} \\
\end{array}
$$

$x - 1\overline{)x^3 + 0x^2 + 0x - 1}$	Insert $0x^2$ and $0x$.
$\underline{x^3 - x^2}$	Multiply x^2 by $(x-1)$.
$x^2 + 0x$	Subtract and bring down $0x$.
$\underline{x^2 - x}$	Multiply x by $(x-1)$.
$x - 1$	Subtract and bring down -1.
$\underline{x - 1}$	Multiply 1 by $(x-1)$.
0	Remainder

So, $x - 1$ divides evenly into $x^3 - 1$ and you can write

$$\frac{x^3 - 1}{x - 1} = x^2 + x + 1, \quad x \neq 1.$$

Example 8   Accounting for Missing Power of x

Use the long division algorithm to perform the division.

$$\frac{2x^4 - 3x^2 + 8x - 1}{x + 2}$$

Solution

$x + 2\overline{)2x^4 + 0x^3 - 3x^2 + 8x - 1}$ quotient $2x^3 - 4x^2 + 5x - 2$	Insert $0x^3$.
$\underline{2x^4 + 4x^3}$	Multiply $2x^3$ by $(x+2)$.
$-4x^3 - 3x^2$	Subtract and bring down $-3x^2$.
$\underline{-4x^3 - 8x^2}$	Multiply $-4x^2$ by $(x+2)$.
$5x^2 + 8x$	Subtract and bring down $8x$.
$\underline{5x^2 + 10x}$	Multiply $5x$ by $(x+2)$.
$-2x - 1$	Subtract and bring down -1.
$\underline{-2x - 4}$	Multiply -2 by $(x+2)$.
3	Remainder

This shows that $\dfrac{2x^4 - 3x^2 + 8x - 1}{x + 2} = 2x^3 - 4x^2 + 5x - 2 + \dfrac{3}{x + 2}.$

5.4 Exercises

Review   *Concepts, Skills, and Problem Solving*

Keep mathematically in shape by doing these exercises *before* the problems of this section.

Properties and Definitions

1. *Writing* ✏ Explain how to write the fraction $\dfrac{24x}{18}$ in simplest form.

Divide both the numerator and denominator by 6. $\dfrac{4x}{3}$

2. *Writing* ✏ The point $(-1, 4)$ lies in what quadrant? Explain.   Quadrant II. Since the x-coordinate is negative, the point lies to the left of the y-axis. Since the y-coordinate is positive, the point lies above the x-axis.

Simplifying Expressions

In Exercises 3–6, simplify the fraction.

3. $\dfrac{-8}{12}$ $-\frac{2}{3}$

4. $\dfrac{18}{-144}$ $-\frac{1}{8}$

5. $\dfrac{-60}{-150}$ $\frac{2}{5}$

6. $\dfrac{-175}{-42}$ $\frac{25}{6}$

In Exercises 7–10, find the product and simplify.

7. $-2x^2(5x^3)$ $-10x^5$ **8.** $(2z + 1)(2z - 1)$ $4z^2 - 1$

9. $(x + 7)^2$ $x^2 + 14x + 49$

10. $(x + 4)(2x - 5)$ $2x^2 + 3x - 20$

Creating a Model and Problem Solving

11. *Number Problem* Write an algebraic expression that represents the product of two consecutive odd integers, the first of which is $2n + 1$. $(2n + 1)(2n + 3) = 4n^2 + 8n + 3$

12. *Average Speed* After traveling for 3 hours, you are still 24 miles from completing a 180-mile trip. It takes you one-half hour to travel the last 24 miles. Find your average speed during the trip.   51.4 miles per hour

Developing Skills

In Exercises 1–52, perform the division and simplify. (Assume that no denominator is zero.) See Examples 1–8.

1. $\dfrac{3z + 3}{3}$ $z + 1$

2. $\dfrac{7x + 7}{7}$ $x + 1$

3. $\dfrac{4z - 12}{4}$ $z - 3$

4. $\dfrac{8u - 24}{8}$ $u - 3$

5. $\dfrac{5x + 5}{-5}$ $-x - 1$

6. $\dfrac{8y + 8}{-8}$ $-y - 1$

7. $\dfrac{6z - 6}{-6}$ $-z + 1$

8. $\dfrac{10x - 10}{-10}$ $-x + 1$

9. $\dfrac{9x - 5}{3}$ $3x - \frac{5}{3}$

10. $\dfrac{3 - 10x}{5}$ $\frac{3}{5} - 2x$

11. $\dfrac{8a + 5}{-4}$ $-2a - \frac{5}{4}$

12. $\dfrac{7x + 12}{-6}$ $-\frac{7}{6}x - 2$

13. $\dfrac{b^2 - 2b}{b}$ $b - 2$

14. $\dfrac{3x + 2x^3}{x}$ $3 + 2x^2$

15. $(5x^2 - 2x) \div x$ $5x - 2$

16. $(16a^2 + 5a) \div a$ $16a + 5$

17. $(3y^2 - 2y) \div (-y)$ $-3y + 2$

18. $(7x^2 + 3x) \div (-x)$ $-7x - 3$

19. $\dfrac{25z^3 + 10z^2}{-5z}$ $-5z^2 - 2z$

20. $\dfrac{12c^4 - 36c}{-6c}$ $-2c^3 + 6$

21. $\dfrac{8z^3 + 3z^2 - 2z}{2z}$ $4z^2 + \frac{3}{2}z - 1$

22. $\dfrac{3x^3 + 5x^2 - 4x}{3x}$ $x^2 + \frac{5}{3}x - \frac{4}{3}$

23. $\dfrac{m^3 + 3m - 4}{m}$ $m^2 + 3 - \frac{4}{m}$

24. $\dfrac{l^2 - 4l + 8}{l}$ $l - 4 + \frac{8}{l}$

25. $\dfrac{4x^2 - 12x}{4x^2}$ $1 - \frac{3}{x}$

26. $\dfrac{5z^2 - 20z}{5z^2}$ $1 - \frac{4}{z}$

27. $\dfrac{8x^2 + 12x^3}{-4x^3}$ $-\frac{2}{x} - 3$

28. $\dfrac{14y^4 + 21y^3}{-7y^3}$ $-2y - 3$

29. $\dfrac{6x^4 - 2x^3 + 3x^2 - x + 4}{2x^3}$ $3x - 1 + \dfrac{3}{2x} - \dfrac{1}{2x^2} + \dfrac{2}{x^3}$

30. $\dfrac{12x^4 + 4x^3 - 3x^2 + 8x - 7}{4x^3}$

$3x + 1 - \dfrac{3}{4x} + \dfrac{2}{x^2} - \dfrac{7}{4x^3}$

31. $\dfrac{15x^5 + 10x^4 - 8x^2 - 3x}{-5x^2}$ $-3x^3 - 2x^2 + \dfrac{8}{5} + \dfrac{3}{5x}$

32. $\dfrac{9x^5 - 12x^3 + 3x^2 - 5x}{-3x^2}$ $-3x^3 + 4x - 1 + \dfrac{5}{3x}$

33. $\dfrac{x^2 - x - 2}{x + 1}$ $x - 2$ **34.** $\dfrac{x^2 - 5x + 6}{x - 2}$ $x - 3$

35. $\dfrac{x^2 + 9x + 20}{x + 4}$ $x + 5$ **36.** $\dfrac{x^2 - 7x - 30}{x - 10}$ $x + 3$

37. $\dfrac{3y^2 + 4y - 4}{3y - 2}$ $y + 2$ **38.** $\dfrac{7t^2 - 10t - 8}{7t + 4}$ $t - 2$

39. $(18t^2 - 21t - 4) \div (3t - 4)$ $6t + 1$

40. $(20t^2 + 32t - 16) \div (2t + 4)$ $10t - 4$

41. $(x^3 - 4x^2 + 9x - 7) \div (x - 2)$

$x^2 - 2x + 5 + \dfrac{3}{x - 2}$

42. $(2x^3 - 2x^2 + 3x + 9) \div (x + 1)$

$2x^2 - 4x + 7 + \dfrac{2}{x + 1}$

43. $(7x + 3) \div (x + 2)$ $7 - \dfrac{11}{x + 2}$

44. $(8x - 5) \div (2x + 1)$ $4 - \dfrac{9}{2x + 1}$

45. $\dfrac{x^3 - 8}{x - 2}$ $x^2 + 2x + 4$ **46.** $\dfrac{x^3 + 27}{x + 3}$ $x^2 - 3x + 9$

47. $\dfrac{x^2 + 9}{x + 3}$ $x - 3 + \dfrac{18}{x + 3}$

48. $\dfrac{y^2 + 3}{y + 3}$ $y - 3 + \dfrac{12}{y + 3}$

49. Divide $9x^2 - 1$ by $3x + 1$. $3x - 1$

50. Divide $25y^2 - 4$ by $5y - 2$. $5y + 2$

51. Divide $x^4 - 1$ by $x - 1$. $x^3 + x^2 + x + 1$

52. Divide x^4 by $x - 1$. $x^3 + x^2 + x + 1 + \dfrac{1}{x - 1}$

In Exercises 53–64, simplify the expression. (Assume that no denominator is zero.)

53. $\dfrac{4x^3}{x^2} - \dfrac{8x}{4}$ $2x$ **54.** $\dfrac{9z^3}{z^2} - \dfrac{6z}{3}$ $7z$

55. $\dfrac{25x^2}{10x} + \dfrac{3x}{2}$ $4x$ **56.** $\dfrac{21y^2}{14y} + \dfrac{5y}{2}$ $4y$

57. $\dfrac{8u^2v}{2u} + \dfrac{(uv)^2}{uv}$ $5uv$ **58.** $\dfrac{18xy^2}{6y} + \dfrac{(xy)^2}{xy}$ $4xy$

59. $\dfrac{9x^5y}{3x^4} - \dfrac{(x^2y)^3}{x^5y^2}$ **60.** $\dfrac{35a^3b^2}{5a^2b} - \dfrac{(a^2b^3)^2}{a^3b^5}$

$2xy$ $6ab$

61. $\dfrac{x^2 + 2x + 1}{x + 1} - (3x - 4)$ $-2x + 5$

62. $\dfrac{x^2 - 2x - 3}{x + 1} - (4x - 1)$ $-3x - 2$

63. $\dfrac{x^2 - 3x + 2}{x - 1} + (4x - 3)$ $5x - 5$

64. $\dfrac{x^2 + x - 6}{x - 2} + (2x - 7)$ $3x - 4$

In Exercises 65–68, determine whether the dividing out is valid.

65. $\dfrac{3 + 4}{3} = \dfrac{3 + \cancel{4}}{\cancel{3}} = 4$ Error; You can divide out only common factors of the numerator and denominator.

66. $\dfrac{4 + 7}{4 + 11} = \dfrac{\cancel{4} + 7}{\cancel{4} + 11} = \dfrac{7}{11}$ Error; You can divide out only common factors of the numerator and denominator.

67. $\dfrac{7 \cdot 12}{19 \cdot 7} = \dfrac{\cancel{7} \cdot 12}{19 \cdot \cancel{7}} = \dfrac{12}{19}$ Valid

68. $\dfrac{24}{43} = \dfrac{2\cancel{4}}{4\cancel{3}} = \dfrac{2}{3}$ Error; You cannot divide out digits of numbers.

Solving Problems

69. ▲ *Geometry* The area of a rectangle is $x^2 + 5x - 6$ and its width is $x - 1$ (see figure). Find the length of the rectangle. $x + 6$

Area:
$x^2 + 5x - 6$
$x - 1$

70. ▲ *Geometry* The area of a rectangle is $x^2 + 2x - 15$ and its length is $x + 5$ (see figure). Find the width of the rectangle. $x - 3$

$x + 5$

Area:
$x^2 + 2x - 15$

71. *Exploration*   Consider the equation
$(x + 3)(x^2 + 2x - 1) = x^3 + 5x^2 + 5x - 3$.

(a) ▦ Use a graphing calculator to verify that the equation is an identity by graphing both the left side and the right side of the equation in the same viewing window. Are the graphs the same?
See Additional Answers.   Yes.

(b) Verify the identity equation by multiplying the polynomials on the left side of the equation.

$(x + 3)(x^2 + 2x - 1)$
$= x(x^2) + x(2x) - x(1) + 3(x^2) + 3(2x) - 3(1)$
$= x^3 + 5x^2 + 5x - 3$

(c) Verify the identity equation by performing the long division

$$\frac{x^3 + 5x^2 + 5x - 3}{x + 3}.$$

$(x^3 + 5x^2 + 5x - 3) \div (x + 3) = x^2 + 2x - 1$

72. *Exploration*   Consider the equation
$2x^3 - 5x^2 + 2x - 5 = (2x - 5)(x^2 + 1)$.

(a) ▦ Use a graphing calculator to verify that the equation is an identity by graphing both the left side and the right side of the equation in the same viewing window. Are the graphs the same?
See Additional Answers. Yes.

(b) Verify the identity equation by multiplying the polynomials on the right side of the equation.

$(2x - 5)(x^2 + 1) = 2x(x^2) + 2x(1) - 5(x^2) - 5(1)$
$= 2x^3 + 2x - 5x^2 - 5$

(c) Verify the identity equation by performing the long division

$$\frac{2x^3 - 5x^2 + 2x - 5}{2x - 5}.$$

$(2x^3 - 5x^2 + 2x - 5) \div (2x - 5) = x^2 + 1$

73. *Comparing Ages*   Your sister has two children: one is 18 years old and the other is 8 years old. In t years, their ages will be $t + 18$ and $t + 8$.

(a) Use long division to rewrite the ratio of the older child's age to the younger child's age.

$$1 + \frac{10}{t + 8}$$

(b) Complete the table.

t	0	10	20	30	40	50	60
$\dfrac{t + 18}{t + 8}$	2.25	1.56	1.36	1.26	1.21	1.17	1.15

(c) What happens to the values of the ratio as t increases? Use the result of part (a) to explain your conclusion.   The values approach 1.

74. *Comparing Salaries*   At a manufacturing company, Employee A and Employee B earn $30,000 and $40,000, respectively. After a company-wide raise of m dollars, their salaries will be $30,000 + m$ dollars and $40,000 + m$ dollars.

(a) Use long division to rewrite the ratio of Employee B's salary to Employee A's salary.

$$1 + \frac{10,000}{30,000 + m}$$

(b) Complete the table.

m	0	1000	2000	3000	4000
$\dfrac{40,000 + m}{30,000 + m}$	1.333	1.323	1.313	1.303	1.294

(c) What happens to the values of the ratio as m increases? Use the result of part (a) to explain your conclusion.   The values approach 1.

Explaining Concepts

75. ⟳ Answer part (g) of Motivating the Chapter on page 264.

76. Label each part of the equation with its name:

$$\frac{x^2 + 2}{x - 3} = x + 3 + \frac{11}{x - 3}.$$

(a) Dividend $x^2 + 2$ (b) Divisor $x - 3$

(c) Quotient $x + 3$ (d) Remainder 11

77. *Writing* ✎   What does it mean when the divisor divides *evenly* into the dividend?   The remainder is 0 and the divisor is a factor of the dividend.

78. *Writing* ✎   Explain how you can check the result of a division problem algebraically *and* graphically.

Algebraically: Multiply the quotient by the divisor and add the remainder; the result should be the dividend.

Graphically: Use a graphing calculator to graph the rational expression and the result of the division process. The graphs should coincide.

What Did You Learn?

Key Terms

negative exponents, *p. 270*
scientific notation, *p. 273*
polynomial, *p. 280*
constant term, *p. 280*

standard form of a
    polynomial, *p. 280*
degree of a polynomial,
    p. 280
leading coefficient, *p. 280*

monomial, *p. 281*
binomial, *p. 281*
trinomial, *p. 281*
FOIL Method, *p. 292*

Key Concepts

5.1 ○ Rules of Exponents

Let m and n be integers, and let a and b be real numbers, variables, or algebraic expressions, such that $a \neq 0$ and $b \neq 0$.

1. $a^m a^n = a^{m+n}$ 2. $\dfrac{a^m}{a^n} = a^{m-n}$

3. $(ab)^m = a^m b^m$ 4. $\left(\dfrac{a}{b}\right)^m = \dfrac{a^m}{b^m}$

5. $(a^m)^n = a^{mn}$ 6. $a^{-n} = \dfrac{1}{a^n}$

7. $a^0 = 1$

5.2 ○ Polynomial in *x*

Let $a_n, a_{n-1}, \ldots, a_2, a_1, a_0$ be real numbers and let n be a nonnegative integer. A polynomial in x is an expression of the form

$$a_n x^n + a_{n-1} x^{n-1} + \cdots + a_2 x^2 + a_1 x + a_0$$

where $a_n \neq 0$. The polynomial is of degree n, and the number a_n is called the leading coefficient. The number a_0 is called the constant term.

5.2 ○ Adding polynomials

To add polynomials, you combine like terms (those having the same degree) by using the Distributive Property.

5.2 ○ Subtracting polynomials

To subtract one polynomial from another, you add the opposite by changing the sign of each term of the polynomial that is being subtracted and then adding the resulting like terms.

5.3 ○ Multiplying polynomials

1. To multiply a polynomial by a monomial, apply the Distributive Property.

2. To multiply two binomials, use the FOIL Method. Combine the product of the **F**irst terms, the product of the **O**uter terms, the product of the **I**nner terms, and the product of the **L**ast terms.

3. To multiply two polynomials, use the Distributive Property to multiply each term of one polynomial by each term of the other polynomial.

5.3 ○ Special Products

Let a and b be real numbers, variables, or algebraic expressions.
Sum and Difference of Two Terms:

$$(a + b)(a - b) = a^2 - b^2$$

Square of a Binomial:

$$(a + b)^2 = a^2 + 2ab + b^2$$

$$(a - b)^2 = a^2 - 2ab + b^2$$

5.4 ○ Dividing a polynomial by a monomial

Let a, b, and c be real numbers, variables, or algebraic expressions, such that $c \neq 0$.

1. $\dfrac{a + b}{c} = \dfrac{a}{c} + \dfrac{b}{c}$ 2. $\dfrac{a - b}{c} = \dfrac{a}{c} - \dfrac{b}{c}$

5.4 ○ Dividing polynomials

Along with the long division algorithm, follow these steps when performing long division of polynomials.

1. Write the dividend and divisor in descending powers of the variable.

2. Insert placeholders with zero coefficients for missing powers of the variable.

3. Perform the long division of the polynomials as you would with integers.

4. Continue the process until the degree of the remainder is less than the degree of the divisor.

Review Exercises

5.1 Negative Exponents and Scientific Notation

1 Apply the rules of exponents to rewrite exponential expressions.

In Exercises 1–20, simplify the expression. (Assume that no denominator is zero.)

1. $x^2 \cdot x \cdot x^4$ x^7

2. $y^2 \cdot y^3 \cdot y$ y^6

3. $(x^3)^2$ x^6

4. $(t^4)^3$ t^{12}

5. $t^4(-2t^2)$ $-2t^6$

6. $u^2(3u^2)$ $3u^4$

7. $(xy)^4(-5x^2y^3)$
$-5x^6y^7$

8. $(3uv)(-2uv^2)^3$
$-24u^4v^7$

9. $(-4x^3y)(8xy)^2$
$-256x^5y^3$

10. $(12a^4b^6)^2(-a^2b)$
$-144a^{10}b^{13}$

11. $\dfrac{x^6}{x}$ x^5

12. $\dfrac{y^4}{y^3}$ y

13. $\dfrac{5a^5}{a^2}$ $5a^3$

14. $\dfrac{b^7}{6b^4}$ $\dfrac{b^3}{6}$

15. $\dfrac{24x^3}{12x^2}$ $2x$

16. $\dfrac{20y^2}{5y^2}$ 4

17. $\dfrac{9z^5}{(-z)^3}$ $-9z^2$

18. $\dfrac{-36v^6}{4v^3}$ $-9v^3$

19. $\dfrac{64u^3v^2}{32uv}$ $2u^2v$

20. $\dfrac{21x^3y^7}{-7x^3y}$ $-3y^6$

2 Use the negative exponent rule to rewrite exponential expressions.

In Exercises 21–26, rewrite the expression using only positive exponents.

21. 4^{-2} $\dfrac{1}{4^2}$

22. 3^{-4} $\dfrac{1}{3^4}$

23. $6x^{-3}$ $\dfrac{6}{x^3}$

24. $5x^{-5}$ $\dfrac{5}{x^5}$

25. $\dfrac{4}{t^{-2}}$ $4t^2$

26. $\dfrac{7}{y^{-3}}$ $7y^3$

In Exercises 27–48, use the rules of exponents to simplify the expression using only positive exponents. (Assume that no variable is zero.)

27. $t^{-2} \cdot t$ $\dfrac{1}{t}$

28. $-u^{-3} \cdot u$ $-\dfrac{1}{u^2}$

29. $t^{-4} \cdot t^2$ $\dfrac{1}{t^2}$

30. $x^5 \cdot x^{-8}$ $\dfrac{1}{x^3}$

31. $\dfrac{a^3}{a^{-4}}$ a^7

32. $\dfrac{b^6}{b^{-1}}$ b^7

33. $\dfrac{x^{-1}}{7x^{-6}}$ $\dfrac{x^5}{7}$

34. $\dfrac{y^{-2}}{2y^{-4}}$ $\dfrac{y^2}{2}$

35. $4x^{-6}y^2 \cdot x^6$ $4y^2$

36. $-2u^5v^{-4} \cdot v^4$ $-2u^5$

37. $(-3a^2)^{-2}(a^2)^0$ $\dfrac{1}{9a^4}$

38. $(3u^{-3})(9u^6)^0$ $\dfrac{3}{u^3}$

39. $(2x^2y^{-3})^2$ $\dfrac{4x^4}{y^6}$

40. $(5x^{-1}y^3)^{-2}$ $\dfrac{x^2}{25y^6}$

41. $\dfrac{t^{-4} \cdot t^5}{t^{-1}}$ t^2

42. $\dfrac{a^3 \cdot a^{-2}}{a^{-1}}$ a^2

43. $\left(\dfrac{y}{5}\right)^{-2}$ $\dfrac{25}{y^2}$

44. $\left(\dfrac{7}{x^4}\right)^{-1}$ $\dfrac{x^4}{7}$

45. $\dfrac{(-3y)^{-4}}{-9y^{-6}}$ $-\dfrac{y^2}{729}$

46. $\dfrac{5z^{-3}}{(-10z^2)^{-2}}$ $500z$

47. $(2u^{-2}v)^3(4u^{-5}v^4)^{-1}$
$\dfrac{2}{uv}$

48. $(3x^2y^4)^3(3x^2y^4)^{-3}$
1

3 Write very large and very small numbers in scientific notation.

In Exercises 49 and 50, write the number in scientific notation.

49. 0.000728
7.28×10^{-4}

50. 39,200,000
3.92×10^7

In Exercises 51 and 52, write the number in decimal notation.

51. 1.809×10^8
180,900,000

52. 3.112×10^{-6}
0.000003112

5.2 Adding and Subtracting Polynomials

1 Identify the degrees and leading coefficients of polynomials.

In Exercises 53–60, write the polynomial in standard form. Then identify its degree and leading coefficient.

53. $10x - 4 - 5x^3$ Standard form: $-5x^3 + 10x - 4$; Degree: 3; Leading coefficient: -5

54. $2x^2 + 9$ Standard form: $2x^2 + 9$; Degree: 2; Leading coefficient: 2

55. $4x^3 - 2x + 5x^4 - 7x^2$
Standard form: $5x^4 + 4x^3 - 7x^2 - 2x$; Degree: 4; Leading coefficient: 5

56. $6 - 3x + 6x^2 - x^3$
Standard form: $-x^3 + 6x^2 - 3x + 6$; Degree: 3; Leading coefficient: -1

57. $7x^4 - 1$ Standard form: $7x^4 - 1$; Degree: 4; Leading coefficient: 7

58. $12x^2 + 2x - 8x^5 + 1$
Standard form: $-8x^5 + 12x^2 + 2x + 1$; Degree: 5; Leading coefficient: -8

59. -2
Standard form: -2; Degree: 0; Leading coefficient: -2

60. $\frac{1}{4}t^2$
Standard form: $\frac{1}{4}t^2$; Degree: 2; Leading coefficient: $\frac{1}{4}$

In Exercises 61–64, give an example of a polynomial in one variable that satisfies the condition. (There are many correct answers.)

61. A trinomial of degree 4 $x^4 + x^2 + 2$

62. A monomial of degree 2 $7t^2$

63. A binomial of degree 1 $-2x + 3$

64. A trinomial of degree 5 $4x^5 - 2x^3 + x$

2 Add polynomials using a horizontal or vertical format.

In Exercises 65–76, find the sum.

65. $(2x + 3) + (x - 4)$ $3x - 1$

66. $(5x + 7) + (x - 2)$ $6x + 5$

67. $\left(\frac{1}{2}x + \frac{2}{3}\right) + \left(4x + \frac{1}{3}\right)$ $\frac{9}{2}x + 1$

68. $\left(\frac{3}{4}y + 2\right) + \left(\frac{1}{2}y - \frac{2}{5}\right)$ $\frac{5}{4}y + \frac{8}{5}$

69. $(2x^3 - 4x^2 + 3) + (x^3 + 4x^2 - 2x)$
$3x^3 - 2x + 3$

70. $(3y^3 + 5y^2 - 9y) + (2y^3 - 3y + 10)$
$5y^3 + 5y^2 - 12y + 10$

71. $-4(6 - x + x^2) + (3x^2 + x)$
$-x^2 + 5x - 24$

72. $(4 - x^2) + 2(x - 2)$ $-x^2 + 2x$

73. $(3u + 4u^2) + 5(u + 1) + 3u^2$
$7u^2 + 8u + 5$

74. $6(u^2 + 2) + 12u + (u^2 - 5u + 2)$
$7u^2 + 7u + 14$

75. $\begin{array}{r} -x^4 - 2x^2 + 3 \\ \underline{3x^4 - 5x^2\phantom{{}+ 3}} \end{array}$ $2x^4 - 7x^2 + 3$

76. $\begin{array}{r} 5z^3 - 4z - 7 \\ \underline{z^2 - 2z\phantom{{}- 7}} \end{array}$ $5z^3 + z^2 - 6z - 7$

77. ▲ *Geometry*   The length of a rectangular wall is x units, and its height is $(x - 3)$ units (see figure). Find an expression for the perimeter of the wall.
$(4x - 6)$ units

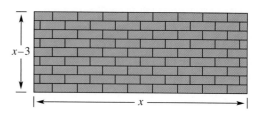

78. ▲ *Geometry*   A rectangular garden has length $(t + 5)$ feet and width $2t$ feet (see figure). Find an expression for the perimeter of the garden.
$(6t + 10)$ feet

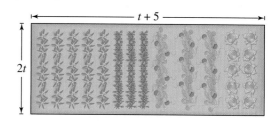

3 Subtract polynomials using a horizontal or vertical format.

In Exercises 79–92, find the difference.

79. $(t - 5) - (3t - 1)$ $-2t - 4$

80. $(y + 3) - (y - 9)$ 12

81. $\left(\frac{1}{2}x + 5\right) - \left(\frac{3}{4}x - \frac{1}{3}\right)$ $-\frac{1}{4}x + \frac{16}{3}$

82. $\left(2x - \frac{1}{5}\right) - \left(\frac{1}{4}x + \frac{1}{4}\right)$ $\frac{7}{4}x - \frac{9}{20}$

83. $(6x^2 - 9x - 5) - (4x^2 - 6x + 1)$ $2x^2 - 3x - 6$

84. $(3y^2 + 2y - 9) - (5y^2 - y + 7)$ $-2y^2 + 3y - 16$

85. $3(2x^2 - 4) - (2x^2 - 5)$ $4x^2 - 7$

86. $(5t^2 + 2) - 2(4t^2 + 1)$ $-3t^2$

87. $(z^2 + 6z) - 3(z^2 + 2z)$ $-2z^2$

88. $(-x^3 - 3x) - 2(2x^3 + x + 1)$ $-5x^3 - 5x - 2$

89. $4y^2 - [y - 3(y^2 + 2)]$ $7y^2 - y + 6$

90. $(6a^3 + 3a) - 2[a - (a^3 + 2)]$ $8a^3 + a + 4$

91. $\begin{array}{r} 5x^2 + 2x - 27 \\ \underline{-(2x^2 - 2x - 13)} \end{array}$ $3x^2 + 4x - 14$

92. $\begin{array}{r} 12y^4 - 15y^2 + 7 \\ \underline{-(18y^4 + 4y^2 - 9)} \end{array}$ $-6y^4 - 19y^2 + 16$

93. *Cost, Revenue, and Profit*   The cost C of producing x units of a product is $C = 15 + 26x$. The revenue R for selling x units is $R = 40x - \frac{1}{2}x^2$, $0 \le x \le 20$. The profit P is the difference between revenue and cost.

(a) Perform the subtraction required to find the polynomial representing profit P.
$-\frac{1}{2}x^2 + 14x - 15$

(b) ▦ Use a graphing calculator to graph the polynomial representing profit.
See Additional Answers.

(c) ▦ Determine the profit when 14 units are produced and sold. Use the graph in part (b) to describe the profit when x is less than or greater than 14. $83; When x is less than or greater than 14, the profit is less than $83.

94. ▦ *Comparing Models*   The table shows population projections (in millions) for the United States for selected years from 2005 to 2030. There are three series of projections: lowest P_L, middle P_M, and highest P_H.   (Source: U.S. Census Bureau)

Year	2005	2010	2015	2020	2025	2030
P_L	284.0	291.4	298.0	303.7	308.2	311.7
P_M	287.7	299.9	312.3	324.9	337.8	351.1
P_H	292.3	310.9	331.6	354.6	380.4	409.6

In the following models for the data, $t = 5$ corresponds to the year 2005.

$P_L = -0.020t^2 + 1.81t + 275.4$

$P_M = 2.53t + 274.6$

$P_H = 0.052t^2 + 2.84t + 277.0$

(a) Use a graphing calculator to plot the data and graph the models in the same viewing window.
See Additional Answers.

(b) Find $(P_L + P_H)/2$. Use a graphing calculator to graph this polynomial and state which graph from part (a) it most resembles. Does this seem reasonable? Explain.

$\dfrac{P_L + P_H}{2} = 0.016t^2 + 2.325t + 276.2$

See Additional Answers.   The graph is most similar to P_M. Yes, because the average of P_L and P_H should be similar to P_M.

(c) Find $P_H - P_L$. Use a graphing calculator to graph this polynomial. Explain why it is increasing.

5.3 Multiplying Polynomials: Special Products

① Find products with monomial multipliers.

In Exercises 95–98, perform the multiplication and simplify.

95. $2x(x + 4)$ $2x^2 + 8x$

96. $3y(y + 1)$ $3y^2 + 3y$

97. $(4x - 2)(-3x^2)$ $-12x^3 + 6x^2$

98. $(5 - 7y)(-6y^2)$ $42y^3 - 30y^2$

② Multiply binomials using the Distributive Property and the FOIL Method.

In Exercises 99–104, perform the multiplication and simplify.

99. $(x - 4)(x + 6)$
$x^2 + 2x - 24$

100. $(u + 5)(u - 2)$
$u^2 + 3u - 10$

101. $(x + 3)(2x - 4)$
$2x^2 + 2x - 12$

102. $(y + 2)(4y - 3)$
$4y^2 + 5y - 6$

103. $(4x - 3)(3x + 4)$
$12x^2 + 7x - 12$

104. $(6 - 2x)(7x + 10)$
$-14x^2 + 22x + 60$

③ Multiply polynomials using a horizontal or vertical format.

In Exercises 105–114, perform the multiplication and simplify.

105. $(x^2 + 5x + 2)(2x + 3)$ $2x^3 + 13x^2 + 19x + 6$

106. $(s^2 + 4s - 3)(s - 3)$ $s^3 + s^2 - 15s + 9$

107. $(2t - 1)(t^2 - 3t + 3)$ $2t^3 - 7t^2 + 9t - 3$

108. $(4x + 2)(x^2 + 6x - 5)$ $4x^3 + 26x^2 - 8x - 10$

109.
$$\begin{array}{r} 3x^2 + x - 2 \\ \times \quad\quad 2x - 1 \end{array}$$ $6x^3 - x^2 - 5x + 2$

110.
$$\begin{array}{r} 5y^2 - 2y + 9 \\ \times \quad\quad 3y + 4 \end{array}$$ $15y^3 + 14y^2 + 19y + 36$

111.
$$\begin{array}{r} y^2 - 4y + 5 \\ \times \ y^2 + 2y - 3 \end{array}$$ $y^4 - 2y^3 - 6y^2 + 22y - 15$

112.
$$\begin{array}{r} x^2 + 8x - 12 \\ \times \ x^2 - 9x + 2 \end{array}$$ $x^4 - x^3 - 82x^2 + 124x - 24$

113. $(2x + 1)^3$ $8x^3 + 12x^2 + 6x + 1$

114. $(3y - 2)^3$ $27y^3 - 54y^2 + 36y - 8$

94. (c) $P_H - P_L = 0.072t^2 + 1.03t + 1.6$

See Additional Answers.   The vertical distance between P_L and P_H is increasing.

115. ▲ *Geometry*   The width of a rectangular window is $(2x + 6)$ inches and its height is $(3x + 10)$ inches (see figure). Find an expression for the area of the window. $(6x^2 + 38x + 60)$ square inches

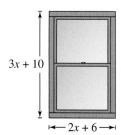

$3x + 10$

$\longmapsto 2x + 6 \longrightarrow$

116. ▲ *Geometry*   The width of a rectangular parking lot is $(x + 25)$ meters and its length is $(x + 30)$ meters (see figure). Find an expression for the area of the parking lot. $(x^2 + 55x + 750)$ square inches

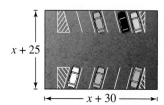

$x + 25$

$\longmapsto x + 30 \longrightarrow$

④ Identify and use special binomial products.

In Exercises 117–128, use a special product pattern to find the product.

117. $(x + 3)^2$
$x^2 + 6x + 9$

118. $(x - 5)^2$
$x^2 - 10x + 25$

119. $(4x - 7)^2$
$16x^2 - 56x + 49$

120. $(9 - 2x)^2$
$81 - 36x + 4x^2$

121. $\left(\frac{1}{2}x - 4\right)^2$
$\frac{1}{4}x^2 - 4x + 16$

122. $(4 + 3b)^2$
$16 + 24b + 9b^2$

123. $(u - 6)(u + 6)$ $u^2 - 36$

124. $(r + 3)(r - 3)$ $r^2 - 9$

125. $(2x - y)^2$ $4x^2 - 4xy + y^2$

126. $(3a + b)^2$ $9a^2 + 6ab + b^2$

127. $(2x - 4y)(2x + 4y)$ $4x^2 - 16y^2$

128. $(4u + 5v)(4u - 5v)$ $16u^2 - 25v^2$

5.4 Dividing Polynomials

① Divide polynomials by monomials.

In Exercises 129–134, perform the division and simplify. (Assume that no denominator is zero.)

129. $\dfrac{8x^3 - 12x}{4x^2}$ $2x - \dfrac{3}{x}$

130. $\dfrac{8u^3 + 4u^2}{2u}$ $4u^2 + 2u$

131. $(18 - 3x + 9x^2) \div (12x^2)$ $\dfrac{3}{2x^2} - \dfrac{1}{4x} + \dfrac{3}{4}$

132. $(5x^2 + 15x - 25) \div (5x)$ $x + 3 - \dfrac{5}{x}$

133. $\dfrac{6y^3 - 4y^2 + 10y}{-2y}$ $-3y^2 + 2y - 5$

134. $\dfrac{6x^4 + 9x^3 - 12x^2}{-3x^2}$ $-2x^2 - 3x + 4$

② Divide polynomials by binomials.

In Exercises 135–140, perform the division. (Assume that no denominator is zero.)

135. $\dfrac{x^2 - x - 6}{x - 3}$ $x + 2$

136. $\dfrac{x^2 + x - 20}{x + 5}$ $x - 4$

137. $\dfrac{24x^2 - x - 8}{3x - 2}$ $8x + 5 + \dfrac{2}{3x - 2}$

138. $\dfrac{21x^2 + 4x + 7}{3x - 2}$ $7x + 6 + \dfrac{19}{3x - 2}$

139. $\dfrac{2x^3 + 2x^2 - x + 2}{x - 1}$ $2x^2 + 4x + 3 + \dfrac{5}{x - 1}$

140. $\dfrac{6x^4 - 4x^3 - 27x^2 + 18x}{3x - 2}$ $2x^3 - 9x$

141. ▲ *Geometry*   The area of a rectangle is $2x^2 - 5x - 12$ and its width is $x - 4$. Find the length of the rectangle. $2x + 3$

142. ▲ *Geometry*   The area of a rectangle is $3x^2 + 5x - 3$ and its length is $x + 3$. Find the width of the rectangle.
$3x - 4 + \dfrac{9}{x + 3}$

Chapter Test

Take this test as you would take a test in class. After you are done, check your work against the answers in the back of the book.

In Exercises 1–6, simplify the expression. (Assume that no variable is zero.)

1. $(3x^2y)(-xy)^2$ $3x^4y^3$

2. $(5x^2y^3)^2(-2xy)^3$ $-200x^7y^9$

3. $\dfrac{-6a^2b}{-9ab}$ $\dfrac{2a}{3}$

4. $(3x^{-2}y^3)^{-2}$ $\dfrac{x^4}{9y^6}$

5. $(4u^{-3}v^2)^{-2}(8u^{-1}v^{-2})^0$ $\dfrac{u^6}{16v^4}$

6. $\dfrac{12x^{-3}y^5}{4x^{-2}y^{-1}}$ $\dfrac{3y^6}{x}$

7. Evaluate each expression *without* using a calculator.

 (a) $\dfrac{2^{-3}}{3^{-1}}$ (b) $(1.5 \times 10^5)^2$ (c) $\dfrac{6.3 \times 10^{-3}}{2.1 \times 10^2}$

 (a) $\frac{3}{8}$ (b) 22,500,000,000   (c) 0.00003

8. (a) Write 0.000032 in scientific notation. 3.2×10^{-5}

 (b) Write 6.04×10^7 in decimal notation.   60,400,000

9. Determine the degree and the leading coefficient of $-3x^4 - 5x^2 + 2x - 10$.
 Degree: 4; Leading coefficient: -3

10. Give an example of a trinomial in one variable of degree 4. $z^4 + 2z^2 - 3$

In Exercises 11–22, perform the indicated operation and simplify. (Assume that no variable or denominator is zero.)

11. $(3z^2 - 3z + 7) + (8 - z^2)$ $2z^2 - 3z + 15$

12. $(8u^3 + 3u^2 - 2u - 1) - (u^3 + 3u^2 - 2u)$ $7u^3 - 1$

13. $6y - [2y - (3 + 4y - y^2)]$ $-y^2 + 8y + 3$

14. $-5(x^2 - 1) + 3(4x + 7) - (x^2 + 26)$ $-6x^2 + 12x$

15. $(5b + 3)(2b - 1)$ $10b^2 + b - 3$ **16.** $(3x - 4)(6x - 5)$ $18x^2 - 39x + 20$

17. $(z + 2)(2z^2 - 3z + 5)$
 $2z^3 + z^2 - z + 10$

18. $(x - 5)^2$
 $x^2 - 10x + 25$

19. $(2x - 3)(2x + 3)$ $4x^2 - 9$

20. $\dfrac{15x - 25}{-5}$ $-3x + 5$

21. $\dfrac{x^3 - x - 6}{x - 2}$ $x^2 + 2x + 3$

22. $\dfrac{4x^3 + 10x^2 - 2x - 5}{2x + 1}$ $2x^2 + 4x - 3 - \dfrac{2}{2x + 1}$

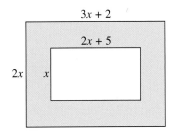

Figure for 23

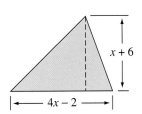

Figure for 24

23. $4x^2 - x$

24. $2x^2 + 11x - 6$

25. $750r^2 + 1500r + 750$

23. Find an expression for the area of the shaded region shown in the figure.

24. Find an expression for the area of the triangle shown in the figure.

25. After 2 years, an investment of $750 compounded annually at interest rate r will yield an amount of $750(1 + r)^2$. Find this product.

26. The area of a rectangle is $x^2 - 2x - 3$ and its length is $x + 1$. Find the width of the rectangle. $x - 3$

Motivating the Chapter

 ## Dimensions of a Potato Storage Bin

A bin used to store potatoes has the form of a rectangular solid with a volume (in cubic feet) given by the polynomial

$$12x^3 + 64x^2 - 48x.$$

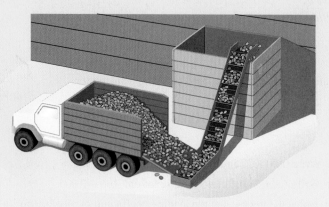

See Section 6.3, Exercise 117.

a. The height of the bin is $4x$ feet. Write an expression for the area of the base of the bin. $3x^2 + 16x - 12$

b. Factor the expression for the area of the base of the bin. Use the result to write expressions for the length and width of the bin.
$3x - 2$ and $x + 6$

See Section 6.5, Exercise 101.

c. The area of the base of the bin is 32 square feet. What are the dimensions of the bin?   4 feet × 8 feet × 8 feet

d. You are told that the bin has a volume of 256 cubic feet. Can you find the dimensions of the bin? Explain your reasoning.   Yes. When the area of the base is 32 square feet, $x = 2$ and the dimensions of the bin are 4 feet × 8 feet × 8 feet.

e. A polynomial that represents the volume of the truck bin in cubic feet is

$$6x^3 + 32x^2 - 24x.$$

How many truckloads does it take to fill the bin? Explain your reasoning.
The volume of the bin is twice the volume of the truck bin. So, it takes two truckloads to fill the bin.

Nik Wheeler/Corbis

Factoring and Solving Equations

6.1 Factoring Polynomials with Common Factors

Dave G. Houser/Corbis

What You Should Learn

1 Find the greatest common factor of two or more expressions.
2 Factor out the greatest common monomial factor from polynomials.
3 Factor polynomials by grouping.

Why You Should Learn It

In some cases, factoring a polynomial enables you to determine unknown quantities. For example, in Exercise 118 on page 327, you will factor the expression for the revenue from selling pool tables to determine an expression for the price of the pool tables.

1 Find the greatest common factor of two or more expressions.

Greatest Common Factor

In Chapter 5, you used the Distributive Property to multiply polynomials. In this chapter, you will study the *reverse* process, which is **factoring.**

Multiplying Polynomials

$2x(7 - 3x)$ ⟹ $14x - 6x^2$
Factor Factor   Product

Factoring Polynomials

$14x - 6x^2$ ⟹ $2x(7 - 3x)$
Product   Factor Factor

To factor an expression efficiently, you need to understand the concept of the *greatest common factor* of two (or more) integers or terms. In Section 1.4, you learned that the **greatest common factor** of two or more integers is the greatest integer that is a factor of each integer. For example, the greatest common factor of $12 = 2 \cdot 2 \cdot 3$ and $30 = 2 \cdot 3 \cdot 5$ is $2 \cdot 3 = 6$.

Example 1 Finding the Greatest Common Factor

To find the greatest common factor of $5x^2y^2$ and $30x^3y$, first factor each term.

$$5x^2y^2 = 5 \cdot x \cdot x \cdot y \cdot y = (5x^2y)(y)$$

$$30x^3y = 2 \cdot 3 \cdot 5 \cdot x \cdot x \cdot x \cdot y = (5x^2y)(6x)$$

So, you can conclude that the greatest common factor is $5x^2y$.

Example 2 Finding the Greatest Common Factor

To find the greatest common factor of $8x^5$, $20x^3$, and $16x^4$, first factor each term.

$$8x^5 = 2 \cdot 2 \cdot 2 \cdot x \cdot x \cdot x \cdot x \cdot x = (4x^3)(2x^2)$$

$$20x^3 = 2 \cdot 2 \cdot 5 \cdot x \cdot x \cdot x = (4x^3)(5)$$

$$16x^4 = 2 \cdot 2 \cdot 2 \cdot 2 \cdot x \cdot x \cdot x \cdot x = (4x^3)(4x)$$

So, you can conclude that the greatest common factor is $4x^3$.

② Factor out the greatest common monomial factor from polynomials.

Common Monomial Factors

Consider the three terms listed in Example 2 as terms of the polynomial

$$8x^5 + 16x^4 + 20x^3.$$

The greatest common factor, $4x^3$, of these terms is the **greatest common monomial factor** of the polynomial. When you use the Distributive Property to remove this factor from each term of the polynomial, you are **factoring out** the greatest common monomial factor.

$$8x^5 + 16x^4 + 20x^3 = 4x^3(2x^2) + 4x^3(4x) + 4x^3(5) \qquad \text{Factor each term.}$$

$$= 4x^3(2x^2 + 4x + 5) \qquad \text{Factor out common monomial factor.}$$

Study Tip

To find the greatest common monomial factor of a polynomial, answer these two questions.

1. What is the greatest integer factor common to each coefficient of the polynomial?

2. What is the highest-power variable factor common to each term of the polynomial?

Example 3   Greatest Common Monomial Factor

Factor out the greatest common monomial factor from $6x - 18$.

Solution

The greatest common integer factor of $6x$ and 18 is 6. There is no common variable factor.

$$6x - 18 = 6(x) - 6(3) \qquad \text{Greatest common monomial factor is 6.}$$

$$= 6(x - 3) \qquad \text{Factor 6 out of each term.}$$

Example 4   Greatest Common Monomial Factor

Factor out the greatest common monomial factor from $10y^3 - 25y^2$.

Solution

For the terms $10y^3$ and $25y^2$, 5 is the greatest common integer factor and y^2 is the highest-power common variable factor.

$$10y^3 - 25y^2 = (5y^2)(2y) - (5y^2)(5) \qquad \text{Greatest common factor is } 5y^2.$$

$$= 5y^2(2y - 5) \qquad \text{Factor } 5y^2 \text{ out of each term.}$$

Example 5   Greatest Common Monomial Factor

Factor out the greatest common monomial factor from $45x^3 - 15x^2 - 15$.

Solution

The greatest common integer factor of $45x^3, 15x^2$, and 15 is 15. There is no common variable factor.

$$45x^3 - 15x^2 - 15 = 15(3x^3) - 15(x^2) - 15(1)$$

$$= 15(3x^3 - x^2 - 1)$$

Example 6   Greatest Common Monomial Factor

Factor out the greatest common monomial factor from $35y^3 - 7y^2 - 14y$.

Solution

$$35y^3 - 7y^2 - 14y = 7y(5y^2) - 7y(y) - 7y(2) \qquad \text{Greatest common factor is } 7y.$$
$$= 7y(5y^2 - y - 2) \qquad \text{Factor } 7y \text{ out of each term.}$$

Example 7   Greatest Common Monomial Factor

Factor out the greatest common monomial factor from $3xy^2 - 15x^2y + 12xy$.

Solution

$$3xy^2 - 15x^2y + 12xy = 3xy(y) - 3xy(5x) + 3xy(4) \qquad \text{Greatest common factor is } 3xy.$$
$$= 3xy(y - 5x + 4) \qquad \text{Factor } 3xy \text{ out of each term.}$$

The greatest common monomial factor of the terms of a polynomial is usually considered to have a positive coefficient. However, sometimes it is convenient to factor a negative number out of a polynomial.

Example 8   A Negative Common Monomial Factor

Factor the polynomial $-2x^2 + 8x - 12$ in two ways.

a. Factor out a common monomial factor of 2.

b. Factor out a common monomial factor of -2.

Solution

a. To factor out the common monomial factor of 2, write the following.

$$-2x^2 + 8x - 12 = 2(-x^2) + 2(4x) + 2(-6) \qquad \text{Factor each term.}$$
$$= 2(-x^2 + 4x - 6) \qquad \text{Factored form}$$

b. To factor -2 out of the polynomial, write the following.

$$-2x^2 + 8x - 12 = -2(x^2) + (-2)(-4x) + (-2)(6) \qquad \text{Factor each term.}$$
$$= -2(x^2 - 4x + 6) \qquad \text{Factored form}$$

Check this result by multiplying $(x^2 - 4x + 6)$ by -2. When you do, you will obtain the original polynomial.

With experience, you should be able to omit writing the first step shown in Examples 6, 7, and 8. For instance, to factor -2 out of $-2x^2 + 8x - 12$, you could simply write

$$-2x^2 + 8x - 12 = -2(x^2 - 4x + 6).$$

③ Factor polynomials by grouping.

Factoring by Grouping

There are occasions when the common factor of an expression is not simply a monomial. For instance, the expression.

$$x^2(x - 2) + 3(x - 2)$$

has the common *binomial* factor $(x - 2)$. Factoring out this common factor produces

$$x^2(x - 2) + 3(x - 2) = (x - 2)(x^2 + 3).$$

This type of factoring is part of a more general procedure called **factoring by grouping.**

Example 9   Common Binomial Factors

Factor each expression.

a. $5x^2(7x - 1) - 3(7x - 1)$ **b.** $2x(3x - 4) + (3x - 4)$
c. $3y^2(y - 3) + 4(3 - y)$

Solution

a. Each of the terms of this expression has a binomial factor of $(7x - 1)$.

$$5x^2(7x - 1) - 3(7x - 1) = (7x - 1)(5x^2 - 3)$$

b. Each of the terms of this expression has a binomial factor of $(3x - 4)$.

$$2x(3x - 4) + (3x - 4) = (3x - 4)(2x + 1)$$

Be sure you see that when $(3x - 4)$ is factored out of itself, you are left with the factor 1. This follows from the fact that $(3x - 4)(1) = (3x - 4)$.

c. $3y^2(y - 3) + 4(3 - y) = 3y^2(y - 3) - 4(y - 3)$ Write $4(3 - y)$ as $-4(y - 3)$.

$$= (y - 3)(3y^2 - 4)$$ Common factor is $(y - 3)$.

Students may find it helpful to write $2x(3x - 4) + (3x - 4)$ as $2x(3x - 4) + 1(3x - 4)$ before factoring it as $(3x - 4)(2x + 1)$.

In Example 9, the polynomials were already grouped so that it was easy to determine the common binomial factors. In practice, you will have to do the grouping as well as the factoring. To see how this works, consider the expression

$$x^3 + 2x^2 + 3x + 6$$

and try to factor it. Note first that there is no common monomial factor to take out of all four terms. But suppose you *group* the first two terms together and the last two terms together.

$$x^3 + 2x^2 + 3x + 6 = (x^3 + 2x^2) + (3x + 6)$$ Group terms.

$$= x^2(x + 2) + 3(x + 2)$$ Factor out common monomial factor in each group.

$$= (x + 2)(x^2 + 3)$$ Factored form

When factoring by grouping, be sure to group terms that have a common monomial factor. For example, in the polynomial above, you should not group the first term x^3 with the fourth term 6.

Additional Examples
Factor each polynomial.

a. $(3y - 1)5y^2 - (3y - 1)6$

b. $2x^3 + 8x^2 + 3x + 12$

Answers:

a. $(3y - 1)(5y^2 - 6)$

b. $(2x^2 + 3)(x + 4)$

Example 10 Factoring by Grouping

Factor $x^3 + 2x^2 + x + 2$.

Solution

$$x^3 + 2x^2 + x + 2 = (x^3 + 2x^2) + (x + 2) \quad \text{Group terms.}$$
$$= x^2(x + 2) + (x + 2) \quad \text{Factor out common monomial factor in each group.}$$
$$= (x + 2)(x^2 + 1) \quad \text{Factored form}$$

Note that in Example 10 the polynomial is factored by grouping the first and second terms and the third and fourth terms. You could just as easily have grouped the first and third terms and the second and fourth terms, as follows.

$$x^3 + 2x^2 + x + 2 = (x^3 + x) + (2x^2 + 2)$$
$$= x(x^2 + 1) + 2(x^2 + 1) = (x^2 + 1)(x + 2)$$

Example 11 Factoring by Grouping

Factor $3x^2 - 12x - 5x + 20$.

Solution

$$3x^2 - 12x - 5x + 20 = (3x^2 - 12x) + (-5x + 20) \quad \text{Group terms.}$$
$$= 3x(x - 4) - 5(x - 4) \quad \text{Factor out common monomial factor in each group.}$$
$$= (x - 4)(3x - 5) \quad \text{Factored form}$$

Note how a -5 is factored out so that the common binomial factor $x - 4$ appears.

You can always check to see that you have factored an expression correctly by multiplying the factors and comparing the result with the original expression. Try using multiplication to check the results of Examples 10 and 11.

Study Tip

Notice in Example 12 that the polynomial is not written in standard form. You could have rewritten the polynomial before factoring and still obtained the same result.

$$2x^3 + 4x - x^2 - 2$$
$$= 2x^3 - x^2 + 4x - 2$$
$$= (2x^3 - x^2) + (4x - 2)$$
$$= x^2(2x - 1) + 2(2x - 1)$$
$$= (2x - 1)(x^2 + 2)$$

Example 12 Geometry: Area of a Rectangle

The area of a rectangle of width $2x - 1$ is given by the polynomial $2x^3 + 4x - x^2 - 2$, as shown in Figure 6.1. Factor this expression to determine the length of the rectangle.

Solution

$$2x^3 + 4x - x^2 - 2 = (2x^3 + 4x) + (-x^2 - 2) \quad \text{Group terms.}$$
$$= 2x(x^2 + 2) - (x^2 + 2) \quad \text{Factor out common monomial factor in each group.}$$
$$= (x^2 + 2)(2x - 1) \quad \text{Factored form}$$

You can see that the length of the rectangle is $x^2 + 2$.

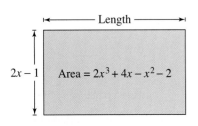

Length

$2x - 1$ Area $= 2x^3 + 4x - x^2 - 2$

Figure 6.1

6.1 Exercises

Review   *Concepts, Skills, and Problem Solving*

Keep mathematically in shape by doing these exercises *before* the problems of this section.

Properties and Definitions

1. *Writing✎*   Find the greatest common factor of 18 and 42. Explain how you arrived at your answer.
6; The greatest common factor is the product of the common prime factors.

2. *Writing✎*   Find the greatest common factor of 30, 45, and 135. Explain how you arrived at your answer.   15; The greatest common factor is the product of the common prime factors.

Simplifying Expressions

In Exercises 3–6, simplify the expression.

3. $2x - (x - 5) + 4(3 - x)$ $-3x + 17$

4. $3(x - 2) - 2(4 - x) - 7x$ $-2x - 14$

5. $\left(\dfrac{3x^2y^3}{2x^5y^2}\right)^2 \quad \dfrac{9y^2}{4x^6}$

6. $\left(\dfrac{-a^3b^{-1}}{4a^{-2}b^3}\right)^2 \quad \dfrac{a^{10}}{16b^8}$

Graphing Equations

In Exercises 7–10, graph the equation and show the coordinates of at least three solution points, including any intercepts.   See Additional Answers.

7. $y = 8 - 4x$

8. $3x - y = 6$

9. $y = -\frac{1}{2}x^2$

10. $y = |x + 2|$

Problem Solving

11. *Commission Rate*   Determine the commission rate for an employee who earned $1620 in commissions on sales of $54,000.   3%

12. *Work Rate*   One person can complete a typing project in 10 hours, and another can complete the same project in 6 hours. Working together, how long will they take to complete the project?   3 hours 45 minutes

Developing Skills

In Exercises 1–16, find the greatest common factor of the expressions. See Examples 1 and 2.

1. $z^2, -z^6$ z^2

2. t^4, t^7 t^4

3. $2x^2, 12x$ $2x$

4. $36x^4, 18x^3$ $18x^3$

5. u^2v, u^3v^2 u^2v

6. $r^6s^4, -rs$ rs

7. $9yz^2, -12y^2z^3$ $3yz^2$

8. $-15x^6y^3, 45xy^3$ $15xy^3$

9. $14x^2, 1, 7x^4$ 1

10. $5y^4, 10x^2y^2, 1$ 1

11. $28a^4b^2, 14a^3b^3, 42a^2b^5$ $14a^2b^2$

12. $16x^2y, 12xy^2, 36x^2y^2$ $4xy$

13. $2(x + 3), 3(x + 3)$ $x + 3$

14. $4(x - 5), 3x(x - 5)$ $x - 5$

15. $x(7x + 5), 7x + 5$ $7x + 5$

16. $x - 4, y(x - 4)$ $x - 4$

In Exercises 17–60, factor the polynomial. (*Note:* Some of the polynomials have no common monomial factor.) See Examples 3–7.

17. $3x + 3$ $3(x + 1)$

18. $5y + 5$ $5(y + 1)$

19. $6z - 6$ $6(z - 1)$

20. $3x - 3$ $3(x - 1)$

21. $8t - 16$ $8(t - 2)$

22. $3u + 12$ $3(u + 4)$

23. $-25x - 10$
$-5(5x + 2)$

24. $-14y - 7$
$-7(2y + 1)$

25. $24y^2 - 18$ $6(4y^2 - 3)$

26. $7z^3 + 21$ $7(z^3 + 3)$

27. $x^2 + x$ $x(x + 1)$

28. $-s^3 - s$ $-s(s^2 + 1)$

29. $25u^2 - 14u$
$u(25u - 14)$

30. $36t^4 + 24t^2$
$12t^2(3t^2 + 2)$

31. $2x^4 + 6x^3$
$2x^3(x + 3)$

32. $9z^6 + 27z^4$
$9z^4(z^2 + 3)$

33. $7s^2 + 9t^2$
No common factor

34. $12x^2 - 5y^3$
No common factor

35. $12x^2 - 2x$
$2x(6x - 1)$

36. $12u + 9u^2$
$3u(4 + 3u)$

37. $-10r^3 - 35r$
$-5r(2r^2 + 7)$

38. $-144a^2 + 24a$
$-24a(6a - 1)$

39. $16a^3b^3 + 24a^4b^3$
$8a^3b^3(2 + 3a)$

40. $6x^4y + 12x^2y$
$6x^2y(x^2 + 2)$

41. $10ab + 10a^2b$
$10ab(1 + a)$

42. $21x^2z - 35xz$
$7xz(3x - 5)$

43. $12x^2 + 16x - 8$
$4(3x^2 + 4x - 2)$

44. $9 - 3y - 15y^2$
$3(3 - y - 5y^2)$

45. $100 + 75z - 50z^2$
$25(4 + 3z - 2z^2)$

46. $42t^3 - 21t^2 + 7$
$7(6t^3 - 3t^2 + 1)$

47. $9x^4 + 6x^3 + 18x^2$
$3x^2(3x^2 + 2x + 6)$

48. $32a^5 - 2a^3 + 6a$
$2a(16a^4 - a^2 + 3)$

49. $5u^2 + 5u^2 + 5u$
$5u(2u + 1)$

50. $11y^3 - 22y^2 + 11y^2$
$11y^2(y - 1)$

51. $x(x - 3) + 5(x - 3)$
$(x - 3)(x + 5)$

52. $x(x + 6) + 3(x + 6)$
$(x + 6)(x + 3)$

53. $t(s + 10) - 8(s + 10)$
$(s + 10)(t - 8)$

54. $y(q - 5) - 10(q - 5)$
$(q - 5)(y - 10)$

55. $a^2(b + 2) - b(b + 2)$
$(b + 2)(a^2 - b)$

56. $x^3(y + 4) + y(y + 4)$
$(y + 4)(x^3 + y)$

57. $z^3(z + 5) + z^2(z + 5)$ $z^2(z + 5)(z + 1)$

58. $x^3(x - 2) + x(x - 2)$ $x(x - 2)(x^2 + 1)$

59. $(a + b)(a - b) + a(a + b)$ $(a + b)(2a - b)$

60. $(x + y)(x - y) - x(x - y)$ $y(x - y)$

In Exercises 61–68, factor a negative real number from the polynomial and then write the polynomial factor in standard form. See Example 8.

61. $5 - 10x$ $-5(2x - 1)$

62. $3 - 6x$ $-3(2x - 1)$

63. $3000 - 3x$
$-3(x - 1000)$

64. $9 - 2x^2$
$-(2x^2 - 9)$

65. $4 + 2x - x^2$
$-(x^2 - 2x - 4)$

66. $18 - 12x - 6x^2$
$-6(x^2 + 2x - 3)$

67. $4 + 12x - 2x^2$
$-2(x^2 - 6x - 2)$

68. $8 - 4x - 12x^2$
$-4(3x^2 + x - 2)$

In Exercises 69–100, factor the polynomial by grouping. See Examples 9–11.

69. $x^2 + 10x + x + 10$ $(x + 10)(x + 1)$

70. $x^2 - 5x + x - 5$ $(x - 5)(x + 1)$

71. $a^2 - 4a + a - 4$ $(a - 4)(a + 1)$

72. $x^2 + 25x + x + 25$ $(x + 25)(x + 1)$

73. $x^2 + 3x + 4x + 12$ $(x + 3)(x + 4)$

74. $x^2 - x + 3x - 3$ $(x - 1)(x + 3)$

75. $x^2 + 2x + 5x + 10$ $(x + 2)(x + 5)$

76. $x^2 - 6x + 5x - 30$ $(x - 6)(x + 5)$

77. $x^2 + 3x - 5x - 15$ $(x + 3)(x - 5)$

78. $x^2 + 4x + x + 4$ $(x + 4)(x + 1)$

79. $4x^2 - 14x + 14x - 49$ $(2x - 7)(2x + 7)$

80. $4x^2 - 6x + 6x - 9$ $(2x - 3)(2x + 3)$

81. $6x^2 + 3x - 2x - 1$ $(2x + 1)(3x - 1)$

82. $5x^2 + 20x - x - 4$ $(x + 4)(5x - 1)$

83. $8x^2 + 32x + x + 4$ $(x + 4)(8x + 1)$

84. $8x^2 - 4x - 2x + 1$ $(2x - 1)(4x - 1)$

85. $3x^2 - 2x + 3x - 2$ $(3x - 2)(x + 1)$

86. $12x^2 + 42x - 10x - 35$ $(2x + 7)(6x - 5)$

87. $2x^2 - 4x - 3x + 6$ $(x - 2)(2x - 3)$

88. $35x^2 - 40x + 21x - 24$ $(7x - 8)(5x + 3)$

89. $ky^2 - 4ky + 2y - 8$ $(y - 4)(ky + 2)$

90. $ay^2 + 3ay + 3y + 9$ $(y + 3)(ay + 3)$

91. $t^3 - 3t^2 + 2t - 6$ $(t - 3)(t^2 + 2)$

92. $3s^3 + 6s^2 + 2s + 4$ $(s + 2)(3s^2 + 2)$

93. $x^3 + 2x^2 + x + 2$ $(x + 2)(x^2 + 1)$

94. $x^3 - 5x^2 + x - 5$ $(x - 5)(x^2 + 1)$

95. $6z^3 + 3z^2 - 2z - 1$ $(2z + 1)(3z^2 - 1)$

96. $4u^3 - 2u^2 - 6u + 3$ $(2u - 1)(2u^2 - 3)$

97. $x^3 - 3x - x^2 + 3$ $(x - 1)(x^2 - 3)$

98. $x^3 + 7x - 3x^2 - 21$ $(x^2 + 7)(x - 3)$

99. $4x^2 - x^3 - 8 + 2x$ $(4 - x)(x^2 - 2)$

100. $5x^2 + 10x^3 + 4 + 8x$ $(2x + 1)(5x^2 + 4)$

In Exercises 101–106, fill in the missing factor.

101. $\frac{1}{4}x + \frac{3}{4} = \frac{1}{4}(\quad x + 3 \quad)$

102. $\frac{5}{6}x - \frac{1}{6} = \frac{1}{6}(\quad 5x - 1 \quad)$

103. $2y - \frac{1}{5} = \frac{1}{5}(\quad 10y - 1 \quad)$

104. $3z + \frac{3}{4} = \frac{1}{4}(\quad 12z + 3 \quad)$

105. $\frac{7}{8}x + \frac{5}{16}y = \frac{1}{16}(\quad 14x + 5y \quad)$

106. $\frac{5}{12}u - \frac{5}{8}v = \frac{1}{24}(\quad 10u - 15v \quad)$

In Exercises 107–110, use a graphing calculator to graph both equations in the same viewing window. What can you conclude? See Additional Answers.

107. $y_1 = 9 - 3x$
$y_2 = -3(x - 3)$
$y_1 = y_2$

108. $y_1 = x^2 - 4x$
$y_2 = x(x - 4)$
$y_1 = y_2$

109. $y_1 = 6x - x^2$
$y_2 = x(6 - x)$ $y_1 = y_2$

110. $y_1 = x(x + 2) - 3(x + 2)$
$y_2 = (x + 2)(x - 3)$ $y_1 = y_2$

Solving Problems

▲ *Geometry*   In Exercises 111 and 112, factor the polynomial to find an expression for the length of the rectangle. See Example 12.

111. Area $= 2x^2 + 2x$ $x + 1$

2x

112. Area $= x^2 + 2x + 10x + 20$ $x + 10$

$x + 2$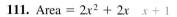

▲ *Geometry*   In Exercises 113 and 114, write an expression for the area of the shaded region and factor the expression if possible.

113.

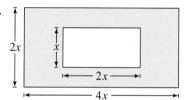

$6x^2$

2x   x

2x

4x

114.

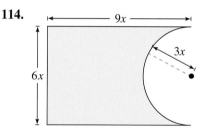

9x

$9x^2\left(6 - \dfrac{\pi}{2}\right)$

3x

6x

115. ▲ *Geometry*   The surface area of a right circular cylinder is given by

$$2\pi r^2 + 2\pi rh$$

where r is the radius of the base of the cylinder and h is the height of the cylinder. Factor the expression for the surface area. $2\pi r(r + h)$

116. *Simple Interest*   The amount after t years when a principal of P dollars is invested at $r\%$ simple interest is given by

$$P + Prt.$$

Factor the expression for simple interest. $P(1 + rt)$

117. *Chemical Reaction*   The rate of change in a chemical reaction is

$$kQx - kx^2$$

where Q is the original amount, x is the new amount, and k is a constant of proportionality. Factor the expression. $kx(Q - x)$

118. *Unit Price*   The revenue R from selling x units of a product at a price of p dollars per unit is given by $R = xp$. For a pool table the revenue is

$$R = 900x - 0.1x^2.$$

Factor the revenue model and determine an expression that represents the price p in terms of x. $R = x(900 - 0.1x); p = 900 - 0.1x$

Explaining Concepts

119. Give an example of a polynomial that is written in factored form. $x^2 + x - 6 = (x - 2)(x + 3)$

120. Give an example of a trinomial whose greatest common monomial factor is $3x$.
$3x^3 + 3x^2 + 3x = 3x(x^2 + x + 1)$

121. *Writing* ✐   How do you check your result when factoring a polynomial?   Multiply the factors.

122. *Writing* ✐   In your own words, describe a method for finding the greatest common factor of a polynomial.   Determine the prime factorization of each term. The greatest common factor contains each common prime factor, repeated the minimum number of times it occurs in any one of the factorizations.

✂ 123. *Writing* ✐   Explain how the word *factor* can be used as a noun and as a verb.   Noun: Any one of the expressions that, when multiplied together, yield the product. Verb: To find the expressions that, when multiplied together, yield the given product.

✂ 124. Give several examples of the use of the Distributive Property in factoring.
$2x + 4 = 2(x + 2)$
$x(x^2 + 1) - 3(x^2 + 1) = (x^2 + 1)(x - 3)$

✂ 125. Give an example of a polynomial with four terms that can be factored by grouping.
$x^3 - 3x^2 - 5x + 15 = (x^3 - 3x^2) + (-5x + 15)$
$= x^2(x - 3) - 5(x - 3)$
$= (x - 3)(x^2 - 5)$

6.2 Factoring Trinomials

What You Should Learn

1. Factor trinomials of the form $x^2 + bx + c$.
2. Factor trinomials in two variables.
3. Factor trinomials completely.

Why You Should Learn It

The techniques for factoring trinomials will help you in solving quadratic equations in Section 6.5.

1. **Factor trinomials of the form $x^2 + bx + c$.**

Factoring Trinomials of the Form $x^2 + bx + c$

From Section 5.3, you know that the product of two binomials is often a trinomial. Here are some examples.

Factored Form   F   O   I   L   *Trinomial Form*

$$(x - 1)(x + 5) = x^2 + 5x - x - 5 = x^2 + 4x - 5$$
$$(x - 3)(x - 3) = x^2 - 3x - 3x + 9 = x^2 - 6x + 9$$
$$(x + 5)(x + 1) = x^2 + x + 5x + 5 = x^2 + 6x + 5$$
$$(x - 2)(x - 4) = x^2 - 4x - 2x + 8 = x^2 - 6x + 8$$

Try covering the factored forms in the left-hand column above. Can you determine the factored forms from the trinomial forms? In this section, you will learn how to factor trinomials of the form $x^2 + bx + c$. To begin, consider the following factorization.

$$(x + m)(x + n) = x^2 + nx + mx + mn$$

$$= x^2 + \underbrace{(n + m)}_{\substack{\text{Sum of} \\ \text{terms}}}x + \underbrace{mn}_{\substack{\text{Product} \\ \text{of terms}}}$$

$$= x^2 + \boxed{b}\,x + \boxed{c}$$

So, to *factor* a trinomial $x^2 + bx + c$ into a product of two binomials, you must find two numbers m and n whose product is c and whose sum is b.

There are many different techniques that can be used to factor trinomials. The most common technique is to use *guess, check, and revise* with mental math. For example, try factoring the trinomial

$$x^2 + 5x + 6.$$

You need to find two numbers whose product is 6 and whose sum is 5. Using mental math, you can determine that the numbers are 2 and 3.

The product of 2 and 3 is 6.

$$x^2 + 5x + 6 = (x + 2)(x + 3)$$

The sum of 2 and 3 is 5.

Example 1   Factoring a Trinomial

Factor the trinomial $x^2 + 5x - 6$.

Solution

You need to find two numbers whose product is -6 and whose sum is 5.

$$\overset{\text{The product of } -1 \text{ and } 6 \text{ is } -6.}{x^2 + 5x - 6 = (x - 1)(x + 6)}$$
$$\underset{\text{The sum of } -1 \text{ and } 6 \text{ is } 5.}{}$$

Example 2   Factoring a Trinomial

Factor the trinomial $x^2 - x - 6$.

Solution

You need to find two numbers whose product is -6 and whose sum is -1.

$$\overset{\text{The product of } -3 \text{ and } 2 \text{ is } -6.}{x^2 - x - 6 = (x - 3)(x + 2)}$$
$$\underset{\text{The sum of } -3 \text{ and } 2 \text{ is } -1.}{}$$

Example 3   Factoring a Trinomial

Factor the trinomial $x^2 - 5x + 6$.

Solution

You need to find two numbers whose product is 6 and whose sum is -5.

$$\overset{\text{The product of } -2 \text{ and } -3 \text{ is } 6.}{x^2 - 5x + 6 = (x - 2)(x - 3)}$$
$$\underset{\text{The sum of } -2 \text{ and } -3 \text{ is } -5.}{}$$

Example 4   Factoring a Trinomial

Factor the trinomial $14 + 5x - x^2$.

Solution

It is helpful first to factor out -1 and write the polynomial factor in standard form.

$$14 + 5x - x^2 = -1(x^2 - 5x - 14)$$

Now you need two numbers -7 and 2 whose product is -14 and whose sum is -5. So,

$$14 + 5x - x^2 = -(x^2 - 5x - 14) = -(x - 7)(x + 2).$$

Study Tip

Use a list to help you find the two numbers with the required product and sum. For Example 2:

Factors of -6	*Sum*
$1, -6$	-5
$-1, 6$	5
$2, -3$	-1
$-2, 3$	1

Because -1 is the required sum, the correct factorization is

$$x^2 - x - 6 = (x - 3)(x + 2).$$

If you have trouble factoring a trinomial, it helps to make a list of all the distinct pairs of factors and then check each sum. For instance, consider the trinomial

$$x^2 - 5x - 24 = (x + \boxed{})(x - \boxed{}).$$ Opposite signs

In this trinomial the constant term is negative, so you need to find two numbers with opposite signs whose product is -24 and whose sum is -5.

Factors of -24	Sum
$1, -24$	-23
$-1, 24$	23
$2, -12$	-10
$-2, 12$	10
$3, -8$	-5 Correct choice
$-3, 8$	5
$4, -6$	-2
$-4, 6$	2

So, $x^2 - 5x - 24 = (x + 3)(x - 8)$.

With experience, you will be able to narrow the list of possible factors *mentally* to only two or three possibilities whose sums can then be tested to determine the correct factorization. Here are some suggestions for narrowing the list.

Guidelines for Factoring $x^2 + bx + c$

To factor $x^2 + bx + c$, you need to find two numbers m and n whose product is c and whose sum is b.

$$x^2 + bx + c = (x + m)(x + n)$$

1. If c is positive, then m and n have like signs that match the sign of b.

2. If c is negative, then m and n have unlike signs.

3. If $|b|$ is small relative to $|c|$, first try those factors of c that are closest to each other in absolute value.

Study Tip

Notice that factors may be written in any order. For example,

$(x - 5)(x + 3) = (x + 3)(x - 5)$

and

$(x + 2)(x + 18) = (x + 18)(x + 2)$

because of the Commutative Property of Multiplication.

Example 5   Factoring a Trinomial

Factor the trinomial $x^2 - 2x - 15$.

Solution

You need to find two numbers whose product is -15 and whose sum is -2.

The product of -5 and 3 is -15.

$$x^2 - 2x - 15 = (x - 5)(x + 3)$$

The sum of -5 and 3 is -2.

Study Tip

Not all trinomials are factorable using integer factors. For instance, $x^2 - 2x - 6$ is not factorable using integer factors because there is no pair of factors of -6 whose sum is -2. Such nonfactorable trinomials are called **prime polynomials.**

② Factor trinomials in two variables.

Example 6 Factoring a Trinomial

Factor the trinomial $x^2 + 7x - 30$.

Solution

You need to find two numbers whose product is -30 and whose sum is 7.

The product of -3 and 10 is -30.

$$x^2 + 7x - 30 = (x - 3)(x + 10)$$

The sum of -3 and 10 is 7.

Factoring Trinomials in Two Variables

So far, the examples in this section have all involved trinomials of the form

$$x^2 + bx + c. \qquad \text{Trinomial in one variable}$$

The next three examples show how to factor trinomials of the form

$$x^2 + bxy + cy^2. \qquad \text{Trinomial in two variables}$$

Note that this trinomial has two variables, x and y. However, from the factorization

$$x^2 + bxy + cy^2 = (x + my)(x + ny) = x^2 + (m + n)xy + mny^2$$

you can see that you still need to find two factors of c whose sum is b.

Study Tip

With *any* factoring problem, remember that you can check your result by multiplying. For instance, in Example 7, you can check the result by multiplying $(x - 4y)$ by $(x + 3y)$ to see that you obtain $x^2 - xy - 12y^2$.

Example 7 Factoring a Trinomial in Two Variables

Factor the trinomial $x^2 - xy - 12y^2$.

Solution

You need to find two numbers whose product is -12 and whose sum is -1.

The product of -4 and 3 is -12.

$$x^2 - xy - 12y^2 = (x - 4y)(x + 3y)$$

The sum of -4 and 3 is -1.

Encourage students to play detective and put together the clues leading to the correct factors. These problems, like other puzzle-solving challenges, can be intriguing.

Example 8 Factoring a Trinomial in Two Variables

Factor the trinomial $y^2 - 6xy + 8x^2$.

Solution

You need to find two numbers whose product is 8 and whose sum is -6.

The product of -2 and -4 is 8.

$$y^2 - 6xy + 8x^2 = (y - 2x)(y - 4x)$$

The sum of -2 and -4 is -6.

Example 9   Factoring a Trinomial in Two Variables

Factor the trinomial $x^2 + 11xy + 10y^2$.

Solution

You need to find two numbers whose product is 10 and whose sum is 11.

The product of 1 and 10 is 10.

$$x^2 + 11xy + 10y^2 = (x + y)(x + 10y)$$

The sum of 1 and 10 is 11.

③ Factor trinomials completely.

Factoring Completely

Some trinomials have a common monomial factor. In such cases you should first factor out the common monomial factor. Then you can try to factor the resulting trinomial by the methods of this section. This "multiple-stage factoring process" is called **factoring completely.** The trinomial below is completely factored.

$$2x^2 - 4x - 6 = 2(x^2 - 2x - 3)$$ Factor out common monomial factor 2.

$$= 2(x - 3)(x + 1)$$ Factor trinomial.

Remind students to include the common monomial factor in the final result.

Example 10   Factoring Completely

Factor the trinomial $2x^2 - 12x + 10$ completely.

Solution

$$2x^2 - 12x + 10 = 2(x^2 - 6x + 5)$$ Factor out common monomial factor 2.

$$= 2(x - 5)(x - 1)$$ Factor trinomial.

Additional Examples:
Factor.

a. $x^2 + 20x + 36$

b. $x^2 + 3xy - 4y^2$

c. $2b^4 - 26b^3 + 84b^2$

Answers:

a. $(x + 2)(x + 18)$

b. $(x + 4y)(x - y)$

c. $2b^2(b - 6)(b - 7)$

Example 11   Factoring Completely

Factor the trinomial $3x^3 - 27x^2 + 54x$ completely.

Solution

$$3x^3 - 27x^2 + 54x = 3x(x^2 - 9x + 18)$$ Factor out common monomial factor 3x.

$$= 3x(x - 3)(x - 6)$$ Factor trinomial.

Example 12   Factoring Completely

Factor the trinomial $4y^4 + 32y^3 + 28y^2$ completely.

Solution

$$4y^4 + 32y^3 + 28y^2 = 4y^2(y^2 + 8y + 7)$$ Factor out common monomial factor $4y^2$.

$$= 4y^2(y + 1)(y + 7)$$ Factor trinomial.

6.2 Exercises

Review   Concepts, Skills, and Problem Solving

Keep mathematically in shape by doing these exercises *before* the problems of this section.

Properties and Definitions

1. *Writing*✏️   Explain what is meant by the intercepts of a graph and explain how to find the intercepts of a graph.   Intercepts are the points at which the graph intersects the *x*- or *y*-axis. To find the *x*-intercept(s), let *y* be zero and solve the equation for *x*. To find the *y*-intercept(s), let *x* be zero and solve the equation for *y*.

2. What is the leading coefficient of the polynomial $3x - 7x^2 + 4x^3 - 4$?   4

Rewriting Algebraic Expressions

In Exercises 3–8, find the product.

3. $y(y + 2)$ $y^2 + 2y$

4. $-a^2(a - 1)$ $-a^3 + a^2$

5. $(x - 2)(x - 5)$ $x^2 - 7x + 10$

6. $(v - 4)(v + 7)$ $v^2 + 3v - 28$

7. $(2x + 5)(2x - 5)$ $4x^2 - 25$

8. $x^2(x + 1) - 5(x^2 - 2)$ $x^3 - 4x^2 + 10$

Problem Solving

9. *Profit*   A consulting company showed a loss of \$2,500,000 during the first 6 months of 2002. The company ended the year with an overall profit of \$1,475,000. What was the profit during the second 6 months of the year?   \$3,975,000

10. *Cost*   Computer printer ink cartridges cost \$11.95 per cartridge. There are 12 cartridges per box, and five boxes were ordered. Determine the total cost of the order.   \$717

11. *Cost, Revenue, and Profit*   The revenue *R* from selling *x* units of a product is $R = 75x$. The cost *C* of producing *x* units is $C = 62.5x + 570$. In order to obtain a profit *P*, the revenue must be greater than the cost. For what values of *x* will this product produce a profit? $x \geq 46$

12. *Distance Traveled*   The minimum and maximum speeds on an interstate highway are 40 miles per hour and 65 miles per hour. You travel nonstop for $3\frac{1}{2}$ hours on this highway. Assuming that you stay within the speed limits, write an inequality for the distance you travel.   140 miles $\leq x \leq$ 227.5 miles

Explaining Concepts

In Exercises 1–8, fill in the missing factor. Then check your answer by multiplying the factors.

1. $x^2 + 4x + 3 = (x + 3)(\ x + 1\ )$

2. $x^2 + 5x + 6 = (x + 3)(\ x + 2\ )$

3. $a^2 + a - 6 = (a + 3)(\ a - 2\ )$

4. $c^2 + 2c - 3 = (c + 3)(\ c - 1\ )$

5. $y^2 - 2y - 15 = (y + 3)(\ y - 5\ )$

6. $y^2 - 4y - 21 = (y + 3)(\ y - 7\ )$

7. $z^2 - 5z + 6 = (z - 3)(\ z - 2\ )$

8. $z^2 - 4z + 3 = (z - 3)(\ z - 1\ )$

In Exercises 9–14, find all possible products of the form $(x + m)(x + n)$ where $m \cdot n$ is the specified product. (Assume that *m* and *n* are integers.)

9. $m \cdot n = 11$ $(x + 1)(x + 11); (x - 1)(x - 11)$

10. $m \cdot n = 5$ $(x + 1)(x + 5); (x - 1)(x - 5)$

11. $m \cdot n = 14$ $(x + 14)(x + 1); (x - 14)(x - 1);$ $(x + 7)(x + 2); (x - 7)(x - 2)$

12. $m \cdot n = 10$ $(x + 1)(x + 10); (x - 1)(x - 10);$ $(x + 2)(x + 5); (x - 2)(x - 5)$

13. $m \cdot n = 12$ $(x + 12)(x + 1); (x - 12)(x - 1);$ $(x + 6)(x + 2); (x - 6)(x - 2);\ (x + 4)(x + 3);$ $(x - 4)(x - 3)$

14. $m \cdot n = 18$ $(x + 18)(x + 1); (x - 18)(x - 1);$ $(x + 9)(x + 2); (x - 9)(x - 2); (x + 6)(x + 3);$ $(x - 6)(x - 3)$

In Exercises 15–44, factor the trinomial. (*Note:* Some of the trinomials may be prime.) See Examples 1–9.

15. $x^2 + 6x + 8$
   $(x + 2)(x + 4)$

16. $x^2 + 13x + 12$
   $(x + 1)(x + 12)$

17. $x^2 - 13x + 40$
   $(x - 5)(x - 8)$

18. $x^2 - 9x + 14$
   $(x - 2)(x - 7)$

19. $z^2 - 7z + 12$
$(z - 3)(z - 4)$

20. $x^2 + 10x + 24$
$(x + 4)(x + 6)$

21. $y^2 + 5y + 11$
Prime

22. $s^2 - 7s - 25$
Prime

23. $x^2 - x - 6$
$(x + 2)(x - 3)$

24. $x^2 + x - 6$
$(x - 2)(x + 3)$

25. $x^2 + 2x - 15$
$(x - 3)(x + 5)$

26. $b^2 - 2b - 15$
$(b - 5)(b + 3)$

27. $y^2 - 6y + 10$
Prime

28. $c^2 - 6c + 10$
Prime

29. $u^2 - 22u - 48$
$(u + 2)(u - 24)$

30. $x^2 - x - 36$
Prime

31. $x^2 + 19x + 60$
$(x + 15)(x + 4)$

32. $x^2 + 3x - 70$
$(x - 7)(x + 10)$

33. $x^2 - 17x + 72$
$(x - 8)(x - 9)$

34. $x^2 + 21x + 108$
$(x + 9)(x + 12)$

35. $x^2 - 8x - 240$
$(x + 12)(x - 20)$

36. $r^2 - 30r + 216$
$(r - 18)(r - 12)$

37. $x^2 + xy - 2y^2$
$(x + 2y)(x - y)$

38. $x^2 - 5xy + 6y^2$
$(x - 2y)(x - 3y)$

39. $x^2 + 8xy + 15y^2$
$(x + 5y)(x + 3y)$

40. $u^2 - 4uv - 5v^2$
$(u - 5v)(u + v)$

41. $x^2 - 7xz - 18z^2$
$(x - 9z)(x + 2z)$

42. $x^2 + 15xy + 50y^2$
$(x + 5y)(x + 10y)$

43. $a^2 + 2ab - 15b^2$
$(a + 5b)(a - 3b)$

44. $y^2 + 4yz - 60z^2$
$(y + 10z)(y - 6z)$

In Exercises 45–64, factor the trinomial completely. (*Note*: Some of the trinomials may be prime.) See Examples 10–12.

45. $3x^2 + 21x + 30$
$3(x + 5)(x + 2)$

46. $4x^2 - 32x + 60$
$4(x - 3)(x - 5)$

47. $4y^2 - 8y - 12$
$4(y - 3)(y + 1)$

48. $5x^2 - 20x - 25$
$5(x + 1)(x - 5)$

49. $3z^2 + 5z + 6$
Prime

50. $7x^2 + 5x + 10$
Prime

51. $9x^2 + 18x - 18$
$9(x^2 + 2x - 2)$

52. $6x^2 - 24x - 6$
$6(x^2 - 4x - 1)$

53. $x^3 - 13x^2 + 30x$
$x(x - 10)(x - 3)$

54. $x^3 + x^2 - 2x$
$x(x + 2)(x - 1)$

55. $x^4 - 5x^3 + 6x^2$
$x^2(x - 2)(x - 3)$

56. $x^4 + 3x^3 - 10x^2$
$x^2(x - 2)(x + 5)$

57. $-3y^2x - 9yx + 54x$ $-3x(y - 3)(y + 6)$

58. $-5x^2z + 15xz + 50z$ $-5z(x - 5)(x + 2)$

59. $x^3 + 5x^2y + 6xy^2$ $x(x + 2y)(x + 3y)$

60. $x^2y - 6xy^2 + y^3$ $y(x^2 - 6xy + y^2)$

61. $2x^3y + 4x^2y^2 - 6xy^3$ $2xy(x + 3y)(x - y)$

62. $2x^3y - 10x^2y^2 + 6xy^3$ $2xy(x^2 - 5xy + 3y^2)$

63. $x^4y^2 + 3x^3y^3 + 2x^2y^4$ $x^2y^2(x + 2y)(x + y)$

64. $x^4y^2 + x^3y^3 - 2x^2y^4$ $x^2y^2(x + 2y)(x - y)$

In Exercises 65–70, find all integers b such that the trinomial can be factored.

65. $x^2 + bx + 18$ $\pm 9, \pm 11, \pm 19$

66. $x^2 + bx + 10$ $\pm 7, \pm 11$

67. $x^2 + bx - 21$ $\pm 4, \pm 20$

68. $x^2 + bx - 7$ ± 6

69. $x^2 + bx + 36$ $\pm 12, \pm 13, \pm 15, \pm 20, \pm 37$

70. $x^2 + bx - 48$ $\pm 2, \pm 8, \pm 13, \pm 22, \pm 47$

In Exercises 71–76, find two integers c such that the trinomial can be factored. (There are many correct answers.)

71. $x^2 + 3x + c$ $2, -10$

72. $x^2 + 5x + c$ $4, -14$

73. $x^2 - 4x + c$ $3, 4$

74. $x^2 - 15x + c$ $14, -16$

75. $x^2 - 9x + c$ $8, -10$

76. $x^2 + 12x + c$ $11, -13$

Graphical Reasoning In Exercises 77–80, use a graphing calculator to graph the two equations in the same viewing window. What can you conclude?

See Additional Answers.

77. $y_1 = x^2 - x - 6$
$y_2 = (x + 2)(x - 3)$ $y_1 = y_2$

78. $y_1 = x^2 - 10x + 16$
$y_2 = (x - 2)(x - 8)$ $y_1 = y_2$

79. $y_1 = x^3 + x^2 - 20x$
$y_2 = x(x - 4)(x + 5)$ $y_1 = y_2$

80. $y_1 = 2x - x^2 - x^3$
$y_2 = x(1 - x)(2 + x)$ $y_1 = y_2$

Solving Problems

81. *Exploration*   An open box is to be made from a four-foot-by-six-foot sheet of metal by cutting equal squares from the corners and turning up the sides (see figure). The volume of the box can be modeled by $V = 4x^3 - 20x^2 + 24x$, $\ 0 < x < 2$.

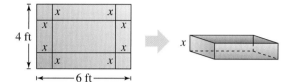

(a) Factor the trinomial that models the volume of the box. Use the factored form to explain how the model was found. $4x(x - 2)(x - 3)$; This is equivalent to $x(4 - 2x)(6 - 2x)$, where x, $4 - 2x$, and $6 - 2x$ are the dimensions of the box. The model was found by expanding this expression.

(b) 🖩 Use a graphing calculator to graph the trinomial over the specified interval. Use the graph to approximate the size of the squares to be cut from the corners so that the volume of the box is greatest.   See Additional Answers.   0.785 foot

82. *Exploration*   If the box in Exercise 81 is to be made from a six-foot-by-eight-foot sheet of metal, the volume of the box would be modeled by

$V = 4x^3 - 28x^2 + 48x$, $\ 0 < x < 3$.

(a) Factor the trinomial that models the volume of the box. Use the factored form to explain how the model was found.

(b) 🖩 Use a graphing calculator to graph the trinomial over the specified interval. Use the graph to approximate the size of the squares to be cut from the corners so that the volume of the box is greatest.   See Additional Answers.   1.131 feet

83. 📐 *Geometry*   The area of the rectangle shown in the figure is $x^2 + 30x + 200$. What is the area of the shaded region?   200 square units

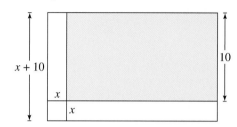

84. 📐 *Geometry*   The area of the rectangle shown in the figure is $x^2 + 17x + 70$. What is the area of the shaded region?   70 square units

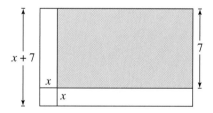

82. (a) $4x(x - 3)(x - 4)$;   This   is   equivalent   to $x(6 - 2x)(8 - 2x)$, where x, $6 - 2x$, and $8 - 2x$ are the dimensions of the box. The model was found by expanding this expression.

Explaining Concepts

85. State which of the following are factorizations of $2x^2 + 6x - 20$. For each correct factorization, state whether or not it is complete.   (a) and (d)

(a) $(2x - 4)(x + 5)$ (b) $(2x - 4)(2x + 10)$

(c) $(x - 2)(x + 5)$ (d) $2(x - 2)(x + 5)$

(a) Not completely factored; (d) Completely factored

86. *Writing*✏   In factoring $x^2 - 4x + 3$, why is it unnecessary to test $(x - 1)(x + 3)$ and $(x + 1)(x - 3)$?   Because the constant term is positive in the polynomial, the signs in the binomial factors must be the same.

87. *Writing*✏   In your own words, explain how to factor a trinomial of the form $x^2 + bx + c$. Give examples with your explanation.

88. *Writing*✏   What is meant by a prime trinomial?   The trinomial is not factorable using factors with integer coefficients.

89. *Writing*✏   Can you completely factor a trinomial into two different sets of prime factors? Explain.   No. The factorization into prime factors is unique.

90. *Writing*✏   In factoring the trinomial $x^2 + bx + c$, is the process easier if c is a prime number such as 5 or a composite number such as 120? Explain.   A prime number, because there are not as many possible factorizations to examine.

87. When attempting to factor $x^2 + bx + c$, find factors of c whose sum is b. $\ x^2 + 7x + 10 = (x + 2)(x + 5)$

6.3 More About Factoring Trinomials

Robert Harding Picture Library Ltd/Alamy

What You Should Learn

① Factor trinomials of the form $ax^2 + bx + c$.

② Factor trinomials completely.

③ Factor trinomials by grouping.

Why You Should Learn It

Trinomials can be used in many geometric applications. For example, in Exercise 112 on page 343, you will factor the expression for the volume of a swimming pool to determine an expression for the width of the swimming pool.

① **Factor trinomials of the form** $ax^2 + bx + c$.

Factoring Trinomials of the Form $ax^2 + bx + c$

In this section, you will learn how to factor a trinomial whose leading coefficient is *not* 1. To see how this works, consider the following.

$$ax^2 + bx + c = (\quad x + \quad)(\quad x + \quad)$$

Factors of a

Factors of c

The goal is to find a combination of factors of a and c such that the outer and inner products add up to the middle term bx.

Example 1 Factoring a Trinomial of the Form $ax^2 + bx + c$

Factor the trinomial $4x^2 - 4x - 3$.

Solution

First, observe that $4x^2 - 4x - 3$ has no common monomial factor. For this trinomial, $a = 4$ and $c = -3$. You need to find a combination of the factors of 4 and -3 such that the outer and inner products add up to $-4x$. The possible combinations are as follows.

Factors	$O + I$	
Inner product = $4x$ $(x + 1)(4x - 3)$ Outer product = $-3x$	$-3x + 4x = x$	x does not equal $-4x$.
$(x - 1)(4x + 3)$	$3x - 4x = -x$	$-x$ does not equal $-4x$.
$(x + 3)(4x - 1)$	$-x + 12x = 11x$	$11x$ does not equal $-4x$.
$(x - 3)(4x + 1)$	$x - 12x = -11x$	$-11x$ does not equal $-4x$.
$(2x + 1)(2x - 3)$	$-6x + 2x = -4x$	$-4x$ equals $-4x$. ✓
$(2x - 1)(2x + 3)$	$6x - 2x = 4x$	$4x$ does not equal $-4x$.

So, the correct factorization is $4x^2 - 4x - 3 = (2x + 1)(2x - 3)$.

Example 2   Factoring a Trinomial of the Form $ax^2 + bx + c$

Factor the trinomial $6x^2 + 5x - 4$.

Solution

First, observe that $6x^2 + 5x - 4$ has no common monomial factor. For this trinomial, $a = 6$ and $c = -4$. You need to find a combination of the factors of 6 and -4 such that the outer and inner products add up to $5x$.

Factors	$O + I$	
$(x + 1)(6x - 4)$	$-4x + 6x = 2x$	$2x$ does not equal $5x$.
$(x - 1)(6x + 4)$	$4x - 6x = -2x$	$-2x$ does not equal $5x$.
$(x + 4)(6x - 1)$	$-x + 24x = 23x$	$23x$ does not equal $5x$.
$(x - 4)(6x + 1)$	$x - 24x = -23x$	$-23x$ does not equal $5x$.
$(x + 2)(6x - 2)$	$-2x + 12x = 10x$	$10x$ does not equal $5x$.
$(x - 2)(6x + 2)$	$2x - 12x = -10x$	$-10x$ does not equal $5x$.
$(2x + 1)(3x - 4)$	$-8x + 3x = -5x$	$-5x$ does not equal $5x$.
$(2x - 1)(3x + 4)$	$8x - 3x = 5x$	$5x$ equals $5x$. ✓
$(2x + 4)(3x - 1)$	$-2x + 12x = 10x$	$10x$ does not equal $5x$.
$(2x - 4)(3x + 1)$	$2x - 12x = -10x$	$-10x$ does not equal $5x$.
$(2x + 2)(3x - 2)$	$-4x + 6x = 2x$	$2x$ does not equal $5x$.
$(2x - 2)(3x + 2)$	$4x - 6x = -2x$	$-2x$ does not equal $5x$.

So, the correct factorization is $6x^2 + 5x - 4 = (2x - 1)(3x + 4)$.

The following guidelines can help shorten the list of possible factorizations.

Guidelines for Factoring $ax^2 + bx + c (a > 0)$

1. If the trinomial has a common monomial factor, you should factor out the common factor before trying to find the binomial factors.

2. Because the resulting trinomial has no common monomial factors, you do not have to test any binomial factors that have a common monomial factor.

3. Switch the signs of the factors of c when the middle term $(O + I)$ is correct except in sign.

Using these guidelines, you can shorten the list in Example 2 to the following.

$(x + 4)(6x - 1) = 6x^2 + 23x - 4$ 23x does not equal 5x.

$(2x + 1)(3x - 4) = 6x^2 - 5x - 4$ Opposite sign

$(2x - 1)(3x + 4) = 6x^2 + 5x - 4$ ⟸   Correct factorization

Technology: Tip

As with other types of factoring, you can use a graphing calculator to check your results. For instance, graph

$y_1 = 2x^2 + x - 15$ and

$y_2 = (2x - 5)(x + 3)$

in the same viewing window, as shown below. Because both graphs are the same, you can conclude that

$2x^2 + x - 15$

$= (2x - 5)(x + 3).$

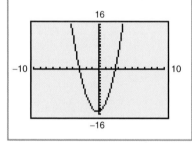

Example 3   Factoring a Trinomial of the Form $ax^2 + bx + c$

Factor the trinomial $2x^2 + x - 15$.

Solution

First, observe that $2x^2 + x - 15$ has no common monomial factor. For this trinomial, $a = 2$, which factors as $(1)(2)$, and $c = -15$, which factors as $(1)(-15), (-1)(15), (3)(-5),$ and $(-3)(5)$.

$$(2x + 1)(x - 15) = 2x^2 - 29x - 15$$

$$(2x + 15)(x - 1) = 2x^2 + 13x - 15$$

$$(2x + 3)(x - 5) = 2x^2 - 7x - 15$$

$$(2x + 5)(x - 3) = 2x^2 - x - 15 \qquad \text{Middle term has opposite sign.}$$

$$(2x - 5)(x + 3) = 2x^2 + x - 15 \qquad \Leftarrow \text{Correct factorization}$$

So, the correct factorization is

$$2x^2 + x - 15 = (2x - 5)(x + 3).$$

Notice in Example 3 that when the middle term has the incorrect sign, you need only to change the signs of the second terms of the two factors.

Factoring Completely

Remember that if a trinomial has a common monomial factor, the common monomial factor should be factored out first. The complete factorization will then show all monomial and binomial factors.

2 Factor trinomials completely.

Additional Examples
Factor completely.

a. $2x^2 + 11x - 21$

b. $24x^3 - 126x^2 + 30x$

Answers:

a. $(2x - 3)(x + 7)$

b. $6x(4x - 1)(x - 5)$

Example 4   Factoring Completely

Factor $4x^3 - 30x^2 + 14x$ completely.

Solution

Begin by factoring out the common monomial factor.

$$4x^3 - 30x^2 + 14x = 2x(2x^2 - 15x + 7)$$

Now, for the new trinomial $2x^2 - 15x + 7$, $a = 2$ and $c = 7$. The possible factorizations of this trinomial are as follows.

$$(2x - 7)(x - 1) = 2x^2 - 9x + 7$$

$$(2x - 1)(x - 7) = 2x^2 - 15x + 7 \qquad \Leftarrow \text{Correct factorization}$$

So, the complete factorization of the original trinomial is

$$4x^3 - 30x^2 + 14x = 2x(2x^2 - 15x + 7)$$

$$= 2x(2x - 1)(x - 7).$$

In factoring a trinomial with a negative leading coefficient, first factor -1 out of the trinomial, as demonstrated in Example 5.

Example 5   A Negative Leading Coefficient

Factor the trinomial $-5x^2 + 7x + 6$.

Solution

This trinomial has a negative leading coefficient, so you should begin by factoring -1 out of the trinomial.

$$-5x^2 + 7x + 6 = (-1)(5x^2 - 7x - 6)$$

Now, for the new trinomial $5x^2 - 7x - 6$, you have $a = 5$ and $c = -6$. After testing the possible factorizations, you can conclude that

$$(x - 2)(5x + 3) = 5x^2 - 7x - 6. \qquad \text{Correct factorization}$$

So, a correct factorization is

$$-5x^2 + 7x + 6 = (-1)(x - 2)(5x + 3)$$

$$= (-x + 2)(5x + 3). \qquad \text{Distributive Property}$$

Another correct factorization is $(x - 2)(-5x - 3)$.

---

3 Factor trinomials by grouping.

Factoring by Grouping

The examples in this and the preceding section have shown how to use the *guess, check, and revise* strategy to factor trinomials. An alternative technique to use is factoring by grouping. Recall from Section 6.1 that the polynomial

$$x^3 + 2x^2 + 3x + 6$$

was factored by first grouping terms and then applying the Distributive Property.

$$x^3 + 2x^2 + 3x + 6 = (x^3 + 2x^2) + (3x + 6) \qquad \text{Group terms.}$$

$$= x^2(x + 2) + 3(x + 2) \qquad \text{Factor out common monomial factor in each group.}$$

$$= (x + 2)(x^2 + 3) \qquad \text{Distributive Property}$$

By rewriting the middle term of the trinomial $2x^2 + x - 15$ as

$$2x^2 + x - 15 = 2x^2 + 6x - 5x - 15$$

you can group the first two terms and the last two terms and factor the trinomial as follows.

$$2x^2 + x - 15 = 2x^2 + 6x - 5x - 15 \qquad \text{Rewrite middle term.}$$

$$= (2x^2 + 6x) + (-5x - 15) \qquad \text{Group terms.}$$

$$= 2x(x + 3) - 5(x + 3) \qquad \text{Factor out common monomial factor in each group.}$$

$$= (x + 3)(2x - 5) \qquad \text{Distributive Property}$$

Guidelines for Factoring $ax^2 + bx + c$ by Grouping

1. If necessary, write the trinomial in standard form.

2. Choose factors of the product ac that add up to b.

3. Use these factors to rewrite the middle term as a sum or difference.

4. Group and remove any common monomial factors from the first two terms and the last two terms.

5. If possible, factor out the common binomial factor.

Example 6   Factoring a Trinomial by Grouping

Use factoring by grouping to factor the trinomial $2x^2 + 5x - 3$.

Solution

In the trinomial $2x^2 + 5x - 3$, $a = 2$ and $c = -3$, which implies that the product ac is -6. Now, because -6 factors as $(6)(-1)$, and $6 - 1 = 5 = b$, you can rewrite the middle term as $5x = 6x - x$. This produces the following.

$$2x^2 + 5x - 3 = 2x^2 + 6x - x - 3 \qquad \text{Rewrite middle term.}$$
$$= (2x^2 + 6x) + (-x - 3) \qquad \text{Group terms.}$$
$$= 2x(x + 3) - (x + 3) \qquad \text{Factor out common monomial factor in each group.}$$
$$= (x + 3)(2x - 1) \qquad \text{Factor out common binomial factor.}$$

So, the trinomial factors as

$$2x^2 + 5x - 3 = (x + 3)(2x - 1).$$

Example 7   Factoring a Trinomial by Grouping

Use factoring by grouping to factor the trinomial $6x^2 - 11x - 10$.

Solution

In the trinomial $6x^2 - 11x - 10$, $a = 6$ and $c = -10$, which implies that the product ac is -60. Now, because -60 factors as $(-15)(4)$ and $-15 + 4 = -11 = b$, you can rewrite the middle term as $-11x = -15x + 4x$. This produces the following.

$$6x^2 - 11x - 10 = 6x^2 - 15x + 4x - 10 \qquad \text{Rewrite middle term.}$$
$$= (6x^2 - 15x) + (4x - 10) \qquad \text{Group terms.}$$
$$= 3x(2x - 5) + 2(2x - 5) \qquad \text{Factor out common monomial factor in each group.}$$
$$= (2x - 5)(3x + 2) \qquad \text{Factor out common binomial factor.}$$

So, the trinomial factors as

$$6x^2 - 11x - 10 = (2x - 5)(3x + 2).$$

6.3 Exercises

Review *Concepts, Skills, and Problem Solving*

Keep mathematically in shape by doing these exercises *before* the problems of this section.

Properties and Definitions

1. Is 29 prime or composite?   Prime

2. Without dividing 255 by 3, how can you tell whether it is divisible by 3?   The sum of the digits is divisible by 3.

Simplifying Expressions

In Exercises 3–6, write the prime factorization.

3. 500 $2^2 \cdot 5^3$

4. 315 $3^2 \cdot 5 \cdot 7$

5. 792 $2^3 \cdot 3^2 \cdot 11$

6. 2275 $5^2 \cdot 7 \cdot 13$

In Exercises 7 and 8, multiply and simplify.

7. $(2x - 5)(x + 7)$
$2x^2 + 9x - 35$

8. $(3x - 2)^2$
$9x^2 - 12x + 4$

Graphing Equations

In Exercises 9 and 10, graph the equation and identify any intercepts.   See Additional Answers.

9. $y = (3 + x)(3 - x)$

10. $3x + 6y - 12 = 0$

11. *Stretching a Spring*   An equation for the distance y (in inches) a spring is stretched from its equilibrium point when a force of x pounds is applied is $y = 0.066x$.

    (a) Graph the model.   See Additional Answers.

    (b) Estimate y when a force of 100 pounds is applied.   6.6 inches

Developing Skills

In Exercises 1–18, fill in the missing factor.

1. $2x^2 + 7x - 4 = (2x - 1)(\; x + 4 \;)$

2. $3x^2 + x - 4 = (3x + 4)(\; x - 1 \;)$

3. $3t^2 + 4t - 15 = (3t - 5)(\; t + 3 \;)$

4. $5t^2 + t - 18 = (5t - 9)(\; t + 2 \;)$

5. $7x^2 + 15x + 2 = (7x + 1)(\; x + 2 \;)$

6. $3x^2 + 4x + 1 = (3x + 1)(\; x + 1 \;)$

7. $5x^2 + 18x + 9 = (x + 3)(\; 5x + 3 \;)$

8. $5x^2 + 19x + 12 = (x + 3)(\; 5x + 4 \;)$

9. $5a^2 + 12a - 9 = (a + 3)(\; 5a - 3 \;)$

10. $5c^2 + 11c - 12 = (c + 3)(\; 5c - 4 \;)$

11. $4z^2 - 13z + 3 = (z - 3)(\; 4z - 1 \;)$

12. $6z^2 - 23z + 15 = (z - 3)(\; 6z - 5 \;)$

13. $6x^2 - 23x + 7 = (3x - 1)(\; 2x - 7 \;)$

14. $6x^2 - 13x + 6 = (2x - 3)(\; 3x - 2 \;)$

15. $9a^2 - 6a - 8 = (3a + 2)(\; 3a - 4 \;)$

16. $4a^2 - 4a - 15 = (2a + 3)(\; 2a - 5 \;)$

17. $18t^2 + 3t - 10 = (6t + 5)(\; 3t - 2 \;)$

18. $12x^2 - 31x + 20 = (3x - 4)(\; 4x - 5 \;)$

In Exercises 19–22, find all possible products of the form $(5x + m)(x + n)$, where $m \cdot n$ is the specified product. (Assume that m and n are integers.)

19. $m \cdot n = 3$ $(5x + 3)(x + 1)$; $(5x - 3)(x - 1)$; $(5x + 1)(x + 3)$; $(5x - 1)(x - 3)$

20. $m \cdot n = 21$ $(5x + 21)(x + 1)$; $(5x - 21)(x - 1)$; $(5x + 1)(x + 21)$; $(5x - 1)(x - 21)$; $(5x + 7)(x + 3)$; $(5x - 7)(x - 3)$; $(5x + 3)(x + 7)$; $(5x - 3)(x - 7)$

21. $m \cdot n = 12$ $(5x + 12)(x + 1)$; $(5x - 12)(x - 1)$; $(5x + 6)(x + 2)$; $(5x - 6)(x - 2)$; $(5x + 4)(x + 3)$; $(5x - 4)(x - 3)$; $(5x + 1)(x + 12)$; $(5x - 1)(x - 12)$; $(5x + 2)(x + 6)$; $(5x - 2)(x - 6)$; $(5x + 3)(x + 4)$; $(5x - 3)(x - 4)$

22. $m \cdot n = 36$ $(5x + 36)(x + 1)$; $(5x - 36)(x - 1)$; $(5x + 1)(x + 36)$; $(5x - 1)(x - 36)$; $(5x + 18)(x + 2)$; $(5x - 18)(x - 2)$; $(5x + 2)(x + 18)$; $(5x - 2)(x - 18)$; $(5x + 12)(x + 3)$; $(5x - 12)(x - 3)$; $(5x + 3)(x + 12)$; $(5x - 3)(x - 12)$; $(5x + 9)(x + 4)$; $(5x - 9)(x - 4)$; $(5x + 4)(x + 9)$; $(5x - 4)(x - 9)$; $(5x + 6)(x + 6)$; $(5x - 6)(x - 6)$

In Exercises 23–50, factor the trinomial. (*Note:* Some of the trinomials may be prime.) See Examples 1–3.

23. $2x^2 + 5x + 3$
$(2x + 3)(x + 1)$

24. $3x^2 + 7x + 2$
$(3x + 1)(x + 2)$

25. $4y^2 + 5y + 1$
$(4y + 1)(y + 1)$

26. $3x^2 + 5x - 2$
$(3x - 1)(x + 2)$

27. $2y^2 - 3y + 1$
$(2y - 1)(y - 1)$

28. $3a^2 - 5a + 2$
$(3a - 2)(a - 1)$

29. $2x^2 - x - 3$
$(2x - 3)(x + 1)$

30. $3z^2 - z - 2$
$(3z + 2)(z - 1)$

31. $5x^2 - 2x + 1$ Prime

32. $4z^2 - 8z + 1$ Prime

33. $2x^2 + x + 3$ Prime

34. $6x^2 - 10x + 5$ Prime

35. $5s^2 - 10s + 6$ Prime

36. $6v^2 + v - 2$ $(3v + 2)(2v - 1)$

37. $4x^2 + 13x - 12$ $(x + 4)(4x - 3)$

38. $6y^2 - 7y - 20$ $(2y - 5)(3y + 4)$

39. $9x^2 - 18x + 8$ $(3x - 2)(3x - 4)$

40. $4a^2 - 16a + 15$ $(2a - 3)(2a - 5)$

41. $18u^2 - 9u - 2$ $(3u - 2)(6u + 1)$

42. $24s^2 + 37s - 5$ $(8s - 1)(3s + 5)$

43. $15a^2 + 14a - 8$ $(5a - 2)(3a + 4)$

44. $12x^2 - 8x - 15$ $(2x - 3)(6x + 5)$

45. $10t^2 - 3t - 18$ $(5t + 6)(2t - 3)$

46. $10t^2 + 43t - 9$ $(5t - 1)(2t + 9)$

47. $15m^2 + 16m - 15$ $(5m - 3)(3m + 5)$

48. $21b^2 - 40b - 21$ $(7b + 3)(3b - 7)$

49. $16z^2 - 34z + 15$ $(8z - 5)(2z - 3)$

50. $12x^2 - 41x + 24$ $(3x - 8)(4x - 3)$

In Exercises 51–60, factor the trinomial. (*Note:* The leading coefficient is negative.) See Example 5.

51. $-2x^2 + x + 3$ $-(2x - 3)(x + 1)$

52. $-5x^2 + x + 4$ $-(5x + 4)(x - 1)$

53. $4 - 4x - 3x^2$ $-(3x - 2)(x + 2)$

54. $-4x^2 + 17x + 15$ $-(4x + 3)(x - 5)$

55. $-6x^2 + 7x + 10$ $-(6x + 5)(x - 2)$

56. $2 + x - 6x^2$ $-(3x - 2)(2x + 1)$

57. $1 - 4x - 60x^2$ $-(10x - 1)(6x + 1)$

58. $2 + 5x - 12x^2$ $-(3x - 2)(4x + 1)$

59. $16 - 8x - 15x^2$ $-(5x - 4)(3x + 4)$

60. $20 + 17x - 10x^2$ $-(2x - 5)(5x + 4)$

In Exercises 61–82, factor the polynomial completely. (*Note:* Some of the polynomials may be prime.) See Examples 4 and 5.

61. $6x^2 - 3x$
$3x(2x - 1)$

62. $3a^4 - 9a^3$
$3a^3(a - 3)$

63. $15y^2 + 18y$
$3y(5y + 6)$

64. $24y^3 - 16y$
$8y(3y^2 - 2)$

65. $u(u - 3) + 9(u - 3)$
$(u - 3)(u + 9)$

66. $x(x - 8) - 2(x - 8)$
$(x - 8)(x - 2)$

67. $2v^2 + 8v - 42$
$2(v + 7)(v - 3)$

68. $4z^2 - 12z - 40$
$4(z + 2)(z - 5)$

69. $-3x^2 - 3x - 60$
$-3(x^2 + x + 20)$

70. $5y^2 + 40y + 35$
$5(y + 1)(y + 7)$

71. $9z^2 - 24z + 15$
$3(z - 1)(3z - 5)$

72. $6x^2 + 8x - 8$
$2(3x - 2)(x + 2)$

73. $4x^2 + 4x + 2$
$2(2x^2 + 2x + 1)$

74. $6x^2 - 6x - 36$
$6(x - 3)(x + 2)$

75. $-15x^4 - 2x^3 + 8x^2$
$-x^2(5x + 4)(3x - 2)$

76. $15y^2 - 7y^3 - 2y^4$
$-y^2(2y - 3)(y + 5)$

77. $3x^3 + 4x^2 + 2x$
$x(3x^2 + 4x + 2)$

78. $5x^3 - 3x^2 - 4x$
$x(5x^2 - 3x - 4)$

79. $6x^3 + 24x^2 - 192x$
$6x(x - 4)(x + 8)$

80. $35x + 28x^2 - 7x^3$
$-7x(x - 5)(x + 1)$

81. $18u^4 + 18u^3 - 27u^2$
$9u^2(2u^2 + 2u - 3)$

82. $12x^5 - 16x^4 + 8x^3$
$4x^3(3x^2 - 4x + 2)$

In Exercises 83–88, find all integers b such that the trinomial can be factored.

83. $3x^2 + bx + 10$
$\pm 11, \pm 13, \pm 17, \pm 31$

84. $4x^2 + bx + 3$
$\pm 7, \pm 8, \pm 13$

85. $2x^2 + bx - 6$
$\pm 1, \pm 4, \pm 11$

86. $5x^2 + bx - 6$
$\pm 1, \pm 7, \pm 13, \pm 29$

87. $6x^2 + bx + 20$
$\pm 22, \pm 23, \pm 26, \pm 29, \pm 34, \pm 43, \pm 62, \pm 121$

88. $8x^2 + bx - 18$
$\pm 7, \pm 10, \pm 18, \pm 32, \pm 45, \pm 70, \pm 143$

In Exercises 89–94, find two integers c such that the trinomial can be factored. (There are many correct answers.)

89. $4x^2 + 3x + c$ $-1, -7$

90. $2x^2 + 5x + c$ $2, -3$

91. $3x^2 - 10x + c$
$-8, 3$

92. $8x^2 - 3x + c$
$-5, -26$

93. $6x^2 - 5x + c$ $-6, -1$

94. $4x^2 - 9x + c$ $-9, 2$

In Exercises 95–110, factor the trinomial by grouping. See Examples 6 and 7.

95. $3x^2 + 7x + 2$
$(3x + 1)(x + 2)$

96. $2x^2 + 5x + 2$
$(x + 2)(2x + 1)$

97. $2x^2 + x - 3$
$(2x + 3)(x - 1)$

98. $5x^2 - 14x - 3$
$(5x + 1)(x - 3)$

99. $6x^2 + 5x - 4$
$(3x + 4)(2x - 1)$

100. $12y^2 + 11y + 2$
$(4y + 1)(3y + 2)$

101. $15x^2 - 11x + 2$
$(5x - 2)(3x - 1)$

102. $12x^2 - 13x + 1$
$(x - 1)(12x - 1)$

103. $3a^2 + 11a + 10$
$(3a + 5)(a + 2)$

104. $3z^2 - 4z - 15$
$(3z + 5)(z - 3)$

105. $16x^2 + 2x - 3$
$(8x - 3)(2x + 1)$

106. $20c^2 + 19c - 1$
$(c + 1)(20c - 1)$

107. $12x^2 - 17x + 6$
$(3x - 2)(4x - 3)$

108. $10y^2 - 13y - 30$
$(2y - 5)(5y + 6)$

109. $6u^2 - 5u - 14$
$(u - 2)(6u + 7)$

110. $12x^2 + 28x + 15$
$(2x + 3)(6x + 5)$

Solving Problems

111. ▲ *Geometry*   The sandbox shown in the figure has a height of x and a width of $x + 2$. The volume of the sandbox is $2x^3 + 7x^2 + 6x$. Find the length l of the sandbox. $l = 2x + 3$

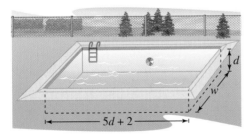

112. ▲ *Geometry*   The swimming pool shown in the figure has a depth of d and a length of $5d + 2$. The volume of the swimming pool is $15d^3 - 14d^2 - 8d$. Find the width w of the swimming pool. $w = 3d - 4$

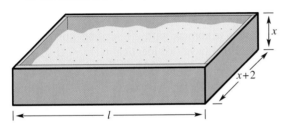

113. ▲ *Geometry*   The area of the rectangle shown in the figure is $2x^2 + 9x + 10$. What is the area of the shaded region? $2x + 10$

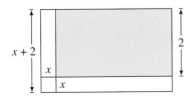

114. ▲ *Geometry*   The area of the rectangle shown in the figure is $3x^2 + 10x + 3$. What is the area of the shaded region? $6x + 3$

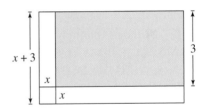

115. *Graphical Exploration*   Consider the equations

$$y_1 = 2x^3 + 3x^2 - 5x$$

and

$$y_2 = x(2x + 5)(x - 1).$$

(a) Factor the trinomial represented by y_1. What is the relationship between y_1 and y_2? $y_1 = y_2$

(b) ▦ Demonstrate your answer to part (a) graphically by using a graphing calculator to graph y_1 and y_2 in the same viewing window.
See Additional Answers.

(c) ▦ Identify the x- and y-intercepts of the graphs of y_1 and y_2. $\left(-\frac{5}{2}, 0\right), (0, 0), (1, 0)$

116. *Beam Deflection*   A cantilever beam of length l is fixed at the origin. A load weighing W pounds is attached to the end of the beam (see figure). The deflection y of the beam x units from the origin is given by

$$y = -\frac{1}{10}x^2 - \frac{1}{120}x^3, \ 0 \le x \le 3.$$

Factor the expression for the deflection. (Write the binomial factor with positive integer coefficients.)

$-\dfrac{x^2}{120}(12 + x)$

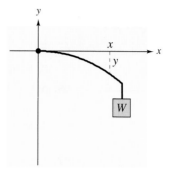

Figure for 116

Explaining Concepts

117. ⊘ Answer parts (a) and (b) of Motivating the Chapter on page 318.

118. *Writing* ✏ Explain the meaning of each letter of FOIL.   First, Outer, Inner, Last

119. *Writing* ✏ Without multiplying the factors, explain why $(2x + 3)(x + 5)$ is not a factorization of $2x^2 + 7x - 15$?   The product of the last terms of the binomials is 15, not -15.

120. *Error Analysis*   Describe the error.

$9x^2 - 9x - 54 = \cancel{(3x + 6)(3x - 9)}$

$\qquad\qquad\qquad = \cancel{3(x + 2)(x - 3)}$

$9x^2 - 9x - 54 = 9(x^2 - x - 6)$

$\qquad\qquad\qquad = 9(x - 3)(x + 2)$

121. *Writing* ✏   In factoring $ax^2 + bx + c$, how many possible factorizations must be tested if a and c are prime? Explain your reasoning.

Four. $(ax + 1)(x + c), (ax + c)(x + 1),$
$(ax - 1)(x - c), (ax - c)(x - 1)$

122. Give an example of a prime trinomial that is of the form $ax^2 + bx + c$. $x^2 + x + 1$

123. Give an example of a trinomial of the form $ax^3 + bx^2 + cx$ that has a common monomial factor of $2x$. $2x^3 + 2x^2 + 2x$

124. Can a trinomial with its leading coefficient not equal to 1 have two identical factors? If so, give an example.   Yes, $9x^2 + 12x + 4 = (3x + 2)^2$

125. *Writing* ✏   Many people think the technique of factoring a trinomial by grouping is more efficient than the *guess, check, and revise* strategy, especially when the coefficients a and c have many factors. Try factoring $6x^2 - 13x + 6$, $2x^2 + 5x - 12$, and $3x^2 + 11x - 4$ using both methods. Which method do you prefer? Explain the advantages and disadvantages of each method.

Factoring by grouping:
$$\begin{aligned}
6x^2 - 13x + 6 &= 6x^2 - (4x + 9x) + 6 \\
&= (6x^2 - 4x) - (9x - 6) \\
&= 2x(3x - 2) - 3(3x - 2) \\
&= (3x - 2)(2x - 3) \\
2x^2 + 5x - 12 &= 2x^2 + (8x - 3x) - 12 \\
&= (2x^2 + 8x) - (3x + 12) \\
&= 2x(x + 4) - 3(x + 4) \\
&= (x + 4)(2x - 3) \\
3x^2 + 11x - 4 &= 3x^2 + (12x - x) - 4 \\
&= (3x^2 + 12x) - (x + 4) \\
&= 3x(x + 4) - (x + 4) \\
&= (x + 4)(3x - 1)
\end{aligned}$$
Preferences, advantages, and disadvantages will vary.

Mid-Chapter Quiz

Take this quiz as you would take a quiz in class. After you are done, check your work against the answers in the back of the book.

In Exercises 1–4, fill in the missing factor.

1. $\frac{2}{3}x - 1 = \frac{1}{3}(\ \boxed{2x - 3}\ )$

2. $x^2y - xy^2 = xy(\ \boxed{x - y}\ )$

3. $y^2 + y - 42 = (y + 7)(\ \boxed{y - 6}\ )$

4. $3y^2 - y - 30 = (3y - 10)(\ \boxed{y + 3}\ )$

In Exercises 5–16, factor the polynomial completely.

5. $10x^2 + 70$ $10(x^2 + 7)$

6. $2a^3b - 4a^2b^2$ $2a^2b(a - 2b)$

7. $x(x + 2) - 3(x + 2)$ $(x + 2)(x - 3)$

8. $t^3 - 3t^2 + t - 3$ $(t - 3)(t^2 + 1)$

9. $y^2 + 11y + 30$ $(y + 6)(y + 5)$ 10. $u^2 + u - 30$ $(u + 6)(u - 5)$

11. $x^3 - x^2 - 30x$ $x(x - 6)(x + 5)$ 12. $2x^2y + 8xy - 64y$ $2y(x + 8)(x - 4)$

13. $2y^2 - 3y - 27$ $(2y - 9)(y + 3)$ 14. $6 - 13z - 5z^2$ $(3 + z)(2 - 5z)$

15. $6x^2 - x - 2$ $(3x - 2)(2x + 1)$ 16. $10s^4 - 14s^3 + 2s^2$ $2s^2(5s^2 - 7s + 1)$

17. Find all integers b such that the trinomial

    $x^2 + bx + 12$

    can be factored. Describe the method you used. $\pm 7, \pm 8, \pm 13$; These integers are the sums of the factors of 12.

18. Find two integers c such that the trinomial

    $x^2 - 10x + c$

    can be factored. Describe the method you used. (There are many correct answers.)   16, 21; The factors of c have a sum of -10.

19. Find all possible products of the form

    $(3x + m)(x + n)$

    such that $m \cdot n = 6$. Describe the method you used.

    19. m and n are factors of 6.
    $(3x + 1)(x + 6)$ $(3x - 1)(x - 6)$
    $(3x + 6)(x + 1)$ $(3x - 6)(x - 1)$
    $(3x + 2)(x + 3)$ $(3x - 2)(x - 3)$
    $(3x + 3)(x + 2)$ $(3x - 3)(x - 2)$

20. The area of the rectangle shown in the figure is $3x^2 + 38x + 80$. What is the area of the shaded region? $10(2x + 8)$

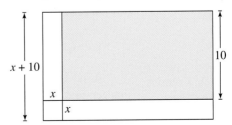

21. Use a graphing calculator to graph $y_1 = -2x^2 + 11x - 12$ and $y_2 = (3 - 2x)(x - 4)$ in the same viewing window. What can you conclude?
    See Additional Answers. $y_1 = y_2$

6.4 Factoring Polynomials with Special Forms

Carol Simowitz Photography

What You Should Learn

① Factor the difference of two squares.

② Recognize repeated factorization.

③ Identify and factor perfect square trinomials.

④ Factor the sum or difference of two cubes.

Why You Should Learn It

You can factor polynomials with special forms that model real-life situations. For instance, in Example 12 on page 351, an expression that models the safe working load for a piano lifted by a rope is factored.

① **Factor the difference of two squares.**

Difference of Two Squares

One of the easiest special polynomial forms to recognize and to factor is the form $a^2 - b^2$. It is called a **difference of two squares**, and it factors according to the following pattern.

Difference of Two Squares

Let a and b be real numbers, variables, or algebraic expressions.

$$a^2 - b^2 = (a + b)(a - b)$$

Difference   Opposite signs

Technology: Discovery

Use your calculator to verify the special polynomial form called the "difference of two squares." To do so, evaluate the equation when $a = 16$ and $b = 9$. Try more values, including negative values. What can you conclude?

See Technology Answers.

This pattern can be illustrated geometrically, as shown in Figure 6.2. The area of the shaded region on the left is represented by $a^2 - b^2$ (the area of the larger square minus the area of the smaller square). On the right, the *same* area is represented by a rectangle whose width is $a + b$ and whose length is $a - b$.

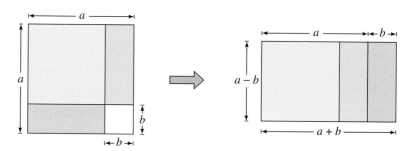

Figure 6.2

Study Tip

Note in the following that x-terms of higher power can be perfect squares.

$$25 - 64x^4 = (5)^2 - (8x^2)^2$$
$$= (5 + 8x^2)(5 - 8x^2)$$

To recognize perfect square terms, look for coefficients that are squares of integers and for variables raised to *even* powers. Here are some examples.

Original Polynomial		Difference of Squares		Factored Form
$x^2 - 1$	⇒	$(x)^2 - (1)^2$	⇒	$(x + 1)(x - 1)$
$4x^2 - 9$	⇒	$(2x)^2 - (3)^2$	⇒	$(2x + 3)(2x - 3)$

Additional Examples

Factor.

a. $y^2 - 121$

b. $9x^2 - 16$

Answers:

a. $(y + 11)(y - 11)$

b. $(3x + 4)(3x - 4)$

Students may find this problem challenging. Compare

$$k^2 - 49 = (k + 7)(k - 7)$$

with

$$(k + m)^2 - 49$$

$$= [(k + m) + 7][(k + m) - 7]$$

$$= (k + m + 7)(k + m - 7).$$

You could also compare

$$a^2 - 64 = (a + 8)(a - 8)$$

with

$$(a + 3)^2 - 64$$

$$= [(a + 3) + 8][(a + 3) - 8]$$

$$= (a + 3 + 8)(a + 3 - 8)$$

$$= (a + 11)(a - 5).$$

Example 1   Factoring the Difference of Two Squares

Factor each polynomial.

a. $x^2 - 36$ **b.** $x^2 - \frac{4}{25}$ **c.** $81x^2 - 49$

Solution

a. $x^2 - 36 = x^2 - 6^2$ Write as difference of two squares.

$\qquad\qquad = (x + 6)(x - 6)$ Factored form

b. $x^2 - \frac{4}{25} = x^2 - \left(\frac{2}{5}\right)^2$ Write as difference of two squares.

$\qquad\qquad = \left(x + \frac{2}{5}\right)\left(x - \frac{2}{5}\right)$ Factored form

c. $81x^2 - 49 = (9x)^2 - 7^2$ Write as difference of two squares.

$\qquad\qquad = (9x + 7)(9x - 7)$ Factored form

Check your results by using the FOIL Method.

The rule $a^2 - b^2 = (a + b)(a - b)$ applies to polynomials or expressions in which a and b are themselves expressions.

Example 2   Factoring the Difference of Two Squares

Factor the polynomial $(x + 1)^2 - 4$.

Solution

$$(x + 1)^2 - 4 = (x + 1)^2 - 2^2$$ Write as difference of two squares.

$$= [(x + 1) + 2][(x + 1) - 2]$$ Factored form

$$= (x + 3)(x - 1)$$ Simplify.

Check your result by using the FOIL Method.

Sometimes the difference of two squares can be hidden by the presence of a common monomial factor. Remember that with all factoring techniques, you should first remove any common monomial factors.

Example 3   Removing a Common Monomial Factor First

Factor the polynomial $20x^3 - 5x$.

Solution

$$20x^3 - 5x = 5x(4x^2 - 1)$$ Factor out common monomial factor $5x$.

$$= 5x[(2x)^2 - 1^2]$$ Write as difference of two squares.

$$= 5x(2x + 1)(2x - 1)$$ Factored form

---

② Recognize repeated factorization.

Repeated Factorization

To factor a polynomial completely, you should always check to see whether the factors obtained might themselves be factorable. That is, can any of the factors be factored? For instance, after factoring the polynomial $(x^4 - 1)$ once as the difference of two squares

$$x^4 - 1 = (x^2)^2 - 1^2 \qquad \text{Write as difference of two squares.}$$

$$= (x^2 + 1)(x^2 - 1) \qquad \text{Factored form}$$

You might explain that $x^2 + 1$ *cannot* be factored (using real numbers). Students frequently attempt to factor the sum of two squares.

you can see that the second factor is itself the difference of two squares. So, to factor the polynomial *completely*, you must continue the factoring process.

$$x^4 - 1 = (x^2 + 1)(x^2 - 1) \qquad \text{Factor as difference of two squares.}$$

$$= (x^2 + 1)(x + 1)(x - 1) \qquad \text{Factor completely.}$$

Another example of repeated factoring is shown in the next example.

Example 4   Factoring Completely

Factor the polynomial $x^4 - 16$ completely.

Solution

Recognizing $x^4 - 16$ as a difference of two squares, you can write

$$x^4 - 16 = (x^2)^2 - 4^2 \qquad \text{Write as difference of two squares.}$$

$$= (x^2 + 4)(x^2 - 4). \qquad \text{Factored form}$$

Note that the second factor $(x^2 - 4)$ is itself a difference of two squares and so

$$x^4 - 16 = (x^2 + 4)(x^2 - 4) \qquad \text{Factor as difference of two squares.}$$

$$= (x^2 + 4)(x + 2)(x - 2). \qquad \text{Factor completely.}$$

Study Tip

Note in Example 4 that no attempt was made to factor the *sum of two squares*. A second-degree polynomial that is the sum of two squares cannot be factored as the product of binomials (using integer coefficients). For instance, the second-degree polynomials

$$x^2 + 4$$

and

$$4x^2 + 9$$

cannot be factored using integer coefficients. In general, *the sum of two squares is not factorable.*

Example 5   Factoring Completely

Factor $48x^4 - 3$ completely.

Solution

Start by removing the common monomial factor.

$$48x^4 - 3 = 3(16x^4 - 1) \qquad \text{Remove common monomial factor 3.}$$

Recognizing $16x^4 - 1$ as the difference of two squares, you can write

$$48x^4 - 3 = 3(16x^4 - 1) \qquad \text{Factor out common monomial.}$$

$$= 3[(4x^2)^2 - 1^2] \qquad \text{Write as difference of two squares.}$$

$$= 3(4x^2 + 1)(4x^2 - 1) \qquad \text{Recognize } 4x^2 - 1 \text{ as a difference of two squares.}$$

$$= 3(4x^2 + 1)[(2x)^2 - 1^2] \qquad \text{Write as difference of two squares.}$$

$$= 3(4x^2 + 1)(2x + 1)(2x - 1). \qquad \text{Factor completely.}$$

③ Identify and factor perfect square trinomials.

Perfect Square Trinomials

A **perfect square trinomial** is the square of a binomial. For instance,

$$x^2 + 4x + 4 = (x + 2)(x + 2)$$
$$= (x + 2)^2$$

is the square of the binomial $(x + 2)$. Perfect square trinomials come in two forms: one in which the middle term is positive and the other in which the middle term is negative. In both cases, the first and last terms are positive perfect squares.

Perfect Square Trinomials

Let a and b be real numbers, variables, or algebraic expressions.

    1. $a^2 + 2ab + b^2 = (a + b)^2$ **2.** $a^2 - 2ab + b^2 = (a - b)^2$

             Same sign   Same sign

Example 6   Identifying Perfect Square Trinomials

Study Tip

To recognize a perfect square trinomial, remember that the first and last terms must be perfect squares and positive, and the middle term must be twice the product of a and b. (The middle term can be positive or negative.) Watch for squares of fractions.

$$4x^2 - \tfrac{4}{3}x + \tfrac{1}{9}$$
$$\downarrow \qquad \downarrow \qquad \downarrow$$
$$(2x)^2 \quad 2(2x)\left(\tfrac{1}{3}\right) \quad \left(\tfrac{1}{3}\right)^2$$

Which of the following are perfect square trinomials?

a. $m^2 - 4m + 4$
b. $4x^2 - 2x + 1$
c. $y^2 + 6y - 9$
d. $x^2 + x + \tfrac{1}{4}$

Solution

a. This polynomial *is* a perfect square trinomial. It factors as $(m - 2)^2$.

b. This polynomial *is not* a perfect square trinomial because the middle term is not twice the product of $2x$ and 1.

c. This polynomial *is not* a perfect square trinomial because the last term, -9, is not positive.

d. This polynomial *is* a perfect square trinomial. The first and last terms are perfect squares, x^2 and $\left(\tfrac{1}{2}\right)^2$, and it factors as $\left(x + \tfrac{1}{2}\right)^2$.

Additional Examples
Factor.

a. $16y^2 + 24y + 9$

b. $9x^2 - 30xy + 25y^2$

Answers:

a. $(4y + 3)^2$

b. $(3x - 5y)^2$

Example 7   Factoring a Perfect Square Trinomial

Factor the trinomial $y^2 - 6y + 9$.

Solution

$$y^2 - 6y + 9 = y^2 - 2(3y) + 3^2 \qquad \text{Recognize the pattern.}$$
$$= (y - 3)^2 \qquad\qquad\quad \text{Write in factored form.}$$

Example 8   Factoring a Perfect Square Trinomial

Factor the trinomial $16x^2 + 40x + 25$.

Solution

$$16x^2 + 40x + 25 = (4x)^2 + 2(4x)(5) + 5^2 \qquad \text{Recognize the pattern.}$$

$$= (4x + 5)^2 \qquad \text{Write in factored form.}$$

Example 9   Factoring a Perfect Square Trinomial

Factor the trinomial $9x^2 - 24xy + 16y^2$.

Solution

$$9x^2 - 24xy + 16y^2 = (3x)^2 - 2(3x)(4y) + (4y)^2 \qquad \text{Recognize the pattern.}$$

$$= (3x - 4y)^2 \qquad \text{Write in factored form.}$$

④ Factor the sum or difference of two cubes.

Sum or Difference of Two Cubes

The last type of special factoring presented in this section is the sum or difference of two *cubes*. The patterns for these two special forms are summarized below.

Study Tip

When using either of the factoring patterns at the right, pay special attention to the signs. Remembering the "like" and "unlike" patterns for the signs is helpful.

Sum or Difference of Two Cubes

Let a and b be real numbers, variables, or algebraic expressions.

Like signs

1. $a^3 + b^3 = (a + b)(a^2 - ab + b^2)$

Unlike signs

Like signs

2. $a^3 - b^3 = (a - b)(a^2 + ab + b^2)$

Unlike signs

Example 10   Factoring a Sum of Two Cubes

Factor the polynomial $y^3 + 27$.

Solution

$$y^3 + 27 = y^3 + 3^3 \qquad \text{Write as sum of two cubes.}$$

$$= (y + 3)[y^2 - (y)(3) + 3^2] \qquad \text{Factored form}$$

$$= (y + 3)(y^2 - 3y + 9) \qquad \text{Simplify.}$$

Additional Examples
Factor.

a. $x^3 - 125$

b. $8y^3 + 1$

Answers:

a. $(x - 5)(x^2 + 5x + 25)$

b. $(2y + 1)(4y^2 - 2y + 1)$

Example 11   Factoring Differences of Two Cubes

Factor each polynomial.

a. $64 - x^3$ **b.** $2x^3 - 16$

Solution

a. $64 - x^3 = 4^3 - x^3$ Write as difference of two cubes.

$= (4 - x)[4^2 + (4)(x) + x^2]$ Factored form

$= (4 - x)(16 + 4x + x^2)$ Simplify.

b. $2x^3 - 16 = 2(x^3 - 8)$ Factor out common monomial factor 2.

$= 2(x^3 - 2^3)$ Write as difference of two cubes.

$= 2(x - 2)[x^2 + (x)(2) + 2^2]$ Factored form

$= 2(x - 2)(x^2 + 2x + 4)$ Simplify.

Example 12   Safe Working Load

An object lifted with a rope should not weigh more than the safe working load for the rope. To lift a 600-pound piano, the safe working load for a natural fiber rope is given by $150c^2 - 600$, where c is the circumference of the rope (in inches). Factor this expression.

Solution

$150c^2 - 600 = 150(c^2 - 4)$ Factor out common monomial factor.

$= 150(c^2 - 2^2)$ Write as difference of two squares.

$= 150(c + 2)(c - 2)$ Factored form

The following guidelines are steps for applying the various procedures involved in factoring polynomials.

Guidelines for Factoring Polynomials

1. Factor out any common factors.

2. Factor according to one of the special polynomial forms: difference of two squares, sum or difference of two cubes, or perfect square trinomials.

3. Factor trinomials, $ax^2 + bx + c$, with $a = 1$ or $a \neq 1$.

4. Factor by grouping—for polynomials with four terms.

5. Check to see whether the factors themselves can be factored.

6. Check the results by multiplying the factors.

6.4 Exercises

Review   *Concepts, Skills, and Problem Solving*

Keep mathematically in shape by doing these exercises *before* the problems of this section.

Properties and Definitions

In Exercises 1 and 2, determine the quadrant or quadrants in which the point must be located.

1. $(-5, 2)$ Quadrant II

2. $(x, 3)$, x is a real number.   Quadrant I or II

3. Find the coordinates of the point on the x-axis and four units to the left of the y-axis. $(-4, 0)$

4. Find the coordinates of the point nine units to the right of the y-axis and six units below the x-axis. $(9, -6)$

Solving Equations

In Exercises 5–10, solve the equation and check your solution.

5. $7 + 5x = 7x - 1$ 4

6. $2 - 5(x - 1) = 2[x + 10(x - 1)]$ 1

7. $2(x + 1) = 0$ -1

8. $\frac{3}{4}(12x - 8) = 10$ $\frac{16}{9}$

9. $\frac{x}{5} + \frac{1}{5} = \frac{7}{10}$ $\frac{5}{2}$

10. $\frac{3x}{4} + \frac{1}{2} = 8$ 10

Problem Solving

11. *Membership Drive*   Because of a membership drive for a public television station, the current membership is 120% of what it was a year ago. The current membership is 8345. How many members did the station have last year?   6954 members

12. *Budget*   You budget 26% of your annual after-tax income for housing. Your after-tax income is $46,750. What amount can you spend on housing? $12,155

Developing Skills

In Exercises 1–22, factor the difference of two squares. See Examples 1 and 2.

1. $x^2 - 36$
   $(x + 6)(x - 6)$

2. $y^2 - 49$
   $(y + 7)(y - 7)$

3. $u^2 - 64$
   $(u + 8)(u - 8)$

4. $x^2 - 4$
   $(x + 2)(x - 2)$

5. $49 - x^2$
   $(7 + x)(7 - x)$

6. $81 - x^2$
   $(9 + x)(9 - x)$

7. $u^2 - \frac{1}{4}$
   $\left(u + \frac{1}{2}\right)\left(u - \frac{1}{2}\right)$

8. $t^2 - \frac{1}{16}$
   $\left(t + \frac{1}{4}\right)\left(t - \frac{1}{4}\right)$

9. $v^2 - \frac{4}{9}$
   $\left(v + \frac{2}{3}\right)\left(v - \frac{2}{3}\right)$

10. $u^2 - \frac{25}{81}$
    $\left(u + \frac{5}{9}\right)\left(u - \frac{5}{9}\right)$

11. $16y^2 - 9$
    $(4y + 3)(4y - 3)$

12. $9z^2 - 25$
    $(3z + 5)(3z - 5)$

13. $100 - 49x^2$
    $(10 + 7x)(10 - 7x)$

14. $16 - 81x^2$
    $(4 + 9x)(4 - 9x)$

15. $(x - 1)^2 - 4$
    $(x + 1)(x - 3)$

16. $(t + 2)^2 - 9$
    $(t + 5)(t - 1)$

17. $25 - (z + 5)^2$
    $-z(10 + z)$

18. $16 - (a + 2)^2$
    $(6 + a)(2 - a)$

19. $x^2 - y^2$
    $(x + y)(x - y)$

20. $x^2 - a^2$
    $(x + a)(x - a)$

21. $9y^2 - 25z^2$
    $(3y + 5z)(3y - 5z)$

22. $100x^2 - 81y^2$
    $(10x + 9y)(10x - 9y)$

In Exercises 23–36, factor the polynomial completely. See Examples 3–5.

23. $2x^2 - 72$
    $2(x + 6)(x - 6)$

24. $3x^2 - 27$
    $3(x + 3)(x - 3)$

25. $4x - 25x^3$
    $x(2 + 5x)(2 - 5x)$

26. $a^3 - 16a$
    $a(a + 4)(a - 4)$

27. $8y^2 - 50z^2$
    $2(2y + 5z)(2y - 5z)$

28. $20x^2 - 180y^2$
    $20(x - 3y)(x + 3y)$

29. $y^4 - 81$
    $(y^2 + 9)(y + 3)(y - 3)$

30. $z^4 - 16$
    $(z^2 + 4)(z + 2)(z - 2)$

31. $1 - x^4$
    $(1 + x^2)(1 + x)(1 - x)$

32. $256 - u^4$
    $(16 + u^2)(4 + u)(4 - u)$

33. $3x^4 - 48$
    $3(x + 2)(x - 2)(x^2 + 4)$

34. $18 - 2x^4$
    $2(3 + x^2)(3 - x^2)$

35. $81x^4 - 16y^4$
    $(9x^2 + 4y^2)(3x + 2y)(3x - 2y)$

36. $81x^4 - z^4$
    $(9x^2 + z^2)(9x^2 - z^2)$

In Exercises 37–54, factor the perfect square trinomial. See Examples 6–9.

37. $x^2 - 4x + 4$
$(x - 2)^2$

38. $x^2 + 10x + 25$
$(x + 5)^2$

39. $z^2 + 6z + 9$
$(z + 3)^2$

40. $a^2 - 12a + 36$
$(a - 6)^2$

41. $4t^2 + 4t + 1$
$(2t + 1)^2$

42. $9x^2 - 12x + 4$
$(3x - 2)^2$

43. $25y^2 - 10y + 1$
$(5y - 1)^2$

44. $16z^2 + 24z + 9$
$(4z + 3)^2$

45. $b^2 + b + \frac{1}{4}$
$\left(b + \frac{1}{2}\right)^2$

46. $x^2 + \frac{2}{5}x + \frac{1}{25}$
$\left(x + \frac{1}{5}\right)^2$

47. $4x^2 - x + \frac{1}{16}$
$\left(2x - \frac{1}{4}\right)^2$

48. $4t^2 - \frac{4}{3}t + \frac{1}{9}$
$\left(2t - \frac{1}{3}\right)^2$

49. $x^2 - 6xy + 9y^2$
$(x - 3y)^2$

50. $16x^2 - 8xy + y^2$
$(4x - y)^2$

51. $4y^2 + 20yz + 25z^2$
$(2y + 5z)^2$

52. $u^2 + 8uv + 16v^2$
$(u + 4v)^2$

53. $9a^2 - 12ab + 4b^2$
$(3a - 2b)^2$

54. $49m^2 - 28mn + 4n^2$
$(7m - 2n)^2$

Think About It   In Exercises 55–60, find two real numbers b such that the expression is a perfect square trinomial.

55. $x^2 + bx + 1$ ± 2

56. $x^2 + bx + 100$ ± 20

57. $x^2 + bx + \frac{16}{25}$ $\pm \frac{8}{5}$

58. $y^2 + by + \frac{1}{9}$ $\pm \frac{2}{3}$

59. $4x^2 + bx + 81$ ± 36

60. $4x^2 + bx + 9$ ± 12

In Exercises 61–64, find a real number c such that the expression is a perfect square trinomial.

61. $x^2 + 6x + c$ 9

62. $x^2 + 10x + c$ 25

63. $y^2 - 4y + c$ 4

64. $z^2 - 14z + c$ 49

In Exercises 65–76, factor the sum or difference of two cubes. See Examples 10 and 11.

65. $x^3 - 8$
$(x - 2)(x^2 + 2x + 4)$

66. $x^3 - 27$
$(x - 3)(x^2 + 3x + 9)$

67. $y^3 + 64$
$(y + 4)(y^2 - 4y + 16)$

68. $z^3 + 125$
$(z + 5)(z^2 - 5z + 25)$

69. $1 + 8t^3$
$(1 + 2t)(1 - 2t + 4t^2)$

70. $27s^3 + 1$
$(3s + 1)(9s^2 - 3s + 1)$

71. $27u^3 - 8$ $(3u - 2)(9u^2 + 6u + 4)$

72. $64v^3 - 125$ $(4v - 5)(16v^2 + 20v + 25)$

73. $x^3 - y^3$ $(x - y)(x^2 + xy + y^2)$

74. $a^3 - b^3$ $(a - b)(a^2 + ab + b^2)$

75. $27x^3 + 64y^3$ $(3x + 4y)(9x^2 - 12xy + 16y^2)$

76. $27y^3 + 125z^3$ $(3y + 5z)(9y^2 - 15yz + 25z^2)$

In Exercises 77–118, factor the polynomial completely. (*Note:* Some of the polynomials may be prime.)

77. $6x - 36$ $6(x - 6)$

78. $8t + 48$ $8(t + 6)$

79. $u^2 + 3u$ $u(u + 3)$

80. $x^3 - 4x^2$ $x^2(x - 4)$

81. $5y^2 - 25y$
$5y(y - 5)$

82. $12a^2 - 24a$
$12a(a - 2)$

83. $5y^2 - 125$
$5(y + 5)(y - 5)$

84. $6x^2 - 54y^2$
$6(x + 3y)(x - 3y)$

85. $y^4 - 25y^2$
$y^2(y + 5)(y - 5)$

86. $y^4 - 49y^2$
$y^2(y + 7)(y - 7)$

87. $x^2 - 4xy + 4y^2$
$(x - 2y)^2$

88. $9y^2 - 6yz + z^2$
$(3y - z)^2$

89. $x^2 - 2x + 1$
$(x - 1)^2$

90. $16 + 6x - x^2$
$(2 + x)(8 - x)$

91. $9x^2 + 10x + 1$
$(9x + 1)(x + 1)$

92. $4x^3 + 3x^2 + x$
$x(4x^2 + 3x + 1)$

93. $2x^3 - 2x^2y - 4xy^2$
$2x(x - 2y)(x + y)$

94. $2y^3 - 7y^2z - 15yz^2$
$y(2y + 3z)(y - 5z)$

95. $9t^2 - 16$
$(3t + 4)(3t - 4)$

96. $16t^2 - 144$
$16(t + 3)(t - 3)$

97. $36 - (z + 6)^2$
$-z(z + 12)$

98. $(t - 4)^2 - 9$
$(t - 7)(t - 1)$

99. $(t - 1)^2 - 121$
$(t + 10)(t - 12)$

100. $(x - 3)^2 - 100$
$(x + 7)(x - 13)$

101. $u^3 + 2u^2 + 3u$
$u(u^2 + 2u + 3)$

102. $u^3 + 2u^2 - 3u$
$u(u + 3)(u - 1)$

103. $x^2 + 81$ Prime

104. $x^2 + 16$ Prime

105. $2t^3 - 16$ $2(t - 2)(t^2 + 2t + 4)$

106. $24x^3 - 3$ $3(2x - 1)(4x^2 + 2x + 1)$

107. $2a^3 - 16b^3$ $2(a - 2b)(a^2 + 2ab + 4b^2)$

108. $54x^3 - 2y^3$ $2(3x - y)(9x^2 + 3xy + y^2)$

109. $x^4 - 81$ $(x^2 + 9)(x + 3)(x - 3)$

110. $2x^4 - 32$ $2(x^2 + 4)(x + 2)(x - 2)$

111. $x^4 - y^4$ $(x^2 + y^2)(x + y)(x - y)$

112. $81y^4 - z^4$ $(9y^2 + z^2)(3y + z)(3y - z)$

113. $x^3 - 4x^2 - x + 4$ $(x + 1)(x - 1)(x - 4)$

114. $y^3 + 3y^2 - 4y - 12$ $(y + 2)(y - 2)(y + 3)$

115. $x^4 + 3x^3 - 16x^2 - 48x$ $x(x + 3)(x + 4)(x - 4)$

116. $36x + 18x^2 - 4x^3 - 2x^4$ $2x(2 + x)(3 + x)(3 - x)$

117. $64 - y^6$ $(2 + y)(2 - y)(y^2 + 2y + 4)(y^2 - 2y + 4)$

118. $1 - y^8$ $(1 + y)(1 - y)(1 + y^2)(1 + y^4)$

Graphical Reasoning In Exercises 119–122, use a graphing calculator to graph the two equations in the same viewing window. What can you conclude?
See Additional Answers.

119. $y_1 = x^2 - 36$
$y_2 = (x + 6)(x - 6)$ $y_1 = y_2$

120. $y_1 = x^2 - 8x + 16$
$y_2 = (x - 4)^2$ $y_1 = y_2$

121. $y_1 = x^3 - 6x^2 + 9x$
$y_2 = x(x - 3)^2$ $y_1 = y_2$

122. $y_1 = x^3 + 27$
$y_2 = (x + 3)(x^2 - 3x + 9)$ $y_1 = y_2$

Mental Math In Exercises 123–126, evaluate the quantity mentally using the two samples as models.

$29^2 = (30 - 1)^2$
$= 30^2 - 2 \cdot 30 \cdot 1 + 1^2$
$= 900 - 60 + 1 = 841$

$48 \cdot 52 = (50 - 2)(50 + 2)$
$= 50^2 - 2^2 = 2496$

123. 21^2 441

124. 49^2 2401

125. $59 \cdot 61$ 3599

126. $28 \cdot 32$ 896

Solving Problems

127. ▲ *Geometry* An annulus is the region between two concentric circles. The area of the annulus shown in the figure is $\pi R^2 - \pi r^2$. Give the complete factorization of the expression for the area.
$\pi(R - r)(R + r)$

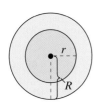

128. *Free-Falling Object* The height of an object that is dropped from the top of the USX Tower in Pittsburgh is given by the expression $-16t^2 + 841$, where t is the time in seconds. Factor this expression. $(29 + 4t)(29 - 4t)$

In Exercises 129 and 130, write the polynomial as the difference of two squares. Use the result to factor the polynomial completely.

129. $x^2 + 6x + 8 = (x^2 + 6x + 9) - 1$
$= (x + 3)^2 - 1^2$
$(x + 4)(x + 2)$

130. $x^2 + 8x + 12 = (x^2 + 8x + 16) - 4$
$= (x + 4)^2 - 2^2$
$(x + 6)(x + 2)$

131. *Writing* The figure below shows two cubes: a large cube whose volume is a^3 and a smaller cube whose volume is b^3. If the smaller cube is removed from the larger, the remaining solid has a volume of $a^3 - b^3$ and is composed of three rectangular boxes, labeled Box 1, Box 2, and Box 3. Find the volume of each box and describe how these results are related to the following special product pattern.
$$a^3 - b^3 = (a - b)(a^2 + ab + b^2)$$
$$= (a - b)a^2 + (a - b)ab + (a - b)b^2$$

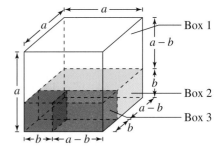

Box 1: $(a - b)a^2$; Box 2: $(a - b)ab$; Box 3: $(a - b)b^2$
The sum of the volumes of boxes 1, 2, and 3 equals the volume of the large cube minus the volume of the small cube, which is the difference of two cubes.

132. ▲ *Geometry*   From the eight vertices of a cube of dimension x, cubes of dimension y are removed (see figure).

(a) Write an expression for the volume of the solid that remains after the eight cubes at the vertices are removed. $x^3 - 8y^3$

(b) Factor the expression for the volume in part (a).
$(x - 2y)(x^2 + 2xy + 4y^2)$

(c) In the context of this problem, y must be less than what multiple of x? Explain your answer geometrically and from the result of part (b).
$y < \frac{1}{2}x$; If $y \geq \frac{1}{2}x$, then $2y \geq x$ and $x - 2y \leq 0$; If $y \geq \frac{1}{2}x$, then $V \leq 0$.

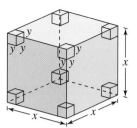

Figure for 132

Explaining Concepts

133. *Writing*   Explain how to identify and factor the difference of two squares.
$a^2 - b^2 = (a + b)(a - b)$

134. *Writing*   Explain how to identify and factor a perfect square trinomial.
$a^2 + 2ab + b^2 = (a + b)^2$ or $a^2 - 2ab + b^2 = (a - b)^2$

135. Is the expression $x(x + 2) - 2(x + 2)$ completely factored? If not, rewrite it in factored form.
No. $(x + 2)(x - 2)$

136. *Writing*   Is $x^2 + 4$ equal to $(x + 2)^2$? Explain.
No. $(x + 2)^2 = x^2 + 4x + 4$

137. *True or False?*   Because the sum of two squares cannot be factored, it follows that the sum of two cubes cannot be factored. Justify your answer.
False. $a^3 + b^3 = (a + b)(a^2 - ab + b^2)$

138. *Writing*   In your own words, state the guidelines for factoring polynomials.

1. Factor out any common factors.
2. Factor according to one of the special polynomial forms: difference of two squares, sum or difference of two cubes, or perfect square trinomials.
3. Factor trinomials, $ax^2 + bx + c$, with $a = 1$ or $a \neq 1$.
4. Factor by grouping—for polynomials with four terms.
5. Check to see whether the factors themselves can be factored.
6. Check the results by multiplying the factors.

6.5 Solving Quadratic Equations by Factoring

Carol Havens/Corbis

What You Should Learn

1. Use the Zero-Factor Property to solve equations.
2. Use factoring to solve quadratic equations.
3. Solve application problems by factoring.

Why You Should Learn It

Quadratic equations can be used to model and solve real-life problems. For instance, Exercise 94 on page 365 shows how a quadratic equation can be used to model the time it takes an object thrown from the Royal Gorge Bridge to reach the ground.

1. **Use the Zero-Factor Property to solve equations.**

The Zero-Factor Property

You have spent nearly two chapters developing skills for *rewriting* (simplifying and factoring) polynomials. You are now ready to use these skills together with the **Zero-Factor Property** to solve **second-degree polynomial equations** (*quadratic* equations).

Zero-Factor Property

Let a and b be real numbers, variables, or algebraic expressions. If a and b are factors such that

$$ab = 0$$

then $a = 0$ or $b = 0$. This property also applies to three or more factors.

Study Tip

The Zero-Factor Property is just another way of saying that the only way the product of two or more factors can equal zero is if one or more of the factors equals zero.

The Zero-Factor Property is the primary property for solving equations in algebra. For instance, to solve the equation

$$(x - 1)(x + 2) = 0 \qquad \text{Original equation}$$

you can use the Zero-Factor Property to conclude that either $(x - 1)$ or $(x + 2)$ equals zero. Setting the first factor equal to zero implies that $x = 1$ is a solution.

$$x - 1 = 0 \quad \Longrightarrow \quad x = 1 \qquad \text{First solution}$$

Similarly, setting the second factor equal to zero implies that $x = -2$ is a solution.

$$x + 2 = 0 \quad \Longrightarrow \quad x = -2 \qquad \text{Second solution}$$

So, the equation $(x - 1)(x + 2) = 0$ has exactly two solutions: $x = 1$ and $x = -2$. Check these solutions by substituting them into the original equation.

$$(x - 1)(x + 2) = 0 \qquad \text{Write original equation.}$$
$$(1 - 1)(1 + 2) \stackrel{?}{=} 0 \qquad \text{Substitute 1 for } x.$$
$$(0)(3) = 0 \qquad \text{First solution checks. } \checkmark$$
$$(-2 - 1)(-2 + 2) \stackrel{?}{=} 0 \qquad \text{Substitute } -2 \text{ for } x.$$
$$(-3)(0) = 0 \qquad \text{Second solution checks. } \checkmark$$

2 Use factoring to solve quadratic equations.

Solving Quadratic Equations by Factoring

Definition of Quadratic Equation

A **quadratic equation** is an equation that can be written in the general form

$$ax^2 + bx + c = 0 \qquad \text{Quadratic equation}$$

where a, b, and c are real numbers with $a \neq 0$.

Here are some examples of quadratic equations.

$$x^2 - 2x - 3 = 0, \quad 2x^2 + x - 1 = 0, \quad \text{and} \quad x^2 - 5x = 0$$

In the next four examples, note how you can combine your factoring skills with the Zero-Factor Property to solve quadratic equations.

Example 1   Using Factoring to Solve a Quadratic Equation

Solve $x^2 - x - 6 = 0$.

Solution

First, check to see that the right side of the equation is zero. Next, factor the left side of the equation. Finally, apply the Zero-Factor Property to find the solutions.

$$x^2 - x - 6 = 0 \qquad \text{Write original equation.}$$
$$(x + 2)(x - 3) = 0 \qquad \text{Factor left side of equation.}$$
$$x + 2 = 0 \implies x = -2 \qquad \text{Set 1st factor equal to 0 and solve for } x.$$
$$x - 3 = 0 \implies x = 3 \qquad \text{Set 2nd factor equal to 0 and solve for } x.$$

The equation has two solutions: $x = -2$ and $x = 3$.

Check

$$(-2)^2 - (-2) - 6 \overset{?}{=} 0 \qquad \text{Substitute } -2 \text{ for } x \text{ in original equation.}$$
$$4 + 2 - 6 \overset{?}{=} 0 \qquad \text{Simplify.}$$
$$0 = 0 \qquad \text{Solution checks.} \checkmark$$
$$(3)^2 - 3 - 6 \overset{?}{=} 0 \qquad \text{Substitute 3 for } x \text{ in original equation.}$$
$$9 - 3 - 6 \overset{?}{=} 0 \qquad \text{Simplify.}$$
$$0 = 0 \qquad \text{Solution checks.} \checkmark$$

Study Tip

In Section 3.1, you learned that the general strategy for solving a linear equation is to *isolate the variable*. Notice in Example 1 that the general strategy for solving a quadratic equation is to factor the equation into linear factors.

Factoring and the Zero-Factor Property allow you to solve a quadratic equation by converting it into two *linear* equations, which you already know how to solve. This is a common strategy of algebra—to break down a given problem into simpler parts, each of which can be solved by previously learned methods.

In order for the Zero-Factor Property to be used, a quadratic equation *must* be written in **general form.** That is, the quadratic must be on one side of the equation and zero must be the only term on the other side of the equation. To write $x^2 - 3x = 10$ in general form, subtract 10 from each side of the equation.

$$x^2 - 3x = 10 \qquad \text{Write original equation.}$$

$$x^2 - 3x - 10 = 10 - 10 \qquad \text{Subtract 10 from each side.}$$

$$x^2 - 3x - 10 = 0 \qquad \text{General form}$$

To solve this equation, factor the left side as $(x - 5)(x + 2)$, then form the linear equations $x - 5 = 0$ and $x + 2 = 0$. The solutions of these two linear equations are $x = 5$ and $x = -2$, respectively. Be sure you see that the Zero-Factor Property can be applied only to a product that is equal to *zero*. For instance, you cannot factor the left side as $x(x - 3) = 10$ and assume that $x = 10$ and $x - 3 = 10$ yield solutions. For instance, if you substitute $x = 10$ into the original equation you obtain the false statement $70 = 10$. Similarly, when $x = 13$ is substituted into the original equation you obtain another false statement, $130 = 10$. The general strategy for solving a quadratic equation by factoring is summarized in the following guidelines.

Guidelines for Solving Quadratic Equations

1. Write the quadratic equation in general form.
2. Factor the left side of the equation.
3. Set each factor with a variable equal to zero.
4. Solve each linear equation.
5. Check each solution in the original equation.

Example 2   Solving a Quadratic Equation by Factoring

Solve $2x^2 + 5x = 12$.

Solution

$$2x^2 + 5x = 12 \qquad \text{Write original equation.}$$
$$2x^2 + 5x - 12 = 0 \qquad \text{Write in general form.}$$
$$(2x - 3)(x + 4) = 0 \qquad \text{Factor left side of equation.}$$
$$2x - 3 = 0 \qquad \text{Set 1st factor equal to 0.}$$
$$x = \tfrac{3}{2} \qquad \text{Solve for } x.$$
$$x + 4 = 0 \qquad \text{Set 2nd factor equal to 0.}$$
$$x = -4 \qquad \text{Solve for } x.$$

You might tell students that Chapter 10 will introduce methods for solving quadratic equations that can't be solved by factoring.

The solutions are $x = \tfrac{3}{2}$ and $x = -4$. Check these solutions in the original equation.

Technology: Discovery

Write the equation in Example 3 in general form. Graph this equation on your graphing calculator.

$$y = x^2 - 8x + 16$$

What are the x-intercepts of the graph of the equation?

Write the equation in Example 4 in general form. Graph this equation on your graphing calculator.

$$y = x^2 + 9x + 14$$

What are the x-intercepts of the graph of the equation?

What can you conclude about the solutions to the equations and the x-intercepts of the graphs of the equations?

See Technology Answers.

In Examples 1 and 2, the original equations each involved a second-degree (quadratic) polynomial and each had *two different* solutions. You will sometimes encounter second-degree polynomial equations that have only one (repeated) solution. This occurs when the left side of the equation is a perfect square trinomial, as shown in Example 3.

Example 3   **A Quadratic Equation with a Repeated Solution**

Solve $x^2 - 2x + 16 = 6x$.

Solution

$x^2 - 2x + 16 = 6x$	Write original equation.
$x^2 - 8x + 16 = 0$	Write in general form.
$(x - 4)^2 = 0$	Factor.
$x - 4 = 0 \text{ or } x - 4 = 0$	Set factors equal to 0.
$x = 4$	Solve for x.

Note that even though the left side of this equation has two factors, the factors are the same. So, the only solution of the equation is $x = 4$. This solution is called a **repeated solution.**

Check

$x^2 - 2x + 16 = 6x$	Write original equation.
$(4)^2 - 2(4) + 16 \overset{?}{=} 6(4)$	Substitute 4 for x.
$16 - 8 + 16 \overset{?}{=} 24$	Simplify.
$24 = 24$	Solution checks. ✓

Example 4   **Solving a Quadratic Equation by Factoring**

Solve $(x + 3)(x + 6) = 4$.

Solution

A common error is setting $x + 3 = 4$ or $x + 6 = 4$. Emphasize the necessity of having a product equal to *zero* before applying the Zero-Factor Property.

Begin by multiplying the factors on the left side.

$(x + 3)(x + 6) = 4$	Write original equation.
$x^2 + 9x + 18 = 4$	Multiply factors.
$x^2 + 9x + 14 = 0$	Write in general form.
$(x + 2)(x + 7) = 0$	Factor.
$x + 2 = 0 \implies x = -2$	Set 1st factor equal to 0 and solve for x.
$x + 7 = 0 \implies x = -7$	Set 2nd factor equal to 0 and solve for x.

The equation has two solutions: $x = -2$ and $x = -7$. Check these in the original equation.

Applications

Example 5 Consecutive Integers

The product of two consecutive positive integers is 56. What are the integers?

Solution

Verbal Model: First integer $\cdot$ Second integer $= 56$

Labels: First integer $= n$
Second integer $= n + 1$

Equation:

$$n(n + 1) = 56 \qquad \text{Original equation}$$

$$n^2 + n - 56 = 0 \qquad \text{Write in general form.}$$

$$(n + 8)(n - 7) = 0 \qquad \text{Factor.}$$

$$n = -8 \text{ or } n = 7 \qquad \text{Solutions using Zero-Factor Property}$$

Because the problem states that the integers are positive, discard -8 as a solution and choose $n = 7$. So, the two integers are $n = 7$ and $n + 1 = 7 + 1 = 8$. Check these in the original statement of the problem.

Example 6 Free-Falling Object

A rock is dropped from the top of a 256-foot river gorge, as shown in Figure 6.3. The height h (in feet) of the rock is modeled by the position equation

$$h = -16t^2 + 256$$

where t is the time measured in seconds. How long will it take the rock to hit the bottom of the gorge?

Solution

From Figure 6.3, note that the bottom of the gorge corresponds to a height of 0 feet. So, substitute a height of 0 for h into the equation and solve for t.

$$0 = -16t^2 + 256 \qquad \text{Substitute 0 for } h.$$

$$16t^2 - 256 = 0 \qquad \text{Write in general form.}$$

$$16(t^2 - 16) = 0 \qquad \text{Factor out common factor.}$$

$$16(t + 4)(t - 4) = 0 \qquad \text{Factor.}$$

$$t + 4 = 0 \implies t = -4 \qquad \text{Set 1st factor equal to 0 and solve for } t.$$

$$t - 4 = 0 \implies t = 4 \qquad \text{Set 2nd factor equal to 0 and solve for } t.$$

256 ft

Height = 0 ft

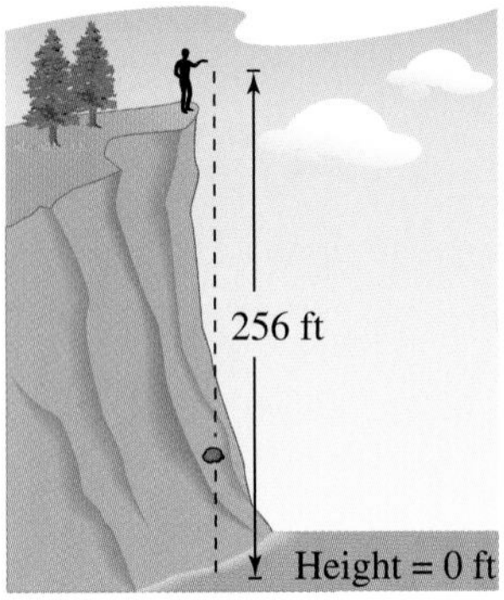

Figure 6.3

Because a time of -4 seconds does not make sense in this problem, choose the positive solution $t = 4$ and conclude that the rock hits the bottom of the gorge 4 seconds after it is dropped. Check this solution in the original statement of the problem.

In Example 6, the equation is a second-degree equation and, as such, cannot have more than two solutions, $t = 4$ and $t = -4$. The factor 16 in the equation

$$16(t + 4)(t - 4) = 0$$

does not yield another solution. Setting this factor equal to zero yields the *false* statement $16 = 0$.

Example 7   An Application from Geometry

A rectangular family room has an area of 160 square feet. The length of the room is 6 feet greater than its width. Find the dimensions of the room.

Solution

To begin, make a sketch of the room, as shown in Figure 6.4. Label the width of the room as x and the length of the room as $x + 6$ because the length is 6 feet greater than the width.

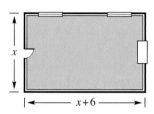

x

$x+6$

Figure 6.4

Verbal Model: Length · Width = Area

Labels:
Width = x (feet)
Length = $x + 6$ (feet)
Area = 160   (square feet)

Equation:
$$x(x + 6) = 160$$ Original equation

$$x^2 + 6x = 160$$ Distributive Property

$$x^2 + 6x - 160 = 0$$ Write in general form.

$$(x + 16)(x - 10) = 0$$ Factor.

$x + 16 = 0$ $x = -16$ Set 1st factor equal to 0 and solve for x.

$x - 10 = 0$ $x = 10$ Set 2nd factor equal to 0 and solve for x.

Because the dimensions must be positive, discard -16 as a solution and use the positive solution $x = 10$. So, the width of the room is 10 feet and the length of the room is

Length = $x + 6$

= $10 + 6$ Substitute 10 for x.

= 16 feet.   Simplify.

To check this solution, go back to the original statement of the problem. Note that a length of 16 feet is 6 feet greater than a width of 10 feet. Moreover, a rectangular room with dimensions 16 feet by 10 feet has an area of 160 square feet. So, the solution checks.

6.5 Exercises

Review   Concepts, Skills, and Problem Solving

Keep mathematically in shape by doing these exercises *before* the problems of this section.

Properties and Definitions

In Exercises 1–4, identify the property of real numbers illustrated by the statement.

1. $2ab - 2ab = 0$ Additive Inverse Property

2. $8t \cdot 1 = 8t$ Multiplicative Identity Property

3. $2x(1 - x) = 2x - 2x^2$ Distributive Property

4. $3x + (2x + 5) = (3x + 2x) + 5$
   Associative Property of Addition

Rewriting Expressions

In Exercises 5–10, perform the required operations and simplify.

5. (a) $2(-3) + 9$ 3   (b) $(-5)^2 + 3$ 28

6. (a) $4 - \dfrac{5}{2}$ $\dfrac{3}{2}$ (b) $\dfrac{|18 - 25|}{6}$ $\dfrac{7}{6}$

7. $\left(-\dfrac{7}{12}\right)\left(\dfrac{3}{28}\right)$ $-\dfrac{1}{16}$

8. $\dfrac{4}{3} \div \dfrac{5}{6}$ $\dfrac{8}{5}$

9. $2t(t - 3) + 4t + 1$ $2t^2 - 2t + 1$

10. $2u - 5(2u - 3)$ $-8u + 15$

Problem Solving

11. *Simple Interest*   Find the interest on a $1000 bond paying an annual percentage rate of 7.5% for 10 years. $750

12. *Speed*   A car leaves a town 1 hour after a fully loaded truck. The speed of the truck is approximately 50 miles per hour. The car overtakes the truck in 2.5 hours. Find the speed of the car.   70 miles per hour

Developing Skills

In Exercises 1–16, use the Zero-Factor Property to solve the equation.

1. $x(x - 5) = 0$ 0, 5

2. $z(z - 3) = 0$ 0, 3

3. $(y - 2)(y - 3) = 0$ 2, 3

4. $(s - 4)(s - 10) = 0$ 4, 10

5. $(a + 1)(a - 2) = 0$ $-1, 2$

6. $(t - 3)(t + 8) = 0$ $-8, 3$

7. $(2t - 5)(3t + 1) = 0$ $-\dfrac{1}{3}, \dfrac{5}{2}$

8. $(2 - 3x)(5 - 2x) = 0$ $\dfrac{2}{3}, \dfrac{5}{2}$

9. $\left(\dfrac{2}{3}x - 4\right)(x + 2) = 0$ $-2, 6$

10. $\left(\dfrac{3}{4}u - 2\right)\left(\dfrac{2}{5}u + \dfrac{1}{2}\right) = 0$ $-\dfrac{5}{4}, \dfrac{8}{3}$

11. $(0.2y - 12)(0.7y + 10) = 0$ $-\dfrac{100}{7}, 60$

12. $(1.5s + 12)(0.75s - 18) = 0$ $-8, 24$

13. $3x(x + 8)(4x - 5) = 0$ $-8, 0, \dfrac{5}{4}$

14. $x(x - 3)(x + 25) = 0$ $-25, 0, 3$

15. $(y - 1)(2y + 3)(y + 12) = 0$ $-12, -\dfrac{3}{2}, 1$

16. $x(5x + 3)(x - 8) = 0$ $-\dfrac{3}{5}, 0, 8$

In Exercises 17–72, solve the equation. See Examples 1–4.

17. $x^2 - 16 = 0$
    $-4, 4$

18. $x^2 - 144 = 0$
    $-12, 12$

19. $100 - v^2 = 0$
    $-10, 10$

20. $4 - x^2 = 0$
    $-2, 2$

21. $3y^2 - 27 = 0$
    $-3, 3$

22. $25z^2 - 100 = 0$
    $-2, 2$

23. $(t - 3)^2 - 25 = 0$
    $-2, 8$

24. $1 - (x + 1)^2 = 0$
    $-2, 0$

25. $81 - (u + 4)^2 = 0$
    $-13, 5$

26. $(s + 5)^2 - 49 = 0$
    $-12, 2$

27. $2x^2 + 4x = 0$
    $-2, 0$

28. $6x^2 + 3x = 0$
    $-\dfrac{1}{2}, 0$

29. $4x^2 - x = 0$
    $0, \dfrac{1}{4}$

30. $x - 3x^2 = 0$
    $0, \dfrac{1}{3}$

31. $2x^2 = 32x$
    $0, 16$

32. $5y^2 = 15y$
    $0, 3$

33. $8x^2 = 5x$
    $0, \dfrac{5}{8}$

34. $3x^2 = 7x$
    $0, \dfrac{7}{3}$

35. $y(y - 4) + 3(y - 4) = 0$ $-3, 4$

36. $u(u + 2) - 3(u + 2) = 0$ $-2, 3$

37. $x(x - 8) + 2(x - 8) = 0$ $-2, 8$

38. $x(x + 2) - 3(x + 2) = 0$ $-2, 3$

39. $m^2 - 2m + 1 = 0$
1

40. $a^2 + 6a + 9 = 0$
-3

41. $x^2 + 14x + 49 = 0$
-7

42. $x^2 - 10x + 25 = 0$
5

43. $4t^2 - 12t + 9 = 0$
$\frac{3}{2}$

44. $16x^2 + 56x + 49 = 0$
$-\frac{7}{4}$

45. $x^2 - 2x - 8 = 0$
$-2, 4$

46. $x^2 - 8x - 9 = 0$
$-1, 9$

47. $3 + 5x - 2x^2 = 0$
$-\frac{1}{2}, 3$

48. $33 + 5y - 2y^2 = 0$
$-3, \frac{11}{2}$

49. $6x^2 + 4x - 10 = 0$
$-\frac{5}{3}, 1$

50. $12x^2 + 7x + 1 = 0$
$-\frac{1}{3}, -\frac{1}{4}$

51. $3y^2 - 2 = -y$ $-1, \frac{2}{3}$

52. $2x^2 - 5 = 3x$ $-1, \frac{5}{2}$

53. $-2x - 15 = -x^2$
$-3, 5$

54. $-9x + 20 = -x^2$
$4, 5$

55. $x^2 - 35 = -2x$
$-7, 5$

56. $x^2 - 26 = -11x$
$-13, 2$

57. $1 + x^2 = 2x$ 1

58. $16 + x^2 = 8x$ 4

59. $x^2 - 2x = 18 + 5x$
$-2, 9$

60. $x^2 + 10x = 3x - 6$
$-6, -1$

61. $3x^2 - 2x = 9 - 8x$
$-3, 1$

62. $6x^2 - 8x = 10 - 4x$
$-1, \frac{5}{3}$

63. $z(z + 2) = 15$
$-5, 3$

64. $x(x - 1) = 6$
$-2, 3$

65. $x(x - 5) = 14$
$-2, 7$

66. $x(x + 4) = -4$
-2

67. $y(2y + 1) = 3$
$-\frac{3}{2}, 1$

68. $x(5x - 14) = 3$
$-\frac{1}{5}, 3$

69. $(x - 3)(x - 6) = 4$
$2, 7$

70. $(x + 1)(x - 2) = 4$
$-2, 3$

71. $(x + 1)(x + 4) = 4$
$-5, 0$

72. $(x - 9)(x + 2) = 12$
$-3, 10$

Graphical Estimation In Exercises 73–76, use the graph to estimate the x-intercepts. Set the quadratic equation equal to zero and solve. What do you notice?

73. $y = x^2 + 2x - 3$

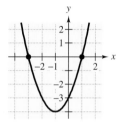

$(-3, 0), (1, 0)$; The number of solutions equals the number of x-intercepts.

74. $y = 2 + x - x^2$

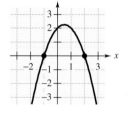

$(-1, 0), (2, 0)$; The number of solutions equals the number of x-intercepts.

75. $y = 12 + x - x^2$

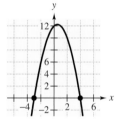

$(-3, 0), (4, 0)$; The number of solutions equals the number of x-intercepts.

76. $y = 2x^2 + x - 3$

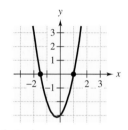

$\left(-\frac{3}{2}, 0\right), (1, 0)$; The number of solutions equals the number of x-intercepts.

Graphical Estimation In Exercises 77–80, use a graphing calculator to graph the equation. Use the graph to estimate the x-intercepts. Check your estimates by substituting them into the equation. See Additional Answers.

77. $y = x^2 - 4$
$(-2, 0), (2, 0)$

78. $y = x^2 - 4x$
$(0, 0), (4, 0)$

79. $y = 2x^2 - 5x - 12$
$\left(-\frac{3}{2}, 0\right), (4, 0)$

80. $y = 4x^2 + 3x - 10$
$(-2, 0), \left(\frac{5}{4}, 0\right)$

Solving Problems

81. *Number Problem* Find two consecutive positive integers whose product is 72. 8, 9

82. *Number Problem* Find two consecutive positive integers whose product is 240. 15, 16

83. *Number Problem* Find two consecutive positive even integers whose product is 440. 20, 22

84. *Number Problem* Find two consecutive positive odd integers whose product is 323. 17, 19

85. ▲ *Geometry* The length of a rectangular picture frame is 3 inches greater than its width. The area of the picture frame is 108 square inches. Find the dimensions of the picture frame. 9 inches × 12 inches

86. ▲ *Geometry* The width of a rectangular garden is 4 feet less than its length. The area of the garden is 320 square feet. Find the dimensions of the garden.
16 feet × 20 feet

87. ▲ *Geometry*   The length of a rectangle is 2 times its width. The area of the rectangle is 450 square inches. Find the dimensions of the rectangle.

15 inches × 30 inches

88. ▲ *Geometry*   The length of a rectangle is $1\frac{1}{2}$ times its width. The area of the rectangle is 600 square inches. Find the dimensions of the rectangle.

20 inches × 30 inches

89. ▲ *Geometry*   An open box is to be made from a square piece of material by cutting two-inch squares from the corners and turning up the sides (see figure).

(a) Show that the volume of the box is given by $V = 2x^2$. $V = lwh,\ V = (x)(x)(2),\ V = 2x^2$

(b) Complete the table.

x	2	4	6	8
V	8	32	72	128

(c) Find the dimensions of the original piece of material when $V = 200$ cubic inches.

14 inches × 14 inches

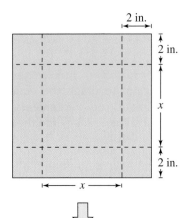

90. ▲ *Geometry*   An open box with a square base is to be constructed from 108 square inches of material. The height of the box is 3 inches. Find the dimensions of the base of the box. (*Hint:* The surface area is given by $S = x^2 + 4xh$.)   6 inches × 6 inches

91. *Free-Falling Object*   An object is dropped from a weather balloon 1600 feet above the ground (see figure). Find the time t (in seconds) for the object to reach the ground by solving the equation

$$-16t^2 + 1600 = 0. \quad t = 10 \text{ seconds}$$

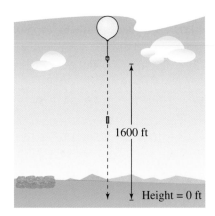

92. *Free-Falling Object*   An object is dropped from a cliff 400 feet above the ground. Find the time t (in seconds) for the object to reach the ground by solving the equation

$$-16t^2 + 400 = 0. \quad 5 \text{ seconds}$$

93. *Height of a Diver*   A diver jumps from a diving board that is 32 feet above the water (see figure). The height h (in feet) of the diver is modeled by the position equation

$$h = -16t^2 + 16t + 32$$

where t is the time measured in seconds. How long will it take for the diver to reach the water?

2 seconds

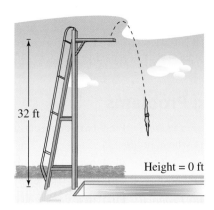

94. *Free-Falling Object*   An object is thrown upward from the Royal Gorge Bridge in Colorado, 1053 feet above the Arkansas River, with an intial velocity of 48 feet per second. The height h (in feet) of the object is modeled by the position equation

$$h = -16t^2 + 48t + 1053 \quad \text{9.75 seconds}$$

where t is the time measured in seconds. How long will it take for the object to reach the ground?

95. *Profit*   The profit P from selling x video games is

$$P = -0.4x^2 + 8x - 10.$$

(a) ⊞ Use a graphing calculator to graph the profit equation.   See Additional Answers.

(b) ⊞ Use the graph to estimate any values of x that yield a profit of $20.   5, 15

(c) Use factorization to find any values of x that yield a profit of $20.   5, 15

96. *Revenue*   The revenue R from selling x lamps is

$$R = 25x - 0.2x^2.$$

(a) ⊞ Use a graphing calculator to graph the revenue equation.   See Additional Answers.

(b) ⊞ Use the graph to estimate the smaller of the two values of x that yield a revenue of $680.   40

(c) Use factorization to find the smaller of the two values of x that yield a revenue of $680.   40

97. *Sum of Natural Numbers*

(a) Find the following sums.

$$1 + 2 + 3 + 4 + 5 = \boxed{15}$$

$$1 + 2 + 3 + 4 + 5 + 6 + 7 + 8 = \boxed{36}$$

$$1 + 2 + 3 + 4 + 5 + 6 +$$
$$7 + 8 + 9 + 10 = \boxed{55}$$

(b) Use the following formula for the sum of the first n natural numbers to verify your answers to part (a).   Verification

$$1 + 2 + 3 + \cdots + n = \frac{1}{2}n(n + 1)$$

(c) Use the formula in part (b) to find n if the sum of the first n natural numbers is 210.   20

98. *Sum of Natural Numbers*   Use the formula in part (b) of Exercise 97 to find n if the sum of the first n natural numbers is 325.   25

99. *Exploration*   If a and b are nonzero real numbers, show that the equation $ax^2 + bx = 0$ must have two different solutions.

$$x(ax + b) = 0, x = 0, -\frac{b}{a}$$

100. *Exploration*   If a is a nonzero real number, find two solutions of the equation $ax^2 - ax = 0$.   0, 1

Explaining Concepts

101. ✿ Answer parts (c)–(e) of Motivating the Chapter on page 318.

102. Use the Zero-Factor Property to complete the statement. If $ab = 0$, then $a = 0$ or $b = 0$.

103. *Writing*✐   In your own words, describe a strategy for solving a quadratic equation by factoring.   The Zero-Factor Property allows you to solve a quadratic equation by factoring and converting it into two linear equations.

104. *Writing*✐   Is it possible for a quadratic equation to have one solution? Explain.
Yes. $x^2 + 2x + 1 = (x + 1)^2 = 0$. There is one solution, $x = -1$.

105. *True or False?*   The only equation with solutions $x = 2$ and $x = -5$ is $(x - 2)(x + 5) = 0$. Justify your answer.   False. $3(x - 2)(x + 5) = 0$ also has solutions $x = 2$ and $x = -5$.

106. *Writing*✐   A student submits the following steps in solving the equation $x^2 + 4x = 12$:

$$x(x + 4) = 12 \qquad \text{Factor.}$$

$$x = 12 \quad \text{and} \quad x + 4 = 12 \qquad \begin{array}{l}\text{Set each factor} \\ \text{equal to 12.}\end{array}$$

$$x = 8$$

Write an explanation of why the method does not work. Solve the equation correctly and check your solutions.   The student should have written the equation in general form before factoring, and then set each factor equal to 0, not 12. Correct solution:

$$x^2 + 4x = 12$$
$$x^2 + 4x - 12 = 0$$
$$(x + 6)(x - 2) = 0$$
$$x + 6 = 0 \implies x = -6$$
$$x - 2 = 0 \implies x = 2$$

What Did You Learn?

Key Terms

factoring, *p. 320*
greatest common factor, *p. 320*
greatest common
  monomial factor, *p. 321*

factoring out, *p. 321*
prime polynomials, *p. 331*
factoring completely, *p. 332*

quadratic equation, *p. 357*
general form, *p. 358*
repeated solution, *p. 359*

Key Concepts

6.1 ◗ Factoring out common monomial factors
Use the Distributive Property to remove the greatest common monomial factor from each term of a polynomial.

6.1 ◗ Factoring polynomials by grouping
For polynomials with four terms, group the first two terms together and the last two terms together. Factor these two groupings and then look for a common binomial factor.

6.2 ◗ Guidelines for factoring $x^2 + bx + c$
To factor $x^2 + bx + c$, you need to find two numbers m and n whose product is c and whose sum is b.

$$x^2 + bx + c = (x + m)(x + n)$$

1. If c is positive, then m and n have like signs that match the sign of b.
2. If c is negative, then m and n have unlike signs.
3. If $|b|$ is small relative to $|c|$, first try those factors of c that are closest to each other in absolute value.

6.3 ◗ Guidelines for factoring $ax^2 + bx + c$ $(a > 0)$
1. If the trinomial has a common monomial factor, you should factor out the common factor before trying to find the binomial factors.
2. You do not have to test any binomial factors that have a common monomial factor.
3. Switch the signs of the factors of c when the middle term (O + I) is correct except in sign.

6.3 ◗ Guidelines for factoring $ax^2 + bx + c$ by grouping
1. If necessary, write the trinomial in standard form.
2. Choose factors of the product ac that add up to b.
3. Use these factors to rewrite the middle term as a sum or difference.
4. Group and remove any common monomial factors from the first two terms and the last two terms.

5. If possible, factor out the common binomial factor.

6.4 ◗ Difference of two squares
Let a and b be real numbers, variables, or algebraic expressions. Then the expression $a^2 - b^2$ can be factored as follows: $a^2 - b^2 = (a + b)(a - b)$.

6.4 ◗ Perfect square trinomials
Let a and b be real numbers, variables, or algebraic expressions. Then the expressions $a^2 \pm 2ab + b^2$ can be factored as follows: $a^2 \pm 2ab + b^2 = (a \pm b)^2$.

6.4 ◗ Sum or difference of two cubes
Let a and b be real numbers, variables, or algebraic expressions. Then the expressions $a^3 \pm b^3$ can be factored as follows: $a^3 \pm b^3 = (a \pm b)(a^2 \mp ab + b^2)$.

6.4 ◗ Guidelines for factoring polynomials
1. Factor out any common factors.
2. Factor according to one of the special polynomial forms: difference of two squares, sum or difference of two cubes, or perfect square trinomials.
3. Factor trinomials, $ax^2 + bx + c$, with $a = 1$ or $a \neq 1$.
4. Factor by grouping—for polynomials with four terms.
5. Check to see whether the factors themselves can be factored.
6. Check the results by multiplying the factors.

6.5 ◗ Zero-Factor Property
Let a and b be real numbers, variables, or algebraic expressions. If a and b are factors such that $ab = 0$, then $a = 0$ or $b = 0$. This property also applies to three or more factors.

6.5 ◗ Solving a quadratic equation
To solve a quadratic equation, write the equation in general form. Factor the quadratic into linear factors and apply the Zero-Factor Property.

Review Exercises

6.1 Factoring Polynomials with Common Factors

 Find the greatest common factor of two or more expressions.

In Exercises 1–8, find the greatest common factor of the expressions.

1. t^2, t^5 t^2

2. $-y^3, y^8$ y^3

3. $3x^4, 21x^2$ $3x^2$

4. $14z^2, 21z$ $7z$

5. $14x^2y^3, -21x^3y^5$
$7x^2y^3$

6. $-15y^2z^2, 5y^2z$
$5y^2z$

7. $8x^2y, 24xy^2, 4xy$
$4xy$

8. $27ab^5, 9ab^6, 18a^2b^3$
$9ab^3$

 Factor out the greatest common monomial factor from polynomials.

In Exercises 9–18, factor the polynomial.

9. $3x - 6$ $3(x - 2)$

10. $7 + 21x$ $7(1 + 3x)$

11. $3t - t^2$ $t(3 - t)$

12. $u^2 - 6u$ $u(u - 6)$

13. $5x^2 + 10x^3$
$5x^2(1 + 2x)$

14. $7y - 21y^4$
$7y(1 - 3y^3)$

15. $8a - 12a^3$
$4a(2 - 3a^2)$

16. $14x - 26x^4$
$2x(7 - 13x^3)$

17. $5x^3 + 5x^2 - 5x$
$5x(x^2 + x - 1)$

18. $6u - 9u^2 + 15u^3$
$3u(2 - 3u + 5u^2)$

Geometry **In Exercises 19 and 20, write an expression for the area of the shaded region and factor the expression.**

19.

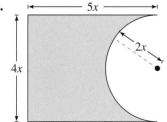

$x(3x + 4)$

20.

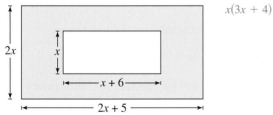

$2x^2(10 - \pi)$

3 Factor polynomials by grouping.

In Exercises 21–30, factor the polynomial by grouping.

21. $x(x + 1) - 3(x + 1)$ $(x + 1)(x - 3)$

22. $5(y - 3) - y(y - 3)$ $(y - 3)(5 - y)$

23. $2u(u - 2) + 5(u - 2)$ $(u - 2)(2u + 5)$

24. $7(x + 8) + 3x(x + 8)$ $(x + 8)(7 + 3x)$

25. $y^3 + 3y^2 + 2y + 6$
$(y + 3)(y^2 + 2)$

26. $z^3 - 5z^2 + z - 5$
$(z - 5)(z^2 + 1)$

27. $x^3 + 2x^2 + x + 2$
$(x^2 + 1)(x + 2)$

28. $x^3 - 5x^2 + 5x - 25$
$(x - 5)(x^2 + 5)$

29. $x^2 - 4x + 3x - 12$
$(x + 3)(x - 4)$

30. $2x^2 + 6x - 5x - 15$
$(x + 3)(2x - 5)$

6.2 Factoring Trinomials

① Factor trinomials of the form $x^2 + bx + c$.

In Exercises 31–38, factor the trinomial.

31. $x^2 - 3x - 28$
$(x - 7)(x + 4)$

32. $x^2 - 3x - 40$
$(x - 8)(x + 5)$

33. $u^2 + 5u - 36$
$(u - 4)(u + 9)$

34. $y^2 + 15y + 56$
$(y + 7)(y + 8)$

35. $x^2 - 2x - 24$
$(x - 6)(x + 4)$

36. $x^2 + 8x + 15$
$(x + 5)(x + 3)$

37. $y^2 + 10y + 21$
$(y + 7)(y + 3)$

38. $a^2 - 7a + 12$
$(a - 4)(a - 3)$

In Exercises 39–42, find all integers b such that the trinomial can be factored.

39. $x^2 + bx + 9$
$\pm6, \pm10$

40. $y^2 + by + 25$
$\pm10, \pm26$

41. $z^2 + bz + 11$
±12

42. $x^2 + bx + 14$
$\pm9, \pm15$

② Factor trinomials in two variables.

In Exercises 43–48, factor the trinomial.

43. $x^2 + 9xy - 10y^2$
$(x - y)(x + 10y)$

44. $u^2 + uv - 5v^2$
Prime

45. $y^2 - 6xy - 27x^2$
$(y + 3x)(y - 9x)$

46. $v^2 + 18uv + 32u^2$
$(v + 2u)(v + 16u)$

47. $x^2 - 2xy - 8y^2$
$(x + 2y)(x - 4y)$

48. $a^2 - ab - 30b^2$
$(a - 6b)(a + 5b)$

③ Factor trinomials completely.

In Exercises 49–54, factor the trinomial completely.

49. $4x^2 - 24x + 32$
$4(x - 2)(x - 4)$

50. $3u^2 - 6u - 72$
$3(u + 4)(u - 6)$

51. $x^3 + 9x^2 + 18x$
$x(x + 3)(x + 6)$

52. $y^3 - 8y^2 + 15y$
$y(y - 5)(y - 3)$

53. $4x^3 + 36x^2 + 56x$
$4x(x + 2)(x + 7)$

54. $2y^3 - 4y^2 - 30y$
$2y(y - 5)(y + 3)$

6.3 More About Factoring Trinomials

① Factor trinomials of the form $ax^2 + bx + c$.

In Exercises 55–68, factor the trinomial.

55. $5 - 2x - 3x^2$
$(1 - x)(5 + 3x)$

56. $8x^2 - 18x + 9$
$(2x - 3)(4x - 3)$

57. $50 - 5x - x^2$
$(10 + x)(5 - x)$

58. $7 + 5x - 2x^2$
$(7 - 2x)(1 + x)$

59. $6x^2 + 7x + 2$
$(3x + 2)(2x + 1)$

60. $16x^2 + 13x - 3$
$(16x - 3)(x + 1)$

61. $4y^2 - 3y - 1$
$(4y + 1)(y - 1)$

62. $5x^2 - 12x + 7$
$(5x - 7)(x - 1)$

63. $3x^2 + 7x - 6$
$(3x - 2)(x + 3)$

64. $45y^2 - 8y - 4$
$(5y - 2)(9y + 2)$

65. $3x^2 + 5x - 2$
$(3x - 1)(x + 2)$

66. $7x^2 - 4x - 3$
$(7x + 3)(x - 1)$

67. $2x^2 - 3x + 1$
$(2x - 1)(x - 1)$

68. $3x^2 + 8x + 4$
$(3x + 2)(x + 2)$

In Exercises 69 and 70, find all integers b such that the trinomial can be factored.

69. $x^2 + bx - 24$
$\pm 2, \pm 5, \pm 10, \pm 23$

70. $2x^2 + bx - 16$
$\pm 4, \pm 14, \pm 31$

In Exercises 71 and 72, find two integers c such that the trinomial can be factored. (There are many correct answers.)

71. $2x^2 - 4x + c$
$2, -6$

72. $5x^2 + 6x + c$
$1, -8$

② Factor trinomials completely.

In Exercises 73–78, factor the polynomial completely.

73. $6u^3 + 3u^2 - 30u$
$3u(2u + 5)(u - 2)$

74. $8x^3 - 8x^2 - 30x$
$2x(2x + 3)(2x - 5)$

75. $8y^3 - 20y^2 + 12y$
$4y(2y - 3)(y - 1)$

76. $14x^3 + 26x^2 - 4x$
$2x(7x - 1)(x + 2)$

77. $6x^3 + 14x^2 - 12x$
$2x(3x - 2)(x + 3)$

78. $12y^3 + 36y^2 + 15y$
$3y(2y + 1)(2y + 5)$

79. ▲ *Geometry* The cake box shown in the figure has a height of x and a width of $x + 1$. The volume of the box is $3x^3 + 4x^2 + x$. Find the length l of the box. $3x + 1$

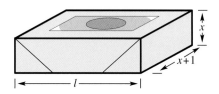

80. ▲ *Geometry* The area of the rectangle shown in the figure is $2x^2 + 5x + 3$. What is the area of the shaded region? $x + 3$

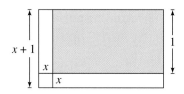

③ Factor trinomials by grouping.

In Exercises 81–86, factor the trinomial by grouping.

81. $2x^2 - 13x + 21$
$(2x - 7)(x - 3)$

82. $3a^2 - 13a - 10$
$(3a + 2)(a - 5)$

83. $4y^2 + y - 3$
$(4y - 3)(y + 1)$

84. $6z^2 - 43z + 7$
$(6z - 1)(z - 7)$

85. $6x^2 + 11x - 10$
$(3x - 2)(2x + 5)$

86. $21x^2 - 25x - 4$
$(3x - 4)(7x + 1)$

6.4 Factoring Polynomials with Special Forms

① Factor the difference of two squares.

In Exercises 87–94, factor the difference of two squares.

87. $a^2 - 100$
$(a + 10)(a - 10)$

88. $36 - b^2$
$(6 + b)(6 - b)$

89. $25 - 4y^2$
$(5 + 2y)(5 - 2y)$

90. $16b^2 - 1$
$(4b + 1)(4b - 1)$

91. $12x^2 - 27$
$3(2x + 3)(2x - 3)$

92. $100x^2 - 64$
$4(5x + 4)(5x - 4)$

93. $(u + 1)^2 - 4$
$(u + 3)(u - 1)$

94. $(y - 2)^2 - 9$
$(y + 1)(y - 5)$

② Recognize repeated factorization.

In Exercises 95 and 96, fill in the missing factors.

95. $x^3 - x = x(\boxed{x+1})(\boxed{x-1})$

96. $u^4 - v^4 = (u^2 + v^2)(\boxed{u+v})(\boxed{u-v})$

In Exercises 97–100, factor the polynomial completely.

97. $s^3t - st^3$
$st(s + t)(s - t)$

98. $5x^3 - 20xy^2$
$5x(x + 2y)(x - 2y)$

99. $x^4 - y^4$
$(x^2 + y^2)(x + y)(x - y)$

100. $2a^4 - 32$
$2(a^2 + 4)(a + 2)(a - 2)$

③ Identify and factor perfect square trinomials.

In Exercises 101–106, factor the perfect square trinomial.

101. $x^2 - 8x + 16$
$(x - 4)^2$

102. $y^2 + 24y + 144$
$(y + 12)^2$

103. $9s^2 + 12s + 4$
$(3s + 2)^2$

104. $16x^2 - 40x + 25$
$(4x - 5)^2$

105. $y^2 + 4yz + 4z^2$
$(y + 2z)^2$

106. $u^2 - 2uv + v^2$
$(u - v)^2$

④ Factor the sum or difference of two cubes.

In Exercises 107–112, factor the sum or difference of two cubes.

107. $a^3 + 1$
$(a + 1)(a^2 - a + 1)$

108. $z^3 + 8$
$(z + 2)(z^2 - 2z + 4)$

109. $27 - 8t^3$
$(3 - 2t)(9 + 6t + 4t^2)$

110. $z^3 - 125$
$(z - 5)(z^2 + 5z + 25)$

111. $8x^3 + y^3$ $\quad (2x + y)(4x^2 - 2xy + y^2)$

112. $125a^3 - 27b^3$ $\quad (5a - 3b)(25a^2 + 15ab + 9b^2)$

6.5 Solving Quadratic Equations by Factoring

① Use the Zero-Factor Property to solve equations.

In Exercises 113–118, use the Zero-Factor Property to solve the equation.

113. $x(2x - 3) = 0$ $\quad 0, \frac{3}{2}$

114. $3x(5x + 1) = 0$ $\quad -\frac{1}{5}, 0$

115. $(x + 3)(x - 2) = 0$ $\quad -3, 2$

116. $(a - 5)(a + 8) = 0$ $\quad -8, 5$

117. $(3x - 7)(2x + 1) = 0$ $\quad -\frac{1}{2}, \frac{7}{3}$

118. $(2y + 3)(4y - 5) = 0$ $\quad -\frac{3}{2}, \frac{5}{4}$

② Use factoring to solve quadratic equations.

In Exercises 119–130, solve the equation.

119. $x^2 - 81 = 0$
$-9, 9$

120. $121 - y^2 = 0$
$-11, 11$

121. $x^2 - 12x + 36 = 0$
6

122. $y^2 - y - 6 = 0$
$-2, 3$

123. $2t^2 - 3t - 2 = 0$
$-\frac{1}{2}, 2$

124. $4s^2 + s - 3 = 0$
$-1, \frac{3}{4}$

125. $(z - 2)^2 - 4 = 0$
$0, 4$

126. $(x + 1)^2 - 16 = 0$
$-5, 3$

127. $x(7 - x) = 12$
$3, 4$

128. $x(x + 5) = 24$
$-8, 3$

129. $(x - 1)(x + 2) = 10$
$-4, 3$

130. $(x + 4)(x - 3) = 18$
$-6, 5$

③ Solve application problems by factoring.

131. *Number Problem* Find two consecutive positive even integers whose product is 168. $\quad 12, 14$

132. *Number Problem* Find two consecutive positive odd integers whose product is 399. $\quad 19, 21$

133. ▲ *Geometry* The height of a rectangular window is $1\frac{1}{2}$ times its width. The area of the window is 2400 square inches. Find the dimensions of the window. $\quad$ 40 inches × 60 inches

134. ▲ *Geometry* A box with a square base has a surface area of 400 square inches. The height of the box is 5 inches. Find the dimensions of the box. (*Hint:* The surface area is given by $S = 2x^2 + 4xh$.)
10 inches × 10 inches × 5 inches

135. *Free-Falling Object* A rock is thrown vertically upward from a height of 48 feet with an initial velocity of 32 feet per second. The height h (in feet) of the rock is modeled by the position equation $h = -16t^2 + 32t + 48$, where t is the time measured in seconds. How long will it take the rock to reach the water? $\quad$ 3 seconds

136. *Revenue* The revenue R from selling x craft kits is $R = 12x - 0.3x^2$.

(a) ▦ Use a graphing calculator to graph the revenue equation. $\quad$ See Additional Answers.

(b) ▦ Use the graph to estimate the value of x that yields a revenue of $120. $\quad$ 20

(c) Use factorization to find the value of x that yields a revenue of $120. $\quad$ 20

Chapter Test

Take this test as you would take a test in class. After you are done, check your work against the answers in the back of the book.

In Exercises 1–10, factor the polynomial completely .

1. $7x^2 - 14x^3$
$7x^2(1 - 2x)$

2. $z(z + 7) - 3(z + 7)$
$(z + 7)(z - 3)$

3. $t^2 - 4t - 5$ $(t - 5)(t + 1)$

4. $6x^2 - 11x + 4$ $(3x - 4)(2x - 1)$

5. $3y^3 + 72y^2 - 75y$
$3y(y - 1)(y + 25)$

6. $4 - 25v^2$
$(2 + 5v)(2 - 5v)$

7. $4x^2 - 20x + 25$
$(2x - 5)^2$

8. $16 - (z + 9)^2$
$(-z - 5)(z + 13)$

9. $x^3 + 2x^2 - 9x - 18$
$(x + 2)(x + 3)(x - 3)$

10. $16 - z^4$
$(4 + z^2)(2 + z)(2 - z)$

11. Fill in the missing factor: $\dfrac{2}{5}x - \dfrac{3}{5} = \dfrac{1}{5}($ $2x - 3$ $)$.

12. Find all integers b such that $x^2 + bx + 5$ can be factored. ± 6

13. Find a real number c such that $x^2 + 12x + c$ is a perfect square trinomial. 36

14. Explain why $(x + 1)(3x - 6)$ is not a complete factorization of $3x^2 - 3x - 6$. $3x^2 - 3x - 6 = 3(x + 1)(x - 2)$

In Exercises 15–18, solve the equation.

15. $(x + 4)(2x - 3) = 0$ $-4, \frac{3}{2}$

16. $3x^2 + 7x - 6 = 0$ $-3, \frac{2}{3}$

17. $y(2y - 1) = 6$ $-\frac{3}{2}, 2$

18. $2x^2 - 3x = 8 + 3x$ $-1, 4$

19. The suitcase shown below has a height of x and a width of $x + 2$. The volume of the suitcase is $x^3 + 6x^2 + 8x$. Find the length l of the suitcase. $x + 4$

20. The width of a rectangle is 5 inches less than its length. The area of the rectangle is 84 square inches. Find the dimensions of the rectangle.
7 inches $\times$ 12 inches

21. An object is thrown upward from the top of the AON Center in Chicago, with an initial velocity of 14 feet per second at a height of 1136 feet. The height h (in feet) of the object is modeled by the position equation

$$h = -16t^2 + 14t + 1136$$

where t is the time measured in seconds. How long will it take for the object to reach the ground? How long will it take the object to fall to a height of 806 feet?   8.875 seconds; 5 seconds

22. Find two consecutive positive even integers whose product is 624.   24, 26

Take this test as you would take a test in class. After you are done, check your work against the answers in the back of the book.

1. Describe how to identify the quadrants in which the points $(-2, y)$ must be located. (y is a real number.)

1. Because $x = -2$, the point must lie in Quadrant II or Quadrant III.

2. Determine whether the ordered pairs are solution points of the equation $9x - 4y + 36 = 0$.
   (a) $(-1, -1)$ (b) $(8, 27)$ (c) $(-4, 0)$ (d) $(3, -2)$
   Not a solution   Solution   Solution   Not a solution

In Exercises 3 and 4, sketch the graph of the equation and determine any intercepts of the graph. See Additional Answers.

3. $y = 2 - |x|$ $(-2, 0), (2, 0), (0, 2)$ **4.** $x + 2y = 8$ $(8, 0), (0, 4)$

5. The slope of a line is $-\frac{1}{4}$ and a point on the line is $(2, 1)$. Find the coordinates of a second point on the line. Explain why there are many correct answers.
   $(-2, 2)$; There are infinitely many points on a line.

6. Find an equation of the line through $\left(0, -\frac{3}{2}\right)$ with slope $m = \frac{5}{6}$. $y = \frac{5}{6}x - \frac{3}{2}$

In Exercises 7 and 8, sketch the lines and determine whether they are parallel, perpendicular, or neither. See Additional Answers.

7. $y_1 = \frac{2}{3}x - 3, y_2 = -\frac{3}{2}x + 1$ **8.** $y_1 = 2 - 0.4x, y_2 = -\frac{2}{5}x$
   Perpendicular   Parallel

9. Subtract: $(x^3 - 3x^2) - (x^3 + 2x^2 - 5)$. $-5x^2 + 5$

10. Multiply: $(6z)(-7z)(z^2)$. $-42z^4$

11. $3x^2 - 7x - 20$

11. Multiply: $(3x + 5)(x - 4)$.   **12.** Multiply: $(5x - 3)(5x + 3)$.

12. $25x^2 - 9$

13. Expand: $(5x + 6)^2$.   **14.** Divide: $(6x^2 + 72x) \div 6x$.

13. $25x^2 + 60x + 36$

14. $x + 12$

15. Divide: $\dfrac{x^2 - 3x - 2}{x - 4}$.   **16.** Simplify: $\dfrac{(3xy^2)^{-2}}{6x^{-3}}$.

15. $x + 1 + \dfrac{2}{x - 4}$

17. Factor: $2u^2 - 6u$.   **18.** Factor and simplify: $(x - 2)^2 - 16$.

16. $\dfrac{x}{54y^4}$

19. Factor completely: $x^3 + 8x^2 + 16x$. $x(x + 4)^2$

20. Factor completely: $x^3 + 2x^2 - 4x - 8$. $(x + 2)^2(x - 2)$

17. $2u(u - 3)$

21. Solve: $u(u - 12) = 0$.   **22.** Solve: $5x^2 - 12x - 9 = 0$.

18. $(x + 2)(x - 6)$

21. $0, 12$

22. $-\frac{3}{5}, 3$

23. Rewrite the expression $\left(\dfrac{x}{2}\right)^{-2}$ using only positive exponents. $\dfrac{4}{x^2}$

24. A sales representative is reimbursed $125 per day for lodging and meals, plus $0.35 per mile driven. Write a linear equation giving the daily cost C to the company in terms of x, the number of miles driven. Explain the reasoning you used to write the model. Find the cost for a day when the representative drives 70 miles. $C = 125 + 0.35x$; $149.50

25. The cost of operating a pizza delivery car is $0.70 per mile after an initial investment of $9000. What mileage on the car will keep the cost at or below $36,400?   39,142 miles

Motivating the Chapter

⟳ Predicting College Enrollment

A college recently changed from an all women's college to a coeducational institution. The admissions office projects that enrollment F of female students for the next 5 years can be approximated by the model

$$F = \frac{18,000 + 2t}{16 - t}$$

where t represents the time in years and $t = 0$ is the year before the school became coeducational. The projected enrollment M of male students for the next 5 years can be approximated by the model

$$M = \frac{500t}{8 - t}$$

where t represents the time in years and $t = 0$ is the year before the school became coeducational.

See Section 7.3, Exercise 107.

a. What was the enrollment of the school the year before it became coeducational?   1125 students

b. Approximate the number of male students for the third year that the school is coeducational. Then, approximate the number of female students for the third year that the school is coeducational. Round your answers to the nearest whole number. What is the projected total enrollment for the third year? $M(3) = 300$, $F(3) \approx 1385$, $M(3) + F(3) \approx 1685$

c. Write an expression for the total number of students in any given year t. Simplify the result.

$$F + M = \frac{-502t^2 - 9984t + 144,000}{(16 - t)(8 - t)}$$

d. Use the expression from part (c) to find the projected total enrollment for the third year. Round your answer to the nearest whole number. Does this agree with your answer in part (b)? Explain.

   1685 students; Yes; You can find the total enrollment either by substituting a value for t in the models for female and male enrollments and then adding or by adding the models for female and male enrollments and then substituting a value for t.

See Section 7.5, Exercise 91.

e. ▦ Use the *table* feature of a graphing calculator to determine if the male enrollment will reach or surpass the female enrollment at any time over the five-year projection period.   It will not.

f. ▦ Write an equation in standard quadratic form ($at^2 + bt + c = 0$) that will predict when the male and female enrollments will be equal. Use a graphing calculator to graph the equation. From the graph, estimate when the male and female enrollments will be equal.
   $249t^2 - 12,992t + 72,000 = 0$; $t \approx 6.3$;   See Additional Answers.

Bob Krist/Corbis

7

Rational Expressions and Equations

7.1 Simplifying Rational Expressions

Lester Lefkowitz/Corbis

What You Should Learn

1. Find the domain of a rational expression.
2. Simplify rational expressions.

Why You Should Learn It

Rational expressions can be used to model and solve real-life problems. For instance, in Exercise 103 on page 382, you will use a rational expression that models the cost of removing air pollutants in the stack emission of a utility company.

1 Find the domain of a rational expression.

Point out that the *numerator* of an algebraic fraction *can* be zero. Do not exclude $x = 5$ from the domain of $\dfrac{x-5}{9}$, and do not exclude $x = 0$ from the domain of $\dfrac{x}{x^2-16}$.

Technology: Discovery

Use a graphing calculator to graph $y = 7/(x+3)$. Then use the *trace* feature to determine the behavior of the graph near $x = -3$, where the denominator is zero. Use the *table* feature to determine the behavior of the graph near $x = -3$. Graph equations that correspond to parts (b), (c), and (d) of Example 1. What conclusions can you draw about the behavior of each graph at certain values of x?   See Technology Answers.

Rational Expressions and Their Domains

In this chapter, you will learn how to use the rules for fractions to simplify, add, subtract, multiply, and divide rational expressions. A **rational expression** is a fraction whose numerator and denominator are polynomials. Some examples are

$$\frac{x-2}{5}, \quad \frac{x^3-4x}{x-4}, \quad \text{and} \quad \frac{3x}{x^2-1}.$$

The **domain** of a rational expression is the set of all real numbers for which it is defined. The domain of the first expression above is the set of all real numbers. Because division by zero is not defined, the domain of the second expression is all real numbers *except* $x = 4$. Similarly, the domain of the third expression is all real numbers *except* $x = \pm 1$. To find the values to *exclude* from the domain, set the denominator equal to zero and find the solution(s) to that equation.

Example 1   Finding the Domain of a Rational Expression

Find the domain of each rational expression.

a. $\dfrac{7}{x+3}$ **b.** $\dfrac{x-5}{9}$ **c.** $\dfrac{x}{x^2-16}$ **d.** $\dfrac{x^2+1}{x^2+4x-5}$

Solution

a. The denominator is zero when $x + 3 = 0$ or $x = -3$. So, the domain is all real values of x such that $x \neq -3$.

b. The denominator, 9, is never zero, and so the domain is the set of *all* real numbers.

c. For this rational expression, the denominator is zero when $x^2 - 16 = 0$.

$$x^2 - 16 = 0 \quad \Longrightarrow \quad (x+4)(x-4) = 0 \qquad \text{Set denominator equal to 0 and factor.}$$

So, the domain is all real values of x such that $x \neq -4$ and $x \neq 4$.

d. For this rational expression, the denominator is zero when $x^2 + 4x - 5 = 0$.

$$x^2 + 4x - 5 = 0 \quad \Longrightarrow \quad (x+5)(x-1) = 0 \qquad \text{Set denominator equal to 0 and factor.}$$

So, the domain is all real values of x such that $x \neq -5$ and $x \neq 1$.

In applications involving rational expressions, it is often necessary to restrict the domain further. In cases where the variable represents a quantity, such as the number of people, the number of hours worked, or the speed of a vehicle, the domain is the set of positive numbers. To indicate such a restriction, you should write the domain to the right of the fraction. For instance, if x represents the number of hours worked, the domain of the expression

$$\frac{x^2 + 10}{x + 2}, \quad x > 0$$

is restricted to the set of *positive* real numbers, as indicated by the inequality $x > 0$.

Example 2   An Application Involving a Restricted Domain

A publisher asks a printing company to print copies of a book. The printing company charges $5000 as a start-up fee, plus $4 per book. The total cost (in dollars) of printing x books is

Total cost = $5000 + 4x$.   Total cost of x books

The average cost per book depends on the number of books printed. For instance, the average cost per book for printing 100 books is

$$\text{Average cost} = \frac{\text{Total cost}}{\text{Number of books}}$$

$$= \frac{5000 + 4(100)}{100}$$

$$= \$54 \text{ per book.}$$ Average cost per book for 100 books

On the other hand, the average cost per book for printing 10,000 books is

$$\frac{5000 + 4(10,000)}{10,000} = \$4.50 \text{ per book.}$$ Average cost per book for 10,000 books

In general, the average cost of printing x books is

$$\frac{5000 + 4x}{x}.$$ Average cost per book for x books

What is the domain of this rational expression?

Solution

If you were simply considering the rational expression $(5000 + 4x)/x$ as a mathematical quantity, you would say that the domain is all real values of x such that $x \neq 0$. However, because this fraction is a mathematical model representing a real-life situation, you must decide which values of x make sense in real life. For this model, the variable x represents the number of books printed. Because the number of books printed cannot be a fraction, you can conclude that the domain is the set of positive integers. That is,

Domain = $\{1, 2, 3, 4, \ldots\}$.

In 2000, more than 2 billion books were sold in the United States. Of these, 37% were hardbacks and 63% were paperbacks.   (Source: Book Industry Study Group, Inc.)

② Simplify rational expressions.

Simplifying Rational Expressions

The rules for operating with rational expressions are like those for numerical fractions (see Section 1.4). As with numerical fractions, a rational expression is in **simplified** (or **reduced**) **form** if its numerator and denominator have no common factors (other than ± 1). To simplify rational expressions, use the rule below.

Simplifying Rational Expressions

Let a, b, and c represent real numbers, variables, or algebraic expressions such that $b \neq 0$ and $c \neq 0$. Then the following is valid.

$$\frac{ac}{bc} = \frac{a\cancel{c}}{b\cancel{c}} = \frac{a}{b}$$

Be sure you divide out only *factors*, not *terms*. For instance, consider the expressions below.

$$\frac{3 \cdot x}{3(x + 4)}$$ You *can* divide out common factor 3.

$$\frac{3 + x}{3 + (x + 4)}$$ You *cannot* divide out common term 3.

Study Tip

Simplifying a rational expression requires two steps.

1. Completely factor the numerator and denominator.

2. Divide out any *factors* that are common to both the numerator and denominator.

Your success in simplifying rational expressions actually lies in your ability to factor completely the polynomials in both the numerator and denominator.

Example 3   Simplifying a Rational Expression by Factoring

$$\frac{15(x^2 - 2x)}{3x^2} = \frac{3 \cdot 5 \cdot x(x - 2)}{3 \cdot x \cdot x}$$ Factor completely.

$$= \frac{3 \cdot 5 \cdot \cancel{x}(x - 2)}{3 \cdot \cancel{x} \cdot x}$$ Divide out common factors.

$$= \frac{5(x - 2)}{x}$$ Simplified form

In simplified form, the domain of the rational expression is the same as that of the original expression—all real values of x such that $x \neq 0$.

Example 4   Simplifying a Rational Expression by Factoring

$$\frac{2x^2 - 4x}{(x - 2)^2} = \frac{2x(x - 2)}{(x - 2)(x - 2)}$$ Factor completely.

$$= \frac{2x\cancel{(x - 2)}}{(x - 2)\cancel{(x - 2)}}$$ Divide out common factors.

$$= \frac{2x}{x - 2}$$ Simplified form

Emphasize the distinction between terms and factors. Dividing out terms is one of the most common algebraic errors.

In simplified form, the domain of the rational expression is the same as that of the original expression—all real values of x such that $x \neq 2$.

In this text, all values of x that must be excluded from the domain of a simplified rational expression in order to make the domains of the simplified and original expressions agree *are listed beside the simplified expression.*

Study Tip

In Example 5, be sure you see that simplifying a rational expression can change its domain. For instance, in part (a) the domain of the original expression is all real values of x such that $x \neq 0$. So, to equate the original expression with the simplified expression, you must restrict the domain of the simplified expression to exclude 0. Similarly, 2 must be excluded in part (b).

Example 5   Adjusting the Domain After Simplifying

Simplify each rational expression.

a. $\dfrac{4x^3 - 8x^2}{4x^2}$ **b.** $\dfrac{x^2 + 4x - 12}{3x - 6}$

Solution

a. $\dfrac{4x^3 - 8x^2}{4x^2} = \dfrac{4x^2(x - 2)}{4x^2}$ Factor numerator.

$= \dfrac{4x^2(x - 2)}{4x^2}$ Divide out common factor $4x^2$.

$= x - 2, \quad x \neq 0$ Simplified form

b. $\dfrac{x^2 + 4x - 12}{3x - 6} = \dfrac{(x + 6)(x - 2)}{3(x - 2)}$ Factor numerator and denominator.

$= \dfrac{(x + 6)(x - 2)}{3(x - 2)}$ Divide out common factor $(x - 2)$.

$= \dfrac{x + 6}{3}, \quad x \neq 2$ Simplified form

Be sure to factor *completely* the numerator and denominator of a rational expression before concluding that there is no common factor. This may involve a change in signs. Remember that the Distributive Property allows you to write $(b - a)$ as $-1(a - b)$. Watch for this in the next example.

Technology: Tip

You can graphically check that you have simplified an expression correctly by using a graphing calculator to graph both the original and simplified expressions in the same viewing window. For instance, try graphing

$y_1 = \dfrac{2x - 6}{12 - x - x^2}$ and

$y_2 = -\dfrac{2}{x + 4}$

in the same viewing window to check the result of Example 6.

See Technology Answers.

Example 6   Simplifying a Rational Expression by a Sign Change

$\dfrac{2x - 6}{12 - x - x^2} = \dfrac{2(x - 3)}{(4 + x)(3 - x)}$ Factor completely.

$= \dfrac{2(x - 3)}{(4 + x)(-1)(x - 3)}$ $(3 - x) = (-1)(x - 3)$

$= \dfrac{2(x - 3)}{(4 + x)(-1)(x - 3)}$ Divide out common factor $(x - 3)$.

$= -\dfrac{2}{x + 4}, x \neq 3$ Simplified form

In the simplified form, this text lists the minus sign in front of the fraction. However, this is a personal preference. All of the following are equivalent.

$$-\dfrac{2}{x + 4} = \dfrac{-2}{x + 4} = \dfrac{2}{-(x + 4)} = \dfrac{2}{-x - 4}$$

2

Additional Examples

Simplify each rational expression.

a. $\dfrac{x^2 + 2x - 48}{2x - 12}$

b. $\dfrac{x^3 - 9x}{x^2 - 10x + 21}$

Answers:

a. $\dfrac{x + 8}{2}, x \neq 6$

b. $\dfrac{x(x + 3)}{x - 7}, x \neq 3$

Notice in Example 6 that the restriction $x \neq 3$ is listed with the simplified expression in order to make the two domains agree. The value of $x = -4$ is excluded from *both* domains, so it is not necessary to list this value.

In the next example, a rational expression that involves more than one variable is simplified.

Example 7   A Rational Expression Involving Two Variables

Simplify the rational expression $\dfrac{x^2 - 2xy + y^2}{5x - 5y}$.

Solution

$$\frac{x^2 - 2xy + y^2}{5x - 5y} = \frac{(x - y)(x - y)}{5(x - y)}$$ Factor numerator and denominator.

$$= \frac{(x - y)(x - y)}{5(x - y)}$$ Divide out common factor $(x - y)$.

$$= \frac{x - y}{5}, \quad x \neq y$$ Simplified form

In simplified form, the domain of the expression is all real values of x such that $x \neq y$.

Example 8   Probability

A parachutist plans to land in the middle of a football stadium during a halftime show. The parachutist is certain of landing in the stadium field, but cannot guarantee a landing in a square area in the center of the field, as shown in Figure 7.1. The probability that the parachutist will land in the center square is equal to the ratio of the area of the center square to the area of the entire field. Find the probability.

Solution

The area of the center square is given by

Area of center square $=$ (Length)(Width) $= x^2$.

The area of the entire field is given by

Area of field $=$ (Length)(Width) $= (6x)(12x + 25)$.

So, the probability of the parachutist landing in the center square is

$$\frac{x^2}{(6x)(12x + 25)} = \frac{x}{6(12x + 25)}.$$

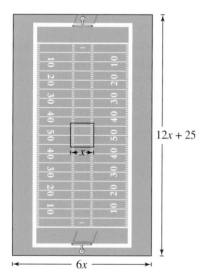

Figure 7.1

$12x + 25$

$6x$

As you study the examples and work the exercises in this and the following two sections, keep in mind that you are *rewriting expressions in simpler forms*. You are not solving equations. Equal signs are used in the steps only to indicate that the new form of the expression (fraction) is equivalent to the original form.

7.1 Exercises

Review Concepts, Skills, and Problem Solving

Keep mathematically in shape by doing these exercises *before* the problems of this section.

Properties and Definitions

1. *Writing*✐ Define the slope of the line through the points (x_1, y_1) and (x_2, y_2).

$$m = \frac{y_2 - y_1}{x_2 - x_1}$$

2. *Writing*✐ Make a statement about the slope m of a line for each condition.

   (a) The line rises from left to right. $m > 0$

   (b) The line falls from left to right. $m < 0$

   (c) The line is horizontal. $m = 0$

   (d) The line is vertical. m is undefined.

Algebraic Operations

In Exercises 3–6, find the slope of the line passing through the two points.

3. $(0, 4), (10, 0)$ $-\frac{2}{5}$ 4. $(0, 0), (5, -3)$ $-\frac{3}{5}$

5. $(-1, 3), (4, 8)$ 1   6. $(2, 6), (5, 1)$ $-\frac{5}{3}$

In Exercises 7–10, solve the equation and check your solution.

7. $14 - 2x = x + 2$ 4

8. $7 - 3(1 - 2p) = 2p + 4$ 0

9. $\dfrac{x}{3} + 5 = 8$ 9   10. $\dfrac{x}{3} + \dfrac{x}{2} = \dfrac{1}{3}$ $\frac{2}{5}$

Problem Solving

11. *Insurance Premiums* The annual insurance premium for a policyholder is normally $645. However, after having an automobile accident, the policyholder is charged an additional 25%. What is the new annual premium? $806.25

12. *Wages* An employee is paid $9.50 per hour for the first 40 hours and $14 for each hour of overtime. During the first week on the job, the employee's gross pay is $478. How many hours of overtime did the employee work? 7 hours

Developing Skills

In Exercises 1–4, determine whether the expression is a rational expression. If not, explain why.

1. $\dfrac{x^2 + 1}{5x - 2}$ Rational

2. $\dfrac{x}{x^{-2} + 2}$ Not rational because the denominator is not a polynomial

3. $\dfrac{x^{1/2} - 2x}{x + 1}$ Not rational because the numerator is not a polynomial

4. $\dfrac{6}{x - 2}$ Rational

In Exercises 5–20, find the domain of the rational expression. See Example 1.

5. $\dfrac{5}{x - 4}$
   All real values of x such that $x \neq 4$

6. $\dfrac{10}{x - 6}$
   All real values of x such that $x \neq 6$

7. $\dfrac{x}{x + 2}$
   All real values of x such that $x \neq -2$

8. $\dfrac{2z}{z + 8}$
   All real values of z such that $z \neq -8$

9. $\dfrac{x^2 - 4}{3}$
   All real numbers

10. $\dfrac{y^2 - 1}{5}$
    All real numbers

11. $\dfrac{3}{x^2 + 4}$
    All real numbers

12. $\dfrac{5}{x^2 + 4}$
    All real numbers

13. $\dfrac{4t}{t^2 - 25}$
    All real values of t such that $t \neq -5$ and $t \neq 5$

14. $\dfrac{z + 2}{z^2 - 4}$
    All real values of z such that $z \neq -2$ and $z \neq 2$

15. $\dfrac{-5(y + 2)}{y^2 - 3y - 28}$
    All real values of y such that $y \neq -4$ and $y \neq 7$

16. $\dfrac{-3(x - 2)}{x^2 + 6x + 8}$
    All real values of x such that $x \neq -2$ and $x \neq -4$

17. $\dfrac{x^2}{x^2 - x - 2}$
All real values of x
such that $x \neq -1$ and $x \neq 2$

18. $\dfrac{x^2}{x^2 + 3x - 10}$
All real values of x
such that $x \neq -5$ and $x \neq 2$

19. $\dfrac{z - 3}{3z^2 - z - 2}$
All real values of z
such that $z \neq -\frac{2}{3}$ and $z \neq 1$

20. $\dfrac{y + 5}{4y^2 + y - 3}$
All real values of y
such that $y \neq -1$ and $y \neq \frac{3}{4}$

In Exercises 21–24, evaluate the rational expression for each specified value. If not possible, state the reason.

Expression	Values

21. $\dfrac{x}{x - 3}$
(a) $x = 0$ (b) $x = 3$
(c) $x = 10$ (d) $x = -3$
(a) 0   (b) Division by zero is undefined.
(c) $\frac{10}{7}$ (d) $\frac{1}{2}$

22. $\dfrac{3y}{y + 6}$
(a) $y = 0$ (b) $y = -3$
(c) $y = -10$ (d) $y = -6$
(a) 0   (b) -3
(c) $\frac{15}{2}$ (d) Division by zero is undefined.

23. $\dfrac{x + 1}{x^2 - 4}$
(a) $x = 2$ (b) $x = 1$
(c) $x = -5$ (d) $x = -2$
(a) Division by zero is undefined.   (b) $-\frac{2}{3}$
(c) $-\frac{4}{21}$ (d) Division by zero is undefined.

24. $\dfrac{2x - 3}{x^2 + 2x - 3}$
(a) $x = 0$ (b) $x = 1$
(c) $x = -\frac{3}{2}$ (d) $x = -3$
(a) 1   (b) Division by zero is undefined.
(c) $\frac{8}{5}$ (d) Division by zero is undefined.

Think About It   In Exercises 25–28, write two equivalent versions of the expression by changing the sign of the numerator, the denominator, and/or the fraction.

25. $\dfrac{x}{12}$ $\dfrac{-x}{12}, \dfrac{-x}{-12}$

26. $-\dfrac{4}{y}$ $\dfrac{-4}{y}, \dfrac{4}{-y}$

27. $-\dfrac{t + 2}{t^2 - 1}$ $\dfrac{t + 2}{1 - t^2}, \dfrac{-t - 2}{t^2 - 1}$

28. $\dfrac{x - 6}{x + 1}$ $\dfrac{6 - x}{x + 1}, \dfrac{6 - x}{-1 - x}$

In Exercises 29–40, fill in the missing factor.

29. $\dfrac{5}{2x} = \dfrac{5\,(3x)}{6x^2}$

30. $\dfrac{3x}{2} = \dfrac{3x\,(2x^2)}{4x^2}$

31. $\dfrac{3}{4} = \dfrac{3\,(x + 1)}{4(x + 1)}$

32. $\dfrac{11}{16} = \dfrac{11\,[2(x - 4)^2]}{32(x - 4)^2}$

33. $\dfrac{-7}{3x} = \dfrac{7\,(-x^2)}{3x^3}$

34. $\dfrac{5x}{-8} = \dfrac{25x^2}{8\,(-5x)}$

35. $\dfrac{x}{2} = \dfrac{x(x + 2)}{2\,(x + 2)}$

36. $\dfrac{2x}{x + 2} = \dfrac{2(x^2 + 2x)}{(x + 2)\,(x + 2)}$

37. $\dfrac{x + 1}{x} = \dfrac{(x + 1)\,(x - 2)}{x(x - 2)}$

38. $\dfrac{3y - 4}{y + 1} = \dfrac{(3y - 4)\,(y - 1)}{y^2 - 1}$

39. $\dfrac{3x}{x - 3} = \dfrac{3x\,(x + 2)}{x^2 - x - 6}$

40. $\dfrac{1 - z}{z^2} = \dfrac{(1 - z)\,(z + 1)}{z^3 + z^2}$

In Exercises 41–90, simplify the rational expression. See Examples 3–7.

41. $\dfrac{4x}{12}$ $\dfrac{x}{3}$

42. $\dfrac{18y}{36}$ $\dfrac{y}{2}$

43. $\dfrac{2y^2}{y}$ $2y, y \neq 0$

44. $\dfrac{5z^3}{z}$ $5z^2, z \neq 0$

45. $\dfrac{15x^2}{10x}$ $\dfrac{3x}{2}, x \neq 0$

46. $\dfrac{18y^2}{60y^5}$ $\dfrac{3}{10y^3}$

47. $\dfrac{60a^2b}{40ab^3}$ $\dfrac{3a}{2b^2}, a \neq 0$

48. $\dfrac{45x^3y^2}{9x^5y}$ $\dfrac{5y}{x^2}, y \neq 0$

49. $\dfrac{75x(x - 1)^2}{15x(x - 1)}$ $5(x - 1), x \neq 0, x \neq 1$

50. $\dfrac{5b(b - 3)}{b(b - 3)^2}$ $\dfrac{5}{b - 3}, b \neq 0$

51. $\dfrac{x^2(x + 1)}{x(x + 1)}$ $x, x \neq 0, x \neq -1$

52. $\dfrac{b(b - 2)}{b^2(b - 2)}$ $\dfrac{1}{b}, b \neq 2$

53. $\dfrac{x - 5}{2x - 10}$ $\dfrac{1}{2}, x \neq 5$

54. $\dfrac{5 - x}{2x - 10}$ $-\dfrac{1}{2}, x \neq 5$

55. $\dfrac{3y + 3}{xy + x}$ $\dfrac{3}{x}, y \neq -1$

56. $\dfrac{x^2y + x}{xy + 1}$ $x, xy \neq -1$

57. $\dfrac{9 - 3t}{t - 3}$ $-3, t \neq 3$

58. $\dfrac{y - 3}{9 - 3y}$ $-\frac{1}{3}, y \neq 3$

59. $\dfrac{x^2 - 25}{5 - x}$
$-(x + 5), x \neq 5$

60. $\dfrac{x^2 - 25}{x - 5}$
$x + 5, x \neq 5$

61. $\dfrac{y^2 - 16}{3y + 12}$
$\dfrac{y - 4}{3}, y \neq -4$

62. $\dfrac{x^2 - 25z^2}{x - 5z}$
$x + 5z, x \neq 5z$

63. $\dfrac{7s^2 - 28t^2}{28t^2 - 7s^2}$
$-1, s^2 \neq 4t^2$

64. $\dfrac{9m^2 - 4n^2}{4n^2 - 9m^2}$
$-1, 4n^2 \neq 9m^2$

65. $\dfrac{x^2 - 2x + 1}{1 - x^2}$
$\dfrac{1 - x}{1 + x}, x \neq 1$

66. $\dfrac{u^2 - 3u - 4}{16 - u^2}$
$-\dfrac{u + 1}{u + 4}, u \neq 4$

67. $\dfrac{a + 2}{a^2 + 4a + 4}$
$\dfrac{1}{a + 2}$

68. $\dfrac{u^2 - 6u + 9}{u - 3}$
$u - 3, u \neq 3$

69. $\dfrac{x^2 - 5x}{x^2 - 10x + 25}$
$\dfrac{x}{x - 5}$

70. $\dfrac{z^2 + 12z + 36}{5z + 30}$
$\dfrac{z + 6}{5}, z \neq -6$

71. $\dfrac{y^2 - 4}{y^2 + 3y - 10}$
$\dfrac{y + 2}{y + 5}, y \neq 2$

72. $\dfrac{x^2 - 7x}{x^2 - 8x + 7}$
$\dfrac{x}{x - 1}, x \neq 7$

73. $\dfrac{x^2 - 4x + 3}{x^2 - 5x + 6}$
$\dfrac{x - 1}{x - 2}, x \neq 3$

74. $\dfrac{y^2 - 7y + 12}{y^2 + 3y - 18}$
$\dfrac{y - 4}{y + 6}, y \neq 3$

75. $\dfrac{x^2 + 8x - 20}{x^2 + 11x + 10}$
$\dfrac{x - 2}{x + 1}, x \neq -10$

76. $\dfrac{z^2 - 3z - 18}{z^2 + 2z - 48}$
$\dfrac{z + 3}{z + 8}, z \neq 6$

77. $\dfrac{5r + 5s}{r^2 + 7rs + 6s^2}$
$\dfrac{5}{r + 6s}, r \neq -s$

78. $\dfrac{6x + 12y}{x^2 - xy - 6y^2}$
$\dfrac{6}{x - 3y}, x \neq -2y$

79. $\dfrac{x^2 + 3xy - 18y^2}{x^2 + 2xy - 15y^2}$
$\dfrac{x + 6y}{x + 5y}, x \neq 3y$

80. $\dfrac{a^2 + 10ab + 21b^2}{a^2 + 11ab + 28b^2}$ $\dfrac{a + 3b}{a + 4b}, a \neq -7b$

81. $\dfrac{x^3 + 5x^2 + 6x}{x^2 - 4}$ $\dfrac{x(x + 3)}{x - 2}, x \neq -2$

82. $\dfrac{t^3 - t}{t^3 + 5t^2 - 6t}$ $\dfrac{t + 1}{t + 6}, t \neq 0, t \neq 1$

83. $\dfrac{x^3 - 2x^2 + x - 2}{x - 2}$ $x^2 + 1, x \neq 2$

84. $\dfrac{x^2 - 9}{x^3 + x^2 - 9x - 9}$ $\dfrac{1}{x + 1}, x \neq \pm 3$

85. $\dfrac{x^3 + 2x^2 + x + 2}{x^2 + 1}$ $x + 2$

86. $\dfrac{z - 3}{z^3 - 3z^2 + z - 3}$ $\dfrac{1}{z^2 + 1}, z \neq 3$

87. $\dfrac{a^3 - 8}{a^2 - 4}$ $\dfrac{a^2 + 2a + 4}{a + 2}, a \neq 2$

88. $\dfrac{1 - y^3}{1 + y + y^2}$ $1 - y$

89. $\dfrac{x^4 - y^4}{(y - x)^4}$ $-\dfrac{(x^2 + y^2)(x + y)}{(y - x)^3}$

90. $\dfrac{16y^4 - x^4}{(x^2 + 4y^2)(x - 2y)}$ $-2y - x, x \neq 2y$

Graphical Reasoning   In Exercises 91–96, simplify the rational expression. Then use a graphing calculator to verify your result.

91. $\dfrac{2x^2 + 4x}{2x}$ $x + 2, x \neq 0$

92. $\dfrac{5x^2 - 6x}{2x}$ $\dfrac{5x - 6}{2}, x \neq 0$

93. $\dfrac{3(x - 3)^2}{x - 3}$ $3(x - 3), x \neq 3$

94. $\dfrac{2(x - 4)^2}{x - 4}$ $2(x - 4), x \neq 4$

95. $\dfrac{x^3 - 2x^2}{x^3 + 2x}$ $\dfrac{x(x - 2)}{x^2 + 2}, x \neq 0$

96. $\dfrac{x^3 - 4x^2}{x^3 + 2x}$ $\dfrac{x(x - 4)}{x^2 + 2}, x \neq 0$

⚡ *Exploration* In Exercises 97 and 98, complete the table. Explain why the values of the expressions agree for all values of x except one.

97.

x	2	2.5	3	3.5	4
$\dfrac{x^3 - 3x^2}{x - 3}$	4	6.25	Undef.	12.25	16
x^2	4	6.25	9	12.25	16

98.

x	0	0.5	1	1.5	2
$\dfrac{x - 1}{x^2 + 2x - 3}$	$\frac{1}{3}$	$\frac{2}{7}$	Undef.	$\frac{2}{9}$	$\frac{1}{5}$
$\dfrac{1}{x + 3}$	$\frac{1}{3}$	$\frac{2}{7}$	$\frac{1}{4}$	$\frac{2}{9}$	$\frac{1}{5}$

97. and **98.** The expressions are equivalent. So, all the values will agree except for the value that is not in the domain of the unsimplified rational expression.

Solving Problems

99. *Average Cost* A machine shop has a setup cost of $3000 for the production of a new alarm clock. The cost of labor and materials for producing each unit is $7.50.

(a) Write a rational expression that models the average cost per unit when x units are produced.
$$\frac{3000 + 7.50x}{x}$$

(b) Find the domain of the expression in part (a).
$\{1, 2, 3, 4, \ldots\}$

(c) Find the average cost per unit when 100 units are produced. $37.50

100. *Average Cost* A machine shop has a setup cost of $5000 for the production of a new computer keyboard. The cost of labor and materials for producing each unit is $12.50.

(a) Write a rational expression that models the average cost per unit when x units are produced.
$$\frac{5000 + 12.50x}{x}$$

(b) Find the domain of the expression in part (a).
$\{1, 2, 3, 4, \ldots\}$

(c) Find the average cost per unit when 200 units are produced. $37.50

101. *Boiling Temperature* As air pressure increases, the temperature at which water boils also increases. (This is the purpose of pressure canners.) A model that relates air pressure x (in pounds per square inch) to boiling temperature B (in degrees Fahrenheit) is

$$B = \frac{156.89x + 7.34x^2}{x + 0.017x^2}, \quad 10 \le x \le 100.$$

(a) Simplify the rational expression.
$$B = \frac{156.89 + 7.34x}{1 + 0.017x}, \ 10 \le x \le 100$$

(b) Use the model to estimate the boiling temperature of water when $x = 14.7$ pounds per square inch (approximate air pressure at sea level). Round your answer to three decimal places.
211.847° F

102. *Depreciation* The value V of an automobile t years after it is purchased is given by

$$V = \frac{50P}{51 + 22t}, \quad 0 \le t \le 5$$

where P is the purchase price.

(a) 🖩 Use a graphing calculator to graph the function V for an automobile for which $P = \$20,000$.   See Additional Answers.

(b) Use the model to estimate the value of an automobile 3 years after it is purchased if the purchase price is $22,500. $9615.38

103. *Environment* A utility company burns coal to produce electricity. The cost C (in dollars) of removing p percent of the air pollutants in the stack emission of the utility company is given by the rational equation

$$C = \frac{80,000p}{100 - p}.$$

(a) Determine the domain of the expression.
$0 \le p < 100$

(b) Create a table showing the costs of removing 20%, 40%, 60%, and 80% of the pollutants in the stack emission.   See Additional Answers.

(c) Use the table to describe the relationship between the cost and the percent. According to this model, can you remove 100% of the pollutants? Explain.   As p increases, C increases. No: the function is not defined for $p = 100$.

104. *Comparing Distances*   You start a trip and drive at an average speed of 50 miles per hour. Two hours later, a friend starts a trip on the same road and drives at an average speed of 60 miles per hour.

(a) Find polynomial expressions that represent the distance d that each of you has driven when your friend has been driving for t hours.
    Friend: $d = 60t$; Yourself: $d = 50(t + 2)$

(b) Use the result of part (a) to determine the ratio of the distance your friend has driven to the distance you have driven.
$$\frac{6t}{5(t + 2)}$$

(c) Evaluate the ratio described in part (b) when $t = 5$ and $t = 10$. $t = 5: \frac{6}{7}; t = 10: 1$

Probability   In Exercises 105–108, the probability of hitting the shaded portion of the region with a dart is equal to the ratio of the shaded area to the total area of the figure. Find the probability.

105. $\dfrac{x - 3}{x + 5}$

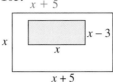

106. $\dfrac{11}{x + 7}$

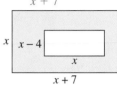

107. $\dfrac{1}{6}$

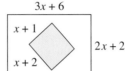

108. $\dfrac{5}{6}$

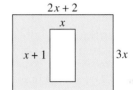

Explaining Concepts

109. *Writing*   Explain what is meant by a *rational expression*.   A rational expression is a fraction whose numerator and denominator are polynomials.

110. *Writing*   Define the *domain* of a rational expression.   The domain of a rational expression consists of all values of the variable for which the denominator is not zero.

111. *Writing*   How can you determine whether a rational expression is in simplified form?   A rational expression is in simplified form if its numerator and denominator have no common factors other than ±1.

112. *True or False?*   To simplify a rational expression, you can divide out common factors of the numerator and denominator. Justify your answer.
True, since $\frac{ac}{bc} = \frac{a}{b}$ when a, b, c are real numbers, variables, or algebraic expressions where $b \neq 0$ and $c \neq 0$.

113. *Error Analysis*   Describe the error.
$$\frac{2x^2 + 3x}{x^2 + 4x} = \frac{2 + 3}{1 + 4} = 1$$
A fraction can be reduced only by dividing out common factors of the numerator and denominator.

114. *Writing*   Is $(4x)/(2x)$ equivalent to 2? Explain.
They are equivalent for $x \neq 0$.

115. Construct a rational expression that cannot be simplified. There are many correct answers.
$$\frac{x}{x^2 + 1}$$

116. Construct a rational expression that can be simplified. There are many correct answers.
$$\frac{x}{x^2 + x}$$

117. *Writing*   Two algebra students hand in the following work.

Student A   *Student B*
$$\frac{\overset{1}{\cancel{x}} - 25}{\overset{5}{\cancel{x}} - 5} = -4 \qquad \frac{3(x + 2)}{3x + 2} = \frac{x + 2}{x + 2} = 1$$

Find all errors or misconceptions in each student's work. In a written, detailed explanation, show the students how to work each problem correctly.
Both students used incorrect simplification techniques—only common factors of the numerator and denominator can be divided out. Neither of these expressions can be simplified any further than their original forms.

7.2 Multiplying and Dividing Rational Expressions

Paul Barton/Corbis

Why You Should Learn It

Multiplication and division of rational expressions can be used to solve real-life applications. For instance, in Exercise 86 on page 392, you will divide a rational expression by another to find the average yearly basic cable TV rate per subscriber.

① Multiply rational expressions and simplify.

What You Should Learn

① Multiply rational expressions and simplify.

② Divide rational expressions and simplify.

Multiplying Rational Expressions

The rule for multiplying rational expressions is the same as the rule for multiplying numerical fractions. That is, you *multiply numerators, multiply denominators, and write the new fraction in simplified form.*

$$\frac{3}{4} \cdot \frac{7}{6} = \frac{21}{24} = \frac{3 \cdot 7}{3 \cdot 8} = \frac{7}{8} \qquad \frac{x}{2} \cdot \frac{5}{x^2} = \frac{5 \cdot \cancel{x}}{2 \cdot \cancel{x} \cdot x} = \frac{5}{2x}$$

Multiplying Rational Expressions

Let a, b, c, and d represent real numbers, variables, or algebraic expressions such that $b \neq 0$ and $d \neq 0$. Then the product of a/b and c/d is

$$\frac{a}{b} \cdot \frac{c}{d} = \frac{ac}{bd}.$$

In order to recognize common factors in the product, write the numerators and denominators in completely factored form, as demonstrated in Example 1.

Example 1   Multiplying Rational Expressions

Multiply and simplify: $\dfrac{3x^2y}{2xy^2} \cdot \dfrac{-10xy^3}{6x^3}$.

Solution

$$\frac{3x^2y}{2xy^2} \cdot \frac{-10xy^3}{6x^3} = \frac{(3x^2y)(-10xy^3)}{(2xy^2)(6x^3)} \qquad \text{Multiply numerators and denominators.}$$

$$= \frac{-3(10)x^3y^4}{2(6)x^4y^2} \qquad \text{Simplify.}$$

$$= \frac{-3(2)(5)(x^3)(y^2)(y^2)}{2(3)(2)(x^3)(x)(y^2)} \qquad \text{Factor and divide out common factors.}$$

$$= -\frac{5y^2}{2x}, \quad y \neq 0 \qquad \text{Simplified form}$$

Example 2   Multiplying Rational Expressions

Multiply and simplify: $\dfrac{4x}{x^2-9} \cdot \dfrac{x-3}{8x^2+12x}$.

Solution

$\dfrac{4x}{x^2-9} \cdot \dfrac{x-3}{8x^2+12x}$

$= \dfrac{4x(x-3)}{(x^2-9)(8x^2+12x)}$ Multiply numerators and denominators.

$= \dfrac{4x(x-3)}{(x+3)(x-3)(4x)(2x+3)}$ Factor.

$= \dfrac{4x(x-3)}{(x+3)(x-3)(4x)(2x+3)}$ Divide out common factors. Factor of 1 remains in numerator.

$= \dfrac{1}{(x+3)(2x+3)}, \quad x \neq 0, \ x \neq 3$ Simplified form

The rule for multiplying rational expressions can be extended to cover products involving expressions that are not in fractional form. To do this, rewrite the (nonfractional) expression as a fraction whose denominator is 1.

Example 3   Multiplying Rational Expressions

a. $\dfrac{x+2}{x+4} \cdot (3x) = \dfrac{x+2}{x+4} \cdot \dfrac{3x}{1}$ Rewrite in fractional form.

$= \dfrac{(x+2)(3x)}{x+4}$ Multiply numerators and denominators.

$= \dfrac{3x(x+2)}{x+4}$ Simplified form

b. $\dfrac{x}{2x^2-x-3} \cdot (2x^2+11x-21)$

$= \dfrac{x}{2x^2-x-3} \cdot \dfrac{2x^2+11x-21}{1}$ Rewrite in fractional form.

$= \dfrac{x(2x^2+11x-21)}{2x^2-x-3}$ Multiply numerators and denominators.

$= \dfrac{x(2x-3)(x+7)}{(2x-3)(x+1)}$ Factor.

$= \dfrac{x(2x-3)(x+7)}{(2x-3)(x+1)}$ Divide out common factor.

$= \dfrac{x(x+7)}{x+1}, \quad x \neq \dfrac{3}{2}$ Simplified form

Additional Examples
Multiply and simplify.

a. $\dfrac{x^2 + 2x - 3}{x^2 - x} \cdot \dfrac{2x + 3}{3x^2 + 5x - 12}$

b. $\dfrac{x + y}{x^2 + 3xy + 2y^2} \cdot \dfrac{x^2 + xy - 2y^2}{2x + 4y}$

Answers:

a. $\dfrac{2x + 3}{x(3x - 4)}, x \neq -3, x \neq 1$

b. $\dfrac{x - y}{2(x + 2y)}, x \neq -y$

In the next example, note how to divide out factors that differ only in sign.

Example 4   Multiplying Rational Expressions

Multiply and simplify: $\dfrac{x - y}{6x + 4y} \cdot \dfrac{3x + 2y}{y^2 - x^2}$.

Solution

$$\dfrac{x - y}{6x + 4y} \cdot \dfrac{3x + 2y}{y^2 - x^2}$$

$$= \dfrac{(x - y)(3x + 2y)}{(6x + 4y)(y^2 - x^2)} \quad \text{Multiply numerators and denominators.}$$

$$= \dfrac{(x - y)(3x + 2y)}{2(3x + 2y)(y + x)(y - x)} \quad \text{Factor.}$$

$$= \dfrac{(x - y)(3x + 2y)}{2(3x + 2y)(y + x)(-1)(x - y)} \quad (y - x) = (-1)(x - y)$$

$$= \dfrac{(x - y)(3x + 2y)}{2(3x + 2y)(y + x)(-1)(x - y)} \quad \text{Divide out common factors.}$$

$$= -\dfrac{1}{2(y + x)}, \quad x \neq y, \ x \neq -\dfrac{2}{3}y \quad \text{Simplified form}$$

In the third step of Example 4, the factor -1 could have been written in the numerator instead of the denominator, with the same results.

$$\dfrac{(x - y)(3x + 2y)}{2(3x + 2y)(y + x)(y - x)} = \dfrac{(-1)(y - x)(3x + 2y)}{2(3x + 2y)(y + x)(y - x)}$$

The rule for multiplying rational expressions can be extended to cover the product of three or more expressions, as shown in Example 5.

Example 5   Multiplying Three Rational Expressions

Multiply and simplify: $\dfrac{3}{x} \cdot \dfrac{x + 1}{x + 2} \cdot \dfrac{x^2 - 2x - 8}{x + 1}$.

Solution

$$\dfrac{3}{x} \cdot \dfrac{x + 1}{x + 2} \cdot \dfrac{x^2 - 2x - 8}{x + 1}$$

$$= \dfrac{3(x + 1)(x + 2)(x - 4)}{x(x + 2)(x + 1)} \quad \text{Multiply and factor.}$$

$$= \dfrac{3(x + 1)(x + 2)(x - 4)}{x(x + 2)(x + 1)} \quad \text{Divide out common factors.}$$

$$= \dfrac{3(x - 4)}{x}, \quad x \neq -2, \ x \neq -1 \quad \text{Simplified form}$$

② Divide rational expressions and simplify.

Dividing Rational Expressions

To divide two rational expressions, multiply the first expression by the *reciprocal* of the second. That is, *invert the divisor and multiply.*

Dividing Rational Expressions

Let *a, b, c,* and *d* represent real numbers, variables, or algebraic expressions such that $b \neq 0$, $c \neq 0$, and $d \neq 0$. Then the quotient of a/b and c/d is

$$\frac{a}{b} \div \frac{c}{d} = \frac{a}{b} \cdot \frac{d}{c} = \frac{ad}{bc}.$$

Example 6   Dividing Rational Expressions

Divide and simplify: $\dfrac{x}{x+4} \div \dfrac{x+3}{x+4}$.

Solution

$$\frac{x}{x+4} \div \frac{x+3}{x+4} = \frac{x}{x+4} \cdot \frac{x+4}{x+3} \qquad \text{Invert divisor and multiply.}$$

$$= \frac{(x)(x+4)}{(x+4)(x+3)} \qquad \text{Multiply numerators and denominators.}$$

You might remind students that only common *factors* can be divided out.

$$= \frac{(x)(x+4)}{(x+4)(x+3)} \qquad \text{Divide out common factor.}$$

$$= \frac{x}{x+3}, \quad x \neq -4 \qquad \text{Simplified form}$$

Example 7   Dividing Rational Expressions

$$\frac{x^2 - 2x}{x^2 - 6x + 8} \div \frac{2x}{3x - 12}$$

$$= \frac{x^2 - 2x}{x^2 - 6x + 8} \cdot \frac{3x - 12}{2x} \qquad \text{Invert divisor and multiply.}$$

$$= \frac{(x^2 - 2x)(3x - 12)}{(x^2 - 6x + 8)(2x)} \qquad \text{Multiply numerators and denominators.}$$

$$= \frac{(x)(x-2)(3)(x-4)}{(x-2)(x-4)(2x)} \qquad \text{Factor.}$$

$$= \frac{(x)(x-2)(3)(x-4)}{(x-2)(x-4)(2x)} \qquad \text{Divide out common factors.}$$

$$= \frac{3}{2}, \quad x \neq 0, \ x \neq 2, \ x \neq 4 \qquad \text{Simplified form}$$

The rule for dividing rational expressions can be extended to cover expressions that are not in fractional form. To do this, rewrite the (non-fractional) expression as a fraction whose denominator is 1.

Example 8   Dividing Rational Expressions

Divide and simplify: $\dfrac{x^2 + 3x - 10}{x^2} \div (x - 2)^2$.

Solution

$$\frac{x^2 + 3x - 10}{x^2} \div (x - 2)^2$$

$$= \frac{x^2 + 3x - 10}{x^2} \div \frac{(x - 2)^2}{1} \qquad \text{Rewrite in fractional form.}$$

$$= \frac{x^2 + 3x - 10}{x^2} \cdot \frac{1}{(x - 2)^2} \qquad \text{Invert divisor and multiply.}$$

$$= \frac{(x + 5)(x - 2)}{(x^2)(x - 2)(x - 2)} \qquad \text{Factor.}$$

$$= \frac{(x + 5)\cancel{(x - 2)}}{(x^2)(x - 2)\cancel{(x - 2)}} \qquad \text{Divide out common factor.}$$

$$= \frac{x + 5}{x^2(x - 2)} \qquad \text{Simplified form}$$

Example 9   Dividing Rational Expressions

Divide and simplify: $\dfrac{16 - x^2}{2x + 8} \div \dfrac{2x^2 - 9x + 4}{6x + 2}$.

Solution

$$\frac{16 - x^2}{2x + 8} \div \frac{2x^2 - 9x + 4}{6x + 2}$$

$$= \frac{16 - x^2}{2x + 8} \cdot \frac{6x + 2}{2x^2 - 9x + 4} \qquad \text{Invert divisor and multiply.}$$

$$= \frac{(-1)(x^2 - 16)(6x + 2)}{(2x + 8)(2x^2 - 9x + 4)} \qquad 16 - x^2 = (-1)(x^2 - 16)$$

$$= \frac{(-1)(x + 4)(x - 4)(2)(3x + 1)}{(2)(x + 4)(x - 4)(2x - 1)} \qquad \text{Factor.}$$

$$= \frac{(-1)\cancel{(x + 4)}\cancel{(x - 4)}(2)(3x + 1)}{(2)\cancel{(x + 4)}\cancel{(x - 4)}(2x - 1)} \qquad \text{Divide out common factors.}$$

$$= -\frac{3x + 1}{2x - 1}, \ x \neq -4, x \neq 4 \qquad \text{Simplified form}$$

Addition Examples
Divide and simplify.

a. $\dfrac{4x}{10x + 5} \div \dfrac{3x^2 - 8x}{2x + 1}$

b. $\dfrac{x^2 - 14x + 49}{x^2 - 49} \div \dfrac{3x - 21}{x^2 + 2x - 35}$

Answers:

a. $\dfrac{4}{5(3x - 8)}, x \neq 0, x \neq -\dfrac{1}{2}$

b. $\dfrac{x - 5}{3}, x \neq 7, x \neq -7$

7.2 Exercises

Review   Concepts, Skills, and Problem Solving

Keep mathematically in shape by doing these exercises *before* the problems of this section.

Properties and Definitions

1. *Writing* ✏ Explain how to factor the difference of two squares $4x^2 - 9$.

   $a^2 - b^2 = (a + b)(a - b)$

   $4x^2 - 9 = (2x + 3)(2x - 3)$

2. *Writing* ✏ Explain how to factor the perfect square trinomial $x^2 - 8x + 16$.

   $a^2 - 2ab + b^2 = (a - b)^2$

   $x^2 - 8x + 16 = (x - 4)^2$

3. *Writing* ✏ Explain how to factor the sum of two cubes $8x^3 + 27$.

   $a^3 + b^3 = (a + b)(a^2 - ab + b^2)$

   $8x^3 + 27 = (2x + 3)(4x^2 - 6x + 9)$

4. *Writing* ✏ Factor $3x^2 + 13x - 10$, and explain how you can check your answer.

   $3x^2 + 13x - 10 = (3x - 2)(x + 5)$

   Check by multiplying $(3x - 2)(x + 5)$.

Algebraic Operations

In Exercises 5–8, factor the polynomial completely.

5. $3x^2 + 7x$

   $x(3x + 7)$

6. $16 - (x - 11)^2$

   $-(x - 7)(x - 15)$

7. $x^2 + 7x - 18$

   $(x + 9)(x - 2)$

8. $10x^2 + 13x - 3$

   $(5x - 1)(2x + 3)$

In Exercises 9 and 10, fill in the missing factor.

9. $\frac{1}{3}x + \frac{5}{9} = \frac{1}{9}(\ 3x + 5\ )$

10. $\frac{5}{8}x - \frac{3}{2} = \frac{1}{8}(\ 5x - 12\ )$

Graphs

In Exercises 11 and 12, sketch the line through the point with each indicated slope on the same set of coordinate axes.   See Additional Answers.

Point	Slopes	
11. $(2, 3)$	(a) 0	(b) 1
	(c) 2	(d) $-\frac{1}{3}$
12. $(-4, 1)$	(a) 3	(b) -3
	(c) $\frac{1}{2}$	(d) Undefined

Developing Skills

In Exercises 1–44, multiply and simplify. See Examples 1–5.

1. $\frac{8x^2}{3} \cdot \frac{9}{16x}$ $\frac{3x}{2}, x \neq 0$

2. $\frac{6x}{5} \cdot \frac{1}{x}$ $\frac{6}{5}, x \neq 0$

3. $\frac{12x^2}{6x} \cdot \frac{12x}{8x^2}$ $3, x \neq 0$

4. $\frac{25x^2}{8x} \cdot \frac{8x}{5x}$ $5x, x \neq 0$

5. $\left(-\frac{5a^4}{6a}\right)\left(-\frac{2}{a}\right)$ $\frac{5a^2}{3}, a \neq 0$

6. $\left(-\frac{10}{t^7}\right)\left(-\frac{3t^2}{25t}\right)$ $\frac{6}{5t^6}$

7. $\frac{10y^3}{6xy} \cdot \frac{4x^2y}{5y^2}$ $\frac{4xy}{3}, x \neq 0, y \neq 0$

8. $\frac{14x^2y^2}{3x^4} \cdot \frac{2x^4}{7y^3}$ $\frac{4x^2}{3y}, x \neq 0$

9. $\frac{y - 1}{5} \cdot \frac{5}{y - 1}$ $1, y \neq 1$

10. $\frac{x + 1}{2} \cdot \frac{2}{x + 1}$ $1, x \neq -1$

11. $\frac{x + 1}{2} \cdot \frac{4x}{x + 1}$ $2x, x \neq -1$

12. $\frac{x - 3}{6x} \cdot \frac{4}{x - 3}$ $\frac{2}{3x}, x \neq 3$

13. $\frac{1 - r}{3} \cdot \frac{3}{r - 1}$ $-1, r \neq 1$

14. $\frac{t - 6}{7} \cdot \frac{7}{6 - t}$ $-1, t \neq 6$

15. $\frac{y + 5}{y - 2} \cdot (2y)$ $\frac{2y(y + 5)}{y - 2}$

16. $\frac{z - 4}{z - 1} \cdot (-2z)$ $\frac{-2z(z - 4)}{z - 1}$

17. $\frac{3x}{5x - 15} \cdot (3 - x)$ $-\frac{3x}{5}, x \neq 3$

18. $\frac{7y}{24 - 6y} \cdot (y - 4)$ $-\frac{7y}{6}, y \neq 4$

19. $\frac{(x - 5)^2}{x + 5} \cdot \frac{x + 5}{x - 5}$ $x - 5, x \neq \pm 5$

20. $\frac{y + 2}{y - 2} \cdot \frac{(y - 2)^2}{y + 2}$ $y - 2, y \neq \pm 2$

21. $\dfrac{5}{x-1} \cdot \dfrac{x-1}{25(x-2)}$

$\dfrac{1}{5(x-2)}, x \neq 1$

22. $\dfrac{8}{r+3} \cdot \dfrac{r+3}{16(r-2)}$

$\dfrac{1}{2(r-2)}, r \neq -3$

23. $\dfrac{2-t}{2+t} \cdot \dfrac{t+2}{t-2}$

$-1, t \neq \pm 2$

24. $\dfrac{1-z}{1+z} \cdot \dfrac{z+1}{z-1}$

$-1, z \neq \pm 1$

25. $\dfrac{9y-15z}{7y+14z} \cdot \dfrac{2y+4z}{3y-5z}$

$\dfrac{6}{7}, y \neq -2z, 3y \neq 5z$

26. $\dfrac{4y-16x}{5y+15x} \cdot \dfrac{2y+6x}{y-4x}$

$\dfrac{8}{5}, y \neq 4x, y \neq -3x$

27. $\dfrac{r}{r-t} \cdot \dfrac{r^2-t^2}{r^2}$

$\dfrac{r+t}{r}, r \neq t$

28. $\dfrac{y^2-16}{2y^3} \cdot \dfrac{4y}{y^2-6y+8}$

$\dfrac{2(y+4)}{y^2(y-2)}, y \neq 4$

29. $(x^2-4) \cdot \dfrac{x}{(x-2)^2}$

$\dfrac{x(x+2)}{x-2}$

30. $(u-2)^2 \cdot \dfrac{u+2}{u-2}$

$u^2-4, u \neq 2$

31. $\dfrac{x-3}{x^2-16} \cdot \dfrac{x+4}{2x^2-6x}$ $\dfrac{1}{2x(x-4)}, x \neq -4, x \neq 3$

32. $\dfrac{x+7}{3x^2-15x} \cdot \dfrac{x-5}{x^2-49}$ $\dfrac{1}{3x(x-7)}, x \neq -7, x \neq 5$

33. $\dfrac{t^2-t-6}{t^2+6t+9} \cdot \dfrac{t+3}{t^2-4}$ $\dfrac{t-3}{(t+3)(t-2)}, t \neq -2$

34. $\dfrac{2x^2-9x+9}{8x-12} \cdot \dfrac{2x}{x^2-3x}$ $\dfrac{1}{2}, x \neq 0, x \neq \dfrac{3}{2}, x \neq 3$

35. $\dfrac{x^2+x-2}{x^3+x^2} \cdot \dfrac{x}{x^2+3x+2}$ $\dfrac{x-1}{x(x+1)^2}, x \neq -2$

36. $\dfrac{x^2-25}{x^2-3x-10} \cdot \dfrac{x+2}{x}$ $\dfrac{x+5}{x}, x \neq -2, x \neq 5$

37. $\dfrac{t^2+2t-3}{t^2+4t-5} \cdot \dfrac{t^2-3t-10}{t^2+5t+6}$

$\dfrac{t-5}{t+5}, t \neq 1, t \neq -3, t \neq -2$

38. $\dfrac{x^2+5x+4}{x^2-6x+8} \cdot \dfrac{x^2+5x-14}{x^2+8x+7}$

$\dfrac{x+4}{x-4}, x \neq 2, x \neq -7, x \neq -1$

39. $\dfrac{4}{x} \cdot \dfrac{x+2}{x+6} \cdot \dfrac{x}{x+2}$ $\dfrac{4}{x+6}, x \neq 0, x \neq -2$

40. $\dfrac{x}{7} \cdot \dfrac{x-7}{x} \cdot \dfrac{x+1}{x-7}$ $\dfrac{x+1}{7}, x \neq 0, x \neq 7$

41. $\dfrac{a+1}{a-1} \cdot \dfrac{a^2-2a+1}{a} \cdot (3a^2+3a)$

$3(a+1)^2(a-1), a \neq 0, a \neq 1$

42. $\dfrac{z^2-z-2}{z} \cdot \dfrac{2z^2+3z}{2z+3} \cdot \dfrac{z}{z-2}$

$z(z+1), z \neq 0, z \neq -\dfrac{3}{2}, z \neq 2$

43. $\dfrac{2}{z+3} \cdot \dfrac{z^2+6z+9}{z-3} \cdot \dfrac{4}{z^2-9}$ $\dfrac{8}{(z-3)^2}, z \neq -3$

44. $(x+5)^2 \cdot \dfrac{x}{x^2-25} \cdot \dfrac{x^2-x-20}{x^2}$

$\dfrac{(x+4)(x+5)}{x}, x \neq \pm 5$

In Exercises 45–50, fill in the missing factor.

45. $\dfrac{x-7}{x+10} \cdot \dfrac{x-3}{x-7} = \dfrac{x-3}{x+10}, x \neq 7$

46. $\dfrac{2x+1}{x-11} \cdot \dfrac{x-11}{x-1} = \dfrac{2x+1}{x-1}, x \neq 11$

47. $\dfrac{x^2-2x-8}{x-5} \cdot \dfrac{x-5}{x-4} = x+2,$

$x \neq 4, x \neq 5$

48. $\dfrac{x-1}{x^2+8x+12} \cdot \dfrac{(x+6)^2(x+2)}{x-1} = x+6,$

$x \neq -6, x \neq -2, x \neq -1$

49. $\dfrac{x+3}{x-4} \cdot \dfrac{(x-4)(x+2)}{x+3} = x+2,$

$x \neq -3, x \neq 4$

50. $\dfrac{x-5}{x+2} \cdot \dfrac{(x+2)(2x-3)}{x-5} = 2x-3,$

$x \neq -2, x \neq 5$

In Exercises 51–56, find the reciprocal of each expression.

51. $\dfrac{3-x}{x^2+4}$ $\dfrac{x^2+4}{3-x}$

52. $\dfrac{x+7}{2x-7}$ $\dfrac{2x-7}{x+7}$

53. $\dfrac{x^2+2x-5}{x^2-4x+7}$ $\dfrac{x^2-4x+7}{x^2+2x-5}$

54. $\dfrac{x^2-3x+1}{x^2+7x-1}$ $\dfrac{x^2+7x-1}{x^2-3x+1}$

55. a^3-8a $\dfrac{1}{a^3-8a}$

56. x^2-4x $\dfrac{1}{x^2-4x}$

In Exercises 57–76, divide and simplify. See Examples 6–9.

57. $\dfrac{x^3}{6} \div \dfrac{x^2}{3}$ $\dfrac{x}{2}, x \neq 0$

58. $\dfrac{5}{x} \div \dfrac{5}{x}$ $1, x \neq 0$

59. $\dfrac{7x^2}{10} \div \dfrac{14x^3}{15}$ $\dfrac{3}{4x}$

60. $\dfrac{2x}{3} \div \dfrac{4x^2}{15}$ $\dfrac{5}{2x}$

61. $\dfrac{a}{a + 1} \div \dfrac{6}{(a + 1)^2}$ $\dfrac{a(a + 1)}{6}, a \neq -1$

62. $\dfrac{z + 3}{6} \div \dfrac{z}{2}$ $\dfrac{z + 3}{3z}$

63. $\dfrac{3(x + 4)}{4} \div \dfrac{x + 4}{2}$ $\dfrac{3}{2}, x \neq -4$

64. $\dfrac{10(x - 3)}{7} \div \dfrac{x - 3}{14}$ $20, x \neq 3$

65. $\dfrac{(2x)^2}{(x + 2)^2} \div \dfrac{2x}{(x + 2)^3}$
$2x(x + 2), x \neq -2, x \neq 0$

66. $\dfrac{3(x + 1)^2}{5x} \div \dfrac{9(x + 1)}{10x^2}$
$\dfrac{2x(x + 1)}{3}, x \neq -1, x \neq 0$

67. $\dfrac{y^2 - 4}{y^2} \div \dfrac{y - 2}{3y}$ $\dfrac{3(y + 2)}{y}, y \neq 2$

68. $\dfrac{5x - 25}{x^2} \div \dfrac{x^2 - 25}{2x}$ $\dfrac{10}{x(x + 5)}, x \neq 5$

69. $\dfrac{x^2 - 7x + 12}{x + 4} \div (3 - x)$ $\dfrac{4 - x}{4 + x}, x \neq 3$

70. $\dfrac{x^2 - x - 2}{x + 2} \div (2 - x)$ $-\dfrac{x + 1}{x + 2}, x \neq 2$

71. $(x - 3) \div \dfrac{x^2 + 3x - 18}{x}$ $\dfrac{x}{x + 6}, x \neq 0, x \neq 3$

72. $(x + 3) \div \dfrac{3x^2 + 18x + 27}{x^2 + 1}$ $\dfrac{x^2 + 1}{3(x + 3)}$

73. $\dfrac{x + 2}{7 - x} \div \dfrac{x^2 - 5x + 6}{x^2 - 9x + 14}$ $-\dfrac{x + 2}{x - 3}, x \neq 7, x \neq 2$

74. $\dfrac{x + 6}{4 - x} \div \dfrac{x^2 + 9x + 18}{x^2 - 3x - 4}$
$-\dfrac{x + 1}{x + 3}, x \neq 4, x \neq -6, x \neq -1$

75. $\dfrac{a^2 - 10a + 25}{a^2 + 7a + 12} \div \dfrac{a^2 - a - 20}{a^2 + 6a + 9}$
$\dfrac{(a - 5)(a + 3)}{(a + 4)^2}, a \neq 5, a \neq -3$

76. $\dfrac{a^2 + 5a + 4}{a^2 - 2a + 1} \div \dfrac{a^2 + 8a + 16}{a^2 - 5a - 6}$
$\dfrac{(a + 1)^2(a - 6)}{(a - 1)^2(a + 4)}, a \neq -1, a \neq 6$

In Exercises 77–82, perform the indicated operations and simplify.

77. $\left(\dfrac{x^2}{5} \cdot \dfrac{x + a}{2}\right) \div \dfrac{x}{30}$ $3x(x + a), x \neq 0$

78. $\left(\dfrac{4u^2}{3} \cdot \dfrac{5}{u}\right) \div \dfrac{6u^2}{4}$ $\dfrac{40}{9u}$

79. $\left[\left(\dfrac{x + 2}{3}\right)^2 \cdot \left(\dfrac{x + 1}{2}\right)^2\right] \div \dfrac{(x + 1)(x + 2)}{36}$
$(x + 2)(x + 1), x \neq -2, x \neq -1$

80. $\left[\left(\dfrac{4}{x - 1}\right)^2 \cdot \left(\dfrac{x + 1}{3}\right)^3\right] \div \dfrac{(x + 1)^2}{27(x - 1)}$
$\dfrac{16(x + 1)}{x - 1}, x \neq -1$

81. $\left[\dfrac{(t + 2)^3}{(t + 1)^3} \div \dfrac{t^2 + 4t + 4}{t^2 + 2t + 1}\right] \cdot \dfrac{t + 1}{t + 2}$
$1, t \neq -1, t \neq -2$

82. $\left(\dfrac{3y^3 + 6y^2}{y^2 - y - 12} \div \dfrac{y^2 - y}{y^2 - 2y - 8}\right) \cdot \dfrac{y^2 + 5y + 6}{y^2}$
$\dfrac{3(y + 2)^3}{y(y - 1)}, y \neq 4, y \neq -3$

Solving Problems

83. *Pump Rate*   A pump in a well can pump water at the rate of 24 gallons per minute. Determine the time required to pump (a) 1 gallon, (b) x gallons, and (c) 120 gallons.

(a) $\dfrac{1}{24}$ minute   (b) $\dfrac{x}{24}$ minutes   (c) 5 minutes

84. *Photocopy Rate*   A photocopier produces copies at a rate of 12 pages per minute. Find the time required to copy (a) 1 page, (b) x pages, and (c) 32 pages.

(a) $\dfrac{1}{12}$ minute   (b) $\dfrac{x}{12}$ minutes   (c) $\dfrac{8}{3}$ minutes

85.   *Geometry*   The base and height of a triangle are given by

$$\dfrac{8}{x^2 + 5x} \quad \text{and} \quad \dfrac{x + 5}{2}$$

respectively. (Assume $x > 0$.)

(a) Write an expression for the area of the triangle in terms of x. Simplify the expression. $\dfrac{2}{x}$

(b) As x increases, determine whether each of the following increases or decreases: (i) base, (ii) height, and (iii) area.   Base: decreases; Height: increases; Area: decreases

86. *Analyzing Data*   The number N (in millions) of subscribers and the revenue R (in millions of dollars) for basic cable TV in the United States for the years 1990 through 2000 can be modeled by

$$N = \frac{-3.190t + 50.63}{0.0027t^2 - 0.099t + 1.00},\ 0 \le t \le 10$$

$$R = \frac{1101.78t + 10{,}003.5}{-0.016t + 1.00},\ 0 \le t \le 10$$

where t represents the year, with $t = 0$ corresponding to 1990.   (Source: Paul Kagan Associates, Inc.)

(a) Find a model for the average yearly basic cable TV rate per subscriber. (Do not simplify.)

$$\frac{1101.78t + 10{,}003.5}{-0.016t + 1.00} \div \frac{-3.190t + 50.63}{0.0027t^2 - 0.099t + 1.00}$$

(b) Use the model from part (a) to complete the table.

Year, t	0	2	4
Yearly rate	197.58	231.64	263.12

Year, t	6	8	10
Yearly rate	293.68	327.27	374.11

Explaining Concepts

87. *Writing*   Explain how to multiply two rational expressions.   To multiply two rational expressions, multiply numerators, multiply denominators, and simplify.

88. *Writing*   Explain how to divide two rational expressions.   To divide two rational expressions, invert the divisor and multiply.

89. *Writing*   Define the reciprocal of a rational expression.

The reciprocal of $\dfrac{a}{b}$ is $\dfrac{b}{a}$.

90. What is the product of an algebraic expression and its reciprocal? What basic rule of algebra does this demonstrate?   1; Multiplicative Inverse Property

91. Give an example of three rational expressions whose product is 1. There are many correct answers.

$$\frac{x + 2}{x} \cdot \frac{x^2}{(x + 2)^3} \cdot \frac{(x + 2)^2}{x}$$

92. *True or False?* $10 \div x = \frac{1}{10} \cdot x$ Justify your answer.

False. $10 \div x = \dfrac{10}{x}$

93. *Error Analysis*   Describe the error.

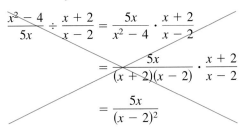

The dividend was mistakenly inverted. Instead, the divisor should have been inverted.

94. The following expressions have the same value when $x = 4$.

$$\frac{3x^2 + 4x - 4}{2x^2 + 5x + 2}, \qquad \frac{x + 6}{2x + 1}, \qquad \frac{3x - 2}{2x + 1}$$

Are they all equivalent? Is there an equivalent pair? Are there any excluded values?

No; $\dfrac{3x^2 + 4x - 4}{2x^2 + 5x + 2} = \dfrac{(3x - 2)(x + 2)}{(2x + 1)(x + 2)} = \dfrac{3x - 2}{2x + 1}$; $x = -\frac{1}{2}$ and $x = -2$ are excluded from the first expression, and $x = -\frac{1}{2}$ is excluded from each of the other expressions.

7.3 Adding and Subtracting Rational Expressions

Dwayne Newton/PhotoEdit

What You Should Learn

1️⃣ Add or subtract rational expressions with like denominators.

2️⃣ Find the least common multiple of two or more polynomials.

3️⃣ Add or subtract rational expressions with unlike denominators.

Why You Should Learn It

Addition and subtraction of rational expressions can be used to solve real-life applications. For instance, in Exercise 106 on page 402, you will add two rational expressions to find the total number of military personnel in the United States.

1️⃣ **Add or subtract rational expressions with like denominators.**

Rational Expressions with Like Denominators

As with numerical fractions, the procedure used to add or subtract two rational expressions depends on whether they have *like* or *unlike* denominators.

Combining Rational Expressions with Like Denominators

Let a, b, and c represent real numbers, variables, or algebraic expressions such that $c \neq 0$.

1. $\dfrac{a}{c} + \dfrac{b}{c} = \dfrac{a+b}{c}$ Add fractions with like denominators.

2. $\dfrac{a}{c} - \dfrac{b}{c} = \dfrac{a-b}{c}$ Subtract fractions with like denominators.

Example 1   Adding and Subtracting with Like Denominators

a. $\dfrac{x}{3} + \dfrac{2-x}{3} = \dfrac{x+(2-x)}{3} = \dfrac{2}{3}$ Add numerators.

b. $\dfrac{5}{x+4} - \dfrac{2x}{x+4} = \dfrac{5-2x}{x+4}$ Subtract numerators.

After adding or subtracting two (or more) rational expressions, check the resulting fraction to see if it can be simplified.

Example 2   Adding Rational Expressions and Simplifying

$$\frac{x}{x^2-4} + \frac{2}{x^2-4} = \frac{x+2}{x^2-4}$$ Add numerators.

$$= \frac{(x+2) \cdot 1}{(x+2)(x-2)}$$ Factor and divide out common factor.

$$= \frac{1}{x-2}, \quad x \neq -2$$ Simplified form

Example 3 **Subtracting Rational Expressions and Simplifying**

Subtract the rational expressions.

$$\frac{x}{x^2 - 2x} - \frac{5x - x^2}{x^2 - 2x}$$

Solution

$$\frac{x}{x^2 - 2x} - \frac{5x - x^2}{x^2 - 2x} = \frac{x - (5x - x^2)}{x^2 - 2x} \qquad \text{Subtract numerators.}$$

$$= \frac{x - 5x + x^2}{x^2 - 2x} \qquad \text{Distributive Property}$$

$$= \frac{x^2 - 4x}{x^2 - 2x} \qquad \text{Combine like terms.}$$

$$= \frac{\cancel{x}(x - 4)}{\cancel{x}(x - 2)} \qquad \begin{array}{l}\text{Factor and divide out}\\ \text{common factor.}\end{array}$$

$$= \frac{x - 4}{x - 2}, x \neq 0 \qquad \text{Simplified form}$$

To add or subtract rational expressions with *opposite* denominators (denominators that differ only in sign), multiply the numerator and denominator of either expression by -1. Then both expressions will have the same denominator. This is demonstrated in Example 4.

Example 4 **Adding Rational Expressions with Opposite Denominators**

Add the rational expressions.

$$\frac{2x}{x - 4} + \frac{5x - 7}{4 - x}$$

Solution

$$\frac{2x}{x - 4} + \frac{5x - 7}{4 - x} = \frac{2x}{x - 4} + \frac{(-1)(5x - 7)}{(-1)(4 - x)} \qquad \begin{array}{l}\text{Multiply numerator and}\\ \text{denominator by } -1.\end{array}$$

$$= \frac{2x}{x - 4} + \frac{(-5x + 7)}{(-4 + x)} \qquad \text{Distributive Property}$$

$$= \frac{2x}{x - 4} + \frac{(-5x + 7)}{x - 4} \qquad \text{Rewrite } -4 + x \text{ as } x - 4.$$

$$= \frac{2x - 5x + 7}{x - 4} \qquad \text{Add numerators.}$$

$$= \frac{7 - 3x}{x - 4} \qquad \text{Combine like terms.}$$

2 Find the least common multiple of two or more polynomials.

Finding least common multiples is an important skill. Here are some additional examples:

Polynomials

a. $6x^2, 9x$

b. $5a, a + 2$

c. $y - 2, y - 6$

d. $k^2 + 6k, 3k^3$

e. $m - 3, 3 - m$

f. $x^2 - 3x - 10, x^2 - 25$

Least Common Multiples

a. $18x^2$

b. $5a(a + 2)$

c. $(y - 2)(y - 6)$

d. $3k^3(k + 6)$

e. $(-1)(m - 3)$

f. $(x - 5)(x + 2)(x + 5)$

Study Tip

Remember that the exponent on a number or variable factor indicates the number of repetitions of that factor. For example, in the term $25x^3$, the factorization $5^2 \cdot x^3$ shows that 5 is repeated twice and x is repeated three times.

Least Common Multiple

To add or subtract rational expressions with *unlike* denominators, you must first rewrite each expression using the **least common multiple (LCM)** of the denominators of the individual expressions. The least common multiple of two (or more) polynomials is the simplest polynomial that is a multiple of each of the original polynomials.

Guidelines for Finding the Least Common Multiple

1. Factor each polynomial completely.

2. The least common multiple must contain all the *different* factors of the polynomials and each such factor must be repeated the maximum number of times it occurs in any of the factorizations.

Example 5   Finding Least Common Multiples

Find the least common multiple of each set of polynomials.

a. $5x, x^2$ **b.** $x^2 - x, x - 1, x$

c. $3x^2 + 6x, x^2 + 4x + 4$ **d.** $9 - x^2, x^2 - x - 6$

Solution

a. These polynomials factor as

$$5x = 5 \cdot x \quad \text{and} \quad x^2 = x \cdot x.$$

The different factors are 5 and x^2. This implies that the least common multiple is $5 \cdot x \cdot x = 5x^2$.

b. These polynomials factor as

$$x^2 - x = x(x - 1), x - 1, \text{ and } x.$$

The different factors are x and $x - 1$. This implies that the least common multiple is $x(x - 1)$.

c. These polynomials factor as

$$3x^2 + 6x = 3x(x + 2) \quad \text{and} \quad x^2 + 4x + 4 = (x + 2) \cdot (x + 2).$$

The different factors are 3, x, and $(x + 2)^2$. This implies that the least common multiple is $3x(x + 2) \cdot (x + 2) = 3x(x + 2)^2$.

d. These polynomials factor as

$$9 - x^2 = (-1) \cdot (x - 3) \cdot (x + 3) \quad \text{and}$$

$$x^2 - x - 6 = (x - 3) \cdot (x + 2).$$

The different factors are -1, $x - 3$, $x + 3$, and $x + 2$. This implies that the least common multiple is $(-1)(x - 3)(x + 3)(x + 2)$.

3 Add or subtract rational expressions with unlike denominators.

Rational Expressions with Unlike Denominators

To add or subtract rational expressions with *unlike* denominators, you must first rewrite the expressions so that they have *like* denominators. The like denominator that you use is the least common multiple of the original denominators and is called the **least common denominator (LCD)** of the original rational expressions.

Example 6 Adding Rational Expressions with Unlike Denominators

Add the rational expressions $\dfrac{5}{3x} + \dfrac{7}{4x}$.

Solution

By factoring the denominators, $3x = 3 \cdot x$ and $4x = 2^2 \cdot x$, you can conclude that the least common denominator of the fractions is $2^2 \cdot 3 \cdot x = 12x$.

$$\frac{5}{3x} + \frac{7}{4x} = \frac{5(4)}{3x(4)} + \frac{7(3)}{4x(3)}$$ Rewrite expressions using LCD of $12x$.

$$= \frac{20}{12x} + \frac{21}{12x}$$ Like denominators

$$= \frac{20 + 21}{12x}$$ Add numerators.

$$= \frac{41}{12x}$$ Simplified form

Advise students to be especially careful with signs when subtracting more than one term.

Example 7 Subtracting Rational Expressions with Unlike Denominators

Subtract the rational expressions $\dfrac{4}{x - 2} - \dfrac{2}{x + 1}$.

Solution

The only factors of the denominators are $x - 2$ and $x + 1$. So, the least common denominator is $(x - 2)(x + 1)$.

$$\frac{4}{x - 2} - \frac{2}{x + 1}$$

$$= \frac{4(x + 1)}{(x - 2)(x + 1)} - \frac{2(x - 2)}{(x - 2)(x + 1)}$$ Rewrite expressions using LCD of $(x - 2)(x + 1)$.

$$= \frac{4(x + 1) - 2(x - 2)}{(x - 2)(x + 1)}$$ Subtract numerators.

$$= \frac{4x + 4 - 2x + 4}{(x - 2)(x + 1)}$$ Distributive Property

$$= \frac{2x + 8}{(x - 2)(x + 1)}$$ Simplified form

Study Tip

Notice that the subtraction of numerators in Example 7 means that the minus sign has to be distributed over the quantity $(2x - 4)$. That is,

$$-(2x - 4) = -2x + 4.$$

Example 8   Adding Rational Expressions with Unlike Denominators

$$\frac{2x}{x^2 - 5x + 6} + \frac{1}{3 - x}$$

$$= \frac{2x}{x^2 - 5x + 6} + \frac{1}{(-1)(x - 3)} \qquad 3 - x = -1(x - 3)$$

$$= \frac{2x}{(x - 2)(x - 3)} - \frac{1}{x - 3} \qquad \text{Factor denominator.}$$

$$= \frac{2x}{(x - 2)(x - 3)} - \frac{x - 2}{(x - 2)(x - 3)} \qquad \begin{array}{l}\text{Rewrite expressions} \\ \text{using LCD of} \\ (x - 2)(x - 3).\end{array}$$

$$= \frac{2x - (x - 2)}{(x - 2)(x - 3)} \qquad \text{Subtract numerators.}$$

$$= \frac{2x - x + 2}{(x - 2)(x - 3)} \qquad \text{Distributive Property}$$

$$= \frac{x + 2}{(x - 2)(x - 3)} \qquad \text{Simplified form}$$

Example 9   Combining Three Rational Expressions

Perform the indicated operations and simplify: $\dfrac{2x - 5}{6x + 9} - \dfrac{4}{2x^2 + 3x} + \dfrac{1}{x}$.

Solution

$$\frac{2x - 5}{6x + 9} - \frac{4}{2x^2 + 3x} + \frac{1}{x}$$

$$= \frac{2x - 5}{3(2x + 3)} - \frac{4}{x(2x + 3)} + \frac{1}{x} \qquad \text{Factor denominators.}$$

$$= \frac{(2x - 5)(x)}{3(2x + 3)(x)} - \frac{(4)(3)}{x(2x + 3)(3)} + \frac{3(2x + 3)}{x(3)(2x + 3)} \qquad \begin{array}{l}\text{Rewrite expressions} \\ \text{using LCD of} \\ 3(2x + 3)(x).\end{array}$$

$$= \frac{(2x - 5)(x) - (4)(3) + 3(2x + 3)}{3x(2x + 3)} \qquad \text{Combine numerators.}$$

$$= \frac{2x^2 - 5x - 12 + 6x + 9}{3x(2x + 3)} \qquad \text{Distributive Property}$$

$$= \frac{2x^2 + x - 3}{3x(2x + 3)} \qquad \text{Combine like terms.}$$

$$= \frac{(x - 1)(2x + 3)}{3x(2x + 3)} \qquad \begin{array}{l}\text{Factor and divide out} \\ \text{common factor.}\end{array}$$

$$= \frac{x - 1}{3x}, \quad x \neq -\frac{3}{2} \qquad \text{Simplified form}$$

Example 10 **Combining Three Rational Expressions**

Perform the indicated operations and simplify.

$$\frac{4x}{x^2 - 16} + \frac{x}{x + 4} - 2$$

Solution

$$\frac{4x}{x^2 - 16} + \frac{x}{x + 4} - 2$$

$$= \frac{4x}{(x + 4)(x - 4)} + \frac{x}{x + 4} - \frac{2}{1} \qquad \text{Factor denominators.}$$

$$= \frac{4x}{(x + 4)(x - 4)} + \frac{x(x - 4)}{(x + 4)(x - 4)} - \frac{2(x + 4)(x - 4)}{(x + 4)(x - 4)}$$

$$= \frac{4x + x(x - 4) - 2(x^2 - 16)}{(x + 4)(x - 4)} \qquad \text{Combine numerators.}$$

$$= \frac{4x + x^2 - 4x - 2x^2 + 32}{(x + 4)(x - 4)} \qquad \text{Distributive Property}$$

$$= \frac{32 - x^2}{(x + 4)(x - 4)} \qquad \text{Combine like terms.}$$

To add or subtract *two* rational expressions, you can use the LCD method or the basic definition

$$\frac{a}{b} \pm \frac{c}{d} = \frac{ad \pm bc}{bd}, b \neq 0, d \neq 0. \qquad \text{Basic definition}$$

This definition provides an efficient way of adding or subtracting two rational expressions that have no common factors in their denominators.

Example 11 **Adding Rational Expressions**

Add the rational expressions.

$$\frac{x}{x - 3} + \frac{2}{3x + 4}$$

Solution

$$\frac{x}{x - 3} + \frac{2}{3x + 4} = \frac{x(3x + 4) + 2(x - 3)}{(x - 3)(3x + 4)} \qquad \text{Basic definition}$$

$$= \frac{3x^2 + 4x + 2x - 6}{(x - 3)(3x + 4)} \qquad \text{Distributive Property}$$

$$= \frac{3x^2 + 6x - 6}{(x - 3)(3x + 4)} \qquad \text{Combine like terms.}$$

7.3 Exercises

Review   Concepts, Skills, and Problem Solving

Keep mathematically in shape by doing these exercises *before* the problems of this section.

Properties and Definitions

1. Write the equation $3y - 7x = 4$ in the following forms.

    (a) Slope-intercept form $y = \frac{7}{3}x + \frac{4}{3}$

    (b) Point-slope form (more than one correct answer)
       $y - 6 = \frac{7}{3}(x - 2)$

    (c) General form $7x - 3y + 4 = 0$

2. *Writing* Explain how you can visually determine the sign of the slope of a line by observing its graph.
If the line rises from left to right, $m > 0$. If the line falls from left to right, $m < 0$.

Solving Equations

In Exercises 3–10, solve the equation.

3. $50 - z = 15$ 35 **4.** $x - 6 = 3x + 10$ -8

5. $\frac{1}{3}x + 5 = 8$ 9 **6.** $\frac{3}{4}x + \frac{1}{2} = 8$ 10

7. $-3(x + 2) = 0$ -2 **8.** $\left(\frac{1}{2}x - 3\right)(x + 1) = 0$ $-1, 6$

9. $4x^2 - 25 = 0$ $-\frac{5}{2}, \frac{5}{2}$ **10.** $2x^2 - 7x - 15 = 0$ $-\frac{3}{2}, 5$

Creating Expressions

▲ *Geometry*   In Exercises 11 and 12, write and simplify expressions for the perimeter and area of the figure.

11.

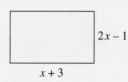

Perimeter: $6x + 4$
Area: $2x^2 + 5x - 3$

12.

Perimeter: $12x$
Area: $6x^2$

Developing Skills

In Exercises 1–16, combine and simplify. See Examples 1–3.

1. $\frac{y}{4} + \frac{3y}{4}$ y **2.** $\frac{2x}{5} - \frac{7x}{5}$ $-x$

3. $\frac{9}{x} - \frac{4}{x}$ $\frac{5}{x}$ **4.** $\frac{7}{z^2} + \frac{10}{z^2}$ $\frac{17}{z^2}$

5. $\frac{5}{3a} + \frac{9}{3a}$ $\frac{14}{3a}$ **6.** $\frac{16}{5z} - \frac{11}{5z}$ $\frac{1}{z}$

7. $\frac{x}{3} + \frac{1-x}{3}$ $\frac{1}{3}$ **8.** $\frac{4z}{3} - \frac{4z-3}{3}$ 1

9. $\frac{-6t}{9} - \frac{12-8t}{9}$ $\frac{2t-12}{9}$ **10.** $\frac{-16u}{7} - \frac{14-16u}{7}$ -2

11. $\frac{5-2x}{x-1} + \frac{x-4}{x-1}$ -1 **12.** $\frac{2x-1}{x+3} + \frac{1-x}{x+3}$ $\frac{x}{x+3}$

13. $\frac{3x-2}{(x+1)^2} + \frac{4x+5}{(x+1)^2}$ $\frac{7x+3}{(x+1)^2}$ **14.** $\frac{2x+1}{(x-5)^2} + \frac{3x-6}{(x-5)^2}$ $\frac{5x-5}{(x-5)^2}$

15. $\frac{5x-2}{x^2+x+1} - \frac{3x-4}{x^2+x+1}$ $\frac{2x+2}{x^2+x+1}$

16. $\frac{7y-10}{y^2-y-1} - \frac{6y+5}{y^2-y-1}$ $\frac{y-15}{y^2-y-1}$

In Exercises 17–24, combine and simplify. See Example 4.

17. $\frac{3}{x-2} + \frac{4}{2-x}$ $-\frac{1}{x-2}$ **18.** $\frac{1}{7-x} + \frac{8}{x-7}$ $\frac{7}{x-7}$

19. $\frac{5x}{2x-3} - \frac{9}{3-2x}$ $\frac{5x+9}{2x-3}$ **20.** $\frac{14}{6x-1} - \frac{8x}{1-6x}$ $\frac{8x+14}{6x-1}$

21. $\dfrac{4x-1}{x-11} + \dfrac{2x}{11-x}$

$\dfrac{2x-1}{x-11}$

22. $\dfrac{10x-3}{x-7} - \dfrac{5x}{7-x}$

$\dfrac{15x-3}{x-7}$

23. $\dfrac{3x+1}{x-15} + \dfrac{x-2}{15-x}$

$\dfrac{2x+3}{x-15}$

24. $\dfrac{5x-1}{12-x} + \dfrac{x+2}{x-12}$

$\dfrac{-4x+3}{x-12}$

In Exercises 25–44, find the least common multiple of the polynomials. See Example 5.

25. $2x, x^3$ $2x^3$

26. $2t^2, 24t$ $24t^2$

27. $9y^2, 12y$ $36y^2$

28. $x^3, 4x$ $4x^3$

29. $2(y-3), 6(y-3)$

$6(y-3)$

30. $4(x-1), 8(x-1)$

$8(x-1)$

31. $16x, 12x(x+2)$

$48x(x+2)$

32. $18y^2, 27y(y-3)$

$54y^2(y-3)$

33. $x, 3(x+5)$

$3x(x+5)$

34. $y, 5(y-2)$

$5y(y-2)$

35. $x-7, x^2, x(x+7)$

$x^2(x+7)(x-7)$

36. $x-1, x, 2(x+3)$

$2x(x-1)(x+3)$

37. $x^2-4, x(x+2)$

$x(x+2)(x-2)$

38. $x^2-25x, x^2(x-5)$

$x^2(x-5)(x-25)$

39. $7x^2-28x, x^2-3x-4$ $7x(x-4)(x+1)$

40. t^2+3t+9, t^2-9 $(t^2+3t+9)(t+3)(t-3)$

41. $x+2, x^2-4, x$

$x(x+2)(x-2)$

42. $x^2-1, x+1, x$

$x(x+1)(x-1)$

43. t^3+4t^2+4t, t^2-4t

$t(t+2)^2(t-4)$

44. y^3-y^2, y^4-y^2

$y^2(y-1)(y+1)$

In Exercises 45–52, fill in the missing factor. Assume all denominators are nonzero.

45. $\dfrac{3}{7x} = \dfrac{15}{7x\ (5)}$

46. $\dfrac{5a}{2} = \dfrac{5a\ (8)}{16}$

47. $\dfrac{3a}{7} = \dfrac{3a\ (a^2)}{7a^2}$

48. $\dfrac{5}{2x} = \dfrac{5x^2}{2x\ (x^2)}$

49. $\dfrac{x}{x+1} = \dfrac{x\ (x+1)}{(x+1)^2}$

50. $\dfrac{x}{x+3} = \dfrac{x\ (x+3)}{(x+3)^2}$

51. $\dfrac{2x}{x+2} = \dfrac{2x\ (2-x)}{4-x^2}$

52. $\dfrac{x^2}{5-x} = \dfrac{x^2\ (-x)}{x^2-5x}$

In Exercises 53–58, find the least common denominator of the expressions and rewrite one or both of the fractions using the least common denominator.

53. $\dfrac{x+5}{3x-6}, \dfrac{10}{x-2}$

$\dfrac{x+5}{3(x-2)}, \dfrac{30}{3(x-2)}$

54. $\dfrac{8x}{(x+2)}, \dfrac{3}{4x+8}$

$\dfrac{32x}{4(x+2)}, \dfrac{3}{4(x+2)}$

55. $\dfrac{2}{(x+3)^2}, \dfrac{5}{x(x+3)}$

$\dfrac{2x}{x(x+3)^2}, \dfrac{5(x+3)}{x(x+3)^2}$

56. $\dfrac{5t}{(t-3)^2}, \dfrac{4}{t(t-3)}$

$\dfrac{5t^2}{t(t-3)^2}, \dfrac{4(t-3)}{t(t-3)^2}$

57. $\dfrac{x-8}{x^2-16}, \dfrac{9x}{x^2-8x+16}$

$\dfrac{(x-8)(x-4)}{(x+4)(x-4)^2}, \dfrac{9x(x+4)}{(x+4)(x-4)^2}$

58. $\dfrac{3y}{y^2-y-6}, \dfrac{y+2}{y^2-3y}$

$\dfrac{3y^2}{y(y-3)(y+2)}, \dfrac{(y+2)^2}{y(y-3)(y+2)}$

In Exercises 59–100, combine and simplify. See Examples 6–11.

59. $\dfrac{3}{2s} - \dfrac{1}{5s}$ $\dfrac{13}{10s}$

60. $\dfrac{5}{6z} + \dfrac{3}{8z}$ $\dfrac{29}{24z}$

61. $\dfrac{1}{5x} - \dfrac{3}{5}$ $\dfrac{1-3x}{5x}$

62. $\dfrac{2}{3} + \dfrac{1}{2x}$ $\dfrac{4x+3}{6x}$

63. $\dfrac{5}{u} + \dfrac{2}{u^2}$ $\dfrac{5u+2}{u^2}$

64. $\dfrac{5}{z} + \dfrac{6}{z^2}$ $\dfrac{5z+6}{z^2}$

65. $\dfrac{3}{2b} + \dfrac{5}{2b^2}$ $\dfrac{3b+5}{2b^2}$

66. $\dfrac{8}{6u^2} - \dfrac{2}{9u}$ $\dfrac{2(6-u)}{9u^2}$

67. $\dfrac{4}{x-3} + \dfrac{4}{3-x}$

$0, x \neq 3$

68. $\dfrac{5}{6-t} - \dfrac{4}{t-6}$

$\dfrac{9}{6-t}$

69. $\dfrac{2x}{x-5} - \dfrac{5}{5-x}$

$\dfrac{2x+5}{x-5}$

70. $\dfrac{3}{x-2} + \dfrac{5}{2-x}$

$\dfrac{2}{2-x}$

71. $6 - \dfrac{5}{x+3}$

$\dfrac{6x+13}{x+3}$

72. $\dfrac{3}{x-1} - 5$

$\dfrac{-5x+8}{x-1}$

73. $7 + \dfrac{2}{2x - 3}$

$\dfrac{14x - 19}{2x - 3}$

74. $\dfrac{3}{2x - 5} + 2$

$\dfrac{4x - 7}{2x - 5}$

75. $\dfrac{3}{x - 5} + \dfrac{2}{x + 3}$

$\dfrac{5x - 1}{(x + 3)(x - 5)}$

76. $\dfrac{6}{x + 4} + \dfrac{3}{x - 1}$

$\dfrac{3(3x + 2)}{(x - 1)(x + 4)}$

77. $\dfrac{1}{x - 1} - \dfrac{1}{x + 2}$

$\dfrac{3}{(x + 2)(x - 1)}$

78. $\dfrac{3}{x - 4} - \dfrac{2}{x + 1}$

$\dfrac{x + 11}{(x - 4)(x + 1)}$

79. $\dfrac{3}{2(x - 4)} - \dfrac{1}{2x}$

$\dfrac{x + 2}{x(x - 4)}$

80. $\dfrac{1}{2x} + \dfrac{1}{2(x - 1)}$

$\dfrac{2x - 1}{2x(x - 1)}$

81. $\dfrac{x}{x^2 - 9} + \dfrac{3}{x + 3}$

$\dfrac{4x - 9}{(x + 3)(x - 3)}$

82. $\dfrac{6}{z + 2} - \dfrac{z - 3}{z^2 - 4}$

$\dfrac{5z - 9}{(z + 2)(z - 2)}$

83. $\dfrac{5v}{v(v + 4)} + \dfrac{2v}{v^2}$

$\dfrac{7v + 8}{v(v + 4)}$

84. $\dfrac{2t}{t^2} - \dfrac{t}{t(t + 1)}$

$\dfrac{t + 2}{t(t + 1)}$

85. $\dfrac{x + 2}{x^2 - 5x + 6} - \dfrac{3}{2 - x}$ $\dfrac{4x - 7}{(x - 2)(x - 3)}$

86. $\dfrac{2}{x + 1} + \dfrac{1 - x}{x^2 - 2x + 3}$ $\dfrac{x^2 - 4x + 7}{(x + 1)(x^2 - 2x + 3)}$

87. $\dfrac{x}{6x^2 + 7x + 2} - \dfrac{5}{2x^2 - 3x - 2}$

$\dfrac{x^2 - 17x - 10}{(3x + 2)(2x + 1)(x - 2)}$

88. $\dfrac{x}{5x^2 - 9x - 2} - \dfrac{2}{3x^2 - 7x + 2}$

$\dfrac{3x^2 - 11x - 2}{(5x + 1)(x - 2)(3x - 1)}$

89. $\dfrac{2x + 1}{x^2 - 16} + \dfrac{4x}{4 - x}$

$\dfrac{4x^2 + 14x - 1}{(4 + x)(4 - x)}$

90. $\dfrac{7x}{x - 5} - \dfrac{3x - 4}{25 - x^2}$

$\dfrac{7x^2 + 38x - 4}{(x + 5)(x - 5)}$

91. $\dfrac{3}{x} - \dfrac{1}{x^2} + \dfrac{1}{x + 1}$

$\dfrac{4x^2 + 2x - 1}{x^2(x + 1)}$

92. $\dfrac{6x}{x^2} - \dfrac{3}{x} + \dfrac{7}{x - 3}$

$\dfrac{10x - 9}{x(x - 3)}$

93. $\dfrac{x + 2}{3(x - 2)^2} + \dfrac{4}{3(x - 2)} + \dfrac{1}{2x}$ $\dfrac{13x^2 - 24x + 12}{6x(x - 2)^2}$

94. $\dfrac{5}{2(x + 1)} - \dfrac{1}{2x} - \dfrac{3}{2(x + 1)^2}$ $\dfrac{4x^2 - 1}{2x(x + 1)^2}$

95. $\dfrac{3x - 4}{4x + 18} + \dfrac{5 - x}{2x^2 + 9x} - \dfrac{3}{x}$ $\dfrac{3x^2 - 18x - 44}{2x(2x + 9)}$

96. $\dfrac{7 - 4x}{6x + 15} - \dfrac{x - 2}{2x^2 + 5x} + \dfrac{4}{x}$ $\dfrac{2(2x^2 - 14x - 33)}{3x(2x + 5)}$

97. $\dfrac{x}{x^2 - 4} + \dfrac{3x}{x + 2} + \dfrac{3x^2 - 5x}{4 - x^2}$ $0, x \neq -2, x \neq 2$

98. $\dfrac{x + 6}{4 - x^2} - \dfrac{x + 3}{x + 2} + \dfrac{x - 3}{2 - x}$ $\dfrac{-2x + 3}{x - 2}$

99. $\dfrac{5x}{x^2 + x - 6} + \dfrac{x}{x + 3} - \dfrac{2}{x - 2}$ $1, x \neq -3, x \neq 2$

100. $\dfrac{1}{x^2 + 3x + 2} - \dfrac{x}{x + 1} + \dfrac{3x}{x + 2}$ $\dfrac{2x^2 + x + 1}{(x + 2)(x + 1)}$

Solving Problems

101. *Work Rate* After two people work together for t hours on a common task, the fractional parts of the task done by each of the workers are $t/8$ and $t/7$. What fractional part of the task has been completed?

$\dfrac{15t}{56}$

102. *Work Rate* After two people work together for t hours on a common task, the fractional parts of the task done by each of the workers are $t/6$ and $t/9$. What fractional part of the task has been completed?

$\dfrac{5t}{18}$

103. ▲ *Geometry* Two angles are complementary (complements of each other) if their sum is 90°. One angle measures $(40 - x)/x$ degrees (see figure). Find an expression for the measure of its complement.

$\left(\dfrac{91x - 40}{x}\right)^{\circ}$

104. 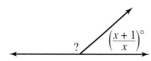 *Geometry*   Two angles are supplementary (supplements of each other) if their sum is 180°. One angle measures $(x + 1)/x$ degrees (see figure). Find an expression for the measure of its supplement. $\left(\dfrac{179x - 1}{x}\right)^\circ$

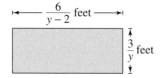

105. 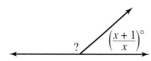 *Geometry*   Find an expression for the perimeter of the rectangle (see figure).
$P = \dfrac{6(3y - 2)}{y(y - 2)}$ feet

$\longleftarrow \dfrac{6}{y - 2}$ feet $\longrightarrow$

$\dfrac{3}{y}$ feet

106. *Analyzing Data*   The number of Army and Navy personnel A (in thousands) and the number of Marine Corps and Air Force personnel M (in thousands) in the United States for the years 1994 through 2002 can be modeled by

$$A = \frac{17.198t^2 + 516.22t + 1707.5}{t}, 4 \le t \le 12$$

$$M = \frac{43.25t - 690.5}{0.0029t^2 + 0.022t - 1}, 4 \le t \le 12$$

where t represents the year, with $t = 4$ corresponding to 1994. (Source: U.S. Department of Defense)

(a) Find a model for the total number of military personnel in the United States. (Do not simplify.)
$\dfrac{17.198t^2 + 516.22t + 1707.5}{t} + \dfrac{43.25t - 690.5}{0.0029t^2 + 0.022t - 1}$

(b) Use the model in part (a) to complete the table.

Year, t	4	5	6	7	8
Personnel	1610	1524	1468	1431	1407

Year, t	9	10	11	12
Personnel	1392	1385	1388	1404

Explaining Concepts

107. ✪ Answer parts (a)–(d) of Motivating the Chapter on page 372.

108. *Writing*   In your own words, explain how to add or subtract rational expressions with like denominators.   Add or subtract the numerators and place the result over the common denominator.

109. *Writing*   In your own words, explain how to add or subtract rational expressions with unlike denominators.   Rewrite each fraction in terms of the lowest common denominator, combine the numerators, and place the result over the lowest common denominator.

110. *Writing*   Explain how to find the least common multiple of two or more polynomials. Give an example.   Determine the prime factorization of each polynomial. The least common multiple contains each prime factor, repeated the maximum number of times it occurs in any one of the factorizations.
$6x + 18 = 2 \cdot 3(x + 3)$
$x^2 - 9 = (x + 3)(x - 3)$ LCM is $12(x + 3)(x - 3)^2$.
$4x^2 - 24x + 36 = 2^2(x - 3)^2$

111. Is it possible for the least common multiple of two polynomials to be the same as one of the polynomials? If so, give an example.   Yes. For example, for $2(x + 2)$ and $x + 2$, the least common multiple is $2(x + 2)$.

112. *Error Analysis*   Describe the error.

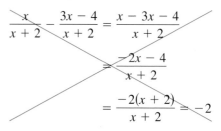

When the numerators are subtracted, the result should be $x - (3x - 4) = x - 3x + 4$.

Mid-Chapter Quiz

Take this quiz as you would take a quiz in class. After you are done, check your work against the answers in the back of the book.

1. The domain of a rational expression is the set of all real numbers for which the denominator of the expression is not equal to zero.
   (a) All real x (b) $x \neq \pm 2$

1. In your own words, explain the meaning of *domain*. Find the domain of
   (a) $\dfrac{x^2}{x^2 + 4}$ and   (b) $\dfrac{x^2}{x^2 - 4}$.

2. (a) $\frac{7}{12}$ (b) 0
   (c) Division by zero is undefined.

2. Evaluate $\dfrac{y - 3}{y + 2}$ for (a) $y = 10$, (b) $y = 3$, and (c) $y = -2$.

In Exercises 3–8, simplify the rational expression.

3. $\dfrac{14z^4}{35z}$ $\dfrac{2z^3}{5}, z \neq 0$

4. $\dfrac{15u(u - 3)^2}{25u^2(u - 3)}$ $\dfrac{3(u - 3)}{5u}, u \neq 3$

5. $\dfrac{24(9 - x)}{15(x - 9)}$ $-\dfrac{8}{5}, x \neq 9$

6. $\dfrac{y^2 - 4}{8 - 4y}$ $-\dfrac{y + 2}{4}, y \neq 2$

7. $\dfrac{1}{b - 1}, b \neq 0, b \neq -3$

7. $\dfrac{b^2 + 3b}{b^3 + 2b^2 - 3b}$

8. $\dfrac{2x - 3}{x + 1}, x \neq \frac{3}{2}$

8. $\dfrac{4x^2 - 12x + 9}{2x^2 - x - 3}$

In Exercises 9–18, perform the specified operation and simplify.

11. $\dfrac{5x^3}{x - 2}, x \neq -4$ **12.** $\dfrac{r - 4}{r(r + 4)}$

9. $\dfrac{3y^3}{5} \cdot \dfrac{25}{9y}$ $\dfrac{5y^2}{3}, y \neq 0$

10. $\dfrac{s - 5}{15} \cdot \dfrac{12s}{25 - s^2}$ $-\dfrac{4s}{5(s + 5)}, s \neq 5$

13. $\dfrac{2}{5(x + 2)^2}, x \neq 0$ **14.** $\dfrac{20}{y^3}, x \neq 0$

11. $(x^3 + 4x^2) \cdot \dfrac{5x}{x^2 + 2x - 8}$

12. $\dfrac{r^2 - 16}{r} \div (r + 4)^2$

15. $\dfrac{5x + 24}{(x + 6)(x - 6)}$

13. $\dfrac{x}{25} \div \dfrac{x^2 + 2x}{10} \cdot \dfrac{1}{x + 2}$

14. $\dfrac{10x^2}{3y} \div \left(\dfrac{y}{x} \cdot \dfrac{x^3y}{6}\right)$

16. $\dfrac{20x}{(x - 5)(x + 5)}$

17. $\dfrac{x^2 + 2x - 15}{(x + 4)^2}$

15. $\dfrac{x}{x^2 - 36} + \dfrac{4}{x - 6}$

16. $\dfrac{x + 5}{x - 5} - \dfrac{x - 5}{x + 5}$

18. $\dfrac{1}{x - 7}, x \neq -2$

17. $\dfrac{x - 3}{x^2 + 8x + 16} + \dfrac{x - 3}{x + 4}$

18. $\dfrac{5x + 10}{x^2 - 5x - 14} - \dfrac{4}{x - 7}$

19. (a) $\dfrac{10,000 + 25x}{x}$

19. A small business has a setup cost of \$10,000 for the production of downhill skis. The cost of labor and materials for producing each unit is \$25.
   (a) Write a rational expression that models the average cost per unit, when x units are produced.
   (b) Complete the table and describe any trends you find.   As the number of units increases, the cost per unit decreases.

x	2000	3000	4000	5000
Average cost	\$30.00	\$28.33	\$27.50	\$27.00

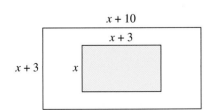

Figure for 20

20. Find the ratio of the area of the shaded region to the total area of the figure.
   $\dfrac{x}{x + 10}, x \neq -3$

7.4 Complex Fractions

EyeWire, Inc./Getty Images

What You Should Learn

① Simplify complex fractions using rules for dividing rational expressions.

② Simplify complex fractions having a sum or difference in the numerator and/or denominator.

Why You Should Learn It

Complex fractions can be used to model real-life situations. For instance, in Exercise 41 on page 410, a complex fraction is used to model the electrical resistance of two resistors.

① **Simplify complex fractions using rules for dividing rational expressions.**

Complex Fractions

Problems involving the division of two rational expressions are sometimes written as **complex fractions.** A complex fraction is a fraction that has a fraction in its numerator or denominator, or both. The rules for dividing rational expressions still apply. For instance, consider the following complex fraction.

$$\dfrac{\left(\dfrac{x+2}{3}\right)}{\left(\dfrac{x-2}{x}\right)}$$

$\longrightarrow$ Numerator fraction

         Main fraction line

         Denominator fraction

To perform the division implied by this complex fraction, invert the denominator fraction and multiply, as follows.

$$\dfrac{\left(\dfrac{x+2}{3}\right)}{\left(\dfrac{x-2}{x}\right)} = \dfrac{x+2}{3} \cdot \dfrac{x}{x-2} = \dfrac{x(x+2)}{3(x-2)}, \quad x \neq 0$$

Note that for complex fractions you make the main fraction line slightly longer than the fraction lines in the numerator and denominator. Example 1 shows how to simplify a complex numerical fraction.

Example 1   Simplifying a Complex Fraction

$$\dfrac{\left(\dfrac{5}{14}\right)}{\left(\dfrac{25}{8}\right)} = \dfrac{5}{14} \cdot \dfrac{8}{25}$$ Invert divisor and multiply.

$$= \dfrac{5 \cdot 2 \cdot 2 \cdot 2}{2 \cdot 7 \cdot 5 \cdot 5}$$ Multiply, factor, and divide out common factors.

$$= \dfrac{4}{35}$$ Simplified form

Example 2   **Simplifying a Complex Fraction**

Simplify the complex fraction.

$$\dfrac{\left(\dfrac{4y^3}{(5x)^2}\right)}{\left(\dfrac{(2y)^2}{10x^3}\right)}$$

Solution

$$\dfrac{\left(\dfrac{4y^3}{(5x)^2}\right)}{\left(\dfrac{(2y)^2}{10x^3}\right)} = \dfrac{4y^3}{25x^2} \cdot \dfrac{10x^3}{4y^2} \qquad \text{Invert divisor and multiply.}$$

$$= \dfrac{4y^2 \cdot y \cdot 2 \cdot 5x^2 \cdot x}{5 \cdot 5x^2 \cdot 4y^2} \qquad \text{Multiply and factor.}$$

$$= \dfrac{4y^2 \cdot y \cdot 2 \cdot 5x^2 \cdot x}{5 \cdot 5x^2 \cdot 4y^2} \qquad \text{Divide out common factors.}$$

$$= \dfrac{2xy}{5}, \quad x \neq 0, \ y \neq 0 \qquad \text{Simplified form}$$

Example 3   **Simplifying a Complex Fraction**

Simplify the complex fraction.

$$\dfrac{\left(\dfrac{x+1}{x+2}\right)}{\left(\dfrac{x+1}{x+5}\right)}$$

Solution

$$\dfrac{\left(\dfrac{x+1}{x+2}\right)}{\left(\dfrac{x+1}{x+5}\right)} = \dfrac{x+1}{x+2} \cdot \dfrac{x+5}{x+1} \qquad \text{Invert divisor and multiply.}$$

$$= \dfrac{(x+1)(x+5)}{(x+2)(x+1)} \qquad \text{Multiply numerators and denominators.}$$

$$= \dfrac{(x+1)(x+5)}{(x+2)(x+1)} \qquad \text{Divide out common factors.}$$

$$= \dfrac{x+5}{x+2}, \quad x \neq -1, \ x \neq -5 \qquad \text{Simplified form}$$

In Example 3, the domain of the complex fraction is restricted by the denominators in the original expression and by the denominators in the expression after the divisor has been inverted. So, the domain of the original expression is all real values of x except $x \neq -2$, $x \neq -5$, and $x \neq -1$.

Example 4   **Simplifying a Complex Fraction**

Simplify the complex fraction.

$$\frac{\left(\dfrac{x^2 + 4x + 3}{x - 2}\right)}{2x + 6}$$

Solution

$$\frac{\left(\dfrac{x^2 + 4x + 3}{x - 2}\right)}{2x + 6} = \frac{\left(\dfrac{x^2 + 4x + 3}{x - 2}\right)}{\left(\dfrac{2x + 6}{1}\right)} \qquad \text{Rewrite denominator.}$$

$$= \frac{x^2 + 4x + 3}{x - 2} \cdot \frac{1}{2x + 6} \qquad \text{Inverse divisor and multiply.}$$

$$= \frac{(x + 1)(x + 3)}{(x - 2)(2)(x + 3)} \qquad \text{Multiply and factor.}$$

$$= \frac{(x + 1)(\cancel{x + 3})}{(x - 2)(2)(\cancel{x + 3})} \qquad \text{Divide out common factor.}$$

$$= \frac{x + 1}{2(x - 2)}, \quad x \neq -3 \qquad \text{Simplified form}$$

---

② Simplify complex fractions having a sum or difference in the numerator and/or denominator.

Complex Fractions with Sums or Differences

Complex fractions can have numerators and/or denominators that are sums or differences of fractions. To simplify a complex fraction, combine its numerator and its denominator into single fractions. Then divide by inverting the denominator and multiplying.

Example 5   **Simplifying a Complex Fraction**

$$\frac{\left(\dfrac{x}{3} + \dfrac{2}{3}\right)}{\left(1 - \dfrac{2}{x}\right)} = \frac{\left(\dfrac{x}{3} + \dfrac{2}{3}\right)}{\left(\dfrac{x}{x} - \dfrac{2}{x}\right)} \qquad \text{Rewrite with least common denominators.}$$

$$= \frac{\left(\dfrac{x + 2}{3}\right)}{\left(\dfrac{x - 2}{x}\right)} \qquad \text{Add fractions.}$$

$$= \frac{x + 2}{3} \cdot \frac{x}{x - 2} \qquad \text{Invert divisor and multiply.}$$

$$= \frac{x(x + 2)}{3(x - 2)}, \quad x \neq 0 \qquad \text{Simplified form}$$

Another way of simplifying the complex fraction in Example 5 is to multiply the numerator and denominator by the least common denominator for all fractions in the numerator and denominator.

$$\frac{\left(\dfrac{x}{3} + \dfrac{2}{3}\right)}{\left(1 - \dfrac{2}{x}\right)} = \frac{\left(\dfrac{x}{3} + \dfrac{2}{3}\right)}{\left(1 - \dfrac{2}{x}\right)} \cdot \frac{3x}{3x}$$ $3x$ is the least common denominator.

$$= \frac{\dfrac{x}{3}(3x) + \dfrac{2}{3}(3x)}{(1)(3x) - \dfrac{2}{x}(3x)}$$ Distributive Property

$$= \frac{x^2 + 2x}{3x - 6} = \frac{x(x + 2)}{3(x - 2)}, \quad x \neq 0$$ Simplify.

As you can see, you obtain the same result as in Example 5.

Example 6   Simplifying a Complex Fraction

Simplify the complex fraction.

$$\frac{\left(\dfrac{2}{x + 2}\right)}{\left(\dfrac{3}{x + 2} + \dfrac{2}{x}\right)}$$

Solution

$$\frac{\left(\dfrac{2}{x + 2}\right)}{\left(\dfrac{3}{x + 2} + \dfrac{2}{x}\right)} = \frac{\left(\dfrac{2}{x + 2}\right)(x)(x + 2)}{\left(\dfrac{3}{x + 2} + \dfrac{2}{x}\right)(x)(x + 2)}$$ $x(x + 2)$ is the least common denominator.

$$= \frac{\left(\dfrac{2}{x + 2}\right)(x)(x + 2)}{\left(\dfrac{3}{x + 2}\right)(x)(x + 2) + \left(\dfrac{2}{x}\right)(x)(x + 2)}$$ Distributive Property

$$= \frac{2x}{3x + 2(x + 2)}$$ Multiply and simplify.

$$= \frac{2x}{3x + 2x + 4}$$ Distributive Property

$$= \frac{2x}{5x + 4}, \quad x \neq -2, x \neq 0$$ Simplify.

Notice that the numerator and denominator of the complex fraction were multiplied by $(x)(x + 2)$, which is the least common denominator of the fractions in the original complex fraction.

7.4 Exercises

Review   Concepts, Skills, and Problem Solving

Keep mathematically in shape by doing these exercises *before* the problems of this section.

Properties and Definitions

1. *Writing* ✎   In your own words, define a polynomial.
   A polynomial is an expression of the form
   $a_n x^n + a_{n-1} x^{n-1} + \cdots + a_2 x^2 + a_1 x + a_0$, where
   $a_n \neq 0$.

2. *Writing* ✎   Explain how to determine the degree of a polynomial.   The degree of a polynomial is the degree of the term with the highest power.

3. *Writing* ✎   Explain how to determine the leading coefficient of a polynomial.   The leading coefficient is the coefficient of the term with the highest power.

4. *Writing* ✎   Explain the FOIL Method of multiplying two binomials.   The FOIL method is a procedure that guarantees that each term of one binomial is multiplied by each term of the other binomial.
   $(ax + b)(cx + d) = ax(cx) + ax(d) + b(cx) + b(d)$.

Algebraic Operations

In Exercises 5–10, perform the indicated operations.

5. $(12x^2 + 4x - 8) + (-x^2 - 6x + 5)$ $11x^2 - 2x - 3$

6. $(-2x^2 + 8x - 7) - (3x^2 + 12)$ $-5x^2 + 8x - 19$

7. $-4x(-6x^2 + 4x - 2)$ $24x^3 - 16x^2 + 8x$

8. $(3x + 6)(-4x + 1)$ $-12x^2 - 21x + 6$

9. $\dfrac{8x^2 + 4x}{x}$ $8x + 4, x \neq 0$

10. $(x^2 + x - 12) \div (x - 3)$ $x + 4, x \neq 3$

Problem Solving

11. *Number Problem*   The sum of three consecutive integers is 39. Find the three integers.   12, 13, 14

12. ▲ *Geometry*   The length of a regulation tennis court is 6 feet more than twice its width. The perimeter of the court is 228 feet. Find the dimensions of the court.   78 feet × 36 feet

Developing Skills

In Exercises 1–20, simplify the complex fraction. See Examples 1–4.

1. $\dfrac{\left(\dfrac{3}{10}\right)}{\left(\dfrac{9}{15}\right)}$ $\dfrac{1}{2}$

2. $\dfrac{\left(\dfrac{7}{16}\right)}{\left(-\dfrac{4}{21}\right)}$ $-\dfrac{147}{64}$

3. $\dfrac{\left(-\dfrac{5}{8}\right)}{\left(\dfrac{4}{15}\right)}$ $-\dfrac{75}{32}$

4. $\dfrac{\left(\dfrac{2}{9}\right)}{\left(\dfrac{8}{45}\right)}$ $\dfrac{5}{4}$

5. $\dfrac{\left(\dfrac{3}{x}\right)}{\left(\dfrac{6}{x^2}\right)}$ $\dfrac{x}{2}, x \neq 0$

6. $\dfrac{\left(\dfrac{2}{3}\right)}{\left(\dfrac{u}{v}\right)}$ $\dfrac{2v}{3u}, v \neq 0$

7. $\dfrac{\left(\dfrac{x^3}{4}\right)}{\left(\dfrac{x}{8}\right)}$ $2x^2, x \neq 0$

8. $\dfrac{\left(\dfrac{y^4}{12}\right)}{\left(\dfrac{y}{16}\right)}$ $\dfrac{4y^3}{3}, y \neq 0$

9. $\dfrac{\left(\dfrac{8x^2 y}{3z^2}\right)}{\left(\dfrac{4xy}{9z^5}\right)}$ $6xz^3, x \neq 0, y \neq 0, z \neq 0$

10. $\dfrac{\left(\dfrac{36x^4}{5y^4 z^5}\right)}{\left(\dfrac{9xy^2}{15z^5}\right)}$ $\dfrac{12x^3}{y^6}, x \neq 0, z \neq 0,$

11. $\dfrac{\left[\dfrac{6x^3}{(5y)^2}\right]}{\left[\dfrac{(3x)^2}{15y^4}\right]}$ $\dfrac{2xy^2}{5}, x \neq 0, y \neq 0$

12. $\dfrac{\left[\dfrac{(3r)^3}{10t^4}\right]}{\left[\dfrac{9r}{(2t)^2}\right]}$ $\dfrac{6r^2}{5t^2}, r \neq 0$

13. $\dfrac{\left(\dfrac{y}{3-y}\right)}{\left(\dfrac{y^2}{y-3}\right)}$ $-\dfrac{1}{y}, y \neq 3$

14. $\dfrac{\left(\dfrac{x}{x-4}\right)}{\left(\dfrac{x}{4-x}\right)}$ $-1, x \neq 4, x \neq 0$

15. $\dfrac{\left(\dfrac{2x-10}{x+1}\right)}{\left(\dfrac{x-5}{x+1}\right)}$ $2, x \neq -1, x \neq 5$

16. $\dfrac{\left(\dfrac{a+5}{6a-15}\right)}{\left(\dfrac{a+5}{2a-5}\right)}$ $\dfrac{1}{3}, a \neq -5, a \neq \dfrac{5}{2}$

17. $\dfrac{\left(\dfrac{x^2+3x-10}{x+4}\right)}{3x-6}$ $\dfrac{x+5}{3(x+4)}, x \neq 2$

18. $\dfrac{\left(\dfrac{x^2-2x-8}{x-1}\right)}{5x-20}$ $\dfrac{x+2}{5(x-1)}, x \neq 4$

19. $\dfrac{\left(\dfrac{6x^2-17x+5}{3x^2+3x}\right)}{\left(\dfrac{3x-1}{3x+1}\right)}$ $\dfrac{(2x-5)(3x+1)}{3x(x+1)}, x \neq \pm\dfrac{1}{3}$

20. $\dfrac{\left(\dfrac{6x^2-13x-5}{5x^2+5x}\right)}{\left(\dfrac{2x-5}{5x+1}\right)}$ $\dfrac{(3x+1)(5x+1)}{5x(x+1)}, x \neq -\dfrac{1}{5}, x \neq \dfrac{5}{2}$

In Exercises 21–36, simplify the complex fraction. See Examples 5 and 6.

21. $\dfrac{\left(3+\dfrac{2}{3}\right)}{\left(1+\dfrac{1}{3}\right)}$ $\dfrac{11}{4}$

22. $\dfrac{\left(3+\dfrac{4}{5}\right)}{\left(1-\dfrac{9}{16}\right)}$ $\dfrac{304}{35}$

23. $\dfrac{\left(1+\dfrac{3}{y}\right)}{y}$ $\dfrac{y+3}{y^2}$

24. $\dfrac{x}{\left(\dfrac{5}{x}+2\right)}$ $\dfrac{x^2}{5+2x}, x \neq 0$

25. $\dfrac{\left(\dfrac{x}{2}\right)}{\left(2+\dfrac{3}{x}\right)}$ $\dfrac{x^2}{2(2x+3)}, x \neq 0$

26. $\dfrac{\left(1-\dfrac{2}{x}\right)}{\left(\dfrac{x}{2}\right)}$ $\dfrac{2(x-2)}{x^2}$

27. $\dfrac{\left(\dfrac{x}{3}-4\right)}{\left(5+\dfrac{1}{x}\right)}$ $\dfrac{x(x-12)}{3(5x+1)}, x \neq 0$

28. $\dfrac{\left(\dfrac{4}{x}+2\right)}{\left(\dfrac{1}{2x}-8\right)}$ $\dfrac{4(x+2)}{1-16x}, x \neq 0$

29. $\dfrac{\left(\dfrac{x}{4}-\dfrac{4}{x}\right)}{(x-4)}$ $\dfrac{x+4}{4x}, x \neq 4$

30. $\dfrac{x-5}{\left(\dfrac{x}{5}-\dfrac{5}{x}\right)}$ $\dfrac{5x}{x+5}, x \neq 0, x \neq 5$

31. $\dfrac{\left(\dfrac{1}{x}+\dfrac{1}{y}\right)}{\left(\dfrac{1}{x}-\dfrac{1}{y}\right)}$ $\dfrac{y+x}{y-x}, x \neq 0, y \neq 0$

32. $\dfrac{\left(\dfrac{4}{x^2}+\dfrac{2}{x}\right)}{\left(\dfrac{4}{x}+\dfrac{2}{x^2}\right)}$ $\dfrac{x+2}{2x+1}, x \neq 0$

33. $\dfrac{\left(\dfrac{10}{x+1}\right)}{\left(\dfrac{1}{2x+2}+\dfrac{3}{x+1}\right)}$ $\dfrac{20}{7}, x \neq -1$

34. $\dfrac{\left(\dfrac{2}{x+5}\right)}{\left(\dfrac{2}{x+5}+\dfrac{1}{4x+20}\right)}$ $\dfrac{8}{9}, x \neq -5$

35. $\dfrac{\left(\dfrac{1}{x}-\dfrac{1}{x+1}\right)}{\left(\dfrac{1}{x+1}\right)}$ $\dfrac{1}{x}, x \neq -1$

36. $\dfrac{\left(\dfrac{5}{y}-\dfrac{6}{2y+1}\right)}{\left(\dfrac{5}{2y+1}\right)}$ $\dfrac{4y+5}{5y}, y \neq -\dfrac{1}{2}$

Solving Problems

37. *Average of Two Numbers*   Determine the average of the two real numbers $x/5$ and $x/6$.
$\dfrac{11x}{60}$

38. *Average of Two Numbers*   Determine the average of the two real numbers $2x/3$ and $3x/5$.
$\dfrac{19x}{30}$

39. *Equal Lengths*   Find expressions for three real numbers that divide the real number line between $x/9$ and $x/6$ into four parts of equal length (see figure).

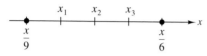

$\dfrac{x}{8}, \dfrac{5x}{36}, \dfrac{11x}{72}$

40. *Equal Lengths*   Find expressions for two real numbers that divide the real number line between $x/3$ and $5x/4$ into three parts of equal length (see figure).

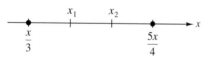

$\dfrac{23x}{36}, \dfrac{17x}{18}$

41. *Electrical Resistance*   When two resistors of resistances R_1 and R_2 are connected in parallel, the total resistance is modeled by

$$\dfrac{1}{\left(\dfrac{1}{R_1} + \dfrac{1}{R_2}\right)}. \quad \dfrac{R_1 R_2}{R_1 + R_2}$$

Simplify this complex fraction.

42. *Monthly Payment*   The approximate annual percent rate r of a monthly installment loan is

$$r = \dfrac{\left[\dfrac{24(MN - P)}{N}\right]}{\left(P + \dfrac{MN}{12}\right)}$$

where N is the total number of payments, M is the monthly payment, and P is the amount financed.

(a) Simplify the expression. $\dfrac{288(MN - P)}{N(MN + 12P)}$

(b) Find the annual percent rate for a four-year car loan of \$10,000 with monthly payments of \$300.
19.6%

Explaining Concepts

43. *Writing*   Define the term "complex fraction." Give an example and show how to simplify the fraction.

44. *Writing*   Of the two methods discussed in this section for simplifying complex fractions, select the method you prefer and explain the method in your own words.   Answers will vary.

43. A complex fraction is a fraction with a fraction in its numerator or denominator, or both.

$$\dfrac{\left(\dfrac{x + 1}{2}\right)}{\left(\dfrac{x + 1}{3}\right)}$$

Simplify by inverting the denominator and multiplying:

$$\left(\dfrac{x + 1}{2}\right) \cdot \left(\dfrac{3}{x + 1}\right) = \dfrac{3}{2}, x \neq -1.$$

45. What are the numerator and denominator of each complex fraction?

(a) $\dfrac{\left(\dfrac{x - 1}{5}\right)}{\left(\dfrac{2}{x^2 + 2x - 35}\right)}$

(b) $\dfrac{\left(\dfrac{1}{2y} + x\right)}{\left(\dfrac{3}{y} + x\right)}$

(a) Numerator: $\dfrac{x - 1}{5}$

Denominator: $\dfrac{2}{x^2 + 2x - 35}$

(b) Numerator: $\dfrac{1}{2y} + x$

Denominator: $\dfrac{3}{y} + x$

7.5 Rational Equations and Applications

Patrick Ward/Getty Images

What You Should Learn

1. Solve rational equations with constant denominators.
2. Solve rational equations with variable denominators.
3. Solve application problems using rational equations with variable denominators.

Why You Should Learn It

Rational equations can be used to model and solve real-life problems related to wildlife management. For instance, in Exercise 89 on page 420, a rational equation is used to model the population of deer in state game lands.

Equations Containing Constant Denominators

In Section 3.2, you learned how to solve equations containing fractions with *constant* denominators. Example 1 reviews that procedure.

1. Solve rational equations with constant denominators.

Study Tip

In Example 1, note that the key to solving an equation that involves fractions is to multiply each side of the equation by the least common denominator (LCD) of the fractions.

Example 1   An Equation Containing Constant Denominators

Solve (a) $\dfrac{x}{5} = 6 - \dfrac{x}{10}$ and (b) $\dfrac{x+6}{9} - \dfrac{x-2}{5} = \dfrac{4}{15}$.

Solution

a.

$$\frac{x}{5} = 6 - \frac{x}{10}$$
     Write original equation.

$$10\left(\frac{x}{5}\right) = 10\left(6 - \frac{x}{10}\right)$$
     Multiply each side by LCD of 10.

$$2x = 60 - x$$
     Distribute and simplify.

$$3x = 60$$
     Add x to each side.

$$x = 20$$
     Divide each side by 3.

The solution is $x = 20$. Check this in the original equation.

b.

$$\frac{x+6}{9} - \frac{x-2}{5} = \frac{4}{15}$$
     Write original equation.

$$45\left(\frac{x+6}{9} - \frac{x-2}{5}\right) = 45\left(\frac{4}{15}\right)$$
     Multiply each side by LCD of 45.

$$5(x+6) - 9(x-2) = 3(4)$$
     Distribute and simplify.

$$5x + 30 - 9x + 18 = 12$$
     Distributive Property

$$-4x + 48 = 12$$
     Combine like terms.

$$-4x = -36$$
     Subtract 48 from each side.

$$x = 9$$
     Divide each side by -4.

The solution is $x = 9$. Check this in the original equation.

Additional Examples

Solve each equation.

a. $\dfrac{x+1}{4} - \dfrac{x-3}{6} = \dfrac{3}{2}$

b. $\dfrac{3x-1}{4} - \dfrac{x}{2} = \dfrac{1}{4}$

Answers:

a. $x = 9$

b. $x = 2$

② Solve rational equations with variable denominators.

Equations Containing Variable Denominators

As stated in Section 7.1, the domain of a rational expression does not contain the values of a variable that make the denominator zero. This is especially critical in solving equations that contain variable denominators. You will see why in the examples that follow.

Example 2   An Equation Containing Variable Denominators

Solve $\dfrac{1}{x} - \dfrac{2}{3} = \dfrac{3}{x}$.

Solution

Begin by multiplying each side by the least common denominator $3x$.

$$\frac{1}{x} - \frac{2}{3} = \frac{3}{x} \qquad \text{Write original equation.}$$

$$3x\left(\frac{1}{x} - \frac{2}{3}\right) = 3x\left(\frac{3}{x}\right) \qquad \text{Multiply each side by LCD of } 3x.$$

$$\frac{3x}{x} - \frac{6x}{3} = \frac{9x}{x} \qquad \text{Distributive Property}$$

$$3 - 2x = 9, \quad x \neq 0 \qquad \text{Simplify.}$$

$$-2x = 6 \qquad \text{Subtract 3 from each side.}$$

$$x = -3 \qquad \text{Divide each side by } -2.$$

The solution is $x = -3$. You can check this in the original equation as follows.

Check

$$\frac{1}{x} - \frac{2}{3} = \frac{3}{x} \qquad \text{Write original equation.}$$

$$\frac{1}{-3} - \frac{2}{3} \overset{?}{=} \frac{3}{-3} \qquad \text{Substitute } -3 \text{ for } x.$$

$$-\frac{3}{3} \overset{?}{=} -1 \qquad \text{Simplify.}$$

$$-1 = -1 \qquad \text{Solution checks.} \checkmark$$

Students could use a graphing calculator to verify that the graph of $y = \dfrac{1}{x} - \dfrac{2}{3} - \dfrac{3}{x}$ has an x-intercept at -3.

In Example 2, notice that the original equation is a *rational equation*—that is, it contains rational expressions. After being multiplied by the least common denominator, the equation is converted into a *linear equation*. As you have seen repeatedly, this is a common strategy in mathematics—*to rewrite complicated problems into simpler forms.*

Notice that during the solution of Example 2, the restriction $x \neq 0$ is placed at the point where the equation becomes linear. The domain of the original equation is all real numbers except $x = 0$. The domain of the equation remains the same throughout the solution even though the form of the equation changes.

Throughout the text, the importance of checking solutions is emphasized. Up to this point, the main reason for checking has been to make sure that you did not make arithmetic errors in the solution process. In the next example, you will see that there is another reason for checking solutions in the *original* equation. That is, even with no mistakes in the solution process, it can happen that a "trial solution" does not satisfy the original equation. This type of solution is called an **extraneous solution.** An extraneous solution of an equation does not, by definition, satisfy the original equation and so *must not* be listed as an actual solution.

Example 3   An Equation with No Solution

Solve $\dfrac{2x}{x+3} = 1 - \dfrac{6}{x+3}$.

Solution

Remind students to check for extraneous solutions.

Begin by multiplying each side by the least common denominator $x + 3$.

$$\frac{2x}{x+3} = 1 - \frac{6}{x+3} \qquad \text{Write original equation.}$$

$$(x+3)\left(\frac{2x}{x+3}\right) = (x+3)\left(1 - \frac{6}{x+3}\right) \qquad \begin{array}{l}\text{Multiply each side by} \\ \text{LCD of } x + 3.\end{array}$$

$$\frac{(x+3)(2x)}{x+3} = (x+3) - \frac{(x+3)6}{x+3} \qquad \text{Distributive Property}$$

$$2x = (x+3) - 6, \quad x \neq -3 \qquad \text{Simplify.}$$

$$2x = x - 3 \qquad \text{Combine like terms.}$$

$$x = -3 \qquad \text{Subtract } x \text{ from each side.}$$

At this point, the solution appears to be $x = -3$. However, by performing a check, you can see that this "trial solution" is extraneous.

Check

Additional Examples
Solve each equation.

a. $\dfrac{2}{x} = \dfrac{3}{x-2} - 1$

b. $\dfrac{2}{x^2+1} + \dfrac{1}{x} = \dfrac{2}{x}$

Answers:

a. $x = 4, x = -1$

b. $x = 1$

$$\frac{2x}{x+3} = 1 - \frac{6}{x+3} \qquad \text{Write original equation.}$$

$$\frac{2(-3)}{-3+3} \overset{?}{=} 1 - \frac{6}{-3+3} \qquad \text{Substitute } -3 \text{ for } x.$$

$$\frac{-6}{0} \overset{?}{=} 1 - \frac{6}{0} \qquad \text{Solution does not check.} \ \times$$

Because the check results in *division by zero,* you can conclude that -3 is extraneous. So, the original equation has no solution.

Looking back at the original equation in Example 3, you can see that $x = -3$ is excluded from the domain of two of the fractions that occur in the equation. When solving rational equations, you may find it helpful to list the domain restrictions before beginning the solution process.

Example 4   An Equation with One Solution

Solve $\dfrac{6}{x-1} + \dfrac{2x}{x-2} = 2$.

Solution

$$\frac{6}{x-1} + \frac{2x}{x-2} = 2$$

$$(x-1)(x-2)\left(\frac{6}{x-1} + \frac{2x}{x-2}\right) = 2(x-1)(x-2)$$

$$6(x-2) + 2x(x-1) = 2(x^2 - 3x + 2), \quad x \neq 1, x \neq 2$$

$$6x - 12 + 2x^2 - 2x = 2x^2 - 6x + 4$$

$$10x = 16$$

$$x = \frac{8}{5} \qquad \text{Solution}$$

Check

$$\frac{6}{\frac{8}{5} - 1} + \frac{2\left(\frac{8}{5}\right)}{\frac{8}{5} - 2} \overset{?}{=} 2 \qquad \text{Substitute } \tfrac{8}{5} \text{ for } x \text{ in original equation.}$$

$$\frac{6}{\frac{3}{5}} + \frac{\frac{16}{5}}{-\frac{2}{5}} \overset{?}{=} 2 \qquad \text{Simplify.}$$

$$10 + (-8) = 2 \qquad \text{Solution checks. } \checkmark$$

Example 5   An Equation with Two Solutions

Solve $\dfrac{2x}{x+2} = \dfrac{1}{x^2 - 4} + 1$.

Solution

$$\frac{2x}{x+2} = \frac{1}{x^2 - 4} + 1 \qquad \text{Write original equation.}$$

$$(x^2 - 4)\left(\frac{2x}{x+2}\right) = (x^2 - 4)\left(\frac{1}{x^2 - 4} + 1\right) \qquad \text{Multiply each side by LCD of } x^2 - 4.$$

$$(x-2)(2x) = 1 + (x^2 - 4), \quad x \neq \pm 2 \qquad \text{Distribute and simplify.}$$

$$2x^2 - 4x = 1 + x^2 - 4 \qquad \text{Distributive Property}$$

$$x^2 - 4x + 3 = 0 \qquad \text{Write in standard form.}$$

$$(x-3)(x-1) = 0 \qquad \text{Factor.}$$

$$x - 3 = 0 \quad \Longrightarrow \quad x = 3 \qquad \text{Set 1st factor equal to 0.}$$

$$x - 1 = 0 \quad \Longrightarrow \quad x = 1 \qquad \text{Set 2nd factor equal to 0.}$$

The solutions are $x = 1$ and $x = 3$. Check these in the original equation.

Technology: Discovery

Use a graphing calculator to graph the equation

$$y = \frac{2x}{x+2} - \frac{1}{x^2 - 4} - 1.$$

Then use the *zero* or *root* feature of the graphing calculator to determine the x-intercepts. How do the x-intercepts compare with the solutions to Example 5? Graph each side of the equation in Example 5 in the same viewing window. Use the *intersect* feature to find the point(s) of intersection. What can you conclude?

See Technology Answers.

Point out that this problem differs from previous examples in this section because it requires solving a *second-degree* equation.

Study Tip

The symbol $\pm$ in Example 5 is read as "plus or minus." For instance, $x \neq \pm 2$ means that x is not equal to 2 or -2.

③ Solve application problems using rational equations with variable denominators.

Applications

In Section 3.5, you studied a formula that relates distance, rate, and time.

$$\text{Distance} = (\text{Rate})(\text{Time})$$

By solving this equation for the rate or the time, you can obtain two other versions of the formula.

$$\text{Rate} = \frac{\text{Distance}}{\text{Time}} \quad \text{and} \quad \text{Time} = \frac{\text{Distance}}{\text{Rate}}$$

The second of these formulas is used in Example 6.

Example 6   Average Speeds

You and your friend travel to separate colleges in the same amount of time. You drive 380 miles and your friend drives 400 miles. Your friend's average speed is 3 miles per hour faster than your average speed. What is your average speed and what is your friend's average speed?

Solution

Begin by setting your time equal to your friend's time. Then use the formula above that gives the time in terms of the distance and the rate.

Verbal Model: Your time $=$ Your friend's time

$$\frac{\text{Your distance}}{\text{Your rate}} = \frac{\text{Friend's distance}}{\text{Friend's rate}}$$

Labels:
Your distance $= 380$ (miles)
Your rate $= r$ (miles per hour)
Friend's distance $= 400$ (miles)
Friend's rate $= r + 3$ (miles per hour)

Equation:
$$\frac{380}{r} = \frac{400}{r + 3}$$

$$380(r + 3) = 400(r), \quad r \neq 0, r \neq -3$$

$$380r + 1140 = 400r$$

$$1140 = 20r \quad \blacktriangleright \quad 57 = r$$

Your average speed is 57 miles per hour and your friend's average speed is $57 + 3 = 60$ miles per hour. Check this solution as follows.

Your friend's average speed is 3 miles per hour faster than yours.

$57 + 3 = 60$ Solution checks. ✓

You drive 380 miles and your friend drives 400 miles in the same time.

$\dfrac{380}{57} = 6\dfrac{2}{3}$ hours and $\dfrac{400}{60} = 6\dfrac{2}{3}$ hours   Solution checks. ✓

Example 7 **Work Rate**

With only the cold water valve open, it takes 7 minutes to fill the tub of a washing machine. With both the hot and cold water valves open, it takes only 4 minutes to fill the tub. How long will it take to fill the tub with only the hot water valve open?

Solution

Verbal Model:

| Rate for cold water | + | Rate for hot water | = | Rate for warm water |

Labels: Rate for warm water = $\frac{1}{4}$ (tub per minute)
Rate for cold water = $\frac{1}{7}$ (tub per minute)
Rate for hot water = $1/t$ (tub per minute)

Equation:
$$\frac{1}{7} + \frac{1}{t} = \frac{1}{4}$$

$$28t\left(\frac{1}{7} + \frac{1}{t}\right) = 28t\left(\frac{1}{4}\right)$$

$$4t + 28 = 7t, \quad t \neq 0$$

$$28 = 3t \quad \Longrightarrow \quad \frac{28}{3} = t$$

So, it will take $9\frac{1}{3}$ minutes to fill the tub with hot water alone. Check this in the original statement of the problem.

After learning to clear an equation of fractions, students may mistakenly attempt to use this technique in an addition or subtraction problem. Ask students to compare an equation with an addition problem that *looks* similar.

$$\frac{4}{x^2 - 4} + \frac{x}{2x - 4} = \frac{1}{2}$$

$$\frac{4}{x^2 - 4} + \frac{x}{2x - 4} + \frac{1}{2}$$

The solution of this equation is $x = -6$; the sum is $\dfrac{x^2 + x + 2}{(x + 2)(x - 2)}$

Example 8 **Batting Average**

In this year's playing season, a baseball player has been up to bat 280 times (excluding walks) and has hit the ball safely 70 times. So, the batting average for the player is 70/280 = .250. How many additional *consecutive* times must the player hit the ball safely to obtain a batting average of .300?

Solution

Verbal Model:

| Batting average | = | Total hits | ÷ | Total times at bat |

Labels: Current times at bat = 280
Current hits = 70
Additional consecutive hits = x

Equation:
$$.300 = \frac{x + 70}{x + 280}$$

$$.300(x + 280) = x + 70, \quad x \neq -280$$

$$0.3x + 84 = x + 70$$

$$14 = 0.7x \quad \Longrightarrow \quad 20 = x$$

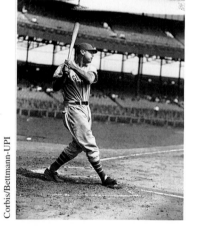

Between 1930 and 2001, only one National League batting champion had a batting average greater than .400: Bill Terry of the New York Giants in 1930. (Source: Major League Baseball)

The player must hit safely for the next 20 times at bat. After that, the player's batting average will be 90/300 = .300.

7.5 Exercises

Review   Concepts, Skills, and Problem Solving

Keep mathematically in shape by doing these exercises *before* the problems of this section.

Properties and Definitions

1. *Writing* Explain what is meant by the domain of a rational expression.   The domain of a rational expression consists of all values of the variable for which the denominator is not zero.

2. *Writing* Explain how you find the least common denominator of two or more rational expressions.   First factor each polynomial completely. Then construct the denominator to contain all the different factors of the polynomials in the denominator.

Solving Equations

In Exercises 3–10, solve the equation.

3. $16x - 3 = 29$ 2

4. $\frac{x}{5} + \frac{1}{5} = \frac{7}{10}$ $\frac{5}{2}$

5. $12(4 + x) = 3(10 - 6x)$ $-\frac{3}{5}$

6. $2x - 8(5 - x) = 9x - 3(x + 2)$ $\frac{17}{2}$

7. $x(x - 8) = 0$ 0, 8

8. $6x\left(\frac{2}{3}x + 1\right) = 0$ $0, -\frac{3}{2}$

9. $x(8 - x) = 16$ 4

10. $x^2 + x - 56 = 0$ $-8, 7$

Models

11. *Depreciation* A business purchases a computer system for $8500. It is estimated that after 4 years the system's depreciated value will be $3000. Assuming straight-line depreciation, write a linear model for the value V of the system in terms of time t. Estimate the value of the computer system after 1 year. $V = 8500 - 1375t$; $7125

12. *Education* A liberal arts college had an enrollment of 3750 students in 2000. During the next 10 years the enrollment increased by approximately 125 students per year.

(a) Write a linear model for the enrollment N in terms of the year t. (Let $t = 0$ correspond to the year 2000.) $N = 3750 + 125t$

(b) *Linear Extrapolation* If this constant rate of growth continues, predict the enrollment in the year 2020.   6250 students

(c) *Linear Interpolation* Use the model to estimate the enrollment in 2008.   4750 students

Developing Skills

In Exercises 1–4, determine whether each value of x is a solution to the equation.

Equation	Values	
1. $\frac{x}{5} - \frac{3}{x} = \frac{1}{10}$	(a) $x = 0$	(b) $x = -1$
	(c) $x = \frac{1}{6}$	(d) $x = 6$

(a) Not a solution   (b) Not a solution
(c) Not a solution   (d) Not a solution

2. $\frac{3x}{5} + \frac{x^2}{2} = \frac{4}{5}$	(a) $x = 0$	(b) $x = -2$
	(c) $x = \frac{4}{5}$	(d) $x = 2$

(a) Not a solution   (b) Solution
(c) Solution   (d) Not a solution

Equation	Values	
3. $\frac{5}{2x} - \frac{4}{x} = 3$	(a) $x = -\frac{1}{2}$	(b) $x = 4$
	(c) $x = 0$	(d) $x = \frac{1}{4}$

(a) Solution   (b) Not a solution
(c) Not a solution   (d) Not a solution

4. $3 + \frac{1}{x + 2} = 4$	(a) $x = -1$	(b) $x = -2$
	(c) $x = 0$	(d) $x = 5$

(a) Solution   (b) Not a solution
(c) Not a solution   (d) Not a solution

Mental Math In Exercises 5–12, solve the rational equation mentally and check your solution.

5. $\frac{1}{2} + \frac{x}{2} = \frac{5}{2}$ 4

6. $\frac{x}{5} + \frac{3}{5} = \frac{9}{5}$ 6

7. $\dfrac{x}{2} + \dfrac{x}{2} = x$

All real numbers

8. $\dfrac{x}{4} + \dfrac{3x}{4} = x$

All real numbers

9. $\dfrac{3}{x-2} = \dfrac{x+4}{x-2}$ -1

10. $\dfrac{x-6}{x+7} = \dfrac{-2}{x+7}$ 4

11. $\dfrac{x-2}{3} + \dfrac{x-2}{3} = \dfrac{x+3}{3}$ 7

12. $\dfrac{x-5}{9} + \dfrac{x-1}{9} = \dfrac{x-10}{9}$ -4

In Exercises 13–60, solve the rational equation. (Check for extraneous solutions.) See Examples 1–5.

13. $\dfrac{z}{3} - \dfrac{2z}{8} = 1$ 12

14. $3 + \dfrac{y}{5} = \dfrac{y}{2}$ 10

15. $\dfrac{t}{3} = 25 - \dfrac{t}{6}$ 50

16. $\dfrac{x}{10} + \dfrac{x}{5} = 20$ $\dfrac{200}{3}$

17. $\dfrac{5x}{7} - \dfrac{2x}{3} = \dfrac{1}{2}$ $\dfrac{21}{2}$

18. $\dfrac{2}{3} - \dfrac{3x}{6} = -\dfrac{4x}{9}$ 12

19. $\dfrac{a+3}{4} - \dfrac{a-1}{6} = \dfrac{4}{3}$ 5

20. $\dfrac{u-5}{10} + \dfrac{u+8}{15} = \dfrac{7}{10}$ 4

21. $\dfrac{x-4}{3} + \dfrac{2x+1}{4} = \dfrac{5}{6}$ $\dfrac{23}{10}$

22. $\dfrac{y+6}{5} + \dfrac{3y-2}{20} = \dfrac{3}{4}$ -1

23. $2 - \dfrac{4}{x} = 1$ 4

24. $\dfrac{2}{b} + 1 = 9$ $\dfrac{1}{4}$

25. $3 - \dfrac{16}{a} = \dfrac{5}{3}$ 12

26. $\dfrac{14}{x} - 4 = 3$ 2

27. $\dfrac{3}{x} + \dfrac{1}{4} = \dfrac{2}{x}$ -4

28. $\dfrac{3}{5} - \dfrac{7}{x} = -\dfrac{4}{x}$ 5

29. $\dfrac{6}{12x} + \dfrac{3}{4} = \dfrac{2}{3x}$ $\dfrac{2}{9}$

30. $\dfrac{5}{2x} + \dfrac{5}{8} = -\dfrac{5}{8x}$ -5

31. $\dfrac{10}{y+3} + \dfrac{10}{3} = 6$ $\dfrac{3}{4}$

32. $\dfrac{5}{2} - \dfrac{12}{x-4} = 6$ $\dfrac{4}{7}$

33. $\dfrac{3}{x} = \dfrac{9}{2(x+2)}$ 4

34. $\dfrac{5}{x+4} = \dfrac{5}{3(x+1)}$ $\dfrac{1}{2}$

35. $\dfrac{7x}{x+1} = \dfrac{5}{x-3} + 7$ $\dfrac{4}{3}$

36. $\dfrac{-3x}{x+4} = \dfrac{2}{x+1} - 3$ $-\dfrac{2}{5}$

37. $\dfrac{4}{x+2} - \dfrac{1}{x} = \dfrac{1}{x}$ 2

38. $\dfrac{3}{x+5} + \dfrac{5}{x} = \dfrac{10}{x}$ $-\dfrac{25}{2}$

39. $10 - \dfrac{13}{x} = 4 + \dfrac{5}{x}$ 3

40. $\dfrac{15}{x} - 4 = \dfrac{6}{x} + 3$ $\dfrac{9}{7}$

41. $\dfrac{3}{x(x-3)} + \dfrac{4}{x} = \dfrac{1}{x-3}$ No solution

42. $\dfrac{3}{z+2} = \dfrac{2}{z} + \dfrac{2}{z(z+2)}$ 6

43. $\dfrac{1}{x+3} + \dfrac{4}{x+4} = -\dfrac{x}{(x+3)(x+4)}$ $-\dfrac{8}{3}$

44. $\dfrac{2x}{(x-4)(x-2)} = \dfrac{1}{x-4} + \dfrac{2}{x-2}$ 10

45. $\dfrac{1}{x-3} + \dfrac{1}{x+3} = \dfrac{10}{x^2-9}$ 5

46. $\dfrac{2}{x-4} - \dfrac{3}{x+4} = \dfrac{9x}{x^2-16}$ 2

47. $\dfrac{1}{x-2} + \dfrac{3}{x+3} = \dfrac{4}{x^2+x-6}$ $\dfrac{7}{4}$

48. $\dfrac{4}{x+5} - \dfrac{2}{x-2} = \dfrac{-6}{x^2+3x-10}$ 6

49. $x + 4 = \dfrac{-4}{x}$ -2

50. $\dfrac{25}{t} = 10 - t$ 5

51. $\dfrac{20-x}{x}=x$ $-5,4$

52. $\dfrac{x+30}{x}=x$ $-5,6$

53. $2y=\dfrac{y+6}{y+1}$
$-2,\tfrac{3}{2}$

54. $\dfrac{3x}{x+1}=\dfrac{2}{x-1}$
$-\tfrac{1}{3},2$

55. $x+\dfrac{1}{x}=\dfrac{5}{2}$ $\tfrac{1}{2},2$

56. $\dfrac{4}{x}-\dfrac{x}{6}=\dfrac{5}{3}$ $-12,2$

57. $\dfrac{x+3}{x^2-9}+\dfrac{4}{x-3}=-2$ $\tfrac{1}{2}$

58. $-1-\dfrac{6}{x-4}=\dfrac{x+2}{x^2-16}$ $-5,-2$

59. $\dfrac{x}{2}=\dfrac{1+\dfrac{3}{x}}{1+\dfrac{1}{x}}$ $-2,3$

60. $\dfrac{2x}{3}=\dfrac{1+\dfrac{1}{x}}{1+\dfrac{2}{x}}$ $-\tfrac{3}{2},1$

In Exercises 61–68, (a) use a graphing calculator to graph the equation and identify any x-intercepts of the graph and (b) set $y=0$ and solve the resulting equation to confirm the result of part (a).

(a) See Additional Answers.

61. $y=\dfrac{5}{x+1}-1$
(b) 4

62. $y=\dfrac{12}{x}-6$
(b) 2

63. $y=\dfrac{6}{x}-\dfrac{3x}{2}$
(b) $2,-2$

64. $y=\dfrac{7}{x+3}-\dfrac{x}{4}$
(b) $4,-7$

65. $y=2\left(\dfrac{3}{x-4}-1\right)$
(b) 7

66. $y=\dfrac{5}{x}+1$
(b) -5

67. $y=x+\dfrac{x-6}{x}$
(b) $2,-3$

68. $y=x-\dfrac{5}{x}-4$
(b) $5,-1$

Solving Problems

69. *Number Problem* Find a number such that the sum of the number and its reciprocal is $\tfrac{10}{3}$. $\tfrac{1}{3},3$

70. *Number Problem* Find a number such that the sum of the number and 3 times its reciprocal is $\tfrac{28}{5}$. $\tfrac{3}{5},5$

71. *Number Problem* Find a number such that the sum of three times the number and 25 times the reciprocal of the number is 20. $\tfrac{5}{3},5$

72. *Number Problem* Find consecutive even integers such that the sum of the first and 3 times the reciprocal of the second is $\tfrac{41}{4}$. 10, 12

73. *Average Speed* One car makes a trip of 440 miles. Another car takes the same amount of time to make a trip of 416 miles. The average speed of the second car is 3 miles per hour lower than the average speed of the first car. What is the average speed of each car? First car: 55 miles per hour; Second car: 52 miles per hour

74. *Average Speed* One car makes a trip of 400 miles. Another car takes the same amount of time to make a trip of 480 miles. The average speed of the second car is 10 miles per hour higher than the average speed of the first car. What is the average speed of each car?

75. *Average Speed* A car leaves a town 20 minutes after a truck. The speed of the truck is approximately 10 miles per hour lower than that of the car. After traveling 100 miles, the car overtakes the truck. Find the average speed of each vehicle.

76. *Average Speed* A car leaves a town 30 minutes after a bus. The speed of the bus is 15 miles per hour lower than that of the car. After traveling 150 miles, the car overtakes the bus. Find the average speed of each vehicle. Car: 75 miles per hour; Bus: 60 miles per hour

77. *Wind Speed* You fly to a meeting in San Francisco 1500 miles away (see figure). After traveling for the same amount of time on the return flight, the pilot states that you still have 300 miles to travel. The plane has a speed of 600 miles per hour in still air. How fast is the wind blowing? (Assume that both flights are made on the same day, the wind direction is parallel to the flight path, and the wind blows at a constant speed all day.) $66\tfrac{2}{3}$ miles per hour

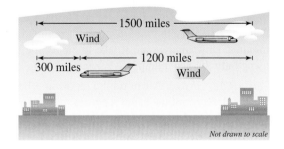

Not drawn to scale

74. First car: 50 miles per hour; Second car: 60 miles per hour
75. Truck: 50 miles per hour; Car: 60 miles per hour

78. *Wind Speed* A small plane has a speed of 170 miles per hour in still air. The plane travels a distance of 400 miles with a tail wind in the same time it takes to travel 280 miles into a head wind. In both cases, the wind has the same speed. Find the speed of the wind. Use a diagram and a verbal model to help solve the problem.
30 miles per hour   See Additional Answers.

Work Rate In Exercises 79 and 80, the first two columns of the table show the times required by each of two people working *alone* to complete a task. Complete the table by finding the time required for each pair of individuals working *together* to complete the task. (Assume that when they work together, their individual rates do not change.)

79.

Person #1	Person #2	Together
4 days	4 days	2 days
4 hours	6 hours	$2\frac{2}{5}$ hours
4 hours	$2\frac{1}{2}$ hours	$1\frac{7}{13}$ hours

80.

Person #1	Person #2	Together
30 minutes	30 minutes	15 minutes
$6\frac{1}{2}$ hours	4 hours	$2\frac{10}{21}$ hours
a days	b days	$\frac{ab}{a+b}$ days

81. *Work Rate* One person can paint a wall in 4 hours. The same person working with a friend can paint a similar wall in 1 hour. Working alone, how long would it take the second person to paint the wall? $\frac{4}{3}$ hours

82. *Work Rate* A pump empties a storage tank in 50 minutes. When a new pump is added to the system, the time to empty the tank using both pumps is 20 minutes. How long would it take to empty the tank using only the new pump? $33\frac{1}{3}$ minutes

83. *Work Rate* Printer A can print a report in 3 hours. Printer B can print the same report in 4 hours. How long would it take both printers working together to print the report? $\frac{12}{7}$ hours

84. *Flow Rate* The flow rate of one pipe is $1\frac{1}{4}$ times that of another pipe. A swimming pool can be filled in 5 hours using both pipes. Find the time required to fill the swimming pool using only the pipe with the lower flow rate. $11\frac{1}{4}$ hours

85. *Batting Average* A softball player has been up to bat 35 times and has hit the ball safely 6 times. How many additional consecutive times must the player hit the ball safely to obtain a batting average of .275? 5 hits

86. *Batting Average* After 50 times at bat, a baseball player has a batting average of .160. How many additional consecutive times must the player hit the ball safely to obtain a batting average of .250? 6 hits

87. *Average Cost* The average cost $\overline{C}$ of printing x books is given by
$$\overline{C} = \frac{1}{2} + \frac{5000}{x}.$$
Determine the number of books that must be produced to obtain an average cost of $2.50 per book. 2500 books

88. *Average Cost* The average cost $\overline{C}$ of producing x picture frames is given by
$$\overline{C} = \frac{1}{4} + \frac{500}{x}.$$
Determine the number of picture frames that must be produced to obtain an average cost of $0.50 per frame. 2000 picture frames

89. *Deer Population* The game commission introduces 50 deer into newly acquired state game lands. The population N of the herd is given by the model
$$N = \frac{250(5 + 3t)}{25 + t}$$
where t is time in years. Find the time required for the herd to increase to 125 deer. 3 years

90. *Environment* A utility company burns oil to generate electricity. The cost C in dollars of removing p percent of the air pollution in the stack emission is given by
$$C = \frac{80,000p}{100 - p}.$$
Determine the percent of the stack emission that can be removed for $240,000. 75%

Explaining Concepts

91. 🔷 Answer parts (e) and (f) of Motivating the Chapter on page 372.

92. *Writing*✏ Explain the difference between the following.

$$\frac{5}{x+3} + \frac{5}{3} = 3, \qquad \frac{5}{x+3} + \frac{5}{3} + 3$$

The first is a rational equation and the second is a rational expression.

93. *Writing*✏ Describe the steps used to solve a rational equation. Multiply each side of the equation by the lowest common denominator, solve the resulting equation, and check the result. It is important to check the result for any errors or extraneous solutions.

94. *Writing*✏ Explain what is meant by an *extraneous* solution. How do you identify an extraneous solution? An extraneous solution is a "trial solution" that does not satisfy the original equation, and so may not be listed as an actual solution.

95. *Writing*✏ Describe the steps that can be used to transform an equation into an equivalent equation. (Section 2.4)

   (i) Simplify each side by removing symbols of grouping, combining like terms, and reducing fractions on one or both sides.

   (ii) Add (or subtract) the same quantity to (from) both sides of the equation.

   (iii) Multiply (or divide) each side of the equation by the same nonzero real number.

   (iv) Interchange the two sides of the equation.

96. *Writing*✏ Which step from your list in Exercise 95 is not followed when an extraneous solution is found for a rational equation? Explain. Step (iii). Extraneous solutions for rational equations can appear when each side of the equation is multiplied or divided by an expression equivalent to zero.

97. *Writing*✏ When can you use cross-multiplication to solve rational equations? Explain. (Section 3.2) When the equation involves only two fractions, one on each side of the equation, the equation can be solved by cross-multiplication.

98. *Writing*✏ A student submits the following argument to prove that 1 is equal to 0. Discuss what is wrong with the argument.

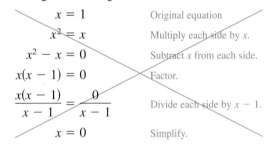

$x = 1$	Original equation
$x^2 = x$	Multiply each side by x.
$x^2 - x = 0$	Subtract x from each side.
$x(x-1) = 0$	Factor.
$\dfrac{x(x-1)}{x-1} = \dfrac{0}{x-1}$	Divide each side by $x - 1$.
$x = 0$	Simplify.

Because you are given $x = 1$ and $x - 1 = 0$, you have division by zero in Step 4, which is undefined.

What Did You Learn?

Key Terms

rational expression, *p. 374*
domain of a rational
   expression, *p. 374*

simplified or reduced form,
   p. 376
least common multiple
   (LCM), *p. 395*

least common denominator
   (LCD), *p. 396*
complex fractions, *p. 404*
extraneous solution, *p. 413*

Key Concepts

7.1 ◔ Finding the domain of a rational expression

To find the values to exclude from the domain of a rational expression, follow these steps.

1. Set the denominator equal to zero.

2. Find the solution(s) to the equation in Step 1.

7.1 ◔ Simplifying rational expressions

Let a, b, and c represent real numbers, variables, or algebraic expressions such that $b \neq 0$ and $c \neq 0$. Then the following is valid.

$$\frac{ac}{bc} = \frac{a\acute{c}}{b\acute{c}} = \frac{a}{b}$$

7.2 ◔ Multiplying rational expressions

Let a, b, c, and d represent real numbers, variables, or algebraic expressions such that $b \neq 0$ and $d \neq 0$. Then the product of a/b and c/d is

$$\frac{a}{b} \cdot \frac{c}{d} = \frac{ac}{bd}.$$

7.2 ◔ Dividing rational expressions

Let a, b, c, and d represent real numbers, variables, or algebraic expressions such that $b \neq 0$, $c \neq 0$, and $d \neq 0$. Then the quotient of a/b and c/d is

$$\frac{a}{b} \div \frac{c}{d} = \frac{a}{b} \cdot \frac{d}{c} = \frac{ad}{bc}.$$

7.3 ◔ Combining rational expressions with like denominators

Let a, b, and c represent real numbers, variables, or algebraic expressions such that $c \neq 0$.

1. $\dfrac{a}{c} + \dfrac{b}{c} = \dfrac{a+b}{c}$

2. $\dfrac{a}{c} - \dfrac{b}{c} = \dfrac{a-b}{c}$

7.3 ◔ Guidelines for finding the least common multiple

1. Factor each polynomial completely.

2. The least common multiple must contain all the different factors of the polynomials and each such factor must be repeated the maximum number of times it occurs in any of the factorizations.

7.3 ◔ Combining rational expressions with unlike denominators

1. Find the least common denominator (LCD) of the rational expressions.

2. Rewrite each rational expression so that it has the LCD in its denominator.

3. Combine these rational expressions with like denominators.

7.4 ◔ Simplifying Complex Fractions

1. Combine the numerator and denominator into single fractions, if necessary.

2. Divide by inverting the denominator and multiplying.

7.5 ◔ Solving rational equations

To solve a rational equation, multiply each side of the equation by the LCD of the expressions, then solve the resulting equation. Check the solutions for any extraneous solutions.

Review Exercises

7.1 Simplifying Rational Expressions

1 Find the domain of a rational expression.

In Exercises 1–4, find the domain of the rational expression.

1. $\dfrac{8x}{x - 5}$
All real numbers x such that $x \neq 5$

2. $\dfrac{y + 1}{y + 3}$
All real numbers y such that $y \neq -3$

3. $\dfrac{t}{t^2 - 3t + 2}$ All real numbers t such that $t \neq 1$ and $t \neq 2$

4. $\dfrac{x - 10}{x(x^2 - 4)}$ All real numbers x such that $x \neq 0$, $x \neq -2$, and $x \neq 2$

In Exercises 5–8, evaluate the rational expression for each specified value. If not possible, state the reason.

Expression & *Values*

5. $\dfrac{2x}{x + 4}$
(a) $x = 0$ (b) $x = 2$
(c) $x = -3$ (d) $x = -4$
(a) 0 (b) $\frac{2}{3}$
(c) -6 (d) Division by zero is undefined.

6. $\dfrac{3}{y - 10}$
(a) $y = 0$ (b) $y = -2$
(c) $y = 10$ (d) $y = -5$
(a) $-\frac{3}{10}$ (b) $-\frac{1}{4}$
(c) Division by zero is undefined. (d) $-\frac{1}{5}$

7. $\dfrac{x - 1}{x^2 + 4}$
(a) $x = 2$ (b) $x = 1$
(c) $x = -2$ (d) $x = -5$
(a) $\frac{1}{8}$ (b) 0 (c) $-\frac{3}{8}$ (d) $-\frac{6}{29}$

8. $\dfrac{3x - 2}{x^2 - x - 6}$
(a) $x = 0$ (b) $x = 6$
(c) $x = \frac{2}{3}$ (d) $x = -2$
(a) $\frac{1}{3}$ (b) $\frac{2}{3}$ (c) 0 (d) Division by zero is undefined.

2 Simplify rational expressions.

In Exercises 9–26, simplify the rational expression.

9. $\dfrac{6t}{18}$ $\dfrac{t}{3}$

10. $\dfrac{45x}{15}$ $3x$

11. $\dfrac{4x^5}{x^2}$ $4x^3, x \neq 0$

12. $\dfrac{88z^2}{33z}$ $\dfrac{8z}{3}, z \neq 0$

13. $\dfrac{7x^2y}{21xy^2}$ $\dfrac{x}{3y}, x \neq 0$

14. $\dfrac{2(yz)^2}{6yz^4}$ $\dfrac{y}{3z^2}, y \neq 0$

15. $\dfrac{3b - 6}{4b - 8}$ $\dfrac{3}{4}, b \neq 2$

16. $\dfrac{2a + 5}{10a + 25}$ $\dfrac{1}{5}, a \neq -\dfrac{5}{2}$

17. $\dfrac{4x - 4y}{y - x}$ $-4, x \neq y$

18. $\dfrac{x - y}{3y - 3x}$ $-\dfrac{1}{3}, x \neq y$

19. $\dfrac{x^2 - 9}{x^2 - x - 6}$ $\dfrac{x + 3}{x + 2}, x \neq 3$

20. $\dfrac{x^2 - 4}{x^2 + x - 6}$ $\dfrac{x + 2}{x + 3}, x \neq 2$

21. $\dfrac{1 - x^3}{x^2 - 1}$ $-\dfrac{x^2 + x + 1}{x + 1}, x \neq 1$

22. $\dfrac{x^2 - 4}{x^3 + 8}$ $\dfrac{x - 2}{x^2 - 2x + 4}, x \neq -2$

23. $\dfrac{x^2 - 3xy - 18y^2}{x^2 + 4xy + 3y^2}$ $\dfrac{x - 6y}{x + y}, x \neq -3y$

24. $\dfrac{x^2 + 2xy + y^2}{x^2 - xy - 2y^2}$ $\dfrac{x + y}{x - 2y}, x \neq -y$

25. $\dfrac{x(x - 4) + 7(x - 4)}{x^2 + 7x}$ $\dfrac{x - 4}{x}, x \neq -7$

26. $\dfrac{x^3 - 5x^2 + 2x - 10}{x^3 + 2x}$ $\dfrac{x - 5}{x}$

In Exercises 27 and 28, complete the table. Explain why the values of the expressions agree for all values of x except one. The expressions are equivalent. So, all the values will agree except for the value that is not in the domain of the unsimplified rational expression.

27.

x	1	1.5	2	2.5	3
$\dfrac{x - 2}{x^2 - 4}$	$\frac{1}{3}$	$\frac{2}{7}$	Undef.	$\frac{2}{9}$	$\frac{1}{5}$
$\dfrac{1}{x + 2}$	$\frac{1}{3}$	$\frac{2}{7}$	$\frac{1}{4}$	$\frac{2}{9}$	$\frac{1}{5}$

28.

x	1	1.5	2	2.5	3
$\dfrac{x - 2}{x^2 - x - 2}$	$\frac{1}{2}$	$\frac{2}{5}$	Undef.	$\frac{2}{7}$	$\frac{1}{4}$
$\dfrac{1}{x + 1}$	$\frac{1}{2}$	$\frac{2}{5}$	$\frac{1}{3}$	$\frac{2}{7}$	$\frac{1}{4}$

Probability In Exercises 29 and 30, the probability of hitting the shaded portion of the figure with a dart is equal to the ratio of the shaded area to the total area of the figure. Find the probability.

29.

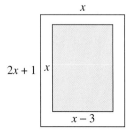

30.

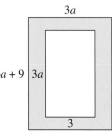

$$\frac{x-3}{2x+1}$$

$$\frac{2(a+1)}{2a+3}$$

7.2 Multiplying and Dividing Rational Expressions

① Multiply rational expressions and simplify.

In Exercises 31–38, multiply and simplify.

31. $\dfrac{x^2}{6} \cdot \dfrac{2x}{x^3}$
$\dfrac{1}{3}, x \neq 0$

32. $\dfrac{5y^4}{8y} \cdot \dfrac{12}{-15y^2}$
$-\dfrac{y}{2}, y \neq 0$

33. $\dfrac{5x^2y}{4} \cdot \dfrac{6x}{10y^3}$
$\dfrac{3x^3}{4y^2}$

34. $\dfrac{x^2y^3}{6} \cdot \dfrac{3}{(xy)^3}$
$\dfrac{1}{2x}, y \neq 0$

35. $\dfrac{2-x}{x+3} \cdot \dfrac{4x+12}{x^2-4}$
$-\dfrac{4}{x+2}, x \neq -3, x \neq 2$

36. $\dfrac{8x-10}{7-x} \cdot \dfrac{x^2-49}{4x-5}$
$-2(x+7), x \neq 7, x \neq \dfrac{5}{4}$

37. $\dfrac{x^2-36}{6} \cdot \dfrac{3}{x^2-12x+36}$
$\dfrac{x+6}{2(x-6)}$

38. $\dfrac{x^2-5x+4}{9} \cdot \dfrac{-18x}{x^2-8x+16}$
$-\dfrac{2x(x-1)}{x-4}$

② Divide rational expressions and simplify.

In Exercises 39–46, divide and simplify.

39. $\dfrac{5}{8} \div \dfrac{u}{v}$
$\dfrac{5v}{8u}, v \neq 0$

40. $\dfrac{x}{9} \div \dfrac{x^2}{3}$
$\dfrac{1}{3x}$

41. $10y^2 \div \dfrac{y}{5}$
$50y, y \neq 0$

42. $\dfrac{5}{z^2} \div 3z^2$
$\dfrac{5}{3z^4}$

43. $\dfrac{u^2}{u^2-9} \div \dfrac{u}{u+3}$
$\dfrac{u}{u-3}, u \neq 0, u \neq -3$

44. $\dfrac{v+5}{v^2} \div \dfrac{v+5}{v^2}$
$1, v \neq 0, v \neq -5$

45. $\dfrac{x^2-8x}{x-1} \div \dfrac{x^2-16x+64}{x^2-1}$
$\dfrac{x(x+1)}{x-8}, x \neq -1, x \neq 1$

46. $\dfrac{x^2-x}{x+1} \div \dfrac{5x-5}{x^2+6x+5}$
$\dfrac{x(x+5)}{5}, x \neq 1, x \neq -1$

7.3 Adding and Subtracting Rational Expressions

① Add or subtract rational expressions with like denominators.

In Exercises 47–52, combine and simplify.

47. $\dfrac{5x}{8} - \dfrac{3x}{8}$
$\dfrac{x}{4}$

48. $\dfrac{4t}{9} + \dfrac{11t}{9}$
$\dfrac{5t}{3}$

49. $\dfrac{4x-5}{x+2} + \dfrac{2x+1}{x+2}$
$\dfrac{6x-4}{x+2}$

50. $\dfrac{3y+4}{2y+1} - \dfrac{y+3}{2y+1}$
$1, y \neq -\dfrac{1}{2}$

51. $\dfrac{5t+1}{2t-3} - \dfrac{2t-7}{2t-3}$
$\dfrac{3t+8}{2t-3}$

52. $\dfrac{3x+11}{x-5} + \dfrac{x-2}{x-5}$
$\dfrac{4x+9}{x-5}$

② Find the least common multiple of two or more polynomials.

In Exercises 53–56, find the least common multiple.

53. $20x^2, 24, 30x^3$ $120x^3$

54. $4y^2, 6y, 18z$ $36y^2z$

55. $x-5, 2x^2, x(x+5)$ $2x^2(x^2-25)$

56. $5(x-1), 2(x^2+x+1)$ $10(x^3-1)$

In Exercises 57–60, fill in the missing factor. Assume all denominators are nonzero.

57. $\dfrac{7}{4x} = \dfrac{7\ (3x^2)}{12x^3}$

58. $\dfrac{5x}{8} = \dfrac{5x\ (x-3)}{8(x-3)}$

59. $\dfrac{x-3}{x-1} = \dfrac{(x-3)\ (x+1)}{x^2-1}$

60. $\dfrac{x+2}{x-2} = \dfrac{(x+2)\ (x+2)}{x^2-4}$

3 Add or subtract rational expressions with unlike denominators.

In Exercises 61–78, combine and simplify.

61. $\dfrac{2x}{x-6} + \dfrac{3x}{6-x} \quad -\dfrac{x}{x-6}$

62. $\dfrac{x}{9-x} + \dfrac{2x}{x-9} \quad \dfrac{x}{x-9}$

63. $\dfrac{4x}{x-1} - \dfrac{3}{1-x} \quad \dfrac{4x+3}{x-1}$

64. $\dfrac{8}{x-10} - \dfrac{4x}{10-x} \quad \dfrac{4x+8}{x-10}$

65. $\dfrac{2x-1}{5-3x} - \dfrac{3x+2}{3x-5} \quad \dfrac{5x+1}{5-3x}$

66. $\dfrac{5x+2}{2x-1} - \dfrac{x-4}{1-2x} \quad \dfrac{6x-2}{2x-1}$

67. $\dfrac{5t}{16} - \dfrac{5t}{24} \quad \dfrac{5t}{48}$

68. $\dfrac{x}{8} + \dfrac{5x}{6} \quad \dfrac{23x}{24}$

69. $\dfrac{1}{x+2} - \dfrac{1}{x+1} \quad -\dfrac{1}{(x+1)(x+2)}$

70. $\dfrac{4}{x-3} + \dfrac{1}{x+4} \quad \dfrac{5x+13}{(x-3)(x+4)}$

71. $\dfrac{1}{x+4} - \dfrac{x-1}{x^2+4x+4} \quad \dfrac{x+8}{(x+4)(x+2)^2}$

72. $\dfrac{1}{x-1} + \dfrac{1-x}{x^2+2x+1} \quad \dfrac{4x}{(x-1)(x+1)^2}$

73. $\dfrac{x-7}{x^2-16} - \dfrac{3}{4-x} \quad \dfrac{4x+5}{(x+4)(x-4)}$

74. $\dfrac{2x-7}{x^2-36} - \dfrac{1}{6-x} \quad \dfrac{3x-1}{(x+6)(x-6)}$

75. $x-1 + \dfrac{1}{x+2} + \dfrac{1}{x-1} \quad \dfrac{x^3-x+3}{(x+2)(x-1)}$

76. $\dfrac{2}{x} - \dfrac{3}{x-1} + \dfrac{4}{x+1} \quad \dfrac{3x^2-7x-2}{x(x-1)(x+1)}$

77. $2x + \dfrac{3}{2(x-4)} - \dfrac{1}{2(x+2)} \quad \dfrac{2x^3-4x^2-15x+5}{(x-4)(x+2)}$

78. $\dfrac{1}{x-2} + \dfrac{1}{(x-2)^2} + \dfrac{1}{x+2} \quad \dfrac{2x^2-3x+2}{(x+2)(x-2)^2}$

7.4 Complex Fractions

1 Simplify complex fractions using rules for dividing rational expressions.

In Exercises 79–82, simplify the complex fraction.

79. $\dfrac{\left(\dfrac{4}{x}\right)}{\left(\dfrac{1}{x^2}\right)} \quad 4x,\ x \neq 0$

80. $\dfrac{5x}{\left(\dfrac{x}{y}\right)} \quad 5y,\ x \neq 0,\ y \neq 0$

81. $\dfrac{\left(\dfrac{6x-21}{x+3}\right)}{\left(\dfrac{2x-7}{x^2-9}\right)} \quad 3(x-3),\ x \neq \pm 3,\ x \neq \dfrac{7}{2}$

82. $\dfrac{\left(\dfrac{x^2+x}{x^2+x-12}\right)}{\left(\dfrac{4x+4}{x^2-6x+9}\right)} \quad \dfrac{x(x-3)}{4(x+4)},\ x \neq -1,\ x \neq 3$

2 Simplify complex fractions having a sum or difference in the numerator and/or denominator.

In Exercises 83–86, simplify the complex fraction.

83. $\dfrac{\left(\dfrac{2}{x}+2\right)}{\left(1-\dfrac{1}{x}\right)} \quad \dfrac{2(x+1)}{x-1},\ x \neq 0$

84. $\dfrac{\left(1+\dfrac{2}{x}\right)}{\left(\dfrac{2}{x}-1\right)} \quad \dfrac{x+2}{2-x},\ x \neq 0$

85. $\dfrac{\left(\dfrac{1}{x+1}-\dfrac{1}{4}\right)}{x-3} \quad \dfrac{-1}{4(x+1)},\ x \neq 3$

86. $\dfrac{\left(\dfrac{x}{x+1}-\dfrac{4}{5}\right)}{x-4} \quad \dfrac{1}{5(x+1)},\ x \neq 4$

7.5 Rational Equations and Applications

1 Solve rational equations with constant denominators.

In Exercises 87 and 88, solve the rational equation.

87. $\dfrac{t+1}{9} = \dfrac{2}{3} - t \quad \dfrac{1}{2}$

88. $\dfrac{2t-5}{14} = \dfrac{5}{7} - 2t \quad \dfrac{1}{2}$

2 Solve rational equations with variable denominators.

In Exercises 89–94, solve the rational equation. (Check for extraneous solutions.)

89. $\dfrac{7}{x} - 2 = \dfrac{3}{x} + 6$ $\dfrac{1}{2}$ **90.** $\dfrac{2}{x} + \dfrac{5}{3} = 1 + \dfrac{4}{x}$ 3

91. $\dfrac{x}{x + 3} - \dfrac{3x}{x^2 - 9} = 1$ **92.** $\dfrac{2x}{x - 1} + \dfrac{4}{x + 4} = 2$
$\dfrac{3}{2}$ $-\dfrac{2}{3}$

93. $\dfrac{t}{t - 4} + \dfrac{3}{t - 2} = 0$ **94.** $\dfrac{2x}{x - 3} - \dfrac{4}{x - 1} = 4$
$-4, 3$ $0, 5$

⊞ **In Exercises 95 and 96, (a) use a graphing calculator to graph the equation and identify the x-intercepts of the graph and (b) set y = 0 and solve the resulting rational equation to confirm the result of part (a).**   (a) See Additional Answers.

95. $y = \dfrac{1}{x + 2} + \dfrac{3}{x - 4}$ **96.** $y = \dfrac{x + 2}{x - 4} + \dfrac{3}{4}$

(b) $-\dfrac{1}{2}$ (b) $\dfrac{4}{7}$

3 Solve application problems using rational equations with variable denominators.

97. *Number Problem* Find a number such that the sum of the number and its reciprocal is $\dfrac{41}{20}$. $\dfrac{5}{4}, \dfrac{4}{5}$

98. *Average Speed* You drive 72 miles one way on a service call for your company. The return trip takes 10 minutes less because you drive an average of 6 miles per hour faster. Find the average speeds going to the service call and on the return trip.
48 miles per hour, 54 miles per hour

99. *Average Speed* You drive 180 miles to pick up supplies for your company. The return trip takes 24 minutes less because you drive an average of 5 miles per hour faster. Find the average speed going to and returning from the supplier.
45 miles per hour, 50 miles per hour

100. *Work Rate* One bricklayer lays $1\frac{1}{4}$ times as many bricks as a second bricklayer in the same amount of time. Find their individual times to complete a task if it takes them 12 hours working together.
21.6 hours, 27 hours

101. *Batting Average* After 40 times at bat, a baseball player has a batting average of .300. How many additional consecutive times must the player hit the ball safely to obtain a batting average of .440?
10 hits

102. *Environment* The cost C (in millions of dollars) for a government project to remove p percent of a river's pollutants is given by

$$C = \dfrac{250p}{100 - p}.$$

(a) What percent of the pollutants can be removed for $375 million?   60%

(b) Complete the table.

p	80	90	95	99	100
C	1,000	2,250	4,750	24,750	Undef.

(c) Using the table from part (b), is it possible to remove 100% of the pollutants from the river?
No

Chapter Test

Take this test as you would take a test in class. After you are done, check your work against the answers in the back of the book.

1. All real numbers x such that $x \neq 10$

1. Find the domain of the rational expression $\dfrac{x}{x - 10}$.

2. Fill in the missing factor: $\dfrac{2x^2}{x + 1} = \dfrac{2x^2\left(\boxed{}\,\right)}{x(x + 1)^2}$.

3. $\dfrac{1}{2y^2(2y - 1)}$ **4.** $\dfrac{x + 8}{x + 5}, x \neq 8$

3. Simplify: $\dfrac{3y(2y - 1)}{6y^3(2y - 1)^2}$

4. Simplify: $\dfrac{x^2 - 64}{x^2 - 3x - 40}$

In Exercises 5–10, perform the indicated operations and simplify.

6. $\dfrac{(x + 2)(x - 2)}{x^2}, x \neq -2$

5. $\dfrac{18x}{5} \cdot \dfrac{15}{3x^3}$ $\dfrac{18}{x^2}$

6. $(x + 2)^2 \cdot \dfrac{x - 2}{x^3 + 2x^2}$

8. $\dfrac{x^3}{(x - 3)^8}, x \neq 0$

7. $\dfrac{3x^2}{4} \div \dfrac{9x^3}{10}$ $\dfrac{5}{6x}$

8. $\left[\left(\dfrac{x}{x - 3}\right)^2 \cdot \dfrac{x^2}{x^2 - 3x}\right] \div (x - 3)^5$

10. $\dfrac{2}{(x + 1)^2}$

9. $\dfrac{3}{x + 2} - 6$ $-\dfrac{6x + 9}{x + 2}$

10. $\dfrac{2}{x + 1} - \dfrac{2x}{x^2 + 2x + 1}$

In Exercises 11–14, simplify the complex fraction.

12. $-\dfrac{6}{5}, x \neq 1, x \neq -1$

11. $\dfrac{4}{\left(\dfrac{2}{x} + 8\right)}$ $\dfrac{2x}{1 + 4x}, x \neq 0$

12. $\dfrac{\left(\dfrac{3}{x + 1} - \dfrac{3}{x - 1}\right)}{\left(\dfrac{5}{x^2 - 1}\right)}$

13. $\dfrac{\left(\dfrac{t}{t - 5}\right)}{\left(\dfrac{t^2}{5 - t}\right)}$ $-\dfrac{1}{t}, t \neq 5$

14. $\dfrac{\left(9x - \dfrac{1}{x}\right)}{\left(\dfrac{1}{x} - 3\right)}$ $-3x - 1, x \neq 0, x \neq \dfrac{1}{3}$

15. (a) Not a solution   (b) Solution
(c) Not a solution   (d) Solution

15. Determine whether each value of x is a solution of $\dfrac{x}{4} + \dfrac{2}{x} = \dfrac{3}{2}$.

(a) $x = 1$ (b) $x = 2$ (c) $x = -\dfrac{1}{2}$ (d) $x = 4$

In Exercises 16–18, solve the rational equation.

16. $5 + \dfrac{t}{3} = t + 2$ $\dfrac{9}{2}$

17. $\dfrac{5}{x + 1} - \dfrac{1}{x} = \dfrac{3}{x}$ 4

18. $\dfrac{x - 3}{x - 2} + \dfrac{x + 1}{x + 3} = \dfrac{2x^2 - 15}{x^2 + x - 6}$ 4

19. Van: 48 miles per hour; Car: 60 miles per hour

19. A car leaves a town 1 hour after a moving van. The speed of the moving van is 12 miles per hour lower than that of the car. After traveling 240 miles, the car overtakes the moving van. Find the average speed of each vehicle.

427

Motivating the Chapter

⊘ Pay Analysis

You have two job offers for sales positions in the technology industry. The first job pays $600 per week plus a sales commission of 4% of your sales. The second job pays $400 per week plus a sales commission of 5% of your sales.

See Section 8.1, Exercise 61.

a. The system of equations that represents this situation is

$$\begin{cases} y = 600 + 0.04x \\ y = 400 + 0.05x \end{cases}$$

where y represents the weekly salary and x represents weekly sales. Graph both equations on the same coordinate system. Estimate the point at which the lines intersect. What does this point of intersection represent in the real-life context of this problem?

See Additional Answers.   (20,000, 1400); Point where the two job offers yield the same weekly salary

See Section 8.2, Exercise 69.

b. If you were given the system of equations from part (a) written as

$$\begin{cases} y - 0.04x = 600 \\ y - 0.05x = 400 \end{cases}$$

which variable would you solve for first using the method of substitution? Explain why you chose that variable.

Solve for y because the coefficient is 1.

c. Solve the system of linear equations given in part (b) using the method of substitution. Do you obtain the same answer as in part (a)?

(20,000, 1400); Yes

See Section 8.3, Exercise 63.

d. To solve the system of equations in part (a), you could multiply each equation by 100 to produce a system with integer coefficients. What are the advantages of this technique when using the method of elimination?   When adding the equations, the addition is easier if the coefficients are integers.

e. Solve the system using the method of elimination. For what level of sales will the second job pay more? Explain.   Sales greater than $20,000; (20,000, 1400) is the solution point of the system. If a value greater than 20,000 is substituted for x, the second equation will yield a greater y-value than the first equation.

Jose Luis Pelaez, Inc./Corbis

8

Systems of Linear Equations and Inequalities

8.1 Solving Systems of Equations by Graphing

What You Should Learn

1 Determine if an ordered pair is a solution to a system of equations.

2 Use a coordinate system to solve systems of linear equations graphically.

Why You Should Learn It

Many businesses use systems of equations to help determine their sales goals. For instance, Example 6 on page 435 shows how to graph a system of equations to determine the break-even point of producing and selling a new energy bar.

1 **Determine if an ordered pair is a solution to a system of equations.**

Systems of Linear Equations

Until this chapter, most problems have involved just one equation in either one or two variables. However, many problems in science, business, health services, and government involve two or more equations in two or more variables. For example, consider the following problem, which will be presented again in Section 8.2.

A total of $12,000 is invested in two funds paying 9% and 11% simple interest. The combined annual interest for the two funds is $1180. How much of the $12,000 is invested at each rate?

Letting x and y denote the amounts (in dollars) in the two funds, you can translate this problem into the following pair of linear equations in two variables.

$$\begin{cases} x + y = 12{,}000 \\ 0.09x + 0.11y = 1180 \end{cases}$$

Taken together, these two equations form a **system of linear equations.** A **solution** of a system of linear equations in two variables x and y is an ordered pair (x, y) that satisfies *both* equations.

Example 1   Checking a Solution

Show that $(2, -1)$ is a solution of the system of equations.

$$\begin{cases} 3x + 2y = 4 & \text{Equation 1} \\ -x + 3y = -5 & \text{Equation 2} \end{cases}$$

Solution

To check that a point is a solution, substitute the coordinates of the point into each equation. In the first equation, substitute 2 for x and -1 for y.

$$3(2) + 2(-1) \stackrel{?}{=} 4 \quad \implies \quad 6 - 2 = 4 \qquad \text{Solution checks in 1st equation. } \checkmark$$

In the second equation, substitute 2 for x and -1 for y.

$$-(2) + 3(-1) \stackrel{?}{=} -5 \quad \implies \quad -2 - 3 = -5 \qquad \text{Solution checks in 2nd equation. } \checkmark$$

Because the solution $(2, -1)$ checks in *both* equations, you can conclude that it is a solution of the original system of linear equations. You can also check by graphing both equations. In Figure 8.1, $(2, -1)$ is the point of intersection.

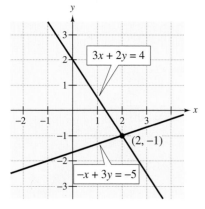

Figure 8.1

② Use a coordinate system to solve systems of linear equations graphically.

Solving a System of Linear Equations by Graphing

In this chapter, you will study three methods for solving a system of two linear equations in two variables. The first method is *solution by graphing*. With this method, you first sketch the lines representing the two equations. Then you try to determine whether the two lines intersect in a point, as illustrated in Example 2.

Remind students that they could sketch the graphs of these lines by considering slopes and y-intercepts instead of making tables of values. The line $y = -x - 2$ has slope $m = -1$ and y-intercept $(0, -2)$. The line $y = \frac{2}{3}x + 3$ has slope $m = \frac{2}{3}$ and y-intercept $(0, 3)$.

Example 2   Solving a System of Linear Equations

Solve the system of equations by graphing each equation and locating the point of intersection.

$$\begin{cases} x + y = -2 & \text{Equation 1} \\ 2x - 3y = -9 & \text{Equation 2} \end{cases}$$

Solution

One way to begin is to write each equation in slope-intercept form.

Equation 1	*Equation 2*
$x + y = -2$	$2x - 3y = -9$
$y = -x - 2$	$-3y = -2x - 9$
	$y = \dfrac{2}{3}x + 3$

Then use a numerical approach by creating two tables of values.

Table of Values for Equation 1

x	-4	-3	-2	-1	0	1	2
$y = -x - 2$	2	1	0	-1	-2	-3	-4

Table of Values for Equation 2

x	-4	-3	-2	-1	0	1	2
$y = \frac{2}{3}x + 3$	$\frac{1}{3}$	1	$\frac{5}{3}$	$\frac{7}{3}$	3	$\frac{11}{3}$	$\frac{13}{3}$

It may happen, as it did in these two tables, that you discover a common solution point, the point $(-3, 1)$. Another way to solve the system is to sketch the graphs of both equations, as shown in Figure 8.2. From the graph, it appears that the lines intersect at the point $(-3, 1)$. To verify this, substitute the coordinates of the point into each of the two original equations.

Substitute into Equation 1	*Substitute into Equation 2*
$x + y = -2$	$2x - 3y = -9$
$-3 + 1 \stackrel{?}{=} -2$	$2(-3) - 3(1) \stackrel{?}{=} -9$
$-2 = -2$ ✓	$-9 = -9$ ✓

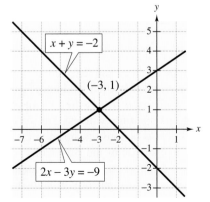

Figure 8.2

Because both equations are satisfied, the point $(-3, 1)$ is the solution of the system.

Technology: Discovery

Rewrite each system of equations in slope-intercept form and graph the equations using a graphing calculator. What is the relationship between the slopes of the two lines and the number of points of intersection?

See Technology Answers.

a. $\begin{cases} 3x + 4y = 12 \\ 2x - 3y = -9 \end{cases}$ **b.** $\begin{cases} -x + 2y = 8 \\ 2x - 4y = 5 \end{cases}$ **c.** $\begin{cases} x + y = 6 \\ 3x + 3y = 18 \end{cases}$

A system of linear equations can have exactly one solution, infinitely many solutions, or no solution. To see why this is true, consider the graphical interpretations of three systems of two linear equations shown below.

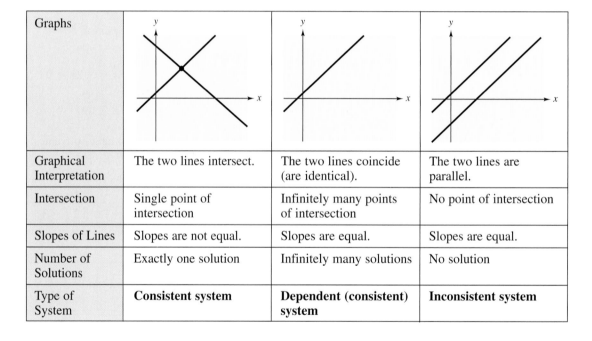

Graphs			
Graphical Interpretation	The two lines intersect.	The two lines coincide (are identical).	The two lines are parallel.
Intersection	Single point of intersection	Infinitely many points of intersection	No point of intersection
Slopes of Lines	Slopes are not equal.	Slopes are equal.	Slopes are equal.
Number of Solutions	Exactly one solution	Infinitely many solutions	No solution
Type of System	**Consistent system**	**Dependent (consistent) system**	**Inconsistent system**

Note that for dependent systems, the slopes of the lines and the *y*-intercepts are equal. For inconsistent systems, the slopes are equal, but the *y*-intercepts of the two lines are different. Also, note that the word *consistent* is used to mean that the system of linear equations has at least one solution, whereas the word *inconsistent* is used to mean that the system of linear equations has no solution.

You can see from the graphs above that a comparison of the slopes of two lines gives useful information about the number of solutions of the corresponding system of equations. So, to solve a system of equations graphically, it helps to begin by writing the equations in slope-intercept form,

$$y = mx + b. \qquad \text{Slope-intercept form}$$

Example 3   A System with No Solution

Solve the system of linear equations.

$$\begin{cases} x - y = 2 & \text{Equation 1} \\ -3x + 3y = 6 & \text{Equation 2} \end{cases}$$

Solution

Begin by writing each equation in slope-intercept form.

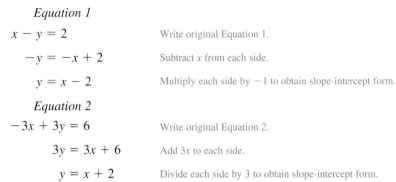

Equation 1

$$x - y = 2 \qquad \text{Write original Equation 1.}$$
$$-y = -x + 2 \qquad \text{Subtract } x \text{ from each side.}$$
$$y = x - 2 \qquad \text{Multiply each side by } -1 \text{ to obtain slope-intercept form.}$$

Equation 2

$$-3x + 3y = 6 \qquad \text{Write original Equation 2.}$$
$$3y = 3x + 6 \qquad \text{Add } 3x \text{ to each side.}$$
$$y = x + 2 \qquad \text{Divide each side by 3 to obtain slope-intercept form.}$$

From these slope-intercept forms, you can see that the lines representing the two equations are parallel (each has a slope of 1), as shown in Figure 8.3. So, the original system of linear equations has no solution and is an inconsistent system. Try constructing tables of values for the two equations. The tables should help convince you that there is no solution.

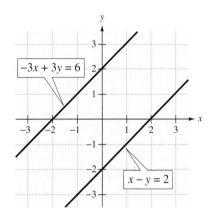

Figure 8.3

Example 4   A System with Infinitely Many Solutions

Solve the system of linear equations.

$$\begin{cases} x - y = 2 & \text{Equation 1} \\ -3x + 3y = -6 & \text{Equation 2} \end{cases}$$

Solution

Begin by writing each equation in slope-intercept form.

$$\begin{cases} y = x - 2 & \text{Slope-intercept form of Equation 1} \\ y = x - 2 & \text{Slope-intercept form of Equation 2} \end{cases}$$

From these forms, you can see that the lines representing the two equations are the same (see Figure 8.4). So, the given system of linear equations is dependent and has infinitely many solutions. You can describe the solution set by saying that each point on the line $y = x - 2$ is a solution of the system of linear equations.

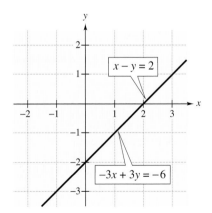

Figure 8.4

Note in Examples 3 and 4 that if the two lines representing a system of linear equations have the same slope, the system must have either no solution or infinitely many solutions. On the other hand, if the two lines have different slopes, they must intersect in a single point and the corresponding system has a single solution.

Example 5   A System with a Single Solution

Solve the system of linear equations.

$$\begin{cases} 2x + y = 4 & \text{Equation 1} \\ 4x + 3y = 9 & \text{Equation 2} \end{cases}$$

Solution

Begin by writing each equation in slope-intercept form.

Equation 1	*Equation 2*
$2x + y = 4$	$4x + 3y = 9$
$y = -2x + 4$	$3y = -4x + 9$
	$y = -\frac{4}{3}x + 3$

The slope-intercept forms of the two equations are as follows.

$$\begin{cases} y = -2x + 4 & \text{Slope-intercept form of Equation 1} \\ y = -\frac{4}{3}x + 3 & \text{Slope-intercept form of Equation 2} \end{cases}$$

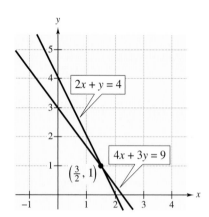

Figure 8.5

Because the lines do not have the same slope, you know that they intersect. To find the point of intersection, sketch both lines on the same rectangular coordinate system, as shown in Figure 8.5. From this sketch, it appears that the solution occurs near the point $\left(\frac{3}{2}, 1\right)$. To check this solution, substitute the coordinates of the point into each of the two original equations.

Substitute into Equation 1	*Substitute into Equation 2*
$2x + y = 4$	$4x + 3y = 9$
$2\left(\frac{3}{2}\right) + 1 \overset{?}{=} 4$	$4\left(\frac{3}{2}\right) + 3(1) \overset{?}{=} 9$
$3 + 1 = 4 \checkmark$	$6 + 3 = 9 \checkmark$

Because *both* equations are satisfied, the point $\left(\frac{3}{2}, 1\right)$ is the solution of the system.

Technology: Tip

The *zoom* and *trace* features of a graphing calculator can be used to approximate the solution point of a system of linear equations. A more accurate result can be obtained by using the *intersect* feature of a graphing calculator. Consult the user's guide of your graphing calculator for the steps in using this feature. Then, use the *intersect* feature to find the solution point in Example 5.

See Technology Answers.

There are two things you should note in Example 5. First, your success in applying the graphical method of solving a system of linear equations depends on sketching accurate graphs. Second, once you have made a graph and "guessed" at the point of intersection, it is critical that you check to see whether the point you have chosen is actually the solution.

As you take other courses in algebra, you will study systems of equations that are *not* linear. When you do that, you will learn that the discussion of the number of solutions on page 432 applies only to systems of *linear* equations. A nonlinear system such as

$$\begin{cases} y = 2x + 3 & \text{Equation 1} \\ y = x^2 & \text{Equation 2} \end{cases}$$

does not have to have zero, one, or infinitely many solutions. Try sketching a graph of this system. How many solutions does it have? Try estimating the solutions from the graphs and then check your estimates in each equation.

A common business application that involves systems of equations is break-even analysis. The total cost C of producing x units of a product usually has two components—the *initial cost* and the *cost per unit*. When enough units have been sold so that the total revenue R equals the total cost C, the sales have reached the **break-even point.** You can find this break-even point by finding the point of intersection of the cost and revenue graphs.

Example 6   Break-Even Analysis

A small business invests $14,000 to produce a new energy bar. Each bar costs $0.80 to produce and is sold for $1.50. The total cost C of producing x units of the bar is given by

$$C = 0.80x + 14,000. \qquad \text{Cost equation}$$

The revenue R from selling x units of the bar is given by

$$R = 1.50x. \qquad \text{Revenue equation}$$

How many energy bars must be sold before the business breaks even?

Solution

The system of equations that represents this problem is

$$\begin{cases} C = 0.80x + 14,000 & \text{Equation 1} \\ R = 1.50x & \text{Equation 2} \end{cases}$$

The two equations are in slope-intercept form and because the lines do not have the same slope, you know that they intersect. So, to find the break-even point, graph both equations and determine the point of intersection of the two graphs, as shown in Figure 8.6. From this graph, it appears that the break-even point occurs near the point (20,000, 30,000). To check this solution, substitute the coordinates of the point into each of the two original equations.

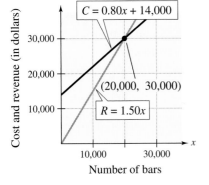

Figure 8.6

Substitute into Equation 1	*Substitute into Equation 2*
$C = 0.80x + 14,000$	$R = 1.50x$
$30,000 \overset{?}{=} 0.80(20,000) + 14,000$	$30,000 \overset{?}{=} 1.50(20,000)$
$30,000 \overset{?}{=} 16,000 + 14,000$	$30,000 = 30,000 \checkmark$
$30,000 = 30,000 \checkmark$	

Because both equations are satisfied, the business must sell 20,000 energy bars before it breaks even.

Profit P (or loss) for the business can be determined by the equation $P = R - C$. Note in Figure 8.6 that sales less than the break-even point correspond to a loss for the business, whereas sales greater than the break-even point correspond to a profit for the business.

8.1 Exercises

Review   *Concepts, Skills, and Problem Solving*

Keep mathematically in shape by doing these exercises *before* the problems of this section.

Properties and Definitions

In Exercises 1–4, use $x^2 + bx + c = (x + m)(x + n)$.

1. $mn = \quad c$

2. If $c > 0$, then what must be true about the signs of m and n?   Like signs

3. If $c < 0$, then what must be true about the signs of m and n?   Unlike signs

4. If m and n have like signs, then $m + n = \quad b$.

Solving Equations

In Exercises 5–10, solve the equation and check your solution.

5. $x - 6 = 5x$ $\quad -\frac{3}{2}$

6. $2 - 3x = 14 + x$ $\quad -3$

7. $y - 3(4y - 2) = 1$ $\quad \frac{5}{11}$

8. $y + 6(3 - 2y) = 4$ $\quad \frac{14}{11}$

9. $\dfrac{x}{2} - \dfrac{x}{5} = 15$ $\quad 50$

10. $\dfrac{x - 4}{10} = 6$ $\quad 64$

Models

In Exercises 11 and 12, translate the phrase into an algebraic expression.

11. The time to travel 250 miles at an average speed of r miles per hour $\dfrac{250}{r}$

12. The perimeter of a rectangle of length L and width $L/2$ $\quad 3L$

Developing Skills

In Exercises 1–6, determine whether each ordered pair is a solution of the system of equations. See Example 1.

System	Ordered Pairs

1. $\begin{cases} x + 3y = 11 \\ -x + 3y = 7 \end{cases}$
    (a) $(2, 3)$ (b) $(5, 4)$
    (a) Solution   (b) Not a solution

2. $\begin{cases} 3x - y = -2 \\ x - 3y = 2 \end{cases}$
    (a) $(0, 2)$ (b) $(-1, -1)$
    (a) Not a solution   (b) Solution

3. $\begin{cases} 2x - 3y = -8 \\ x + y = 1 \end{cases}$
    (a) $(5, -3)$ (b) $(-1, 2)$
    (a) Not a solution   (b) Solution

4. $\begin{cases} 5x - 3y = -12 \\ x - 4y = 1 \end{cases}$
    (a) $(-3, -1)$ (b) $(3, 1)$
    (a) Solution   (b) Not a solution

5. $\begin{cases} 5x - 6y = -2 \\ 7x + y = -31 \end{cases}$
    (a) $(-4, -3)$ (b) $(-3, -4)$
    (a) Solution   (b) Not a solution

6. $\begin{cases} -x - y = 6 \\ -5x - 2y = 3 \end{cases}$
    (a) $(7, -13)$ (b) $(3, -9)$
    (a) Not a solution   (b) Solution

7. $(2, 0)$ **8.** $(-1, 1)$ **9.** $(-1, -1)$ **10.** $(3, 4)$

In Exercises 7–16, use the graphs of the equations to determine the solution (if any) of the system of linear equations. Check your solution.

7. $\begin{cases} 2x + y = 4 \\ x - y = 2 \end{cases}$

8. $\begin{cases} x + 3y = 2 \\ -x + 2y = 3 \end{cases}$

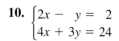

9. $\begin{cases} x - y = 0 \\ 3x - 2y = -1 \end{cases}$

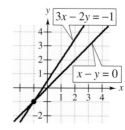

10. $\begin{cases} 2x - y = 2 \\ 4x + 3y = 24 \end{cases}$

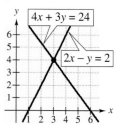
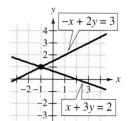

11. $\begin{cases} x - 2y = -4 \\ -0.5x + y = 2 \end{cases}$

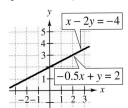

Infinitely many solutions

12. $\begin{cases} x - 3y = 3 \\ 2x - y = 6 \end{cases}$

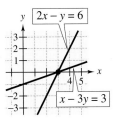

$(3, 0)$

13. $\begin{cases} 2x - 3y = 6 \\ 4x + 3y = 12 \end{cases}$

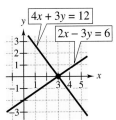

$(3, 0)$

14. $\begin{cases} 2x - 5y = 10 \\ 6x - 15y = 75 \end{cases}$

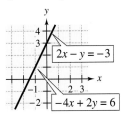

No solution

15. $\begin{cases} x + 4y = 8 \\ 3x + 12y = 12 \end{cases}$

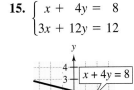

No solution

16. $\begin{cases} 2x - y = -3 \\ -4x + 2y = 6 \end{cases}$

Infinitely many solutions

In Exercises 17–42, sketch the graphs of the equations and approximate any solutions of the system of linear equations. See Examples 2–5.   See Additional Answers.

17. $\begin{cases} y = -x + 3 \quad (1, 2) \\ y = x + 1 \end{cases}$

18. $\begin{cases} y = 2x - 1 \quad (2, 3) \\ y = x + 1 \end{cases}$

19. $\begin{cases} y = 2x - 4 \quad (2, 0) \\ y = -\frac{1}{2}x + 1 \end{cases}$

20. $\begin{cases} y = \frac{1}{2}x + 2 \quad (4, 4) \\ y = -x + 8 \end{cases}$

21. $\begin{cases} x - y = 2 \quad (2, 0) \\ x + y = 2 \end{cases}$

22. $\begin{cases} x - y = 0 \quad (2, 2) \\ x + y = 4 \end{cases}$

23. $\begin{cases} -x + 2y = 4 \\ x - 2y = 4 \end{cases}$

No solution

24. $\begin{cases} 3x - y = 1 \\ -3x + y = 1 \end{cases}$

No solution

25. $\begin{cases} 4x - 5y = 0 \quad (5, 4) \\ 6x - 5y = 10 \end{cases}$

26. $\begin{cases} 3x + 2y = -6 \quad (0, -3) \\ 3x - 2y = 6 \end{cases}$

27. $\begin{cases} x - 2y = 4 \quad \text{Infinitely many solutions} \\ 2x - 4y = 8 \end{cases}$

28. $\begin{cases} 2x + 3y = 6 \quad \text{Infinitely many solutions} \\ 4x + 6y = 12 \end{cases}$

29. $\begin{cases} -2x + y = -1 \quad (2, 3) \\ x - 2y = -4 \end{cases}$

30. $\begin{cases} 2x + y = -4 \quad (0, -4) \\ 4x - 2y = 8 \end{cases}$

31. $\begin{cases} 4x - 3y = 3 \quad \text{No solution} \\ 4x - 3y = 0 \end{cases}$

32. $\begin{cases} 2x + 5y = 5 \quad \text{Infinitely many solutions} \\ -2x - 5y = -5 \end{cases}$

33. $\begin{cases} x + 2y = 3 \quad (7, -2) \\ x - 3y = 13 \end{cases}$

34. $\begin{cases} -x + 10y = 30 \quad (-10, 2) \\ x + 10y = 10 \end{cases}$

35. $\begin{cases} x + 7y = -5 \quad (2, -1) \\ 3x - 2y = 8 \end{cases}$

36. $\begin{cases} x + 2y = 4 \quad \left(1, \frac{3}{2}\right) \\ 2x - 2y = -1 \end{cases}$

37. $\begin{cases} -3x + 10y = 15 \quad \text{No solution} \\ 3x - 10y = 15 \end{cases}$

38. $\begin{cases} 4x - 9y = 12 \quad \text{No solution} \\ -4x + 9y = 12 \end{cases}$

39. $\begin{cases} 4x + 5y = 20 \quad \text{Infinitely many solutions} \\ \frac{4}{5}x + y = 4 \end{cases}$

40. $\begin{cases} 3x + 7y = 15 \quad \text{Infinitely many solutions} \\ x + \frac{7}{3}y = 5 \end{cases}$

41. $\begin{cases} 8x - 6y = -12 \quad \text{No solution} \\ x - \frac{3}{4}y = -2 \end{cases}$

42. $\begin{cases} -x + \frac{2}{3}y = 5 \quad \text{No solution} \\ 9x - 6y = 6 \end{cases}$

⊞ In Exercises 43–46, use a graphing calculator to graph the equations and approximate any solutions of the system of linear equations. Check your solution. See Additional Answers.

43. $\begin{cases} y = 2x - 1 \\ y = -3x + 9 \end{cases}$ $(2, 3)$

44. $\begin{cases} y = \frac{3}{4}x + 2 \\ y = x + 1 \end{cases}$ $(4, 5)$

45. $\begin{cases} y = x - 1 \\ y = -2x + 8 \end{cases}$ $(3, 2)$

46. $\begin{cases} y = 2x + 3 \\ y = -x - 3 \end{cases}$ $(-2, -1)$

In Exercises 47–54, write the equations of the lines in slope-intercept form. What can you conclude about the number of solutions of the system?

47. $\begin{cases} 2x - 3y = -12 \\ -8x + 12y = -12 \end{cases}$ $y = \frac{2}{3}x + 4, \ y = \frac{2}{3}x - 1;$ No solution

48. $\begin{cases} -5x + 8y = 8 \\ 7x - 4y = 14 \end{cases}$ $y = \frac{5}{8}x + 1, \ y = \frac{7}{4}x - \frac{7}{2};$ One solution

49. $\begin{cases} -x + 4y = 7 \\ 3x - 12y = -21 \end{cases}$ $y = \frac{1}{4}x + \frac{7}{4}, \ y = \frac{1}{4}x + \frac{7}{4};$ Infinitely many solutions

50. $\begin{cases} 3x + 8y = 28 \\ -4x + 9y = 1 \end{cases}$ $y = -\frac{3}{8}x + \frac{7}{2}, \ y = \frac{4}{9}x + \frac{1}{9};$ One solution

51. $\begin{cases} -2x + 3y = 4 \\ 2x + 3y = 8 \end{cases}$ $y = \frac{2}{3}x + \frac{4}{3}, \ y = -\frac{2}{3}x + \frac{8}{3};$ One solution

52. $\begin{cases} 2x + 5y = 15 \\ 2x - 5y = 5 \end{cases}$ $y = -\frac{2}{5}x + 3, \ y = \frac{2}{5}x - 1;$ One solution

53. $\begin{cases} -6x + 8y = 9 \\ 3x - 4y = -6 \end{cases}$ $y = \frac{3}{4}x + \frac{9}{8}, \ y = \frac{3}{4}x + \frac{3}{2};$ No solution

54. $\begin{cases} -6x + 8y = 9 \\ 3x - 4y = -4.5 \end{cases}$ $y = \frac{3}{4}x + \frac{9}{8}, \ y = \frac{3}{4}x + \frac{9}{8};$ Infinitely many solutions

Solving Problems

55. *Number Problem*   The sum of two numbers x and y is 20 and the difference of the two numbers is 2. The system of equations that represents this problem is

$$\begin{cases} x + y = 20 \\ x - y = 2 \end{cases}.$$

Solve the system graphically to find the two numbers. See Additional Answers. $(11, 9)$

56. *Number Problem*   The sum of two numbers x and y is 35 and the difference of the two numbers is 11. The system of equations that represents this problem is

$$\begin{cases} x + y = 35 \\ x - y = 11 \end{cases}.$$

Solve the system graphically to find the two numbers. See Additional Answers. $(23, 12)$

57. *Break-Even Analysis*   A small company produces bird feeders that sell for $23 per unit. The cost of producing each unit is $16.75, and the company has fixed costs of $400.

(a) Use a verbal model to show that the cost C of producing x units is $C = 16.75x + 400$ and the revenue R from selling x units is $R = 23x$. See Additional Answers.

(b) ⊞ Use a graphing calculator to graph the cost and revenue functions in the same viewing window. Approximate the point of intersection of the graphs and interpret the result. $x = 64$ units, $C = R = \$1472$; This means that the company must sell 64 feeders to cover their cost. Sales over 64 feeders will generate profit. See Additional Answers.

58. ⊞ *Supply and Demand*   The Law of Supply and Demand states that as the price of a product increases, the demand for the product decreases and the supply increases. The demand and supply equations for a tool set are $p = 90 - x$ and $p = 2x - 48$, respectively, where p is the price in dollars and x represents the number of units. Market equilibrium is the point of intersection of the two equations. Use a graphing calculator to graph the equations in the same viewing window and determine the price of the tool set that yields market equilibrium. See Additional Answers. $p = \$44$

📊 *Think About It*   In Exercises 59 and 60, the graphs of the two equations appear parallel. Are the two lines actually parallel? Does the system have a solution? If so, find the solution.

59. $\begin{cases} x - 200y = -200 \\ x - 199y = 198 \end{cases}$

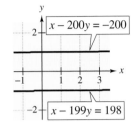

Because the slopes of the two lines are not equal, the lines intersect and the system has one solution: $(79{,}400,\ 398)$.

60. $\begin{cases} 25x - 24y = 0 \\ 13x - 12y = 24 \end{cases}$

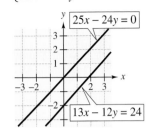

Because the slopes of the two lines are not equal, the lines intersect and the system has one solution: $(48, 50)$.

Explaining Concepts

61. 🔄   Answer part (a) of Motivating the Chapter on page 428.

62. *Writing*   Give geometric descriptions of the three cases for a system of linear equations in two variables.

63. *Writing*   In your own words, explain what is meant by a dependent system of linear equations.
A system that has an infinite number of solutions

64. *Writing*   In your own words, explain what is meant by an inconsistent system of linear equations.
A system that has no solution

65. *True or False?*   It is possible for a consistent system of linear equations to have exactly two solutions. Justify your answer.   False. It may have one solution or infinitely many solutions.

62. • Two lines that intersect in one point; the system has a unique solution.
   • Two lines that coincide; the system has an infinite number of solutions.
   • Two parallel lines; the system has no solution.

66. *Creating a System*   Write a system of linear equations with integer coefficients that has the unique solution $(3, 1)$. (There are many correct answers.)
$\begin{cases} x + 2y = 5 \\ -x + 3y = 0 \end{cases}$

📊 **67.** *Creating a System*   Write a system of linear equations that has no solution. (There are many correct answers.)
$\begin{cases} x + y = 0 \\ x + y = 1 \end{cases}$

📊 **68.** *Creating a System*   Write a system of linear equations that has infinitely many solutions. (There are many correct answers.)
$\begin{cases} x + y = 3 \\ 2x + 2y = 6 \end{cases}$

8.2 Solving Systems of Equations by Substitution

What You Should Learn

1. Use the method of substitution to solve systems of equations algebraically.
2. Use the method of substitution to solve systems with no solution or infinitely many solutions.
3. Use the method of substitution to solve application problems.

Why You Should Learn It

The method of substitution is one method of solving a system of linear equations. For instance, in Exercise 66 on page 448, a system of linear equations is used to find the cost of heating a home.

The Method of Substitution

Solving systems of equations by graphing is useful but less accurate than algebraic methods. In this section, you will study an algebraic method called the **method of substitution.** The goal of the method of substitution is to *reduce a system of two linear equations in two variables to a single equation in one variable.* Examples 1 and 2 illustrate the basic steps of this method.

1. Use the method of substitution to solve systems of equations algebraically.

Example 1   The Method of Substitution

Solve the system of linear equations.

$$\begin{cases} -x + y = 1 & \text{Equation 1} \\ 2x + y = -2 & \text{Equation 2} \end{cases}$$

Solution

Begin by solving for y in Equation 1.

$$y = x + 1 \qquad \text{Revised Equation 1}$$

Next, substitute this expression for y into Equation 2.

$$2x + y = -2 \qquad \text{Equation 2}$$
$$2x + (x + 1) = -2 \qquad \text{Substitute } x + 1 \text{ for } y.$$
$$3x + 1 = -2 \qquad \text{Combine like terms.}$$
$$3x = -3 \qquad \text{Subtract 1 from each side.}$$
$$x = -1 \qquad \text{Divide each side by 3.}$$

> You might point out that the solution obtained by substitution is the same solution that would be found by graphing. The two lines would intersect at $(-1, 0)$.

At this point, you know that the x-coordinate of the solution is -1. To find the y-coordinate, *back-substitute* the x-value into the revised Equation 1.

$$y = x + 1 \qquad \text{Revised Equation 1}$$
$$y = -1 + 1 = 0 \qquad \text{Substitute } -1 \text{ for } x.$$

So, the solution is $(-1, 0)$. Check this solution by substituting $x = -1$ and $y = 0$ into both of the original equations.

Study Tip

The term **back-substitute** implies that you work backwards. After solving for one of the variables, substitute that value back into one of the equations in the original (or revised) system to find the value of the other variable.

Example 2   The Method of Substitution

Solve the system of linear equations.

$$\begin{cases} 5x + 7y = 1 & \text{Equation 1} \\ x + 4y = -5 & \text{Equation 2} \end{cases}$$

Solution

For this system, it is convenient to begin by solving for x in the second equation.

$x + 4y = -5$ Original Equation 2

$x = -4y - 5$ Revised Equation 2

Substituting this expression for x into the first equation produces the following.

$5(-4y - 5) + 7y = 1$ Substitute $-4y - 5$ for x in Equation 1.

$-20y - 25 + 7y = 1$ Distributive Property

$-13y - 25 = 1$ Combine like terms.

$-13y = 26$ Add 25 to each side.

$y = -2$ Divide each side by -13.

Finally, back-substitute this y-value into the revised second equation.

$x = -4(-2) - 5 = 3$ Substitute -2 for y in revised Equation 2.

The solution is $(3, -2)$. Check this by substituting $x = 3$ and $y = -2$ into both of the original equations, as follows.

Substitute into Equation 1	*Substitute into Equation 2*
$5x + 7y = 1$	$x + 4y = -5$
$5(3) + 7(-2) \overset{?}{=} 1$	$(3) + 4(-2) \overset{?}{=} -5$
$15 - 14 = 1 \checkmark$	$3 - 8 = -5 \checkmark$

The steps for using the method of substitution to solve a system of equations involving two variables are summarized as follows.

The Method of Substitution

1. Solve one of the equations for one variable in terms of the other.

2. Substitute the expression obtained in Step 1 into the other equation to obtain an equation in one variable.

3. Solve the equation obtained in Step 2.

4. Back-substitute the solution from Step 3 into the expression obtained in Step 1 to find the value of the other variable.

5. Check the solution to see that it satisfies *both* of the original equations.

Study Tip

When you use the method of substitution, it does not matter which variable you choose to solve for first. You should choose the variable and equation that are easier to work with. For instance, in the system

$$\begin{cases} 3x - 2y = 1 & \text{Equation 1} \\ x + 4y = 3 & \text{Equation 2} \end{cases}$$

it is easier first to solve for x in the second equation. On the other hand, in the system

$$\begin{cases} 2x + y = 5 & \text{Equation 1} \\ 3x - 2y = 11 & \text{Equation 2} \end{cases}$$

it is easier first to solve for y in the first equation.

If neither variable has a coefficient of 1 in a system of linear equations, you can still use the method of substitution. However, you may have to work with some fractions in the solution steps.

Example 3   The Method of Substitution

Solve the system of linear equations.

$$\begin{cases} 5x + 3y = 18 & \text{Equation 1} \\ 2x - 7y = -1 & \text{Equation 2} \end{cases}$$

Solution

Step 1 ▷ Because neither variable has a coefficient of 1, you can choose to solve for either variable. For instance, you can begin by solving for x in Equation 1.

$$5x + 3y = 18 \qquad \text{Original Equation 1}$$

$$x = -\frac{3}{5}y + \frac{18}{5} \qquad \text{Revised Equation 1}$$

Step 2 ▷ Substitute for x in Equation 2.

$$2x - 7y = -1 \qquad \text{Equation 2}$$

$$2\left(-\frac{3}{5}y + \frac{18}{5}\right) - 7y = -1 \qquad \text{Substitute } -\frac{3}{5}y + \frac{18}{5} \text{ for } x.$$

Step 3 ▷ Solve for y.

$$-\frac{6}{5}y + \frac{36}{5} - 7y = -1 \qquad \text{Distributive Property}$$

$$-6y + 36 - 35y = -5 \qquad \text{Multiply each side by 5.}$$

$$36 - 41y = -5 \qquad \text{Combine like terms.}$$

$$-41y = -41 \qquad \text{Subtract 36 from each side.}$$

$$y = 1 \qquad \text{Divide each side by } -41.$$

Step 4 ▷ Back-substitute for y in the revised first equation.

$$x = -\frac{3}{5}y + \frac{18}{5} \qquad \text{Revised Equation 1}$$

$$x = -\frac{3}{5}(1) + \frac{18}{5} \qquad \text{Substitute 1 for } y.$$

$$x = 3 \qquad \text{Simplify.}$$

Step 5 ▷ The solution is $(3, 1)$. Check this in the original system.

In Example 3, notice that you can begin the solution by solving for x or y in either the first or second equation. You will obtain the same solution no matter which variable you solve for first.

② Use the method of substitution to solve systems with no solution or infinitely many solutions.

The No-Solution and Many-Solution Cases

The next two examples show how the method of substitution identifies systems of equations that have no solution or infinitely many solutions.

Example 4   The Method of Substitution: No-Solution Case

Solve the system of linear equations.

$$\begin{cases} x - 3y = 2 & \text{Equation 1} \\ -2x + 6y = 2 & \text{Equation 2} \end{cases}$$

Solution

Begin by solving for x in Equation 1 to obtain $x = 3y + 2$. Then substitute this expression for x in Equation 2.

$$-2x + 6y = 2 \qquad \text{Equation 2}$$
$$-2(3y + 2) + 6y = 2 \qquad \text{Substitute } 3y + 2 \text{ for } x.$$
$$-6y - 4 + 6y = 2 \qquad \text{Distributive Property}$$
$$-4 \neq 2 \qquad \text{Simplify.}$$

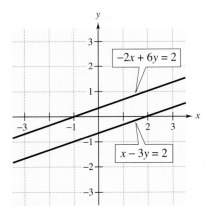

Figure 8.7

Because -4 does not equal 2, you can conclude that the original system is inconsistent and has no solution. The graphs in Figure 8.7 confirm this result.

Example 5   The Method of Substitution: Many-Solution Case

Solve the system of linear equations.

$$\begin{cases} 9x + 3y = 15 & \text{Equation 1} \\ 3x + y = 5 & \text{Equation 2} \end{cases}$$

Solution

Begin by solving for y in Equation 2 to obtain $y = -3x + 5$. Then substitute this expression for y in Equation 1.

$$9x + 3y = 15 \qquad \text{Equation 1}$$
$$9x + 3(-3x + 5) = 15 \qquad \text{Substitute } -3x + 5 \text{ for } y.$$
$$9x - 9x + 15 = 15 \qquad \text{Distributive Property}$$
$$15 = 15 \qquad \text{Simplify.}$$

The equation $15 = 15$ is true for any value of x. This implies that any solution of Equation 2 is also a solution of Equation 1. In other words, the original system of linear equations is *dependent* and has infinitely many solutions. The solutions consist of all ordered pairs (x, y) lying on the line $3x + y = 5$. Some sample solutions are $(-1, 8)$, $(0, 5)$, and $(1, 2)$.

Additional Examples
Solve each system of linear equations using the method of substitution.

a. $\begin{cases} -x + y = 3 \\ 3x + y = -1 \end{cases}$

b. $\begin{cases} 2x - 2y = 0 \\ x - y = 1 \end{cases}$

Answers:

a. $(-1, 2)$

b. No solution

Study Tip

By writing both equations in Example 5 in slope-intercept form, you will get identical equations. This means that the lines coincide and the system has infinitely many solutions.

③ Use the method of substitution to solve application problems.

Application

Example 6 uses the method of substitution to solve the interest rate problem introduced on page 430.

Example 6   An Application of a System of Linear Equations

Solve the system of linear equations.

$$\begin{cases} x + y = 12{,}000 & \text{Equation 1} \\ 0.09x + 0.11y = 1180 & \text{Equation 2} \end{cases}$$

Solution

To begin, it is convenient to multiply each side of the second equation by 100. This eliminates the need to work with decimals.

$$9x + 11y = 118{,}000 \qquad \text{Revised Equation 2}$$

Then solve for x in Equation 1.

$$x + y = 12{,}000 \qquad\qquad \text{Equation 1}$$
$$x = 12{,}000 - y \qquad\qquad \text{Subtract } y \text{ from each side.}$$

Next, substitute this expression for x into the revised Equation 2 and solve for y.

$$9x + 11y = 118{,}000 \qquad\qquad \text{Revised Equation 2}$$
$$9(12{,}000 - y) + 11y = 118{,}000 \qquad \text{Substitute } 12{,}000 - y \text{ for } x.$$
$$108{,}000 - 9y + 11y = 118{,}000 \qquad \text{Distributive Property}$$
$$2y = 10{,}000 \qquad\qquad \text{Simplify.}$$
$$y = 5000 \qquad\qquad \text{Divide each side by 2.}$$

Back-substitute the value $y = 5000$ into Equation 1 to obtain $x = 12{,}000 - 5000 = 7000$. The solution is $(7000, 5000)$. So, $7000 was invested at 9% simple interest and $5000 was invested at 11% simple interest. Check this in the original statement of the problem given on page 430.

Technology: Tip

The general solution of the linear system

$$\begin{cases} ax + by = c \\ dx + ey = f \end{cases}$$

is $x = (ce - bf)/(ae - bd)$ and $y = (af - cd)/(ae - bd)$. If $ae - db = 0$, the system does not have a unique solution. Graphing calculator programs for solving such a system can be found at our website *math.college.hmco.com/students*. Try using the program for your graphing calculator to solve the system in Example 6. See Technology Answers.

8.2 Exercises

Review   Concepts, Skills, and Problem Solving

Keep mathematically in shape by doing these exercises *before* the problems of this section.

Properties and Definitions

1. How many solutions does a linear equation of the form $2x + 8 = 7$ have?   One

2. What is a helpful usual first step when solving an equation such as
$$\frac{x}{6} + \frac{3}{2} = \frac{7}{4}?$$ Multiply each side of the equation by the lowest common denominator.

Factoring and Solving Equations

In Exercises 3–6, factor the expression.

3. $x(3 - x) - 2(3 - x)$
$(3 - x)(x - 2)$

4. $4t^2 - 9$
$(2t + 3)(2t - 3)$

5. $4y^2 - 20y + 25$
$(2y - 5)^2$

6. $6u^2 - 5u - 21$
$(3u - 7)(2u + 3)$

In Exercises 7–10, solve the equation.

7. $14 - 2x = x + 2$ 4

8. $\dfrac{9 + x}{3} = 15$ 36

9. $z^2 - 4z - 12 = 0$ $-2, 6$

10. $t^3 + t^2 - 4t - 4 = 0$ $-2, -1, 2$

Graphs and Models

11. ▲ *Geometry*   The length of each edge of a cube is x inches. Write an equation for the surface area A of the cube in terms of x. Graph the model.
$A = 6x^2$ See Additional Answers.

12. *Distance*   The speed of an airplane is 475 miles per hour. Write an equation for the distance d the plane travels in terms of the flight time t. Graph the model.
$d = 475t$ See Additional Answers.

Developing Skills

In Exercises 1–10, solve the system by the method of substitution. Use the graph to check the solution.

1. $\begin{cases} x - y = 0 \\ x + y = 2 \end{cases}$ $(1, 1)$

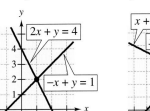

2. $\begin{cases} x + y = 1 \\ 2x - y = 2 \end{cases}$ $(1, 0)$

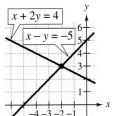

3. $\begin{cases} 2x + y = 4 \\ -x + y = 1 \end{cases}$ $(1, 2)$

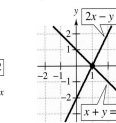

4. $\begin{cases} x - y = -5 \\ x + 2y = 4 \end{cases}$ $(-2, 3)$

5. $\begin{cases} -x + y = 1 \\ x - y = 1 \end{cases}$

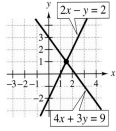

No solution

6. $\begin{cases} 4x + 3y = 8 \\ -4x + y = 8 \end{cases}$

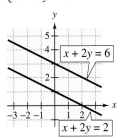

$(-1, 4)$

7. $\begin{cases} 2x - y = 2 \\ 4x + 3y = 9 \end{cases}$

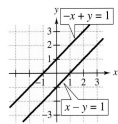

$\left(\frac{3}{2}, 1\right)$

8. $\begin{cases} x + 2y = 6 \\ x + 2y = 2 \end{cases}$

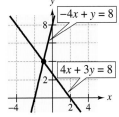

No solution

9. $\begin{cases} 2x + y = 3 \\ 4x + 2y = 6 \end{cases}$

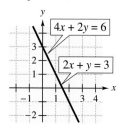

Infinitely many solutions

10. $\begin{cases} -4x + 3y = 6 \\ 8x - 6y = -12 \end{cases}$

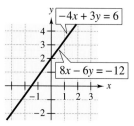

Infinitely many solutions

33. $\begin{cases} 4x - y = 2 \\ 2x - \frac{1}{2}y = 1 \end{cases}$
Infinitely many solutions

34. $\begin{cases} 3x - y = 6 \\ 4x - \frac{2}{3}y = -4 \end{cases}$
$(-4, -18)$

35. $\begin{cases} \frac{1}{5}x + \frac{1}{2}y = 8 \\ 2x + y = 20 \end{cases}$
$\left(\frac{5}{2}, 15\right)$

36. $\begin{cases} \frac{1}{2}x + \frac{3}{4}y = 10 \\ 4x - y = 4 \end{cases}$
$\left(\frac{26}{7}, \frac{76}{7}\right)$

37. $\begin{cases} -5x + 4y = 14 \\ 5x - 4y = 4 \end{cases}$
No solution

38. $\begin{cases} 3x - 2y = 3 \\ -6x + 4y = -6 \end{cases}$
Infinitely many solutions

In Exercises 11–48, solve the system by the method of substitution. See Examples 1–5.

11. $\begin{cases} y = 2x - 1 \\ y = -x + 5 \end{cases}$ $(2, 3)$

12. $\begin{cases} y = -2x + 9 \\ y = 3x - 1 \end{cases}$ $(2, 5)$

13. $\begin{cases} x = 4y - 5 \\ x = 3y \end{cases}$
$(15, 5)$

14. $\begin{cases} x = -5y - 2 \\ x = 2y - 23 \end{cases}$
$(-17, 3)$

15. $\begin{cases} 2x = 8 \\ x + y = 1 \end{cases}$ $(4, -3)$

16. $\begin{cases} 3x - y = 0 \\ 3y = 6 \end{cases}$ $\left(\frac{2}{3}, 2\right)$

17. $\begin{cases} x - y = 2 \\ x - y = 1 \end{cases}$
No solution

18. $\begin{cases} x + y = 8 \\ x + y = -1 \end{cases}$
No solution

19. $\begin{cases} x - y = 0 \\ 2x + y = 0 \end{cases}$ $(0, 0)$

20. $\begin{cases} x - y = 0 \\ 5x - 3y = 10 \end{cases}$ $(5, 5)$

21. $\begin{cases} x - 2y = -10 \\ 3x - y = 0 \end{cases}$
$(2, 6)$

22. $\begin{cases} x - 2y = 0 \\ 3x - y = 0 \end{cases}$
$(0, 0)$

23. $\begin{cases} 2x - y = -2 \\ 4x + y = 5 \end{cases}$ $\left(\frac{1}{2}, 3\right)$

24. $\begin{cases} x + 6y = 7 \\ -x + 4y = -2 \end{cases}$ $\left(4, \frac{1}{2}\right)$

25. $\begin{cases} x + 2y = 1 \\ 5x - 4y = -23 \end{cases}$
$(-3, 2)$

26. $\begin{cases} -3x + 6y = 4 \\ 2x + y = 4 \end{cases}$
$\left(\frac{4}{3}, \frac{4}{3}\right)$

27. $\begin{cases} 5x + 3y = 11 \\ x - 5y = 5 \end{cases}$
$\left(\frac{5}{2}, -\frac{1}{2}\right)$

28. $\begin{cases} -3x + y = 4 \\ -9x + 5y = 10 \end{cases}$
$\left(-\frac{5}{3}, -1\right)$

29. $\begin{cases} 5x + 2y = 0 \\ x - 3y = 0 \end{cases}$ $(0, 0)$

30. $\begin{cases} 4x + 3y = 0 \\ 2x - y = 0 \end{cases}$ $(0, 0)$

31. $\begin{cases} 2x + 5y = -4 \\ 3x - y = 11 \end{cases}$
$(3, -2)$

32. $\begin{cases} 2x + 5y = 1 \\ -x + 6y = 8 \end{cases}$
$(-2, 1)$

39. $\begin{cases} 2x + y = 8 \\ 5x + 2.5y = 10 \end{cases}$
No solution

40. $\begin{cases} 0.5x + 0.5y = 4 \\ x + y = -1 \end{cases}$
No solution

41. $\begin{cases} -6x + 1.5y = 6 \\ 8x - 2y = -8 \end{cases}$
Infinitely many solutions

42. $\begin{cases} 0.3x - 0.3y = 0 \\ x - y = 4 \end{cases}$
No solution

43. $\begin{cases} \dfrac{x}{3} - \dfrac{y}{4} = 2 \\ \dfrac{x}{2} + \dfrac{y}{6} = 3 \end{cases}$
$(6, 0)$

44. $\begin{cases} -\dfrac{x}{5} + \dfrac{y}{2} = -3 \\ \dfrac{x}{4} - \dfrac{y}{4} = 0 \end{cases}$
$(-10, -10)$

45. $\begin{cases} \dfrac{x}{4} + \dfrac{y}{2} = 1 \\ \dfrac{x}{2} - \dfrac{y}{3} = 1 \end{cases}$
$\left(\frac{5}{2}, \frac{3}{4}\right)$

46. $\begin{cases} -\dfrac{x}{6} + \dfrac{y}{12} = 1 \\ \dfrac{x}{2} + \dfrac{y}{8} = 1 \end{cases}$
$\left(-\frac{2}{3}, \frac{32}{3}\right)$

47. $\begin{cases} 2(x - 5) = y + 2 \\ 3x = 4(y + 2) \end{cases}$ $(8, 4)$

48. $\begin{cases} 3(x - 2) + 5 = 4(y + 3) - 2 \\ 2x + 7 = 2y + 8 \end{cases}$ $\left(-9, -\frac{19}{2}\right)$

In Exercises 49–54, solve the system by the method of substitution. Use a graphing calculator to verify the solution graphically.

49. $\begin{cases} y = -2x + 10 \\ y = x + 4 \end{cases}$ $(2, 6)$

50. $\begin{cases} y = \frac{5}{4}x + 3 \\ y = \frac{1}{2}x + 6 \end{cases}$ $(4, 8)$

51. $\begin{cases} 3x + 2y = 12 \\ x - y = 3 \end{cases}$ $\left(\frac{18}{5}, \frac{3}{5}\right)$

52. $\begin{cases} 2x - y = 1 \\ x - y = -2 \end{cases}$ $(3, 5)$

53. $\begin{cases} 5x + 3y = 15 \\ 2x - 3y = 6 \end{cases}$
$(3, 0)$

54. $\begin{cases} 4x - 5y = 0 \\ 2x - 5y = -10 \end{cases}$
$(5, 4)$

Think About It   In Exercises 55–58, find a system of linear equations that has the given solution. (There are many correct answers.)

55. $(2, 1)$

$$\begin{cases} x - 2y = 0 \\ x + y = 3 \end{cases}$$

56. $(4, -3)$

$$\begin{cases} 2x + y = 5 \\ -x + y = -7 \end{cases}$$

57. $\left(\frac{7}{2}, -3\right)$

$$\begin{cases} 2x - y = 10 \\ 4x + 3y = 5 \end{cases}$$

58. $\left(-\frac{1}{2}, 1\right)$

$$\begin{cases} 4x + 5y = 3 \\ 2x + 3y = 2 \end{cases}$$

Solving Problems

59. *Number Problem*   The sum of two numbers x and y is 40 and the difference of the two numbers is 10. The system of equations that represents this problem is

$$\begin{cases} x + y = 40 \\ x - y = 10 \end{cases}.$$

Solve this system to find the two numbers.   (25, 15)

60. *Number Problem*   The sum of two numbers x and y is 50 and the difference of the two numbers is 20. The system of equations that represents this problem is

$$\begin{cases} x + y = 50 \\ x - y = 20 \end{cases}.$$

Solve this system to find the two numbers.   (35, 15)

61. *Investment*   A total of $15,000 is invested in two funds paying 5% and 8% simple interest. The combined annual interest for the two funds is $900. The system of equations that represents this situation is

$$\begin{cases} x + y = 15{,}000 \\ 0.05x + 0.08y = 900 \end{cases}$$

where x represents the amount invested in the 5% fund and y represents the amount invested in the 8% fund. Solve this system to determine how much of the $15,000 is invested at each rate.
5%: $10,000; 8%: $5000

62. *Investment*   A total of $10,000 is invested in two funds paying 7% and 10% simple interest. The combined annual interest for the two funds is $775. The system of equations that represents this situation is

$$\begin{cases} x + y = 10{,}000 \\ 0.07x + 0.10y = 775 \end{cases}$$

where x represents the amount invested in the 7% fund and y represents the amount invested in the 10% fund. Solve this system to determine how much of the $10,000 is invested at each rate.
7%: $7500; 10%: $2500

63. *Dinner Price*   Six people ate dinner for $63.90. The price for adults was $16.95 and the price for children was $7.50. The system of equations that represents this situation is

$$\begin{cases} x + y = 6 \\ 16.95x + 7.50y = 63.90 \end{cases}$$

where x represents the number of adults and y represents the number of children. Solve this system to determine how many adults attended the dinner.
2 adults

64. *Ticket Sales*   You are selling football tickets. Student tickets cost $2 and general admission tickets cost $3. You sell 1957 tickets and collect $5035. The system of equations that represents this situation is

$$\begin{cases} x + y = 1957 \\ 2x + 3y = 5035 \end{cases}$$

where x represents the number of student tickets sold and y represents the number of general admission tickets sold. Solve this system to determine how many of each type of ticket were sold.
Student tickets: 836; General admission tickets: 1121

65. *Comparing Costs*   Car model ES costs $16,000 and costs an average of $0.26 per mile to maintain. Car model LS costs $18,000 and costs an average of $0.22 per mile to maintain. The system of equations that represents this situation is

$$\begin{cases} y = 16{,}000 + 0.26x & \text{Model ES} \\ y = 18{,}000 + 0.22x & \text{Model LS} \end{cases}$$

where y represents the total cost of the car and x represents the number of miles driven. Solve this system to determine after how many miles the total costs of the two models will be the same (one of each model is driven the same number of miles).   50,000 miles

66. *Comparing Costs*   Heating a three-bedroom home using a solar heating system costs $28,500 for installation and $125 per year to operate. Heating the same home using an electric heating system costs $5750 for installation and $1000 per year to operate. The system of equations that represents this situation is

$$\begin{cases} y = 28{,}500 + 125x \quad \text{Solar heating} \\ y = 5750 + 1000x \quad \text{Electric heating} \end{cases}$$

where y represents the total cost of heating the home and x is the number of years. Solve this system to determine after how many years the total costs for solar heating and electric heating will be the same. What will be the cost at that time?   26 years; $31,750

67. ▲ *Geometry*   Find an equation of the line with slope $m = 2$ passing through the intersection of the lines $x - 2y = 3$ and $3x + y = 16$.
$2x - y - 9 = 0$

68. ▲ *Geometry*   Find an equation of the line with slope $m = -3$ passing through the intersection of the lines $4x + 6y = 26$ and $5x - 2y = -15$.
$3x + y - 2 = 0$

Explaining Concepts

69. 🌀 Answer parts (b) and (c) of Motivating the Chapter on page 428.

70. *Writing*   In your own words, explain the basic steps in solving a system of linear equations by the method of substitution.

71. *Writing*   When solving a system of linear equations by the method of substitution, how do you recognize that it has no solution?   Solve one of the equations for one variable in terms of the other variable. Substitute that expression into the other equation. If a false statement results, the system has no solution.

72. *Writing*   When solving a system of linear equations by the method of substitution, how do you recognize that it has infinitely many solutions?
Solve one of the equations for one variable in terms of the other variable. Substitute that expression into the other equation. If an identity statement results, the system has infinitely many solutions.

73. *Writing*   Describe any advantages of the method of substitution over the graphical method of solving a system of linear equations.   The substitution method yields exact solutions.

70. (a) Solve one of the equations for one variable in terms of the other.
(b) Substitute the expression found in Step (a) into the other equation to obtain an equation in one variable.
(c) Solve the equation obtained in Step (b).
(d) Back-substitute the solution from Step (c) into the expression obtained in Step (a) to find the value of the other variable.
(e) Check the solution to see that it satisfies both of the original equations.

74. *Writing*   Explain what is meant by a consistent system of linear equations.   A consistent system of linear equations is a system that has at least one solution.

75. *Writing*   Explain how you can check the solution of a system of linear equations algebraically and graphically.   Algebraically: Substitute the solution into each equation of the original system. Graphically: Graph the two lines and verify that the solution is the point of intersection of the lines.

76. Your instructor says, "An equation (not in standard form) such as $2x - 3 = 5x - 9$ can be considered a system of equations." Create the system, and find the solution point. How many solution points does the "system" $x^2 - 1 = 2x - 1$ have? Illustrate your results with a graphing calculator.
$y = 2x - 3$, $y = 5x - 9$; Solution point: $(2, 1)$;
The "system" $x^2 - 1 = 2x - 1$ has two solution points.
See Additional Answers.

Think About It   In Exercises 77–80, find the value of *a* or *b* such that the system of linear equations is inconsistent.

77. $\begin{cases} x + by = 1 \quad b = 2 \\ x + 2y = 2 \end{cases}$

78. $\begin{cases} ax + 3y = 6 \quad a = -3 \\ 5x - 5y = 2 \end{cases}$

79. $\begin{cases} -6x + y = 4 \quad b = -\frac{1}{3} \\ 2x + by = 3 \end{cases}$

80. $\begin{cases} 6x - 3y = 4 \quad a = 2 \\ ax - y = -2 \end{cases}$

8.3 Solving Systems of Equations by Elimination

Staffan Widstrand/Corbis

What You Should Learn

① Solve systems of linear equations algebraically using the method of elimination.

② Choose a method for solving systems of equations.

Why You Should Learn It

The method of elimination is one method of solving a system of linear equations. For instance, in Exercise 60 on page 458, this method is convenient for solving a system of linear equations used to find the focal length of a camera.

① Solve systems of linear equations algebraically using the method of elimination.

The Method of Elimination

In this section, you will study another way to solve a system of linear equations algebraically—the **method of elimination.** The key step in this method is to obtain opposite coefficients for one of the variables so that *adding* the two equations eliminates this variable. For instance, by adding the equations

$$\begin{cases} 3x + 5y = 7 & \text{Equation 1} \\ -3x - 2y = -1 & \text{Equation 2} \\ \hline 3y = 6 & \text{Add equations.} \end{cases}$$

you eliminate the variable x and obtain a single equation in one variable, y.

Example 1   The Method of Elimination

Solve the system of linear equations.

$$\begin{cases} 4x + 3y = 1 & \text{Equation 1} \\ 2x - 3y = 5 & \text{Equation 2} \end{cases}$$

Solution

Begin by noting that the coefficients of y are opposites. So, by adding the two equations, you can eliminate y.

$$\begin{cases} 4x + 3y = 1 & \text{Equation 1} \\ 2x - 3y = 5 & \text{Equation 2} \\ \hline 6x = 6 & \text{Add equations.} \end{cases}$$

So, $x = 1$. By back-substituting this value into the first equation, you can solve for y, as follows.

$$4(1) + 3y = 1 \qquad \text{Substitute 1 for } x \text{ in Equation 1.}$$

$$3y = -3 \qquad \text{Subtract 4 from each side.}$$

$$y = -1 \qquad \text{Divide each side by 3.}$$

The solution is $(1, -1)$. Check this in both of the original equations.

Study Tip

Try solving the system in Example 1 by substitution. Which method do you think is easier? Many people find that the method of elimination is more efficient.

To obtain opposite coefficients for one of the variables, you often need to multiply one or both of the equations by a suitable constant. This is demonstrated in the following example.

Example 2   The Method of Elimination

Solve the system of linear equations.

$$\begin{cases} 2x - 3y = -7 & \text{Equation 1} \\ 3x + y = -5 & \text{Equation 2} \end{cases}$$

Solution

For this system, you can obtain opposite coefficients of y by multiplying the second equation by 3.

$$\begin{cases} 2x - 3y = -7 \\ 3x + y = -5 \end{cases} \implies \begin{array}{ll} 2x - 3y = -7 & \text{Equation 1} \\ \underline{9x + 3y = -15} & \text{Multiply Equation 2 by 3.} \\ 11x \qquad = -22 & \text{Add equations.} \end{array}$$

So, $x = -2$. By back-substituting this value of x into the second equation, you can solve for y.

$$\begin{array}{ll} 3x + y = -5 & \text{Equation 2} \\ 3(-2) + y = -5 & \text{Substitute } -2 \text{ for } x. \\ -6 + y = -5 & \text{Simplify.} \\ y = 1 & \text{Add 6 to each side.} \end{array}$$

The solution is $(-2, 1)$. Check this in the original equations, as follows.

Substitute into Equation 1

$$2x - 3y = -7$$
$$2(-2) - 3(1) \overset{?}{=} -7$$
$$-4 - 3 = -7 \checkmark$$

Substitute into Equation 2

$$3x + y = -5$$
$$3(-2) + 1 \overset{?}{=} -5$$
$$-6 + 1 = -5 \checkmark$$

This method is called "elimination" because the first step in the process is to "eliminate" one of the variables. This method is summarized as follows.

The Method of Elimination

1. Obtain opposite coefficients of x (or y) by multiplying all terms of one or both equations by suitable constants.

2. Add the equations to eliminate one variable and solve the resulting equation.

3. Back-substitute the value obtained in Step 2 into either of the original equations and solve for the other variable.

4. Check your solution in *both* of the original equations.

Example 3   The Method of Elimination

Solve the system of linear equations.

$$\begin{cases} 5x + 3y = 6 \\ 2x - 4y = 5 \end{cases}$$ Equation 1

    Equation 2

Solution

You can obtain opposite coefficients of y by multiplying the first equation by 4 and the second equation by 3.

$$\begin{cases} 5x + 3y = 6 \\ 2x - 4y = 5 \end{cases}$$ $\Longrightarrow$ $20x + 12y = 24$ Multiply Equation 1 by 4.

$\Longrightarrow$ $\underline{6x - 12y = 15}$ Multiply Equation 2 by 3.

$26x \qquad = 39$ Add equations.

From this equation, you can see that $x = \frac{3}{2}$. By back-substituting this value of x into the second equation, you can solve for y, as follows.

$$2x - 4y = 5$$ Equation 2

$$2\left(\frac{3}{2}\right) - 4y = 5$$ Substitute $\frac{3}{2}$ for x.

$$3 - 4y = 5$$ Simplify.

$$-4y = 2$$ Subtract 3 from each side.

$$y = -\frac{1}{2}$$ Divide each side by -4.

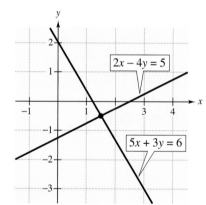

Figure 8.8

The solution is $\left(\frac{3}{2}, -\frac{1}{2}\right)$. You can check this as follows.

Substitute into Equation 1   *Substitute into Equation 2*

$$5x + 3y = 6 \qquad\qquad 2x - 4y = 5$$

$$5\left(\frac{3}{2}\right) + 3\left(-\frac{1}{2}\right) \overset{?}{=} 6 \qquad 2\left(\frac{3}{2}\right) - 4\left(-\frac{1}{2}\right) \overset{?}{=} 5$$

$$\frac{15}{2} - \frac{3}{2} = 6 \checkmark \qquad\qquad 3 + 2 = 5 \checkmark$$

The graph of this system is shown in Figure 8.8. From the graph it appears that the solution $\left(\frac{3}{2}, -\frac{1}{2}\right)$ is reasonable.

In Example 3, the y-variable was eliminated first. You could just as easily have solved the system by eliminating the x-variable first, as follows.

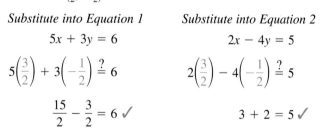

$$\begin{cases} 5x + 3y = 6 \\ 2x - 4y = 5 \end{cases}$$ $\Longrightarrow$ $10x + 6y = 12$ Multiply Equation 1 by 2.

$\Longrightarrow$ $\underline{-10x + 20y = -25}$ Multiply Equation 2 by -5.

$26y = -13$ Add equations.

From this equation, $y = -\frac{1}{2}$. By back-substituting this value of y into the second equation, you can solve for x to obtain $x = \frac{3}{2}$.

In the next example, note how the method of elimination can be used to determine that a system of linear equations has no solution. As with substitution, notice that the key is recognizing the occurrence of a *false statement*.

Example 4 The Method of Elimination: No-Solution Case

Solve the system of linear equations.

$$\begin{cases} 2x - 6y = 5 & \text{Equation 1} \\ 3x - 9y = 2 & \text{Equation 2} \end{cases}$$

Solution

To obtain coefficients that differ only in sign, multiply the first equation by 3 and multiply the second equation by -2.

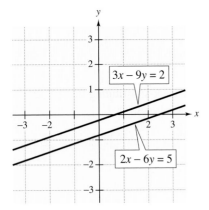

Figure 8.9

$$\begin{cases} 2x - 6y = 5 \\ 3x - 9y = 2 \end{cases} \Longrightarrow \begin{array}{ll} 6x - 18y = 15 & \text{Multiply Equation 1 by 3.} \\ \underline{-6x + 18y = -4} & \text{Multiply Equation 2 by } -2. \\ 0 \neq 11 & \text{Add equations.} \end{array}$$

Because 0 does not equal 11, you can conclude that the system is inconsistent and has no solution. The lines corresponding to the two equations of this system are shown in Figure 8.9. Note that the two lines are parallel and so have no point of intersection.

Example 5 shows how the method of elimination works with a system that has infinitely many solutions. Notice that you can recognize this case by the occurrence of an equation that is true for all real values of x and y.

Example 5 The Method of Elimination: Many-Solution Case

Solve the system of linear equations.

$$\begin{cases} 2x - 6y = -5 & \text{Equation 1} \\ -4x + 12y = 10 & \text{Equation 2} \end{cases}$$

Solution

To obtain the coefficients of x that differ only in sign, multiply the first equation by 2.

$$\begin{cases} 2x - 6y = -5 \\ -4x + 12y = 10 \end{cases} \Longrightarrow \begin{array}{ll} 4x - 12y = -10 & \text{Multiply Equation 1 by 2.} \\ \underline{-4x + 12y = 10} & \text{Equation 2} \\ 0 = 0 & \text{Add equations.} \end{array}$$

Study Tip

By writing both equations in Example 5 in slope-intercept form, you will obtain identical equations. This shows that the system has infinitely many solutions.

Because $0 = 0$ is a true statement, you can conclude that the system is dependent and has infinitely many solutions. The solution set consists of all ordered pairs (x, y) lying on the line $2x - 6y = -5$.

The next example shows how the method of elimination works with a system of linear equations with decimal coefficients.

Example 6   Solving a System with Decimal Coefficients

Solve the system of linear equations.

$$\begin{cases} 0.02x - 0.05y = -0.38 & \text{Equation 1} \\ 0.03x + 0.04y = 1.04 & \text{Equation 2} \end{cases}$$

Solution

Because the coefficients in this system have two decimal places, begin by multiplying each equation by 100. This produces a system in which the coefficients are all integers.

$$\begin{cases} 2x - 5y = -38 & \text{Revised Equation 1} \\ 3x + 4y = 104 & \text{Revised Equation 2} \end{cases}$$

Now, to obtain coefficients of x that differ only in sign, multiply the first equation by 3 and multiply the second equation by -2.

$$\begin{cases} 2x - 5y = -38 \\ 3x + 4y = 104 \end{cases} \implies \begin{array}{ll} 6x - 15y = -114 & \text{Multiply Equation 1 by 3.} \\ \underline{-6x - 8y = -208} & \text{Multiply Equation 2 by } -2. \\ -23y = -322 & \text{Add equations.} \end{array}$$

So, the y-coordinate of the solution is

$$y = \frac{-322}{-23} = 14.$$

Back-substituting this value into revised Equation 2 produces the following.

$$3x + 4(14) = 104 \qquad \text{Substitute 14 for } y \text{ in revised Equation 2.}$$
$$3x + 56 = 104 \qquad \text{Simplify.}$$
$$3x = 48 \qquad \text{Subtract 56 from each side.}$$
$$x = 16 \qquad \text{Divide each side by 3.}$$

So, the solution is $(16, 14)$. You can check this solution as follows.

Substitute into Equation 1

$$0.02x - 0.05y = -0.38 \qquad \text{Equation 1}$$
$$0.02(16) - 0.05(14) \stackrel{?}{=} -0.38 \qquad \text{Substitute 16 for } x \text{ and 14 for } y.$$
$$0.32 - 0.70 = -0.38 \checkmark \qquad \text{Solution checks.}$$

Substitute into Equation 2

$$0.03x + 0.04y = 1.04 \qquad \text{Equation 2}$$
$$0.03(16) + 0.04(14) \stackrel{?}{=} 1.04 \qquad \text{Substitute 16 for } x \text{ and 14 for } y.$$
$$0.48 + 0.56 = 1.04 \checkmark \qquad \text{Solution checks.}$$

Study Tip

When multiplying an equation by a negative number, be sure to distribute the negative sign to each term of the equation. For instance, in Example 6 the second equation is multiplied by -2.

Additional Examples
Solve each system using the method of elimination.

a. $\begin{cases} 5x + 3y = 8 \\ 2x - 4y = 11 \end{cases}$

b. $\begin{cases} -2x + 6y = 3 \\ 4x - 12y = -6 \end{cases}$

Answers:

a. $\left(\frac{5}{2}, -\frac{3}{2}\right)$

b. Infinitely many solutions

Example 7 An Application of a System of Linear Equations

A fundraising dinner was held on two consecutive nights. On the first night, 100 adult tickets and 175 children's tickets were sold, for a total of $1225. On the second night, 200 adult tickets and 316 children's tickets were sold, for a total of $2348. The system of linear equations that represents this problem is

$$\begin{cases} 100x + 175y = 1225 & \text{Equation 1} \\ 200x + 316y = 2348 & \text{Equation 2} \end{cases}$$

where x represents the price of the adult tickets and y represents the price of the children's tickets. Solve this system to find the price of each type of ticket.

Solution

To obtain coefficients of x that differ only in sign, multiply Equation 1 by -2.

$$\begin{cases} 100x + 175y = 1225 \\ 200x + 316y = 2348 \end{cases} \Rightarrow \begin{array}{l} -200x - 350y = -2450 \quad \text{Multiply Equation 1 by } -2. \\ \underline{200x + 316y = 2348} \quad \text{Equation 2} \\ -34y = -102 \quad \text{Add equations.} \end{array}$$

So, the y-coordinate of the solution is $y = -102/-34 = 3$. Back-substituting this value into Equation 2 produces the following.

$$200x + 316(3) = 2348 \qquad \text{Substitute 3 for } y \text{ in Equation 2.}$$
$$200x = 1400 \qquad \text{Simplify.}$$
$$x = 7 \qquad \text{Divide each side by 200.}$$

The solution is $(7, 3)$. So the price of the adult tickets was $7 and the price of the children's tickets was $3. Check this solution in both of the original equations.

---

2 Choose a method for solving systems of equations.

Choosing Methods

To decide which of the three methods (graphing, substitution, or elimination) to use to solve a system of two linear equations, use the following guidelines.

These guidelines could be the basis for a classroom discussion of the advantages and disadvantages of each of the three methods.

Guidelines for Solving a System of Linear Equations

To decide whether to use graphing, substitution, or elimination, consider the following.

1. The graphing method is useful for approximating the solution and for giving an overall picture of how one variable changes with respect to the other.

Mention to students that when they go on to more advanced algebra courses, they will find that these methods can be generalized to systems that contain more than two variables.

2. To find exact solutions, use either substitution or elimination.

3. For systems of equations in which one variable has a coefficient of 1, substitution may be more efficient than elimination.

4. Elimination is usually more efficient. This is especially true when the coefficients of one of the variables are opposites.

8.3 Exercises

Review   Concepts, Skills, and Problem Solving

Keep mathematically in shape by doing these exercises *before* the problems of this section.

Properties and Definitions

In Exercises 1–4, identify the property of real numbers illustrated by the statement.

1. $2ab \cdot \dfrac{1}{2ab} = 1$ Multiplicative Inverse Property

2. $8t + 0 = 8t$ Additive Identity Property

3. $2yx = 2xy$ Commutative Property of Multiplication

4. $3(2x) = (3 \cdot 2)x$ Associative Property of Multiplication

Algebraic Operations

In Exercises 5–10, plot the points on a rectangular coordinate system. Find the slope of the line passing through the points. If not possible, state why.
See Additional Answers.

5. $(-6, 4), (-3, -4)$ $-\frac{8}{3}$

6. $(4, 6), (8, -2)$ -2

7. $\left(\frac{7}{2}, \frac{9}{2}\right), \left(\frac{4}{3}, -3\right)$ $\frac{45}{13}$

8. $\left(-\frac{3}{4}, -\frac{7}{4}\right), \left(-1, \frac{5}{2}\right)$ -17

9. $(-3, 6), (-3, 2)$ Undefined

10. $(6, 2), (10, 2)$ 0

Problem Solving

11. *Quality Control*   A quality control engineer for a buyer found three defective units in a sample of 100. At this rate, what is the expected number of defective units in a shipment of 5000 units?   150 defective units

12. *Consumer Awareness*   The cost of a long distance telephone call is $0.70 for the first minute and $0.42 for each additional minute. The total cost of the call cannot exceed $8. Find the interval of time that is available for the call. Round the maximum value to one decimal place. $0 < t \le 18.4$

Developing Skills

In Exercises 1–4, solve the system by the method of elimination. Use the graph to check your solution.

1. $\begin{cases} 2x + y = 4 \\ x - y = 2 \end{cases}$

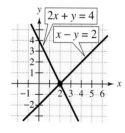

2. $\begin{cases} x + 3y = 2 \\ -x + 2y = 3 \end{cases}$

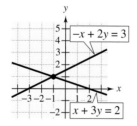

3. $\begin{cases} x - y = 0 \\ 3x - 2y = -1 \end{cases}$

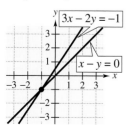

4. $\begin{cases} 2x - y = 2 \\ 4x + 3y = 24 \end{cases}$

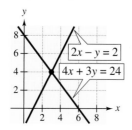

1. $(2, 0)$ **2.** $(-1, 1)$ **3.** $(-1, -1)$ **4.** $(3, 4)$

In Exercises 5–32, solve the system by the method of elimination. See Examples 1–6.

5. $\begin{cases} x - y = 4 \\ x + y = 12 \end{cases}$ $(8, 4)$

6. $\begin{cases} x + y = 7 \\ x - y = 3 \end{cases}$ $(5, 2)$

7. $\begin{cases} -x + 2y = 12 \\ x + 6y = 20 \end{cases}$ $(-4, 4)$

8. $\begin{cases} x + 2y = 14 \\ x - 2y = 10 \end{cases}$ $(12, 1)$

9. $\begin{cases} 3x - 5y = 1 \\ 2x + 5y = 9 \end{cases}$ $(2, 1)$

10. $\begin{cases} -2x + 3y = -4 \\ 2x - 4y = 6 \end{cases}$ $(-1, -2)$

11. $\begin{cases} 3a + 3b = 7 \\ 3a + 5b = 3 \end{cases}$ $\left(\frac{13}{3}, -2\right)$

12. $\begin{cases} 4a + 5b = 9 \\ 2a + 5b = 7 \end{cases}$ $(1, 1)$

13. $\begin{cases} -x + 2y = 12 \\ 3x - 6y = 10 \end{cases}$ No solution

14. $\begin{cases} -6x + 3y = 18 \\ 2x - y = 11 \end{cases}$ No solution

15. $\begin{cases} 3x - 4y = 11 \\ 2x + 3y = -4 \end{cases}$ $(1, -2)$

16. $\begin{cases} 2x + 3y = 16 \\ 5x - 10y = 30 \end{cases}$ $\left(\frac{50}{7}, \frac{4}{7}\right)$

17. $\begin{cases} 3x + 2y = -1 \\ -2x + 7y = 9 \end{cases}$
$(-1, 1)$

18. $\begin{cases} 5x + 3y = 27 \\ 7x - 2y = 13 \end{cases}$
$(3, 4)$

19. $\begin{cases} 3x - 4y = 1 \\ 4x + 3y = 1 \end{cases}$
$\left(\frac{7}{25}, -\frac{1}{25}\right)$

20. $\begin{cases} 2x - 5y = -1 \\ 2x - y = 1 \end{cases}$
$\left(\frac{3}{4}, \frac{1}{2}\right)$

21. $\begin{cases} 3x + 2y = 10 \\ 2x + 5y = 3 \end{cases}$
$(4, -1)$

22. $\begin{cases} 4x + 5y = 7 \\ 6x - 2y = -18 \end{cases}$
$(-2, 3)$

23. $\begin{cases} 5u + 6v = 14 \\ 3u + 5v = 7 \end{cases}$
$(4, -1)$

24. $\begin{cases} 5x + 3y = 18 \\ 2x - 7y = -1 \end{cases}$
$(3, 1)$

25. $\begin{cases} 6r + 5s = 3 \\ \frac{3}{2}r - \frac{5}{4}s = \frac{3}{4} \end{cases}$
$\left(\frac{1}{2}, 0\right)$

26. $\begin{cases} \frac{2}{3}x + \frac{1}{6}y = \frac{2}{3} \\ 4x + y = 4 \end{cases}$
Infinitely many solutions

27. $\begin{cases} \frac{1}{2}s - t = \frac{3}{2} \\ 4s + 2t = 27 \end{cases}$ $\left(6, \frac{3}{2}\right)$

28. $\begin{cases} 3u + 4v = 14 \\ \frac{1}{6}u - v = -2 \end{cases}$ $\left(\frac{18}{11}, \frac{25}{11}\right)$

29. $\begin{cases} 0.4a + 0.7b = 3 \\ 0.2a + 0.6b = 5 \end{cases}$ $(-17, 14)$

30. $\begin{cases} 0.2u - 0.1v = 1 \\ -0.8u + 0.4v = 3 \end{cases}$ No solution

31. $\begin{cases} 0.02x - 0.05y = -0.19 \\ 0.03x + 0.04y = 0.52 \end{cases}$ $(8, 7)$

32. $\begin{cases} 0.05x - 0.03y = 0.21 \\ 0.01x + 0.01y = 0.09 \end{cases}$ $(6, 3)$

In Exercises 33–38, solve the system by the method of elimination. Use a graphing calculator to verify your solution.

33. $\begin{cases} x + 2y = 3 \\ -x - y = -1 \end{cases}$
$(-1, 2)$

34. $\begin{cases} -2x + 2y = 7 \\ 2x + y = 8 \end{cases}$
$\left(\frac{3}{2}, 5\right)$

35. $\begin{cases} 7x + 8y = 6 \\ 3x - 4y = 10 \end{cases}$
$(2, -1)$

36. $\begin{cases} 10x - 11y = 7 \\ 2x - y = 5 \end{cases}$
$(4, 3)$

37. $\begin{cases} 5x + 2y = 7 \\ 3x - 6y = -3 \end{cases}$
$(1, 1)$

38. $\begin{cases} -4x + 5y = 8 \\ 2x + 3y = 18 \end{cases}$
$(3, 4)$

In Exercises 39–52, use the most convenient method (graphing, substitution, or elimination) to solve the system of linear equations. State which method you used.

39. $\begin{cases} x - y = 2 \\ y = 3 \end{cases}$ $(5, 3)$

40. $\begin{cases} y = 7 \\ x - 3y = 0 \end{cases}$ $(21, 7)$

41. $\begin{cases} 6x + 21y = 132 \\ 6x - 4y = 32 \end{cases}$ $(8, 4)$

42. $\begin{cases} -2x + y = 12 \\ 2x + 3y = 20 \end{cases}$ $(-2, 8)$

43. $\begin{cases} y = 2x - 1 \\ y = x + 1 \end{cases}$ $(2, 3)$

44. $\begin{cases} 2x - y = 4 \\ y = x \end{cases}$ $(4, 4)$

45. $\begin{cases} -4x + 3y = 11 \\ 3x - 10y = 15 \end{cases}$ $(-5, -3)$

46. $\begin{cases} -3x + 5y = -11 \\ 5x - 9y = 19 \end{cases}$ $(2, -1)$

47. $\begin{cases} x + y = 0 \\ 8x + 3y = 15 \end{cases}$ $(3, -3)$

48. $\begin{cases} x - 2y = 0 \\ 0.2x + 0.8y = 2.4 \end{cases}$ $(4, 2)$

49. $\begin{cases} -\dfrac{x}{4} + y = 1 \\ \dfrac{x}{4} + \dfrac{y}{2} = 1 \end{cases}$ $\left(\frac{4}{3}, \frac{4}{3}\right)$

50. $\begin{cases} \dfrac{x}{3} - \dfrac{y}{5} = 1 \\ \dfrac{x}{12} + \dfrac{y}{40} = 1 \end{cases}$ $(9, 10)$

51. $\begin{cases} 3(x + 5) - 7 = 2(3 - 2y) \\ 2x + 1 = 4(y + 2) \end{cases}$ $\left(1, -\frac{5}{4}\right)$

52. $\begin{cases} \frac{1}{2}(x - 4) + 9 = y - 10 \\ -5(x + 3) = 8 - 2(y - 3) \end{cases}$ $\left(\frac{5}{4}, \frac{141}{8}\right)$

Solving Problems

53. *Ticket Sales*   Ticket sales for a play were $3799 on the first night and $4905 on the second night. On the first night, 213 student tickets were sold and 632 general admission tickets were sold. On the second night, 275 student tickets were sold and 816 general admission tickets were sold. The system of equations that represents this situation is

$$\begin{cases} 213x + 632y = 3799 \\ 275x + 816y = 4905 \end{cases}$$

where x represents the price of the student tickets and y represents the price of the general admission tickets. Solve this system to determine the price of each type of ticket.

Student ticket: $3, General admission ticket: $5

54. *Ticket Sales*   Ticket sales for an annual variety show were $540 the first night and $850 the second night. On the first night, 150 student tickets were sold and 80 general admission tickets were sold. On the second night, 200 student tickets were sold and 150 general admission tickets were sold. The system of equations that represents this situation is

$$\begin{cases} 150x + 80y = 540 \\ 200x + 150y = 850 \end{cases}$$

where x represents the price of the student tickets and y represents the price of the general admission tickets. Solve this system to determine the price of each type of ticket.

Student ticket: $2, General admission ticket: $3

55. *Investment*   You invest a total of $10,000 in two investments earning 7.5% and 10% simple interest. (There is more risk in the 10% fund.) Your goal is to have a total annual interest income of $850. The system of equations that represents this situation is

$$\begin{cases} x + y = 10{,}000 \quad \$4000 \\ 0.075x + 0.10y = 850 \end{cases}$$

where x is the amount invested in the 7.5% fund and y is the amount invested in the 10% fund. Solve this system to determine the smallest amount that you can invest at 10% in order to meet your objective.

56. *Investment*   You invest a total of $12,000 in two investments earning 8% and 11.5% simple interest. (There is more risk in the 11.5% fund.) Your goal is to have a total annual interest income of $1065. The system of equations that represents this situation is

$$\begin{cases} x + y = 12{,}000 \quad \$3000 \\ 0.08x + 0.115y = 1065 \end{cases}$$

where x is the amount invested in the 8% fund and y is the amount invested in the 11.5% fund. Solve this system to determine the smallest amount that you can invest at 11.5% in order to meet your objective.

57. *Number Problem*   The sum of two numbers x and y is 82 and the difference of the numbers is 14. The system of equations that represents this problem is

$$\begin{cases} x + y = 82 \\ x - y = 14 \end{cases}$$

Solve this system to find the two numbers.   (48, 34)

58. *Number Problem*   The sum of two numbers x and y is 154 and the difference of the numbers is 38. The system of equations that represents this problem is

$$\begin{cases} x + y = 154 \\ x - y = 38 \end{cases}$$

Solve this system to find the two numbers.   (96, 58)

59. *Jewelry*   A bracelet that is supposed to be 18-karat gold weighs 277.92 grams. The volume of the bracelet is 18.52 cubic centimeters. The bracelet is made of gold and copper. Gold weighs 19.3 grams per cubic centimeter and copper weighs 9 grams per cubic centimeter. The system of equations that represents this situation is

$$\begin{cases} x + y = 18.52 \\ 19.3x + 9y = 277.92 \end{cases}$$

where x represents the volume of gold and y represents the volume of copper. Solve this system to determine whether or not the bracelet is 18-karat gold.   Yes, it is.

18K = 3/4 gold by weight

60. *Focal Length*   When parallel rays of light pass through a convex lens, they are bent inward and meet at a *focus* (see figure). The distance from the center of the lens to the focus is called the *focal length.* The equations of the lines containing the two bent rays in the camera are

$$\begin{cases} x + 3y = 1 \\ -x + 3y = -1 \end{cases}$$

where x and y are measured in inches. Which equation is the upper ray? What is the focal length?

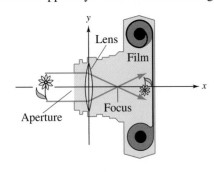

61. ▲ *Geometry*   Find an equation of the line of slope $m = 3$ passing through the intersection of the lines

$$3x + 4y = 7 \quad \text{and} \quad 5x - 4y = 1.$$
$3x - y - 2 = 0$

62. ▲ *Geometry*   Find an equation of the line of slope $m = -2$ passing through the intersection of the lines

$$2x + 5y = 11 \quad \text{and} \quad 4x - y = 11.$$
$2x + y - 7 = 0$

60. $x + 3y = 1$ is the equation of the upper ray because the slope of the line is negative. The focal length is 1.

67. $\begin{cases} 3x - 2y = 9 \\ 7x + 2y = 11 \end{cases}$ **68.** $\begin{cases} x - 4y = 3 \\ 7x + 9y = 11 \end{cases}$

Explaining Concepts

63. 🌀 Answer parts (d) and (e) of Motivating the Chapter on page 428.

64. *Writing* ✎   When solving a system by the method of elimination, how do you recognize that it has no solution?   When you add the equations to eliminate one variable, both are eliminated, yielding a contradiction. For example, adding the equations in the system $x - y = 3$ and $-x + y = 8$ yields $0 = 11$.

65. *Writing* ✎   When solving a system by the method of elimination, how do you recognize that it has infinitely many solutions?   When you add the equations to eliminate one variable, both are eliminated, yielding an identity. For example, adding the equations in the system $x - y = 3$ and $-x + y = -3$ yields $0 = 0$.

66. *Creating an Example*   Write an example of "clearing" a system of decimals. There are many correct answers.

$$\begin{cases} 0.3x - 0.2y = 0.9 \\ 0.7x + 0.2y = 1.1 \end{cases}$$

Multiply each side of each equation by 10 to clear the decimals.

$$\begin{cases} 3x - 2y = 9 \\ 7x + 2y = 11 \end{cases}$$

67. *Creating a System*   Write a system of linear equations that is better solved by the method of elimination than by the method of substitution. There are many correct answers.

68. *Creating a System*   Write a system of linear equations that is better solved by the method of substitution than by the method of elimination. There are many correct answers.

⚡ 69. *Writing* ✎   Both $(-2, 3)$ and $(8, 1)$ are solutions to a system of linear equations. How many solutions does the system have? Explain.   Infinitely many solutions. Since two solutions are given, the system is dependent.

⚡ 70. Consider the system of linear equations.

$$\begin{cases} x + y = 8 \\ 2x + 2y = k \end{cases}$$

(a) Find the value(s) of k for which the system has an infinite number of solutions. $k = 16$

(b) Find one value of k for which the system has no solution. There are many correct answers. $k = 1$

(c) Can the system have a single solution for some value of k? Why or why not?   No. Both variables are eliminated when the second equation is subtracted from 2 times the first equation.

Mid-Chapter Quiz

Take this quiz as you would take a quiz in class. After you are done, check your work against the answers in the back of the book.

1. Not a solution because substituting $x = 4$ and $y = -2$ into $5x - 3y = 14$ yields $26 = 14$, which is a contradiction.

2. Solution because substituting $x = 2$ and $y = -1$ into the equations yields true equalities.

1. Is $(4, -2)$ a solution of $3x + 4y = 4$ *and* $5x - 3y = 14$? Explain.

2. Is $(2, -1)$ a solution of $2x - 3y = 7$ *and* $3x + 5y = 1$? Explain.

In Exercises 3–5, use the given graphs to solve the system of linear equations.

3. $\begin{cases} x + y = 5 \\ x - 3y = -3 \end{cases}$ $(3, 2)$

4. $\begin{cases} x + 2y = 6 \\ 3x - 4y = 8 \end{cases}$ $(4, 1)$

5. $\begin{cases} x + 2y = 2 \\ x - 2y = 6 \end{cases}$ $(4, -1)$

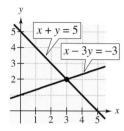

Figure for 3

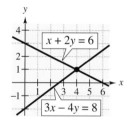

Figure for 4

In Exercises 6–8, sketch the graphs of the equations and approximate any solutions of the system of linear equations. Check your solution.

See Additional Answers.

6. $\begin{cases} x = 6 \\ x + y = 8 \end{cases}$ $(6, 2)$

7. $\begin{cases} y = \frac{3}{2}x - 1 \\ y = -x + 4 \end{cases}$ $(2, 2)$

8. $\begin{cases} 4x + y = 0 \\ -x + y = 5 \end{cases}$ $(-1, 4)$

In Exercises 9–11, solve the system by the method of substitution.

9. $\begin{cases} x - y = 4 \\ y = 2 \end{cases}$

10. $\begin{cases} y = -\frac{2}{3}x + 5 \\ y = 2x - 3 \end{cases}$

11. $\begin{cases} 2x - y = -7 \\ 4x + 3y = 16 \end{cases}$

In Exercises 12–14, solve the system by the method of elimination.

12. $\begin{cases} x + y = -2 \\ x - y = 4 \end{cases}$

13. $\begin{cases} 2x + y = 1 \\ 6x + 5y = 13 \end{cases}$

14. $\begin{cases} -x + 3y = 10 \\ 9x - 4y = 5 \end{cases}$

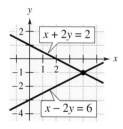

Figure for 5

9. $(6, 2)$ **10.** $(3, 3)$

11. $\left(-\frac{1}{2}, 6\right)$ **12.** $(1, -3)$

13. $(-2, 5)$ **14.** $\left(\frac{55}{23}, \frac{95}{23}\right)$

In Exercises 15 and 16, find a system of linear equations that has the ordered pair as its only solution. (There are many correct answers.)

15. $(0, 0)$ $\begin{cases} x + y = 0 \\ 3x - y = 0 \end{cases}$

16. $(6, -8)$ $\begin{cases} 4x + 3y = 0 \\ x + y = -2 \end{cases}$

In Exercises 17 and 18, find the value of k such that the system of linear equations is inconsistent.

17. $\begin{cases} 5x + ky = 3 \\ 10x - 4y = 1 \end{cases}$ $k = -2$

18. $\begin{cases} 8x - 5y = 16 \\ kx - 0.5y = 3 \end{cases}$ $k = 0.8$

19. The sum of two numbers x and y is 50 and their difference is 22. Solve the following system of equations that represents this problem.

$\begin{cases} x + y = 50 \\ x - y = 22 \end{cases}$ $(36, 14)$

$\begin{cases} x + y = 32 \\ x = 4y + 2 \end{cases}$

System for 20

20. A student spent a total of $32 for a book and a calendar. The price of the book was $2 more than four times the price of the calendar. The system of equations that represents this situation is shown at the left, where x is the price of the book and y is the price of the calendar. Solve this system to determine the price of each item. book: $26, calendar: $6

8.4 Applications of Systems of Linear Equations

Spencer Grant/PhotoEdit

What You Should Learn

1. Construct a system of linear equations from an application problem.
2. Solve real-life applications using systems of linear equations.

Why You Should Learn It

Systems of linear equations can be used to model and solve real-life problems. For instance, in Exercise 57 on page 470, you will use a system of linear equations to determine the time spent walking on a treadmill.

---

1. **Construct a system of linear equations from an application problem.**

Constructing Systems of Linear Equations

At this point, you may be wondering how you can tell which applications can be solved using systems of linear equations. The answer comes from the following considerations.

1. Does the problem involve more than one unknown quantity?
2. Are there two (or more) equations or conditions to be satisfied?

If one or both of these conditions occur, then the appropriate mathematical model for the problem may be a system of linear equations. You can construct a model for the problem below as follows.

A total of $12,000 is invested in two funds paying 9% and 11% simple interest. The combined annual interest for the two funds is $1180. How much of the $12,000 is invested at each rate?

Notice that this problem has two unknowns: the amount (in dollars) invested at 9% and the amount (in dollars) invested at 11%.

Remind students of the importance of verbal models and labels.

Verbal Model:

$$\boxed{\text{Amount in 9\% fund}} + \boxed{\text{Amount in 11\% fund}} = \boxed{\text{Total amount}}$$

$$\boxed{\text{Interest from 9\% fund}} + \boxed{\text{Interest from 11\% fund}} = \boxed{\text{Total interest}}$$

Labels:
Amount in 9% fund $= x$	(dollars)
Amount in 11% fund $= y$	(dollars)
Total amount $= 12{,}000$	(dollars)
Interest from 9% fund $= 0.09x$	(dollars)
Interest from 11% fund $= 0.11y$	(dollars)
Total interest $= 1180$	(dollars)

System:
$$\begin{cases} x + y = 12{,}000 & \text{Equation 1} \\ 0.09x + 0.11y = 1180 & \text{Equation 2} \end{cases}$$

Study Tip

When solving application problems, make sure your answers make sense. For instance, a negative result for the x- or y-value in Example 1 would not make sense.

This system was solved in Example 6 on page 444. There you found that $x = 7000$ and $y = 5000$, which means that $7000 was invested at 9% simple interest and $5000 was invested at 11% simple interest.

② Solve real-life applications using systems of linear equations.

Applications of Systems of Linear Equations

Many of the problems solved in Chapters 3 and 4 using one variable can now be solved using a system of linear equations in two variables.

Example 1 A Coin Mixture Problem

A cash register contains $20.90 in dimes and quarters. There are 119 coins in the register. How many of each type of coin are in the cash register?

Solution

Note that this problem involves two unknowns: the number of dimes and the number of quarters.

Verbal Model:

$$\boxed{\text{Number of dimes}} + \boxed{\text{Number of quarters}} = \boxed{\text{Number of coins}}$$

$$\boxed{\text{Value of dimes}} + \boxed{\text{Value of quarters}} = \boxed{\text{Value of coins}}$$

Labels:

Number of dimes $= x$	(dimes)
Number of quarters $= y$	(quarters)
Number of coins $= 119$	(coins)
Value of dimes $= 0.10x$	(dollars)
Value of quarters $= 0.25y$	(dollars)
Value of coins $= 20.90$	(dollars)

System:

$$\begin{cases} x + y = 119 & \text{Equation 1} \\ 0.10x + 0.25y = 20.90 & \text{Equation 2} \end{cases}$$

To solve this equation, use the method of elimination. Begin by multiplying the second equation by 100 to eliminate the decimal points. Then, to obtain opposite coefficients, multiply the first equation by -10.

$$\begin{cases} x + y = 119 \\ 0.10x + 0.25y = 20.90 \end{cases} \implies \begin{array}{r} -10x - 10y = -1190 \\ 10x + 25y = 2090 \\ \hline 15y = 900 \end{array}$$

So, $y = 60$. To solve for x, back-substitute this value into Equation 1.

$$x + y = 119 \qquad\qquad \text{Equation 1}$$

$$x + 60 = 119 \qquad\qquad \text{Substitute 60 for } y.$$

$$x = 59 \qquad\qquad \text{Subtract 60 from each side.}$$

So, the cash register contains 59 dimes and 60 quarters. Check this in the original statement of the problem, as follows.

The cash register contains $20.90 in dimes and quarters.

$$0.10(59) + 0.25(60) = 5.90 + 15.00 = \$20.90 \qquad \text{Solution checks. ✓}$$

There are 119 coins in the register.

$$59 + 60 = 119 \qquad\qquad\qquad\qquad\qquad \text{Solution checks. ✓}$$

Example 2   Geometry: Dimensions of a Rectangle

A rectangle is twice as long as it is wide and its perimeter is 132 inches. Find the dimensions of the rectangle using a system of linear equations.

Solution

The two unknowns in this problem are the length and width of the rectangle, as shown in Figure 8.10.

Figure 8.10

Verbal Model:
$$2\left(\begin{array}{c}\text{Width of}\\\text{rectangle}\end{array}\right) + 2\left(\begin{array}{c}\text{Length of}\\\text{rectangle}\end{array}\right) = \begin{array}{c}\text{Perimeter}\\\text{of rectangle}\end{array}$$

$$\begin{array}{c}\text{Length of}\\\text{rectangle}\end{array} = 2 \cdot \begin{array}{c}\text{Width of}\\\text{rectangle}\end{array}$$

Labels: Width of rectangle $= w$ (inches)
           Length of rectangle $= l$ (inches)
           Perimeter of rectangle $= 132$ (inches)

System:
$$\begin{cases} 2w + 2l = 132 & \text{Equation 1} \\ \qquad\quad l = 2w & \text{Equation 2} \end{cases}$$

To solve this system, use the method of substitution.

$$2w + 2l = 132 \qquad \text{Equation 1}$$

$$2w + 2(2w) = 132 \qquad \text{Substitute } 2w \text{ for } l.$$

$$6w = 132 \qquad \text{Combine like terms.}$$

So, $w = 22$ inches. To solve for l, back-substitute into Equation 2.

$$l = 2(22) \qquad \text{Substitute 22 for } w \text{ in Equation 2.}$$

$$l = 44 \qquad \text{Simplify.}$$

This implies that the rectangle has a width of 22 inches and a length of 44 inches. Check this in the original statement of the problem, as follows.

> *The rectangle is twice as long as it is wide.*

Length $= 2(\text{width}) = 2(22) = 44$ inches   Solution checks. ✓

> *The rectangle's perimeter is 132 inches.*

$2(\text{width}) + 2(\text{length}) = 2(22) + 2(44) = 132$ inches   Solution checks. ✓

When solving real-life problems, remember that there are several effective problem-solving strategies that you can use. For instance, in Example 2, you could use *Guess, Check,* and *Revise* to find the dimensions, as shown in the following table.

Width	15	16	17	18	19	20	21	22
Length	30	32	34	36	38	40	42	44
Perimeter	90	96	102	108	114	120	126	132 ✓

You might want to review the three techniques for solving systems of linear equations and discuss which method to use in various situations.

Example 3   Selling Price and Wholesale Cost

The selling price of a pair of ski boots is $99.20. The markup rate is 55% of the wholesale cost. What is the wholesale cost?

Solution

You could solve this problem using only one unknown (the wholesale cost). However, for the sake of illustration, use two unknowns: the wholesale cost and the markup (in dollars).

Verbal Model:

$$\boxed{\text{Wholesale cost}} + \boxed{\text{Markup}} = \boxed{\text{Selling price}}$$

$$\boxed{\text{Markup}} = \boxed{\text{Markup rate}} \cdot \boxed{\text{Wholesale cost}}$$

Labels:

Wholesale cost $= C$	(dollars)
Markup $= M$	(dollars)
Selling price $= 99.20$	(dollars)
Markup rate $= 0.55$	(percent in decimal form)

System:

$$\begin{cases} C + M = 99.20 & \text{Equation 1} \\ M = 0.55C & \text{Equation 2} \end{cases}$$

To solve this system of linear equations, use the method of substitution.

$$C + M = 99.20 \qquad \text{Equation 1}$$

$$C + 0.55C = 99.20 \qquad \text{Substitute } 0.55C \text{ for } M.$$

$$1.55C = 99.20 \qquad \text{Combine like terms.}$$

$$C = 64 \qquad \text{Divide each side by 1.55.}$$

The wholesale cost of the pair of boots is $64. (You weren't asked to find the markup, so it is unnecessary to back-substitute to find the value of M.)

The problem in Example 3 could also have been modeled with a single equation in one variable. Using the verbal model

$$\boxed{\text{Wholesale cost}} + \boxed{\text{Markup}} = \boxed{\text{Selling price}}$$

with C representing the wholesale cost, $0.55C$ representing the markup, and 99.20 representing the selling price, you obtain the equation and solution.

$$C + 0.55C = 99.20 \qquad \text{Original equation}$$

$$1.55C = 99.20 \qquad \text{Combine like terms.}$$

$$C = 64 \qquad \text{Divide each side by 1.55.}$$

Which of these two solution techniques do you prefer: the one using a system of equations or the one using a single equation?

Example 4   A Mixture Problem

A company with two stores buys six large delivery vans and five small delivery vans. The first store receives four large vans and two small vans for a total cost of $160,000. The second store receives two large vans and three small vans for a total cost of $128,000. What is the cost of each type of van?

Solution

The two unknowns in this problem are the costs of the two types of vans.

Verbal
Model: $4\left(\begin{array}{c} \text{Cost of} \\ \text{large van} \end{array} \right) + 2\left(\begin{array}{c} \text{Cost of} \\ \text{small van} \end{array} \right) = \begin{array}{c} \text{Total cost for} \\ \text{first store} \end{array}$

$2\left(\begin{array}{c} \text{Cost of} \\ \text{large van} \end{array} \right) + 3\left(\begin{array}{c} \text{Cost of} \\ \text{small van} \end{array} \right) = \begin{array}{c} \text{Total cost for} \\ \text{second store} \end{array}$

Labels: Cost of large van $= x$ (dollars)
Cost of small van $= y$ (dollars)
Total cost for first store $= 160,000$ (dollars)
Total cost for second store $= 128,000$ (dollars)

System: $\begin{cases} 4x + 2y = 160,000 & \text{Equation 1} \\ 2x + 3y = 128,000 & \text{Equation 2} \end{cases}$

To solve this system of linear equations, use the method of elimination. To obtain coefficients of x that differ only in sign, multiply the second equation by -2.

$$\begin{cases} 4x + 2y = 160,000 \\ 2x + 3y = 128,000 \end{cases} \implies \begin{array}{rcl} 4x + 2y &=& 160,000 \\ -4x - 6y &=& -256,000 \\ \hline -4y &=& -96,000 \end{array}$$

So, the cost of each small van is $y = \$24,000$. By back-substituting this value into Equation 1, you can find the cost of each large van, as follows.

$4x + 2(24,000) = 160,000$ Substitute 24,000 for y in Equation 1.

$4x = 112,000$ Subtract 48,000 from each side.

$x = 28,000$ Divide each side by 4.

The cost of each large van is $x = \$28,000$, and the cost of each small van is $y = \$24,000$. Check this solution, as follows.

The first store receives four large vans and two small vans for $160,000.

$4(28,000) + 2(24,000) = 112,000 + 48,000 = \$160,000$ Solution checks. ✓

The second store receives two large vans and three small vans for $128,000.

$2(28,000) + 3(24,000) = 56,000 + 72,000 = \$128,000$ Solution checks. ✓

When assigning labels to a verbal model, remember to check that the units of measure make sense in the problem. For instance, each of the terms in the equations in Example 4 is measured in dollars. If, in the same equation, the terms had different units, you would know that an error had occurred.

Example 5   An Application Involving Two Speeds

You are taking a motorboat trip on a river—18 miles upstream and 18 miles back downstream. You run the motor at the same speed going up and down the river, but because of the current of the river, the trip upstream takes longer than the trip downstream. You don't know the speed of the river's current, but you know that the trip upstream takes $1\frac{1}{2}$ hours and the trip downstream takes only 1 hour. From this information, determine the speed of the current.

Solution

One unknown in this problem is the speed of the current. The other unknown is the speed of the boat in still water. To set up a model, use the fact that the effective speed of the boat (relative to the land) going upstream is

$$\frac{18}{1.5} = 12 \text{ miles per hour} \qquad \text{Rate} = \text{distance} \div \text{time}$$

and the speed going downstream is

$$\frac{18}{1} = 18 \text{ miles per hour} \qquad \text{Rate} = \text{distance} \div \text{time}$$

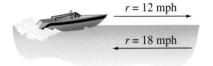

Figure 8.11

as shown in Figure 8.11. In the following verbal model, note that the current fights against you going upstream but helps you going downstream.

Verbal Model:

| Boat speed (still water) | − | Speed of current | = | Upstream speed |

| Boat speed (still water) | + | Speed of current | = | Downstream speed |

Labels:

Boat speed in still water $= x$	(miles per hour)
Current speed $= y$	(miles per hour)
Upstream speed $= 12$	(miles per hour)
Downstream speed $= 18$	(miles per hour)

System:
$$\begin{cases} x - y = 12 & \text{Equation 1} \\ x + y = 18 & \text{Equation 2} \end{cases}$$

To solve this system of linear equations, use the method of elimination.

$$
\begin{aligned}
x - y &= 12 && \text{Equation 1} \\
\underline{x + y} &= \underline{18} && \text{Equation 2} \\
2x &= 30 && \text{Add equations.}
\end{aligned}
$$

So, the speed of the boat in still water is $x = 15$ miles per hour. To find the speed of the current, back-substitute this value into Equation 2.

$$15 + y = 18 \qquad \text{Substitute 15 for } x \text{ in Equation 2.}$$

$$y = 3 \qquad \text{Subtract 15 from each side.}$$

So, the speed of the current is 3 miles per hour. Check this solution in the original statement of the problem.

In 2000, there were 153,000 chemists employed in the United States. Of these, 30.3% were women. (Source: U.S. Bureau of Labor Statistics)

Greg Pease/Getty Images

Example 6 A Mixture Problem

A chemist has two different solutions. One is 50% alcohol and 50% water, and the other is 75% alcohol and 25% water. How much of each type of solution should be mixed to obtain 8 liters of a solution comprised of 60% alcohol and 40% water?

Solution

The two unknowns in this problem are the amounts of each type of solution.

Verbal Model:

$$\boxed{\text{Liters of 50\% solution}} + \boxed{\text{Liters of 75\% solution}} = \boxed{\text{Liters of 60\% solution}}$$

$$\boxed{\text{Alcohol in 50\% solution}} + \boxed{\text{Alcohol in 75\% solution}} = \boxed{\text{Alcohol in 60\% solution}}$$

Labels:

Liters of 50% solution $= x$	(liters)
Liters of 75% solution $= y$	(liters)
Liters of 60% solution $= 8$	(liters)
Alcohol in 50% solution $= 0.50x$	(liters)
Alcohol in 75% solution $= 0.75y$	(liters)
Alcohol in 60% solution $= 0.60(8) = 4.8$	(liters)

System:
$$\begin{cases} x + y = 8 & \text{Equation 1} \\ 0.50x + 0.75y = 4.8 & \text{Equation 2} \end{cases}$$

To solve this system of linear equations, use the method of elimination by multiplying the first equation by -50 and the second equation by 100.

$$\begin{cases} x + y = 8 \\ 0.50x + 0.75y = 4.8 \end{cases} \implies \begin{aligned} -50x - 50y &= -400 \\ \underline{50x + 75y} &= \underline{480} \\ 25y &= 80 \end{aligned}$$

So, the y-coordinate of the solution is

$$y = \frac{80}{25} = 3.2.$$

To solve for x, back-substitute the value of y into the first equation.

$$x + y = 8 \qquad \text{Equation 1}$$
$$x + 3.2 = 8 \qquad \text{Substitute 3.2 for } y.$$
$$x = 4.8 \qquad \text{Subtract 3.2 from each side.}$$

So, the chemist should use 4.8 liters of the 50% solution and 3.2 liters of the 75% solution. Check this solution in the original system, as follows.

Substitute into Equation 1	*Substitute into Equation 2*
$x + y = 8$	$0.50x + 0.75y = 4.8$
$4.8 + 3.2 \overset{?}{=} 8$	$0.50(4.8) + 0.75(3.2) \overset{?}{=} 4.8$
$8 = 8 \checkmark$	$4.8 = 4.8 \checkmark$

Similar mixture problems were introduced in Section 3.5, where they were modeled with *one* equation. You might ask students to discuss which method they prefer.

8.4 Exercises

Review   Concepts, Skills, and Problem Solving

Keep mathematically in shape by doing these exercises *before* the problems of this section.

Properties and Definitions

1. *Writing* ✍   Describe the procedure for finding the x- and y-intercepts of the graph of $y = 4 - (x - 1)^2$.

To find the x-intercepts, let $y = 0$ and solve the equation for x. To find the y-intercept, let $x = 0$ and solve the equation for y.

2. Find the domain of the rational expression

$$\frac{3x^2 - 7x + 2}{x^2 - 16}.$$ All real values of x such that $x \neq \pm 4$

Algebraic Operations

In Exercises 3–10, perform the operation and simplify.

3. $(3x^2 - 2x) - (x^2 + 10)$ $2x^2 - 2x - 10$

4. $7x - [(3x + 2) + 8x]$ $-4x - 2$

5. $2t(t^2 - 2t + 3)$
$2t^3 - 4t^2 + 6t$

6. $(u + 2)(u^2 + u - 7)$
$u^3 + 3u^2 - 5u - 14$

7. $(5z - 3)(5z + 3)$
$25z^2 - 9$

8. $(2y - 11)^2$
$4y^2 - 44y + 121$

9. $\dfrac{10x^2 - 12x}{2x}$

$5x - 6,\ x \neq 0$

10. $\dfrac{x^2 + 3x + 6}{x + 1}$

$x + 2 + \dfrac{4}{x + 1}$

Graphs and Models

11. *Depreciation*   A business purchases a new machine for \$32,000. It is estimated that in 4 years its depreciated value will be \$8000. Assume that the depreciation of the machine can be approximated by a straight line.

(a) Write a linear equation for the value y of the machine in terms of time t. $y = -6000t + 32,000$

(b) ⊞ Use a graphing calculator to graph the equation in part (a).   See Additional Answers.

(c) ⊞ Use the *trace* feature of the graphing calculator to approximate the value of the machine after 3 years.   \$14,000

12. *Reimbursed Expenses*   A sales representative is reimbursed \$130 per day for lodging and meals plus \$0.35 per mile driven. Write a linear equation giving the daily cost C to the company in terms of x, the number of miles driven. $C = 130 + 0.35x$

Solving Problems

Modeling   In Exercises 1 and 2, construct and solve a system of linear equations.

1. The total cost of 15 gallons of regular gasoline and 10 gallons of premium gasoline is \$35.50. Premium costs \$0.20 more per gallon than regular. What is the cost per gallon of each type of gasoline?

(a) Write a verbal model for this problem.
15(Price of regular) + 10(Price of premium)
= 35.50
(Price of premium) = (Price of regular) + 0.20

(b) Assign labels to the verbal model.
x = Price per gallon of regular
y = Price per gallon of premium

(c) Use the labels to write a linear system.
$$\begin{cases} 15x + 10y = 35.50 \\ \qquad\quad y = x + 0.20 \end{cases}$$

(d) Solve the system and answer the question.
Regular: \$1.34 per gallon; Premium: \$1.54 per gallon

2. A total of \$12,000 is invested in two bonds that pay 10.5% and 12% simple interest. The combined annual interest is \$1380. How much is invested in each bond?

(a) Write a verbal model for this problem.
(Amount at 10.5%) + (Amount at 12%) = 12,000
(Interest from 10.5% bond) + (Interest from 12% bond)
= 1380

(b) Assign labels to the verbal model.
x = Amount at 10.5%; y = Amount at 12%

(c) Use the labels to write a linear system.
$$\begin{cases} x + \qquad y = 12,000 \\ 0.105x + 0.12y = \quad 1,380 \end{cases}$$

(d) Solve the system and answer the question.
\$4000 in 10.5% fund; \$8000 in 12% fund

Number Problems   In Exercises 3–8, use a system of linear equations to find two numbers that satisfy the requirements.

3. The sum of the numbers is 67, and their difference is 17.   42, 25

4. The sum of the numbers is 75, and their difference is 15.   45, 30

5. The sum of the numbers is 132, and the larger number is 6 more than twice the smaller number.   90, 42

6. The sum of the numbers is 46, and the larger number is 2 less than twice the smaller number.   30, 16

7. The sum of the larger number and twice the smaller number is 100, and their difference is 10.   40, 30

8. The sum of three times the smaller number and four times the larger number is 225. Nine times the smaller number plus two times the larger number gives the same sum.   15, 45

Coin Problems   In Exercises 9–14, use a system of linear equations to determine the number of each type of coin. See Example 1.

	Number	Types of coins	Value
9.	21	Dimes and quarters	$4.05
	8 dimes, 13 quarters		
10.	21	Dimes and quarters	$2.70
	17 dimes, 4 quarters		
11.	35	Nickels and quarters	$5.75
	15 nickels, 20 quarters		
12.	35	Nickels and quarters	$7.75
	5 nickels, 30 quarters		
13.	44	Nickels and dimes	$3.00
	28 nickels, 16 dimes		
14.	28	Nickels and dimes	$2.40
	8 nickels, 20 dimes		

▲ *Geometry*   In Exercises 15–20, use a system of linear equations to find the dimensions of the rectangle that meet the specified conditions. See Example 2.

Perimeter	Relationship Between Length and Width
15. 40 feet	The length is 4 feet greater than the width. 8 feet × 12 feet
16. 220 inches	The width is 10 inches less than the length. 50 inches × 60 inches
17. 16 yards	The width is one-third of the length. 2 yards × 6 yards
18. 48 meters	The length is twice the width. 8 meters × 16 meters
19. 35.2 meters	The length is 120% of the width. 8 meters × 9.6 meters
20. 35 feet	The width is 75% of the length. 7.5 feet × 10 feet

In Exercises 21–26, use a system of linear equations to solve the problem. See Example 3.

21. *Wholesale Cost*   The selling price of a watch is $108.75. The markup rate is 45% of the wholesale cost. Find the wholesale cost. $75

22. *Wholesale Cost*   The selling price of a cordless phone is $119.91. The markup rate is 40% of the wholesale cost. Find the wholesale cost. $85.65

23. *Wholesale Cost*   The selling price of an air conditioner is $359. The markup rate is 30% of the wholesale cost. Find the wholesale cost. $276.15

24. *List Price*   The sale price of a microwave oven is $275.00. The discount is 20% of the list price. Find the list price. $343.75

25. *List Price*   The sale price of a watch is $35.98. The discount is 30% of the list price. Find the list price. $51.40

26. *List Price*   The sale price of a stereo system is $716. The discount is 20% of the list price. Find the list price. $895

In Exercises 27–58, use a system of linear equations to solve the problem. See Examples 4–6.

27. *Ticket Sales*   Five hundred tickets were sold for a fundraising dinner. The receipts totaled $3312.50. Adult tickets were $7.50 each and children's tickets were $4.00 each. How many tickets of each type were sold?   375 adult tickets, 125 children's tickets

28. *Ticket Sales*   A fundraising dinner was held on two consecutive nights. On the first night, 425 adult tickets and 316 children's tickets were sold, for a total of $2915.00. On the second night, 542 adult tickets and 345 children's tickets were sold, for a total of $3572.50. Find the price of each type of ticket.   Adult ticket: $5.00; Child's ticket: $2.50

29. *Truck Cost*   A bakery with two stores buys three large delivery trucks and six small delivery trucks. One store receives one large delivery truck and four small delivery trucks for a total cost of $118,000. The second store receives two large delivery trucks and two small delivery trucks for a total cost of $107,000. What is the cost of each type of delivery truck?   Large truck: $32,000; Small truck: $21,500

30. *Truck Cost*   A furniture company with two stores buys three large delivery trucks and four small delivery trucks. One store receives one large delivery truck and three small delivery trucks for a total cost of $157,000. The second store receives two large delivery trucks and one small delivery truck for a total cost of $139,000. What is the cost of each type of delivery truck?

Large truck: $52,000; Small truck: $35,000

31. *Gasoline Mixture*   The total cost of 8 gallons of regular gasoline and 12 gallons of premium gasoline is $27.84. Premium gasoline costs $0.17 more per gallon than regular gasoline. Find the price per gallon for each grade of gasoline.

Regular: $1.29 per gallon; Premium: $1.46 per gallon

32. *Gasoline Mixture*   The total cost of 6 gallons of regular gasoline and 11 gallons of premium gasoline is $24.03. Premium gasoline costs $0.50 more per gallon than regular gasoline. Find the price per gallon for each type of gasoline.

Regular: $1.09 per gallon; Premium: $1.59 per gallon

33. *Food Costs*   You and a friend go to a Mexican restaurant. You order two tacos and three enchiladas, and your friend orders three tacos and five enchiladas. Your bill is $7.80 plus tax, and your friend's bill is $12.70 plus tax. How much does each taco and each enchilada cost?   Taco: $0.90; Enchilada: $2.00

34. *Food Costs*   You and a friend go to a fast-food restaurant. You order three deluxe hamburgers and two small fries and pay $5.35. Your friend orders two deluxe hamburgers and two small fries and pays $4.16. How much does a deluxe burger cost? $1.19

35. *Current Speed*   You travel 10 miles upstream and 10 miles downstream on a motorboat trip. You run the motor at the same speed going up and down the river, but because of the speed of the current the trip upstream takes $\frac{1}{2}$ hour and the trip downstream takes $\frac{1}{3}$ hour. Determine the speed of the current.

5 miles per hour

36. *Boat Speed*   You travel 16 miles upstream and 16 miles downstream on a motorboat trip. You run the motor at the same speed going up and down the river, but because of the speed of the current the trip upstream takes 2 hours and the trip downstream takes $1\frac{1}{3}$ hours. Determine the speed of the boat in still water.   10 miles per hour

37. *Airplane Speed*   An airplane flying into a headwind travels 2100 miles in $3\frac{1}{2}$ hours. On the return flight, the same distance is traveled in 3 hours. Find the speed of the plane in still air and the speed of the wind, assuming that both remain constant through out the round trip.   Speed of the plane in still air: 650 miles per hour; Speed of the wind: 50 miles per hour

38. *Airplane Speed*   An airplane flying into a headwind travels 1800 miles in 3 hours and 36 minutes. On the return flight, the same distance is traveled in 3 hours. Find the speed of the plane in still air and the speed of the wind, assuming that both remain constant throughout the round trip.   Speed of the plane in still air: 550 miles per hour; Speed of the wind: 50 miles per hour

39. *Average Speed*   A van travels for 2 hours at an average speed of 40 miles per hour. How much longer must the van travel at an average speed of 55 miles per hour so that the average speed for the entire trip will be 45 miles per hour?   1 hour

40. *Average Speed*   A van travels for 3 hours at an average speed of 40 miles per hour. How much longer must the van travel at an average speed of 55 miles per hour so that the average speed for the entire trip will be 50 miles per hour?   6 hours

41. *Relay Race*   The total time for a two-member team in a 5160-meter relay race is 16 minutes. The first runner on the team averages 300 meters per minute and the second runner averages 360 meters per minute. For how many minutes did the first runner keep the baton before passing it to the second runner?   10 minutes

42. *Driving Distances*   In a trip of 450 kilometers, two people drive. One person drives two times as far as the other. Find the distance that each person drives.

150 miles; 300 miles

43. *Mixture Problem*   How many liters of a 35% alcohol solution must be mixed with a 60% solution to obtain 10 liters of a 50% solution?

35% solution: 4 liters; 60% solution: 6 liters

44. *Mixture Problem*   Ten gallons of a 30% acid solution is obtained by mixing a 20% solution with a 50% solution. How many gallons of each solution must be used to obtain the desired mixture?

20% solution: $6\frac{2}{3}$ gallons; 50% solution: $3\frac{1}{3}$ gallons

45. *Nut Mixture*   Ten pounds of mixed nuts sells for $5.86 per pound. The mixture is obtained from two kinds of nuts: walnuts at $4.25 per pound and cashews at $6.55 per pound. How many pounds of each variety of nut are used in the mixture?

Walnuts: 3 pounds; Cashews: 7 pounds

46. *Nut Mixture*   A grocer mixes two kinds of nuts priced at $3.25 and $5.85 per pound to obtain 15 pounds of mixed nuts that sell for $4.29 per pound. How many pounds of each variety should the grocer use? $3.25 nuts: 9 pounds; $5.85 nuts: 6 pounds

47. *Feed Mixture*   How many tons of hay at $110 per ton and $60 per ton must be purchased to have 100 tons of hay with an average value of $75 per ton?

$110 hay: 30 tons; $60 hay: 70 tons

48. *Feed Mixture*   How many pounds of bird feed at $1.68 per pound and $0.83 per pound must be purchased to have 100 pounds of bird feed at a value of $1.34 per pound? $1.68 bird feed: 60 pounds; $0.83 bird feed: 40 pounds

49. *Investment*   You invest a total of $24,000 in two bonds that pay 6% and 9.5% simple interest. (There is more risk in the 9.5% bond.) Your goal is to have a total annual interest income of $2000. What is the smallest amount you can invest at 9.5% in order to meet your objective? $16,000

50. *Investment*   You invest a total of $12,000 in two bonds that pay 8% and 10% simple interest. (There is more risk in the 10% bond.) Your goal is to have a total annual interest income of $1000. To meet your objective, what is the smallest amount you can invest at 10%? $2000

51. *Sales*   An athletic sportswear store sold 28 pairs of cross-training shoes for a total of $2220. Style A sells for $70 per pair and style B sells for $90 per pair. How many of each style were sold?

Style A: 15 pairs; Style B: 13 pairs

52. *Sales*   A craft store sold 34 candle jars for a total of $261.50. Large candle jars sell for $8.50 and small candle jars sell for $6. How many of each type of candle jar were sold?

23 large candle jars; 11 small candle jars

53. *Job Offer*   You are offered two different sales jobs. Job A offers an annual salary of $30,000 plus a year-end bonus of 1% of your total sales. Job B offers an annual salary of $24,000 plus a year-end bonus of 2% of your total sales. How much would you have to sell to earn the same amount at each job? $600,000

54. *Job Offer*   You are offered two different sales jobs. Job A offers an annual salary of $40,000 plus a year-end bonus of 2% of your total sales. Job B offers an annual salary of $36,000 plus a year-end bonus of 3% of your total sales. How much would you have to sell to earn the same amount at each job? $400,000

55. *Phone Plan*   You are choosing between two different cellular phone plans that give you 1500 minutes free each month. Plan A charges $30 per month plus $0.80 for each minute over the free 1500 minutes. Plan B charges $35 per month plus $0.30 for each minute over the free 1500 minutes. How many minutes over the free 1500 minutes would you have to use in order for each plan to cost the same?   10 minutes

56. *Membership Plan*   You are choosing between two different membership plans at a local gym. Plan A charges a monthly fee of $26 plus $1 for each locker rental. Plan B charges a monthly fee of $19 plus $2 for each locker rental. How many times can you rent a locker in order for each plan to cost the same?

7 times

57. *Exercise*   For 30 minutes you do a combination of walking and jogging on a treadmill. At the end of your workout, the pedometer displays a total of 2.5 miles. You know that you walk 0.05 mile per minute and jog 0.1 mile per minute. How much time were you walking?   10 minutes

58. *Exercise*   For 40 minutes you do a combination of walking and running on a treadmill. At the end of your workout, the pedometer displays 3.5 miles. You know that you walk 0.07 mile per minute and run 0.12 mile per minute. How much time were you running?   14 minutes

Break-Even Analysis   In Exercises 59 and 60, use a graphing calculator to graph the cost and revenue equations in the same viewing window. Find the sales x necessary to break even ($R = C$) and the corresponding revenue R obtained by selling x units. (Round x to the nearest whole unit.)   See Additional Answers.

Cost	Revenue
59. $C = 7650x + 125,000$	$R = 8950x$

96 units; $R(96) = \$859,200$

60. $C = 0.25x + 25,000$	$R = 0.45x$

125,000 units; $R(125,000) = \$56,250$

In Exercises 61–64, find m and b such that $y = mx + b$ is the equation of the line through the points. (*Hint:* Generate a linear system in m and b by substituting the coordinates of the specified points into $y = mx + b$.)

61. $(2, -1), (6, 1)$ $m = \frac{1}{2}, b = -2$

62. $(1, 3), (4, 9)$ $m = 2, b = 1$

63. $(-3, 6), (5, 2)$ $m = -\frac{1}{2}, b = \frac{9}{2}$

64. $(0, 2), (4, -8)$ $m = -\frac{5}{2}, b = 2$

Explaining Concepts

65. *Writing*   How can you determine whether a real-life problem may be modeled with a system of linear equations?   If one or both of the following conditions occur, then a real life problem may be modeled with a system of linear equations. (a) Does the problem involve more than one unknown quantity? (b) Are there two or more equations or conditions to be satisfied?

66. *Writing*   List the three methods for solving systems of linear equations.   Graphing, substitution, elimination

67. *Writing*   What is meant by a verbal model of a real-life problem?   Translation of the verbal description of a problem into the form of an equation using words

68. *Writing*   Compare the one-variable method with the two-variable method for modeling real-life problems.

The one-variable method for modeling real-life problems involves the use of one unknown quantity and the writing of one equation.

The two-variable method involves the use of two unknowns and the writing of two equations.

8.5 Systems of Linear Inequalities

Jonathan Nourok/PhotoEdit

What You Should Learn

(1) Solve systems of linear inequalities in two variables.

(2) Use systems of linear inequalities to model and solve real-life problems.

Why You Should Learn It

Systems of linear inequalities can be used to model and solve real-life problems. For instance, in Exercise 58 on page 478, a system of linear inequalities can be used to analyze the compositions of dietary supplements.

(1) **Solve systems of linear inequalities in two variables.**

Systems of Linear Inequalities in Two Variables

You have already graphed linear inequalities in two variables. However, many practical problems in business, science, and engineering involve **systems of linear inequalities.** This type of system arises in problems that have *constraint* statements that contain phrases such as "more than," "less than," "at least," "no more than," "a minimum of," and "a maximum of." A **solution** of a system of linear inequalities in x and y is a point (x, y) that satisfies each inequality in the system.

To sketch the graph of a system of inequalities in two variables, first sketch (on the same coordinate system) the graph of each individual inequality. The **solution set** is the region that is *common* to every graph in the system.

Example 1   Graphing a System of Linear Inequalities

Sketch the graph of the system of linear inequalities: $\begin{cases} 2x - y \le 5 \\ x + 2y \ge 2 \end{cases}$.

Solution

Begin by rewriting each inequality in slope-intercept form. Then sketch the line for each corresponding equation. See Figures 8.12-8.14.

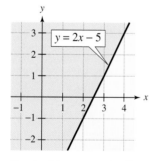

Graph of $2x - y \le 5$ is all points on and above $y = 2x - 5$.
Figure 8.12

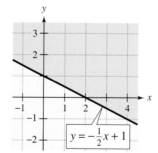

Graph of $x + 2y \ge 2$ is all points on and above $y = -\frac{1}{2}x + 1$.
Figure 8.13

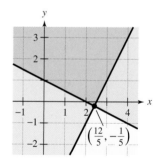

Graph of system is the purple wedge-shaped region.
Figure 8.14

In Figure 8.14, note that the two borderlines of the region

$$y = 2x - 5 \quad \text{and} \quad y = -\tfrac{1}{2}x + 1$$

intersect at the point $\left(\tfrac{12}{5}, -\tfrac{1}{5}\right)$. Such a point is called a **vertex** of the region. The region shown in the figure has only one vertex. Some regions, however, have several vertices. When you are sketching the graph of a system of linear inequalities, it is helpful to find and label any vertices of the region.

Graphing a System of Linear Inequalities

1. Sketch the line that corresponds to each inequality. (Use dashed lines for inequalities with $<$ or $>$ and use solid lines for inequalities with $\leq$ or $\geq$.)

2. Lightly shade the half-plane that is the graph of each linear inequality. (Colored pencils may help distinguish different half-planes.)

3. The graph of the system is the intersection of the half-planes. (If you use colored pencils, it is the region that is shaded with *every* color.)

Example 2   Graphing a System of Linear Inequalities

Sketch the graph of the system of linear inequalities: $\begin{cases} y < 4 \\ y > 1 \end{cases}$.

Solution

The graph of the first inequality is the half-plane *below* the horizontal line

$$y = 4. \qquad \text{Upper boundary}$$

The graph of the second inequality is the half-plane *above* the horizontal line

$$y = 1. \qquad \text{Lower boundary}$$

The graph of the system is the horizontal band that lies *between* the two horizontal lines (where $y < 4$ *and* $y > 1$), as shown in Figure 8.15.

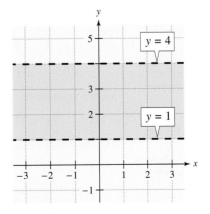

Figure 8.15

Additional Example

Sketch the graph of the system of linear inequalities.

$$\begin{cases} x - y < 2 \\ x > -2 \\ y \leq 3 \end{cases}$$

Answer:

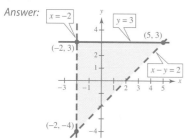

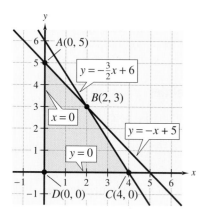

Figure 8.16

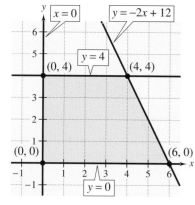

Figure 8.17

Example 3 Graphing a System of Linear Inequalities

Sketch the graph of the system of linear inequalities, and label the vertices.

$$\begin{cases} x + y \leq 5 \\ 3x + 2y \leq 12 \\ x \geq 0 \\ y \geq 0 \end{cases}$$

Solution

Begin by sketching the half-planes represented by the four linear inequalities. The graph of

$$x + y \leq 5$$

is the half-plane lying on and below the line $y = -x + 5$. The graph of

$$3x + 2y \leq 12$$

is the half-plane lying on and below the line $y = -\frac{3}{2}x + 6$. The graph of $x \geq 0$ is the half-plane lying on and to the right of the y-axis, and the graph of $y \geq 0$ is the half-plane lying on and above the x-axis. As shown in Figure 8.16, the region that is common to all four of these half-planes is a four-sided polygon. The vertices of the region are found as follows.

Vertex A: $(0, 5)$

Solution of the system

$$\begin{cases} x + y = 5 \\ x = 0 \end{cases}$$

Vertex B: $(2, 3)$

Solution of the system

$$\begin{cases} x + y = 5 \\ 3x + 2y = 12 \end{cases}$$

Vertex C: $(4, 0)$

Solution of the system

$$\begin{cases} 3x + 2y = 12 \\ y = 0 \end{cases}$$

Vertex D: $(0, 0)$

Solution of the system

$$\begin{cases} x = 0 \\ y = 0 \end{cases}$$

Example 4 Finding the Boundaries of a Region

Find a system of inequalities that defines the region shown in Figure 8.17.

Solution

Three of the boundaries of the region are horizontal or vertical—they are easy to find. To find the diagonal boundary line, use the techniques from Section 4.4 to find the equation of the line passing through the points $(4, 4)$ and $(6, 0)$. You can use the formula for slope to find $m = -2$ and then use the point-slope form with point $(6, 0)$ and $m = -2$ to obtain

$$y - 0 = -2(x - 6).$$

So, the equation is

$$y = -2x + 12.$$

The system of linear inequalities that describes the region is as follows.

$$\begin{cases} y \leq 4 & \text{Region lies on and below line } y = 4. \\ y \geq 0 & \text{Region lies on and above } x\text{-axis.} \\ x \geq 0 & \text{Region lies on and to the right of } y\text{-axis.} \\ y \leq -2x + 12 & \text{Region lies on and below line } y = -2x + 12. \end{cases}$$

Technology: Tip

A graphing calculator can be used to graph a system of linear inequalities. The graph of

$$\begin{cases} 4y < 2x - 6 \\ x + y \geq 7 \end{cases}$$

is shown at the left. The shaded region, in which all points satisfy both inequalities, is the solution of the system.

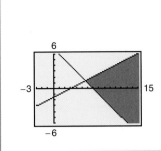

② Use systems of linear inequalities to model and solve real-life problems.

Application

Example 5   Target Heart Rate

A person's maximum heart rate is $220 - x$, where x is the person's age in years ($20 \leq x \leq 70$). When a person exercises, it is recommended that the person strive for a heart rate that is at least 50% of the maximum and at most 75% of the maximum. Write a system of linear inequalities that describes the exercise target zone and sketch a graph of the system.   (Source: American Heart Association)

Solution

Let y represent the person's heart rate. From the given information, you can write the following system of inequalities.

$$\begin{cases} x \geq 20 & \text{Person's age must be at least 20.} \\ x \leq 70 & \text{Person's age can be at most 70.} \\ y \geq 0.50(220 - x) & \text{Target rate is at least 50\% of maximum rate.} \\ y \leq 0.75(220 - x) & \text{Target rate is at most 75\% of maximum rate.} \end{cases}$$

To graph this system, sketch the half-planes represented by the four linear inequalities. The graph of $x \geq 20$ is the half-plane lying on and to the right of the line $x = 20$. The graph of $x \leq 70$ is the half-plane lying on and to the left of the line $x = 70$. The graph of $y \geq 0.50(220 - x)$ is the half-plane lying on and above the line $y = 0.50(220 - x)$. The graph of $y \leq 0.75(220 - x)$ is the half-plane lying on and below the line $y = 0.75(220 - x)$. As shown in Figure 8.18, the region that is common to all four of these half-planes is a four-sided polygon. The vertices of the region are found as follows.

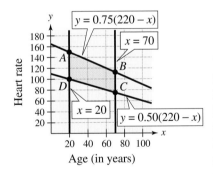

Figure 8.18

Vertex A: $(20, 150)$
Solution of the system
$$\begin{cases} x = 20 \\ y = 0.75(220 - x) \end{cases}$$

Vertex B: $(70, 112.5)$
Solution of the system
$$\begin{cases} x = 70 \\ y = 0.75(220 - x) \end{cases}$$

Vertex C: $(70, 75)$
Solution of the system
$$\begin{cases} x = 70 \\ y = 0.50(220 - x) \end{cases}$$

Vertex D: $(20, 100)$
Solution of the system
$$\begin{cases} x = 20 \\ y = 0.50(220 - x) \end{cases}$$

8.5 Exercises

Review   *Concepts, Skills, and Problem Solving*

Keep mathematically in shape by doing these exercises *before* the problems of this section.

Properties and Definitions

1. *Writing* 🖉   Explain the Multiplication Property of Inequalities in the case of multiplying by a negative quantity.   When multiplying an inequality by a negative number, the inequality must be reversed.

2. *Writing* 🖉   Explain how to use a test point to help determine which half-plane to shade when graphing the solution to a linear inequality in two variables.   The half-plane in which a test point resides should be shaded if the test point satisfies the equality.

Graphing Equations and Inequalities

In Exercises 3–6, sketch a graph of the equation.
See Additional Answers.

3. $2x + 4y = 8$

4. $y = 4(x - 1) + 3$

5. $0.3x - 0.2y = 0.8$

6. $x = 6$

In Exercises 7–10, sketch a graph of the inequality.
See Additional Answers.

7. $3x - 5y > -5$

8. $x + 4y \leq 5$

9. $y \leq -\frac{3}{4}(x + 4) - 3$

10. $y > -5$

Problem Solving

11. *Driving Distances*   Two people share the driving for a trip of 300 miles. One person drives three times as far as the other. Find the distance that each person drives.   225 miles; 75 miles

12. *Rope Length*   You must cut a rope that is 25 feet long into two pieces so that one piece is 15 feet longer than the other piece. Find the length of each piece.   20 feet; 5 feet

Developing Skills

In Exercises 1–6, match the system of linear inequalities with its graph. [The graphs are labeled (a), (b), (c), (d), (e), and (f).]

(a)

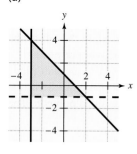

(b)

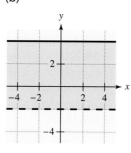

(c)

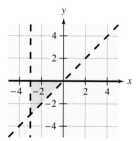

(d)

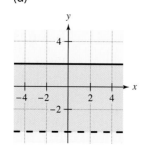

(e)

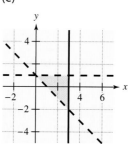

(f)

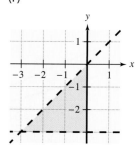

1. $\begin{cases} y > x \quad \text{c} \\ x > -3 \\ y \leq 0 \end{cases}$

2. $\begin{cases} y \leq 4 \quad \text{b} \\ y > -2 \end{cases}$

3. $\begin{cases} y < x \quad \text{f} \\ y > -3 \\ x \leq 0 \end{cases}$

4. $\begin{cases} x \leq 3 \quad \text{e} \\ y < 1 \\ y > -x + 1 \end{cases}$

5. $\begin{cases} y > -1 \quad \text{a} \\ x \geq -3 \\ y \leq -x + 1 \end{cases}$

6. $\begin{cases} y > -4 \quad \text{d} \\ y \leq 2 \end{cases}$

In Exercises 7–44, sketch a graph of the solution of the system of linear inequalities. See Examples 1–3.

See Additional Answers.

7. $\begin{cases} x < 3 \\ x > -2 \end{cases}$

8. $\begin{cases} y > -1 \\ y \le 2 \end{cases}$

9. $\begin{cases} x + y \le 3 \\ x - 1 \le 1 \end{cases}$

10. $\begin{cases} x + y \ge 2 \\ x - y \le 2 \end{cases}$

11. $\begin{cases} 2x - 4y \le 6 \\ x + y \ge 2 \end{cases}$

12. $\begin{cases} 4x + 10y \ge 5 \\ x - y \le 4 \end{cases}$

13. $\begin{cases} x + 2y \le 6 \\ x - 2y \le 0 \end{cases}$

14. $\begin{cases} 2x + y \le 0 \\ x - y \le 8 \end{cases}$

15. $\begin{cases} x - 2y > 4 \\ 2x + y > 8 \end{cases}$

16. $\begin{cases} 3x + y < 9 \\ x + 2y > 2 \end{cases}$

17. $\begin{cases} x + y > -1 \\ x + y < 3 \end{cases}$

18. $\begin{cases} x - y > 2 \\ x - y < -4 \end{cases}$

19. $\begin{cases} y \ge \frac{4}{3}x + 1 \\ y \le 7 \end{cases}$

20. $\begin{cases} y \ge \frac{1}{2}x + \frac{1}{2} \\ y \le 3 \end{cases}$

21. $\begin{cases} y > x - 2 \\ x > 5 \end{cases}$

22. $\begin{cases} y > x - 4 \\ x > -1 \end{cases}$

23. $\begin{cases} y \ge 3x - 3 \\ y \le -x + 1 \end{cases}$

24. $\begin{cases} y \ge 2x - 3 \\ y \le 3x + 1 \end{cases}$

25. $\begin{cases} y > 2x \\ y > -x + 4 \end{cases}$

26. $\begin{cases} y \le -x \\ y \le x + 1 \end{cases}$

27. $\begin{cases} x + 2y \le -4 \\ y \ge x + 5 \end{cases}$

28. $\begin{cases} x + y \le -3 \\ y \ge 3x - 4 \end{cases}$

29. $\begin{cases} x + y \le 4 \\ x \ge 0 \\ y \ge 0 \end{cases}$

30. $\begin{cases} 2x + y \le 6 \\ x \ge 0 \\ y \ge 0 \end{cases}$

31. $\begin{cases} 4x - 2y > 8 \\ x \ge 0 \\ y \le 0 \end{cases}$

32. $\begin{cases} 2x - 6y > 6 \\ x \le 0 \\ y \le 0 \end{cases}$

33. $\begin{cases} y > -5 \\ x \le 2 \\ y \le x + 2 \end{cases}$

34. $\begin{cases} y \ge -1 \\ x \le 2 \\ y \le x + 2 \end{cases}$

35. $\begin{cases} x + y \le 1 \\ -x + y \le 1 \\ y \ge 0 \end{cases}$

36. $\begin{cases} 3x + 2y < 6 \\ x \ge 0 \\ y \ge 0 \end{cases}$

37. $\begin{cases} x + y \le 5 \\ x \ge 2 \\ y \ge 0 \end{cases}$

38. $\begin{cases} 2x + y \ge 2 \\ x \le 2 \\ y \le 1 \end{cases}$

39. $\begin{cases} -3x + 2y < 6 \\ x - 4y > -2 \\ 2x + y < 3 \end{cases}$

40. $\begin{cases} x - 7y > -36 \\ 5x + 2y > 5 \\ 6x + 5y > 6 \end{cases}$

41. $\begin{cases} 2x + y < 2 \\ 6x + 3y > 2 \end{cases}$

42. $\begin{cases} x - 2y < -6 \\ 5x - 3y > -9 \end{cases}$

43. $\begin{cases} x \ge 1 \\ x - 2y \le 3 \\ 3x + 2y \ge 9 \\ x + y \le 6 \end{cases}$

44. $\begin{cases} x + y \le 4 \\ x + y \ge -1 \\ x - y \ge -2 \\ x - y \le 2 \end{cases}$

In Exercises 45–50, use a graphing calculator to graph the solution of the system of linear inequalities.

See Additional Answers.

45. $\begin{cases} 2x - 3y \le 6 \\ y \le 4 \end{cases}$

46. $\begin{cases} 6x + 3y \ge 12 \\ y \le 4 \end{cases}$

47. $\begin{cases} 2x - 2y \le 5 \\ y \le 6 \end{cases}$

48. $\begin{cases} 2x + 3y \ge 12 \\ y \ge 2 \end{cases}$

49. $\begin{cases} 2x + y \le 2 \\ y \ge -4 \end{cases}$

50. $\begin{cases} x - 2y \ge -6 \\ y \le 6 \end{cases}$

In Exercises 51–54, write a system of linear inequalities that describes the shaded region. See Example 4.

51.

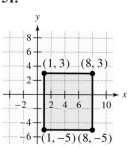

$$\begin{cases} x \geq & 1 \\ x \leq & 8 \\ y \geq & -5 \\ y \leq & 3 \end{cases}$$

52.

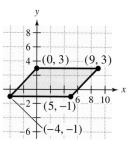

$$\begin{cases} y \leq 3 \\ y \geq -1 \\ y \geq x - 6 \\ y \leq x + 3 \end{cases}$$

53.

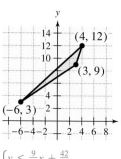

$$\begin{cases} y \leq \frac{9}{10}x + \frac{42}{5} \\ y \geq 3x \\ y \geq \frac{2}{3}x + 7 \end{cases}$$

54.

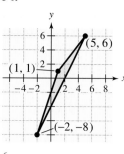

$$\begin{cases} y \geq 2x - 4 \\ y \leq 3x - 2 \\ y \leq \frac{5}{4}x - \frac{1}{4} \end{cases}$$

Solving Problems

55. *Investment*   A person plans to invest up to $20,000 in two different interest-bearing accounts, account X and account Y. Account X is to contain at least $5000. Moreover, account Y should have at least twice the amount in account X. Write a system of linear inequalities describing the various amounts that can be deposited in each account, and sketch the graph of the system.

$$\begin{cases} x + y \leq 20{,}000 \quad \text{See Additional Answers.} \\ \, x \geq 5{,}000 \\ \, y \geq 2x \end{cases}$$

56. *Investment*   A person plans to invest up to $10,000 in two different interest-bearing accounts, account X and account Y. Account Y is to contain at most $3000. Moreover, account X should have at least three times the amount in account Y. Write a system of linear inequalities describing the various amounts that can be deposited in each account, and sketch the graph of the system.

$$\begin{cases} x + y \leq 10{,}000 \quad \text{See Additional Answers.} \\ \, y \leq 3{,}000 \\ \, x \geq 3y \end{cases}$$

57. 🖩 *Ticket Sales*   Two types of tickets are to be sold for a concert. One type costs $15 per ticket and the other type costs $25 per ticket. The promoter of the concert must sell at least 15,000 tickets, including at least 8000 of the $15 tickets and at least 4000 of the $25 tickets. Moreover, the gross receipts must total at least $275,000 in order for the concert to be held. Write a system of linear inequalities describing the different numbers of tickets that can be sold. Use a graphing calculator to graph the system.

$$\begin{cases} x + y \geq 15{,}000 \quad \text{See Additional Answers.} \\ 15x + 25y \geq 275{,}000 \\ \, x \geq 8000 \\ \, y \geq 4000 \end{cases}$$

58. 🖩 *Nutrition*   A dietitian is asked to design a special diet supplement using two different foods. Each ounce of food X contains 20 units of calcium, 15 units of iron, and 10 units of vitamin B. Each ounce of food Y contains 10 units of calcium, 10 units of iron, and 20 units of vitamin B. The minimum daily requirements in the diet are 280 units of calcium, 160 units of iron, and 180 units of vitamin B. Write a system of linear inequalities describing the different amounts of food X and food Y that can be used in the diet. Use a graphing calculator to graph the system.

$$\begin{cases} 20x + 10y \geq 280 \quad \text{See Additional Answers.} \\ 15x + 10y \geq 160 \\ 10x + 20y \geq 180 \\ \, x \geq 0 \\ \, y \geq 0 \end{cases}$$

59. ▲ *Geometry*   The figure shows a cross section of a roped-off swimming area at a beach. Write a system of linear inequalities describing the cross section. (Each unit in the coordinate system represents 1 foot.)

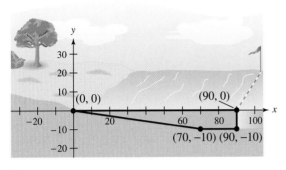

$$\begin{cases} x \leq 90 \\ y \leq 0 \\ y \geq -10 \\ y \geq -\frac{1}{7}x \end{cases}$$

60. ▲ *Geometry*   The figure shows the chorus platform on a stage. Write a system of linear inequalities describing the part of the audience that can see the full chorus. (Each unit in the coordinate system represents 1 meter.)

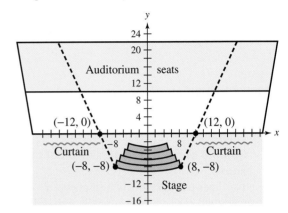

$$\begin{cases} y \leq 22 \\ y \geq 10 \\ y \geq 2x - 24 \\ y \geq -2x - 24 \end{cases}$$

Explaining Concepts

61. *Writing*   Explain the meaning of the term *half-plane*. Give an example of an inequality whose graph is a half-plane.   The graph of a linear equation splits the *xy*-plane into two parts, each of which is a half-plane. $y < 5$ is a half-plane.

62. *Writing*   Explain how you can check any single point (x_1, y_1) to determine whether or not it is a solution to a system of inequalities.   Check to see if (x_1, y_1) satisfies each inequality in the system.

63. *Writing*   Explain how to determine the vertices of the solution region for a system of linear inequalities.   Find all intersection points of the lines corresponding to the inequalities.

64. *Writing*   Is it possible for a system of linear inequalities to have no solution? If so, write an example.

Yes; Example: $\begin{cases} x + y > 7 \\ x + y < 3 \end{cases}$

What Did You Learn?

Key Terms

system of linear equations, *p. 430*

solution of a system of linear equations, *p. 430*

consistent system, *p. 432*

dependent system, *p. 432*

inconsistent system, *p. 432*

method of substitution, *p. 440*

back-substitute, *p. 440*

method of elimination, *p. 449*

system of linear inequalities, *p. 472*

solution of a system of linear inequalities, *p. 472*

vertex, *p. 473*

Key Concepts

8.1 ○ Number of points of intersection of two lines

1. The two lines can intersect in a single point (slopes are not equal). The corresponding system of linear equations has a single solution and is called consistent.

2. The two lines can coincide (slopes are equal) and have infinitely many points of intersection. The corresponding consistent system of linear equations has infinitely many solutions and is called dependent.

3. The two lines can be parallel (slopes are equal) and have no point of intersection. The corresponding system of linear equations has no solution and is called inconsistent.

8.2 ○ The method of substitution

1. Solve one of the equations for one variable in terms of the other.

2. Substitute the expression obtained in Step 1 into the other equation to obtain an equation in one variable.

3. Solve the equation obtained in Step 2.

4. Back-substitute the solution from Step 3 into the expression obtained in Step 1 to find the value of the other variable.

5. Check the solution to see that it satisfies both of the original equations.

8.3 ○ The method of elimination

1. Obtain opposite coefficients of x (or y) by multiplying all terms of one or both equations by suitable constants.

2. Add the equations to eliminate one variable and solve the resulting equation.

3. Back-substitute the value obtained in Step 2 into either of the original equations and solve for the other variable.

4. Check your solution in both of the original equations.

8.3 ○ Guidelines for solving a system of linear equations

To decide whether to use graphing, substitution, or elimination, consider the following.

1. The graphing method is useful for approximating the solution and for giving an overall picture of how one variable changes with respect to the other.

2. To find exact solutions, use either substitution or elimination.

3. For systems of equations in which one variable has a coefficient of 1, substitution may be more efficient than elimination.

4. Elimination is usually more efficient. This is especially true when the coefficients of one of the variables are opposites.

8.5 ○ Graphing a system of linear inequalities

1. Sketch the line that corresponds to each inequality. (Use dashed lines for inequalities with < or > and use solid lines for inequalities with ≤ or ≥.)

2. Lightly shade the half-plane that is the graph of each linear inequality.

3. The graph of the system is the intersection of the half-planes.

Review Exercises

8.1 Solving Systems of Equations by Graphing

1 Determine if an ordered pair is a solution to a system of equations.

In Exercises 1–4, determine whether each ordered pair is a solution to the system of equations.

System	Ordered Pairs

1. $\begin{cases} 3x - 5y = 11 \\ -x + 2y = -4 \end{cases}$ (a) $(2, -1)$ (b) $(3, -2)$

(a) Solution   (b) Not a solution

2. $\begin{cases} 10x + 8y = -2 \\ 2x - 5y = 26 \end{cases}$ (a) $(4, -4)$ (b) $(3, -4)$

(a) Not a solution   (b) Solution

3. $\begin{cases} 0.2x + 0.4y = 5 \\ x + 3y = 30 \end{cases}$ (a) $(0.5, -0.7)$ (b) $(15, 5)$

(a) Not a solution   (b) Solution

4. $\begin{cases} -\frac{1}{2}x - \frac{2}{3}y = \frac{1}{2} \\ x + y = 1 \end{cases}$ (a) $(-5, 6)$ (b) $(7, -3)$

(a) Not a solution   (b) Not a solution

2 Use a coordinate system to solve systems of linear equations graphically.

In Exercises 5–8, use the graphs of the equations to determine the solution (if any) of the system of linear equations. Check your solution.

5. $\begin{cases} 2x + y = 4 \\ 2x - y = 0 \end{cases}$ $(1, 2)$

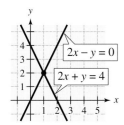

6. $\begin{cases} -x + y = 1 \\ x + y = 5 \end{cases}$ $(2, 3)$

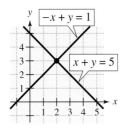

7. $\begin{cases} 3x - 2y = 6 \\ 2x - y = 3 \end{cases}$ $(0, -3)$

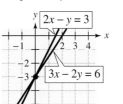

8. $\begin{cases} x + 2y = 6 \\ x + 2y = 2 \end{cases}$ No solution

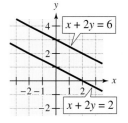

In Exercises 9–14, sketch the graphs of the equations and approximate any solutions of the system of linear equations.   See Additional Answers.

9. $\begin{cases} y = x - 4 \\ y = 2x - 9 \end{cases}$
$(5, 1)$

10. $\begin{cases} y = -\frac{5}{3}x + 6 \\ y = x - 10 \end{cases}$
$(6, -4)$

11. $\begin{cases} x + y = 2 \\ x - y = 0 \end{cases}$
$(1, 1)$

12. $\begin{cases} x - y = 9 \\ -x + y = 1 \end{cases}$
No solution

13. $\begin{cases} 2x + 3 = 3y \\ y = \frac{2}{3}x \end{cases}$
No solution

14. $\begin{cases} x + y = -1 \\ 3x + 2y = 0 \end{cases}$
$(2, -3)$

8.2 Solving Systems of Equations by Substitution

1 Use the method of substitution to solve systems of equations algebraically.

In Exercises 15–22, solve the system by the method of substitution.

15. $\begin{cases} y = 2x \\ y = x + 4 \end{cases}$ $(4, 8)$

16. $\begin{cases} x = -2y + 13 \\ x = \frac{y}{2} + 3 \end{cases}$ $(5, 4)$

17. $\begin{cases} x = 3y - 2 \\ x = 6 - y \end{cases}$
$(4, 2)$

18. $\begin{cases} y = -4x + 1 \\ y = x - 4 \end{cases}$
$(1, -3)$

19. $\begin{cases} x - 2y = 6 \\ 3x + 2y = 10 \end{cases}$
$(4, -1)$

20. $\begin{cases} 5x + y = 20 \\ 7x - 5y = -4 \end{cases}$
$(3, 5)$

21. $\begin{cases} 2x - y = 2 \\ 6x + 8y = 39 \end{cases}$ $\left(\frac{5}{2}, 3\right)$

22. $\begin{cases} 3x + 4y = 1 \\ x - 7y = -3 \end{cases}$ $\left(-\frac{1}{5}, \frac{2}{5}\right)$

② Use the method of substitution to solve systems with no solution or infinitely many solutions.

In Exercises 23–26, solve the system by the method of substitution.

23. $\begin{cases} x = y + 3 \\ x = y + 1 \end{cases}$
No solution

24. $\begin{cases} y = 3x + 4 \\ 9x = 3y - 12 \end{cases}$
Infinitely many solutions

25. $\begin{cases} -6x + y = -3 \\ 12x - 2y = 6 \end{cases}$
Infinitely many solutions

26. $\begin{cases} 3x + 4y = 7 \\ 6x + 8y = 10 \end{cases}$
No solution

③ Use the method of substitution to solve application problems.

27. *Investment*   A total of $12,000 is invested in two funds paying 5% and 10% simple interest. The combined annual interest for the two funds is $800. The system of equations that represents this situation is

$\begin{cases} x + y = 12{,}000 \\ 0.05x + 0.10y = 800 \end{cases}$

where x represents the amount invested in the 5% fund and y represents the amount invested in the 10% fund. Solve this system to determine how much of the $12,000 is invested at each rate.
5%: $8000; 10%: $4000

28. *Ticket Sales*   You are selling tickets to your school musical. Adult tickets cost $5 and children's tickets cost $3. You sell 1510 tickets and collect $6138. The system of equations that represents this situation is

$\begin{cases} x + y = 1510 \\ 5x + 3y = 6138 \end{cases}$

where x represents the number of adult tickets sold and y represents the number of children's tickets sold. Solve this system to determine how many of each type of ticket were sold.
804 adult tickets, 706 children's tickets

8.3 Solving Systems of Equations by Elimination

① Solve systems of linear equations algebraically using the method of elimination.

In Exercises 29–36, solve the system by the method of elimination.

29. $\begin{cases} 3x - y = 5 \\ 2x + y = 5 \end{cases}$
(2, 1)

30. $\begin{cases} 2x + 4y = 2 \\ -2x - 7y = 4 \end{cases}$
(5, −2)

31. $\begin{cases} 5x + 4y = 2 \\ -x + y = -22 \end{cases}$
(10, −12)

32. $\begin{cases} 3x - 2y = 9 \\ x + y = 3 \end{cases}$
(3, 0)

33. $\begin{cases} 8x - 6y = 4 \\ -4x + 3y = -2 \end{cases}$
Infinitely many solutions

34. $\begin{cases} 2x - 5y = 2 \\ 3x - 7y = 1 \end{cases}$
(−9, −4)

35. $\begin{cases} 0.2x + 0.1y = 0.03 \\ 0.3x - 0.1y = -0.13 \end{cases}$ (−0.2, 0.7)

36. $\begin{cases} 0.2x - 0.1y = 0.07 \\ 0.4x - 0.5y = -0.01 \end{cases}$ (0.6, 0.5)

② Choose a method for solving systems of equations.

In Exercises 37–54, use the most convenient method to solve the system of linear equations. State which method you used.

37. $\begin{cases} 6x - 5y = 0 \\ y = 6 \end{cases}$
(5, 6)

38. $\begin{cases} -x + 2y = 2 \\ x = 4 \end{cases}$
(4, 3)

39. $\begin{cases} -x + 4y = 4 \\ x + y = 6 \end{cases}$
(4, 2)

40. $\begin{cases} -x + y = 4 \\ x + y = 4 \end{cases}$
(0, 4)

41. $\begin{cases} x - y = 0 \\ x - 6y = 5 \end{cases}$
(−1, −1)

42. $\begin{cases} x + 2y = 2 \\ x - 4y = 20 \end{cases}$
(8, −3)

43. $\begin{cases} 5x + 8y = 8 \\ x - 8y = 16 \end{cases}$
$\left(4, -\frac{3}{2}\right)$

44. $\begin{cases} -7x + 9y = 9 \\ 2x + 9y = -18 \end{cases}$
$\left(-3, -\frac{4}{3}\right)$

45. $\begin{cases} 6x - 3y = 27 \\ -2x + y = -9 \end{cases}$ Infinitely many solutions

46. $\begin{cases} -5x + 2y = -4 \\ x - 6y = 4 \end{cases}$ $\left(\frac{4}{7}, -\frac{4}{7}\right)$

47. $\begin{cases} \frac{1}{5}x + \frac{3}{2}y = 2 \\ 2x + 13y = 20 \end{cases}$
(10, 0)

48. $\begin{cases} -\frac{1}{4}x + \frac{2}{3}y = 1 \\ 3x - 8y = 1 \end{cases}$
No solution

49. $\begin{cases} x + y = 0 \\ 2x + y = 0 \end{cases}$
(0, 0)

50. $\begin{cases} 2x + 6y = 16 \\ 2x + 3y = 7 \end{cases}$
(−1, 3)

51. $\begin{cases} \frac{1}{3}x + \frac{4}{7}y = 3 \\ 2x + 3y = 15 \end{cases}$
(−3, 7)

52. $\begin{cases} \frac{1}{2}x - \frac{1}{3}y = 0 \\ 3x + 2y = 0 \end{cases}$
(0, 0)

53. $\begin{cases} 1.2s + 4.2t = -1.7 \\ 3.0s - 1.8t = 1.9 \end{cases}$
$\left(\frac{1}{3}, -\frac{1}{2}\right)$

54. $\begin{cases} 0.2u + 0.3v = 0.14 \\ 0.4u + 0.5v = 0.20 \end{cases}$
(−0.5, 0.8)

8.4 Applications of Systems of Linear Equations

1 Construct a system of linear equations from an application problem.

In Exercises 55 and 56, construct a system of linear equations to represent the situation, and then solve the system.

55. *Coin Problem*   A cash register has 15 coins consisting of dimes and quarters. The total value of the coins is $2.85. Find the number of each type of coin.
6 dimes, 9 quarters
$$\begin{cases} x + y = 15 \\ 0.10x + 0.25y = 2.85 \end{cases}$$

56. *Video Rental*   You go to the video store to rent five movies for the weekend. Videos rent for $2 and $3. You spend $13. How many $2 videos did you rent? 2
$$\begin{cases} x + y = 5 \\ 2x + 3y = 13 \end{cases}$$

2 Solve real-life applications using systems of linear equations.

57. ▲ *Geometry*   A rectangular sign has a perimeter of 120 inches. The height of the sign is two-thirds of its width. Find the dimensions of the sign.
24 inches by 36 inches

58. ▲ *Geometry*   The perimeter of a table tennis top is 28 feet. The difference between 4 times the length and 3 times the width is 21 feet. Find the dimensions of the table tennis top.   5 feet by 9 feet

59. *Wholesale Cost*   The selling price of a DVD player is $234. The markup rate is 40% of the wholesale cost. Find the wholesale cost. $167.14

60. *List Price*   The sale price of a computer system is $952. The discount is 30% of the list price. Find the list price. $1360

61. *Fuel Costs*   You buy 2 gallons of gasoline for your lawn mower and 5 gallons of diesel fuel for your garden tractor. The total bill is $8.59. Diesel fuel costs $0.08 more per gallon than gasoline. Find the price per gallon of each type of fuel.   Gasoline: $1.17 per gallon; Diesel fuel: $1.25 per gallon

62. *Seed Mixture*   Ten pounds of mixed bird seed sells for $6.97 per pound. The mixture is obtained from two kinds of bird seed, with one variety priced at $5.65 per pound and the other at $8.95 per pound. How many pounds of each variety of bird seed are used in the mixture?
$5.65 seed: 6 pounds; $8.95 seed: 4 pounds

63. *Average Speed*   A car travels for 3 hours at an average speed of 40 miles per hour. How much longer must the car travel at an average speed of 55 miles per hour so that the average speed for the entire trip will be 49 miles per hour? $\frac{9}{2}$ hours

64. *Average Speed*   A car travels for 4 hours at an average speed of 50 miles per hour. How much longer must the car travel at an average speed of 65 miles per hour so that the average speed for the entire trip will be 55 miles per hour? 2 hours

▦ *Break-Even Analysis*   **In Exercises 65 and 66, use a graphing calculator to graph the cost and revenue equations in the same viewing window. Find the sales x necessary to break even ($R = C$) and the corresponding revenue R obtained by selling x units. (Round x to the nearest whole unit.)**   See Additional Answers.

Cost	*Revenue*
65. $C = 650x + 12{,}500$	$R = 800x$

83 units; $R(83) = \$66{,}400$

66. $C = 3.30x + 1200$ $R = 4.75x$
828 units; $R(828) = \$3933$

8.5 Systems of Linear Inequalities

1 Solve systems of linear inequalities in two variables.

In Exercises 67–74, sketch a graph of the solution of the system of linear inequalities.
See Additional Answers.

67. $\begin{cases} y < 2x - 2 \\ x \geq 3 \end{cases}$ **68.** $\begin{cases} 4x + 6y \leq 24 \\ y > 2 \end{cases}$

69. $\begin{cases} 3x - y \leq 6 \\ x + y \geq 2 \end{cases}$ **70.** $\begin{cases} x + y < 4 \\ x - y < 4 \end{cases}$

71. $\begin{cases} x + y < 5 \\ x > 2 \\ y \geq 0 \end{cases}$ **72.** $\begin{cases} 2x + y > 2 \\ x < 2 \\ y < 1 \end{cases}$

73. $\begin{cases} x + 2y \leq 160 \\ 3x + y \leq 180 \\ x \geq 0 \\ y \geq 0 \end{cases}$

74. $\begin{cases} 2x + 3y \leq 24 \\ 2x + y \leq 16 \\ x \geq 0 \\ y \geq 0 \end{cases}$

In Exercises 75 and 76, write a system of linear inequalities to describe the region.

75. Parallelogram with vertices at $(1, 5)$, $(3, 1)$, $(6, 10)$, and $(8, 6)$

76. Triangle with vertices at $(1, 2)$, $(6, 7)$, and $(8, 1)$

② Use systems of linear inequalities to model and solve real-life problems.

In Exercises 77–80, determine a system of linear inequalities that models the description, and sketch a graph of the solution of the system.
See Additional Answers.

77. *Fruit Distribution*   A Pennsylvania fruit grower has up to 1500 bushels of apples that are to be divided between markets in Harrisburg and Philadelphia. These two markets need at least 400 bushels and 600 bushels, respectively.

78. *Inventory Costs*   A warehouse operator has up to 24,000 square feet of floor space in which to store two products. Each unit of product I requires 20 square feet of floor space and costs $12 per day to store. Each unit of product II requires 30 square feet of floor space and costs $8 per day to store. The total storage cost per day cannot exceed $12,400.

79. *Production*   A company manufactures two types of hedge trimmers, a cordless model and a corded model. The corded trimmer requires 2 hours to assemble and the cordless trimmer requires 4 hours to assemble. The company has no more than 800 work hours to use in assembly each day, and the packing department can package no more than 300 trimmers per day.

80. *Production*   A company manufactures two types of wood chippers, a standard model and a deluxe model. The deluxe model requires 3 hours to assemble and $\frac{1}{2}$ hour to paint. The standard model requires 2 hours to assemble and 1 hour to paint. The assembly line cannot be used more than 24 hours each day. The painting line cannot be used more than 8 hours each day.

79. $\begin{cases} 2x + 4y \leq 800 \\ \ x + \ y \leq 300 \\ \ x \qquad \geq \quad 0 \\ \qquad \ y \geq \quad 0 \end{cases}$ **80.** $\begin{cases} 3x + 2y \leq 24 \\ \frac{1}{2}x + \ y \leq \ 8 \\ \ x \qquad \geq \ 0 \\ \qquad \ y \geq \ 0 \end{cases}$

75. $\begin{cases} 2x + y \geq \quad 7 \\ \ x - y \leq \quad 2 \\ 2x + y \leq \ 22 \\ \ x - y \geq -4 \end{cases}$

76. $\begin{cases} \ x - \ y \geq -1 \\ 3x + \ y \leq \ 25 \\ \ x + 7y \geq \ 15 \end{cases}$

77. $\begin{cases} x + y \leq 1500 \\ x \qquad \geq \ 400 \\ \qquad y \geq \ 600 \end{cases}$

78. $\begin{cases} 20x + 30y \leq 24{,}000 \\ 12x + \ 8y \leq 12{,}400 \\ \quad x \qquad \geq \qquad 0 \\ \qquad \ y \geq \qquad 0 \end{cases}$

Chapter Test

Take this test as you would take a test in class. After you are done, check your work against the answers in the back of the book.

1. Which is the solution of the system $x - 6y = -19$ and $4x - 5y = 0$: $(3, -2)$ or $(5, 4)$? Explain your reasoning.

In Exercises 2–4, determine the number of solutions of the system.

2. $\begin{cases} 3x + 4y = 16 \\ 3x - 4y = 8 \end{cases}$

3. $\begin{cases} x - 2y = -4 \\ x - 2y = 2 \end{cases}$

4. $\begin{cases} 2x - y = 5 \\ -4x + 2y = -10 \end{cases}$

In Exercises 5–8, solve the system of equations graphically. See Additional Answers.

5. $\begin{cases} x - 2y = -2 \\ x + y = 4 \end{cases}$ $(2, 2)$

6. $\begin{cases} 2x + y = 4 \\ x - 2y = -3 \end{cases}$ $(1, 2)$

7. $\begin{cases} x - 3y = -2 \\ 2x + y = 10 \end{cases}$ $(4, 2)$

8. $\begin{cases} 2x = 3 \\ 2x + 3y = 9 \end{cases}$ $\left(\frac{3}{2}, 2\right)$

In Exercises 9–11, solve the system of equations by the method of substitution.

9. $\begin{cases} x + 5y = 10 \\ 4x - 5y = 15 \end{cases}$

10. $\begin{cases} x + 3y = 15 \\ -2x + 5y = 14 \end{cases}$

11. $\begin{cases} y = 14 - 5x \\ x = y - 2 \end{cases}$

In Exercises 12–14, solve the system of equations by the method of elimination.

12. $\begin{cases} x + y = 8 \\ 2x - y = -2 \end{cases}$

13. $\begin{cases} 7x + 6y = 36 \\ 5x - 4y = 5 \end{cases}$

14. $\begin{cases} \frac{1}{2}x - \frac{1}{4}y = 1 \\ 4x + 5y = 22 \end{cases}$

In Exercises 15 and 16, sketch a graph of the solution of the system of linear inequalities. See Additional Answers.

15. $\begin{cases} 3x - y < 4 \\ x > 0 \\ y > 0 \end{cases}$

16. $\begin{cases} x + y < 6 \\ 2x + 3y > 9 \\ x \geq 0 \\ y \geq 0 \end{cases}$

17. Twenty liters of a 20% acid solution is obtained by mixing a 30% solution and a 5% solution. How many liters of each solution are needed to obtain the specified mixture?

18. You have decided to invest some money in certificates of deposit. You have up to $1000 to divide between two different programs. Program A requires a minimum deposit of $300 and program B requires a minimum deposit of $400. Write a system of linear inequalities describing the various amounts that can be deposited in each certificate of deposit, and sketch a graph of the system. See Additional Answers.

---

1. $(5, 4)$ because it satisfies both equations while $(3, -2)$ does not.

2. One solution   **3.** No solution

4. Infinitely many solutions

9. $(5, 1)$ **10.** $(3, 4)$

11. $(2, 4)$ **12.** $(2, 6)$

13. $\left(3, \frac{5}{2}\right)$ **14.** $(3, 2)$

17. 12 liters of 30% solution
8 liters of 5% solution

18. $\begin{cases} x + y \leq 1000 \\ x \geq 300 \\ y \geq 400 \end{cases}$

Motivating the Chapter

Constructing a Walkway

On a college campus, a walkway from the dormitory at point *A* to the library at point *B* has a path that contains a right angle, as shown at the right. The original plan was to make a *diagonal* walkway, but several trees block a direct diagonal path from the dormitory to the library. Two options are shown at the right for placement of the new walkway.

See Section 9.3, Exercise 141.

a. Can the expression for the distance from the dormitory to point *C* in option 1 be simplified further? If so, write the simplified expression.
Yes. $72\sqrt{2}$ feet

b. The distance b_2 in option 2 is one-half the distance from the dormitory to point *C* shown in option 1. What is the length of b_2? Use a calculator and round your answer to three decimal places. $36\sqrt{2} \approx 50.912$ feet

c. Find the area of the right triangle in option 1 given $a_1 = 72\sqrt{2}$.
5184 square feet

d. Find the area of the right triangle in option 2 given $a_2 = 36\sqrt{2}$.
1296 square feet

See Section 9.4, Exercise 93.

A coordinate plane is placed over the original plan in a manner such that point *A* is located at $(100, 100)$, point *B* is located at $(-72, 128)$, and point *C* is located at $(28, 28)$.

e. Use the Distance Formula to find the distance from the dormitory to the library. Round your answer to three decimal places. 174.264 feet

f. Use the Distance Formula to find the distance between point *C* and the library. Round your answer to three decimal places. 141.421 feet

g. Find the distance you would walk from the dormitory to the library in option 1 if you use the diagonal walkway. How much farther is that than a direct diagonal path from the dormitory to the library? 183.598 feet, 9.334 feet

h. Find the distance you would walk from the dormitory to the library in option 2 if you use the diagonal walkway. How much farther is that than a direct diagonal path from the dormitory to the library? How much farther is that than the distance in option 1? 213.421 feet, 39.157 feet, 29.823 feet

Original

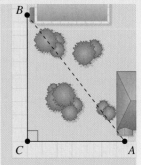

Option 1

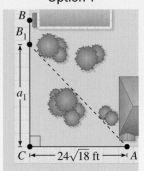

Option 2

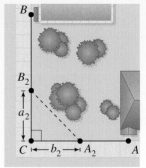

Terry Wild Studio, Inc.

Radical Expressions and Equations

487

9.1 Roots and Radicals

What You Should Learn

1. Find the *n*th roots of real numbers.
2. Use the radical symbol to denote the *n*th roots of numbers.
3. Approximate the values of expressions involving square roots using a calculator.

Why You Should Learn It

Algebraic expressions often involve radicals. For instance, in Exercise 120 on page 496, you will use an expression involving a radical to find the maximum walking speed for a person.

Roots

You already know how to find the square of a number (by multiplying the number by itself). For instance, the square of 3 is $3^2 = 9$. In this section, you will study the reverse problem, which is finding a *square root* of a number.

The **square root** of a number is defined as one of its two equal factors. For example, 3 is a square root of 9 because 3 is one of the two equal factors of 9. In a similar way, a **cube root** of a number is one of its three equal factors.

1. Find the *n*th roots of real numbers.

Number	Equal Factors	Root	Type
$25 = (-5)^2$	$(-5)(-5)$	-5	Square root
$-27 = (-3)^3$	$(-3)(-3)(-3)$	-3	Cube root
$16 = 2^4$	$2 \cdot 2 \cdot 2 \cdot 2$	2	Fourth root

Definition of *n*th Root of a Number

Let a and b be real numbers and let n be an integer such that $n \geq 2$. If

$$a = b^n$$

then b is an ***n*th root of *a*.** If $n = 2$, the root is a **square root.** If $n = 3$, the root is a **cube root.**

Technology: Discovery

When a positive number is raised to an even or odd power, the answer is positive. For instance $2^2 = 4$ and $2^3 = 8$. What happens when a negative number is raised to a power? Evaluate each expression. Use a calculator to verify your result.

a. $(-4)^2$ b. $(-4)^3$
c. $(-2)^4$ d. $(-2)^5$

When a negative number is raised to an even power, is the answer positive or negative? When a negative number is raised to an odd power, is the answer positive or negative?

See Technology Answers.

Example 1   Finding Square Roots of Numbers

Find all square roots of each number.

a. 81   b. $\frac{4}{9}$ c. 0   d. -4

Solution

a. The positive number 81 has two square roots, 9 and -9, because $9^2 = (9)(9) = 81$ and $(-9)^2 = (-9)(-9) = 81$.

b. The positive number $\frac{4}{9}$ has two square roots, $\frac{2}{3}$ and $-\frac{2}{3}$, because $\left(\frac{2}{3}\right)^2 = \left(\frac{2}{3}\right)\left(\frac{2}{3}\right) = \frac{4}{9}$ and $\left(-\frac{2}{3}\right)^2 = \left(-\frac{2}{3}\right)\left(-\frac{2}{3}\right) = \frac{4}{9}$.

c. The number 0 has only one square root: 0.

d. The negative number -4 has no square root because there is no real number that can be multiplied by itself to obtain -4.

9.2 Simplifying Radicals

Dewitt Jones/Corbis

Why You Should Learn It

In real-life applications, you often have to simplify algebraic expressions involving radicals. For instance, in Exercise 135 on page 506, you will simplify a radical expression to find how long it will take a performer to swing through one period of a trapeze.

What You Should Learn

1. Simplify radical expressions involving constants.
2. Simplify radical expressions involving variables.
3. Simplify radical expressions by rationalizing the denominator.

Simplifying Radicals with Constant Factors

You already know that simplifying algebraic expressions is one of the primary tasks in algebra. In this and the following section, you will learn how to simplify **radical expressions.**

The first step in rewriting a radical in simplest form is to evaluate any perfect nth factors. To do this, you need the property of radicals demonstrated by the following two equations.

$$\sqrt{4 \cdot 25} = \sqrt{100} = 10 \quad \text{and} \quad \sqrt{4} \cdot \sqrt{25} = 2 \cdot 5 = 10$$

Note that in these two equations you obtain the same result whether you first multiply 4 and 25 and then take the square root, or first take the square roots of the factors 4 and 25 and then multiply. So, $\sqrt{4 \cdot 25} = \sqrt{4} \cdot \sqrt{25}$. This rule is generalized as follows.

1 Simplify radical expressions involving constants.

Product Rule for Radicals

Let a and b be real numbers, variables, or algebraic expressions. If the nth roots of a and b are real, the following property is true.

$$\sqrt[n]{ab} = \sqrt[n]{a}\,\sqrt[n]{b} \qquad \text{Product Rule for Radicals}$$

Example 1 Simplifying Radical Expressions

Students may factor $\sqrt{18} = \sqrt{6} \cdot \sqrt{3}$. Point out that this is a true statement, but because neither factor is a perfect square, this factoring does not help to simplify the radical.

a. $\sqrt{18} = \sqrt{9 \cdot 2}$ 9 is perfect square factor.

$= \sqrt{9} \cdot \sqrt{2}$ Product Rule for Radicals

$= 3\sqrt{2}$ Simplest form

b. $\sqrt[3]{54} = \sqrt[3]{27 \cdot 2}$ 27 is perfect cube factor.

$= \sqrt[3]{27} \cdot \sqrt[3]{2}$ Product Rule for Radicals

$= 3\sqrt[3]{2}$ Simplest form

c. $\sqrt[4]{80} = \sqrt[4]{16 \cdot 5}$ 16 is 2 raised to the fourth power.

$= \sqrt[4]{16} \cdot \sqrt[4]{5}$ Product Rule for Radicals

$= 2\sqrt[4]{5}$ Simplest form

Some radicands have more than one perfect nth factor. In such cases, choose the *largest perfect nth factor* because this will minimize the number of steps needed to write the radical in simplest form. Compare the following two versions of the same problem.

First Solution

$$\sqrt{72} = \sqrt{36 \cdot 2}$$ 36 is largest perfect square factor of 72.

$$= \sqrt{36} \cdot \sqrt{2}$$ Product Rule for Radicals

$$= 6\sqrt{2}$$ Simplest form

Second Solution

$$\sqrt{72} = \sqrt{9 \cdot 8}$$ 9 is a perfect square factor of 72.

$$= \sqrt{9} \cdot \sqrt{8}$$ Product Rule for Radicals

$$= 3\sqrt{8}$$ Simplify.

$$= 3\sqrt{4 \cdot 2}$$ 4 is perfect square factor of 8.

$$= 3\sqrt{4} \cdot \sqrt{2}$$ Product Rule for Radicals

$$= 3 \cdot 2 \cdot \sqrt{2}$$ Simplify.

$$= 6\sqrt{2}$$ Simplest form

By finding the largest perfect square factor of 72, you can save several steps.

Example 2   Simplifying Radical Expressions

Simplify each radical expression.

a. $\sqrt{96}$ **b.** $\sqrt{108}$

Solution

a. The largest perfect square factor of 96 is 16.

$$\sqrt{96} = \sqrt{16 \cdot 6} = \sqrt{16} \cdot \sqrt{6} = 4\sqrt{6}$$

b. The largest perfect square factor of 108 is 36.

$$\sqrt{108} = \sqrt{36 \cdot 3} = \sqrt{36} \cdot \sqrt{3} = 6\sqrt{3}$$

In the same way that perfect squares can be removed from square root radicals, perfect nth powers can be removed from nth root radicals.

Example 3   Simplifying Radical Expressions

a. In the radical expression $\sqrt[3]{-192}$, the largest perfect cube factor is -64.

$$\sqrt[3]{-192} = \sqrt[3]{-64 \cdot 3} = \sqrt[3]{-64} \cdot \sqrt[3]{3} = -4\sqrt[3]{3}$$

b. In the radical expression $\sqrt[4]{512}$, the largest perfect fourth powered factor is 256.

$$\sqrt[4]{512} = \sqrt[4]{256 \cdot 2} = \sqrt[4]{256} \cdot \sqrt[4]{2} = 4\sqrt[4]{2}$$

Study Tip

The two solutions at the right show that some simplification strategies are more efficient than others. *In spite of this,* don't get hung up on always finding the easiest way to solve a problem. Remember that the primary goal is to solve the problem—efficiency is only a secondary goal.

Encourage students to find the largest perfect nth factor, but assure them that any perfect nth factor can be used as a first step in simplifying a radical.

② Simplify radical expressions involving variables.

Simplifying Radicals with Variable Factors

Simplifying radicals that involve *variable* radicands is trickier than simplifying radicals involving only constant radicands. The reason for this can be seen by considering the radical $\sqrt{x^2}$. At first glance, it would appear that this radical simplifies as x. However, doing so overlooks the possibility that x might be negative. For instance, consider the following.

$$\text{If } x = 2, \text{ then } \sqrt{x^2} = \sqrt{2^2} = \sqrt{4} = 2 = x.$$

$$\text{If } x = -2, \text{ then } \sqrt{x^2} = \sqrt{(-2)^2} = \sqrt{4} = 2 = |x|.$$

In both of these cases, you can conclude that

$$\sqrt{x^2} = |x|$$

but without knowing whether x is positive, zero, or negative, you *cannot* conclude that $\sqrt{x^2} = x$.

The Square Root of x^2

If x is a real number, then

$$\sqrt{x^2} = |x|. \qquad\qquad \text{Restricted by absolute value signs}$$

For the special case in which you know that x is a *nonnegative* real number, you can write $\sqrt{x^2} = x$.

Example 4   Simplifying Square Roots Involving Even Powers

Simplify each radical expression.

a. $\sqrt{25x^2}$ **b.** $\sqrt{18a^2}, a \geq 0$ **c.** $\sqrt{x^4}$

Solution

a. $\sqrt{25x^2} = \sqrt{5^2 x^2}$ Factor radicand.

$\qquad = \sqrt{5^2} \cdot \sqrt{x^2}$ Product Rule for Radicals

$\qquad = 5|x|$ $\sqrt{x^2} = |x|$

b. $\sqrt{18a^2} = \sqrt{3^2 \cdot a^2 \cdot 2}$ Factor radicand.

$\qquad = \sqrt{3^2} \cdot \sqrt{a^2} \cdot \sqrt{2}$ Product Rule for Radicals

$\qquad = 3a\sqrt{2}$ $\sqrt{a^2} = a, a \geq 0$

c. This problem is different. Note that the absolute value signs are not necessary in the final simplified version because you know that x^2 cannot be negative.

$$\sqrt{x^4} = \sqrt{(x^2)^2}$$

$$= |x^2|$$

$$= x^2$$

Additional Examples
Simplify each radical expression.

a. $\sqrt{72y^2}$ b. $\sqrt{9x^6}$

Answers:

a. $6|y|\sqrt{2}$

b. $3|x^3|$

Study the following simplifications of square roots of powers of x. Try to see why absolute values are necessary in some cases and not necessary in others.

$$\sqrt{x^2} = |x|$$ In $\sqrt{x^2}$, x can be positive or negative.

$$\sqrt{x^3} = \sqrt{x^2 \cdot x} = x\sqrt{x}$$ In $\sqrt{x^3}$, x cannot be negative.

$$\sqrt{x^4} = \sqrt{(x^2)^2} = |x^2| = x^2$$ In $\sqrt{x^4}$, x can be positive or negative.

$$\sqrt{x^5} = \sqrt{x^4 \cdot x} = x^2\sqrt{x}$$ In $\sqrt{x^5}$, x cannot be negative.

$$\sqrt{x^6} = \sqrt{(x^3)^2} = |x^3|$$ In $\sqrt{x^6}$, x can be positive or negative.

The reason you don't need to use absolute values in simplifying $\sqrt{x^3} = x\sqrt{x}$ is that the original radical would be undefined if x were negative.

Example 5   Simplifying Square Roots Involving Odd Powers

Simplify each radical expression.

a. $\sqrt{16x^3}$ **b.** $\sqrt{9a^5}$

Solution

a. $\sqrt{16x^3} = \sqrt{4^2 x^2 x}$ Factor radicand.

$\quad = \sqrt{4^2} \cdot \sqrt{x^2} \cdot \sqrt{x}$ Product Rule for Radicals

$\quad = 4x\sqrt{x}$ Simplified form

b. $\sqrt{9a^5} = \sqrt{3^2 a^4 a}$ Factor radicand.

$\quad = \sqrt{3^2} \cdot \sqrt{(a^2)^2} \cdot \sqrt{a}$ Product Rule for Radicals

$\quad = 3a^2\sqrt{a}$ Simplified form

Example 6   Simplifying Radicals Involving *n*th Roots

Simplify each radical expression.

a. $\sqrt[3]{54a^3}$ **b.** $\sqrt[5]{x^6}$ **c.** $\sqrt[4]{32x^7}$

Solution

a. $\sqrt[3]{54a^3} = \sqrt[3]{27 \cdot 2 \cdot a^3}$ Factor radicand.

$\quad = \sqrt[3]{3^3} \cdot \sqrt[3]{2} \cdot \sqrt[3]{a^3}$ Product Rule for Radicals

$\quad = 3a\sqrt[3]{2}$ Simplify.

b. $\sqrt[5]{x^6} = \sqrt[5]{x^5 \cdot x}$ Factor radicand.

$\quad = \sqrt[5]{x^5} \cdot \sqrt[5]{x}$ Product Rule for Radicals

$\quad = x\sqrt[5]{x}$ Simplify.

c. $\sqrt[4]{32x^7} = \sqrt[4]{16 \cdot 2 \cdot x^4 \cdot x^3}$ Factor radicand.

$\quad = \sqrt[4]{2^4} \cdot \sqrt[4]{2} \cdot \sqrt[4]{x^4} \cdot \sqrt[4]{x^3}$ Product Rule for Radicals

$\quad = 2x\sqrt[4]{2x^3}$ Simplify.

③ Simplify radical expressions by rationalizing the denominator.

Rationalizing Denominators

To simplify a square root having a fractional radicand, you can use the following Quotient Rule for Radicals.

Quotient Rule for Radicals

Let a and b be real numbers, variables, or algebraic expressions. If the nth roots of a and b are real, the following property is true.

$$\sqrt[n]{\frac{a}{b}} = \frac{\sqrt[n]{a}}{\sqrt[n]{b}}, \qquad b \neq 0 \qquad \text{Quotient Rule for Radicals}$$

Note how this rule is used to simplify radicals in Examples 7 and 8.

Example 7   Simplifying a Radical Expression Involving a Fraction

Simplify $\sqrt{\frac{21}{4}}$.

Solution

$$\sqrt{\frac{21}{4}} = \frac{\sqrt{21}}{\sqrt{4}} = \frac{\sqrt{21}}{2} \qquad \text{Quotient Rule for Radicals}$$

Example 8   Simplifying Radical Expressions Involving Fractions

a.
$$\sqrt{\frac{48x^4}{3}} = \sqrt{16x^4} \qquad \text{Simplify fraction.}$$

$$= 4x^2 \qquad \text{Simplify.}$$

b.
$$\sqrt{\frac{3x^2}{12y^4}} = \sqrt{\frac{x^2}{4y^4}} \qquad \text{Simplify fraction.}$$

$$= \frac{\sqrt{x^2}}{\sqrt{4y^4}} \qquad \text{Quotient Rule for Radicals}$$

$$= \frac{|x|}{2y^2} \qquad \text{Simplify.}$$

In Examples 7 and 8, note that the denominators are free of radicals. This simplifying process is called **rationalizing the denominator.** In these examples, the rationalizing was easy because the denominators were perfect squares. For more general fractions, rationalizing the denominator can require a little more work. The goal is to find a *rationalizing* factor that creates (in the denominator) a perfect square radicand for square roots, a perfect cube radicand for cube roots, and so on.

You might show students that there can be more than one approach to finding the simplified forms of these radicals. Here are some "less efficient" ways of simplifying Example 9(b).

$$\sqrt{\frac{7}{20}} = \frac{\sqrt{7}}{\sqrt{20}} = \frac{\sqrt{7}}{\sqrt{4}\sqrt{5}}$$

$$= \frac{\sqrt{7}}{2\sqrt{5}} \cdot \frac{\sqrt{5}}{\sqrt{5}} = \frac{\sqrt{35}}{10}$$

$$\sqrt{\frac{7}{20}} = \frac{\sqrt{7}}{\sqrt{20}} \cdot \frac{\sqrt{20}}{\sqrt{20}} = \frac{\sqrt{140}}{20}$$

$$= \frac{\sqrt{4}\sqrt{35}}{20} = \frac{2\sqrt{35}}{20} = \frac{\sqrt{35}}{10}$$

Study Tip

When rationalizing a denominator, remember that for square roots you want a perfect square in the denominator, for cube roots you want a perfect cube, and so on. For instance, to find the radical factor needed to create a perfect square in the denominator of Example 9(b) you can write the prime factorization of 20.

$$20 = 2 \cdot 2 \cdot 5$$
$$= 2^2 \cdot 5$$

From its prime factorization you can see that 2^2 is a square root factor of 20 and you need one more factor of 5 to create a perfect square in the denominator.

Compare two examples such as $\frac{\sqrt{10}}{\sqrt{5}}$ and $\frac{10}{\sqrt{5}}$.

$$\frac{\sqrt{10}}{\sqrt{5}} = \sqrt{\frac{10}{5}} = \sqrt{2}$$

$$\frac{10}{\sqrt{5}} = \frac{10}{\sqrt{5}} \cdot \frac{\sqrt{5}}{\sqrt{5}} = \frac{10\sqrt{5}}{5} = 2\sqrt{5}$$

When rationalizing the denominator of a fraction, you must multiply both the numerator and denominator by the rationalizing factor, as shown in Examples 9, 10, and 11.

Example 9   Rationalizing Denominators

Rationalize the denominator in each expression.

a. $\sqrt{\dfrac{13}{3}}$ **b.** $\sqrt{\dfrac{7}{20}}$ **c.** $\dfrac{12}{\sqrt{18}}$

Solution

a. $\sqrt{\dfrac{13}{3}} = \dfrac{\sqrt{13}}{\sqrt{3}} \cdot \dfrac{\sqrt{3}}{\sqrt{3}} = \dfrac{\sqrt{39}}{\sqrt{3^2}} = \dfrac{\sqrt{39}}{3}$ Multiply by $\sqrt{3}/\sqrt{3}$ to create a perfect square in the denominator.

b. $\sqrt{\dfrac{7}{20}} = \dfrac{\sqrt{7}}{\sqrt{20}} \cdot \dfrac{\sqrt{5}}{\sqrt{5}} = \dfrac{\sqrt{35}}{\sqrt{10^2}} = \dfrac{\sqrt{35}}{10}$ Multiply by $\sqrt{5}/\sqrt{5}$ to create a perfect square in the denominator.

c. $\dfrac{12}{\sqrt{18}} = \dfrac{12}{\sqrt{18}} \cdot \dfrac{\sqrt{2}}{\sqrt{2}} = \dfrac{12\sqrt{2}}{\sqrt{36}} = \dfrac{12\sqrt{2}}{\sqrt{6^2}} = \dfrac{12\sqrt{2}}{6} = 2\sqrt{2}$

The three criteria for a radical expression to be in simplest form are summarized as follows.

Simplifying Radical Expressions

A radical expression is said to be in simplest form if all three of the statements below are true.

1. All possible nth powered factors have been removed from each radical.

2. No radical contains a fraction.

3. No denominator of a fraction contains a radical.

The next two examples show how to rationalize denominators that contain variable factors.

Example 10   Rationalizing Denominators

a. $\sqrt{\dfrac{3}{a}} = \dfrac{\sqrt{3}}{\sqrt{a}} \cdot \dfrac{\sqrt{a}}{\sqrt{a}} = \dfrac{\sqrt{3a}}{\sqrt{a^2}} = \dfrac{\sqrt{3a}}{a}$

b. $\sqrt{\dfrac{1}{4x^3}} = \dfrac{\sqrt{1}}{\sqrt{4x^3}} \cdot \dfrac{\sqrt{x}}{\sqrt{x}} = \dfrac{\sqrt{x}}{\sqrt{4x^4}} = \dfrac{\sqrt{x}}{2x^2}$

c. $\sqrt{\dfrac{10x}{8y^5}} = \sqrt{\dfrac{5x}{4y^5}} = \dfrac{\sqrt{5x}}{\sqrt{4y^5}} \cdot \dfrac{\sqrt{y}}{\sqrt{y}} = \dfrac{\sqrt{5xy}}{\sqrt{4y^6}} = \dfrac{\sqrt{5xy}}{2y^3}$

Additional Examples
Rationalize the denominator of each expression.

a. $\sqrt{\dfrac{2}{12y^3}}$ b. $\sqrt[3]{\dfrac{1}{8x^4}}$

Answers:

a. $\dfrac{\sqrt{6y}}{6y^2}$

b. $\dfrac{\sqrt[3]{x^2}}{2x^2}$

When the denominator of an expression contains a cube root, multiply the numerator and denominator by a factor that will create a perfect cube in the denominator.

Example 11   Rationalizing Denominators with Cube Roots

a. $\dfrac{5}{\sqrt[3]{9}} = \dfrac{5}{\sqrt[3]{9}} \cdot \dfrac{\sqrt[3]{3}}{\sqrt[3]{3}}$ Multiply by $\sqrt[3]{3}/\sqrt[3]{3}$ to create a perfect cube in the denominator.

$= \dfrac{5\sqrt[3]{3}}{\sqrt[3]{27}} = \dfrac{5\sqrt[3]{3}}{3}$ Multiply and simplify.

b. $\sqrt[3]{\dfrac{5}{x}} = \dfrac{\sqrt[3]{5}}{\sqrt[3]{x}}$ Quotient Rule for Radicals

$= \dfrac{\sqrt[3]{5}}{\sqrt[3]{x}} \cdot \dfrac{\sqrt[3]{x^2}}{\sqrt[3]{x^2}}$ Multiply by $\sqrt[3]{x^2}/\sqrt[3]{x^2}$ to create a perfect cube in the denominator.

$= \dfrac{\sqrt[3]{5} \cdot \sqrt[3]{x^2}}{\sqrt[3]{x^3}} = \dfrac{\sqrt[3]{5x^2}}{x}$ Multiply and simplify.

Radicals are also used in many geometric applications, such as applications involving right circular cones. The lateral surface of a cone consists of all segments that connect the vertex with points on the edge of the base, as shown in Figure 9.1. The lateral surface area S of a right circular cone is given by

$$S = \pi r \sqrt{r^2 + h^2}$$

where r is the radius of the base of the cone and h is the height.

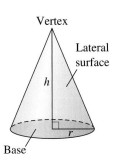

Vertex

Lateral surface

h

r

Base

Figure 9.1

Example 12   Geometry: Lateral Surface Area

The radius of a traffic cone is 14 centimeters and the height of the cone is 34 centimeters (see Figure 9.2). What is the lateral surface area of the traffic cone?

Solution

You can use the formula for the lateral surface area of a cone as follows.

$S = \pi r \sqrt{r^2 + h^2}$ Formula for lateral surface area

$= \pi(14)\sqrt{14^2 + 34^2}$ Substitute 14 for r and 34 for h.

$= 14\pi\sqrt{1352}$ Simplify.

$= 14\pi\sqrt{676 \cdot 2}$ 676 is a perfect square factor of 1352.

$= 14 \cdot 26\pi\sqrt{2}$ Simplify.

$= 364\pi\sqrt{2}$ Simplify.

≈ 1617.2 Use a calculator.

So, the lateral surface area of the cone is about 1617.2 square centimeters.

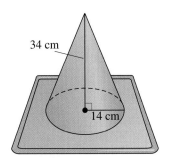

34 cm

14 cm

Figure 9.2

9.2 Exercises

Review   *Concepts, Skills, and Problem Solving*

Keep mathematically in shape by doing these exercises *before* the problems of this section.

Properties and Definitions

1. *Writing* ✏ Explain how to determine the half-plane satisfying $x - y < 2$.   Test one point in each of the half-planes formed by graphing $x - y = 2$. If the point satisfies the inequality, then the half-plane containing the point satisfies the inequality.

2. *Writing* ✏ Describe the difference between the graphs of $2x - 3y \leq 6$ and $2x - 3y < 6$.   In the first inequality, the points on the boundary (solid line) are solutions of the inequality. In the second inequality, the points on the boundary (dashed line) are not solutions.

Graphing Inequalities

In Exercises 3–8, graph the inequality.   See Additional Answers.

3. $3x - 2y > 6$ 4. $y \leq 2 - \frac{1}{2}x$
5. $y \leq 3$ 6. $x > 5$

7. $5x + 10y < 30$ 8. $4x + y - 3 \geq 2$

Solving Equations

In Exercises 9–12, solve the equation.

9. $0.60x = 24$ 40

10. $\dfrac{5}{8} = \dfrac{x}{4}$ $\dfrac{5}{2}$

11. $(2x + 3)(x - 9) = 0$ $-\frac{3}{2}, 9$ 12. $t^2 - 5t = 0$ $0, 5$

Problem Solving

13. *Distance*   A car travels for t hours at an average speed of 65 miles per hour. Write an equation for the distance d in terms of time t. Use the equation to determine the time required to travel 160 miles. $d = 65t$; 2 hours 28 minutes

14. *Mixture Problem*   How many liters of a 65% alcohol solution must be mixed with a 40% solution to obtain 100 liters of a 50% solution?   65% solution: 40 liters; 40% solution: 60 liters

Developing Skills

In Exercises 1–10, write as a single radical.

1. $\sqrt{3} \cdot \sqrt{5}$ $\sqrt{15}$ 2. $\sqrt{3} \cdot \sqrt{2}$ $\sqrt{6}$
3. $\sqrt{10} \cdot \sqrt{7}$ $\sqrt{70}$ 4. $\sqrt{13} \cdot \sqrt{2}$ $\sqrt{26}$
5. $\sqrt{5} \cdot \sqrt{6}$ $\sqrt{30}$ 6. $\sqrt{3} \cdot \sqrt{11}$ $\sqrt{33}$
7. $\sqrt{2} \cdot \sqrt{x}$ $\sqrt{2x}$ 8. $\sqrt{3} \cdot \sqrt{y}$ $\sqrt{3y}$
9. $\sqrt{2x} \cdot \sqrt{3y}$ $\sqrt{6xy}$ 10. $\sqrt{5a} \cdot \sqrt{3b}$ $\sqrt{15ab}$

In Exercises 11–14, write the expression as a product of two radicals and simplify.

11. $\sqrt{4 \cdot 15}$ $2\sqrt{15}$ 12. $\sqrt{16 \cdot 3}$ $4\sqrt{3}$
13. $\sqrt{64 \cdot 11}$ $8\sqrt{11}$ 14. $\sqrt{100 \cdot 3}$ $10\sqrt{3}$

In Exercises 15–34, simplify the radical. See Examples 1–3.

15. $\sqrt{8}$ $2\sqrt{2}$ 16. $\sqrt{12}$ $2\sqrt{3}$
17. $\sqrt{45}$ $3\sqrt{5}$ 18. $\sqrt{28}$ $2\sqrt{7}$
19. $\sqrt{32}$ $4\sqrt{2}$ 20. $\sqrt{20}$ $2\sqrt{5}$

21. $\sqrt{180}$ $6\sqrt{5}$ 22. $\sqrt{128}$ $8\sqrt{2}$
23. $\sqrt{300}$ $10\sqrt{3}$ 24. $\sqrt{432}$ $12\sqrt{3}$
25. $\sqrt{500}$ $10\sqrt{5}$ 26. $\sqrt{800}$ $20\sqrt{2}$
27. $\sqrt[3]{24}$ $2\sqrt[3]{3}$ 28. $\sqrt[3]{16}$ $2\sqrt[3]{2}$
29. $\sqrt[3]{250}$ $5\sqrt[3]{2}$ 30. $\sqrt[3]{54}$ $3\sqrt[3]{2}$
31. $\sqrt[4]{48}$ $2\sqrt[4]{3}$ 32. $\sqrt[4]{32}$ $2\sqrt[4]{2}$
33. $\sqrt[4]{162}$ $3\sqrt[4]{2}$ 34. $\sqrt[4]{512}$ $4\sqrt[4]{2}$

In Exercises 35–58, simplify the radical. Use absolute value signs if appropriate. See Examples 4–6.

35. $\sqrt{4x^2}$ $2|x|$ 36. $\sqrt{9x^4}$ $3x^2$
37. $\sqrt{64x^3}$ $8x\sqrt{x}$ 38. $\sqrt{49z^5}$ $7z^2\sqrt{z}$
39. $\sqrt{x^6}$ $|x^3|$ 40. $\sqrt{y^8}$ y^4
41. $\sqrt{u^7}$ $u^3\sqrt{u}$ 42. $\sqrt{v^5}$ $v^2\sqrt{v}$
43. $\sqrt{20x^6}$ $2|x^3|\sqrt{5}$ 44. $\sqrt{28x^8}$ $2x^4\sqrt{7}$
45. $\sqrt{x^2y^3}$ $|x|y\sqrt{y}$ 46. $\sqrt{a^5b^4}$ $a^2b^2\sqrt{a}$
47. $\sqrt{u^4v^8}$ u^2v^4 48. $\sqrt{x^2y^6}$ $|xy^3|$

49. $\sqrt{200x^2y^9}$ $10|x|y^4\sqrt{2y}$ **50.** $\sqrt{128u^4v^7}$ $8u^2v^3\sqrt{2v}$

51. $\sqrt[3]{27a^4}$ $3a\sqrt[3]{a}$ **52.** $\sqrt[3]{8x^5}$ $2x\sqrt[3]{x^2}$

53. $\sqrt[3]{2y^4}$ $y\sqrt[3]{2y}$ **54.** $\sqrt[3]{6r^5}$ $r\sqrt[3]{6r^2}$

55. $\sqrt[4]{t^7}$ $t\sqrt[4]{t^3}$ **56.** $\sqrt[4]{x^5}$ $x\sqrt[4]{x}$

57. $\sqrt[4]{16y^5}$ $2y\sqrt[4]{y}$ **58.** $\sqrt[4]{80t^7}$ $2t\sqrt[4]{5t^3}$

In Exercises 59–74, write as a single radical and simplify.

59. $\dfrac{\sqrt{6}}{\sqrt{3}}$ $\sqrt{2}$ **60.** $\dfrac{\sqrt{21}}{\sqrt{7}}$ $\sqrt{3}$

61. $\dfrac{\sqrt{54}}{\sqrt{6}}$ 3 **62.** $\dfrac{\sqrt{48}}{\sqrt{8}}$ $\sqrt{6}$

63. $\dfrac{\sqrt{39}}{\sqrt{13}}$ $\sqrt{3}$ **64.** $\dfrac{\sqrt{84}}{\sqrt{12}}$ $\sqrt{7}$

65. $\dfrac{\sqrt{24}}{\sqrt{2}}$ $2\sqrt{3}$ **66.** $\dfrac{\sqrt{60}}{\sqrt{3}}$ $2\sqrt{5}$

67. $\dfrac{\sqrt{156}}{\sqrt{3}}$ $2\sqrt{13}$ **68.** $\dfrac{\sqrt{160}}{\sqrt{5}}$ $4\sqrt{2}$

69. $\dfrac{\sqrt{15x}}{\sqrt{5x}}$ $\sqrt{3}, x\neq 0$ **70.** $\dfrac{\sqrt{10u}}{\sqrt{2u}}$ $\sqrt{5}, u\neq 0$

71. $\dfrac{\sqrt{18a^3}}{\sqrt{a}}$ $3a\sqrt{2}, a\neq 0$ **72.** $\dfrac{\sqrt{16y^4}}{\sqrt{y^2}}$ $4|y|, y\neq 0$

73. $\dfrac{\sqrt{54b^4}}{\sqrt{2b^2}}$ $3|b|\sqrt{3}, b\neq 0$ **74.** $\dfrac{\sqrt{45x^7}}{\sqrt{5x}}$ $3x^3, x\neq 0$

In Exercises 75–90, simplify the expression. See Examples 7 and 8.

75. $\dfrac{\sqrt{35}}{\sqrt{16}}$ $\dfrac{\sqrt{35}}{4}$ **76.** $\dfrac{\sqrt{11}}{\sqrt{25}}$ $\dfrac{\sqrt{11}}{5}$

77. $\dfrac{\sqrt{48}}{\sqrt{64}}$ $\dfrac{\sqrt{3}}{2}$ **78.** $\dfrac{\sqrt{72}}{\sqrt{9}}$ $2\sqrt{2}$

79. $\sqrt{\dfrac{32}{4}}$ $2\sqrt{2}$ **80.** $\sqrt{\dfrac{24}{36}}$ $\dfrac{\sqrt{6}}{3}$

81. $\sqrt{\dfrac{12x^2}{25}}$ $\dfrac{2|x|\sqrt{3}}{5}$ **82.** $\dfrac{\sqrt{5u^2}}{\sqrt{4u^4}}$ $\dfrac{\sqrt{5}}{2|u|}$

83. $\sqrt{\dfrac{3x^2}{27}}$ $\dfrac{|x|}{3}$ **84.** $\sqrt{\dfrac{5x^4}{20}}$ $\dfrac{x^2}{2}$

85. $\sqrt{\dfrac{56y^3}{14}}$ $2y\sqrt{y}$ **86.** $\sqrt{\dfrac{50x^5}{2}}$ $5x^2\sqrt{x}$

87. $\sqrt{\dfrac{x^6}{16y^2}}$ $\dfrac{|x^3|}{4|y|}$ **88.** $\sqrt{\dfrac{u^4}{36v^4}}$ $\dfrac{u^2}{6v^2}$

89. $\sqrt{\dfrac{9u^5}{48u^7}}$ $\dfrac{\sqrt{3}}{4|u|}$ **90.** $\sqrt{\dfrac{6x}{27x^5}}$ $\dfrac{\sqrt{2}}{3x^2}$

In Exercises 91–114, rationalize the denominator of the expression and simplify. (Assume all variables are positive.) See Examples 9 and 10.

91. $\sqrt{\dfrac{1}{3}}$ $\dfrac{\sqrt{3}}{3}$ **92.** $\sqrt{\dfrac{1}{5}}$ $\dfrac{\sqrt{5}}{5}$

93. $\dfrac{1}{\sqrt{7}}$ $\dfrac{\sqrt{7}}{7}$ **94.** $\dfrac{1}{\sqrt{10}}$ $\dfrac{\sqrt{10}}{10}$

95. $\dfrac{5}{\sqrt{10}}$ $\dfrac{\sqrt{10}}{2}$ **96.** $\dfrac{7}{\sqrt{14}}$ $\dfrac{\sqrt{14}}{2}$

97. $\dfrac{\sqrt{2}}{\sqrt{3}}$ $\dfrac{\sqrt{6}}{3}$ **98.** $\dfrac{\sqrt{3}}{\sqrt{8}}$ $\dfrac{\sqrt{6}}{4}$

99. $\sqrt{\dfrac{11}{8}}$ $\dfrac{\sqrt{22}}{4}$

100. $\sqrt{\dfrac{7}{18}}$ $\dfrac{\sqrt{14}}{6}$

101. $\dfrac{\sqrt{6}}{\sqrt{12}}$ $\dfrac{\sqrt{2}}{2}$

102. $\dfrac{\sqrt{10}}{\sqrt{32}}$ $\dfrac{\sqrt{5}}{4}$

103. $\sqrt{\dfrac{100}{11}}$ $\dfrac{10\sqrt{11}}{11}$

104. $\sqrt{\dfrac{169}{2}}$ $\dfrac{13\sqrt{2}}{2}$

105. $\dfrac{1}{\sqrt{y}}$ $\dfrac{\sqrt{y}}{y}$

106. $\dfrac{1}{\sqrt{z}}$ $\dfrac{\sqrt{z}}{z}$

107. $\sqrt{\dfrac{5}{x}}$ $\dfrac{\sqrt{5x}}{x}$

108. $\sqrt{\dfrac{3}{a}}$ $\dfrac{\sqrt{3a}}{a}$

109. $\sqrt{\dfrac{3}{16x^5}}$ $\dfrac{\sqrt{3x}}{4x^3}$

110. $\sqrt{\dfrac{6}{25u^3}}$ $\dfrac{\sqrt{6u}}{5u^2}$

111. $\dfrac{\sqrt{2t}}{\sqrt{8r}}$ $\dfrac{\sqrt{rt}}{2r}$

112. $\dfrac{\sqrt{2x}}{\sqrt{50y}}$ $\dfrac{\sqrt{xy}}{5y}$

113. $\dfrac{\sqrt{12x^3}}{\sqrt{3y}}$ $\dfrac{2x\sqrt{xy}}{y}$

114. $\dfrac{\sqrt{20x^2}}{\sqrt{5y^2}}$ $\dfrac{2x}{y}$

In Exercises 115–122, rationalize the denominator of the expression and simplify. (Assume all variables are positive.) See Example 11.

115. $\dfrac{4}{\sqrt[3]{9}}$ $\dfrac{4\sqrt[3]{3}}{3}$

116. $\dfrac{9}{\sqrt[3]{4}}$ $\dfrac{9\sqrt[3]{2}}{2}$

117. $\dfrac{7}{\sqrt[3]{3}}$ $\dfrac{7\sqrt[3]{9}}{3}$

118. $\dfrac{5}{\sqrt[3]{2}}$ $\dfrac{5\sqrt[3]{4}}{2}$

119. $\sqrt[3]{\dfrac{1}{x^2}}$ $\dfrac{\sqrt[3]{x}}{x}$

120. $\sqrt[3]{\dfrac{3}{x}}$ $\dfrac{\sqrt[3]{3x^2}}{x}$

121. $\sqrt[3]{\dfrac{1}{8y^2}}$ $\dfrac{\sqrt[3]{y}}{2y}$

122. $\sqrt[3]{\dfrac{1}{27a}}$ $\dfrac{\sqrt[3]{a^2}}{3a}$

Graphical Reasoning   In Exercises 123–126, use a graphing calculator to graph the equations in the same viewing window. What can you conclude from the graphs?   See Additional Answers.

123. $y_1 = \sqrt{2} \cdot \sqrt{x}$

$y_2 = \sqrt{2x}$

The expressions are equivalent.

124. $y_1 = \dfrac{\sqrt{x}}{\sqrt{4}}$

$y_2 = \tfrac{1}{2}\sqrt{x}$

The expressions are equivalent.

125. $y_1 = \dfrac{\sqrt{x}}{\sqrt{8}}$

$y_2 = \tfrac{1}{4}\sqrt{2x}$

The expressions are equivalent.

126. $y_1 = \dfrac{\sqrt{3x}}{\sqrt{27}}$

$y_2 = \tfrac{1}{3}\sqrt{x}$

The expressions are equivalent.

In Exercises 127–130, place the correct inequality symbol (< or >) between the real numbers. Do not use a calculator.

127. $4\sqrt{2}$ >   5

128. $\sqrt{300}$ >   12

129. $5\sqrt{6}$ < $6\sqrt{5}$

130. 4 < $\sqrt{5} + \sqrt{10}$

Solving Problems

Geometry   In Exercises 131–134, find the area of the figure. Round your answer to two decimal places.

131. $10\sqrt{2} \approx 14.14$

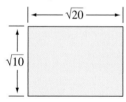

132. 98

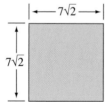

133. $12\sqrt{6} \approx 29.39$

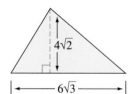

134. $\dfrac{9\sqrt{2}}{4} \approx 3.18$

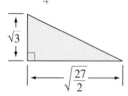

Period of a Pendulum   In Exercises 135 and 136, use the formula

$$t = 2\pi\sqrt{\dfrac{L}{32}}$$

which gives the time t (in seconds) for a pendulum of length L (in feet) to go through one complete cycle, both forward and back (its period).

135. How long will it take the performer in the figure to swing through one period of the 12-foot trapeze?

$\dfrac{\pi\sqrt{6}}{2} \approx 3.85$ seconds

136. Find the period of the pendulum clock in the figure.

$\dfrac{1.5\pi\sqrt{2}}{4} = \dfrac{3\pi\sqrt{2}}{8} \approx 1.67$ seconds

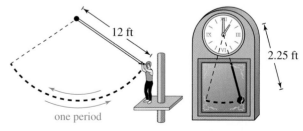

Figure for 135   Figure for 136

Geometry   In Exercises 137–140, find the lateral surface area of the cone. See Example 12. Round your answer to one decimal place.

137.

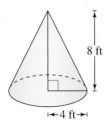

8 ft

4 ft→|

$16\pi\sqrt{5} \approx 112.4$ square feet

138.

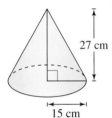

27 cm

|← 15 cm →|

$45\pi\sqrt{106} \approx 1455.5$ square centimeters

139.

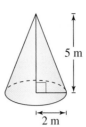

5 m

|←→ 2 m →|

$2\pi\sqrt{29} \approx 33.8$ square meters

140.

12 in.

|←16 in.→|

$320\pi \approx 1005.3$ square inches

Explaining Concepts

141. State the Product Rule for Radicals.   If the nth roots of a and b are real, then $\sqrt[n]{ab} = \sqrt[n]{a}\,\sqrt[n]{b}$.

142. State the Quotient Rule for Radicals.

If the nth roots of a and b are real, then

$$\sqrt[n]{\frac{a}{b}} = \frac{\sqrt[n]{a}}{\sqrt[n]{b}},\ b \neq 0.$$

143. *Writing*   In your own words, describe the three conditions that must be true for a radical expression to be in simplest form.

   (i)   All possible nth powered factors have been removed from each radical.

   (ii) No radical contains a fraction.

   (iii) No denominator of a fraction contains a radical.

144. *Writing*   Show how the Product Rule for Radicals can be used to simplify $\sqrt{28}$.

$\sqrt{28} = \sqrt{4 \cdot 7} = \sqrt{4} \cdot \sqrt{7} = 2\sqrt{7}$

145. *Writing*   Show how to simplify $1/\sqrt{3}$.

$\dfrac{1}{\sqrt{3}} = \dfrac{1}{\sqrt{3}} \cdot \dfrac{\sqrt{3}}{\sqrt{3}} = \dfrac{\sqrt{3}}{3}$

146. Determine which of the statements is true for any real number x. Give an example to show any false statements. Verify your conclusions by using a graphing calculator to graph

$y_1 = \sqrt{x^2}$, $y_2 = x$, $y_3 = |x|$, and $y_4 = \sqrt[3]{x^3}$.

  (a) $\sqrt{x^2} = x$ False. $\sqrt{(-2)^2} \neq -2$

  (b) $\sqrt{x^2} = |x|$ True

  (c) $\sqrt[3]{x^3} = |x|$ False. $\sqrt[3]{(-3)^3} \neq |-3|$

  (d) $\sqrt[3]{x^3} = x$ True

True or False?   In Exercises 147–149, decide whether the statement is true or false. Justify your answer.

147. $\sqrt{3x^2} = x\sqrt{3}$ False. $\sqrt{3x^2} = |x|\sqrt{3}$

148. $\dfrac{\sqrt{50}}{\sqrt{2}} = 25$ False. $\dfrac{\sqrt{50}}{\sqrt{2}} = \sqrt{25} = 5$

149. $\sqrt{x^2 + 16} = x + 4$ False. $(x + 4)^2 = x^2 + 8x + 16$

Mid-Chapter Quiz

Take this quiz as you would take a quiz in class. After you are done, check your work against the answers in the back of the book.

In Exercises 1–6, evaluate the expression, if possible.

1. $\sqrt{121}$ 11 **2.** $-\sqrt{0.25}$ -0.5 **3.** $\sqrt[3]{-8}$ -2

4. $\sqrt[4]{-16}$ Not possible   **5.** $\sqrt{-\frac{1}{49}}$ Not possible   **6.** $-\sqrt{1.44}$ -1.2

In Exercises 7–9, classify the number as rational or irrational.

7. $\sqrt{5}$ Irrational   **8.** $\sqrt{\frac{3}{4}}$ Irrational   **9.** $\sqrt{900}$ Rational

In Exercises 10–12, use a calculator to approximate the value of the expression. Round your answer to three decimal places.

15. $5\sqrt{2}$ **16.** $5\sqrt{2}$

17. $6|x|\sqrt{2}$ **18.** $3b^4\sqrt{5}$

19. $4x$ **20.** $x^2\sqrt[4]{x}$

21. $3|b|\sqrt{5}, \; b \neq 0$

22. $3u^2|v|\sqrt{2u}$

10. $\sqrt{15.8}$ 3.975 **11.** $35 - \sqrt{8.9}$ 32.017 **12.** $\dfrac{-5 + 3\sqrt{20}}{10}$ 0.842

In Exercises 13 and 14, write as a single radical.

13. $\sqrt{5} \cdot \sqrt{19}$ $\sqrt{95}$ **14.** $\dfrac{\sqrt{42}}{\sqrt{6}}$ $\sqrt{7}$

In Exercises 15–22, simplify the expression. Use absolute value signs if appropriate.

15. $\sqrt{50}$ **16.** $\dfrac{\sqrt{600}}{\sqrt{12}}$ **17.** $\sqrt{72x^2}$ **18.** $\sqrt{45b^8}$

19. $\sqrt[3]{64x^3}$ **20.** $\sqrt[4]{x^9}$ **21.** $\sqrt{\dfrac{90b^4}{2b^2}}$ **22.** $\sqrt{18u^5v^2}$

In Exercises 23–28, rationalize the denominator and simplify. (Assume that the variables are positive.)

23. $\sqrt{\dfrac{3}{2}}$ $\dfrac{\sqrt{6}}{2}$ **24.** $\dfrac{2}{\sqrt{12}}$ $\dfrac{\sqrt{3}}{3}$ **25.** $\sqrt[3]{\dfrac{1}{9}}$ $\dfrac{\sqrt[3]{3}}{3}$

26. $\dfrac{4a}{\sqrt{2a}}$ $2\sqrt{2a}, a \neq 0$ **27.** $\sqrt{\dfrac{5a^3}{4a}}$ $\dfrac{a\sqrt{5}}{2}, a \neq 0$ **28.** $\dfrac{\sqrt[3]{x}}{\sqrt[3]{27x^4}}$ $\dfrac{1}{3x}$

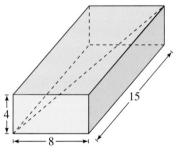

Figure for 29

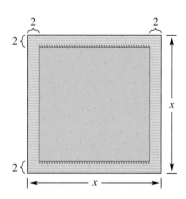

Figure for 30

29. The length of a diagonal of a rectangular solid of length l, width w, and height h is

$$\sqrt{l^2 + w^2 + h^2}.$$

Approximate to two decimal places the length of the diagonal of the solid shown in the figure. 17.46

30. A square room has 361 square feet of floor space. An area rug covers all of the floor space except for a two-foot border all around the room (see figure). How many square feet are in the rug? How many square yards are in the rug?

225 square feet; 25 square yards

9.3 Operations with Radical Expressions

Stuart Cohen/The Image Works

Why You Should Learn It

Many real-life applications can be modeled by equations containing radicals. For instance, in Exercise 140 on page 518, an equation containing a radical is used to determine the wind chill.

What You Should Learn

① Add and subtract like radical expressions.

② Multiply radical expressions using the Distributive Property, the FOIL Method, or a special product pattern.

③ Determine the conjugate of an expression and find the product of the expression and its conjugate.

④ Simplify quotients involving radicals by rationalizing the denominators.

Adding and Subtracting Radical Expressions

Two radical expressions are called **like radicals** if they have the same index and the same radicand. For instance, the expressions $3\sqrt{2}$ and $5\sqrt{2}$ are like radicals, whereas the expressions $2\sqrt{3}$ and $2\sqrt{5}$ are not like radicals. You can combine like radicals by the Distributive Property. For instance, you can combine the radical expression $2\sqrt{5} + 7\sqrt{5}$ as follows.

$$2\sqrt{5} + 7\sqrt{5} = (2 + 7)\sqrt{5} \qquad \text{Distributive Property}$$

$$= 9\sqrt{5} \qquad \text{Simplify.}$$

Note the similarity between this procedure and the procedure used to combine like polynomial terms such as $2x + 7x = (2 + 7)x = 9x$.

① Add and subtract like radical expressions.

Example 1   Combining Radical Expressions

Simplify each expression by combining like terms.

a. $\sqrt{7} + 5\sqrt{7} - 2\sqrt{7}$ **b.** $6\sqrt{6} - \sqrt{3} - 5\sqrt{6} + 2\sqrt{3}$

c. $3\sqrt[4]{5} + 7\sqrt[4]{5}$ **d.** $\sqrt[5]{2} + 6\sqrt[5]{2} - 2\sqrt[5]{2}$

Solution

a. $\sqrt{7} + 5\sqrt{7} - 2\sqrt{7} = (1 + 5 - 2)\sqrt{7}$ Distributive Property

$$= 4\sqrt{7} \qquad \text{Simplify.}$$

b. $6\sqrt{6} - \sqrt{3} - 5\sqrt{6} + 2\sqrt{3}$

$$= (6\sqrt{6} - 5\sqrt{6}) + (-\sqrt{3} + 2\sqrt{3}) \qquad \text{Group like terms.}$$

$$= (6 - 5)\sqrt{6} + (-1 + 2)\sqrt{3} \qquad \text{Distributive Property}$$

$$= \sqrt{6} + \sqrt{3} \qquad \text{Simplify.}$$

c. $3\sqrt[4]{5} + 7\sqrt[4]{5} = (3 + 7)\sqrt[4]{5}$ Distributive Property

$$= 10\sqrt[4]{5} \qquad \text{Simplify.}$$

d. $\sqrt[5]{2} + 6\sqrt[5]{2} - 2\sqrt[5]{2} = (1 + 6 - 2)\sqrt[5]{2}$ Distributive Property

$$= 5\sqrt[5]{2} \qquad \text{Simplify.}$$

Study Tip

It is important to realize that the expression $\sqrt{a} + \sqrt{b}$ is not equal to $\sqrt{a + b}$. For instance, in Example 1(b), you may have been tempted to add $\sqrt{6} + \sqrt{3}$ and get $\sqrt{9}$ or 3. But remember, you cannot add unlike radicals. So, $\sqrt{6} + \sqrt{3}$ cannot be simplified further.

Example 2   **Combining Radicals with Variable Radicands**

Simplify each expression.

a. $3 + 3\sqrt{x} - \sqrt{x} + 2$ **b.** $2\sqrt[3]{2} + 2\sqrt[3]{y^2} - 4\sqrt[3]{2} + 5\sqrt[3]{y^2}$

Solution

a. $3 + 3\sqrt{x} - \sqrt{x} + 2$

$$= (3 + 2) + \left(3\sqrt{x} - \sqrt{x}\right) \qquad \text{Group like terms.}$$

$$= (3 + 2) + (3 - 1)\sqrt{x} \qquad \text{Distributive Property}$$

$$= 5 + 2\sqrt{x} \qquad \text{Simplify.}$$

b. $2\sqrt[3]{2} + 2\sqrt[3]{y^2} - 4\sqrt[3]{2} + 5\sqrt[3]{y^2}$

$$= \left(2\sqrt[3]{2} - 4\sqrt[3]{2}\right) + \left(2\sqrt[3]{y^2} + 5\sqrt[3]{y^2}\right) \qquad \text{Group like terms.}$$

$$= (2 - 4)\sqrt[3]{2} + (2 + 5)\sqrt[3]{y^2} \qquad \text{Distributive Property}$$

$$= -2\sqrt[3]{2} + 7\sqrt[3]{y^2} \qquad \text{Simplify.}$$

Sometimes you may have to simplify individual radicals before combining like radicals. This is demonstrated in Examples 3 and 4.

Study Tip

Remember that the square root of a negative number is not a real number. Because of this, you can assume for $\sqrt{x^3}$ that the implied domain of the expression is the set of nonnegative real numbers. This assumption allows you to write $\sqrt{x^3} = x\sqrt{x}$.

Example 3   **Simplifying a Radical Expression**

$$2\sqrt{72} - 2\sqrt{32} = 2\sqrt{36 \cdot 2} - 2\sqrt{16 \cdot 2} \qquad \text{Factor radicands.}$$

$$= 12\sqrt{2} - 8\sqrt{2} \qquad \text{Simplify radicals.}$$

$$= 4\sqrt{2} \qquad \text{Combine like radicals.}$$

Example 4   **Simplifying Radical Expressions**

a. $3\sqrt{x} + \sqrt{4x} = 3\sqrt{x} + \sqrt{4 \cdot x} \qquad \text{Factor radicand.}$

$$= 3\sqrt{x} + 2\sqrt{x} \qquad \text{Simplify radical.}$$

$$= 5\sqrt{x} \qquad \text{Combine like radicals.}$$

b. $\sqrt{45x} + 2\sqrt{20x} = \sqrt{9 \cdot 5x} + 2\sqrt{4 \cdot 5x} \qquad \text{Factor radicands.}$

$$= 3\sqrt{5x} + 4\sqrt{5x} \qquad \text{Simplify radicals.}$$

$$= 7\sqrt{5x} \qquad \text{Combine like radicals.}$$

c. $5\sqrt{2x^3} - x\sqrt{8x} = 5\sqrt{2 \cdot x^2 \cdot x} - x\sqrt{4 \cdot 2 \cdot x} \qquad \text{Factor radicands.}$

$$= 5x\sqrt{2x} - 2x\sqrt{2x} \qquad \text{Simplify radicals.}$$

$$= (5x - 2x)\sqrt{2x} \qquad \text{Distributive Property}$$

$$= 3x\sqrt{2x} \qquad \text{Combine like radicals.}$$

2 Multiply radical expressions using the Distributive Property, the FOIL Method, or a special product pattern.

Multiplying Radical Expressions

You can multiply radical expressions by using the Distributive Property, the FOIL Method, or a special product formula. These procedures also use the Product Rule for Radicals.

Example 5   Multiplying Radical Expressions

Find each product and simplify.

a. $\sqrt{6} \cdot \sqrt{3}$ **b.** $\sqrt[3]{4} \cdot \sqrt[3]{12}$

Solution

a. $\sqrt{6} \cdot \sqrt{3} = \sqrt{6 \cdot 3} = \sqrt{18} = \sqrt{9 \cdot 2} = 3\sqrt{2}$

b. $\sqrt[3]{4} \cdot \sqrt[3]{12} = \sqrt[3]{4 \cdot 12} = \sqrt[3]{48} = \sqrt[3]{8 \cdot 6} = 2\sqrt[3]{6}$

Example 6   Multiplying Radical Expressions

Find each product and simplify.

a. $\sqrt{5}(\sqrt{15} - \sqrt{5})$ **b.** $\sqrt[3]{4}(3 + \sqrt[3]{2})$

Solution

a. $\sqrt{5}(\sqrt{15} - \sqrt{5}) = \sqrt{5}\sqrt{15} - \sqrt{5}\sqrt{5}$ Distributive Property

$\qquad = \sqrt{75} - \sqrt{25}$ Product Rule for Radicals

$\qquad = \sqrt{25 \cdot 3} - 5$ Factor radicand.

$\qquad = 5\sqrt{3} - 5$ Simplify.

b. $\sqrt[3]{4}(3 + \sqrt[3]{2}) = \sqrt[3]{4} \cdot 3 + \sqrt[3]{4}\sqrt[3]{2}$ Distributive Property

$\qquad = 3\sqrt[3]{4} + \sqrt[3]{8}$ Product Rule for Radicals

$\qquad = 3\sqrt[3]{4} + 2$ Simplify.

Additional Examples
Find each product and simplify.
a. $\sqrt{2}(1 + \sqrt{x})$
b. $(\sqrt[3]{4} + 5)(\sqrt[3]{2} + 6)$

Answers:

a. $\sqrt{2} + \sqrt{2x}$

b. $32 + 6\sqrt[3]{4} + 5\sqrt[3]{2}$

Point out the analogies with the multiplication of polynomials.

Example 7   Multiplying Radical Expressions

Find each product and simplify.

a. $(\sqrt{x} - 1)(\sqrt{x} + 3)$ **b.** $(2 - \sqrt{x})(2 + \sqrt{x})$

Solution

$\qquad\qquad\qquad\quad$ F $\quad$ O $\quad$ I $\quad$ L

a. $(\sqrt{x} - 1)(\sqrt{x} + 3) = \sqrt{x \cdot x} + 3\sqrt{x} - \sqrt{x} - 3$ FOIL Method

$\qquad = x + (3 - 1)\sqrt{x} - 3$ Combine like radicals.

$\qquad = x - 3 + 2\sqrt{x}$ Simplify.

b. $(2 - \sqrt{x})(2 + \sqrt{x}) = 2^2 - (\sqrt{x})^2$ Special product formula

$\qquad = 4 - x$ Simplify.

Example 8   Geometry: Perimeter and Area of a Triangle

Write and simplify expressions for the perimeter and area of the triangle shown in Figure 9.3.

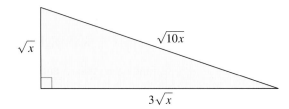

Figure 9.3

Solution

$$P = a + b + c$$ Formula for perimeter of a triangle

$$= \sqrt{x} + 3\sqrt{x} + \sqrt{10x}$$ Substitute

$$= (1 + 3)\sqrt{x} + \sqrt{10x}$$ Distributive Property

$$= 4\sqrt{x} + \sqrt{10x}$$ Simplify.

$$A = \tfrac{1}{2}bh$$ Formula for area of a triangle

$$= \tfrac{1}{2}\left(3\sqrt{x}\right)\left(\sqrt{x}\right)$$ Substitute.

$$= \tfrac{3}{2}\sqrt{x^2}$$ Product Rule for Radicals

$$= \tfrac{3}{2}x$$ Simplify.

③ Determine the conjugate of an expression and find the product of the expression and its conjugate.

Conjugates

Recall from Section 5.3 the special product formula, *the product of the sum and difference of two terms.*

$$(a + b)(a - b) = a^2 - ab + ab - b^2$$ Sum and difference of two terms

$$= a^2 - b^2$$ Difference of two squares

The resulting product is the difference of two squares. Notice in Example 7(b) that this special product formula was used to find the product of $2 - \sqrt{x}$ and $2 + \sqrt{x}$. These expressions are called **conjugates** of each other and they differ only in sign between the terms. Here are some other examples.

Expression	*Conjugate*	*Product*
$1 - \sqrt{3}$	$1 + \sqrt{3}$	$(1)^2 - \left(\sqrt{3}\right)^2 = 1 - 3 = -2$
$\sqrt{5} + \sqrt{2}$	$\sqrt{5} - \sqrt{2}$	$\left(\sqrt{5}\right)^2 - \left(\sqrt{2}\right)^2 = 5 - 2 = 3$
$\sqrt{x} + 2$	$\sqrt{x} - 2$	$\left(\sqrt{x}\right)^2 - (2)^2 = x - 4, x \geq 0$

Notice that when you multiply a binomial with radical terms by its conjugate, the result is free of radicals.

<u>Example 9</u>   **Multiplying Conjugates**

Find the conjugate of the expression. Then find the product of the expression and its conjugate.

a. $\sqrt{3} + \sqrt{7}$

b. $\sqrt{t} - \sqrt{8}$

Solution

a. The conjugate of $\sqrt{3} + \sqrt{7}$ is $\sqrt{3} - \sqrt{7}$. The product of $\sqrt{3} + \sqrt{7}$ and $\sqrt{3} - \sqrt{7}$ is

$$\left(\sqrt{3} + \sqrt{7}\right)\left(\sqrt{3} - \sqrt{7}\right) = \left(\sqrt{3}\right)^2 - \left(\sqrt{7}\right)^2 \qquad \text{Special product formula}$$

$$= 3 - 7 \qquad \text{Simplify.}$$

$$= -4. \qquad \text{Simplify.}$$

b. The conjugate of $\sqrt{t} - \sqrt{8}$ is $\sqrt{t} + \sqrt{8}$. The product of $\sqrt{t} - \sqrt{8}$ and $\sqrt{t} + \sqrt{8}$ is

$$\left(\sqrt{t} - \sqrt{8}\right)\left(\sqrt{t} + \sqrt{8}\right) = \left(\sqrt{t}\right)^2 - \left(\sqrt{8}\right)^2 \qquad \text{Special product formula}$$

$$= t - 8. \qquad \text{Simplify.}$$

④ Simplify quotients involving radicals by rationalizing the denominators.

Simplifying Quotients Involving Radicals

To simplify a *quotient* involving radicals, you can rationalize the denominator by multiplying both the numerator and denominator by the conjugate of the denominator. Note how this is done in Examples 10 and 11.

<u>Example 10</u>   **Simplifying Quotients Involving Radicals**

Simplify the quotient.

$$\frac{2}{3 - \sqrt{5}}$$

Solution

$$\frac{2}{3 - \sqrt{5}} = \frac{2}{3 - \sqrt{5}} \cdot \frac{3 + \sqrt{5}}{3 + \sqrt{5}} \qquad \text{Multiply numerator and denominator by conjugate of denominator.}$$

$$= \frac{2\left(3 + \sqrt{5}\right)}{(3)^2 - \left(\sqrt{5}\right)^2} \qquad \text{Special product formula}$$

$$= \frac{2\left(3 + \sqrt{5}\right)}{4} \qquad \text{Simplify.}$$

$$= \frac{3 + \sqrt{5}}{2} \qquad \text{Divide out common factors.}$$

Example 11 Simplifying Quotients Involving Radicals

a. $\dfrac{5\sqrt{2}}{\sqrt{3}+\sqrt{2}} = \dfrac{5\sqrt{2}}{\sqrt{3}+\sqrt{2}} \cdot \dfrac{\sqrt{3}-\sqrt{2}}{\sqrt{3}-\sqrt{2}}$ Multiply numerator and denominator by conjugate of denominator.

$\phantom{\dfrac{5\sqrt{2}}{\sqrt{3}+\sqrt{2}}} = \dfrac{5\sqrt{6}-5\sqrt{4}}{\left(\sqrt{3}\right)^2-\left(\sqrt{2}\right)^2}$ Special product formula

$\phantom{\dfrac{5\sqrt{2}}{\sqrt{3}+\sqrt{2}}} = \dfrac{5\sqrt{6}-10}{3-2}$ Simplify.

$\phantom{\dfrac{5\sqrt{2}}{\sqrt{3}+\sqrt{2}}} = 5\sqrt{6}-10$ Simplify.

b. $\dfrac{2-\sqrt{3}}{\sqrt{6}+\sqrt{2}} = \dfrac{2-\sqrt{3}}{\sqrt{6}+\sqrt{2}} \cdot \dfrac{\sqrt{6}-\sqrt{2}}{\sqrt{6}-\sqrt{2}}$ Multiply numerator and denominator by conjugate of denominator.

$\phantom{\dfrac{2-\sqrt{3}}{\sqrt{6}+\sqrt{2}}} = \dfrac{2\sqrt{6}-2\sqrt{2}-\sqrt{18}+\sqrt{6}}{\left(\sqrt{6}\right)^2-\left(\sqrt{2}\right)^2}$ FOIL Method and special product formula

$\phantom{\dfrac{2-\sqrt{3}}{\sqrt{6}+\sqrt{2}}} = \dfrac{3\sqrt{6}-2\sqrt{2}-3\sqrt{2}}{6-2}$ Combine like radicals.

$\phantom{\dfrac{2-\sqrt{3}}{\sqrt{6}+\sqrt{2}}} = \dfrac{3\sqrt{6}-5\sqrt{2}}{4}$ Simplify.

c. $\dfrac{6}{\sqrt{x}-2} = \dfrac{6}{\sqrt{x}-2} \cdot \dfrac{\sqrt{x}+2}{\sqrt{x}+2}$ Multiply numerator and denominator by conjugate of denominator.

$\phantom{\dfrac{6}{\sqrt{x}-2}} = \dfrac{6\left(\sqrt{x}+2\right)}{\left(\sqrt{x}\right)^2-(2)^2}$ Special product formula

$\phantom{\dfrac{6}{\sqrt{x}-2}} = \dfrac{6\sqrt{x}+12}{x-4}$ Simplify.

Example 12 Simplifying a Mixed Radical Expression

Write $3 - \dfrac{2}{\sqrt{5}}$ as a single fraction in simplest form.

Solution

First, write the expression as a single fraction, then rationalize the denominator.

$3 - \dfrac{2}{\sqrt{5}} = \dfrac{3}{1} \cdot \dfrac{\sqrt{5}}{\sqrt{5}} - \dfrac{2}{\sqrt{5}}$ Multiply by LCD.

$\phantom{3 - \dfrac{2}{\sqrt{5}}} = \dfrac{3\sqrt{5}-2}{\sqrt{5}}$ Subtract fractions.

$\phantom{3 - \dfrac{2}{\sqrt{5}}} = \dfrac{3\sqrt{5}-2}{\sqrt{5}} \cdot \dfrac{\sqrt{5}}{\sqrt{5}}$ Multiply by $\sqrt{5}/\sqrt{5}$ to create a perfect square in the denominator.

$\phantom{3 - \dfrac{2}{\sqrt{5}}} = \dfrac{3\sqrt{25}-2\sqrt{5}}{\sqrt{25}} = \dfrac{15-2\sqrt{5}}{5}$ Simplify.

9.3 Exercises

Review   Concepts, Skills, and Problem Solving

Keep mathematically in shape by doing these exercises *before* the problems of this section.

Properties and Definitions

1. *Writing*✐   Is it possible for a system of equations to have exactly two solutions? Explain.
   No. There is exactly one solution, an infinite number of solutions, or no solution.

2. *Writing*✐   Explain why the following system of equations has no solution.

$$\begin{cases} 4x - 2y = 5 \\ -2x + y = 1 \end{cases}$$ The graphs of the equations are distinct parallel lines.

Solving Equations and Systems of Equations

In Exercises 3–6, solve the equation.

3. $4[2x - 3(x + 2)] = 5(x - 6)$ $\frac{2}{3}$

4. $2 - 5(x - 1) = 2[x + 10(x - 1)]$ 1

5. $\dfrac{x}{6} + \dfrac{x}{3} = 1$ 2

6. $\dfrac{x + 1}{2} + \dfrac{x}{4} = 1$ $\frac{2}{3}$

In Exercises 7–10, solve the system of linear equations.

7. $\begin{cases} x - 3y = 2 \\ x + y = 6 \end{cases}$ (5, 1)

8. $\begin{cases} 5x + y = 20 \\ 2x - y = 1 \end{cases}$ (3, 5)

9. $\begin{cases} -x + 7y = -1 \\ 2x - 10y = 6 \end{cases}$ (8, 1)

10. $\begin{cases} 9x - 7y = 39 \\ 4x + 3y = -1 \end{cases}$ (2, -3)

Problem Solving

11. *Harvest Rate*   One combine harvests a field of wheat in r hours and a second combine requires $\frac{5}{4}r$ hours. Find the individual times to harvest the field if it takes 5 hours using both machines.   9 hours, 11.25 hours

12. *Wind Speed*   A plane has a speed of 300 miles per hour in still air. Find the speed of the wind if the plane traveled a distance of 700 miles with a tail wind in the same time it took to travel 500 miles into a head wind.   50 miles per hour

Developing Skills

In Exercises 1–48, simplify the expression. See Examples 1–4.

1. $3\sqrt{5} - \sqrt{5}$ $2\sqrt{5}$

2. $5\sqrt{6} - 10\sqrt{6}$ $-5\sqrt{6}$

3. $10\sqrt{11} + 8\sqrt{11}$ $18\sqrt{11}$

4. $\sqrt{15} + 7\sqrt{15}$ $8\sqrt{15}$

5. $\frac{2}{5}\sqrt{3} - \frac{6}{5}\sqrt{3}$ $-\frac{4}{5}\sqrt{3}$

6. $\frac{2}{3}\sqrt{6} + \frac{4}{3}\sqrt{6}$ $2\sqrt{6}$

7. $\sqrt{3} - 5\sqrt{7} - 12\sqrt{3}$ $-11\sqrt{3} - 5\sqrt{7}$

8. $9\sqrt{17} + 7\sqrt{2} - 11\sqrt{17} + \sqrt{2}$ $-2\sqrt{17} + 8\sqrt{2}$

9. $2\sqrt{2} - 3\sqrt{5} + 8\sqrt{2}$ $10\sqrt{2} - 3\sqrt{5}$

10. $\sqrt{6} + 5\sqrt{3} - 7\sqrt{3}$ $\sqrt{6} - 2\sqrt{3}$

11. $4\sqrt[3]{5} + 2\sqrt[3]{5}$ $6\sqrt[3]{5}$

12. $7\sqrt[5]{4} + 3\sqrt[5]{4}$ $10\sqrt[5]{4}$

13. $9\sqrt[4]{8} - 4\sqrt[4]{8}$ $5\sqrt[4]{8}$

14. $6\sqrt[3]{3} - 5\sqrt[3]{3}$ $\sqrt[3]{3}$

15. $9\sqrt[3]{7} + 3\sqrt[3]{7} - 4\sqrt[3]{7}$ $8\sqrt[3]{7}$

16. $8\sqrt[4]{6} - 3\sqrt[4]{6} + 5\sqrt[4]{6}$ $10\sqrt[4]{6}$

17. $4\sqrt[3]{5} + 8\sqrt[3]{5} + \sqrt[3]{5}$ $13\sqrt[3]{5}$

18. $5\sqrt[5]{2} + 3\sqrt[5]{2} - \sqrt[5]{2}$ $7\sqrt[5]{2}$

19. $5\sqrt{x} - 3\sqrt{x}$ $2\sqrt{x}$

20. $2\sqrt{y} + 6\sqrt{y}$ $8\sqrt{y}$

21. $4\sqrt{u} - 3 + \sqrt{u} + 8$ $5\sqrt{u} + 5$

22. $12\sqrt{v} + 6 - 5\sqrt{v} - 9$ $7\sqrt{v} - 3$

23. $3 + \sqrt[3]{x} + 5 + 4\sqrt[3]{x}$ $5\sqrt[3]{x} + 8$

24. $7\sqrt[4]{a} + 6 + 3\sqrt[4]{a} + 2$ $10\sqrt[4]{a} + 8$

25. $9 + 3\sqrt[5]{y} - \sqrt[5]{y} - 6$ $2\sqrt[5]{y} + 3$

26. $8\sqrt[3]{x^2} + 3 - 3\sqrt[3]{x^2} + 2$ $5\sqrt[3]{x^2} + 5$

27. $8\sqrt[5]{a^3} - 3\sqrt[5]{a^3} + 5 + 6\sqrt[5]{a^3}$ $11\sqrt[5]{a^3} + 5$

28. $4\sqrt[3]{b^2} - 8 + 4\sqrt[3]{b^2} - \sqrt[3]{b^2}$ $7\sqrt[3]{b^2} - 8$

29. $12\sqrt{8} - 3\sqrt{8}$ $18\sqrt{2}$

30. $4\sqrt{32} + 2\sqrt{32}$ $24\sqrt{2}$

31. $2\sqrt{50} + 12\sqrt{8}$ $34\sqrt{2}$

32. $4\sqrt{27} - \sqrt{75}$ $7\sqrt{3}$

33. $8\sqrt{75} + \sqrt{50} + \sqrt{2}$ $40\sqrt{3} + 6\sqrt{2}$

34. $3\sqrt{18} - \sqrt{12} - \sqrt{8}$ $7\sqrt{2} - 2\sqrt{3}$

35. $\sqrt{9x} + \sqrt{36x}$ $9\sqrt{x}$

36. $\sqrt{64t} - \sqrt{16t}$ $4\sqrt{t}$

37. $\sqrt{16b} + \sqrt{b}$
$5\sqrt{b}$

38. $\sqrt{x} - \sqrt{25x}$
$-4\sqrt{x}$

39. $\sqrt{45z} - \sqrt{125z}$
$-2\sqrt{5z}$

40. $\sqrt{18u} + 3\sqrt{8u}$
$9\sqrt{2u}$

41. $3\sqrt{3y} - \sqrt{27y} + \sqrt{y}$
$\sqrt{y}$

42. $\sqrt{49v} + 2\sqrt{7v} + \sqrt{v}$
$8\sqrt{v} + 2\sqrt{7v}$

43. $\sqrt{32x^3} - 2\sqrt{8x^3}$
0

44. $3\sqrt{45u^5} + 5\sqrt{48u^5}$
$9u^2\sqrt{5u} + 20u^2\sqrt{3u}$

45. $\sqrt{x^3y} + 4\sqrt{xy}$
$(|x| + 4)\sqrt{xy}$

46. $3t\sqrt{st^3} - s\sqrt{s^3t}$
$(3t^2 - s^2)\sqrt{st}$

47. $\sqrt{\dfrac{a}{4}} - \sqrt{\dfrac{a}{9}}$
$\frac{1}{6}\sqrt{a}$

48. $\sqrt{\dfrac{v}{36}} - \sqrt{\dfrac{v}{9}}$
$-\frac{1}{6}\sqrt{v}$

In Exercises 49–90, multiply and simplify. See Examples 5–7.

49. $\sqrt{2} \cdot \sqrt{8}$ 4

50. $\sqrt{3} \cdot \sqrt{12}$ 6

51. $\sqrt{3} \cdot \sqrt{27}$ 9

52. $\sqrt{5} \cdot \sqrt{15}$ $5\sqrt{3}$

53. $\sqrt{10} \cdot \sqrt{6}$ $2\sqrt{15}$

54. $\sqrt{7} \cdot \sqrt{21}$ $7\sqrt{3}$

55. $\sqrt[3]{4} \cdot \sqrt[3]{2}$ 2

56. $\sqrt[3]{9} \cdot \sqrt[3]{3}$ 3

57. $\sqrt[4]{2} \cdot \sqrt[4]{8}$ 2

58. $\sqrt[4]{9} \cdot \sqrt[4]{9}$ 3

59. $\sqrt{7}(1 - \sqrt{2})$
$\sqrt{7} - \sqrt{14}$

60. $\sqrt{3}(\sqrt{5} - 3)$
$\sqrt{15} - 3\sqrt{3}$

61. $\sqrt{6}(\sqrt{12} + 8)$
$6\sqrt{2} + 8\sqrt{6}$

62. $\sqrt{2}(\sqrt{14} + 3)$
$2\sqrt{7} + 3\sqrt{2}$

63. $\sqrt[3]{2}(\sqrt[3]{4} + 5)$
$2 + 5\sqrt[3]{2}$

64. $\sqrt[3]{4}(\sqrt[3]{2} - 3)$
$2 - 3\sqrt[3]{4}$

65. $\sqrt[4]{2}(6 + \sqrt[4]{8})$
$6\sqrt[4]{2} + 2$

66. $\sqrt[4]{4}(9 - \sqrt[4]{4})$
$9\sqrt[4]{4} - 2$

67. $(\sqrt{2} - 1)(\sqrt{2} + 3)$ $-1 + 2\sqrt{2}$

68. $(\sqrt{5} + \sqrt{2})(\sqrt{5} - \sqrt{3})$ $5 - \sqrt{15} + \sqrt{10} - \sqrt{6}$

69. $(\sqrt{2} + 1)(\sqrt{3} - 5)$
$\sqrt{6} - 5\sqrt{2} + \sqrt{3} - 5$

70. $(\sqrt{6} + 5)(\sqrt{6} + 3)$
$21 + 8\sqrt{6}$

71. $(1 + \sqrt{11})(1 - \sqrt{11})$
-10

72. $(\sqrt{7} + 3)(\sqrt{7} - 3)$
-2

73. $(\sqrt{10} + \sqrt{5})(\sqrt{10} - \sqrt{5})$ 5

74. $(\sqrt{2} + \sqrt{3})(\sqrt{2} - \sqrt{3})$ -1

75. $(\sqrt[3]{4} + 5)(\sqrt[3]{3} + 2)$
$\sqrt[3]{12} + 2\sqrt[3]{4} + 5\sqrt[3]{3} + 10$

76. $(\sqrt[4]{6} - 3)(\sqrt[4]{2} + 7)$
$\sqrt[4]{12} + 7\sqrt[4]{6} - 3\sqrt[4]{2} - 21$

77. $(\sqrt[3]{2} - 1)^2$ $\sqrt[3]{4} - 2\sqrt[3]{2} + 1$

78. $(\sqrt[5]{6} + 3)^2$ $\sqrt[5]{36} + 6\sqrt[5]{6} + 9$

79. $(\sqrt{13} + 2)^2$
$4\sqrt{13} + 17$

80. $(\sqrt{7} + 3)^2$
$6\sqrt{7} + 16$

81. $(3 - \sqrt{8})^2$
$17 - 12\sqrt{2}$

82. $(4 - \sqrt{12})^2$
$28 - 16\sqrt{3}$

83. $\sqrt{x}(\sqrt{x} + 5)$
$x + 5\sqrt{x}$

84. $\sqrt{x}(3 - \sqrt{x})$
$3\sqrt{x} - x$

85. $(\sqrt{x} + 1)(\sqrt{x} - 3)$
$x - 2\sqrt{x} - 3$

86. $(\sqrt{u} - 3)(\sqrt{u} - 4)$
$u - 7\sqrt{u} + 12$

87. $(3 + \sqrt{x})^2$
$x + 6\sqrt{x} + 9$

88. $(5 - \sqrt{v})^2$
$25 - 10\sqrt{v} + v$

89. $(2\sqrt{x} - 3)(2\sqrt{x} + 3)$
$4x - 9$

90. $(4 - 3\sqrt{t})(4 + 3\sqrt{t})$
$16 - 9t$

In Exercises 91–98, find the conjugate of the expression. Then find the product of the expression and its conjugate. See Example 9.

91. $4 + \sqrt{5}$
$4 - \sqrt{5}, 11$

92. $\sqrt{7} - 3$
$\sqrt{7} + 3, -2$

93. $\sqrt{t} - 5$
$\sqrt{t} + 5, t - 25$

94. $5 + \sqrt{y}$
$5 - \sqrt{y}, 25 - y$

95. $\sqrt{15} - \sqrt{7}$
$\sqrt{15} + \sqrt{7}, 8$

96. $\sqrt{10} + \sqrt{2}$
$\sqrt{10} - \sqrt{2}, 8$

97. $\sqrt{u} - \sqrt{2}$
$\sqrt{u} + \sqrt{2}, u - 2$

98. $\sqrt{a} + \sqrt{3}$
$\sqrt{a} - \sqrt{3}, a - 3$

In Exercises 99–120, rationalize the denominator of the expression and simplify. See Examples 10 and 11.

99. $\dfrac{5}{\sqrt{14} - 2}$ $\dfrac{\sqrt{14} + 2}{2}$

100. $\dfrac{5}{\sqrt{10} - 5}$ $-\dfrac{\sqrt{10} + 5}{3}$

101. $\dfrac{16}{\sqrt{11} + 3}$ $8\sqrt{11} - 24$

102. $\dfrac{15}{\sqrt{7} + 2}$ $5\sqrt{7} - 10$

103. $\dfrac{4}{\sqrt{7} - \sqrt{3}}$ $\sqrt{7} + \sqrt{3}$

104. $\dfrac{3}{\sqrt{5} + \sqrt{6}}$ $-3\sqrt{5} + 3\sqrt{6}$

105. $\dfrac{3\sqrt{6}}{\sqrt{5} + \sqrt{6}}$ $-3\sqrt{30} + 18$

106. $\dfrac{2\sqrt{3}}{\sqrt{3} - \sqrt{7}}$ $\dfrac{3 + \sqrt{21}}{2}$

107. $\dfrac{\sqrt{5} + 1}{\sqrt{7} + 1}$ $\dfrac{\sqrt{35} - \sqrt{5} + \sqrt{7} - 1}{6}$

108. $\dfrac{2 - \sqrt{7}}{1 + \sqrt{7}}$ $\dfrac{\sqrt{7} - 3}{2}$

109. $\dfrac{2}{5 - \sqrt{y}}$ $\dfrac{10 + 2\sqrt{y}}{25 - y}$

110. $\dfrac{6}{\sqrt{x} - 1}$ $\dfrac{6\sqrt{x} + 6}{x - 1}$

111. $\dfrac{9}{\sqrt{x} + 2}$ $\dfrac{9\sqrt{x} - 18}{x - 4}$

112. $\dfrac{5}{6 + \sqrt{t}}$ $\dfrac{30 - 5\sqrt{t}}{36 - t}$

113. $\dfrac{\sqrt{y}}{7 - \sqrt{y}}$ $\dfrac{7\sqrt{y} + y}{49 - y}$

114. $\dfrac{\sqrt{a}}{4 + \sqrt{a}}$ $\dfrac{4\sqrt{a} - a}{16 - a}$

115. $\dfrac{-\sqrt{x}}{\sqrt{x} - \sqrt{2}}$ $\dfrac{-x - \sqrt{2x}}{x - 2}$

116. $\dfrac{6\sqrt{x}}{2\sqrt{x} - \sqrt{5}}$ $\dfrac{12x + 6\sqrt{5x}}{4x - 5}$

117. $\dfrac{\sqrt{x} - 5}{\sqrt{x} - 1}$ $\dfrac{x - 4\sqrt{x} - 5}{x - 1}$

118. $\dfrac{\sqrt{t} + 1}{\sqrt{t} - 1}$ $\dfrac{t + 2\sqrt{t} + 1}{t - 1}$

119. $\dfrac{x}{\sqrt{x} + \sqrt{y}}$ $\dfrac{x\sqrt{x} - x\sqrt{y}}{x - y}$

120. $\dfrac{b}{\sqrt{a} - \sqrt{b}}$ $\dfrac{b\sqrt{a} + b\sqrt{b}}{a - b}$

In Exercises 121–126, rewrite the expression as a single fraction and simplify. See Example 12.

121. $3 - \dfrac{1}{\sqrt{3}}$ $\dfrac{9 - \sqrt{3}}{3}$

122. $\dfrac{1}{\sqrt{5}} - 5$ $\dfrac{\sqrt{5} - 25}{5}$

123. $\sqrt{50} - \dfrac{6}{\sqrt{2}}$ $2\sqrt{2}$

124. $\dfrac{7}{\sqrt{3}} + \sqrt{12}$ $\dfrac{13\sqrt{3}}{3}$

125. $\dfrac{4}{\sqrt{3}} + 2$ $\dfrac{4\sqrt{3} + 6}{3}$

126. $9 + \dfrac{3}{\sqrt{2}}$ $\dfrac{18 + 3\sqrt{2}}{2}$

Graphical Interpretation In Exercises 127–130, use a graphing calculator to graph the two equations in the same viewing window. Use the graphs to verify that the expressions are equivalent. Verify your results algebraically.   See Additional Answers.

127. $y_1 = \sqrt{8x} + \sqrt{2x}$
$y_2 = 3\sqrt{2x}$

128. $y_1 = \sqrt{3} \cdot \sqrt{12x}$
$y_2 = 6\sqrt{x}$

129. $y_1 = \dfrac{2}{\sqrt{x + 1}}$
$y_2 = \dfrac{2\sqrt{x + 1}}{x + 1}$

130. $y_1 = \dfrac{x}{\sqrt{2} - 1}$
$y_2 = \left(\sqrt{2} + 1\right)x$

In Exercises 131–134, insert the correct symbol ($<$, $>$, or $=$) between the two real numbers.

131. $\sqrt{5} + \sqrt{3}$ $>$ $\sqrt{5 + 3}$

132. $\sqrt{5} - \sqrt{3}$ $<$ $\sqrt{5 - 3}$

133. 5 $>$ $\sqrt{3^2 + 2^2}$

134. 5 $=$ $\sqrt{3^2 + 4^2}$

Solving Problems

Geometry In Exercises 135–138, write expressions for the perimeter and area of the rectangle. Then simplify the expressions.

135.

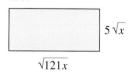

$5\sqrt{x}$

$\sqrt{121x}$

136.

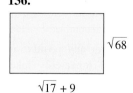

$\sqrt{68}$

$\sqrt{17} + 9$

137.

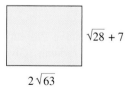

$\sqrt{28} + 7$

$2\sqrt{63}$

138.

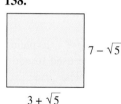

$7 - \sqrt{5}$

$3 + \sqrt{5}$

135. Perimeter: $32\sqrt{x}$; Area: $55x$

136. Perimeter: $6\sqrt{17} + 18$; Area: $18\sqrt{17} + 34$

137. Perimeter: $16\sqrt{7} + 14$; Area: $42\sqrt{7} + 84$

138. Perimeter: 20; Area: $4\sqrt{5} + 16$

139. *The Golden Section*   The ratio of the width of the Temple of Hephaestus to its height (see figure) is

$$\frac{w}{h} \approx \frac{2}{\sqrt{5}-1}. \qquad \frac{\sqrt{5}+1}{2} \approx 1.62$$

This number is called the **golden section.** Early Greeks believed that the most aesthetically pleasing rectangles were those whose sides had this ratio. Rationalize the denominator for this number. Approximate your answer, rounded to two decimal places.

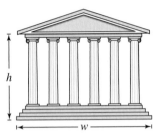

140. *Wind Chill*   A formula that can be used to determine "wind chill" is given by

$$T_{wc} = 0.0817\left(3.71\sqrt{v} + 5.81 - 0.25v\right)\\(T - 91.4) + 91.4$$

where T_{wc} is the wind chill, T is the air temperature in degrees Fahrenheit, and v is the wind speed in miles per hour. Use the formula to determine the wind chill for each combination of air temperature and wind speed in the table.

v \ T	0	5	10
10	$-20.9°$	$-14.8°$	$-8.6°$
20	$-38.5°$	$-31.4°$	$-24.3°$
30	$-47.7°$	$-40.1°$	$-32.5°$
40	$-52.5°$	$-44.7°$	$-36.8°$

v \ T	15	20	25
10	$-2.5°$	$3.7°$	$9.8°$
20	$-17.2°$	$-10.1°$	$-3.0°$
30	$-24.9°$	$-17.3°$	$-9.7°$
40	$-28.9°$	$-21.0°$	$-13.2°$

Explaining Concepts

141. 🔄 Answer parts (a)–(d) of Motivating the Chapter on page 486.

142. *Writing* ✐   In your own words, explain what it means for two square root radicals to be like radicals.   They have the same radicand.

143. Give an example of two like radical expressions. Give an example of two unlike radical expressions.   Like: $3\sqrt{2}, -5\sqrt{2}$; Unlike: $3\sqrt{2}, -5\sqrt{7}$

144. *Writing* ✐   Explain how the Distributive Property can be used to add or subtract like radicals. Give examples.   Add (or subtract) the numbers in front of the radical and keep the same radicand. $9\sqrt{7} - 3\sqrt{7} = (9 - 3)\sqrt{7} = 6\sqrt{7}$

145. *Writing* ✐   Is $\sqrt{2} + \sqrt{18}$ in simplest form? Explain.   No. $\sqrt{2} + \sqrt{18} = \sqrt{2} + 3\sqrt{2} = 4\sqrt{2}$

146. *Writing* ✐   Explain the relationship between $3 - \sqrt{2}$ and $3 + \sqrt{2}$.   These expressions are called conjugates of each other and they differ only in sign between the terms.

147. *Writing* ✐   Is the number $3/(1 + \sqrt{5})$ in simplest form? If not, explain the steps for writing it in simplest form.   No. Multiply the numerator and denominator by the conjugate of the denominator.

148. *Writing* ✐   Square the real number $3/\sqrt{2}$. Is this equivalent to rationalizing the denominator? Explain.   No. Squaring yields $\frac{9}{2}$, whereas rationalizing the denominator yields $3\sqrt{2}/2$.

149. Enter any positive real number in your calculator and repeatedly take the square root. What real number does the display appear to be approaching?   1

9.4 Radical Equations and Applications

Tony Freeman/PhotoEdit

What You Should Learn

① Use the Squaring Property of Equality to solve radical equations.

② Solve application problems using radical equations.

Why You Should Learn It

Radical equations can be used to model and solve real-life problems. For instance, in Exercise 90 on page 528, a radical equation is used to estimate the total monthly cost of daily flights between Chicago and Denver.

① **Use the Squaring Property of Equality to solve radical equations.**

Solving Radical Equations

In this section, you will study techniques for solving equations *and* applications involving radicals. This chapter gives, in short, a picture of algebra. In other words, the algebra of radicals consists of the following five types of problems.

Definitions of roots and radicals, and techniques for evaluating radicals	Section 9.1
Simplifying radicals and radical expressions	Section 9.2
Operations with radicals	Section 9.3
Solving equations involving radicals	Section 9.4
Applications involving radicals	Section 9.4

Remember that the algebra of *linear equations* has the same five types of problems as do the algebra of *polynomials* and the algebra of *rational expressions*.

A **radical equation** is an equation that contains one or more radicals with variable radicands. Here are some examples.

$$\sqrt{x} = 5, \quad \sqrt{3x - 2} = 7, \quad \text{and} \quad \sqrt{x + 3} = \sqrt{7 - x}$$

Solving radical equations is somewhat like solving equations that contain fractions—first try to get rid of the radicals and obtain a linear or quadratic equation. Then, solve the equation using the standard procedures. For square root radicals, the following property plays a key role.

Squaring Property of Equality

Let a and b be real numbers, variables, or algebraic expressions. If $a = b$, then it follows that $a^2 = b^2$. This operation is called **squaring each side of an equation.**

Note how the squaring property is used in the following illustration.

$$\sqrt{x} = 5 \qquad \text{Original equation}$$
$$\left(\sqrt{x}\right)^2 = 5^2 \qquad \text{Square each side.}$$
$$x = 25 \qquad \text{Simplify.}$$

To use the Squaring Property of Equality to solve a radical equation, first try to isolate the radical on one side of the equation.

Example 1   Solving a Radical Equation with One Radical

Solve $\sqrt{2x+1} - 2 = 3$.

Solution

$$\sqrt{2x+1} - 2 = 3 \qquad \text{Write original equation.}$$
$$\sqrt{2x+1} = 5 \qquad \text{Isolate the radical.}$$
$$\left(\sqrt{2x+1}\right)^2 = (5)^2 \qquad \text{Square each side.}$$
$$2x+1 = 25 \qquad \text{Simplify.}$$
$$2x = 24 \qquad \text{Subtract 1 from each side.}$$
$$x = 12 \qquad \text{Divide each side by 2.}$$

Check

$$\sqrt{2x+1} - 2 = 3 \qquad \text{Write original equation.}$$
$$\sqrt{2(12)+1} - 2 \overset{?}{=} 3 \qquad \text{Substitute 12 for } x.$$
$$\sqrt{25} - 2 \overset{?}{=} 3 \qquad \text{Simplify.}$$
$$5 - 2 = 3 \qquad \text{Solution checks. } \checkmark$$

So, the solution is $x = 12$.

Study Tip

You will see in this section that squaring each side of an equation often introduces *extraneous solutions*. When you use this procedure, it is critical that you check each solution in the original equation.

Some radical equations have no solution, as shown in the next example.

Example 2   A Radical with No Solution

Solve $\sqrt{3x} = -9$.

Solution

$$\sqrt{3x} = -9 \qquad \text{Write original equation.}$$
$$\left(\sqrt{3x}\right)^2 = (-9)^2 \qquad \text{Square each side.}$$
$$3x = 81 \qquad \text{Simplify.}$$
$$x = 27 \qquad \text{Divide each side by 3.}$$

Check

$$\sqrt{3x} = -9 \qquad \text{Write original equation.}$$
$$\sqrt{3(27)} \overset{?}{=} -9 \qquad \text{Substitute 27 for } x.$$
$$9 \neq -9 \qquad \text{Solution does not check. } ✗$$

So, $x = 27$ is an *extraneous* solution, which means that the original equation has no solution.

Study Tip

Another way to see that the equation in Example 2 has no solution is to observe that the right side of the original equation is negative but the left side *cannot* be negative.

Example 3   A Radical Equation with Two Radicals

Solve the radical equation.

$$\sqrt{5x + 3} = \sqrt{x + 11}$$

Solution

$\sqrt{5x + 3} = \sqrt{x + 11}$	Write original equation.
$\left(\sqrt{5x + 3}\right)^2 = \left(\sqrt{x + 11}\right)^2$	Square each side.
$5x + 3 = x + 11$	Simplify.
$5x = x + 8$	Subtract 3 from each side.
$4x = 8$	Subtract x from each side.
$x = 2$	Divide each side by 4.

Check

$\sqrt{5x + 3} = \sqrt{x + 11}$	Write original equation.
$\sqrt{5(2) + 3} \stackrel{?}{=} \sqrt{2 + 11}$	Substitute 2 for x.
$\sqrt{13} = \sqrt{13}$	Solution checks. ✓

So, the solution is $x = 2$.

Study Tip

In Example 4 it is necessary to isolate the radicals on each side of the equal sign. If both sides of the equation were squared before isolating the radicals, the resulting equation would still contain a radical, which would require isolating the radical and squaring each side a second time.

Example 4   A Radical Equation with Two Radicals

Solve the radical equation.

$$\sqrt{6x - 4} - 2\sqrt{4 - x} = 0$$

Solution

$\sqrt{6x - 4} - 2\sqrt{4 - x} = 0$	Write original equation.
$\sqrt{6x - 4} = 2\sqrt{4 - x}$	Isolate radicals.
$6x - 4 = 2^2(4 - x)$	Square each side.
$6x - 4 = 16 - 4x$	Distributive Property
$6x = 20 - 4x$	Add 4 to each side.
$10x = 20$	Add $4x$ to each side.
$x = 2$	Divide each side by 10.

Emphasize the necessity of checking any solution obtained by squaring each side of an equation.

Additional Example

Solve $\sqrt{3x} + \sqrt{2x - 5} = 0$.

Answer:

No solution

Check

$\sqrt{6x - 4} - 2\sqrt{4 - x} = 0$	Write original equation.
$\sqrt{6(2) - 4} - 2\sqrt{4 - 2} \stackrel{?}{=} 0$	Substitute 2 for x.
$\sqrt{8} - 2\sqrt{2} \stackrel{?}{=} 0$	Simplify.
$2\sqrt{2} - 2\sqrt{2} = 0$	Solution checks. ✓

So, the solution is $x = 2$.

Remind students that $(a + b)^2 = a^2 + 2ab + b^2$. Squaring a binomial incorrectly is a common cause of errors in equations involving radicals.

The next two examples show how squaring each side of an equation can yield a *nonlinear* equation.

Example 5   Solving a Radical Equation

$1 - 2x = \sqrt{x}$	Original equation
$(1 - 2x)^2 = \left(\sqrt{x}\right)^2$	Square each side.
$1 - 4x + 4x^2 = x$	Square of a binomial
$4x^2 - 5x + 1 = 0$	Write in standard form.
$(4x - 1)(x - 1) = 0$	Factor.
$4x - 1 = 0 \quad\Longrightarrow\quad x = \frac{1}{4}$	Set 1st factor equal to 0.
$x - 1 = 0 \quad\Longrightarrow\quad x = 1$	Set 2nd factor equal to 0.

Check First Solution

$$1 - 2x = \sqrt{x}$$
$$1 - 2\left(\tfrac{1}{4}\right) \overset{?}{=} \sqrt{\tfrac{1}{4}}$$
$$1 - \tfrac{1}{2} \overset{?}{=} \tfrac{1}{2}$$
$$\tfrac{1}{2} = \tfrac{1}{2} \checkmark$$

Check Second Solution

$$1 - 2x = \sqrt{x}$$
$$1 - 2(1) \overset{?}{=} \sqrt{1}$$
$$1 - 2 \overset{?}{=} 1$$
$$-1 \neq 1 ✗$$

From the check, you can see that $x = 1$ is an extraneous solution. So, the only solution is $x = \frac{1}{4}$.

Example 6   Solving a Radical Equation

$\sqrt{6x + 1} = x - 1$	Original equation
$\left(\sqrt{6x + 1}\right)^2 = (x - 1)^2$	Square each side.
$6x + 1 = x^2 - 2x + 1$	Square of a binomial
$0 = x^2 - 8x$	Write in standard form.
$0 = x(x - 8)$	Factor.
$x = 0 \quad\Longrightarrow\quad x = 0$	Set 1st factor equal to 0.
$x - 8 = 0 \quad\Longrightarrow\quad x = 8$	Set 2nd factor equal to 0.

Check First Solution

$$\sqrt{6x + 1} = x - 1$$
$$\sqrt{6(0) + 1} \overset{?}{=} 0 - 1$$
$$\sqrt{1} \overset{?}{=} -1$$
$$1 \neq -1 ✗$$

Check Second Solution

$$\sqrt{6x + 1} = x - 1$$
$$\sqrt{6(8) + 1} \overset{?}{=} 8 - 1$$
$$\sqrt{49} \overset{?}{=} 7$$
$$7 = 7 \checkmark$$

From the check, you can see that $x = 0$ is an extraneous solution. So, the only solution is $x = 8$.

Technology: Tip

You can use a graphing approach to determine the number of solutions of a radical equation.

Using the equation in Example 5, set each side of the equation equal to y.

$$y_1 = 1 - 2x$$
$$y_2 = \sqrt{x}$$

Graph the equations with your graphing calculator. The number of intersections of the two graphs is the number of solutions of the equation. The x-value of the intersection is the solution.

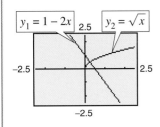

From the graph you can see that there is one intersection point at $\left(\frac{1}{4}, \frac{1}{2}\right)$. So, there is one solution, $x = \frac{1}{4}$.

2 Solve application problems using radical equations.

Applications of Radicals

A common use of radicals occurs in applications involving right triangles. Recall that a right triangle is one that contains a right (or 90°) angle, as shown in Figure 9.4. The relationship among the three sides of a right triangle is described by the **Pythagorean Theorem,** which says that if a and b are the lengths of the legs (the two sides that form the right angle) and c is the length of the hypotenuse (the side across from the right angle), then

$$c^2 = a^2 + b^2 \qquad \text{Pythagorean Theorem}$$

$$c = \sqrt{a^2 + b^2}. \qquad \text{Take square root of each side.}$$

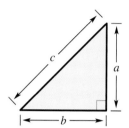

Figure 9.4

Example 7   Dimensions of a Softball Diamond

A softball diamond has the shape of a square with 60-foot sides, as shown in Figure 9.5. The catcher is 4 feet behind home plate. How far does the catcher have to throw to reach second base?

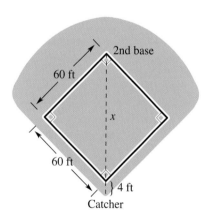

Figure 9.5

Study Tip

The Pythagorean Theorem can also be used to find the lengths of the legs of a right triangle.

$$a = \sqrt{c^2 - b^2}$$

$$b = \sqrt{c^2 - a^2}$$

Solution

In Figure 9.5, let x be the hypotenuse of a right triangle with 60-foot legs. So, by the Pythagorean Theorem, you have the following.

$$x = \sqrt{60^2 + 60^2} \qquad \text{Pythagorean Theorem}$$

$$= \sqrt{7200} \qquad \text{Simplify.}$$

$$\approx 84.9 \text{ feet} \qquad \text{Use a calculator.}$$

The distance from home plate to second base is approximately 84.9 feet. Because the catcher is 4 feet behind home plate, the catcher must make a throw of

$$x + 4 \approx 84.9 + 4$$

$$= 88.9 \text{ feet.}$$

Check this solution in the original statement of the problem.

The Pythagorean Theorem can be used to establish the Distance Formula for finding the distance between two points in the coordinate plane.

The Distance Formula

The distance d between the two points (x_1, y_1) and (x_2, y_2) in a coordinate plane is

$$d = \sqrt{(x_2 - x_1)^2 + (y_2 - y_1)^2}.$$

Example 8   The Distance Formula

Find the distance between the points $(-1, -5)$ and $(2, -2)$.

Solution

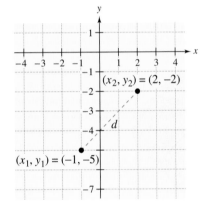

Let $(x_1, y_1) = (-1, -5)$ and $(x_2, y_2) = (2, -2)$, as shown in Figure 9.6. Then apply the Distance Formula as follows.

$d = \sqrt{(x_2 - x_1)^2 + (y_2 - y_1)^2}$	Distance Formula
$\quad = \sqrt{[2 - (-1)]^2 + [-2 - (-5)]^2}$	Substitute for x_1, y_1, x_2, and y_2.
$\quad = \sqrt{3^2 + 3^2}$	Simplify.
$\quad = \sqrt{18}$	Simplify.
$\quad \approx 4.24$	Use a calculator.

So, the distance between the two points is about 4.24 units.

Figure 9.6

Notice in Example 8 that you could have chosen (x_1, y_1) to be the point $(2, -2)$ and (x_2, y_2) to be the point $(-1, -5)$. The results would have been the same.

$d = \sqrt{(-1 - 2)^2 + [-5 - (-2)]^2}$	Distance Formula
$\quad = \sqrt{(-3)^2 + (-3)^2}$	Simplify.
$\quad = \sqrt{18}$	Simplify.
$\quad \approx 4.24$	Use a calculator.

Technology: Discovery

Try solving the equation

$$\sqrt{x} - \sqrt{2x + 1} = 0$$

graphically. First isolate the radicals, then use a graphing calculator to graph each side of the equation in the same viewing window. Do the graphs intersect? Does the equation have a solution? What does an extraneous solution mean graphically?   See Technology Answers.

9.4 Exercises

Review   *Concepts, Skills, and Problem Solving*

Keep mathematically in shape by doing these exercises *before* the problems of this section.

Properties and Definitions

1. *Writing* Explain how to determine the domain of the following rational expression.

$$\frac{10}{x+2}.$$

2. *Writing* Explain the excluded value $(x \neq -2)$ in the following.

$$\frac{x^2 - x - 6}{x^2 - 4} = \frac{(x+2)(x-3)}{(x+2)(x-2)} = \frac{x-3}{x-2}, \quad x \neq -2$$

Simplifying Expressions

In Exercises 3–8, perform the operation and/or simplify.

3. $\dfrac{15x^2}{10x} \quad \dfrac{3x}{2}, x \neq 0$

4. $\dfrac{8x^2 + 4x}{2x + 1} \quad 4x, x \neq -\dfrac{1}{2}$

5. $\dfrac{5}{x+1} \cdot \dfrac{x+1}{25}$
$\dfrac{1}{5}, x \neq -1$

6. $\dfrac{r}{r-1} \div \dfrac{r^2}{r^2-1}$
$\dfrac{r+1}{r}, r \neq 1$

7. $6 - \dfrac{5}{x+3} \quad \dfrac{6x+13}{x+3}$

8. $\dfrac{\left(\dfrac{x}{2} - 1\right)}{(x-2)} \quad \dfrac{1}{2}, x \neq 2$

Graphing Equations

In Exercises 9 and 10, graph the equation and identify any intercepts. Use a graphing calculator to verify your results.   See Additional Answers.

9. $y = 3 - \frac{2}{3}x$

10. $y + 5 = 2x$

1. The domain is the set of all real numbers for which the expression is defined. Because $10/(x+2)$ is not defined for $x = -2$, the domain is all real numbers except -2.

2. $x = -2$ is in the domain of the reduced fraction, but not in the domain of the original fraction.

Developing Skills

In Exercises 1–4, determine whether each value of x is a solution of the equation.

Equation	Values of x
1. $\sqrt{x} - 6 = 0$	(a) $x = -1$ (b) $x = -36$
(a) Not a solution	(c) $x = 36$ (d) $x = 6$
(b) Not a solution (c) Solution (d) Not a solution	
2. $\sqrt{2x} - 3 = 0$	(a) $x = \frac{9}{2}$ (b) $x = -\frac{9}{2}$
(a) Solution	(c) $x = 0$ (d) $x = \frac{3}{2}$
(b) Not a solution (c) Not a solution (d) Not a solution	
3. $x = \sqrt{2x+3}$	(a) $x = -1$ (b) $x = 2$
(a) Not a solution	(c) $x = 8$ (d) $x = 3$
(b) Not a solution (c) Not a solution (d) Solution	
4. $\sqrt{3x+4} = 2\sqrt{x}$	(a) $x = -2$ (b) $x = 2$
(a) Not a solution	(c) $x = 4$ (d) $x = 0$
(b) Not a solution (c) Solution (d) Not a solution	

In Exercises 5–56, solve the equation. (Some of the equations have no solution.) See Examples 1–6.

5. $\sqrt{x} = 4 \quad 16$

6. $\sqrt{t} = 9 \quad 81$

7. $\sqrt{x} = 10 \quad 100$

8. $\sqrt{x} = 3 \quad 9$

9. $\sqrt{y} - 5 = 0 \quad 25$

10. $\sqrt{t} - 12 = 0 \quad 144$

11. $\sqrt{u} + 3 = 0$
No solution

12. $\sqrt{y} + 10 = 0$
No solution

13. $8 - \sqrt{t} = 3$
25

14. $12 - \sqrt{s} = 25$
No solution

15. $\sqrt{x+4} = 3 \quad 5$

16. $\sqrt{x-2} = 5 \quad 27$

17. $\sqrt{10x} = 100 \quad 1000$

18. $\sqrt{8x} = 4 \quad 2$

19. $\sqrt{3x} - 4 = -7$
No solution

20. $\sqrt{2x} + 5 = 1$
No solution

21. $\sqrt{3x-2} = 4 \quad 6$

22. $\sqrt{5x-1} = 8 \quad 13$

23. $\sqrt{3y+5} = 2 \quad -\frac{1}{3}$

24. $\sqrt{5z-2} = 6 \quad \frac{38}{5}$

25. $\sqrt{x-2} + 1 = 7 \quad 38$

26. $\sqrt{3x+4} - 2 = 3 \quad 7$

27. $\sqrt{4x+3} - 6 = -5 \quad -\frac{1}{2}$

28. $\sqrt{5x-4} - 2 = 4 \quad 8$

29. $\sqrt{1-4x} - 3 = 2$
-6

30. $\sqrt{3-2x} + 1 = 8$
-23

31. $5\sqrt{x+1} = 6 \quad \frac{11}{25}$

32. $2\sqrt{x+3} = 5 \quad \frac{13}{4}$

33. $\sqrt{3x+4} = \sqrt{5x-2} \quad 3$

34. $\sqrt{6-7x} = \sqrt{4x+17} \quad -1$

35. $\sqrt{x+3} = \sqrt{4x-3} \quad 2$

36. $\sqrt{x+8} = \sqrt{5x-4} \quad 3$

37. $\sqrt{5x - 1} = 3\sqrt{x}$
No solution

38. $\sqrt{2x + 3} = 2\sqrt{x}$
$\frac{3}{2}$

39. $\sqrt{3x - 4} = 2\sqrt{x}$
No solution

40. $\sqrt{u - 4} = \sqrt{2u}$
No solution

41. $\sqrt{3t + 11} = 5\sqrt{t}$ $\frac{1}{2}$

42. $\sqrt{5 - 4u} = 4\sqrt{u}$ $\frac{1}{4}$

43. $2\sqrt{y + 1} = \sqrt{3y + 6}$ 2

44. $2\sqrt{x + 4} = 3\sqrt{x - 1}$ 5

45. $\sqrt{x^2 + 5} = x + 1$ 2

46. $\sqrt{x^2 - 2} = x - 1$ $\frac{3}{2}$

47. $\sqrt{x} = 2 - x$ 1

48. $6 - x = \sqrt{x}$ 4

49. $x - 5 = 4\sqrt{x}$ 25

50. $5\sqrt{x} = x + 4$ 1, 16

51. $x = \sqrt{20 - x}$ 4

52. $x = \sqrt{6 - x}$ 2

53. $x = \sqrt{18 - 3x}$ 3

54. $x = \sqrt{77 - 4x}$ 7

55. $\sqrt{6x + 7} = x + 2$
$-1, 3$

56. $\sqrt{4x + 17} = x + 3$
2

In Exercises 57–60, use a graphing calculator to graph the equation and estimate the *x*-intercept(s). Set $y = 0$ and solve the resulting radical equation. Compare the result with the *x*-intercept(s) of the graph.
See Additional Answers. They are the same.

57. $y = \sqrt{x - 1} - 2$ 5

58. $y = 6 - \sqrt{3x}$ 12

59. $y = x - \sqrt{4x + 5}$ 5

60. $y = 2\sqrt{x} - x + 3$ 9

In Exercises 61–66, use the Pythagorean Theorem to solve for *x*. Round your answer to two decimal places if necessary.

61.

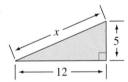

62.

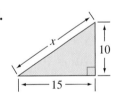

63.

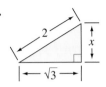

64.

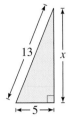

65.

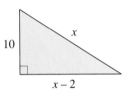

66.

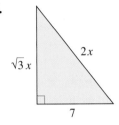

In Exercises 67–74, find the distance between the two points. Round your answer to two decimal places if necessary. See Example 8.

67. $(1, 2), (5, 5)$
5

68. $(2, 5), (7, 1)$
$\sqrt{41} \approx 6.40$

69. $(-5, 4), (3, -2)$
10

70. $(1, -2), (-4, 8)$
$5\sqrt{5} \approx 11.18$

71. $(3, -2), (4, 6)$
$\sqrt{65} \approx 8.06$

72. $(1, 5), (2, -6)$
$\sqrt{122} \approx 11.05$

73. $(-5, 2), (-2, 6)$
5

74. $(-1, 2), (7, -2)$
$4\sqrt{5} \approx 8.94$

61. 13   **62.** $5\sqrt{13} \approx 18.03$

63. 1   **64.** 12

65. 26   **66.** 7

Solving Problems

Height In Exercises 75 and 76, the time *t* in seconds for a free-falling object to fall *d* feet is given by

$$t = \sqrt{\frac{d}{16}}.$$

75. A construction worker drops a nail from a building and observes it strike a water puddle after approximately 3 seconds. Estimate the height from which the nail was dropped.   144 feet

76. A construction worker drops a nail from a building and observes it strike a water puddle after approximately 5 seconds. Estimate the height from which the nail was dropped.   400 feet

Free-Falling Object In Exercises 77 and 78, use the equation for the velocity of a free-falling object

$$v = \sqrt{2gh}$$

where *v* is measured in feet per second, $g = 32$ feet per second squared, and *h* is height in feet.

77. An object strikes the ground with a velocity of 75 feet per second. Estimate the height from which it was dropped.   87.89 feet

78. An object strikes the ground with a velocity of 100 feet per second. Estimate the height from which it was dropped.   156.25 feet

Pendulum Length   In Exercises 79 and 80, the time *t* (in seconds) for a pendulum of length *L* (in feet) to go through one complete cycle, both forward and back (its period), is given by

$$t = 2\pi\sqrt{\frac{L}{32}}.$$

79. How long is the pendulum of a grandfather clock with a period of 2 seconds (see figure)?   3.24 feet

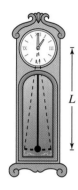

L

80. How long is the pendulum of a mantel clock with a period of 0.8 second?   0.52 foot

81. ▲ *Geometry*   A ladder is 15 feet long, and the bottom of the ladder is 3 feet from the side of a house (see figure). How far does the ladder reach up the side of the house? $6\sqrt{6} \approx 14.70$ feet

15 ft

3 ft

82. ▲ *Geometry*   A 39-foot guy wire on a sailboat is attached to the top of the mast and to the deck 15 feet from the base of the mast (see figure). How tall is the mast?   36 feet

39 ft

15 ft

83. ▲ *Geometry*   A 12-foot plank is used to brace a basement wall during construction of a home. The plank is nailed to the wall 4 feet above the floor (see figure). Find the slope of the plank. $-\dfrac{\sqrt{2}}{4}$

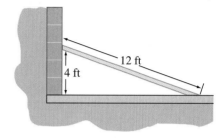

4 ft

12 ft

84. ▲ *Geometry*   A volleyball court is 30 feet wide and 60 feet long. Find the length of the diagonal of the court. $30\sqrt{5} \approx 67.08$ feet

85. ▲ *Geometry*   A baseball diamond is a square that is 90 feet on a side (see figure). Determine the distance between first base and third base. $90\sqrt{2} \approx 127.28$ feet

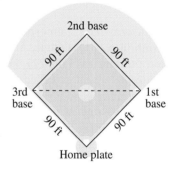

2nd base

90 ft   90 ft

3rd base   1st base

90 ft   90 ft

Home plate

86. ▲ *Geometry*   The distance between Memphis and New Orleans is 410 miles. The distance between Memphis and Chattanooga is 317 miles (see figure). Approximate the distance between Chattanooga and New Orleans. $\sqrt{268{,}589} \approx 518.3$ miles

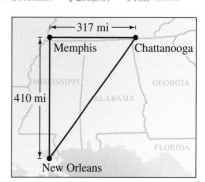

317 mi

Memphis   Chattanooga

MISSISSIPPI   GEORGIA

410 mi   ALABAMA

FLORIDA

New Orleans

87. ▲ *Geometry* For a television set with a 25-inch screen, it is actually the diagonal of the screen that is 25 inches. Find the dimensions of the square screen.

88. ▲ *Geometry* For a television set with a 32-inch screen, it is actually the diagonal of the screen that is 32 inches. Find the dimensions of the square screen.

89. *Demand* The demand equation for a video game is

$$p = 40 - \sqrt{x - 1}$$

where x is the number of units demanded per day and p is the price per game. Find the demand when the price is set at \$34.70. 29 units

90. *Airline Passengers* An airline offers daily flights between Chicago and Denver. The total monthly cost C (in millions of dollars) of these flights is modeled by

$$C = \sqrt{0.2x + 1}, \, x \geq 0$$

where x is measured in thousands of passengers. The total cost of the flights for June is 2.5 million dollars. Approximately how many passengers flew in June? 26.25 thousand passengers

91. *Analyzing Data* The atmospheric pressure decreases with increasing altitude. At sea level, the average air pressure is 1.033227 kilograms per square centimeter, which is called 1 atmosphere. Variations in weather conditions cause changes in the atmospheric pressure of up to ±5%. The table shows the pressure p (in atmospheres) at different altitudes h (in kilometers).

h	0	5	10	15	20	25
p	1	0.55	0.25	0.12	0.06	0.02

A model for this data is $p = 1 - 0.206\sqrt{h}$.

(a) ▦ Use a graphing calculator to plot the data and graph the model in the same viewing window. See Additional Answers.

(b) Approximate the altitude when the atmospheric pressure is 0.4 atmosphere. 8.5 kilometers

92. *Exploration* Pythagorean triples are sets of positive integers a, b, and c that satisfy the Pythagorean Theorem $a^2 + b^2 = c^2$. Consider the positive integers u and v where $u > v$, one is even and the other is odd, and their only common factor is 1. Select some values of u and v that satisfy these criteria, and then generate the numbers

$$a = u^2 - v^2$$
$$b = 2uv$$
$$c = u^2 + v^2.$$

In each case, verify that a, b, and c are a Pythagorean triple.

87. $\dfrac{25\sqrt{2}}{2} \approx 17.68$ inches   **88.** $16\sqrt{2} \approx 22.63$ inches

92. Answers will vary. Sample answer:
   $u = 2, v = 1; a = 3, b = 4, c = 5;$
   $3^2 + 4^2 = 5^2$
   $9 + 16 = 25$
   $25 = 25$

Explaining Concepts

93. ⊘ Answer parts (e)–(h) of Motivating the Chapter on page 486.

94. *Writing* In your own words, describe a radical equation. A radical equation is an equation that contains one or more radicals with variable radicands.

95. *Writing* In your own words, describe the steps that can be used to solve a radical equation.

96. *Writing* Is $x = 25$ a solution of $\sqrt{x} = -5$? Explain. No. The principal square root of a number is positive.

95. Isolate a radical on one side of the equation and then square each side of the equation. If there are more radicals, continue the process. When the radicals have been eliminated, solve the resulting equation and check your results.

97. *Writing* Give two reasons why it is important to check solutions of a radical equation. To check for any errors and to identify extraneous solutions

98. State the Pythagorean Theorem. If a and b are the lengths of the legs of a right triangle and c is the length of the hypotenuse, then $a^2 + b^2 = c^2$.

99. *Writing* Can a right triangle be isosceles (have two sides of the same length)? Explain. Yes. The two legs can be of the same length l and the hypotenuse of length $\sqrt{2}\,l$.

What Did You Learn?

Key Terms

square root, *p. 488*
cube root, *p. 488*
*n*th root of *a*, *p. 488*
radical symbol, *p. 489*
index, *p. 489*
radicand, *p. 489*

perfect square, *p. 490*
radical expressions, *p. 497*
rationalizing the
   denominator, *p. 501*
like radicals, *p. 509*

conjugates, *p. 512*
radical equation, *p. 519*
squaring each side of an
   equation, *p. 519*
Pythagorean Theorem, *p. 523*

Key Concepts

9.1 ○ Definition of *n*th root of a number

Let a and b be real numbers and let n be an integer such that $n \geq 2$. If

$$a = b^n$$

then b is an *n*th root of a. If $n = 2$, the root is a square root. If $n = 3$, the root is a cube root.

9.1 ○ Principal *n*th root of a number

Let a be a real number that has at least one (real number) *n*th root. The principal *n*th root of a is the *n*th root that has the same sign as a, and it is denoted by the radical symbol

$$\sqrt[n]{a}.$$

The positive integer n is the index of the radical, and the number a is the radicand. If $n = 2$, omit the index and write $\sqrt{a}$ rather than $\sqrt[2]{a}$.

9.2 ○ Product Rule for Radicals

Let a and b be real numbers, variables, or algebraic expressions. If the *n*th roots of a and b are real, the following property is true.

$$\sqrt[n]{ab} = \sqrt[n]{a}\,\sqrt[n]{b}$$

9.2 ○ The square root of x^2

If x is a real number, then

$$\sqrt{x^2} = |x|.$$

For the special case in which you know that x is a nonnegative real number, you can write $\sqrt{x^2} = x$.

9.2 ○ Quotient Rule for Radicals

Let a and b be real numbers, variables, or algebraic expressions. If the *n*th roots of a and b are real, the following property is true.

$$\sqrt[n]{\frac{a}{b}} = \frac{\sqrt[n]{a}}{\sqrt[n]{b}}, \quad b \neq 0$$

9.2 ○ Simplifying radical expressions

A radical expression is said to be in simplest form if all three of the statements below are true.

1. All possible *n*th powered factors have been removed from each radical.

2. No radical contains a fraction.

3. No denominator of a fraction contains a radical.

9.4 ○ Squaring Property of Equality

Let a and b be real numbers, variables, or algebraic expressions. If $a = b$, then it follows that

$$a^2 = b^2.$$

This operation is called squaring each side of an equation.

9.4 ○ The Distance Formula

The distance d between the two points (x_1, y_1) and (x_2, y_2) in a coordinate plane is

$$d = \sqrt{(x_2 - x_1)^2 + (y_2 - y_1)^2}.$$

Review Exercises

9.1 Roots and Radicals

1 Find the nth roots of real numbers.

In Exercises 1–4, find the positive and negative square roots of the real number, if possible. (Do not use a calculator.)

1. 49 $7, -7$

2. 100 $10, -10$

3. -4 Not possible

4. -9 Not possible

In Exercises 5–8, find the nth root. (Do not use a calculator.)

5. The cube root of -125 -5

6. The cube root of -1 -1

7. The fourth root of 16 $2, -2$

8. The fourth root of 256 $4, -4$

2 Use the radical symbol to denote the nth roots of numbers.

In Exercises 9–18, find the principal nth root of the number, if possible. (Do not use a calculator.)

9. $\sqrt{121}$ 11

10. $\sqrt{-25}$ Not possible

11. $\sqrt{1.44}$ 1.2

12. $\sqrt{0.09}$ 0.3

13. $\sqrt{-\frac{1}{16}}$ Not possible

14. $-\sqrt{\frac{64}{9}}$ $-\frac{8}{3}$

15. $\sqrt[3]{-27}$ -3

16. $\sqrt[4]{16}$ 2

17. $\sqrt[3]{\frac{8}{125}}$ $\frac{2}{5}$

18. $\sqrt[4]{\frac{1}{81}}$ $\frac{1}{3}$

3 Approximate the values of expressions involving square roots using a calculator.

In Exercises 19–26, use a calculator to approximate the value of the expression. Round your answer to three decimal places.

19. $\sqrt{53}$ 7.280

20. $\sqrt{5335}$ 73.041

21. $\sqrt{\frac{7}{8}}$ 0.935

22. $-\sqrt{\frac{45}{8}}$ -2.372

23. $3 + 2\sqrt{6}$ 7.899

24. $-4 - 3\sqrt{10}$ -13.487

25. $\dfrac{5 - 3\sqrt{3}}{2}$ -0.098

26. $\dfrac{7 - 4\sqrt{2}}{4}$ 0.336

▲ *Geometry* **In Exercises 27 and 28, evaluate the expression when $x = -2$ and $y = 3$. Round your answer to two decimal places.**

27. $\sqrt{x^2 y}$

28. $\sqrt{y^2 - 5x}$

9.2 Simplifying Radicals

1 Simplify radical expressions involving constants.

In Exercises 29–40, simplify the radical.

29. $\sqrt{48}$ $4\sqrt{3}$

30. $\sqrt{72}$ $6\sqrt{2}$

31. $\sqrt{160}$ $4\sqrt{10}$

32. $\sqrt{45}$ $3\sqrt{5}$

33. $\sqrt{\frac{23}{9}}$ $\frac{\sqrt{23}}{3}$

34. $\sqrt{\frac{26}{16}}$ $\frac{\sqrt{26}}{4}$

35. $\sqrt{\frac{20}{9}}$ $\frac{2\sqrt{5}}{3}$

36. $\sqrt{\frac{27}{16}}$ $\frac{3\sqrt{3}}{4}$

37. $\sqrt[3]{24}$ $2\sqrt[3]{3}$

38. $\sqrt[4]{48}$ $2\sqrt[4]{3}$

39. $\sqrt[4]{96}$ $2\sqrt[4]{6}$

40. $\sqrt[3]{54}$ $3\sqrt[3]{2}$

2 Simplify radical expressions involving variables.

In Exercises 41–52, simplify the radical expression. Use absolute value signs if appropriate.

41. $\sqrt{36x^4}$ $6x^2$

42. $\sqrt{81z^2}$ $9|z|$

43. $\sqrt{4y^3}$ $2y\sqrt{y}$

44. $\sqrt{100u^5}$ $10u^2\sqrt{u}$

45. $\sqrt{32a^3b}$ $4a\sqrt{2ab}$

46. $\sqrt{75u^4v^2}$ $5u^2|v|\sqrt{3}$

47. $\sqrt{0.04x^2y}$ $0.2|x|\sqrt{y}$

48. $\sqrt{1.44x^2y^3}$ $1.2|x|y\sqrt{y}$

49. $\sqrt[3]{8x^6}$ $2x^2$

50. $\sqrt[4]{16y^5}$ $2y\sqrt[4]{y}$

51. $\sqrt[3]{64a^5}$ $4a\sqrt[3]{a^2}$

52. $\sqrt[4]{a^8b^{11}}$ $a^2b^2\sqrt[4]{b^3}$

3 Simplify radical expressions by rationalizing the denominator.

In Exercises 53–68, rationalize the denominator of the expression and simplify. (Assume all variables are positive.)

53. $\sqrt{\frac{3}{5}}$ $\frac{\sqrt{15}}{5}$

54. $\sqrt{\frac{7}{10}}$ $\frac{\sqrt{70}}{10}$

55. $\dfrac{6}{\sqrt{3}}$ $2\sqrt{3}$

56. $\dfrac{15}{\sqrt{5}}$ $3\sqrt{5}$

57. $\sqrt{\frac{5}{12}}$ $\frac{\sqrt{15}}{6}$

58. $\sqrt{\frac{13}{32}}$ $\frac{\sqrt{26}}{8}$

59. $\dfrac{3}{\sqrt[3]{2}}$ $\frac{3\sqrt[3]{4}}{2}$

60. $\dfrac{5}{\sqrt[3]{4}}$ $\frac{5\sqrt[3]{2}}{2}$

61. $\dfrac{3}{\sqrt{x}}$ $\frac{3\sqrt{x}}{x}$

62. $\dfrac{7}{\sqrt{t}}$ $\frac{7\sqrt{t}}{t}$

27. $\sqrt{12} \approx 3.46$

28. $\sqrt{19} \approx 4.36$

63. $\sqrt{\dfrac{11a}{b}}$ $\dfrac{\sqrt{11ab}}{b}$

64. $\sqrt{\dfrac{4y}{z}}$ $\dfrac{2\sqrt{yz}}{z}$

65. $\dfrac{\sqrt{6x^2}}{\sqrt{27y^3}}$ $\dfrac{x\sqrt{2y}}{3y^2}$

66. $\dfrac{\sqrt{10a^2}}{\sqrt{8b^2}}$ $\dfrac{a\sqrt{5}}{2b}$

67. $\sqrt[3]{\dfrac{4}{x^2}}$ $\dfrac{\sqrt[3]{4x}}{x}$

68. $\sqrt[3]{\dfrac{7}{y^5}}$ $\dfrac{\sqrt[3]{7y}}{y^2}$

9.3 Operations with Radical Expressions

① Add and subtract like radical expressions.

In Exercises 69–84, simplify the expression.

69. $7\sqrt{2} + 5\sqrt{2}$ $12\sqrt{2}$

70. $15\sqrt{15} - 7\sqrt{15}$ $8\sqrt{15}$

71. $3\sqrt{5} - 7\sqrt{3} + 11\sqrt{3}$ $3\sqrt{5} + 4\sqrt{3}$

72. $5\sqrt{11} + 6\sqrt{2} - 8\sqrt{11}$ $6\sqrt{2} - 3\sqrt{11}$

73. $3\sqrt{20} - 10\sqrt{20}$ $-14\sqrt{5}$

74. $25\sqrt{98} + 2\sqrt{98}$ $189\sqrt{2}$

75. $4\sqrt{48} + 2\sqrt{3} - 5\sqrt{12}$ $8\sqrt{3}$

76. $12\sqrt{50} - 3\sqrt{8} + \sqrt{32}$ $58\sqrt{2}$

77. $\sqrt[4]{4} + 5\sqrt[4]{4}$ $6\sqrt[4]{4}$

78. $2\sqrt[3]{7} - 6\sqrt[3]{7} + 9\sqrt[3]{7}$ $5\sqrt[3]{7}$

79. $\sqrt[5]{x} + 8\sqrt[5]{x}$ $9\sqrt[5]{x}$

80. $5\sqrt[4]{x^3} - 3\sqrt[4]{y^3} + 4\sqrt[4]{x^3}$ $9\sqrt[4]{x^3} - 3\sqrt[4]{y^3}$

81. $\sqrt{36y} - \sqrt{16y}$ $2\sqrt{y}$

82. $\sqrt{25x} + \sqrt{49x}$ $12\sqrt{x}$

83. $\sqrt{18x^3} - 3x\sqrt{2x}$ 0

84. $\sqrt{28y^5} + 4y\sqrt{7y^3}$ $6y^2\sqrt{7y}$

② Multiply radical expressions using the Distributive Property, the FOIL Method, or a special product pattern.

In Exercises 85–96, multiply and simplify.

85. $\sqrt{3}\left(\sqrt{6} - 1\right)$ $3\sqrt{2} - \sqrt{3}$

86. $\sqrt{7}\left(10 - \sqrt{7}\right)$ $10\sqrt{7} - 7$

87. $\sqrt[4]{6}\left(\sqrt[4]{2} - 1\right)$ $\sqrt[4]{12} - \sqrt[4]{6}$

88. $\sqrt[3]{5}\left(\sqrt[3]{4} + 2\right)$ $\sqrt[3]{20} + 2\sqrt[3]{5}$

89. $\left(\sqrt{3} - \sqrt{5}\right)\left(\sqrt{3} + \sqrt{5}\right)$ -2

90. $\left(\sqrt{7} - 2\right)\left(\sqrt{7} + 2\right)$ 3

91. $\left(\sqrt{5} - 2\right)^2$ $9 - 4\sqrt{5}$

92. $\left(\sqrt{3} + 1\right)^2$ $2\sqrt{3} + 4$

93. $\left(\sqrt{8} + 2\right)\left(3\sqrt{2} - 1\right)$ $10 + 4\sqrt{2}$

94. $\left(2\sqrt{3} + 10\right)\left(\sqrt{2} - 3\right)$ $2\sqrt{6} - 6\sqrt{3} + 10\sqrt{2} - 30$

95. $\left(\sqrt[5]{2} + 3\right)\left(\sqrt[5]{3} + 5\right)$ $\sqrt[5]{6} + 5\sqrt[5]{2} + 3\sqrt[5]{3} + 15$

96. $\left(\sqrt[4]{3} - 1\right)\left(\sqrt[4]{3} - 1\right)$ $\sqrt[4]{9} - 2\sqrt[4]{3} + 1$

③ Determine the conjugate of an expression and find the product of the expression and its conjugate.

In Exercises 97–100, find the conjugate of the expression. Then find the product of the expression and its conjugate.

97. $\sqrt{x} + 9$
$\sqrt{x} - 9,\, x - 81$

98. $\sqrt{y} - 15$
$\sqrt{y} + 15,\, y - 225$

99. $12 - \sqrt{t}$
$12 + \sqrt{t},\, 144 - t$

100. $20 + \sqrt{z}$
$20 - \sqrt{z},\, 400 - z$

④ Simplify quotients involving radicals by rationalizing the denominators.

In Exercises 101–104, rationalize the denominator of the expression and simplify.

101. $\dfrac{3}{\sqrt{12} - 3}$
$2\sqrt{3} + 3$

102. $\dfrac{9}{\sqrt{7} - 4}$
$-\left(\sqrt{7} + 4\right)$

103. $\dfrac{\sqrt{x} - 3}{\sqrt{x} + 3}$
$\dfrac{x - 6\sqrt{x} + 9}{x - 9}$

104. $\dfrac{3\sqrt{s} + 2}{\sqrt{s} + 1}$
$\dfrac{3s - \sqrt{s} - 2}{s - 1}$

9.4 Radical Equations and Applications

① Use the Squaring Property of Equality to solve radical equations.

In Exercises 105–114, solve the equation. (Some of the equations have no solution.)

105. $\sqrt{y} = 13$ 169

106. $\sqrt{z} = 25$ 625

107. $\sqrt{x} + 2 = 0$
No solution

108. $\sqrt{x} - 10 = 0$
100

109. $\sqrt{2t + 3} = 5$ 11

110. $\sqrt{2a - 7} = 15$ 116

111. $\sqrt{4x - 3} = \sqrt{x + 6}$ 3

112. $\sqrt{5x + 3} = \sqrt{x + 15}$ 3

113. $\sqrt{x - 4} = x - 6$
8

114. $\sqrt{2x + 7} = x + 4$
-3

⊞ **In Exercises 115 and 116, use a graphing calculator to graph the equation and estimate the x-intercept(s). Set y = 0 and solve the resulting radical equation. Compare the result with the x-intercept(s) of the graph.**   See Additional Answers. They are the same.

115. $y = \sqrt{x + 3} - 4\sqrt{x}$ $\frac{1}{5}$

116. $y = x + 2 - \sqrt{x + 8}$ 1

② Solve application problems using radical equations.

▲ *Geometry*   **In Exercises 117 and 118, use the Pythagorean Theorem to solve for *x*. Round your answer to two decimal places if necessary.**

117.   **118.**

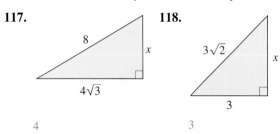

4   3

In Exercises 119–126, find the distance between the two points. Round your answer to two decimal places if necessary.

119. $(1, 4), (-2, 0)$ **120.** $(3, 8), (-3, 5)$
   5                                      $3\sqrt{5} \approx 6.71$

121. $(-5, -2), (-1, 9)$ **122.** $(6, 0), (1, -5)$
   $\sqrt{137} \approx 11.70$ $5\sqrt{2} \approx 7.07$

123. $(-4, -3), (2, -4)$ **124.** $(8, 10), (-1, -1)$
   $\sqrt{37} \approx 6.08$ $\sqrt{202} \approx 14.21$

125. $(7, 3), (2, -8)$ **126.** $(-6, 5), (9, -2)$
   $\sqrt{146} \approx 12.08$ $\sqrt{274} \approx 16.55$

Free-Falling Object   **In Exercises 127 and 128, use the equation for the velocity of a free-falling object $v = \sqrt{2gh}$, where *v* is measured in feet per second, $g = 32$ feet per second squared, and *h* is height in feet.**

127. An object strikes the ground with a velocity of 80 feet per second. Estimate the height from which it was dropped.   100 feet

128. An object strikes the ground with a velocity of 50 feet per second. Estimate the height from which it was dropped. $\frac{625}{16} \approx 39.1$ feet

129. *Period of a Pendulum*   The time *t* (in seconds) for a pendulum of length *L* (in feet) to go through one complete cycle (its period) is

$$t = 2\pi\sqrt{\frac{L}{32}}.$$

How long is the pendulum of a grandfather clock with a period of 1.75 seconds?   2.48 feet

130. *Height*   The time *t* in seconds for a free-falling object to fall *d* feet is given by

$$t = \sqrt{\frac{d}{16}}.$$

A child drops a rock from a bridge and sees it strike the water after approximately 1.5 seconds. Estimate the height of the bridge.   36 feet

131. ▲ *Geometry*   A guy wire on a radio tower is attached to the top of the tower and to an anchor 60 feet from the base of the tower (see figure). The tower is 100 feet high. How long is the wire?
   $20\sqrt{34} \approx 116.6$ feet

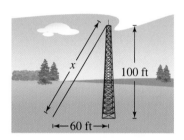

132. ▲ *Geometry*   A baseball diamond is a square that is 90 feet on a side (see figure). The right fielder catches a fly ball on the first-base line approximately 70 feet beyond first base. He then throws a runner out at third base. Determine the distance *d* between the right fielder and third base. $10\sqrt{337} \approx 183.6$ feet

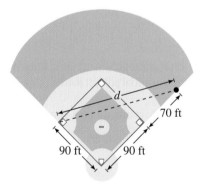

133. ▲ *Geometry*   Determine the length and width of a rectangle that has a perimeter of 70 inches and a diagonal length of 25 inches.
   20 inches × 15 inches or 15 inches × 20 inches

134. *Demand*   The demand equation for a disposable camera is $p = 12 - \sqrt{0.75x - 9}$, where *x* is the number of units demanded per day and *p* is the price per camera. Find the demand when the price is set at $6.00.   60 units per day

Chapter Test

Take this test as you would take a test in class. After you are done, check your work against the answers in the back of the book.

In Exercises 1 and 2, if possible, evaluate each expression without a calculator. If it is not possible, explain why.

1. (b) Not possible; There is no real number that can be multiplied by itself to obtain -36.

1. (a) $\sqrt{121}$ 11   (b) $\sqrt{-36}$

2. (a) $\sqrt[4]{81}$ 3   (b) $\sqrt[3]{-64}$ -4

In Exercises 3–8, simplify the radical.

3. $\sqrt{48}$ $4\sqrt{3}$

4. $\sqrt[3]{54}$ $3\sqrt[3]{2}$

5. $\sqrt{32x^2y^3}$ $4|x|y\sqrt{2y}$

6. $\sqrt{\dfrac{3x^3}{y^4}}$ $\dfrac{x\sqrt{3x}}{y^2}$

7. $\dfrac{5}{\sqrt{15}}$ $\dfrac{\sqrt{15}}{3}$

8. $\dfrac{2}{\sqrt[3]{4}}$ $\sqrt[3]{2}$

In Exercises 9–16, perform the indicated operation.

9. $10\sqrt{27} - 7\sqrt{12}$ $16\sqrt{3}$

10. $5\sqrt{3x} + 3\sqrt{75x}$ $20\sqrt{3x}$

11. $7\sqrt[3]{5} - 6\sqrt[3]{4} + \sqrt[3]{5}$
   $8\sqrt[3]{5} - 6\sqrt[3]{4}$

12. $4\sqrt{2x} - 6\sqrt{32x} + \sqrt{2x^2}$
   $\sqrt{2}x - 20\sqrt{2x}$

13. $\sqrt{3}\left(2 - \sqrt{12}\right)$ $2\sqrt{3} - 6$

14. $\sqrt[3]{5}\left(\sqrt[3]{2} + 3\right)$ $\sqrt[3]{10} + 3\sqrt[3]{5}$

15. $\left(\sqrt{6} - 3\right)\left(\sqrt{6} + 5\right)$ $2\sqrt{6} - 9$

16. $\left(1 - 2\sqrt{x}\right)\left(-2\sqrt{x}\right)$ $4x - 2\sqrt{x}$

17. In your own words, explain what *conjugate* means. Find the conjugate of the expression $\sqrt{3} - 5$. Then find the product of the expression and its conjugate.   Expressions that differ only in the sign between them are conjugates. $\sqrt{3} + 5; -22$

18. Rationalize the denominator: $\dfrac{10}{\sqrt{6} + 1}$. $2\left(\sqrt{6} - 1\right)$

In Exercises 19–22, solve the equation.

19. $\sqrt{y} = 4$ 16

20. $2\sqrt{x + 3} = 5$ $\frac{13}{4}$

21. $\sqrt{5x - 4} = \sqrt{3x + 6}$ 5

22. $2\sqrt{2x} = x + 2$ 2

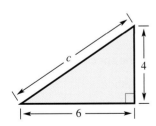

Figure for 23

23. Use the Pythagorean Theorem to solve for c in the figure. Round your answer to two decimal places. $2\sqrt{13} \approx 7.21$

24. Find the distance between the points $(-3, 8)$ and $(5, 2)$. Round your answer to two decimal places.   10

25. The demand equation for a VCR is

$$p = 100 - \sqrt{x - 25}$$

where x is the number of units demanded per day and p is the price per VCR. Find the demand when the price is set at \$90.   125 units per day

Take this test as you would take a test in class. After you are done, check your work against the answers in the back of the book.

1. Find the domain of the rational expression $\dfrac{x-5}{x+2}$.

All real numbers x such that $x \neq -2$

2. Fill in the missing factor: $\dfrac{7}{3x} = \dfrac{7\,(4x)}{12x^2}$.

In Exercises 3 and 4, simplify the expression.

3. $\dfrac{8-2x}{x^2-16}$ $-\dfrac{2}{x+4}, \; x \neq 4$

4. $\dfrac{x^2-3x-10}{x^2-4}$ $\dfrac{x-5}{x-2}, \; x \neq -2$

In Exercises 5–8, perform the indicated operation and simplify.

5. $\dfrac{c+10}{c^2}, \; c \neq 1$

6. $\dfrac{3c^2}{4(c-1)}, \; c \neq 0$

7. $\dfrac{2(x+3)}{(x+2)(x-2)}$

8. $\dfrac{3x+2}{x(x-1)}$

5. $\dfrac{c}{c-1} \cdot \dfrac{c^2+9c-10}{c^3}$

6. $\dfrac{6}{(c-1)^2} \div \dfrac{8}{c^3-c^2}$

7. $\dfrac{3}{x-2} + \dfrac{x}{4-x^2}$

8. $\dfrac{5}{x-1} - \dfrac{2}{x}$

In Exercises 9 and 10, simplify the complex fraction.

9. $\dfrac{\left(\dfrac{9}{x}\right)}{\left(\dfrac{6}{x}+2\right)}$ $\dfrac{9}{6+2x}, \; x \neq 0$

10. $\dfrac{\left(a-\dfrac{1}{a}\right)}{\left(\dfrac{1}{2}+\dfrac{1}{a}\right)}$ $\dfrac{2a^2-2}{a+2}, \; a \neq 0$

In Exercises 11–14, solve the equation.

11. $\dfrac{5}{x} + \dfrac{3}{x} = 24$ $\frac{1}{3}$

12. $\dfrac{x}{5} - \dfrac{x}{2} = 3$ -10

13. $\dfrac{1}{x} + \dfrac{2}{x-5} = 0$ $\frac{5}{3}$

14. $\dfrac{2x-1}{2x+1} = \dfrac{4}{5}$ $\frac{9}{2}$

In Exercises 15–20, solve the system of equations by the specified method.

15. Graphical method:
$$\begin{cases} x + 5y = 0 \quad (5,-1) \\ 7x + 5y = 30 \end{cases}$$

16. Graphical method:
$$\begin{cases} 3x + y = 2 \quad (1,-1) \\ 5x - y = 6 \end{cases}$$

17. Substitution:
$$\begin{cases} x - y = 0 \quad (5,5) \\ 5x - 3y = 10 \end{cases}$$

18. Substitution:
$$\begin{cases} x + 8y = 6 \quad \left(-4, \frac{5}{4}\right) \\ 2x + 4y = -3 \end{cases}$$

19. Elimination:
$$\begin{cases} 4x + 3y = 15 \quad (3,1) \\ 2x - 5y = 1 \end{cases}$$

20. Elimination:
$$\begin{cases} 2x + y = 4 \quad \left(\frac{3}{2}, 1\right) \\ 4x - 3y = 3 \end{cases}$$

21.

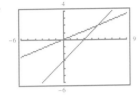

$(5, 2)$

22. Answers will vary.
  Sample answers:

  (a) $\begin{cases} x + y = 1 \\ x + y = -1 \end{cases}$

  (b) $\begin{cases} x + y = 1 \\ 2x + 2y = 2 \end{cases}$

21. ⊞ Use a graphing calculator to solve the system.

$$\begin{cases} 2x - 5y = 0 \\ x - y = 3 \end{cases}$$

22. Give examples of systems of linear equations that have (a) no solution and (b) an infinite number of solutions.

In Exercises 23–26, evaluate the expression, if possible.

23. $\sqrt{-\frac{4}{9}}$ Not possible

24. $-\sqrt{\frac{4}{9}}$ $-\frac{2}{3}$

25. $5\sqrt{144}$ 60

26. $\sqrt[3]{-125}$ -5

In Exercises 27–30, simplify the radical.

27. $-\sqrt{54}$ $-3\sqrt{6}$

28. $\sqrt{50x^3}$ $5x\sqrt{2x}$

29. $3\sqrt[3]{32u^4v^6}$ $6uv^2\sqrt[3]{4u}$

30. $\sqrt{\dfrac{32y}{9y^3}}$ $\dfrac{4\sqrt{2}}{3|y|}$

In Exercises 31–34, perform the indicated operation and simplify.

31. $5\sqrt{x} - 3\sqrt{x}$ $2\sqrt{x}$

32. $\sqrt{7}(\sqrt{7} + 2)$ $7 + 2\sqrt{7}$

33. $(4 - \sqrt{8})^2$ $24 - 16\sqrt{2}$

34. $\dfrac{8y}{\sqrt{5} - 1}$ $2y(\sqrt{5} + 1)$

In Exercises 35–38, solve the equation.

35. $2\sqrt{x} = 10$ 25

36. $\sqrt{a - 4} = 5$ 29

37. $\sqrt{2x + 7} = 3\sqrt{x}$ 1

38. $x(\sqrt{x} - 2) = 0$ 0, 4

39. On the second half of a 200-mile trip, you average 10 more miles per hour than on the first half. What is your average speed on the second half of the trip if the total time for the trip is $4\frac{1}{2}$ hours?   50 miles per hour

40. A new employee takes twice as long as an experienced employee to complete a task. Together they can complete the task in 3 hours. Determine the time it takes each employee to do the task individually.   Experienced employee: $4\frac{1}{2}$ hours; New employee: 9 hours

41. The total cost of 10 gallons of regular gasoline and 12 gallons of premium gasoline is \$31.88. Premium costs \$0.20 more per gallon than regular. Find the price per gallon of each grade of gasoline.   Regular: \$1.34, Premium: \$1.54

42. Use the Pythagorean Theorem to solve for x in the figure at the left. Round your answer to two decimal places if necessary.   8

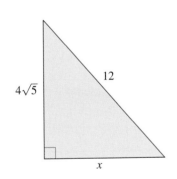

$4\sqrt{5}$ 12 x

Figure for 42

In Exercises 43 and 44, find the distance between the two points. Round your answer to two decimal places, if necessary.

43. $(-2, -1), (1, 3)$ 5

44. $(-2, 3), (5, 0)$ $\sqrt{58} \approx 7.62$

Motivating the Chapter

⚡ Vertical Motion

The mathematical model for the height h in feet of an object moving in a vertical path is given by the quadratic equation

$$h = -16t^2 + bt + c, \quad t \geq 0$$

where t is the time in seconds. This equation is also called the *position equation*. The coefficient b represents the velocity in feet per second at which the object is initially propelled upward $(b > 0)$ or downward $(b < 0)$, and the constant c represents the initial height (when $t = 0$) of the object.

See Section 10.1, Exercise 103.

a. Write the quadratic equation for an object that is dropped from a height of 40 feet. $h = -16t^2 + 40$

b. An object has a position equation of $h = -16t^2 + 48t$. What is the initial velocity of the object? What is its initial height? How long will it take for the object to hit the ground?   48 feet per second, 0 feet, 3 seconds

c. An object has a position equation of $h = -16t^2 + 80$. What is the initial velocity of the object? What is its initial height? How long will it take for the object to hit the ground?   0 feet per second, 80 feet, $\sqrt{5} \approx 2.2$ seconds

d. Suppose the object in part (b) had an initial height of 64 feet. How long would it take for the object to hit the ground? How long would it take for the object to hit the ground if its initial velocity were -48 feet per second?   4 seconds, 1 second

See Section 10.3, Exercise 73.

e. For an object whose position equation is $h = -16t^2 + 48t + 64$, find the time(s) at which the height of the object is 80 feet.
$\dfrac{3 - \sqrt{5}}{2} \approx 0.38$ second, $\dfrac{3 + \sqrt{5}}{2} \approx 2.62$ seconds

f. Use the conditions in the table at the right to identify which method—the Square Root Property, factoring, or the Quadratic Formula—is most practical for solving for t in the position equation $h = -16t^2 + bt + c$ when $h = 0$. $b = 0$: Square Root Property; $b \neq 0, c = 0$: factoring; $b \neq 0, c \neq 0$: Quadratic Formula

Condition	Method
$b = 0$	?
$b \neq 0, c = 0$	?
$b \neq 0, c \neq 0$	?

See Section 10.4, Exercise 99.

g. An object has a position equation of $h = -16t^2 + 48t + 64$. Use the *table* feature of a graphing calculator to estimate the object's maximum height. Verify this height by using the formula for locating the vertex of a parabola.   100 feet

Syracuse Newspapers/David Lassman/The Image Works

Quadratic Equations and Functions

10.1 Solution by the Square Root Property

What You Should Learn

1. Solve quadratic equations by factoring.
2. Solve quadratic equations by the Square Root Property.

Why You Should Learn It

Quadratic equations can be used in applications related to the environment. For instance, in Exercise 93 on page 542, you will use a quadratic equation to estimate the diameter of an oil spill.

Solving Quadratic Equations by Factoring

Recall that a quadratic equation in x is an equation that can be written in the general form $ax^2 + bx + c = 0$, where a, b, and c are real numbers with $a \neq 0$. In Section 6.5, you learned how to solve a quadratic equation by factoring. Example 1 reviews this procedure.

---

1. Solve quadratic equations by factoring.

Remind students that they can verify solutions by using graphing calculators to check the x-intercepts of the graph of $y = ax^2 + bx + c$.

Discuss these two equations as a comparison of two common monomial factors, one a variable and one a constant:

$x^2 - 5x = 0$

$x(x - 5) = 0$

$x = 0$

$x - 5 = 0 \rightarrow x = 5$

and

$2x^2 - 14x + 24 = 0$

$2(x^2 - 7x + 12) = 0$

$2(x - 3)(x - 4) = 0$

$2 \neq 0$

$x - 3 = 0 \rightarrow x = 3$

$x - 4 = 0 \rightarrow x = 4.$

The common monomial factor x in the first example gives us a solution, $x = 0$. The common monomial factor 2 (a constant) in the second example does not provide a solution.

Example 1   Solving Quadratic Equations by Factoring

a.

$x^2 + 5x = 14$	Original equation
$x^2 + 5x - 14 = 0$	Write in general form.
$(x + 7)(x - 2) = 0$	Factor.
$x + 7 = 0 \implies x = -7$	Set 1st factor equal to 0.
$x - 2 = 0 \implies x = 2$	Set 2nd factor equal to 0.

The solutions are $x = -7$ and $x = 2$. Check these solutions in the original equation.

b.

$2x^2 = 32$	Original equation
$2x^2 - 32 = 0$	Write in general form.
$2(x^2 - 16) = 0$	Factor out common monomial factor.
$2(x + 4)(x - 4) = 0$	Factor as difference of squares.
$x + 4 = 0 \implies x = -4$	Set 1st factor equal to 0.
$x - 4 = 0 \implies x = 4$	Set 2nd factor equal to 0.

The solutions are $x = -4$ and $x = 4$. Check these solutions in the original equation.

In Example 1(b), it is not necessary to set the *constant* factor 2 equal to zero. Because of this, when an equation has a constant factor, you can divide each side by the factor. This is valid because dividing each side of an equation by a constant factor produces an equivalent equation.

② Solve quadratic equations by the Square Root Property.

Solving Quadratic Equations by the Square Root Property

In Example 1(b), suppose you had written the equation $2x^2 = 32$ in the form $x^2 = 16$ by first dividing each side by 2. In this form, you could have concluded that the solutions are simply the two square roots of 16–namely, $x = 4$ and $x = -4$. This is a nice shortcut for solving quadratic equations that can be written in the form $u^2 = d$ where $d > 0$ and u is an algebraic expression.

Study Tip

Not all quadratic equations have real number solutions. For instance, there is no real number that is the solution of the equation

$$x^2 + 4 = 0$$

because this would imply that there is a real number x such that $x^2 = -4$. Keep this possibility in mind as you do the exercises for this section.

Square Root Property

Let u be a real number, a variable, or an algebraic expression, and let d be a positive real number; then the equation $u^2 = d$ has exactly two solutions.

If $u^2 = d$, then $u = \sqrt{d}$ and $u = -\sqrt{d}$.

These solutions can also be written as $u = \pm\sqrt{d}$. This form of the solution is read as "u is equal to plus or minus the square root of d." Solving an equation of the form $u^2 = d$ by the Square Root Property is also called **extracting square roots.**

Example 2   Using the Square Root Property

Solve $4x^2 = 12$ by the Square Root Property.

Solution

$$4x^2 = 12 \qquad \text{Write original equation.}$$

$$x^2 = 3 \qquad \text{Divide each side by 4.}$$

$$x = \pm\sqrt{3} \qquad \text{Square Root Property}$$

The solutions are $x = \sqrt{3}$ and $x = -\sqrt{3}$. Check these solutions as follows.

Check

Substitute $\sqrt{3}$ *into Equation*	*Substitute* $-\sqrt{3}$ *into Equation*
$4\left(\sqrt{3}\right)^2 \overset{?}{=} 12$	$4\left(-\sqrt{3}\right)^2 \overset{?}{=} 12$
$4(3) = 12 \checkmark$	$4(3) = 12 \checkmark$

Example 3   Using the Square Root Property

Solve $(x - 3)^2 = 7$ by the Square Root Property.

Solution

$$(x - 3)^2 = 7 \qquad \text{Write original equation.}$$

$$x - 3 = \pm\sqrt{7} \qquad \text{Square Root Property}$$

$$x = 3 \pm \sqrt{7} \qquad \text{Add 3 to each side.}$$

Checking these solutions is easier than it may first appear. You might want to discuss this with students.

Caution students about the common error of forgetting the $\pm$ sign.

The solutions are $x = 3 + \sqrt{7}$ and $x = 3 - \sqrt{7}$. Check these in the original equation.

Example 4   Using the Square Root Property

Solve the quadratic equation by the Square Root Property.

$$9(3x + 5)^2 - 16 = 0$$

Solution

$9(3x + 5)^2 - 16 = 0$	Write original equation.
$9(3x + 5)^2 = 16$	Add 16 to each side.
$(3x + 5)^2 = \frac{16}{9}$	Divide each side by 9.
$3x + 5 = \pm\frac{4}{3}$	Square Root Property
$3x = -5 \pm \frac{4}{3}$	Subtract 5 from each side.
$x = -\frac{5}{3} \pm \frac{4}{9}$	Divide each side by 3.
$x = -\frac{5}{3} + \frac{4}{9} = -\frac{11}{9}$	First solution
$x = -\frac{5}{3} - \frac{4}{9} = -\frac{19}{9}$	Second solution

The solutions are $x = -\frac{11}{9}$ and $x = -\frac{19}{9}$. Check these in the original equation.

Example 5   Calculating a Compound Interest Rate

Five hundred dollars is deposited in an account. At the end of 2 years, the balance in the account is $561.80. The interest earned on the account is compounded annually. What is the interest rate?

Solution

The mathematical model used to find the interest earned on an account is

$$A = P(1 + r)^t$$

where A is the balance in the account, P is the amount deposited, r is the annual interest rate (in decimal form), and t is the number of years.

$A = P(1 + r)^t$	Model for compound interest
$561.80 = 500(1 + r)^2$	Substitute 561.80 for A, 500 for P, and 2 for t.
$1.1236 = (1 + r)^2$	Divide each side by 500.
$\sqrt{1.1236} = 1 + r$	Square Root Property
$\sqrt{1.1236} - 1 = r$	Subtract 1 from each side.
$0.06 = r$	Interest rate in decimal form

So, the interest rate is 6%. Check this in the original statement of the problem.

In Example 5, only the positive square root is used in the solution. The negative square root would have resulted in a negative value for r. In the context of this real-life application, the value of r must be positive. Watch for other such restrictions in the exercises for this chapter.

10.1 Exercises

Review   *Concepts, Skills, and Problem Solving*

Keep mathematically in shape by doing these exercises *before* the problems of this section.

Properties and Definitions

1. *Writing* Identify the leading coefficient in $11x - 2x^4 + 5x^2$. Explain.
   -2. The coefficient of the term of highest degree.

2. *Writing* State the degree of the product $(x^3 - 1)(x^2 + 1)$. Explain. 5. The degree of the term that is the product of the terms of highest degree in the factors.

3. *Writing* Explain how to use the FOIL Method to multiply two binomials.
   Add the products of the first, outer, inner, and last terms.

4. *Writing* Explain how to divide a polynomial by a binomial. Follow the long division pattern used for dividing whole numbers.

In Exercises 5–10, completely factor the expression.

5. $4b^3 - 12b^2$
   $4b^2(b - 3)$

6. $t^3 + 4t^2 - 4t - 16$
   $(t + 4)(t + 2)(t - 2)$

7. $12y^2 - 75$
   $3(2y + 5)(2y - 5)$

8. $4x^2 - 28x + 49$
   $(2x - 7)^2$

9. $2u^2 + 12u - 54$
   $2(u + 9)(u - 3)$

10. $6x^2 - 11x - 35$
    $(3x + 5)(2x - 7)$

Problem Solving

11. *Cost* The selling price of a jacket is $234. The markup rate is 30% of the cost. Find the cost. $180

12. *Mixture Problem* How many liters of a 20% alcohol solution must be mixed with a 50% solution to obtain 12 liters of a 40% solution?
    4 liters

Developing Skills

13. $2, 3$ 14. $3, 4$ 15. -2 16. 5 17. $\frac{5}{4}$ 18. $\frac{2}{3}$
19. $-\frac{4}{5}, 4$ 20. $-\frac{4}{3}, 6$ 21. $-\frac{7}{2}, \frac{4}{3}$ 22. $\frac{1}{2}, 2$ 23. $-2, 4$ 24. $-4, -1$

In Exercises 1–24, solve the quadratic equation by factoring. See Example 1.

1. $y^2 - 3y = 0$ $0, 3$
2. $t^2 + 5t = 0$ $-5, 0$
3. $4x^2 - 8x = 0$ $0, 2$
4. $25y^2 - 100y = 0$ $0, 4$
5. $a^2 - 25 = 0$ $-5, 5$
6. $16 - v^2 = 0$ $-4, 4$
7. $9m^2 = 64$ $-\frac{8}{3}, \frac{8}{3}$
8. $16y^2 = 81$ $-\frac{9}{4}, \frac{9}{4}$
9. $u(u - 10) - 6(u - 10) = 0$ $6, 10$
10. $2y(y + 3) - 5(y + 3) = 0$ $-3, \frac{5}{2}$
11. $3z(z + 20) + 12(z + 20) = 0$ $-20, -4$
12. $16x(x - 3) - 4(x - 3) = 0$ $\frac{1}{4}, 3$
13. $x^2 - 5x + 6 = 0$
14. $x^2 - 7x + 12 = 0$
15. $x^2 + 4x + 4 = 0$
16. $x^2 - 10x + 25 = 0$
17. $16x^2 - 40x + 25 = 0$
18. $9x^2 - 12x + 4 = 0$
19. $5x^2 - 16x - 16 = 0$
20. $3x^2 - 14x - 24 = 0$
21. $6x^2 = -13x + 28$
22. $2x^2 = 5x - 2$
23. $(x - 3)(x + 1) = 5$
24. $(6 + x)(1 - x) = 10$

In Exercises 25–78, solve the quadratic equation by the Square Root Property. (Some equations have no real solutions.) See Examples 2–4.

25. $x^2 = 64$ $-8, 8$
26. $x^2 = 144$ $-12, 12$

27. $h^2 = 169$ $-13, 13$
28. $z^2 = 16$ $-4, 4$
29. $x^2 = -9$
    No real solution
30. $r^2 = -49$
    No real solution
31. $6x^2 = 30$ $-\sqrt{5}, \sqrt{5}$
32. $5x^2 = 35$ $-\sqrt{7}, \sqrt{7}$
33. $7x^2 = 42$
    $-\sqrt{6}, \sqrt{6}$
34. $3x^2 = 33$
    $-\sqrt{11}, \sqrt{11}$
35. $9x^2 = 49$ $-\frac{7}{3}, \frac{7}{3}$
36. $16z^2 = 121$ $-\frac{11}{4}, \frac{11}{4}$
37. $16y^2 = 25$ $-\frac{5}{4}, \frac{5}{4}$
38. $36v^2 = 4$ $-\frac{1}{3}, \frac{1}{3}$
39. $u^2 - 100 = 0$ $-10, 10$
40. $v^2 - 25 = 0$ $-5, 5$
41. $9u^2 - 100 = 0$
    $-\frac{10}{3}, \frac{10}{3}$
42. $16v^2 - 25 = 0$
    $-\frac{5}{4}, \frac{5}{4}$
43. $x^2 + 1 = 0$
    No real solution
44. $a^2 + 9 = 0$
    No real solution
45. $2s^2 - 5 = 27$
    $-4, 4$
46. $81x^2 - 5 = 20$
    $-\frac{5}{9}, \frac{5}{9}$
47. $\frac{1}{2}x^2 - 1 = 3$
    $-2\sqrt{2}, 2\sqrt{2}$
48. $\frac{1}{5}x^2 + 4 = 5$
    $-\sqrt{5}, \sqrt{5}$
49. $\frac{1}{3}t^2 - 14 = 2$
    $-4\sqrt{3}, 4\sqrt{3}$
50. $\frac{1}{4}z^2 - 4 = 4$
    $-4\sqrt{2}, 4\sqrt{2}$
51. $\frac{1}{4}x^2 + 6 = 2$
    No real solution
52. $\frac{1}{3}y^2 - 5 = -6$
    No real solution
53. $(x - 3)^2 = 16$
    $-1, 7$
54. $(x - 2)^2 = 25$
    $-3, 7$

55. $(y - 7)^2 = 6$
$7 - \sqrt{6}, 7 + \sqrt{6}$

56. $(t + 1)^2 = 10$
$-1 - \sqrt{10}, -1 + \sqrt{10}$

57. $(x + 4)^2 = 144$
$-16, 8$

58. $(y - 7)^2 = 121$
$-4, 18$

59. $(x - 3)^2 = 3$
$3 - \sqrt{3}, 3 + \sqrt{3}$

60. $(x - 4)^2 = 5$
$4 - \sqrt{5}, 4 + \sqrt{5}$

61. $(y + 2)^2 = 12$
$-2 - 2\sqrt{3}, -2 + 2\sqrt{3}$

62. $(a - 5)^2 = 8$
$5 - 2\sqrt{2}, 5 + 2\sqrt{2}$

63. $(x - 3)^2 = 18$
$3 - 3\sqrt{2}, 3 + 3\sqrt{2}$

64. $(x + 6)^2 = 24$
$-6 - 2\sqrt{6}, -6 + 2\sqrt{6}$

65. $(3x + 2)^2 = 9$
$-\frac{5}{3}, \frac{1}{3}$

66. $(2x + 1)^2 = 49$
$-4, 3$

67. $(5x + 2)^2 = 5$
$\dfrac{-2 - \sqrt{5}}{5}, \dfrac{-2 + \sqrt{5}}{5}$

68. $(4x + 5)^2 = 10$
$\dfrac{-5 - \sqrt{10}}{4}, \dfrac{-5 + \sqrt{10}}{4}$

69. $(3x - 4)^2 = 27$
$\dfrac{4 - 3\sqrt{3}}{3}, \dfrac{4 + 3\sqrt{3}}{3}$

70. $(5x - 2)^2 = 20$
$\dfrac{2 - 2\sqrt{5}}{5}, \dfrac{2 + 2\sqrt{5}}{5}$

71. $(2x + 1)^2 = -4$
No real solution

72. $(3x + 2)^2 = -7$
No real solution

73. $4(x + 3)^2 = 25$
$-\frac{11}{2}, -\frac{1}{2}$

74. $9(x - 1)^2 = 16$
$-\frac{1}{3}, \frac{7}{3}$

75. $16(x - 5)^2 = 49$
$\frac{13}{4}, \frac{27}{4}$

76. $25(y + 3)^2 = 81$
$-\frac{24}{5}, -\frac{6}{5}$

77. $8(4x + 3)^2 = 14$
$\dfrac{-6 - \sqrt{7}}{8}, \dfrac{-6 + \sqrt{7}}{8}$

78. $12(3x - 7)^2 = 15$
$\dfrac{14 - \sqrt{5}}{6}, \dfrac{14 + \sqrt{5}}{6}$

In Exercises 79–82, factor the left side of the equation and solve the resulting equation.

79. $x^2 - 4x + 4 = 9$
$-1, 5$

80. $x^2 + 6x + 9 = 4$
$-5, -1$

81. $4x^2 + 4x + 1 = 25$
$-3, 2$

82. $4x^2 - 12x + 9 = 16$
$-\frac{1}{2}, \frac{7}{2}$

In Exercises 83–88, use a graphing calculator to graph the equation. Use the graph to estimate the x-intercepts of the graph. Set $y = 0$ and solve the resulting equation. Compare the results with the x-intercepts of the graph.   See Additional Answers.

83. $y = x^2 - 9$ $-3, 3$

84. $y = 1 - x^2$ $-1, 1$

85. $y = 4 - (x - 1)^2$
$-1, 3$

86. $y = (x + 3)^2 - 9$
$-6, 0$

87. $y = (x + 2)^2 - 1$
$-3, -1$

88. $y = 25 - (x - 3)^2$
$-2, 8$

Solving Problems

△ *Geometry*   In Exercises 89–92, solve for x.

89. Area = 16 square centimeters -2

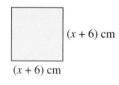

$(x + 6)$ cm
$(x + 6)$ cm
Figure for 89

$(x - 4)$ ft
$(x - 4)$ ft
Figure for 90

90. Area = 64 square feet 12

91. Area = 12 square inches 3

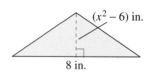

$(x^2 - 6)$ in.
8 in.
Figure for 91

$(x^2 + 3)$ ft
20 ft
Figure for 92

92. Area = 120 square feet 3

93. *Environment*   An oil spill from an offshore drilling platform covers a circular region of approximately 10 square miles. Approximate the diameter of the region. (Use $\pi \approx 3.14$.)   3.57 miles

94. *Storage Bin*   The circular base of a chemical storage bin has an area of 706 square inches. Approximate the diameter of the circular base. (Use $\pi \approx 3.14$.)   29.99 inches

95. *Compound Interest*   After 2 years, a $1000 investment, compounded annually at interest rate r, will yield an amount of $1000(1 + r)^2$. If this amount is $1166.40, find the rate r. $0.08 = 8\%$

96. *Compound Interest*   Four hundred dollars is deposited in an account. At the end of 2 years, the balance in the account is $462.25. The interest earned on the account is compounded annually. What is the interest rate? $0.075 = 7.5\%$

97. *Revenue*   The revenue R (in dollars) when x units of a photo album are sold is modeled by

$$R = x\left(5 - \frac{1}{10}x\right), \quad 0 < x < 25.$$

Determine the number of photo albums that must be sold to produce a revenue of $60.   20 units

98. *Revenue*   The revenue R (in dollars) when x units of a sewing machine are sold is given by

$$R = x\left(100 - \frac{1}{2}x\right), \quad 0 < x < 100.$$

Determine the number of sewing machines that must be sold to produce a revenue of $4200.   60 units

99. *Free-Falling Object*   The height h (in feet) of an object dropped from a tower 64 feet high is modeled by $h = 64 - 16t^2$, where t measures the time in seconds. How long does it take for the object to reach the ground?   2 seconds

100. *Free-Falling Object*   The height h (in feet) of an object dropped from the Brooklyn Bridge, which is 272 feet high, is modeled by

$$h = 272 - 16t^2$$

where t measures the time in seconds. How long does it take for the object to reach the ground?

$\sqrt{17} \approx 4.12$ seconds

101. *The Hammer and the Feather*   In 1971, astronaut David Scott demonstrated that a feather and a hammer fall at the same rate on the moon because the moon has no atmosphere (and hence no air resistance). The height h (in feet) of a falling object on the moon is modeled by $h = -2.7t^2 + s$, where t is time in seconds and s is the height from which the object is dropped. A hammer and a feather are dropped 5 feet from the surface of the moon. How long will it take for each to hit the surface?

1.36 seconds

102. *Analyzing a Free-Falling Object*   An object is dropped from Chase Tower in Houston, Texas, which is 1000 feet tall. Its height h (in feet) after t seconds of falling is modeled by $h = 1000 - 16t^2$.

(a) Complete the table. Round the times to two decimal places.

h	1000	950	900	850	800	750	700
t	0	1.77	2.50	3.06	3.54	3.95	4.33

(b) Consecutive heights in the table differ by a constant amount of 50 feet. Do the corresponding times differ by a constant amount? Explain.

No; the difference in times decreases, because the velocity of the object is increasing.

Explaining Concepts

103. ⊘ Answer parts (a)–(d) of Motivating the Chapter on page 536.

104. *Writing✍*   The general form of a quadratic equation in x is $ax^2 + bx + c = 0$, where a, b, and c are real numbers with $a \neq 0$. Explain why b and c can equal 0, but a cannot.

If $a = 0$, the equation would not be quadratic.

105. *Writing✍*   Explain the Zero-Factor Property. How can it be used to solve a quadratic equation?

Factoring and the Zero-Factor Property allow you to solve a quadratic equation by converting it into two linear equations, which you know how to solve.

106. *True or False?*   The only solution of the equation $x^2 = 16$ is $x = 4$. Justify your answer.   False. The solutions are $x = \pm 4$, because $4^2 = 16$ and $(-4)^2 = 16$.

107. *Writing✍*   In your own words, explain how to solve a quadratic equation by the Square Root Property.   Write the equation in the form $u^2 = d$, where u is an algebraic expression and d is a positive constant. Take the square roots of both sides to obtain the solutions $u = \pm\sqrt{d}$.

108. *Writing✍*   State whether factoring or the Square Root Property is usually the easier method of solving the quadratic equation $ax^2 + bx + c = 0$ when (a) $b = 0$ and (b) $c = 0$. Explain.

(a) Square Root Property   (b) Factoring

109. *Writing✍*   Without attempting to solve the equations, state which of the equations do not have real solutions. Explain.

(a) $(x - 2)^2 = 0$ (b) $(x - 2)^2 = 36$

(c) $(x - 2)^2 + 36 = 0$ (d) $(x - 2)^2 - 5 = 0$

Equation (c). $(x - 2)^2 + 36 \geq 36$ for all x.

110. *Writing✍*   Write a short paragraph explaining how to solve an equation of the form $ax^2 + c = 0$ by the Square Root Property. Use the described procedure to solve the equation $2x^2 + 6 = 0$. Can you suggest any revisions to make the procedure valid?

See Additional Answers.

10.2 Solution by Completing the Square

Michael Newman/PhotoEdit, Inc.

What You Should Learn

1 Construct perfect square trinomials.

2 Solve quadratic equations by completing the square.

Why You Should Learn It

Quadratic equations can be used in many business applications. For instance, in Exercise 98 on page 551, you will use a quadratic equation to model the revenue for an infant stroller.

1 Construct perfect square trinomials.

Constructing Perfect Square Trinomials

You know from Example 3 of the preceding section that the equation $(x - 3)^2 = 7$ has two solutions: $x = 3 + \sqrt{7}$ and $x = 3 - \sqrt{7}$. Suppose you were given the equation in its general form

$$(x - 3)^2 = 7 \qquad \text{Completed square form}$$

$$x^2 - 6x + 9 = 7 \qquad \text{Multiply binomials.}$$

$$x^2 - 6x + 2 = 0. \qquad \text{Write in general form.}$$

How could you solve this form of the quadratic equation? You could try factoring, but you would find that the left side of the equation is not factorable using integer coefficients. You need a procedure for reversing the steps shown above.

In this section, you will study a technique for rewriting an equation in a completed square form. This technique is called **completing the square.** To complete the square, you must realize that all perfect square trinomials with leading coefficients of 1 have a similar form.

Perfect Square Trinomial = Square of Binomial

$$x^2 + bx + \left(\frac{b}{2}\right)^2 = \left(x + \frac{b}{2}\right)^2$$

$$\text{(half of } b)^2$$

Note that the constant term of the perfect square trinomial is the square of half of the coefficient of the x-term. So, to complete the square for an expression of the form $x^2 + bx$, you must add $(b/2)^2$ to the expression.

Completing the Square

To **complete the square** for the expression

$$x^2 + bx$$

add $(b/2)^2$, which is the square of half the coefficient of x. Consequently,

$$x^2 + bx + \left(\frac{b}{2}\right)^2 = \left(x + \frac{b}{2}\right)^2.$$

Example 1   Constructing a Perfect Square Trinomial

What term should be added to the expression

$$x^2 - 4x$$

so that it becomes a perfect square trinomial? Write the new expression as the square of a binomial.

Solution

For this expression, the coefficient of the x-term is -4. By taking half of this coefficient and squaring the result, you obtain $\left(-\frac{4}{2}\right)^2 = 4$. This is the term that should be added to the expression to make it a perfect square trinomial.

$$x^2 - 4x + \left(-\frac{4}{2}\right)^2 = x^2 - 4x + 4 \qquad \text{Add } \left(-\frac{4}{2}\right)^2 = 4 \text{ to the expression.}$$

$$= (x - 2)^2 \qquad \text{Completed square form}$$

In Example 1, don't make the mistake of concluding that $x^2 - 4x$ is equal to $(x - 2)^2$. The point of the example is that $x^2 - 4x$ is not a perfect square trinomial. If, however, you add 4 to the expression, then the *new* expression is a perfect square trinomial.

Example 2   Constructing Perfect Square Trinomials

Here are some additional examples. What term should be added to each expression so that it becomes a perfect square trinomial?

$y^2 - 16y$:
$y^2 - 16y + 64 = (y - 8)^2$

$b^2 + 11b$:
$b^2 + 11b + \frac{121}{4} = \left(b + \frac{11}{2}\right)^2$

$k^2 - \frac{8}{9}k$:
$k^2 - \frac{8}{9}k + \frac{16}{81} = \left(k - \frac{4}{9}\right)^2$

$w^2 + \frac{1}{6}w$:
$w^2 + \frac{1}{6}w + \frac{1}{144} = \left(w + \frac{1}{12}\right)^2$

What terms should be added to each expression so that it becomes a perfect square trinomial? Write each new expression as the square of a binomial.

a. $x^2 + 12x$

b. $x^2 - 7x$

Solution

a. For this expression, the coefficient of the x-term is 12. By taking half of this coefficient and squaring the result, you obtain $\left(\frac{12}{2}\right)^2 = 36$. This is the term that should be added to the expression to make it a perfect square trinomial.

$$x^2 + 12x + \left(\frac{12}{2}\right)^2 = x^2 + 12x + 36 \qquad \text{Add } \left(\frac{12}{2}\right)^2 = 36 \text{ to the expression.}$$

$$= (x + 6)^2 \qquad \text{Completed square form}$$

b. For this expression, the coefficient of the x-term is -7. By taking half of this coefficient and squaring the result, you obtain $\left(-\frac{7}{2}\right)^2 = \frac{49}{4}$. This is the term that should be added to the expression to make it a perfect square trinomial.

$$x^2 - 7x + \left(-\frac{7}{2}\right)^2 = x^2 - 7x + \frac{49}{4} \qquad \text{Add } \left(-\frac{7}{2}\right)^2 = \frac{49}{4} \text{ to the expression.}$$

$$= \left(x - \frac{7}{2}\right)^2 \qquad \text{Completed square form}$$

② Solve quadratic equations by completing the square.

Completing the Square

Completing the square can be used to solve quadratic equations. When using this procedure, remember to *preserve the equality* by adding the same constant to each side of the equation.

Study Tip

In Example 3, completing the square is used for the sake of illustration. This particular equation would be easier to solve by factoring. Try reworking the problem by factoring to see that you obtain the same two solutions.

Example 3   Completing the Square: Leading Coefficient Is 1

Solve $x^2 + 10x = 0$ by completing the square.

Solution

$$x^2 + 10x = 0 \qquad \text{Write original equation.}$$

$$x^2 + 10x + 5^2 = 25 \qquad \text{Add } 5^2 = 25 \text{ to each side.}$$

(half of 10)²

$$(x + 5)^2 = 25 \qquad \text{Completed square form}$$

$$x + 5 = \pm\sqrt{25} \qquad \text{Square Root Property}$$

$$x = -5 \pm 5 \qquad \text{Subtract 5 from each side.}$$

$$x = 0 \text{ or } x = -10 \qquad \text{Solutions}$$

The solutions are $x = 0$ and $x = -10$. You can check these solutions as follows.

Check

Substitute 0 into Equation	*Substitute −10 into Equation*
$x^2 + 10x = 0$	$x^2 + 10x = 0$
$0^2 + 10(0) \overset{?}{=} 0$	$(-10)^2 + 10(-10) \overset{?}{=} 0$
$0 = 0 \checkmark$	$100 - 100 = 0 \checkmark$

Example 4   Completing the Square: Leading Coefficient Is 1

Solve $x^2 - 4x + 1 = 0$ by completing the square.

Solution

$$x^2 - 4x + 1 = 0 \qquad \text{Write original equation.}$$

$$x^2 - 4x = -1 \qquad \text{Subtract 1 from each side.}$$

$$x^2 - 4x + (-2)^2 = -1 + 4 \qquad \text{Add } (-2)^2 = 4 \text{ to each side.}$$

(half of −4)²

$$(x - 2)^2 = 3 \qquad \text{Completed square form}$$

$$x - 2 = \pm\sqrt{3} \qquad \text{Square Root Property}$$

$$x = 2 \pm \sqrt{3} \qquad \text{Add 2 to each side.}$$

Remind students that the statement $x = 2 \pm \sqrt{3}$ indicates two distinct solutions.

The solutions are $x = 2 + \sqrt{3}$ and $x = 2 - \sqrt{3}$. Use your graphing calculator to check these solutions in the original equation.

If the leading coefficient of a quadratic expression is not 1, you must divide each side of the equation by this coefficient *before* completing the square. This process is demonstrated in Examples 5 and 6.

Example 5   Completing the Square: Leading Coefficient Is Not 1

Solve $2x^2 + 5x = 3$ by completing the square.

Solution

$$2x^2 + 5x = 3 \qquad \text{Write original equation.}$$

$$x^2 + \frac{5}{2}x = \frac{3}{2} \qquad \text{Divide each side by 2.}$$

$$x^2 + \frac{5}{2}x + \left(\frac{5}{4}\right)^2 = \frac{3}{2} + \frac{25}{16} \qquad \text{Add } \left(\frac{5}{4}\right)^2 = \frac{25}{16} \text{ to each side.}$$

$$\left(\text{half of } \tfrac{5}{2}\right)^2$$

$$\left(x + \frac{5}{4}\right)^2 = \frac{49}{16} \qquad \text{Completed square form}$$

$$x + \frac{5}{4} = \pm\frac{7}{4} \qquad \text{Square Root Property}$$

$$x = -\frac{5}{4} \pm \frac{7}{4} \qquad \text{Subtract } \tfrac{5}{4} \text{ from each side.}$$

The solutions are

$$x = -\frac{5}{4} + \frac{7}{4} = \frac{2}{4} = \frac{1}{2} \quad \text{and} \quad x = -\frac{5}{4} - \frac{7}{4} = -\frac{12}{4} = -3.$$

You can check these solutions in the original equation as follows.

Check

Substitute $\frac{1}{2}$ into Equation	Substitute -3 into Equation
$2\left(\frac{1}{2}\right)^2 + 5\left(\frac{1}{2}\right) \stackrel{?}{=} 3$	$2(-3)^2 + 5(-3) \stackrel{?}{=} 3$
$2\left(\frac{1}{4}\right) + \frac{5}{2} \stackrel{?}{=} 3$	$2(9) - 15 \stackrel{?}{=} 3$
$\frac{1}{2} + \frac{5}{2} = 3 \ \checkmark$	$18 - 15 = 3 \ \checkmark$

If you solve a quadratic equation by completing the square and obtain solutions that do not involve radicals, you could have solved the equation by factoring. For instance, in Example 5, the equation could have been factored as follows.

$$2x^2 + 5x - 3 = 0 \qquad \text{Write in general form.}$$

$$(2x - 1)(x + 3) = 0 \qquad \text{Factor.}$$

$$2x - 1 = 0 \implies x = \frac{1}{2} \qquad \text{Set 1st factor equal to 0.}$$

$$x + 3 = 0 \implies x = -3 \qquad \text{Set 2nd factor equal to 0.}$$

You might point out that the procedure for completing the square is also used in subsequent algebra courses for purposes other than solving quadratic equations.

Example 6   Completing the Square: Leading Coefficient Is Not 1

Solve $3x^2 - 2x - 4 = 0$ by completing the square. Use a calculator to approximate the solutions to two decimal places.

Solution

$$3x^2 - 2x - 4 = 0 \qquad \text{Write original equation.}$$

$$3x^2 - 2x = 4 \qquad \text{Add 4 to each side.}$$

$$x^2 - \frac{2}{3}x = \frac{4}{3} \qquad \text{Divide each side by 3.}$$

$$x^2 - \frac{2}{3}x + \left(-\frac{1}{3}\right)^2 = \frac{4}{3} + \frac{1}{9} \qquad \text{Add } \left(-\frac{1}{3}\right)^2 = \frac{1}{9} \text{ to each side.}$$

$$\left(\text{half of } -\tfrac{2}{3}\right)^2$$

$$\left(x - \frac{1}{3}\right)^2 = \frac{13}{9} \qquad \text{Completed square form}$$

$$x - \frac{1}{3} = \pm\frac{\sqrt{13}}{3} \qquad \text{Square Root Property}$$

$$x = \frac{1}{3} \pm \frac{\sqrt{13}}{3} \qquad \text{Add } \tfrac{1}{3} \text{ to each side.}$$

The solutions are

$$x = \frac{1}{3} + \frac{\sqrt{13}}{3} \approx 1.54 \quad \text{and} \quad x = \frac{1}{3} - \frac{\sqrt{13}}{3} \approx -0.87.$$

Use your graphing calculator to check these solutions in the original equation.

Additional Examples
Solve each equation by completing the square.

a. $x^2 + 8x + 14 = 0$

b. $3x^2 - 6x + 1 = 0$

Answers:

a. $x = -4 \pm \sqrt{2}$

b. $x = \dfrac{3 \pm \sqrt{6}}{3}$

The method of completing the square can be used to solve *any* quadratic equation. Moreover, this method will identify those quadratic equations that have no real solutions, as demonstrated in the next example.

Example 7   A Quadratic Equation with No Real Solution

Show that $x^2 - 4x + 7 = 0$ has no real solution by completing the square.

Solution

$$x^2 - 4x + 7 = 0 \qquad \text{Write original equation.}$$

$$x^2 - 4x = -7 \qquad \text{Subtract 7 from each side.}$$

$$x^2 - 4x + 4 = -7 + 4 \qquad \text{Add } (-2)^2 = 4 \text{ to each side.}$$

$$(x - 2)^2 = -3 \qquad \text{Completed square form}$$

Because the square of a real number cannot be negative, you can conclude that this equation has no real solution.

Technology: Tip

Use a graphing calculator to graph the quadratic equation

$$y = x^2 - 4x + 7.$$

How can you tell from the graph that the quadratic equation has no real solution? What do you think the graph of a quadratic equation having one (repeated) solution looks like? How about a quadratic equation having two solutions?

See Technology Answers.

10.2 Exercises

Review   Concepts, Skills, and Problem Solving

Keep mathematically in shape by doing these exercises *before* the problems of this section.

Properties and Definitions

In Exercises 1–4, consider the equation of the line $4x - 3y - 12 = 0$.

1. Is the point $(6, 4)$ on the line? Explain.
   Yes. The equation is true when $x = 6$ and $y = 4$.

2. Find two points on the line and use the two points to find the slope of the line. Does it matter which two points you use? Explain. $m = \frac{4}{3}$. Any two points may be used because the rate of change remains the same.

3. Write the equation of the line in point-slope form.
   There are many correct answers. One example is $y - 4 = \frac{4}{3}(x - 6)$.

4. Write the equation of the line in slope-intercept form. $y = \frac{4}{3}x - 4$

Linear Equations

In Exercises 5–8, write an equation of the line that passes through the two points.

5. $(-3, 2), (5, 0)$ $x + 4y - 5 = 0$

6. $(-10, -5), (4, 10)$ $15x - 14y + 80 = 0$

7. $(-4, -4), (-4, 6)$ $x + 4 = 0$

8. $\left(\frac{2}{3}, \frac{1}{2}\right), \left(\frac{5}{6}, \frac{3}{4}\right)$ $3x - 2y - 1 = 0$

In Exercises 9 and 10, solve the system of equations.

9. $\begin{cases} 3x - 5y = 8 \\ x + 2y = 10 \end{cases}$ $(6, 2)$

10. $\begin{cases} 6x + 2y = -22 \\ 5x - 3y = -37 \end{cases}$ $(-5, 4)$

Problem Solving

11. *Reimbursed Expenses*   A sales representative is reimbursed $125 per day for lodging and meals plus $0.36 per mile driven. Write a linear equation giving the daily cost C to the company in terms of x, where x is the number of miles driven.
    $C = 125 + 0.36x$

12. ▲ *Geometry*   A tennis court is a rectangle that is 27 feet wide and 78 feet long. Find the length of the diagonal of the court. $3\sqrt{757} \approx 82.5$ feet

Developing Skills

In Exercises 1–20, what term should be added to the expression to make it a perfect square trinomial? See Examples 1 and 2.

1. $x^2 + 10x$ 25
2. $x^2 + 14x$ 49
3. $y^2 - 24y$ 144
4. $y^2 - 8y$ 16
5. $x^2 + 16x$ 64
6. $t^2 - 18t$ 81
7. $h^2 - 42h$ 441
8. $x^2 + 100x$ 2500
9. $t^2 + 3t$ $\frac{9}{4}$
10. $u^2 + 9u$ $\frac{81}{4}$
11. $y^2 - 7y$ $\frac{49}{4}$
12. $x^2 - 11x$ $\frac{121}{4}$
13. $x^2 - x$ $\frac{1}{4}$
14. $y^2 + 5y$ $\frac{25}{4}$
15. $x^2 + 21x$ $\frac{441}{4}$
16. $t^2 - 33t$ $\frac{1089}{4}$
17. $x^2 + \frac{1}{2}x$ $\frac{1}{16}$
18. $y^2 - \frac{1}{3}y$ $\frac{1}{36}$
19. $t^2 - \frac{3}{4}t$ $\frac{9}{64}$
20. $u^2 + \frac{4}{5}u$ $\frac{4}{25}$

In Exercises 21–60, solve the quadratic equation by completing the square. (Some equations have no real solutions.) See Examples 3–5.

21. $x^2 - 8x = 0$
    $0, 8$
22. $x^2 + 12x = 0$
    $-12, 0$
23. $y^2 + 20y = 0$
    $-20, 0$
24. $u^2 - 16u = 0$
    $0, 16$
25. $x^2 - 2x - 1 = 0$
    $1 \pm \sqrt{2}$
26. $x^2 - 6x + 7 = 0$
    $3 \pm \sqrt{2}$
27. $u^2 - 4u - 1 = 0$
    $2 \pm \sqrt{5}$
28. $a^2 - 10a + 15 = 0$
    $5 \pm \sqrt{10}$
29. $x^2 - 2x + 3 = 0$
    No real solution
30. $x^2 - 6x + 14 = 0$
    No real solution
31. $x^2 - 8x - 2 = 0$
    $4 \pm 3\sqrt{2}$
32. $x^2 + 6x - 3 = 0$
    $-3 \pm 2\sqrt{3}$

33. $y^2 + 14y + 17 = 0$
$-7 \pm 4\sqrt{2}$

34. $y^2 + 2y - 26 = 0$
$-1 \pm 3\sqrt{3}$

35. $x^2 + 2x - 35 = 0$
$-7, 5$

36. $x^2 - 6x - 27 = 0$
$-3, 9$

37. $x^2 - x - 3 = 0$
$\dfrac{1 \pm \sqrt{13}}{2}$

38. $x^2 + 3x + 1 = 0$
$\dfrac{-3 \pm \sqrt{5}}{2}$

39. $t^2 + 5t + 2 = 0$
$\dfrac{-5 \pm \sqrt{17}}{2}$

40. $u^2 - 9u - 5 = 0$
$\dfrac{9 \pm \sqrt{101}}{2}$

41. $x^2 + 3x - 4 = 0$
$-4, 1$

42. $x^2 + 5x + 6 = 0$
$-3, -2$

43. $y^2 - 9y + 14 = 0$
$2, 7$

44. $t^2 - 5t + 4 = 0$
$1, 4$

45. $u^2 + 9u + 21 = 0$
No real solution

46. $x^2 - 11x + 31 = 0$
No real solution

47. $3x^2 - 6x + 9 = 0$
No real solution

48. $2x^2 - 4x + 6 = 0$
No real solution

49. $2x^2 + 6x - 5 = 0$
$\dfrac{-3 \pm \sqrt{19}}{2}$

50. $3x^2 - 12x + 7 = 0$
$\dfrac{6 \pm \sqrt{15}}{3}$

51. $3x^2 + 4x + 5 = 0$
No real solution

52. $2z^2 - z + 1 = 0$
No real solution

53. $2y^2 + 3y - 1 = 0$
$\dfrac{-3 \pm \sqrt{17}}{4}$

54. $4z^2 - 3z - 2 = 0$
$\dfrac{3 \pm \sqrt{41}}{8}$

55. $6x^2 - 10x - 9 = 0$
$\dfrac{5 \pm \sqrt{79}}{6}$

56. $10x^2 - 8x + 15 = 0$
No real solution

57. $\frac{1}{3}x^2 + \frac{1}{3}x - 4 = 0$
$-4, 3$

58. $\frac{1}{5}x^2 + \frac{3}{5}x - 2 = 0$
$-5, 2$

59. $\frac{1}{2}x^2 + x - 1 = 0$
$-1 \pm \sqrt{3}$

60. $\frac{2}{3}x^2 - 4x + 1 = 0$
$\dfrac{6 \pm \sqrt{30}}{2}$

In Exercises 61–68, solve the quadratic equation (a) by completing the square and (b) by factoring.

61. $x^2 - 4x = 0$ $0, 4$

62. $x^2 - 2x = 0$ $0, 2$

63. $t^2 + 6t + 5 = 0$
$-5, -1$

64. $x^2 + 2x - 15 = 0$
$-5, 3$

65. $x^2 + 5x + 6 = 0$
$-3, -2$

66. $x^2 - 7x - 8 = 0$
$-1, 8$

67. $2x^2 - 5x + 2 = 0$
$\frac{1}{2}, 2$

68. $4x^2 - 4x - 3 = 0$
$-\frac{1}{2}, \frac{3}{2}$

In Exercises 69–76, solve the quadratic equation by completing the square. Use a calculator to approximate the solution. Round your answer to two decimal places. See Example 6.

69. $x^2 + x - 3 = 0$
$\dfrac{-1 + \sqrt{13}}{2} \approx 1.30; \dfrac{-1 - \sqrt{13}}{2} \approx -2.30$

70. $c^2 - 3c + 1 = 0$
$\dfrac{3 + \sqrt{5}}{2} \approx 2.62; \dfrac{3 - \sqrt{5}}{2} \approx 0.38$

71. $x^2 - 6x + 7 = 0$
$3 + \sqrt{2} \approx 4.41; 3 - \sqrt{2} \approx 1.59$

72. $y^2 + 4y - 1 = 0$
$-2 + \sqrt{5} \approx 0.24; -2 - \sqrt{5} \approx -4.24$

73. $4z^2 - 4z - 3 = 0$
$-0.50, 1.50$

74. $2x^2 + 6x + 1 = 0$
$\dfrac{-3 + \sqrt{7}}{2} \approx -0.18; \dfrac{-3 - \sqrt{7}}{2} \approx -2.82$

75. $3y^2 - y - 1 = 0$
$\dfrac{1 + \sqrt{13}}{6} \approx 0.77; \dfrac{1 - \sqrt{13}}{6} \approx -0.43$

76. $2x^2 + 2x - 7 = 0$
$\dfrac{-1 + \sqrt{15}}{2} \approx 1.44; \dfrac{-1 - \sqrt{15}}{2} \approx -2.44$

In Exercises 77–86, solve the equation.

77. $\dfrac{x}{2} + \dfrac{1}{x} = 2$ $2 \pm \sqrt{2}$

78. $\dfrac{x}{3} + \dfrac{2}{x} = 4$ $6 \pm \sqrt{30}$

79. $\dfrac{3}{x - 2} = 2x$ $\dfrac{2 \pm \sqrt{10}}{2}$

80. $\dfrac{4}{x} + \dfrac{2}{x - 2} = 1$ $4 \pm 2\sqrt{2}$

81. $\sqrt{2x + 3} = x - 2$ $3 + 2\sqrt{2}$

82. $\sqrt{4x + 5} = x - 6$ $8 + \sqrt{33}$

83. $2\sqrt{x - 1} = x - 4$ 10

84. $3\sqrt{x + 1} = x - 3$ 15

85. $\sqrt{x^2 + 3} - 2\sqrt{x} = 0$ $1, 3$

86. $3\sqrt{x} = \sqrt{x^2 - 10}$ 10

In Exercises 87–90, use a graphing calculator to graph the equation. Use the graph to estimate the x-intercepts of the graph. Set $y = 0$ and solve the resulting equation. Compare the results with the x-intercepts of the graph.   See Additional Answers.

87. $y = x^2 - 4x + 2$ $2 + \sqrt{2} \approx 3.41; 2 - \sqrt{2} \approx 0.59$

88. $y = x^2 + 2x - 2$
$-1 + \sqrt{3} \approx 0.73; -1 - \sqrt{3} \approx -2.73$

89. $y = \sqrt{2x + 1} - x$ $1 + \sqrt{2} \approx 2.41$

90. $y = \sqrt{x} - x + 3$ $\dfrac{7 + \sqrt{13}}{2} \approx 5.30$

Solving Problems

△ *Geometry*   In Exercises 91–94, solve for x.

91. Area = 12 square centimeters   4

92. Area = 31 square feet $-1 + 4\sqrt{2}$

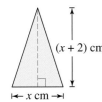

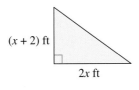

Figure for 91   Figure for 92

93. Area = 16 square millimeters $2 + 2\sqrt{5}$

94. Area = 160 square feet   20

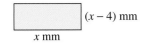

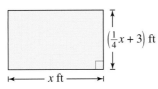

Figure for 93   Figure for 94

95. *Number Problem*   Find two consecutive positive integers such that the sum of their squares is 85.   6, 7

96. *Number Problem*   Find two consecutive positive integers such that the sum of their squares is 41.   4, 5

97. *Revenue*   The revenue R (in dollars) for selling x units of a cordless telephone is modeled by

$$R = x\left(25 - \frac{1}{2}x\right), \quad 0 < x < 25.$$

Find the number of telephones that must be sold to produce a revenue of $304.50.   21 units

98. *Revenue*   The revenue (in dollars) for selling x units of an infant stroller is modeled by

$$R = x\left(50 - \frac{1}{4}x\right), \quad 0 < x < 100.$$

Find the number of strollers that must be sold to produce a revenue of $2059.   58 units

99. △ *Geometry*   The sum of the base and height of a triangle is 100 centimeters. The area of the triangle is 1200 square centimeters.

(a) Draw a figure that gives a visual representation of the problem.   See Additional Answers.

(b) The height of the triangle is h. Write an expression for the base in terms of h. $100 - h$

(c) Write an expression for the area of the triangle in terms of h. Use the result to find the dimensions of the triangle. $A = \frac{1}{2}(100 - h)h$
Height: 40 centimeters; Base: 60 centimeters or
Height: 60 centimeters; Base: 40 centimeters

100. △ *Geometry*   The sum of the base and height of a triangle is 120 centimeters. The area of the triangle is 1792 square centimeters.

(a) Draw a figure that gives a visual representation of the problem.   See Additional Answers.

(b) The height of the triangle is h. Write an expression for the base in terms of h. $120 - h$

(c) Write an expression for the area of the triangle in terms of h. Use the result to find the dimensions of the triangle. $A = \frac{1}{2}(120 - h)h$
Height: 56 centimeters; Base: 64 centimeters or
Height: 64 centimeters; Base: 56 centimeters

101. *Think About It*   Add a term to the expression to make a perfect square trinomial.

$$x^2 + \boxed{24x} + 144$$

Explain how you obtained your answer.
$(x + 12)^2 = x^2 + 24x + 144$

102. *Think About It*   Add a term to the expression to make a perfect square trinomial.

$$x^2 - \boxed{28x} + 196$$

Explain how you obtained your answer.
$(x - 14)^2 = x^2 - 28x + 196$

Explaining Concepts

103. *Writing* ✏️   What is a perfect square trinomial? Give an example.   A perfect square trinomial is one that can be written in the form $(x + k)^2$. Example $x^2 - 4x + 4$

104. *Writing* ✏️   Describe how to complete the square for a quadratic equation when the leading coefficient is 1.

Divide the coefficient of the first-degree term by 2, square, and add the result to both sides of the equation. Write the side containing the variable as a perfect square trinomial, and solve the equation by extracting square roots.

105. *Writing* ✏️   Describe how to complete the square for a quadratic equation when the leading coefficient *is not* 1.   Divide each side of the equation by the leading coefficient and then follow the steps listed in the answer to Exercise 104.

106. *Writing* ✏️   Would you solve $x^2 - 2x - 3 = 0$ by factoring or by completing the square? Explain.

Factoring, because it is easily factored.

107. *Writing* ✏️   Would you solve $x^2 - 2x - 2 = 0$ by factoring or by completing the square? Explain.

Completing the square, because it is not easily factored.

108. *Error Analysis*   Describe the error.

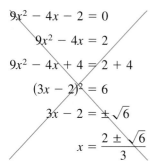

$$9x^2 - 4x - 2 = 0$$
$$9x^2 - 4x = 2$$
$$9x^2 - 4x + 4 = 2 + 4$$
$$(3x - 2)^2 = 6$$
$$3x - 2 = \pm\sqrt{6}$$
$$x = \frac{2 \pm \sqrt{6}}{3}$$

$9x^2 - 4x + 4 \neq (3x - 2)^2$

The leading coefficient should be 1 when completing the square.

109. *Writing* ✏️   Is it possible for a quadratic equation to have no real number solution? If so, give an example.   Yes. $x^2 + 1 = 0$

110. *True or False?*   There exist quadratic equations with real solutions that cannot be solved by completing the square. Justify your answer.

False. Any quadratic equation with real solutions can be solved by completing the square.

111.   The expression $x^2 + 2x$ represents the area of the figure on the left. The expression $(x + 1)^2$ represents the area of the figure on the right. Find the area of the part that was added to the left figure to "complete the square." Draw two such figures for the expression $x^2 + 5x$. What was added to complete the square?   Area required to complete the square: 1   Figures for $x^2 + 5x$:

See Additional Answers.

The value $(2.5)^2$ was added to complete the square.

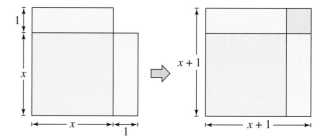

10.3 Solution by the Quadratic Formula

Robert Glusic/Photodisc/Getty Images

What You Should Learn

1. Use the completing the square technique to develop the Quadratic Formula.

2. Solve quadratic equations using the Quadratic Formula.

Why You Should Learn It

Quadratic equations can be used to model and solve real-life problems. For instance, in Exercise 69 on page 560, you will use a quadratic equation to model the time it takes for an object to fall from the Sky Pod of the CN Tower.

The Quadratic Formula

Another technique for solving a quadratic equation involves the **Quadratic Formula.** This formula is obtained by completing the square for a general quadratic equation.

$$ax^2 + bx + c = 0 \qquad \text{General form, } a \neq 0$$

$$ax^2 + bx = -c \qquad \text{Subtract } c \text{ from each side.}$$

$$x^2 + \frac{b}{a}x = -\frac{c}{a} \qquad \text{Divide each side by } a.$$

$$x^2 + \frac{b}{a}x + \left(\frac{b}{2a}\right)^2 = -\frac{c}{a} + \left(\frac{b}{2a}\right)^2 \qquad \text{Add } \left(\frac{b}{2a}\right)^2 \text{ to each side.}$$

$$\left(x + \frac{b}{2a}\right)^2 = \frac{b^2 - 4ac}{4a^2} \qquad \text{Simplify.}$$

$$x + \frac{b}{2a} = \pm\sqrt{\frac{b^2 - 4ac}{4a^2}} \qquad \text{Square Root Property}$$

$$x = -\frac{b}{2a} \pm \frac{\sqrt{b^2 - 4ac}}{2|a|} \qquad \text{Subtract } \frac{b}{2a} \text{ from each side.}$$

$$x = \frac{-b \pm \sqrt{b^2 - 4ac}}{2a} \qquad \text{Simplify.}$$

1. **Use the completing the square technique to develop the Quadratic Formula.**

You might first present this formula, illustrate its use, and then, when students are somewhat familiar with the formula, discuss its derivation.

Study Tip

The Quadratic Formula is one of the most important formulas in algebra, and you should memorize it. It helps to try to memorize a verbal statement of the rule. For instance, you might try to remember the following verbal statement of the Quadratic Formula: "The opposite of b, plus or minus the square root of b squared minus $4ac$, all divided by $2a$."

The Quadratic Formula

The solutions of $ax^2 + bx + c = 0$, $a \neq 0$, are given by the **Quadratic Formula**

$$x = \frac{-b \pm \sqrt{b^2 - 4ac}}{2a}.$$

The expression inside the radical, $b^2 - 4ac$, is called the **discriminant.**

1. If $b^2 - 4ac > 0$, the equation has two real solutions.

2. If $b^2 - 4ac = 0$, the equation has one (repeated) real solution.

3. If $b^2 - 4ac < 0$, the equation has no real solution.

Example 1   Determining the Number of Solutions

Use the discriminant to determine the number of real solutions of $x^2 + 5x - 14 = 0$.

Solution

For this equation, $a = 1$, $b = 5$, and $c = -14$. So, the discriminant is

$$b^2 - 4ac = 5^2 - 4(1)(-14) = 25 + 56 = 81.$$

Because $81 > 0$, there are two real solutions to the equation.

② Solve quadratic equations using the Quadratic Formula.

Solving Equations by the Quadratic Formula

When using the Quadratic Formula, remember that *before* the formula can be applied, you must first write the quadratic equation in general form in order to determine the values of a, b, and c.

Remind students to write the equation in general form before determining the values of $a, b,$ and c.

Example 2   The Quadratic Formula: Two Distinct Solutions

Use the Quadratic Formula to solve $x^2 + 5x = 14$.

Solution

To begin, write the equation in general form, $ax^2 + bx + c = 0$.

$$x^2 + 5x = 14 \qquad \text{Write original equation.}$$

$$x^2 + 5x - 14 = 0 \qquad \text{Write in general form.}$$

From Example 1, you know that this equation has two solutions. Use the values $a = 1$, $b = 5$, and $c = -14$ from the general form to substitute into the Quadratic Formula and obtain the solution.

Students may neglect to write the entire quantity $-b \pm \sqrt{b^2 - 4ac}$ over the denominator $2a$. This is a common cause of errors.

$$x = \frac{-b \pm \sqrt{b^2 - 4ac}}{2a} \qquad \text{Quadratic Formula}$$

$$x = \frac{-5 \pm \sqrt{5^2 - 4(1)(-14)}}{2(1)} \qquad \text{Substitute 1 for } a, 5 \text{ for } b, \text{ and } -14 \text{ for } c.$$

$$x = \frac{-5 \pm \sqrt{25 + 56}}{2} \qquad \text{Simplify.}$$

$$x = \frac{-5 \pm \sqrt{81}}{2} \qquad \text{Simplify.}$$

$$x = \frac{-5 \pm 9}{2} \qquad \text{Simplify.}$$

$$x = \frac{-5 + 9}{2} = \frac{4}{2} = 2 \quad \text{and} \quad x = \frac{-5 - 9}{2} = -\frac{14}{2} = -7.$$

The solutions are $x = 2$ and $x = -7$. Check these in the original equation.

Study Tip

The equation in Example 2 could have been solved by factoring. Try doing so, and compare your results with those obtained in Example 2.

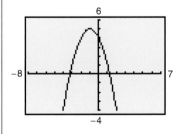
Study Tip

Notice in Example 4 that 2 is factored out of the original equation. To use the Quadratic Formula, you do not have to factor out a common numerical factor; however, it makes the arithmetic easier.

Point out that rational solutions indicate a quadratic equation that can be solved by factoring.

Example 3   The Quadratic Formula: Two Distinct Solutions

Use the Quadratic Formula to solve $-x^2 - 2x = -4$.

Solution

$-x^2 - 2x = -4$	Write original equation.
$-x^2 - 2x + 4 = 0$	Write in general form with $a = -1, b = -2, c = 4$.
$x = \dfrac{-b \pm \sqrt{b^2 - 4ac}}{2a}$	Quadratic Formula
$x = \dfrac{-(-2) \pm \sqrt{(-2)^2 - 4(-1)(4)}}{2(-1)}$	Substitute -1 for a, -2 for b, and 4 for c.
$x = \dfrac{2 \pm \sqrt{4 + 16}}{-2}$	Simplify.
$x = \dfrac{2 \pm \sqrt{20}}{-2}$	Simplify.
$x = \dfrac{2 \pm 2\sqrt{5}}{-2}$	Simplify radical.
$x = \dfrac{(-2)(-1 \pm \sqrt{5})}{(-2)}$	Divide out common factor.
$x = -1 \pm \sqrt{5}$	Simplify.

So, the solutions are $x = -1 + \sqrt{5} \approx 1.24$ and $x = -1 - \sqrt{5} \approx -3.24$. Check these in the original equation.

Example 4   The Quadratic Formula: One Repeated Solution

Use the Quadratic Formula to solve $8x^2 - 24x + 18 = 0$.

Solution

$8x^2 - 24x + 18 = 0$	Write original equation.
$4x^2 - 12x + 9 = 0$	Divide each side by 2.
$x = \dfrac{-b \pm \sqrt{b^2 - 4ac}}{2a}$	Quadratic Formula
$x = \dfrac{-(-12) \pm \sqrt{(-12)^2 - 4(4)(9)}}{2(4)}$	Substitute 4 for a, -12 for b, and 9 for c.
$x = \dfrac{12 \pm \sqrt{144 - 144}}{8}$	Simplify.
$x = \dfrac{12 \pm \sqrt{0}}{8} = \dfrac{3}{2}$	Simplify.

This equation has only one (repeated) solution, $x = \frac{3}{2}$. Check this in the original equation.

In the next example, note how the Quadratic Formula can be used to determine that a quadratic equation has no real solution.

Students can use graphing calculators to verify that the graph of $y = 2x^2 - 4x + 5$ has *no x-intercept*, and therefore has no real solution.

Example 5   The Quadratic Formula: No Real Solution

Use the Quadratic Formula to solve

$$2x^2 - 4x + 5 = 0.$$

Solution

Additional Examples
Use the Quadratic Formula to solve each equation.

a. $-x^2 - 4x + 8 = 0$

b. $3x^2 + 4x = 8$

Answers:

a. $x = -2 \pm 2\sqrt{3}$

b. $x = \dfrac{-2 \pm 2\sqrt{7}}{3}$

$$2x^2 - 4x + 5 = 0$$
General form with $a = 2, b = -4, c = 5$

$$x = \frac{-b \pm \sqrt{b^2 - 4ac}}{2a}$$
Quadratic Formula

$$x = \frac{-(-4) \pm \sqrt{(-4)^2 - 4(2)(5)}}{2(2)}$$
Substitute 2 for a, -4 for b, and 5 for c.

$$x = \frac{4 \pm \sqrt{16 - 40}}{4}$$
Simplify.

$$x = \frac{4 \pm \sqrt{-24}}{4}$$
Simplify.

Because $\sqrt{-24}$ is not a real number, you can conclude that the original equation has no real solution. Notice that -24 is the discriminant, $b^2 - 4ac$, which could have been calculated first to show that the original equation has no real solution.

You have now studied four ways to solve quadratic equations.

1. Square Root Property   Section 10.1
2. Factoring   Sections 6.5 and 10.1
3. Completing the square   Section 10.2
4. The Quadratic Formula   Section 10.3

The following guidelines may help you decide which method best applies to an equation. Note that *completing the square* is not recommended as a practical technique—it is used more as a theoretical technique.

Guidelines for Solving Quadratic Equations

1. First check to see whether you can solve the equation by the Square Root Property.

2. If you can't use the Square Root Property, write the equation in general form and try factoring.

3. If you can't factor the quadratic equation in general form, apply the Quadratic Formula.

Remember that the Quadratic Formula can be used to solve any quadratic equation.

Example 6   Mountain Biker's Speed

A mountain biker spends a total of 5 hours going up a 25-mile mountain trail and coming back down. The biker's speed up the trail is 4 miles per hour slower than the speed down the trail. What is the biker's speed coming down the trail?

Solution

Form a verbal model for the total time. Remember that *Distance = Rate × Time* so,

$$Time = Distance \div Rate.$$

Let x represent the rate coming down the trail and let $x - 4$ represent the rate going up the trail.

Verbal Model:	Total time = Time up + Time down	

Labels:	Total time = 5	(hours)
	Time up $= \dfrac{25}{x - 4}$	(hours)
	Time down $= \dfrac{25}{x}$	(hours)

> **Technology: Tip**
>
> Graphing calculator programs for solving a quadratic equation in general form can be found at our website *math.college.hmco.com/students.* Try using this program to solve the equation in Example 6.

Equation:

$$5 = \frac{25}{x - 4} + \frac{25}{x} \qquad \text{Original equation}$$

$$x(x - 4)(5) = x(x - 4)\left(\frac{25}{x - 4} + \frac{25}{x}\right) \qquad \text{Multiply each side by LCD of } x(x - 4).$$

$$5x^2 - 20x = 25x + 25x - 100 \qquad \text{Distribute and simplify.}$$

$$5x^2 - 70x + 100 = 0 \qquad \text{Quadratic equation in general form}$$

$$x^2 - 14x + 20 = 0 \qquad \text{Divide each side by 5.}$$

This equation does not factor, so use the Quadratic Formula to solve the equation.

$$x = \frac{-b \pm \sqrt{b^2 - 4ac}}{2a} \qquad \text{Quadratic Formula}$$

$$x = \frac{-(-14) \pm \sqrt{(-14)^2 - 4(1)(20)}}{2(1)} \qquad \text{Substitute 1 for } a, -14 \text{ for } b, \text{ and } 20 \text{ for } c.$$

$$x = \frac{14 \pm \sqrt{116}}{2} \qquad \text{Simplify.}$$

$$x = \frac{14 \pm 2\sqrt{29}}{2} \qquad \text{Simplify radical.}$$

$$x = 7 \pm \sqrt{29} \qquad \text{Solution to quadratic equation}$$

The biker's speed coming down the trail is

$$7 + \sqrt{29} \approx 12.4 \text{ miles per hour.}$$

The solution $7 - \sqrt{29} \approx 1.6$ is excluded because the uphill rate $x - 4$ would be negative. Check the solution in the original statement of the problem.

10.3 Exercises

Review   Concepts, Skills, and Problem Solving

Keep mathematically in shape by doing these exercises *before* the problems of this section.

Properties and Definitions

1. *Writing* ✏ Explain how to simplify the expression

$\dfrac{x^2 - 4}{x + 2}$.

Factor completely and divide out the common factor $x + 2$.

2. *Writing* ✏ Explain how to divide $\dfrac{4x^2}{3y}$ by $\dfrac{15y^2}{6x}$.

Invert the divisor and multiply.

Rational Expressions and Equations

In Exercises 3–6, perform the operation and simplify.

3. $\dfrac{5}{x - 1} \cdot \dfrac{x - 1}{25(x - 2)}$

$\dfrac{1}{5(x - 2)}, \; x \neq 1$

4. $\dfrac{x + 2}{5(x - 3)} \div \dfrac{x - 2}{5(x - 3)}$

$\dfrac{x + 2}{x - 2}, \; x \neq 3$

5. $\dfrac{6}{x - 3} + \dfrac{x}{x - 3}$ $\dfrac{x + 6}{x - 3}$

6. $\dfrac{3}{x - 1} - 5$ $-\dfrac{5x - 8}{x - 1}$

In Exercises 7 and 8, solve the equation.

7. $\dfrac{5x - 4}{5x + 4} = \dfrac{2}{3}$ 4

8. $\dfrac{15}{x} - 4 = \dfrac{6}{x} + 3$ $\dfrac{9}{7}$

Problem Solving

9. *Mixture Problem*   Determine the number of gallons of a 25% solution that must be mixed with a 50% solution to obtain 10 gallons of a 40% solution.   4 gallons

10. *Test Scores*   To get an A in a course, a student must have an average of at least 90 on four tests of 100 points each. A student scores 83, 92, and 88 on the first three tests. What must the student score on the fourth test to earn a 90% average for the course?   97

Developing Skills

In Exercises 1–6, write the quadratic equation in general form.

1. $3x^2 - x = 7$ $3x^2 - x - 7 = 0$

2. $x^2 + 6x = -2$ $x^2 + 6x + 2 = 0$

3. $x^2 = 3 - 2x$ $x^2 + 2x - 3 = 0$

4. $2x^2 + 3x = 5$ $2x^2 + 3x - 5 = 0$

5. $x(4 - x) = 10$ $-x^2 + 4x - 10 = 0$

6. $x(8x + 3) = 2$ $8x^2 + 3x - 2 = 0$

In Exercises 7–14, use the discriminant to determine the number of real solutions of the quadratic equation. See Example 1.

7. $2x^2 - 3x - 1 = 0$ Two real solutions

8. $4x^2 + 4x + 1 = 0$ One real solution

9. $x^2 + 4x + 5 = 0$ No real solution

10. $x^2 - 2x + 5 = 0$ No real solution

11. $x^2 + 6x + 1 = 0$ Two real solutions

12. $x^2 + 6x - 10 = 0$ Two real solutions

13. $9x^2 - 12x + 4 = 0$ One real solution

14. $2x^2 + 5x + 3 = 0$ Two real solutions

In Exercises 15–44, use the Quadratic Formula to solve the quadratic equation. (Some equations have no real solutions.) See Examples 2–5.

15. $x^2 - 3x - 18 = 0$ $-3, 6$

16. $x^2 - 3x - 10 = 0$ $-2, 5$

17. $x^2 + 8x + 15 = 0$ $-5, -3$

18. $x^2 + 8x + 12 = 0$ $-6, -2$

19. $t^2 + 5t + 6 = 0$ $-3, -2$

20. $y^2 + y + 1 = 0$ No real solution

21. $x^2 - 6x + 7 = 0$ $3 \pm \sqrt{2}$

22. $x^2 - 10x + 22 = 0$ $5 \pm \sqrt{3}$

23. $t^2 + t + 3 = 0$ No real solution

24. $u^2 + 5u + 2 = 0$ $\dfrac{-5 \pm \sqrt{17}}{2}$

25. $x^2 = 3x + 1$ $\dfrac{3 \pm \sqrt{13}}{2}$

26. $x^2 = 12x - 20$ $2, 10$

27. $4x^2 - 20x + 25 = 0$ $\dfrac{5}{2}$

28. $9x^2 + 6x + 1 = 0$ $-\dfrac{1}{3}$

29. $8x^2 - 10x + 3 = 0$ $\dfrac{1}{2}, \dfrac{3}{4}$

30. $4x^2 - 13x + 3 = 0$ $\dfrac{1}{4}, 3$

31. $3z^2 + 4z + 4 = 0$ No real solution

32. $9z^2 + 10z + 4 = 0$ No real solution

33. $5x^2 + 2x - 2 = 0$ $\dfrac{-1 \pm \sqrt{11}}{5}$

34. $2x^2 - 7x - 9 = 0$ $-1, \dfrac{9}{2}$

35. $4x^2 - 4x - 1 = 0$ $\dfrac{1 \pm \sqrt{2}}{2}$

36. $3x^2 + 4x - 1 = 0$ $\dfrac{-2 \pm \sqrt{7}}{3}$

37. $\dfrac{1}{2}x^2 + 2x - 3 = 0$ $-2 \pm \sqrt{10}$

38. $\dfrac{3}{2}z^2 + 1 = 2z$ No real solution

39. $0.5x^2 - 0.8x + 0.3 = 0$ $\dfrac{3}{5}, 1$

40. $0.3x^2 + 0.7x - 0.4 = 0$ $\dfrac{-7 \pm \sqrt{97}}{6}$

41. $0.2y^2 + y + 6 = 0$ No real solution

42. $0.1x^2 - x - 1 = 0$ $5 \pm \sqrt{35}$

43. $0.36s^2 - 0.12s + 0.01 = 0$ $\dfrac{1}{6}$

44. $0.06t^2 + 0.05t - 0.10 = 0$ $\dfrac{-5 \pm \sqrt{265}}{12}$

In Exercises 45–58, solve the quadratic equation by the most convenient method.

45. $x^2 = 18$ $\pm 3\sqrt{2}$

46. $t^2 = 27$ $\pm 3\sqrt{3}$

47. $y^2 + 8y = 0$ $-8, 0$

48. $7u^2 - 49u = 0$ $0, 7$

49. $2y(y - 12) + 3(y - 12) = 0$ $-\dfrac{3}{2}, 12$

50. $x(x + 2) - 5(x + 2) = 0$ $-2, 5$

51. $(x - 3)^2 - 75 = 0$ $3 \pm 5\sqrt{3}$

52. $(y - 8)^2 - 20 = 0$ $8 \pm 2\sqrt{5}$

53. $x^2 - 6x + 3 = 0$ $3 \pm \sqrt{6}$

54. $x^2 + 14x + 49 = 0$ -7

55. $-2x^2 + 6x + 1 = 0$ $\dfrac{3 \pm \sqrt{11}}{2}$

56. $6x^2 + 20x + 5 = 0$ $\dfrac{-10 \pm \sqrt{70}}{6}$

57. $10x^2 + x - 3 = 0$ $-\dfrac{3}{5}, \dfrac{1}{2}$

58. $4a^2 - 12a + 9 = 0$ $\dfrac{3}{2}$

In Exercises 59–62, solve the quadratic equation using the Quadratic Formula. Use a calculator to approximate your solution to three decimal places.

59. $3x^2 - 14x + 4 = 0$

$\dfrac{7 + \sqrt{37}}{3} \approx 4.361$

$\dfrac{7 - \sqrt{37}}{3} \approx 0.306$

60. $7x^2 + x - 35 = 0$

$\dfrac{-1 + 3\sqrt{109}}{14} \approx 2.166$

$\dfrac{-1 - 3\sqrt{109}}{14} \approx -2.309$

61. $-0.03x^2 + 2x - 0.5 = 0$

$\dfrac{100 + 5\sqrt{394}}{3} \approx 66.416$

$\dfrac{100 - 5\sqrt{394}}{3} \approx 0.251$

62. $1.7x^2 - 4.2x + 2.1 = 0$

$\dfrac{21 + 2\sqrt{21}}{17} \approx 1.774$

$\dfrac{21 - 2\sqrt{21}}{17} \approx 0.696$

In Exercises 63–66, solve the equation.

63. $\dfrac{x + 3}{2} - \dfrac{4}{x} = 2$

$\dfrac{1 \pm \sqrt{33}}{2}$

64. $\dfrac{2}{r + 1} + \dfrac{2}{r} = 1$

$\dfrac{3 \pm \sqrt{17}}{2}$

65. $\sqrt{4x + 3} = x - 1$

$3 + \sqrt{11}$

66. $\sqrt{3x - 2} = x - 2$ 6

Solving Problems

67. *Biker's Speed* A mountain biker spends a total of 4 hours going up a 20-mile mountain trail and coming back down. The biker's speed up the trail is 5 miles per hour lower than the speed down the trail. What is the biker's speed coming down the trail?

13.1 miles per hour

68. *Hiker's Speed* A hiker spends a total of 4 hours going up a six-mile trail and coming back down. The hiker's speed up the trail is 1 mile per hour lower than the speed down the trail. What is the hiker's speed coming down the trail? 3.6 miles per hour

69. *Free-Falling Object* An object is thrown upward with an initial velocity of 20 feet per second from the Sky Pod of the CN Tower 1465 feet above the ground. The height h (in feet) of the object t seconds after it is thrown is modeled by

$$h = -16t^2 + 20t + 1465.$$

(a) Find the two times when the object is 1465 feet above the ground. 0 seconds; $\frac{5}{4}$ seconds

(b) Find the time when the object strikes the ground.
10.21 seconds

70. *Free-Falling Object* An object is thrown upward with an initial velocity of 24 feet per second from the Sydney Harbor Bridge 161 feet above the water. The height h (in feet) of the object t seconds after it is thrown is modeled by $h = -16t^2 + 24t + 161.$

(a) Find the two times when the object is 169 feet above the water level. $\frac{1}{2}$ second; 1 second

(b) Find the time when the object strikes the water.
4.01 seconds

71. ▲ *Geometry* The area of the rectangle shown in the figure is 58.14 square inches. Use the Quadratic Formula to find its dimensions.
11.4 inches × 5.1 inches

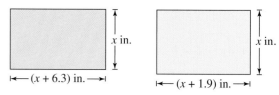

Figure for 71   Figure for 72

72. ▲ *Geometry* The area of the rectangle shown in the figure is 26.66 square inches. Use the Quadratic Formula to find its dimensions.
6.2 inches × 4.3 inches

Explaining Concepts

73. ⊘ Answer parts (e) and (f) of Motivating the Chapter on page 536.

74. *Writing* State the Quadratic Formula in words.

Compute $-b$ plus or minus the square root of the quantity b squared minus $4ac$. This quantity divided by the quantity $2a$ is the Quadratic Formula.

75. *Writing* Can the Quadratic Formula be used to solve the equation $(x - 2)(x - 3) = 0$? If it can, would it be the simplest method? Explain. After the binomial factors are multiplied, the Quadratic Formula could be used. This would not be the most efficient method, because the quadratic equation is already factored.

76. State the four methods used to solve quadratic equations. The four methods are factoring, Square Root Property, completing the square, and the Quadratic Formula.

77. Use the Quadratic Formula to show that the sum of the solutions of a quadratic equation is $-b/a$ and the product of the solutions is c/a. Proof

78. *Writing* Explain how the discriminant of a quadratic equation can be used to determine the number of real solutions of the equation. Create a quadratic equation for each case.

When the discriminant is positive, the equation has two real solutions. When the discriminant is zero, the equation has one (repeated) real solution. When the discriminant is negative, the equation has no real solution. Examples will vary.

Mid-Chapter Quiz

Take this quiz as you would take a quiz in class. After you are done, check your work against the answers in the back of the book.

In Exercises 1–12, solve the quadratic equation by the specified method.

1. Factoring:
   $x^2 - 5x + 6 = 0$ $3, 2$

2. Square Root Property:
   $x^2 = 400$ ± 20

3. Factoring:
   $2x^2 + 9x - 35 = 0$ $-7, \frac{5}{2}$

4. Factoring:
   $x(x - 4) + 3(x - 4) = 0$ $-3, 4$

5. Square Root Property:
   $x^2 - 2500 = 0$ ± 50

6. Square Root Property:
   $(z - 4)^2 - 81 = 0$ $-5, 13$

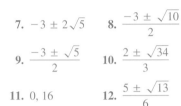

7. $-3 \pm 2\sqrt{5}$

8. $\dfrac{-3 \pm \sqrt{10}}{2}$

9. $\dfrac{-3 \pm \sqrt{5}}{2}$

10. $\dfrac{2 \pm \sqrt{34}}{3}$

11. $0, 16$

12. $\dfrac{5 \pm \sqrt{13}}{6}$

7. Completing the square:
   $y^2 + 6y - 11 = 0$

8. Completing the square:
   $4u^2 + 12u - 1 = 0$

9. Quadratic Formula:
   $x^2 + 3x + 1 = 0$

10. Quadratic Formula:
    $3x^2 - 4x - 10 = 0$

11. Quadratic Formula
    $3x^2 - 48x = 0$

12. Quadratic Formula
    $5x = 3x^2 + 1$

In Exercises 13–16, use the discriminant to determine the number of real solutions of the equation.

13. $x^2 + x + \dfrac{9}{4} = 0$ No real solution

14. $y^2 - 7y - 1 = 0$ Two real solutions

15. $3x^2 - 4x - 4 = 0$ Two real solutions

16. $9x^2 + 6x + 1 = 0$ One real solution

17. Fifteen hundred dollars is deposited in an account. After 2 years, the balance in the account is $1669.54. The interest earned on the account is compounded annually. What is the interest rate?   5.5%

18. On August 7, 1998, Stig Günther from Denmark dove 343 feet into a 39.4-by-49.2-by-14.8-foot air bag. His height h (in feet) after t seconds into the fall is modeled by $h = 343 - 16t^2$. How long was Günther in the air? (Source: Guinness Book of World Records)
    $\dfrac{7\sqrt{7}}{4} \approx 4.63$ seconds

19. The area of the rectangle shown in the figure is 153.92 square inches. Find the dimensions of the rectangle.   7.4 inches × 20.8 inches

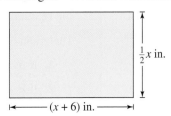

10.4 Graphing Quadratic Equations

Brooklyn Production/Corbis

Why You Should Learn It

Quadratic equations can be used to model data for analysis of consumer behavior. For instance, in Exercise 97 on page 572, a quadratic equation is used to model the number of cellular phone subscribers in the United States.

① Determine whether parabolas open upward or downward using the Leading Coefficient Test.

What You Should Learn

① Determine whether parabolas open upward or downward using the Leading Coefficient Test.

② Sketch graphs of quadratic equations using the point-plotting method and the vertex of a parabola.

Quadratic Equations in Two Variables

In this section, you will study the graphs of quadratic equations. A **quadratic equation in two variables** can be written in the form

$$y = ax^2 + bx + c, \quad a \neq 0.$$

The graph of a quadratic equation is called a **parabola.** Parabolas are U-shaped, like the reflective part of a flashlight or like the flight path of a baseball. If the coefficient of x^2 (the leading coefficient) is positive, the parabola opens upward, as shown in Figure 10.1. The lowest point (minimum) on a parabola that opens upward is the **vertex** of the parabola. If the leading coefficient is negative, the parabola opens downward, and its vertex is the highest point (maximum) on the parabola.

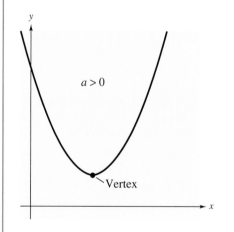

Parabola opens upward.
Figure 10.1

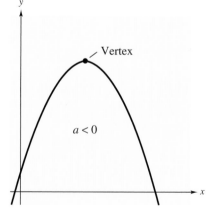

Parabola opens downward.

Technology: Discovery

You can use a graphing calculator to discover a rule for determining the appearance of a parabola. Graph each equation using a graphing calculator.

a. $y_1 = x^2 + 4x + 3$

b. $y_2 = 5 - 2x^2$

c. $y_3 = -7 + 3x^2$

d. $y_4 = -x^2 + 6x$

In your own words, write a rule for determining whether the graph of a parabola opens upward or downward by just looking at the equation. Does $y = 9 - 4x - 2x^2$ open upward or downward?

See Technology Answers.

The Leading Coefficient Test for Parabolas

The graph of the quadratic equation $y = ax^2 + bx + c$ is a **parabola.**

1. If $a > 0$, the parabola opens upward.

2. If $a < 0$, the parabola opens downward.

Example 1 shows how to use the Leading Coefficient Test to determine whether the graph of a quadratic equation opens upward or downward.

Example 1   Using the Leading Coefficient Test

Determine whether the parabola opens upward or downward.

a. $y = -x^2 + 2x + 3$ **b.** $y = 4x^2 - 1$

c. $y = 2 - 5x - 3x^2$ **d.** $y = 4 - 2x(3 - x)$

Solution

a. In general form, you can see that the leading coefficient is negative.

$$y = ax^2 + bx + c \qquad \text{General form}$$

$$y = -x^2 + 2x + 3 \quad \Longrightarrow \quad a = -1 \qquad \text{Leading coefficient is negative.}$$

The parabola opens downward, as shown in Figure 10.2.

b. In general form, you can see that the leading coefficient is positive.

$$y = ax^2 + bx + c \qquad \text{General form}$$

$$y = 4x^2 - 1 \quad \Longrightarrow \quad a = 4 \qquad \text{Leading coefficient is positive.}$$

The parabola opens upward, as shown in Figure 10.3.

c. In general form, you can see that the leading coefficient is negative.

$$y = ax^2 + bx + c \qquad \text{General form}$$

$$y = -3x^2 - 5x + 2 \quad \Longrightarrow \quad a = -3 \qquad \text{Leading coefficient is negative.}$$

The parabola opens downward, as shown in Figure 10.4.

d. In general form, you can see that the leading coefficient is positive.

$$y = ax^2 + bx + c \qquad \text{General form}$$

$$y = 4 - 6x + 2x^2 \qquad \text{Distributive Property}$$

$$y = 2x^2 - 6x + 4 \quad \Longrightarrow \quad a = 2 \qquad \text{Leading coefficient is positive.}$$

The parabola opens upward, as shown in Figure 10.5.

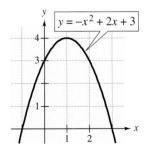

Figure 10.2

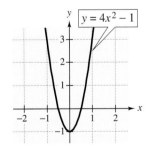

Figure 10.3

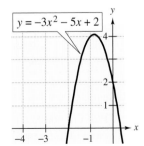

Figure 10.4

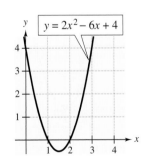

Figure 10.5

Verify the graphs shown in Example 1 using a graphing calculator.

② Sketch graphs of quadratic equations using the point-plotting method and the vertex of a parabola.

Sketching the Graph of a Quadratic Equation

There are three basic approaches to sketching the graph of a quadratic equation.

1. *Numerical Approach* You can create a table of values and use the point-plotting method.

2. *Graphing Calculator* You can use a graphing calculator to graph the equation.

3. *Analytic Approach* You can analyze the characteristics of the graph and use the results to draw the graph.

You will probably find that a combination of these approaches is most efficient.

Review the techniques for determining intercepts in Section 4.2.

Example 2 Using the Point-Plotting Method

Find the intercepts of the graph of $y = x^2 - 4$. Then sketch the graph of the equation and label the intercepts.

Solution

To find the x-intercepts, let y equal zero and solve the resulting equation for x.

$$x^2 - 4 = 0 \qquad \text{Let } y = 0 \text{ and solve for } x.$$

$$(x + 2)(x - 2) = 0 \qquad \text{Factor.}$$

$$x + 2 = 0 \quad\Longrightarrow\quad x = -2 \qquad \text{Set 1st factor equal to 0.}$$

$$x - 2 = 0 \quad\Longrightarrow\quad x = 2 \qquad \text{Set 2nd factor equal to 0.}$$

From these two solutions, you can see that the graph has two x-intercepts: $(-2, 0)$ and $(2, 0)$. To find the y-intercept, let x equal zero in the original equation and solve for y, as follows.

$$y = (0)^2 - 4 \qquad \text{Substitute 0 for } x.$$

$$= -4 \qquad \text{Simplify.}$$

So, the y-intercept is $(0, -4)$. To sketch the graph of the equation, create a table of values. (Note that the three intercepts are included in the table.)

x	-3	-2	-1	0	1	2	3
$y = x^2 - 4$	5	0	-3	-4	-3	0	5
Solution point	$(-3, 5)$	$(-2, 0)$	$(-1, -3)$	$(0, -4)$	$(1, -3)$	$(2, 0)$	$(3, 5)$

Plot the points and connect them with a smooth curve, as shown in Figure 10.6. Note that the parabola opens upward because the leading coefficient is positive.

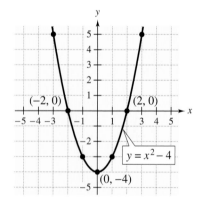

Figure 10.6

When creating a table of values, you should include any intercepts you have found. You should also include points to the left and right of the intercepts.

Vertex of a Parabola

The vertex of a parabola given by $y = ax^2 + bx + c$ occurs at the point whose x-coordinate is

$$x = -\frac{b}{2a}.$$

To find the y-coordinate of the vertex, substitute the x-coordinate in the equation $y = ax^2 + bx + c$.

Example 3 Finding the Vertex of a Parabola

Find the vertex of each parabola.

a. $y = x^2 + 2x - 1$ **b.** $y = -x^2 + 3x$

Solution

a. For this equation, $a = 1$ and $b = 2$. So, the x-coordinate of the vertex is

$$x = -\frac{b}{2a} = -\frac{2}{2(1)} = -1. \qquad \text{Substitute 1 for } a \text{ and 2 for } b.$$

The y-coordinate of the vertex is

$$y = (-1)^2 + 2(-1) - 1 = -2. \qquad \text{Substitute } -1 \text{ for } x.$$

So, the vertex occurs at $(-1, -2)$, as shown in Figure 10.7.

b. For this equation, $a = -1$ and $b = 3$. So, the x-coordinate of the vertex is

$$x = -\frac{b}{2a} = -\frac{3}{2(-1)} = \frac{3}{2}. \qquad \text{Substitute } -1 \text{ for } a \text{ and 3 for } b.$$

The y-coordinate of the vertex is

$$y = -\left(\tfrac{3}{2}\right)^2 + 3\left(\tfrac{3}{2}\right) = \tfrac{9}{4}. \qquad \text{Substitute } \tfrac{3}{2} \text{ for } x.$$

So, the vertex occurs at $\left(\tfrac{3}{2}, \tfrac{9}{4}\right)$, as shown in Figure 10.8.

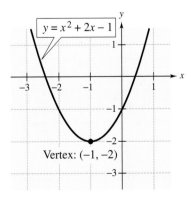

Figure 10.7

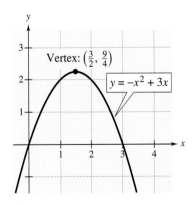

Figure 10.8

Guidelines for Sketching a Parabola

1. Use the Leading Coefficient Test to determine whether the parabola opens upward or downward.

2. Find and plot the x-intercepts (if any) and the y-intercept.

3. Find and plot the vertex.

4. Create a table of values that includes a few additional points.

5. Complete the graph with a smooth, U-shaped curve.

The graph of every quadratic equation has exactly one y-intercept. The number of x-intercepts, however, can vary. In Examples 4, 5, and 6, notice that parabolas can have two x-intercepts, one x-intercept, or no x-intercept.

Example 4   Sketching a Parabola: Two x-Intercepts

Sketch the graph of $y = -2x^2 - x + 6$.

Solution

From the Leading Coefficient Test, the parabola opens downward. The y-intercept is $(0, 6)$ and the x-intercepts are $\left(\frac{3}{2}, 0\right)$ and $(-2, 0)$. The vertex is $\left(-\frac{1}{4}, \frac{49}{8}\right)$.

When finding these x-intercepts, you could multiply each side of the equation by -1.

$$0(-1) = (-2x^2 - x + 6)(-1)$$
$$0 = 2x^2 + x - 6$$
$$0 = (2x - 3)(x + 2)$$
$$2x - 3 = 0 \rightarrow x = \tfrac{3}{2}$$
$$x + 2 = 0 \rightarrow x = -2$$

The x-intercepts are $\left(\frac{3}{2}, 0\right)$ and $(-2, 0)$.

x	-3	-1	1	2
$y = -2x^2 - x + 6$	-9	5	3	-4
Solution point	$(-3, -9)$	$(-1, 5)$	$(1, 3)$	$(2, -4)$

Plot the intercepts, the vertex, and the additional points shown in the table of values. Then connect the points with a smooth curve, as shown in Figure 10.9.

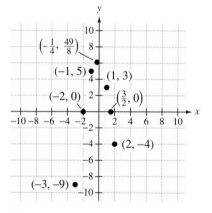
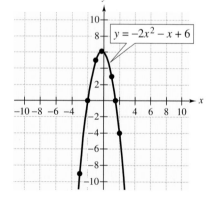

Figure 10.9

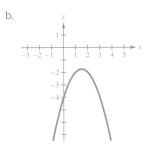
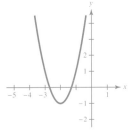
Example 5   **Sketching a Parabola: One x-Intercept**

Sketch the graph of $y = x^2 - 4x + 4$.

Solution

From the Leading Coefficient Test, the parabola opens upward. The y-intercept is $(0, 4)$ and the x-intercept is $(2, 0)$. The vertex is $(2, 0)$. Plot the intercepts, the vertex, and some additional points. Connect the points, as shown in Figure 10.10.

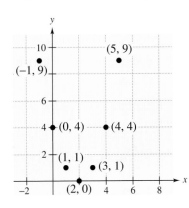
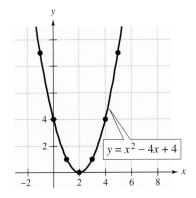

Figure 10.10

Example 6   **Sketching a Parabola: No x-Intercept**

Sketch the graph of $y = x^2 - 6x + 10$.

Solution

From the Leading Coefficient Test, the parabola opens upward. The y-intercept is $(0, 10)$, and there is no x-intercept. The vertex is $(3, 1)$. Plot the vertex and some additional points. Then connect the points, as shown in Figure 10.11.

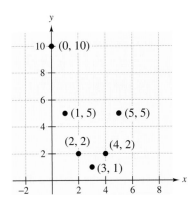
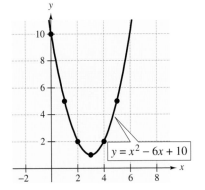

Figure 10.11

In Example 6, notice the relationship between the graph of the equation $y = x^2 - 6x + 10$ and the solution of the equation $x^2 - 6x + 10 = 0$. The discriminant is $(-6)^2 - 4(1)(10) = 36 - 40 = -4$. Because the discriminant is less than 0, the graph has no x-intercept and the equation has no real solution.

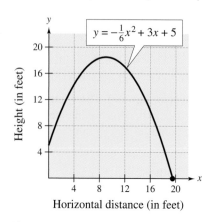

Figure 10.12

Example 7 **Analyzing the Path of a Ball**

The height y (in feet) of a ball thrown on a parabolic path is modeled by

$$y = -\frac{1}{6}x^2 + 3x + 5$$

where x is the horizontal distance (in feet) from where the ball is thrown (see Figure 10.12).

a. From what height is the ball thrown?

b. What is the maximum height reached by the ball?

c. How far does the ball travel horizontally through the air?

Solution

a. The height from which the ball is thrown occurs when $x = 0$. When $x = 0$, the height is

$$y = -\frac{1}{6}(0)^2 + 3(0) + 5 \qquad \text{Substitute 0 for } x.$$

$$= 5 \text{ feet.}$$

b. The maximum height occurs at the vertex of the parabolic path. The x-value of this vertex is

$$x = -\frac{b}{2a} = -\frac{3}{2\left(-\frac{1}{6}\right)} = 9. \qquad \text{Substitute } -\frac{1}{6} \text{ for } a \text{ and 3 for } b.$$

At this x-value, the height is

$$y = -\frac{1}{6}(9)^2 + 3(9) + 5 \qquad \text{Substitute 9 for } x.$$

$$= 18.5 \text{ feet.}$$

c. The distance that the ball travels horizontally through the air corresponds to the x-intercept of the parabolic path. This intercept can be found by letting $y = 0$ and using the Quadratic Formula to solve the resulting equation.

$$x = \frac{-3 \pm \sqrt{3^2 - 4\left(-\frac{1}{6}\right)(5)}}{2\left(-\frac{1}{6}\right)} \qquad \begin{array}{l} \text{Substitute } -\frac{1}{6} \text{ for } a, 3 \text{ for } b, \text{ and} \\ 5 \text{ for } c \text{ in the Quadratic Formula.} \end{array}$$

$$= \frac{-3 \pm \sqrt{\frac{37}{3}}}{-\frac{1}{3}} \qquad \text{Simplify.}$$

$$= 9 \pm \sqrt{111} \qquad \text{Simplify.}$$

Because you want the positive distance, the ball travels

$$x = 9 + \sqrt{111}$$

$$\approx 9 + 10.5$$

$$= 19.5 \text{ feet}$$

horizontally through the air.

10.4 Exercises

Review   Concepts, Skills, and Problem Solving

Keep mathematically in shape by doing these exercises *before* the problems of this section.

Properties and Definitions

In Exercises 1 and 2, rewrite the expression, where a and b are nonnegative real numbers, using the specified property.

1. Multiplication Property: $\sqrt{ab} = $ $\sqrt{a}\ \sqrt{b}$

2. Division Property: $\sqrt{\dfrac{a}{b}} = $ $\sqrt{a}/\sqrt{b}$

3. *Writing* Is $\sqrt{80}$ in simplest form? Explain.
   No. $\sqrt{80} = 4\sqrt{5}$

4. *Writing* Is $10\sqrt{5}$ in simplest form? Explain.
   Yes. All possible factors have been removed from the radical.

Simplifying Expressions

In Exercises 5–10, perform the operation and simplify the expression.

5. $8\sqrt{15} - 6\sqrt{15}$ $2\sqrt{15}$

6. $\sqrt{10}\left(1 - \sqrt{2}\right)$ $\sqrt{10} - 2\sqrt{5}$

7. $\left(\sqrt{3} + 2\right)\left(\sqrt{3} - 2\right)$ -1

8. $\dfrac{4}{5 - \sqrt{2}}$ $\dfrac{4\left(5 + \sqrt{2}\right)}{23}$

9. $\sqrt{\dfrac{6}{5}}$ $\dfrac{\sqrt{30}}{5}$

10. $\dfrac{2 + \sqrt{8}}{\frac{1}{2}}$ $4\left(1 + \sqrt{2}\right)$

Problem Solving

11. ▲ *Geometry* A 20-foot board leans against the side of a house. The bottom is 5 feet from the house. How far does the board reach up the side of the house? $5\sqrt{15} \approx 19.36$ feet

12. *Quality Control* A quality control engineer for a manufacturer found one defective unit in a sample of 125. At that rate, what is the expected number of defective units in a shipment of 150,000?
    1200 defective units

Developing Skills

In Exercises 1–6, match the equation with its graph. [The graphs are labeled (a), (b), (c), (d), (e), and (f).]

(a)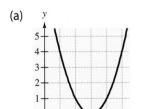

(b)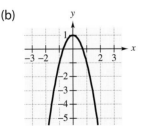

(e)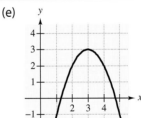

(f)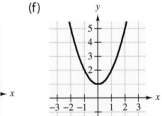

(c)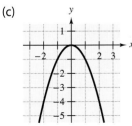

(d)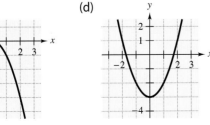

1. $y = -x^2$ c
2. $y = x^2 - 3$ d
3. $y = x^2 + 1$ f
4. $y = -2x^2 + 1$ b
5. $y = x^2 - 6x + 9$ a
6. $y = -(x^2 - 6x + 6)$ e

In Exercises 7–14, determine whether the parabola opens upward or downward. See Example 1.

7. $y = x^2 - 4x + 3$
   Upward
8. $y = 5 - 6x + x^2$
   Upward
9. $y = 6 + x - 2x^2$
10. $y = -2x^2 - 4x + 6$
11. $y = 3 + x(3 - x)$
12. $y = 6 - 2x(2 - x)$
13. $y = -(x + 1)^2 - 1$
14. $y = (x - 3)^2 - 2$

9. Downward   10. Downward   11. Downward
12. Upward   13. Downward   14. Upward

In Exercises 15–28, find the intercepts of the graph. See Example 2.

15. $y = 16 - x^2$
$(-4, 0), (4, 0), (0, 16)$

16. $y = x^2 - 36$
$(-6, 0), (6, 0), (0, -36)$

17. $y = x^2 - 2x$
$(0, 0), (2, 0)$

18. $y = 4x - x^2$
$(0, 0), (4, 0)$

19. $y = x^2 - 4x + 3$
$(1, 0), (3, 0), (0, 3)$

20. $y = x^2 - 6x + 5$
$(1, 0), (5, 0), (0, 5)$

21. $y = -x^2 + 8x - 12$
$(2, 0), (6, 0), (0, -12)$

22. $y = -x^2 - 2x + 3$
$(-3, 0), (1, 0), (0, 3)$

23. $y = 3x^2 + 4x - 4$
$(-2, 0), \left(\frac{2}{3}, 0\right), (0, -4)$

24. $y = 2x^2 - 7x + 3$
$\left(\frac{1}{2}, 0\right), (3, 0), (0, 3)$

25. $y = 3x^2 + 5x + 4$
$(0, 4)$

26. $y = 4x^2 - 6x + 7$
$(0, 7)$

27. $y = \frac{1}{2}x^2 - 2x + 1$
$\left(2 - \sqrt{2}, 0\right), \left(2 + \sqrt{2}, 0\right), (0, 1)$

28. $y = \frac{1}{3}x^2 + 2x - 1$
$\left(-3 - 2\sqrt{3}, 0\right), \left(-3 + 2\sqrt{3}, 0\right), (0, -1)$

In Exercises 29–40, find the vertex of the parabola. See Example 3.

29. $y = -x^2 + 2$ $(0, 2)$

30. $y = 3x^2 - 3$ $(0, -3)$

31. $y = x^2 - 4x + 7$
$(2, 3)$

32. $y = 1 - 2x - x^2$
$(-1, 2)$

33. $y = 6 + 10x - x^2$
$(5, 31)$

34. $y = x^2 - 12x + 9$
$(6, -27)$

35. $y = x^2 + 5x - 3$
$\left(-\frac{5}{2}, -\frac{37}{4}\right)$

36. $y = x^2 + 3x + 4$
$\left(-\frac{3}{2}, \frac{7}{4}\right)$

37. $y = 2x^2 + 4x - 6$
$(-1, -8)$

38. $y = -2x^2 - 4x - 2$
$(-1, 0)$

39. $y = 8 - 9x - 3x^2$
$\left(-\frac{3}{2}, \frac{59}{4}\right)$

40. $y = 3x^2 + 4x - 1$
$\left(-\frac{2}{3}, -\frac{7}{3}\right)$

In Exercises 41–62, sketch the parabola. Label the vertex and any intercepts. See Examples 4–6.
See Additional Answers.

41. $y = x^2 - 1$

42. $y = x^2 - 9$

43. $y = -x^2 + 1$

44. $y = -x^2 + 9$

45. $y = x^2 - 4x$

46. $y = x^2 - 6x$

47. $y = -x^2 + 4x$

48. $y = -x^2 + 6x$

49. $y = x^2 - 4x + 4$

50. $y = -(x^2 + 4x + 4)$

51. $y = x^2 + 6x + 8$

52. $y = x^2 - 6x + 8$

53. $y = -(x^2 + 2x - 3)$

54. $y = -(x^2 + 4x + 2)$

55. $y = -(x^2 + 4x + 8)$

56. $y = x^2 - 6x + 12$

57. $y = x^2 - 4x + 1$

58. $y = x^2 + 6x + 11$

59. $y = 8 + 4x - 3x^2$

60. $y = 2x^2 + 2x - 3$

61. $y = \frac{2}{3}x^2 - x - 3$

62. $y = \frac{3}{2}x^2 + 2x - 2$

In Exercises 63–70, use a graphing calculator to graph the equation. Approximate the vertex from the graph.   See Additional Answers.

63. $y = -x^2 + 6x$

64. $y = x^2 - x$

65. $y = x^2 + 4x + 3$

66. $y = x^2 - 4x + 3$

67. $y = -4x^2 + 4x + 1$

68. $y = -2(x^2 - x - 2)$

69. $y = \frac{1}{2}x^2 + x - 4$

70. $y = -\frac{1}{2}x^2 + 4x - 2$

In Exercises 71–74, sketch the parabola and the horizontal line on the same set of coordinate axes. Find any points of intersection.   See Additional Answers.

71. $y = -x^2 + 3$
$y = 2$

72. $y = x^2 - 6x + 6$
$y = 1$

73. $y = \frac{1}{2}x^2 - 4x + 10$
$y = 3$

74. $y = -2x^2 + 12x - 14$
$y = 5$

Exploration   In Exercises 75–78, use a graphing calculator to graph $y = x^2$ and the following equations in the same viewing window. Describe the relationship between the graph of $y = x^2$ and each of the other graphs.   See Additional Answers.

75. (a) $y = \frac{1}{8}x^2$ (b) $y = -2x^2$
(a) The graph is wider; y is $\frac{1}{8}$ what it is in $y = x^2$.
(b) The parabola opens downward. The graph is not as wide.

76. (a) $y = x^2 + 3$ (b) $y = x^2 - 3$
(a) The graph is shifted 3 units upward from $y = x^2$.
(b) The graph is shifted 3 units downward from $y = x^2$.

77. (a) $y = (x - 2)^2$ (b) $y = (x + 4)^2$
(a) The graph is shifted 2 units to the right of $y = x^2$.
(b) The graph is shifted 4 units to the left of $y = x^2$.

78. (a) $y = -4 + (x + 1)^2$ (b) $y = 1 - (x - 2)^2$
(a) The graph is shifted 1 unit to the left and 4 units downward.
(b) The parabola opens downward and is shifted 2 units to the right and 1 unit upward.

Exploration In Exercises 79–82, find two quadratic equations—one opening upward and one opening downward—whose graphs have the given *x*-intercepts. (The answers are not unique.)

79. $(-2, 0), (2, 0)$ $y = x^2 - 4; y = 4 - x^2$

80. $(-4, 0), (4, 0)$ $y = x^2 - 16; y = 16 - x^2$

81. $(-3, 0), (1, 0)$ $y = x^2 + 2x - 3; y = -x^2 - 2x + 3$

82. $(2, 0), (5, 0)$ $y = x^2 - 7x + 10; y = -x^2 + 7x - 10$

Exploration In Exercises 83–86, graph the equation. Then describe how the vertex can be determined from the completed square form of the equation.

See Additional Answers.

83. $y = (x - 2)^2 - 2$ The vertex is (h, k) when the equation is in the form $y = (x - h)^2 + k$. So, the vertex is $(2, -2)$.

84. $y = -(x - 3)^2 + 1$ The vertex is (h, k) when the equation is in the form $y = (x - h)^2 + k$. So, the vertex is $(3, 1)$.

85. $y = -(x + 1)^2 + 3$ The vertex is (h, k) when the equation is in the form $y = (x - h)^2 + k$. So, the vertex is $(-1, 3)$.

86. $y = (x + 2)^2 - 4$ The vertex is (h, k) when the equation is in the form $y = (x - h)^2 + k$. So, the vertex is $(-2, -4)$.

In Exercises 87–90, complete the square for the right side of the equation and write it in the form of Exercises 83–86. What is the vertex of the parabola?

87. $y = x^2 - 10x + 26$ $y = (x - 5)^2 + 1; (5, 1)$

88. $y = x^2 + 8x + 14$ $y = (x + 4)^2 - 2; (-4, -2)$

89. $y = x^2 - 4x + 3$ $y = (x - 2)^2 - 1; (2, -1)$

90. $y = x^2 - 6x + 5$ $y = (x - 3)^2 - 4; (3, -4)$

Solving Problems

91. *Path of a Ball* The height y (in feet) of a ball thrown by a child is modeled by $y = -0.1x^2 + 2x + 4$, where x is the horizontal distance (in feet) from where the ball is thrown.

(a) How high is the ball when it leaves the child's hand?   4 feet

(b) What is the maximum height of the ball?   14 feet

(c) How far from the child does the ball strike the ground? $10 + 2\sqrt{35} \approx 21.8$ feet

92. *Path of a Ball* The height y (in feet) of a ball thrown by a child is modeled by $y = -0.05x^2 + 2x + 4$, where x is the horizontal distance (in feet) from where the ball is thrown.

(a) How high is the ball when it leaves the child's hand?   4 feet

(b) What is the maximum height of the ball?   24 feet

(c) How far from the child does the ball strike the ground? $20 + 4\sqrt{30} \approx 41.9$ feet

93. *Suspension Bridge* The suspension cables on the Golden Gate Bridge can be modeled by

$$y = \frac{1}{9000}x^2 + 5$$

where x and y are measured in feet. The roadbed of the bridge lies on the *x*-axis, and the *y*-axis is midway between the towers.

(a) What are the coordinates of the lowest points on the cables? $(0, 5)$

(b) The span between the two towers is 4200 feet. How high above the roadbed are the cables connected to the towers?   495 feet

94. *Path of a Football* A football player kicks a 41-yard punt. The path of the football is modeled by $y = -0.035x^2 + 1.4x + 1$, where y is the height (in yards) and x is the horizontal distance (in yards) from where the football was kicked.

(a) What is the maximum height of the football?   15 yards

(b) The player kicks the football toward midfield from the 18-yard line. Over which yard line is the football at its maximum height?   38

95. ▲ *Geometry* The perimeter of a rectangle is 36 meters.

(a) The length of the rectangle is x. Find an expression for its width in terms of x. $18 - x$

(b) Use the result of part (a) to write the area of the rectangle in terms of x. $A = x(18 - x)$

(c) ⊞ Use a graphing calculator to graph the area equation in part (b).   See Additional Answers.

(d) ⊞ Use the graph in part (c) to approximate the dimensions of the rectangle of maximum area.   9 meters × 9 meters

96. *Conjecture* Use the results of Exercise 95 to make a conjecture about the dimensions of a rectangle of fixed perimeter that has a maximum area.
The rectangle is a square.

97. ▦ *Data Analysis* The table shows the numbers S (in thousands) of cellular phone subscribers in the United States for the years 1998 through 2001.
(Source: Cellular Telecommunications & Internet Association)

Year	1998	1999	2000	2001
Subscribers, S	69,209	86,047	109,478	128,375

The data can be approximated by the model $S = 1178.29t^2 - 2816.5t + 17,457$, where t represents the year, with $t = 8$ corresponding to 1998.

(a) Use a graphing calculator to graph the equation and the data in the same viewing window.
See Additional Answers.

(b) Use the graph to approximate the year in which there were 70 million subscribers.   1998

(c) Verify your answer from part (b) algebraically.
$1178.29(8)^2 - 2816.5(8) + 17,457 = 70,335.56$

98. ▦ *Data Analysis* The table shows the numbers D of dalmatians registered with the American Kennel Club for the years 1997 through 2001.
(Source: American Kennel Club)

Year	1997	1998	1999	2000	2001
Dalmatians, D	22,726	9,722	4,652	3,084	2,139

The data can be approximated by the model $D = 1972.86t^2 - 40,292.6t + 207,351$, where t represents the year, with $t = 7$ corresponding to 1997.

(a) Use a graphing calculator to graph the equation and the data in the same viewing window.
See Additional Answers.

(b) Use the graph to approximate the year in which there were 5000 dalmations registered.   1999

(c) Verify your answer from part (b) algebraically.
$1972.86(9)^2 - 40,292.6(9) + 207,351 = 4519.26$

Explaining Concepts

99. ◔ Answer part (g) of Motivating the Chapter on page 536.

100. *Writing* Describe the general shape of the graph of a quadratic equation.   Parabola

101. *Writing* Explain what is meant by the intercepts of a graph of an equation. Explain the method for finding any x- or y-intercepts of a graph.
The intercepts are the points where the graph intersects the x- and y-axes. Find the x-intercepts by setting $y = 0$ and solving the resulting equation. Find the y-intercepts by setting $x = 0$ and solving the resulting equation.

102. *Writing* In your own words, describe the use of the Leading Coefficient Test for parabolas.
The graph of $y = ax^2 + bx + c$ is a parabola opening upward if $a > 0$ and downward if $a < 0$.

103. *Writing* Explain the relationship between the number of x-intercepts of the graph of $y = ax^2 + bx + c$ and the discriminant of $ax^2 + bx + c = 0$.
When the discriminant is positive, there are two x-intercepts. When the discriminant is 0, there is one x-intercept. When the discriminant is negative, there are no x-intercepts.

104. *Writing* What is the relationship between the x-coordinate of the vertex of a parabola and the x-intercepts? Check your conjecture for the graphs in Exercises 83–86.   The x-coordinate of the vertex is the average of the x-coordinates of the x-intercepts.

105. *Writing* Is it possible for the graph of a quadratic equation to have two y-intercepts? Explain.
No. To find the y-intercept of the graph of $y = ax^2 + bx + c$, set $x = 0$ and solve for y. The equation $y = a(0) + b(0) + c$ has only one solution: $y = c$. So, there is only one y-intercept: $(0, c)$.

106. The parabola given by $y = (x - 6)(x - 2)$ has x-intercepts at $(2, 0)$ and $(6, 0)$. The vertex of this parabola is $(4, -4)$. Find equations of parabolas that have the same x-intercepts but that have vertices at the points $(4, 4)$, $(4, 8)$, and $(4, -8)$. Use a graphing calculator to verify your answers.
$= -(x - 2)(x - 6) = -x^2 + 8x - 12$
$= -2(x - 2)(x - 6) = -2x^2 + 16x - 24$
$= 2(x - 2)(x - 6) = 2x^2 - 16x + 24$

10.5 Applications of Quadratic Equations

Why You Should Learn It

Quadratic equations can be used to model and solve real-life problems. For instance, in Exercise 57 on page 580, a quadratic equation is used to model the number of households in the United States with Internet access.

① Solve application problems that can be modeled by quadratic equations.

What You Should Learn

① Solve application problems that can be modeled by quadratic equations.

Applications of Quadratic Equations

In this section, you will study real-life problems that can be modeled by quadratic equations. For such problems, the verbal model may be the *product* of two variable quantities, or the model may be a previously known formula that describes the situation. Watch for these variations in the following examples.

Example 1   Geometry: Dimensions of a Picture

A picture is 3 inches taller than it is wide and has an area of 108 square inches. What are the dimensions of the picture?

Solution

Begin by drawing a diagram, as shown in Figure 10.13.

Figure 10.13

Verbal Model:

$$\boxed{\text{Area of picture}} = \boxed{\text{Width}} \cdot \boxed{\text{Height}}$$

Labels:
Picture width = w	(inches)
Picture height = $w + 3$	(inches)
Area = 108	(square inches)

Equation:

$$108 = w(w + 3)$$

$$0 = w^2 + 3w - 108$$

$$0 = (w + 12)(w - 9)$$

$$w + 12 = 0 \implies w = -12$$

$$w - 9 = 0 \implies w = 9$$

Of the two possible solutions, choose the positive value of w and conclude that the width of the picture is 9 inches. Because the picture is 3 inches taller than it is wide, you can conclude that the height of the picture is $9 + 3 = 12$ inches. Check these dimensions in the original statement of the problem.

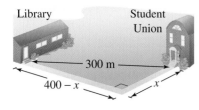

Figure 10.14

Example 2   Using the Pythagorean Theorem

An L-shaped sidewalk from the library to the Student Union on a college campus is 400 meters long, as shown in Figure 10.14. By cutting diagonally across the grass, students shorten the walking distance to 300 meters. What are the lengths of the two legs of the sidewalk?

Solution

From Figure 10.14, you can see that the L-shaped sidewalk and the diagonal form a right triangle. So, to find the lengths of the two legs of the sidewalk, you can use the Pythagorean Theorem.

Common
Formula: $a^2 + b^2 = c^2$ Pythagorean Theorem

Labels: a = length of one leg = x (meters)
               b = length of other leg = $400 - x$ (meters)
               c = length of diagonal = 300 (meters)

Equation: $x^2 + (400 - x)^2 = 300^2$

$$2x^2 - 800x + 160,000 = 90,000$$

$$2x^2 - 800x + 70,000 = 0$$

$$x^2 - 400x + 35,000 = 0$$

Using the Quadratic Formula, you can find the solutions of this equation as follows.

$$x = \frac{-(-400) \pm \sqrt{(-400)^2 - 4(1)(35,000)}}{2(1)}$$ Substitute 1 for a, -400 for b, and 35,000 for c.

$$= \frac{400 \pm \sqrt{160,000 - 140,000}}{2(1)}$$

$$= \frac{400 \pm \sqrt{20,000}}{2}$$

$$= \frac{400 \pm 100\sqrt{2}}{2}$$

$$= 200 \pm 50\sqrt{2}$$

Both solutions are positive, so it does not matter which you choose. If you let

$$x = 200 + 50\sqrt{2} \approx 270.7 \text{ meters}$$

the length of the other leg is

$$400 - x \approx 400 - 270.7 \approx 129.3 \text{ meters}.$$

Try choosing the other value of x to see that you obtain the same two lengths.

Be aware that in real-life applications involving measurements, units of production, or time, you can exclude any negative solution. In Example 2, both solutions are positive and yield the same two dimensions.

Technology: Tip

Use a graphing calculator to graph the quadratic equation $y = x^2 - 400x + 35,000$ shown in Example 2 using the window setting below.

Xmin = 0
Xmax = 300
Xscl = 25
Ymin = -6000
Ymax = 1000
Yscl = 50

Use the *zero* or *root* feature of the graphing calculator to find the x-intercepts of the graph. How do they compare with the solutions of the equation shown in the example?

See Technology Answers.

In Examples 1 and 2, the verbal model or common formula leads directly to a quadratic equation. In many real-life applications, however, the verbal model produces a rational equation, which in turn leads indirectly to a quadratic equation. Notice how this occurs in Examples 3 and 4.

Example 3   Work-Rate Problem

An office has two copy machines. Machine B is known to take 12 minutes longer than machine A to copy the company's monthly report. Using both machines together, it takes 8 minutes to reproduce the report. How long would it take each machine alone to reproduce the report?

Solution

Verbal Model:
$$\boxed{\text{Rate for A}} + \boxed{\text{Rate for B}} = \boxed{\text{Rate for both}}$$

Labels:

Time for both machines $= 8$	(minutes)
Rate for both machines $= \frac{1}{8}$	(job per minute)
Time for machine A $= t$	(minutes)
Rate for machine A $= 1/t$	(job per minute)
Time for machine B $= t + 12$	(minutes)
Rate for machine B $= 1/(t + 12)$	(job per minute)

Equation:

$$\frac{1}{t} + \frac{1}{t + 12} = \frac{1}{8}$$

$$\left(\frac{1}{t} + \frac{1}{t + 12}\right)(8t)(t + 12) = \frac{1}{8}(8t)(t + 12)$$

$$8(t + 12) + 8t = t(t + 12)$$

$$16t + 96 = t^2 + 12t$$

$$0 = t^2 - 4t - 96$$

$$0 = (t - 12)(t + 8)$$

$$t - 12 = 0 \quad \Longrightarrow \quad t = 12$$

$$t + 8 = 0 \quad \Longrightarrow \quad t = -8$$

By choosing the positive value for t, you can conclude that the times for the two machines are

Time for machine A $= t = 12$ minutes

Time for machine B $= t + 12 = 12 + 12 = 24$ minutes.

Check

$$\frac{1}{12} + \frac{1}{12 + 12} \stackrel{?}{=} \frac{1}{8} \qquad \text{Substitute 12 for } t \text{ in original equation.}$$

$$\frac{2}{24} + \frac{1}{24} \stackrel{?}{=} \frac{1}{8} \qquad \text{Rewrite using LCD of 24.}$$

$$\frac{3}{24} = \frac{1}{8} \qquad \text{Solution checks. } \checkmark$$

Example 4 Reduced Rates

A ski club chartered a bus for a ski trip at a cost of $480. In an attempt to lower the bus fare per skier, the club invited nonmembers to go along. When five nonmembers agreed to go on the trip, the fare per skier decreased by $4.80. How many club members are going on the trip?

Solution

Verbal Model: | Cost per skier · Number of skiers | = | 480 |

Labels: Number of ski club members $= x$ (people)
Number of skiers $= x + 5$ (people)
Original cost $= \dfrac{480}{x}$ (dollars per person)
New cost $= \dfrac{480}{x} - 4.80$ (dollars per person)

Equation:
$$\left(\frac{480}{x} - 4.80\right)(x + 5) = 480$$

$$\left(\frac{480 - 4.8x}{x}\right)(x + 5) = 480$$

$$(480 - 4.8x)(x + 5) = 480x, \quad x \neq 0$$

$$480x + 2400 - 4.8x^2 - 24x = 480x$$

$$-4.8x^2 - 24x + 2400 = 0$$

$$x^2 + 5x - 500 = 0$$

$$(x + 25)(x - 20) = 0$$

$$x + 25 = 0 \quad \Longrightarrow \quad x = -25$$

$$x - 20 = 0 \quad \Longrightarrow \quad x = 20$$

Both sides of this equation were divided by -4.8. If -4.8 had not been a factor of each term, we might have multiplied each side by -10 to clear the equation of decimals and then solved the resulting equation.

Choosing the positive value of x, you can conclude that 20 ski club members are going on the trip. Check this solution in the original statement of the problem, as follows.

Original cost for 20 ski club members

$$\frac{480}{20} = \$24$$

New cost with 5 nonmembers

$$\frac{480}{25} = \$19.20$$

Decrease in fare with 5 nonmembers

$$24 - 19.20 = \$4.80 \qquad \text{Solution checks. } \checkmark$$

10.5 Exercises

Review   Concepts, Skills, and Problem Solving

Keep mathematically in shape by doing these exercises *before* the problems of this section.

Properties and Definitions

1. *Writing* ✍   Define the slope of the line through the points (x_1, y_1) and (x_2, y_2). $m = \dfrac{y_2 - y_1}{x_2 - x_1}$

2. Write each form of an equation of a line.
   (a) General form $Ax + By + C = 0$
   (b) Slope-intercept form $y = mx + b$
   (c) Vertical line $x = a$
   (d) Point-slope form $y - y_1 = m(x - x_1)$

Simplifying Expressions

In Exercises 3–6, find the product and simplify.

3. $-2x^2(5x^3)$
   $-10x^5$

4. $(x + 4)(2x - 5)$
   $2x^2 + 3x - 20$

5. $(x + 7)^2$
   $x^2 + 14x + 49$

6. $(x + 1)(x^2 - x + 1)$
   $x^3 + 1$

In Exercises 7–10, factor the expression completely.

7. $9x^2 - 4y^2$
   $(3x + 2y)(3x - 2y)$

8. $10x^4 + 3x^3 - 4x^2$
   $x^2(2x - 1)(5x + 4)$

9. $15x^2 - 11x - 14$
   $(3x + 2)(5x - 7)$

10. $4x^2 - 28x + 49$
    $(2x - 7)^2$

Graphing Equations

In Exercises 11 and 12, graph the equation.
See Additional Answers.

11. $y = \frac{3}{2}t - \frac{1}{2}$

12. $y = -\frac{1}{2}x^2 + 2x - 1$

Solving Problems

1. 15, 16   **2.** 23, 24   **3.** 17, 19
4. 11, 13   **5.** 16, 18   **6.** 12, 14

Number Problem   In Exercises 1–8, find two positive integers satisfying the requirement.

1. The product of two consecutive integers is 240.

2. The product of two consecutive integers is 552.

3. The product of two consecutive odd integers is 323.

4. The product of two consecutive odd integers is 143.

5. The product of two consecutive even integers is 288.

6. The product of two consecutive even integers is 168.

7. The sum of the squares of two consecutive integers is 113.   7, 8

8. The sum of the squares of two consecutive integers is 421.   14, 15

Free-Falling Object   In Exercises 9–14, the height h (in feet) of a falling object at any time t (in seconds) is modeled by $h = h_0 - 16t^2$, where h_0 is the initial height. Use the model to find the time it takes for an object to fall to the ground given h_0.

9. $h_0 = 1600$ 10 seconds

10. $h_0 = 400$ 5 seconds

11. $h_0 = 1532$
    9.79 seconds

12. $h_0 = 2320$
    12.04 seconds

13. $h_0 = 555$ (height of the Washington Monument)
    5.89 seconds

14. $h_0 = 630$ (height of the Gateway Arch)
    6.27 seconds

🔺 *Geometry*   In Exercises 15–24, use the perimeter or area to find the length and width of the rectangle. Then find the area or perimeter, as indicated.

	Width	Length	Perimeter	Area
15.	$0.6l$	l	64 in.	240 in.²
	Width: 12 inches; Length: 20 inches			
16.	w	$1.5w$	75 m	337.5 m²
	Width: 15 meters; Length: 22.5 meters			
17.	w	$2w$	30 ft	50 ft²
	Width: 5 feet; Length: 10 feet			
18.	w	$1.2w$	152.42 cm	1440 cm²
	Width: 34.64 centimeters; Length: 41.57 centimeters			
19.	$\frac{1}{4}l$	l	50 in.	100 in.²
	Width: 5 inches; Length: 20 inches			
20.	$\frac{2}{3}l$	l	20 in.	24 in.²
	Width: 4 inches; Length: 6 inches			

Width	Length	Perimeter	Area
21. w	$w + 4$	56 km	192 km²

Width: 12 kilometers; Length: 16 kilometers

22. $l - 5$	l	18 ft	14 ft²

Width: 2 feet; Length: 7 feet

23. $l - 10$	l	40 m	75 m²

Width: 5 meters; Length: 15 meters

24. w	$w + 6$	52 ft	160 ft²

Width: 10 feet; Length: 16 feet

In Exercises 25–28, write a quadratic equation to solve the problem. See Example 1.

25. ▲ *Geometry*   A picture is 6 inches longer than it is wide and has an area of 187 square inches. What are the dimensions of the picture?

$x(x + 6) = 187$; Width: 11 inches; Length: 17 inches

26. ▲ *Geometry*   A picture whose width is 7 inches less than its length has an area of 144 square inches. What are the dimensions of the picture?

$x(x - 7) = 144$; Width: 9 inches; Length: 16 inches

27. ▲ *Geometry*   A rectangular region in a lumber yard is to be fenced for storage. The region will be fenced on three sides with 175 feet of fence, and the fourth side will be bounded by a building (see figure). The area of the fenced region is 3750 square feet. Find the dimensions of the region.

$w(175 - 2w) = 3750$; 50 feet × 75 feet or 37.5 feet × 100 feet

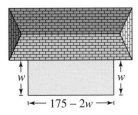

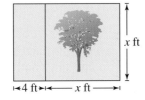

Figure for 27   Figure for 28

28. ▲ *Geometry*   To add more space to your yard, you purchase an additional 4 feet along the side of the property (see figure). The area of the lot is now 9600 square feet. What are the dimensions of the new lot?

$x(x + 4) = 9600$; Length: 100 feet; Width: 96 feet

29. ▲ *Geometry*   A radio station advertises that its broadcasts are heard over a circular region covering approximately 10,000 square miles. Approximate the distance between the station and the listeners farthest from the station.   56.4 miles

30. ▲ *Geometry*   A bag of lawn fertilizer covers 10,000 square feet. Determine the diameter of the circular region that can be fertilized with one bag.   112.8 feet

31. ▲ *Geometry*   The height of a triangle is one-third its base, and the area of the triangle is 24 square inches. Find the dimensions of the triangle.   Base: 12 inches; Height: 4 inches

32. ▲ *Geometry*   The height of a triangle is three times its base, and the area of the triangle is 864 square inches. Find the dimensions of the triangle.   Base: 24 inches; Height: 72 inches

33. *Architecture*   A brick walkway of uniform width surrounds a 20-foot-by-30-foot rectangular garden. The brick walkway covers 336 square feet. Find its width.

3 feet

34. *Architecture*   A decorative tile border of uniform width is to be constructed around a 30-foot-by-40-foot rectangular swimming pool. There is enough tile to cover 296 square feet. Find the width of the tile border.   2 feet

In Exercises 35–40, use the Pythagorean Theorem to solve the problem. See Example 2.

35. *Distance*   A windlass is used to pull a boat to the dock (see figure). The rope is attached to the boat at a point 15 feet below the level of the windlass. Find the distance from the boat to the dock when the length of the rope is 75 feet.   73.48 feet

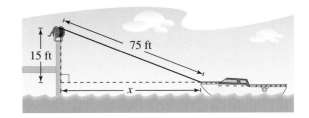

36. *Length*   A 20-foot ladder is leaning against a building (see figure). The ladder must reach a point 19 feet above the ground. Determine the distance from the base of the ladder to the building.   6.24 feet

37. ▲ *Geometry*   A corner lot has sidewalks on two adjacent sides for a total length of 90 feet. The diagonal path across the lot is 64 feet. What are the lengths of the two sides of the sidewalk?
49.80 feet, 40.20 feet

38. ▲ *Geometry*   You are delivering pizzas to an apartment complex and a furniture store (see figure). You are required to keep a log of all mileages between stops. You forget to look at the odometer at the apartment complex, but after getting to the furniture store you record the total distance traveled from the pizza shop as 14 miles. The return distance on the beltway from the furniture store to the pizza shop is 10 miles. The route forms a right triangle. Find the distance from the pizza shop to the apartment complex.
8 miles or 6 miles

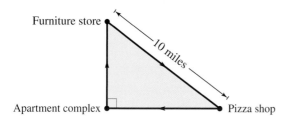

39. ▲ *Geometry*   The perimeter of a rectangle is 68 inches, and the length of the diagonal is 26 inches. Find the dimensions of the rectangle.
10 inches × 24 inches

40. ▲ *Geometry*   The perimeter of a rectangle is 84 centimeters, and the length of the diagonal is 30 centimeters. Find the dimensions of the rectangle.
18 inches × 24 inches

41. ▲ *Geometry*   The floor plan of a building is shown in the figure. Find x if the building has 2400 square feet of floor space.   30 feet

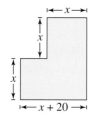 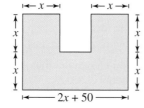

Figure for 41   Figure for 42

42. ▲ *Geometry*   The floor plan of a building is shown in the figure. Find x if the building has 26,250 square feet of floor space.   75 feet

In Exercises 43–46, begin by writing a rational equation to solve the problem. See Example 3.

43. *Work Rate*   Working together, two people can complete a task in 4 hours. Working alone, one person takes 6 hours longer than the other. Working alone, how long would it take each person to complete the task?   6 hours; 12 hours

44. *Work Rate*   Working together, two people can complete a task in $3\frac{1}{2}$ hours. Working alone, one person takes 2 hours longer than the other. Working alone, how long would it take each person to complete the task?   6.14 hours; 8.14 hours

45. *Work Rate*   A farmer has two combines. Combine B is known to take 2 hours longer than combine A to harvest a field. Using both machines, it takes 4 hours to harvest the field. How long would it take to harvest the field using each machine individually?   Combine A: 7.12 hours; Combine B: 9.12 hours

46. *Work Rate*   An office contains two printers. Printer B is known to take 5 minutes longer than printer A to print the company's monthly financial report. With the two printers working together, it takes 3 minutes to print the report. How long would it take each machine individually to produce the report?   Printer A: 4.4 minutes; Printer B: 9.4 minutes

In Exercises 47–50, begin by writing a rational equation to solve the problem. See Example 4.

47. *Reduced Rates*   A service organization paid $100 for a block of tickets to a ball game. The block contained five more tickets than the organization needed for its members. By inviting five more people to attend (and share in the cost), the organization lowered the price per ticket by $1. How many people are going to the game?   25 people

48. *Reduced Rates*   A service organization paid $150 for a block of tickets to a ball game. The block contained four more tickets than the organization needed for its members. By inviting four more people to attend (and share in the cost), the organization lowered the price per ticket by $1.25. How many people are going to the game?   24 people

49. *Reduced Fares*   A science club chartered a bus for $360 to attend a science fair. In order to lower the bus fare per member, the club invited nonmembers to go along. When 10 nonmembers agreed to go on the trip, the fare per person decreased by $3. How many people are going to the science fair?   40 people

50. *Reduced Fares* A literary club chartered a bus for $600 to attend a Shakespearean festival. In order to lower the bus fare per member, the club invited nonmembers to go along. When 10 nonmembers agreed to go on the trip, the fare per person decreased by $3. How many people are going to the festival? 50 people

51. *Average Speed* A truck traveled the first 200 miles of a trip at one speed and the last 225 miles at an average speed of 5 miles per hour less. The entire trip took 10 hours. What were the two average speeds?
45.3 miles per hour; 40.3 miles per hour

52. *Average Speed* A truck traveled the first 150 miles of a trip at one speed and the last 200 miles at an average speed of 10 miles per hour less. The entire trip took 8 hours. What were the two average speeds?
50 miles per hour; 40 miles per hour

Compound Interest In Exercises 53–56, find the interest rate r. The amount A in an account earning r percent (in decimal form) compounded annually for 2 years is given by $A = P(1 + r)^2$, where P is the original investment.

53. $P = \$1000$
$A = \$1123.60$ 6%

54. $P = \$2500$
$A = \$2862.25$ 7%

55. $P = \$200$
$A = \$235.44$ ≈8.5%

56. $P = \$10,000$
$A = \$11,990$ ≈9.5%

57. Data Analysis The numbers H (in millions) of U.S. households with Internet access for the years 1996 through 2001 can be approximated by the model $H = -0.295t^2 + 11.65t - 51.2$,   for $6 \le t \le 11$, where t represents the year, with $t = 6$ corresponding to 1996. (Source: U.S. Department of Commerce)

(a) Use a graphing calculator to graph the model.
See Additional Answers.

(b) Use the graph to approximate the year in which there were 36 million households with Internet access. 2000

(c) Use the graph to complete the table and verify your answer from part (b).

t	6	7	8	9	10	11
H	8.08	15.895	23.12	29.755	35.80	41.255

(d) Verify your answer to part (c) algebraically.
See Additional Answers.

58. Data Analysis Experimental data is obtained in a lab for the breaking strength y (in tons) of a steel cable of diameter x (in inches). The data is shown in the table.

x	0.50	0.75	1.0	1.25	1.50	1.75
y	9.9	21.8	38.3	59.2	84.4	114.0

The data can be approximated by the model

$y = 35.229x^2 + 4.07x - 1.0$.

(a) Use a graphing calculator to plot the data and graph the model in the same viewing window.
See Additional Answers.

(b) Use the graph in part (a) to estimate the diameter of a cable that would support 75 tons.
1.41 inches

(c) Use the model and the Quadratic Formula to estimate the diameter of a cable that would support 30 tons.
≈ 0.88 inches

Explaining Concepts

59. *Writing* In your own words, describe what is meant by the term *equivalent equations*. Give an example. Equivalent equations have the same solution(s). The equations $x + 3 = 10$ and $2x + 6 = 20$ are equivalent.

60. *Writing* Describe the steps that can be used to transform an equation into an equivalent equation.
See Additional Answers.

61. The phrase *increased by* indicates what operation?
Addition

62. The word *per* indicates what operation? Division

63. *Writing* State the guidelines for solving word problems. See Additional Answers.

64. *Writing* State the Pythagorean Theorem and draw a diagram that illustrates this theorem. If a and b are the lengths of the legs of a right triangle and c is the length of the hypotenuse, then $a^2 + b^2 = c^2$. See Additional Answers.

10.6 Complex Numbers

What You Should Learn

① Write square roots of negative numbers in i-form and perform operations on numbers in i-form.
② Determine the equality of two complex numbers.
③ Add, subtract, and multiply complex numbers.
④ Use complex conjugates to write the quotient of two complex numbers in standard form.
⑤ Solve quadratic equations with complex solutions.

Why You Should Learn It

Understanding complex numbers can help you to identify quadratic equations that have no real solutions, as in Example 12 on page 587.

① Write square roots of negative numbers in i-form and perform operations on numbers in i-form.

The Imaginary Unit i

In Section 9.1, you learned that a negative number has no *real* square root. For instance, $\sqrt{-1}$ is not real because there is no real number x such that $x^2 = -1$. So, as long as you are dealing only with real numbers, the equation $x^2 = -1$ has no solution. To overcome this deficiency, mathematicians have expanded the set of numbers by including the **imaginary unit i,** defined as

$$i = \sqrt{-1}. \qquad \text{Imaginary unit}$$

This number has the property that $i^2 = -1$. So, the imaginary unit i is a solution of the equation $x^2 = -1$.

The Square Root of a Negative Number

Let c be a positive real number. Then the square root of $-c$ is given by

$$\sqrt{-c} = \sqrt{c(-1)} = \sqrt{c}\sqrt{-1} = \sqrt{c}\,i.$$

When writing $\sqrt{-c}$ in the **i-form,** $\sqrt{c}\,i$, note that i is outside the radical.

Example 1   Writing Numbers in i-Form

Write each number in i-form.

a. $\sqrt{-36}$ **b.** $\sqrt{-\dfrac{16}{25}}$ **c.** $\sqrt{-54}$ **d.** $\dfrac{\sqrt{-48}}{\sqrt{-3}}$

Solution

a. $\sqrt{-36} = \sqrt{36(-1)} = \sqrt{36}\sqrt{-1} = 6i$

b. $\sqrt{-\dfrac{16}{25}} = \sqrt{\dfrac{16}{25}(-1)} = \sqrt{\dfrac{16}{25}}\sqrt{-1} = \dfrac{4}{5}i$

c. $\sqrt{-54} = \sqrt{54(-1)} = \sqrt{54}\sqrt{-1} = 3\sqrt{6}\,i$

d. $\dfrac{\sqrt{-48}}{\sqrt{-3}} = \dfrac{\sqrt{48}\sqrt{-1}}{\sqrt{3}\sqrt{-1}} = \dfrac{\sqrt{48}\,i}{\sqrt{3}\,i} = \sqrt{\dfrac{48}{3}} = \sqrt{16} = 4$

Technology: Discovery

Use a calculator to evaluate each radical. Does one result in an error message? Explain why.

See Technology Answers.

a. $\sqrt{121}$

b. $\sqrt{-121}$

c. $-\sqrt{121}$

To perform operations with square roots of negative numbers, you must *first* write the numbers in *i*-form. Once the numbers have been written in *i*-form, you add, subtract, and multiply as follows.

$$ai + bi = (a + b)i \qquad \qquad \text{Addition}$$

$$ai - bi = (a - b)i \qquad \qquad \text{Subtraction}$$

$$(ai)(bi) = ab(i^2) = ab(-1) = -ab \qquad \qquad \text{Multiplication}$$

Study Tip

When performing operations with numbers in *i*-form, you sometimes need to be able to evaluate powers of the imaginary unit *i*. The first several powers of *i* are as follows.

$$i^1 = i$$

$$i^2 = -1$$

$$i^3 = i(i^2) = i(-1) = -i$$

$$i^4 = (i^2)(i^2) = (-1)(-1) = 1$$

$$i^5 = i(i^4) = i(1) = i$$

$$i^6 = (i^2)(i^4) = (-1)(1) = -1$$

$$i^7 = (i^3)(i^4) = (-i)(1) = -i$$

$$i^8 = (i^4)(i^4) = (1)(1) = 1$$

Note how the pattern of values $i, -1, -i$, and 1 repeats itself for powers greater than 4.

Example 2   Operations with Square Roots of Negative Numbers

Perform each operation.

a. $\sqrt{-9} + \sqrt{-49}$ **b.** $\sqrt{-32} - 2\sqrt{-2}$

Solution

a. $\sqrt{-9} + \sqrt{-49} = \sqrt{9}\sqrt{-1} + \sqrt{49}\sqrt{-1}$ Product Rule for Radicals

$$= 3i + 7i \qquad \qquad \text{Write in } i\text{-form.}$$

$$= 10i \qquad \qquad \text{Simplify.}$$

b. $\sqrt{-32} - 2\sqrt{-2} = \sqrt{32}\sqrt{-1} - 2\sqrt{2}\sqrt{-1}$ Product Rule for Radicals

$$= 4\sqrt{2}i - 2\sqrt{2}i \qquad \qquad \text{Write in } i\text{-form.}$$

$$= 2\sqrt{2}i \qquad \qquad \text{Simplify.}$$

Example 3   Multiplying Square Roots of Negative Numbers

Find each product.

a. $\sqrt{-15}\sqrt{-15}$ **b.** $\sqrt{-5}\left(\sqrt{-45} - \sqrt{-4}\right)$

Solution

a. $\sqrt{-15}\sqrt{-15} = \left(\sqrt{15}i\right)\left(\sqrt{15}i\right)$ Write in *i*-form.

$$= \left(\sqrt{15}\right)^2 i^2 \qquad \qquad \text{Multiply.}$$

$$= 15(-1) \qquad \qquad i^2 = -1$$

$$= -15 \qquad \qquad \text{Simplify.}$$

b. $\sqrt{-5}\left(\sqrt{-45} - \sqrt{-4}\right) = \sqrt{5}i\left(3\sqrt{5}i - 2i\right)$ Write in *i*-form.

$$= \left(\sqrt{5}i\right)\left(3\sqrt{5}i\right) - \left(\sqrt{5}i\right)(2i) \quad \text{Distributive Property}$$

$$= 3(5)(-1) - 2\sqrt{5}(-1) \qquad \text{Multiply.}$$

$$= -15 + 2\sqrt{5} \qquad \qquad \text{Simplify.}$$

When multiplying square roots of negative numbers, be sure to write them in *i*-form *before multiplying*. If you do not do this, you can obtain incorrect answers. For instance, in Example 3(a) be sure you see that

$$\sqrt{-15}\sqrt{-15} \neq \sqrt{(-15)(-15)} = \sqrt{225} = 15.$$

② Determine the equality of two complex numbers.

Complex Numbers

A number of the form $a + bi$, where a and b are real numbers, is called a **complex number.** The real number a is called the **real part** of the complex number $a + bi$, and the number bi is called the **imaginary part.**

> ### Definition of Complex Number
>
> If a and b are real numbers, the number $a + bi$ is a **complex number,** and it is said to be written in **standard form.** If $b = 0$, the number $a + bi = a$ is a real number. If $b \neq 0$, the number $a + bi$ is called an **imaginary number.** A number of the form bi, where $b \neq 0$, is called a **pure imaginary number.**

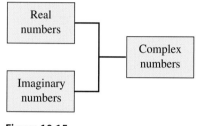

Figure 10.15

A number cannot be both real and imaginary. For instance, the numbers -2, 0, 1, $\frac{1}{2}$, and $\sqrt{2}$ are real numbers (but they are *not* imaginary numbers), and the numbers $-3i$, $2 + 4i$, and $-1 + i$ are imaginary numbers (but they are *not* real numbers). The diagram shown in Figure 10.15 further illustrates the relationship among real, complex, and imaginary numbers.

Two complex numbers $a + bi$ and $c + di$, in standard form, are equal if and only if $a = c$ and $b = d$.

Example 4   Equality of Two Complex Numbers

To determine whether the complex numbers $\sqrt{9} + \sqrt{-48}$ and $3 - 4\sqrt{3}i$ are equal, begin by writing the first number in standard form.

$$\sqrt{9} + \sqrt{-48} = \sqrt{3^2} + \sqrt{4^2(3)(-1)} = 3 + 4\sqrt{3}i$$

From this form, you can see that the two numbers are not equal because they have imaginary parts that differ in sign.

Example 5   Equality of Two Complex Numbers

Find values of x and y that satisfy the equation $3x - \sqrt{-25} = -6 + 3yi$.

Solution

Begin by writing the left side of the equation in standard form.

$$3x - 5i = -6 + 3yi \qquad \text{Each side is in standard form.}$$

For these two numbers to be equal, their real parts must be equal to each other and their imaginary parts must be equal to each other.

Real Parts	*Imaginary Parts*
$3x = -6$	$3yi = -5i$
$x = -2$	$3y = -5$
	$y = -\frac{5}{3}$

So, $x = -2$ and $y = -\frac{5}{3}$.

③ Add, subtract, and multiply complex numbers.

Operations with Complex Numbers

To add or subtract two complex numbers, you add (or subtract) the real and imaginary parts separately. This is similar to combining like terms of a polynomial.

$$(a + bi) + (c + di) = (a + c) + (b + d)i \qquad \text{Addition of complex numbers}$$

$$(a + bi) - (c + di) = (a - c) + (b - d)i \qquad \text{Subtraction of complex numbers}$$

Example 6   Adding and Subtracting Complex Numbers

a. $(3 - i) + (-2 + 4i) = (3 - 2) + (-1 + 4)i = 1 + 3i$

b. $3i + (5 - 3i) = 5 + (3 - 3)i = 5$

c. $4 - (-1 + 5i) + (7 + 2i) = [4 - (-1) + 7] + (-5 + 2)i = 12 - 3i$

d. $(6 + 3i) + (2 - \sqrt{-8}) - \sqrt{-4} = (6 + 3i) + (2 - 2\sqrt{2}i) - 2i$

$$= (6 + 2) + (3 - 2\sqrt{2} - 2)i$$

$$= 8 + (1 - 2\sqrt{2})i$$

Study Tip

Note in part (b) of Example 6 that the sum of two complex numbers can be a real number.

The Commutative, Associative, and Distributive Properties of real numbers are also valid for complex numbers, as is the FOIL Method.

Example 7   Multiplying Complex Numbers

Perform each operation and write the result in standard form.

a. $(7i)(-3i)$ **b.** $(1 - i)(\sqrt{-9})$

c. $(2 - i)(4 + 3i)$ **d.** $(3 + 2i)(3 - 2i)$

Solution

a. $(7i)(-3i) = -21i^2$ Multiply.

$$= -21(-1) = 21 \qquad i^2 = -1$$

b. $(1 - i)(\sqrt{-9}) = (1 - i)(3i)$ Write in i-form.

$$= 3i - 3(i^2) \qquad \text{Distributive Property}$$

$$= 3i - 3(-1) = 3 + 3i \qquad i^2 = -1$$

c. $(2 - i)(4 + 3i) = 8 + 6i - 4i - 3i^2$ FOIL Method

$$= 8 + 6i - 4i - 3(-1) \qquad i^2 = -1$$

$$= 11 + 2i \qquad \text{Combine like terms.}$$

d. $(3 + 2i)(3 - 2i) = 3^2 - (2i)^2$ Special product formula

$$= 9 - 4i^2 \qquad \text{Simplify.}$$

$$= 9 - 4(-1) = 13 \qquad i^2 = -1$$

Additional Examples
Perform the indicated operations and write the result in standard form.

a. $(7 + 4i) - (3 + \sqrt{-12})$

b. $(2 + 5i)^2$

Answers:

a. $4 + 2(2 - \sqrt{3})i$

b. $-21 + 20i$

④ Use complex conjugates to write the quotient of two complex numbers in standard form.

Complex Conjugates

In Example 7(d), note that the product of two complex numbers can be a real number. This occurs with pairs of complex numbers of the form $a + bi$ and $a - bi$, called **complex conjugates.** In general, the product of complex conjugates has the following form.

$$(a + bi)(a - bi) = a^2 - (bi)^2 = a^2 - b^2i^2 = a^2 - b^2(-1) = a^2 + b^2$$

Here are some examples.

Complex Number	Complex Conjugate	Product
$4 - 5i$	$4 + 5i$	$4^2 + 5^2 = 41$
$3 + 2i$	$3 - 2i$	$3^2 + 2^2 = 13$
$-2 = -2 + 0i$	$-2 = -2 - 0i$	$(-2)^2 + 0^2 = 4$
$i = 0 + i$	$-i = 0 - i$	$0^2 + 1^2 = 1$

To write the quotient of $a + bi$ and $c + di$ in standard form, where c and d are not both zero, multiply the numerator and denominator by the *complex conjugate of the denominator*, as shown in Example 8.

Example 8   Writing Quotients of Complex Numbers in Standard Form

a. $\dfrac{2 - i}{4i} = \dfrac{2 - i}{4i} \cdot \dfrac{(-4i)}{(-4i)}$ Multiply numerator and denominator by complex conjugate of denominator.

$= \dfrac{-8i + 4i^2}{-16i^2}$ Multiply fractions.

$= \dfrac{-8i + 4(-1)}{-16(-1)}$ $i^2 = -1$

$= \dfrac{-8i - 4}{16}$ Simplify.

$= -\dfrac{1}{4} - \dfrac{1}{2}i$ Write in standard form.

b. $\dfrac{5}{3 - 2i} = \dfrac{5}{3 - 2i} \cdot \dfrac{3 + 2i}{3 + 2i}$ Multiply numerator and denominator by complex conjugate of denominator.

$= \dfrac{5(3 + 2i)}{(3 - 2i)(3 + 2i)}$ Multiply fractions.

$= \dfrac{5(3 + 2i)}{3^2 + 2^2}$ Product of complex conjugates

$= \dfrac{15 + 10i}{13}$ Simplify.

$= \dfrac{15}{13} + \dfrac{10}{13}i$ Write in standard form.

Example 9   Writing a Quotient of Complex Numbers in Standard Form

$$\frac{8 - i}{8 + i} = \frac{8 - i}{8 + i} \cdot \frac{8 - i}{8 - i}$$ Multiply numerator and denominator by complex conjugate of denominator.

$$= \frac{64 - 16i + i^2}{8^2 + 1^2}$$ Multiply fractions.

$$= \frac{64 - 16i + (-1)}{8^2 + 1^2}$$ $i^2 = -1$

$$= \frac{63 - 16i}{65}$$ Simplify.

$$= \frac{63}{65} - \frac{16}{65}i$$ Write in standard form.

Example 10   Writing a Quotient of Complex Numbers in Standard Form

$$\frac{2 + 3i}{4 - 2i} = \frac{2 + 3i}{4 - 2i} \cdot \frac{4 + 2i}{4 + 2i}$$ Multiply numerator and denominator by complex conjugate of denominator.

$$= \frac{8 + 16i + 6i^2}{4^2 + 2^2}$$ Multiply fractions.

$$= \frac{8 + 16i + 6(-1)}{4^2 + 2^2}$$ $i^2 = -1$

$$= \frac{2 + 16i}{20}$$ Simplify.

$$= \frac{1}{10} + \frac{4}{5}i$$ Write in standard form.

Some students incorrectly rewrite $(2 + 16i)/20$ as $(1 + 16i)/10$ or as $(2 + 4i)/5$.

Example 11   Verifying a Complex Solution of an Equation

Show that $x = 2 + i$ is a solution of the equation

$$x^2 - 4x + 5 = 0.$$

Solution

$$x^2 - 4x + 5 = 0$$ Write original equation.

$$(2 + i)^2 - 4(2 + i) + 5 \overset{?}{=} 0$$ Substitute $2 + i$ for x.

$$4 + 4i + i^2 - 8 - 4i + 5 \overset{?}{=} 0$$ Expand.

$$i^2 + 1 \overset{?}{=} 0$$ Combine like terms.

$$(-1) + 1 \overset{?}{=} 0$$ $i^2 = -1$

$$0 = 0$$ Solution checks. ✓

So, $x = 2 + i$ is a solution of the original equation.

5 Solve quadratic equations with complex solutions.

Quadratic Equations with Complex Solutions

Prior to this section, the only solutions to find were real numbers. But now that you have studied complex numbers, it makes sense to look for other types of solutions. For instance, although the quadratic equation $x^2 + 1 = 0$ has no solutions that are real numbers, it does have two solutions that are complex numbers: i and $-i$. To check this, substitute i and $-i$ for x.

$$(i)^2 + 1 = -1 + 1 = 0 \qquad \text{Solution checks. } \checkmark$$
$$(-i)^2 + 1 = -1 + 1 = 0 \qquad \text{Solution checks. } \checkmark$$

One way to find complex solutions of a quadratic equation is to extend the Square Root Property to cover the case in which d is a negative number.

> **Square Root Property (Complex Square Roots)**
>
> The equation $u^2 = d$, where $d < 0$, has exactly two solutions:
> $$u = \sqrt{|d|}\,i \quad \text{and} \quad u = -\sqrt{|d|}\,i.$$
> These solutions can also be written as $u = \pm\sqrt{|d|}\,i.$

Technology: Discovery

Solve each quadratic equation below algebraically. Then use a graphing calculator to check the solutions. Which equations have real solutions? Which equations have complex solutions? Which graphs have x-intercepts? Which graphs have no x-intercepts? Compare the type(s) of solution(s) of a quadratic equation with the x-intercept(s) of the graph of the equation.

See Technology Answers.

a. $y = 2x^2 + 3x - 5$
b. $y = 2x^2 + 4x + 2$
c. $y = x^2 + 4$
d. $y = (x + 7)^2 + 2$

Example 12   Square Root Property

a. $x^2 + 8 = 0$ \hfill Original equation

$\qquad x^2 = -8$ \hfill Subtract 8 from each side.

$\qquad x = \pm\sqrt{8}\,i = \pm 2\sqrt{2}\,i$ \hfill Square Root Property

The solutions are $x = 2\sqrt{2}\,i$ and $x = -2\sqrt{2}\,i$. Check these in the original equation.

b. $(x - 4)^2 = -3$ \hfill Original equation

$\qquad x - 4 = \pm\sqrt{3}\,i$ \hfill Square Root Property

$\qquad x = 4 \pm \sqrt{3}\,i$ \hfill Add 4 to each side.

The solutions are $x = 4 + \sqrt{3}\,i$ and $x = 4 - \sqrt{3}\,i$. Check these in the original equation.

c. $2(3x - 5)^2 + 32 = 0$ \hfill Original equation

$\qquad 2(3x - 5)^2 = -32$ \hfill Subtract 32 from each side.

$\qquad (3x - 5)^2 = -16$ \hfill Divide each side by 2.

$\qquad 3x - 5 = \pm 4i$ \hfill Square Root Property

$\qquad 3x = 5 \pm 4i$ \hfill Add 5 to each side.

$\qquad x = \dfrac{5 \pm 4i}{3}$ \hfill Divide each side by 3.

The solutions are $x = (5 + 4i)/3$ and $x = (5 - 4i)/3$. Check these in the original equation.

Example 13   A Quadratic Equation with Complex Solutions

Solve $x^2 - 4x + 8 = 0$ by completing the square.

Solution

$$x^2 - 4x + 8 = 0 \qquad \text{Write original equation.}$$

$$x^2 - 4x = -8 \qquad \text{Subtract 8 from each side.}$$

$$x^2 - 4x + (-2)^2 = -8 + 4 \qquad \text{Add } (-2)^2 = 4 \text{ to each side.}$$

$$(x - 2)^2 = -4 \qquad \text{Completed square form}$$

$$x - 2 = \pm 2i \qquad \text{Square Root Property}$$

$$x = 2 \pm 2i \qquad \text{Add 2 to each side.}$$

The solutions are $x = 2 + 2i$ and $x = 2 - 2i$. Check these in the original equation.

———

Note in the next example how the Quadratic Formula can be used to solve a quadratic equation that has complex solutions.

Additional Examples
Solve each equation using the Quadratic Formula.

a. $3x^2 + 4x - 8 = 0$

b. $-x^2 - 2x - 4 = 0$

Answers:

a. $x = \dfrac{-2 \pm 2\sqrt{7}}{3}$

b. $x = -1 \pm \sqrt{3}i$

Example 14   The Quadratic Formula: Complex Solutions

Solve $x^2 - 4x + 5 = 0$ by using the Quadratic Formula.

Solution

$$x^2 - 4x + 5 = 0 \qquad \text{Write original equation.}$$

$$x = \frac{-b \pm \sqrt{b^2 - 4ac}}{2a} \qquad \text{Quadratic Formula}$$

$$x = \frac{-(-4) \pm \sqrt{(-4)^2 - 4(1)(5)}}{2(1)} \qquad \text{Substitute 1 for } a, -4 \text{ for } b, \text{ and 5 for } c.$$

$$x = \frac{4 \pm \sqrt{-4}}{2} \qquad \text{Simplify.}$$

$$x = \frac{4 \pm 2i}{2} \qquad \text{Write in } i\text{-form.}$$

$$x = \frac{2(2 \pm i)}{2} \qquad \text{Factor numerator.}$$

$$x = \frac{2(2 \pm i)}{2} \qquad \text{Divide out common factor.}$$

$$x = 2 \pm i \qquad \text{Solutions}$$

The solutions are $x = 2 + i$ and $x = 2 - i$. Check these in the original equation.

———

10.6 Exercises

Review   Concepts, Skills, and Problem Solving

Keep mathematically in shape by doing these exercises *before* the problems of this section.

Properties and Definitions

1. *Writing*✏   In your own words, describe how to multiply $\dfrac{3t}{5} \cdot \dfrac{8t^2}{15}$. Use the rule $\dfrac{u}{v} \cdot \dfrac{w}{z} = \dfrac{uw}{vz}$.
   That is, you multiply the numerators, multiply the denominators, and write the new fraction in simplified form.

2. *Writing*✏   In your own words, describe how to divide $\dfrac{3t}{5} \div \dfrac{8t^2}{15}$. Use the rule $\dfrac{u}{v} \div \dfrac{w}{z} = \dfrac{u}{v} \cdot \dfrac{z}{w}$.
   That is, you invert the divisor and multiply.

3. *Writing*✏   In your own words, describe how to add $\dfrac{3t}{5} + \dfrac{8t^2}{15}$. Rewrite the fractions so they have common denominators and then use the rule
   $$\dfrac{u}{w} + \dfrac{v}{w} = \dfrac{u+v}{w}.$$

4. *Writing*✏   What is the value of $\dfrac{t-5}{5-t}$? Explain.
   $$\dfrac{t-5}{5-t} = \dfrac{-1(5-t)}{5-t} = -1$$

Simplifying Expressions

In Exercises 5–10, simplify the expression.
See Additional Answers.

5. $\dfrac{x^2}{2x+3} \div \dfrac{5x}{2x+3}$

6. $\dfrac{x-y}{5x} \div \dfrac{x^2-y^2}{x^2}$

7. $\dfrac{\dfrac{9}{x}}{\left(\dfrac{6}{x}+2\right)}$

8. $\dfrac{\left(1+\dfrac{2}{x}\right)}{\left(x-\dfrac{4}{x}\right)}$

9. $\dfrac{\left(\dfrac{4}{x^2-9}\right)}{\left(\dfrac{1}{x-3}\right)}$

10. $\dfrac{\left(\dfrac{1}{x+1}\right)}{\left(\dfrac{3}{2x^2+4x+2}\right)}$

Problem Solving

11. *Number Problem*   Find two real numbers that divide the real number line between $x/2$ and $4x/3$ into three equal parts.
    $\dfrac{7x}{9}, \dfrac{19x}{18}$

12. *Capacitance*   When two capacitors with capacitances C_1 and C_2 are connected in series, the equivalent capacitance is given by
    $$\dfrac{1}{\left(\dfrac{1}{C_1}+\dfrac{1}{C_2}\right)}.$$
    Simplify this complex fraction.
    $\dfrac{C_1 C_2}{C_1 + C_2}$

Developing Skills

In Exercises 1–16, write the number in *i*-form. See Example 1.

1. $\sqrt{-4}$ $2i$
2. $\sqrt{-9}$ $3i$
3. $-\sqrt{-144}$ $-12i$
4. $\sqrt{-49}$ $7i$
5. $\sqrt{-\dfrac{4}{25}}$ $\dfrac{2}{5}i$
6. $-\sqrt{-\dfrac{36}{121}}$ $-\dfrac{6}{11}i$
7. $\sqrt{-0.09}$ $0.3i$
8. $\sqrt{-0.0004}$ $0.02i$
9. $\sqrt{-8}$ $2\sqrt{2}i$
10. $\sqrt{-75}$ $5\sqrt{3}i$
11. $\sqrt{-7}$ $\sqrt{7}i$
12. $\sqrt{-15}$ $\sqrt{15}i$
13. $\dfrac{\sqrt{-12}}{\sqrt{-3}}$ 2
14. $\dfrac{\sqrt{-45}}{\sqrt{-5}}$ 3
15. $\sqrt{-\dfrac{18}{64}}$ $\dfrac{3\sqrt{2}}{8}i$
16. $\sqrt{-\dfrac{8}{25}}$ $\dfrac{2\sqrt{2}}{5}i$

In Exercises 17–38, perform the operation(s) and write the result in standard form. See Examples 2 and 3.

17. $\sqrt{-16} + \sqrt{-36}$ $10i$
18. $\sqrt{-25} - \sqrt{-9}$ $2i$
19. $\sqrt{-50} - \sqrt{-8}$ $3\sqrt{2}i$
20. $\sqrt{-500} + \sqrt{-45}$ $13\sqrt{5}i$
21. $\sqrt{-48} + \sqrt{-12} - \sqrt{-27}$ $3\sqrt{3}i$
22. $\sqrt{-32} - \sqrt{-18} + \sqrt{-50}$ $6\sqrt{2}i$
23. $\sqrt{-8}\sqrt{-2}$ -4
24. $\sqrt{-25}\sqrt{-6}$ $-5\sqrt{6}$
25. $\sqrt{-18}\sqrt{-3}$ $-3\sqrt{6}$
26. $\sqrt{-7}\sqrt{-7}$ -7
27. $\sqrt{-0.16}\sqrt{-1.21}$ -0.44
28. $\sqrt{-0.49}\sqrt{-1.44}$ -0.84
29. $\sqrt{-3}\left(\sqrt{-3} + \sqrt{-4}\right)$ $-2\sqrt{3} - 3$

30. $\sqrt{-12}\left(\sqrt{-3}-\sqrt{-12}\right)$ 6

31. $\sqrt{-5}\left(\sqrt{-16}-\sqrt{-10}\right)$ $5\sqrt{2}-4\sqrt{5}$

32. $\sqrt{-24}\left(\sqrt{-9}+\sqrt{-4}\right)$ $-10\sqrt{6}$

33. $\sqrt{-2}\left(3-\sqrt{-8}\right)$ **34.** $\sqrt{-9}\left(1+\sqrt{-16}\right)$
   $4+3\sqrt{2}\,i$ $-12+3i$

35. $\left(\sqrt{-16}\right)^2$ -16 **36.** $\left(\sqrt{-2}\right)^2$ -2

37. $\left(\sqrt{-4}\right)^3$ $-8i$ **38.** $\left(\sqrt{-5}\right)^3$ $-5\sqrt{5}\,i$

In Exercises 39–44, determine the values of a and b that satisfy the equation. See Examples 4 and 5.

39. $3-4i=a+bi$ $a=3, b=-4$

40. $-8+6i=a+bi$ $a=-8, b=6$

41. $5-4i=(a+3)+(b-1)i$ $a=2, b=-3$

42. $-10+12i=2a+(5b-3)i$ $a=-5, b=3$

43. $-4-\sqrt{-8}=a+bi$ $a=-4, b=-2\sqrt{2}$

44. $\sqrt{-36}-3=a+bi$ $a=-3, b=6$

In Exercises 45–56, perform the operation(s) and write the result in standard form. See Example 6.

45. $(4-3i)+(6+7i)$ $10+4i$

46. $(-10+2i)+(4-7i)$ $-6-5i$

47. $(-4-7i)+(-10-33i)$ $-14-40i$

48. $(15+10i)-(2+10i)$ 13

49. $13i-(14-7i)$ $-14+20i$

50. $(-21-50i)+(21-20i)$ $-70i$

51. $(30-i)-(18+6i)+3i^2$ $9-7i$

52. $(4+6i)+(15+24i)-(1-i)$ $18+31i$

53. $6-(3-4i)+2i$ $3+6i$

54. $22+(-5+8i)+10i$ $17+18i$

55. $15i-(3-25i)+\sqrt{-81}$ $-3+49i$

56. $(-1+i)-\sqrt{2}-\sqrt{-2}$ $\left(-1-\sqrt{2}\right)+\left(1-\sqrt{2}\right)i$

In Exercises 57–76, perform the operation and write the result in standard form. See Example 7.

57. $(3i)(12i)$ -36 **58.** $(-5i)(4i)$ 20

59. $(3i)(-8i)$ 24 **60.** $(-2i)(-10i)$ -20

61. $(-6i)(-i)(6i)$ **62.** $(10i)(12i)(-3i)$
   $-36i$ $360i$

63. $(-3i)^3$ $27i$ **64.** $(8i)^2$ -64

65. $(-3i)^2$ -9 **66.** $(2i)^4$ 16

67. $-5(13+2i)$ **68.** $10(8-6i)$
   $-65-10i$ $80-60i$

69. $4i(-3-5i)$ **70.** $-3i(10-15i)$
   $20-12i$ $-45-30i$

71. $(9-2i)\left(\sqrt{-4}\right)$ **72.** $(11+3i)\left(\sqrt{-25}\right)$
   $4+18i$ $-15+55i$

73. $(4+3i)(-7+4i)$ **74.** $(3+5i)(2+15i)$
   $-40-5i$ $-69+55i$

75. $(-7+7i)(4-2i)$ **76.** $(3+5i)(2-15i)$
   $-14+42i$ $81-35i$

In Exercises 77–86, simplify the expression.

77. i^7 $-i$ **78.** i^{11} $-i$

79. i^{24} 1 **80.** i^{35} $-i$

81. i^{42} -1 **82.** i^{64} 1

83. i^9 i **84.** i^{71} $-i$

85. $(-i)^6$ -1 **86.** $(-i)^4$ 1

In Exercises 87–96, multiply the number by its complex conjugate and simplify.

87. $2+i$ 5 **88.** $3+2i$ 13

89. $-2-8i$ 68 **90.** $10-3i$ 109

91. $5-\sqrt{6}i$ 31 **92.** $-4+\sqrt{2}i$ 18

93. $10i$ 100 **94.** 20 400

95. $1+\sqrt{-3}$ 4 **96.** $-3-\sqrt{-5}$ 14

In Exercises 97–110, write the quotient in standard form. See Examples 8–10.

97. $\dfrac{20}{2i}$ $-10i$ **98.** $\dfrac{-5}{-3i}$ $-\frac{5}{3}i$

99. $\dfrac{2+i}{-5i}$ $-\frac{1}{5}+\frac{2}{5}i$ **100.** $\dfrac{1+i}{3i}$ $\frac{1}{3}-\frac{1}{3}i$

101. $\dfrac{4}{1-i}$ $2+2i$ **102.** $\dfrac{20}{3+i}$ $6-2i$

103. $\dfrac{7i+14}{7i}$ $1-2i$ **104.** $\dfrac{6i+3}{3i}$ $2-i$

105. $\dfrac{-12}{2+7i}$ $-\frac{24}{53}+\frac{84}{53}i$ **106.** $\dfrac{15}{2(1-i)}$ $\frac{15}{4}+\frac{15}{4}i$

107. $\dfrac{3i}{5+2i}$ $\frac{6}{29}+\frac{15}{29}i$ **108.** $\dfrac{4i}{5-3i}$ $-\frac{6}{17}+\frac{10}{17}i$

109. $\dfrac{4+5i}{3-7i}$ $-\frac{23}{58}+\frac{43}{58}i$ **110.** $\dfrac{5+3i}{7-4i}$ $\frac{23}{65}+\frac{41}{65}i$

In Exercises 111–116, perform the operation and write the result in standard form.

111. $\dfrac{5}{3+i} + \dfrac{1}{3-i}$

$\dfrac{9}{5} - \dfrac{2}{5}i$

112. $\dfrac{1}{1-2i} + \dfrac{4}{1+2i}$

$1 - \dfrac{6}{5}i$

113. $\dfrac{3i}{1+i} + \dfrac{2}{2+3i}$

$\dfrac{47}{26} + \dfrac{27}{26}i$

114. $\dfrac{i}{4-3i} - \dfrac{5}{2+i}$

$-\dfrac{53}{25} + \dfrac{29}{25}i$

115. $\dfrac{1+i}{i} - \dfrac{3}{5-2i}$

$\dfrac{14}{29} - \dfrac{35}{29}i$

116. $\dfrac{3-2i}{i} - \dfrac{1}{7+i}$

$-\dfrac{107}{50} - \dfrac{149}{50}i$

In Exercises 117–120, determine whether each number is a solution of the equation. See Example 11.

117–120. (a) Solution and (b) Solution

117. $x^2 + 2x + 5 = 0$

   (a) $x = -1 + 2i$ (b) $x = -1 - 2i$

118. $x^2 - 4x + 13 = 0$

   (a) $x = 2 - 3i$ (b) $x = 2 + 3i$

119. $x^3 + 4x^2 + 9x + 36 = 0$

   (a) $x = -4$ (b) $x = -3i$

120. $x^3 - 8x^2 + 25x - 26 = 0$

   (a) $x = 2$ (b) $x = 3 - 2i$

In Exercises 121–142, solve the equation by using the Square Root Property. See Example 12.

121. $z^2 = -36$ $\pm 6i$

122. $x^2 = -9$ $\pm 3i$

123. $x^2 + 4 = 0$ $\pm 2i$

124. $y^2 + 16 = 0$ $\pm 4i$

125. $9u^2 + 17 = 0$ $\pm \dfrac{\sqrt{17}}{3}i$

126. $4v^2 + 9 = 0$ $\pm \dfrac{3}{2}i$

127. $(t - 3)^2 = -25$ $3 \pm 5i$

128. $(x + 5)^2 = -81$ $-5 \pm 9i$

129. $(3z + 4)^2 + 144 = 0$ $-\dfrac{4}{3} \pm 4i$

130. $(2y - 3)^2 + 25 = 0$ $\dfrac{3}{2} \pm \dfrac{5}{2}i$

131. $(2x - 5)^2 = -54$ $\dfrac{5}{2} \pm \dfrac{3\sqrt{6}}{2}i$

132. $(6y - 5)^2 = -8$ $\dfrac{5}{6} \pm \dfrac{\sqrt{2}}{3}i$

133. $9(x + 6)^2 = -121$ $-6 \pm \dfrac{11}{3}i$

134. $4(x - 4)^2 = -169$ $4 \pm \dfrac{13}{2}i$

135. $(x - 1)^2 = -27$ $1 \pm 3\sqrt{3}i$

136. $(2x + 3)^2 = -54$ $-\dfrac{3}{2} \pm \dfrac{3}{2}\sqrt{6}i$

137. $(x + 1)^2 + 0.04 = 0$ $-1 \pm 0.2i$

138. $(x - 3)^2 + 2.25 = 0$ $3 \pm 1.5i$

139. $x^2 + 900 = 0$ $\pm 30i$

140. $y^2 + 225 = 0$ $\pm 15i$

141. $(x - 5)^2 + 100 = 0$ $5 \pm 10i$

142. $(y + 12)^2 + 400 = 0$ $-12 \pm 20i$

In Exercises 143–152, solve the equation by completing the square. Give the solutions in exact form and in decimal form rounded to two decimal places. See Example 13.

143. $x^2 + 10 = 6x$ $3 \pm i$

144. $z^2 + 4z + 13 = 0$ $-2 \pm 3i$

145. $-x^2 + x - 1 = 0$

$\dfrac{1}{2} + \dfrac{\sqrt{3}}{2}i \approx 0.50 + 0.87i$

$\dfrac{1}{2} - \dfrac{\sqrt{3}}{2}i \approx 0.50 - 0.87i$

146. $y^2 + 5y + 9 = 0$

$-\dfrac{5}{2} + \dfrac{\sqrt{11}}{2}i \approx -2.50 + 1.66i$

$-\dfrac{5}{2} - \dfrac{\sqrt{11}}{2}i \approx -2.50 - 1.66i$

147. $u^2 - \dfrac{2}{3}u + 5 = 0$

$\dfrac{1}{3} + \dfrac{2\sqrt{11}}{3}i \approx 0.33 + 2.21i$

$\dfrac{1}{3} - \dfrac{2\sqrt{11}}{3}i \approx 0.33 - 2.21i$

148. $4z^2 - 3z + 2 = 0$

$\dfrac{3}{8} + \dfrac{\sqrt{23}}{8}i \approx 0.38 + 0.60i$

$\dfrac{3}{8} - \dfrac{\sqrt{23}}{8}i \approx 0.38 - 0.60i$

149. $5x^2 - 3x + 10 = 0$

$\dfrac{3}{10} + \dfrac{\sqrt{191}}{10}i \approx 0.30 + 1.38i$

$\dfrac{3}{10} - \dfrac{\sqrt{191}}{10}i \approx 0.30 - 1.38i$

150. $7x^2 + 4x + 3 = 0$

$-\dfrac{2}{7} + \dfrac{\sqrt{17}}{7}i \approx -0.29 + 0.59i$

$-\dfrac{2}{7} - \dfrac{\sqrt{17}}{7}i \approx -0.29 - 0.59i$

151. $0.5t^2 + t + 2 = 0$

$-1 + \sqrt{3}i \approx -1 + 1.73i$
$-1 - \sqrt{3}i \approx -1 - 1.73i$

152. $0.1x^2 + 0.2x + 0.5 = 0$ $-1 \pm 2i$

In Exercises 153–156, solve the equation by using the Quadratic Formula. See Example 14.

153. $x^2 - 8x + 19$

$4 \pm \sqrt{3}i$

154. $2x^2 - x + 1 = 0$

$\dfrac{1}{4} \pm \dfrac{\sqrt{7}}{4}i$

155. $2x^2 + 3x + 3 = 0$

$-\dfrac{3}{4} \pm \dfrac{\sqrt{15}}{4}i$

156. $8x^2 - 6x + 2 = 0$

$\dfrac{3}{8} \pm \dfrac{\sqrt{7}}{8}i$

157. *Cube Roots*   The principal cube root of 125, $\sqrt[3]{125}$, is 5. Evaluate the expression x^3 for each value of x.

(a) $x = \dfrac{-5 + 5\sqrt{3}i}{2}$ $\left(\dfrac{-5 + 5\sqrt{3}i}{2}\right)^3 = 125$

(b) $x = \dfrac{-5 - 5\sqrt{3}i}{2}$ $\left(\dfrac{-5 - 5\sqrt{3}i}{2}\right)^3 = 125$

158. *Cube Roots*   The principal cube root of 27, $\sqrt[3]{27}$, is 3. Evaluate the expression x^3 for each value of x.

(a) $x = \dfrac{-3 + 3\sqrt{3}i}{2}$ $\left(\dfrac{-3 + 3\sqrt{3}i}{2}\right)^3 = 27$

(b) $x = \dfrac{-3 - 3\sqrt{3}i}{2}$ $\left(\dfrac{-3 - 3\sqrt{3}i}{2}\right)^3 = 27$

159. *Pattern Recognition*   Compare the results of Exercises 157 and 158. Use the results to list possible cube roots of (a) 1, (b) 8, and (c) 64. Verify your results algebraically. See Additional Answers.

160. *Algebraic Properties*   Consider the complex number $1 + 5i$.

(a) Find the additive inverse of the number.
$-(1 + 5i)$

(b) Find the multiplicative inverse of the number.
$\dfrac{1}{1 + 5i} = \dfrac{1}{26} - \dfrac{5}{26}i$

In Exercises 161–164, perform the operations.

161. $(a + bi) + (a - bi)$ $2a$

162. $(a + bi)(a - bi)$ $a^2 + b^2$

163. $(a + bi) - (a - bi)$ $2bi$

164. $(a + bi)^2 + (a - bi)^2$ $2a^2 - 2b^2$

Explaining Concepts

165. *Writing*   Define the imaginary unit i.
$i = \sqrt{-1}$

166. *Writing*   Explain why the equation $x^2 = -1$ does not have real number solutions.
The square of any real number is nonnegative.

167. *Writing*   Describe the error.

$\sqrt{-3}\sqrt{-3} = \sqrt{(-3)(-3)} = \sqrt{9} = 3$
$\sqrt{-3}\sqrt{-3} = \left(\sqrt{3}i\right)\left(\sqrt{3}i\right) = 3i^2 = -3$

168. *True or False?*   Some numbers are both real and imaginary. Justify your answer.
False. A number cannot be both. The number 2 is real but not imaginary. The number $2i$ is imaginary but not real.

169. The polynomial $x^2 + 1$ is prime *with respect to the integers*. It is not, however, prime *with respect to the complex numbers*. Show how $x^2 + 1$ can be factored using complex numbers.
$x^2 + 1 = (x + i)(x - i)$

10.7 Relations, Functions, and Graphs

Robert Grubbs/Photo Network

What You Should Learn

1. Identify the domain and range of a relation.
2. Determine if relations are functions by inspection or by using the Vertical Line Test.
3. Use function notation and evaluate functions.
4. Identify the domain of a function.

Why You Should Learn It

Relations and functions can be used to describe real-life situations. For instance, in Exercise 71 on page 602, a relation is used to model the length of time between sunrise and sunset for Erie, Pennsylvania.

1. Identify the domain and range of a relation.

Relations

Many everyday occurrences involve pairs of quantities that are matched with each other by some rule of correspondence. For instance, each person is matched with a birth month (person, month); the number of hours worked is matched with a paycheck (hours, pay); an instructor is matched with a course (instructor, course); and the time of day is matched with the outside temperature (time, temperature). In each instance, sets of ordered pairs can be formed. Such sets of ordered pairs are called **relations.**

Definition of Relation

A **relation** is any set of ordered pairs. The set of first components in the ordered pairs is the **domain** of the relation. The set of second components is the **range** of the relation.

In mathematics, relations are commonly described by ordered pairs of *numbers.* The set of x-coordinates is the domain, and the set of y-coordinates is the range. In the relation $\{(3, 5), (1, 2), (4, 4), (0, 3)\}$, the domain D and range R are the sets $D = \{3, 1, 4, 0\}$ and $R = \{5, 2, 4, 3\}$.

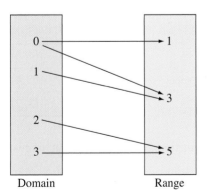

Domain   Range

Figure 10.16

Example 1   Analyzing a Relation

Find the domain and range of the relation $\{(0, 1), (1, 3), (2, 5), (3, 5), (0, 3)\}$. Then sketch a graphical representation of the relation.

Solution

The domain is the set of all first components of the relation, and the range is the set of all second components.

$$D = \{0, 1, 2, 3\} \quad \text{and} \quad R = \{1, 3, 5\}$$

A graphical representation is shown in Figure 10.16.

You should note that it is not necessary to list repeated components of the domain and range of a relation.

2 Determine if relations are functions by inspection or by using the Vertical Line Test.

Functions

In the study of mathematics and its applications, the focus is mainly on a special type of relation, called a **function.**

Definition of Function

A **function** is a relation in which no two ordered pairs have the same first component and different second components.

This definition means that a given first component cannot be paired with two different second components. For instance, the pairs $(1, 3)$ and $(1, -1)$ could not be ordered pairs of a function.

Consider the relations described at the beginning of this section.

This discussion of functions introduces students to an important mathematical concept. You might ask students to define some relations and then decide whether each relation is a function.

Relation	Ordered Pairs	Sample Relation
1	(person, month)	$\{(A, May), (B, Dec), (C, Oct), . . .\}$
2	(hours, pay)	$\{(12, 84), (4, 28), (6, 42), (15, 105), . . .\}$
3	(instructor, course)	$\{(A, MATH001), (A, MATH002), . . .\}$
4	(time, temperature)	$\{(8, 70°), (10, 78°), (12, 78°), . . .\}$

The first relation *is* a function because each person has only one birth month. The second relation *is* a function because the number of hours worked at a particular job can yield only *one* paycheck amount. The third relation *is not* a function because an instructor can teach more than one course. The fourth relation *is* a function. Note that the ordered pairs $(10, 78°)$ and $(12, 78°)$ do not violate the definition of a function.

Study Tip

The ordered pairs of a relation can be thought of in the form (input, output). For a *function*, a given input cannot yield two different outputs. For instance, if the input is a person's name and the output is that person's month of birth, then your name as the input can yield only your month of birth as the output.

Example 2 Testing Whether a Relation Is a Function

Decide whether the relation represents a function.

a. Input: a, b, c

Output: 2, 3, 4

$\{(a, 2), (b, 3), (c, 4)\}$

b.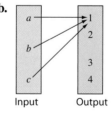

Input   Output

c.

Input x	Output y	(x, y)
3	1	(3, 1)
4	3	(4, 3)
5	4	(5, 4)
3	2	(3, 2)

Solution

a. This set of ordered pairs *does* represent a function. No first component has two different second components.

b. This diagram *does* represent a function. No first component has two different second components.

c. This table *does not* represent a function. The first component 3 is paired with two different second components, 1 and 2.

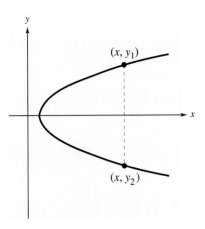

Figure 10.17

In algebra, it is common to represent functions by equations in two variables rather than by ordered pairs. For instance, the equation $y = x^2$ represents the variable y as a function of x. The variable x is the **independent variable** (the input) and y is the **dependent variable** (the output). In this context, the domain of the function is the set of all *allowable* values for x, and the range is the *resulting* set of all values taken on by the dependent variable y.

From the graph of an equation, it is easy to determine whether the equation represents y as a function of x. The graph in Figure 10.17 *does not* represent a function of x because the indicated value of x is paired with two y-values. Graphically, this means that a vertical line intersects the graph more than once.

Vertical Line Test

A set of points on a rectangular coordinate system is the graph of y as a function of x if and only if no vertical line intersects the graph at more than one point.

Example 3   Using the Vertical Line Test for Functions

Use the Vertical Line Test to determine whether y is a function of x.

a.
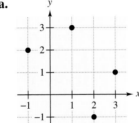

b.
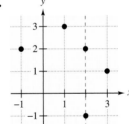

c.
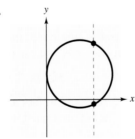

d.
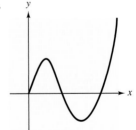

Solution

a. From the graph, you can see that no vertical line intersects more than one point on the graph. So, the relation *does* represent y as a function of x.

b. From the graph, you can see that a vertical line intersects more than one point on the graph. So, the relation *does not* represent y as a function of x.

c. From the graph, you can see that a vertical line intersects more than one point on the graph. So, the relation *does not* represent y as a function of x.

d. From the graph, you can see that no vertical line intersects more than one point on the graph. So, the relation *does* represent y as a function of x.

③ Use function notation and evaluate functions.

Function Notation

To discuss functions represented by equations, it is common practice to give them names using **function notation.** For instance, the function

$$y = 2x - 6$$

can be given the name "f" and written in function notation as

$$f(x) = 2x - 6.$$

Function Notation

In the notation $f(x)$:

> f is the **name** of the function.
> x is a **domain** (or input) value.
> $f(x)$ is a **range** (or output) value y for a given x.

The symbol $f(x)$ is read as **the value of f at x** or simply **f of x.**

The process of finding the value of $f(x)$ for a given value of x is called **evaluating a function.** This is accomplished by substituting a given x-value (input) into the equation to obtain the value of $f(x)$ (output). Here is an example.

Function	*x-Values (input)*	*Function Values (output)*
$f(x) = 4 - 3x$	$x = -2$	$f(-2) = 4 - 3(-2) = 4 + 6 = 10$
	$x = -1$	$f(-1) = 4 - 3(-1) = 4 + 3 = 7$
	$x = 0$	$f(0) = 4 - 3(0) = 4 - 0 = 4$
	$x = 2$	$f(2) = 4 - 3(2) = 4 - 6 = -2$
	$x = 3$	$f(3) = 4 - 3(3) = 4 - 9 = -5$

Although f and x are often used as a convenient function name and independent (input) variable, you can use other letters. For instance, the equations

$$f(x) = x^2 - 3x + 5, \quad f(t) = t^2 - 3t + 5, \quad \text{and} \quad g(s) = s^2 - 3s + 5$$

all define the same function. In fact, the letters used are just "placeholders" and this same function is well described by the form

$$f() = ()^2 - 3() + 5$$

where the parentheses are used in place of a letter. To evaluate $f(-2)$, simply place -2 in each set of parentheses, as follows.

$$f(-2) = (-2)^2 - 3(-2) + 5$$
$$= 4 + 6 + 5$$
$$= 15$$

Remind students to use the order of operations as they evaluate functions.

It is important to put parentheses around the x-value (input) and then simplify the result.

Example 4   Evaluating a Function

Let $f(x) = x^2 + 1$. Find each value of the function.

a. $f(-2)$ **b.** $f(0)$

Solution

a. $f(x) = x^2 + 1$ Write original function.

   $f(-2) = (-2)^2 + 1$ Substitute -2 for x.

   $= 4 + 1 = 5$ Simplify.

b. $f(x) = x^2 + 1$ Write original function.

   $f(0) = (0)^2 + 1$ Substitute 0 for x.

   $= 0 + 1 = 1$ Simplify.

Example 5   Evaluating a Function

Let $g(x) = 3x - x^2$. Find each value of the function.

a. $g(2)$ **b.** $g(0)$

Solution

a. Substituting 2 for x produces $g(2) = 3(2) - (2)^2 = 6 - 4 = 2$.

b. Substituting 0 for x produces $g(0) = 3(0) - (0)^2 = 0 - 0 = 0$.

④ Identify the domain of a function.

Finding the Domain of a Function

The domain of a function may be explicitly described along with the function, or it may be *implied* by the context in which the function is used. For instance, if weekly pay is a function of hours worked (for a 40-hour work week), the implied domain is typically the interval $0 \le x \le 40$. Certainly x cannot be negative in this context.

Example 6   Finding the Domain of a Function

Find the domain of each function.

a. $f:\{(-3, 0), (-1, 2), (0, 4), (2, 4), (4, -1)\}$
b. Area of a square: $A = s^2$

Solution

a. The domain of f consists of all first components in the set of ordered pairs. So, the domain is $\{-3, -1, 0, 2, 4\}$.
b. For the area of a square, you must choose positive values for the side s. So, the domain is the set of all real numbers s such that $s > 0$.

10.7 Exercises

Review   *Concepts, Skills, and Problem Solving*

Keep mathematically in shape by doing these exercises *before* the problems of this section.

Properties and Definitions

1. If $a < b$ and $b < c$, then what is the relationship between a and c? Name this property. $a < c$
Transitive Property

2. Demonstrate the Multiplication Property of Equality for the equation $7x = 21$. $\frac{7x}{7} = \frac{21}{7}$; $x = 3$

Simplifying Expressions

In Exercises 3–6, simplify the expression.

3. $4s - 6t + 7s + t$
$11s - 5t$

4. $2x^2 - 4 + 5 - 3x^2$
$-x^2 + 1$

5. $\frac{5}{3}x - \frac{2}{3}x - 4$ $x - 4$

6. $3x^2y + xy - xy^2 - 6xy$ $3x^2y - xy^2 - 5xy$

Solving Equations

In Exercises 7–10, solve the equation.

7. $3x^2 + 9x - 12 = 0$
$-4, 1$

8. $3 - \frac{4}{x} = \frac{5}{2}$
8

9. $\frac{6}{x + 3} = \frac{5}{x - 2}$ 27

10. $\sqrt{x^2 + 3} = x + 1$ 1

Problem Solving

11. *Simple Interest*   An inheritance of $7500 is invested in a mutual fund, and at the end of 1 year the value of the investment is $8190. What is the annual interest rate for this fund?   9.2%

12. *Number Problem*   One number is 10 times larger than a second number. When the reciprocal of the first number is subtracted from the reciprocal of the second number, the result is 3. Find the two numbers. $\frac{3}{10}, 3$

Developing Skills

In Exercises 1–6, find the domain and range of the relation. See Example 1.

1. $\{(-4, 3), (2, 5), (1, 2), (4, -3)\}$
Domain: $\{-4, 1, 2, 4\}$; Range: $\{-3, 2, 3, 5\}$

2. $\{(-1, 5), (8, 3), (4, 6), (-5, -2)\}$
Domain: $\{-5, -1, 4, 8\}$; Range: $\{-2, 3, 5, 6\}$

3. $\left\{(2, 16), (-9, -10), \left(\frac{1}{2}, 0\right)\right\}$
Domain: $\left\{-9, \frac{1}{2}, 2\right\}$; Range: $\{-10, 0, 16\}$

4. $\left\{\left(\frac{2}{3}, -4\right), \left(-6, \frac{1}{4}\right), (0, 0)\right\}$
Domain: $\left\{-6, 0, \frac{2}{3}\right\}$; Range: $\left\{-4, 0, \frac{1}{4}\right\}$

5. $\{(-1, 3), (5, -7), (-1, 4), (8, -2), (1, -7)\}$
Domain: $\{-1, 1, 5, 8\}$; Range: $\{-7, -2, 3, 4\}$

6. $\{(1, 1), (2, 4), (3, 9), (-2, 4), (-1, 1)\}$
Domain: $\{-2, -1, 1, 2, 3\}$; Range: $\{1, 4, 9\}$

11. Function   **12.** Function

In Exercises 7–26, determine whether the relation represents a function. See Example 2.

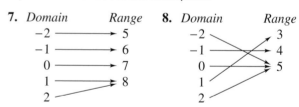

7. Domain   Range   **8.** Domain   Range

Function   Function

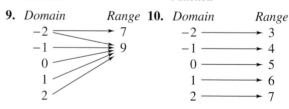

9. Domain   Range   **10.** Domain   Range

Not a function   Function

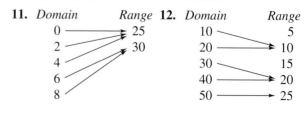

11. Domain   Range   **12.** Domain   Range

13.

Domain	Range
0	1
1	2
2	5
3	9
4	

Not a function

14.

Domain	Range
−4	3
−3	4
−2	
−1	

Not a function

15.

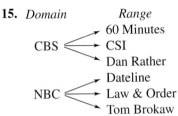

Domain   Range

CBS → 60 Minutes, CSI, Dan Rather

NBC → Dateline, Law & Order, Tom Brokaw

Not a function

16.

Domain   Range

60 Minutes, CSI, Dan Rather → CBS

Dateline, Law & Order, Tom Brokaw → NBC

Function

17.

Domain	Range
Year	Single women in the labor force (in percent)
1997	67.9
1998	68.5
1999	68.7
2000	69.0

Function

(Source: U.S. Bureau of Labor Statistics)

18.

Domain: Percent daily value of vitamin C per serving

Range: Cereal

10% → Corn Flakes, Wheaties

100% → Cheerios, Total

Not a function

19.

Input x	Output y	(x, y)
0	2	$(0, 2)$
1	4	$(1, 4)$
2	6	$(2, 6)$
3	8	$(3, 8)$
4	10	$(4, 10)$

Function

20.

Input x	Output y	(x, y)
0	2	$(0, 2)$
1	4	$(1, 4)$
2	6	$(2, 6)$
1	8	$(1, 8)$
0	10	$(0, 10)$

Not a function

21.

Input x	Output y	(x, y)
1	1	$(1, 1)$
3	2	$(3, 2)$
5	3	$(5, 3)$
3	4	$(3, 4)$
1	5	$(1, 5)$

Not a function

22.

Input x	Output y	(x, y)
2	1	$(2, 1)$
4	1	$(4, 1)$
6	1	$(6, 1)$
8	1	$(8, 1)$
10	1	$(10, 1)$

Function

23. $\{(0, 25), (2, 25), (4, 30), (6, 30), (8, 30)\}$
Function

24. $\{(10, 5), (20, 10), (30, 15), (40, 20), (50, 25)\}$
Function

25. Input: a, b, c

Output: 0, 1, 2

$\{(a, 0), (b, 1), (c, 2)\}$
Function

26. Input: 3, 5, 7

Output: d, e, f

$\{(3, d) (5, e), (7, f), (7, d)\}$
Not a function

In Exercises 27–36, use the Vertical Line Test to determine whether y is a function of x. See Example 3.

27.

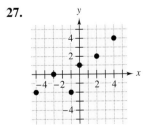

Function

28.

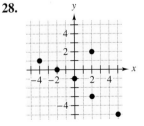

Not a function

29.

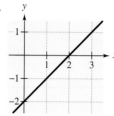

Function

30.

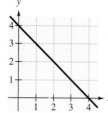

Function

31.

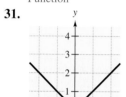

Function

32.

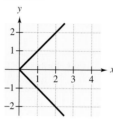

Not a function

33.

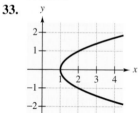

Not a function

34.

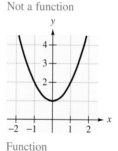

Function

35.

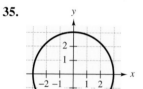

Not a function

36.

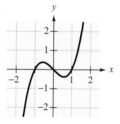

Function

In Exercises 37–52, evaluate the function as indicated, and simplify. See Examples 4 and 5.

37. $f(x) = \frac{1}{2}x$ (a) $f(2)$ (b) $f(5)$
(a) 1   (b) $\frac{5}{2}$ (c) $f(-4)$ (d) $f\left(-\frac{2}{3}\right)$
(c) -2 (d) $-\frac{1}{3}$

38. $g(x) = -\frac{4}{5}x$ (a) $g(5)$ (b) $g(0)$
(a) -4 (b) 0   (c) $g(-3)$ (d) $g\left(-\frac{5}{4}\right)$
(c) $\frac{12}{5}$ (d) 1

39. $f(x) = 2x - 1$ (a) $f(0)$ (b) $f(3)$
(a) -1 (b) 5   (c) $f(-3)$ (d) $f\left(-\frac{1}{2}\right)$
(c) -7 (d) -2

40. $f(t) = 3 - 4t$ (a) $f(0)$ (b) $f(1)$
(a) 3   (b) -1 (c) $f(-2)$ (d) $f\left(\frac{3}{4}\right)$
(c) 11   (d) 0

41. $f(x) = 4x + 1$ (a) $f(1)$ (b) $f(-1)$
(a) 5   (b) -3 (c) $f(-4)$ (d) $f\left(-\frac{4}{3}\right)$
(c) -15 (d) $-\frac{13}{3}$

42. $g(t) = 5 - 2t$ (a) $g\left(\frac{5}{2}\right)$ (b) $g(-10)$
(a) 0   (b) 25   (c) $g(0)$ (d) $g\left(\frac{3}{4}\right)$
(c) 5   (d) $\frac{7}{2}$

43. $h(t) = \frac{1}{4}t - 1$ (a) $h(200)$ (b) $h(-12)$
(a) 49   (b) -4 (c) $h(8)$ (d) $h\left(-\frac{5}{2}\right)$
(c) 1   (d) $-\frac{13}{8}$

44. $f(s) = 4 - \frac{2}{3}s$ (a) $f(60)$ (b) $f(-15)$
(a) -36 (b) 14   (c) $f(-18)$ (d) $f\left(\frac{1}{2}\right)$
(c) 16   (d) $\frac{11}{3}$

45. $f(v) = \frac{1}{2}v^2$ (a) $f(-4)$ (b) $f(4)$
(a) 8   (b) 8   (c) $f(0)$ (d) $f(2)$
(c) 0   (d) 2

46. $g(u) = -2u^2$ (a) $g(0)$ (b) $g(2)$
(a) 0   (b) -8 (c) $g(3)$ (d) $g(-4)$
(c) -18 (d) -32

47. $g(x) = 2x^2 - 3x + 1$ (a) $g(0)$ (b) $g(-2)$
(a) 1   (b) 15   (c) $g(1)$ (d) $g\left(\frac{1}{2}\right)$
(c) 0   (d) 0

48. $h(x) = x^2 + 4x - 1$ (a) $h(0)$ (b) $h(-4)$
(a) -1 (b) -1 (c) $h(10)$ (d) $h\left(\frac{3}{2}\right)$
(c) 139   (d) $\frac{29}{4}$

49. $g(u) = |u + 2|$ (a) $g(2)$ (b) $g(-2)$
(a) 4   (b) 0   (c) $g(10)$ (d) $g\left(-\frac{5}{2}\right)$
(c) 12   (d) $\frac{1}{2}$

50. $h(s) = |s| + 2$ (a) $h(4)$ (b) $h(-10)$
(a) 6   (b) 12   (c) $h(-2)$ (d) $h\left(\frac{3}{2}\right)$
(c) 4   (d) $\frac{7}{2}$

51. $h(x) = x^3 - 1$ (a) $h(0)$ (b) $h(1)$
(a) -1 (b) 0   (c) $h(3)$ (d) $h\left(\frac{1}{2}\right)$
(c) 26   (d) $-\frac{7}{8}$

52. $f(x) = 16 - x^4$ (a) $f(-2)$ (b) $f(2)$
(a) 0   (b) 0   (c) $f(1)$ (d) $f(3)$
(c) 15   (d) -65

In Exercises 53–60, find the domain of the function. See Example 6.

53. $f:\{(0, 4), (1, 3), (2, 2), (3, 1), (4, 0)\}$
$D = \{0, 1, 2, 3, 4\}$

54. $f:\{(-2, -1), (-1, 0), (0, 1), (1, 2), (2, 3)\}$
$D = \{-2, -1, 0, 1, 2\}$

55. $g:\{(-2, 4), (-1, 1), (0, 0), (1, 1), (2, 4)\}$
$D = \{-2, -1, 0, 1, 2\}$

56. $g:\{(0, 7), (1, 6), (2, 6), (3, 7), (4, 8)\}$
$D = \{0, 1, 2, 3, 4\}$

57. $h:\{(-5, 2), (-4, 2), (-3, 2), (-2, 2), (-1, 2)\}$
$D = \{-5, -4, -3, -2, -1\}$

58. $h:\{(10, 100), (20, 200), (30, 300), (40, 400)\}$
$D = \{10, 20, 30, 40\}$

59. Area of a circle: $A = \pi r^2$
The set of all real numbers r such that $r > 0$.

60. Circumference of a circle: $C = 2\pi r$
The set of all real numbers r such that $r > 0$.

Solving Problems

61. *Demand*   The demand for a product is a function of its price. Consider the demand function

$$f(p) = 20 - 0.5p$$

where p is the price in dollars.

(a) Find $f(10)$ and $f(15)$. $f(10) = 15$, $f(15) = 12.5$

(b) Describe the effect a price increase has on demand.   Demand decreases.

62. *Maximum Load*   The maximum safe load L (in pounds) for a wooden beam 2 inches wide and d inches high is $L(d) = 100d^2$.

(a) Complete the table.

d	2	4	6	8
$L(d)$	400	1600	3600	6400

(b) Describe the effect of an increase in height on the maximum safe load.
Maximum safe load increases.

63. *Distance*   The function $d(t) = 50t$ gives the distance (in miles) that a car will travel in t hours at an average speed of 50 miles per hour. Find the distance traveled for (a) $t = 2$, (b) $t = 4$, and (c) $t = 10$.
(a) 100 miles   (b) 200 miles   (c) 500 miles

64. *Speed of Sound*   The function $S(h) = 1116 - 4.04h$ approximates the speed of sound (in feet per second) at altitude h (in thousands of feet). Use the function to approximate the speed of sound for (a) $h = 0$, (b) $h = 10$, and (c) $h = 30$.   (a) 1116 feet per second
(b) 1075.6 feet per second   (c) 994.8 feet per second

Interpreting a Graph   In Exercises 65–68, use the information in the graph. (Source: U.S. National Center for Education Statistics)

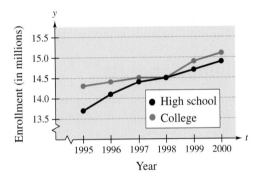

65. Is the high school enrollment a function of the year?
High school enrollment is a function of the year.

66. Is the college enrollment a function of the year?
College enrollment is a function of the year.

67. Let $f(t)$ represent the number of high school students in year t. Find $f(1996)$.
$f(1996) \approx 14,100,000$

68. Let $g(t)$ represent the number of college students in year t. Find $g(2000)$.
$g(2000) \approx 15,100,000$

69. ▲ *Geometry*   Write the formula for the perimeter P of a square with sides of length s. Is P a function of s? Explain. $P = 4s$; P is a function of s.

70. ▲ *Geometry*   Write the formula for the volume V of a cube with sides of length t. Is V a function of t? Explain. $V = t^3$; V is a function of t.

71. *Sunrise and Sunset*   The graph approximates the length of time L (in hours) between sunrise and sunset in Erie, Pennsylvania for the year 2002. The variable t represents the day of the year.   (Source: Fly-By-Day Consulting, Inc.)

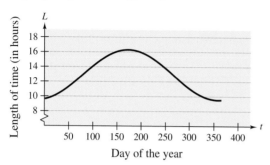

Day of the year

(a) Is the length of time L a function of the day of the year t? L is a function of t.

(b) Estimate the range for this relation.
   $9.5 \le L \le 16.5$

72. *SAT Scores and Grade-Point Average*   The graph shows the SAT score x and the grade-point average (GPA) y for 12 students.

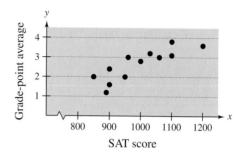

SAT score

(a) Is the GPA y a function of the SAT score x?
   GPA is not a function of the SAT score.

(b) Estimate the range for this relation. $1.2 \le y \le 3.8$

Explaining Concepts

73. *Writing*   Explain the difference between a relation and a function. Give an example of a relation that is not a function.   A relation is any set of ordered pairs. A function is a relation in which no two ordered pairs have the same first component and different second components.   See Additional Answers.

74. Is it possible to find a function that is not a relation? If it is, find one.   No.

75. *Writing*   Explain the meaning of the terms *domain* and *range* in the context of a function.   The domain is the set of inputs of the function, and the range is the set of outputs of the function.

76. *Writing*   In your own woods, explain how to use the Vertical Line Test.   Check to see that no vertical line intersects the graph at two (or more) points. If this is true, then the equation represents y as a function of x.

77. *Writing*   Describe some advantages of using function notation.   You can name the functions $(f, g, \ldots)$, which is convenient when there is more than one function used in solving a problem. The values of the independent and dependent variables are easily seen in function notation.

78. *Writing*   Is it possible for the number of elements in the domain of a relation to be greater than the number of elements in the range of the relation? Explain.   Yes. Example: $f(x) = 10$

79. *Writing*   Determine whether the statement uses the word *function* in a way that is mathematically correct. Explain your reasoning.

(a) The amount of money in your savings account is a function of your salary.   No, your savings account will vary while your salary is constant.

(b) The speed at which a free-falling baseball strikes the ground is a function of the height from which it is dropped.   Yes, each height will be associated with only one speed.

What Did You Learn?

Key Terms

extracting square roots, *p. 539*

completing the square, *p. 544*

Quadratic Formula, *p. 553*

discriminant, *p. 553*

quadratic equation in two variables, *p. 562*

parabola, *p. 562*

vertex, *p. 562*

imaginary unit *i*, *p. 581*

complex number, *p. 583*

real part, *p. 583*

imaginary part, *p. 583*

imaginary number, *p. 583*

complex conjugates, *p. 585*

relation, *p. 593*

domain, *p. 593*

range, *p. 593*

function, *p. 594*

independent variable, *p. 595*

dependent variable, *p. 595*

function notation, *p. 596*

Key Concepts

10.1 ◯ Square Root Property

Let u be a real number, a variable, or an algebraic expression, and let d be a positive real number; then the equation $u^2 = d$ has exactly two solutions.

If $u^2 = d$, then $u = \sqrt{d}$ and $u = -\sqrt{d}$.

These solutions can also be written as $u = \pm\sqrt{d}$. This form of the solution is read as "*u* is equal to plus or minus the square root of *d*."

10.2 ◯ Completing the square

To complete the square for the expression

$$x^2 + bx$$

add $(b/2)^2$, which is the square of half the coefficient of x. Consequently,

$$x^2 + bx + \left(\frac{b}{2}\right)^2 = \left(x + \frac{b}{2}\right)^2.$$

10.3 ◯ The Quadratic Formula

The solutions of $ax^2 + bx + c = 0$, $a \neq 0$, are given by the Quadratic Formula

$$x = \frac{-b \pm \sqrt{b^2 - 4ac}}{2a}.$$

The expression inside the radical, $b^2 - 4ac$, is called the discriminant.

1. If $b^2 - 4ac > 0$, the equation has two real solutions.

2. If $b^2 - 4ac = 0$, the equation has one (repeated) real solution.

3. If $b^2 - 4ac < 0$, the equation has no real solution.

10.3 ◯ Guidelines for solving quadratic equations

See the guidelines on page 556.

10.4 ◯ The Leading Coefficient Test for parabolas

The graph of the quadratic equation $y = ax^2 + bx + c$ is a parabola.

1. If $a > 0$, the parabola opens upward.

2. If $a < 0$, the parabola opens downward.

10.4 ◯ Vertex of a parabola

The vertex of a parabola given by $y = ax^2 + bx + c$ occurs at the point whose x-coordinate is $x = -b/(2a)$. To find the y-coordinate of the vertex, substitute the x-coordinate in the equation $y = ax^2 + bx + c$.

10.4 ◯ Guidelines for sketching a parabola

See the guidelines on page 566.

10.6 ◯ The square root of a negative number

Let c be a positive real number. Then the square root of $-c$ is given by

$$\sqrt{-c} = \sqrt{c(-1)} = \sqrt{c}\sqrt{-1} = \sqrt{c}\,i.$$

When writing $\sqrt{-c}$ in the *i*-form, $\sqrt{c}\,i$, note that i is outside the radical.

10.6 ◯ Square Root Property (complex square roots)

The equation $u^2 = d$, where $d < 0$, has exactly two solutions:

$$u = \sqrt{|d|}\,i \quad \text{and} \quad u = -\sqrt{|d|}\,i.$$

10.7 ◯ Vertical Line Test

A set of points on a rectangular coordinate system is the graph of y as a function of x if and only if no vertical line intersects the graph at more than one point.

Review Exercises

10.1 Solution by the Square Root Property

① Solve quadratic equations by factoring.

In Exercises 1–6, solve the quadratic equation by factoring.

1. $x^2 + 10x = 0$ $-10, 0$ **2.** $u^2 - 12u = 0$ $0, 12$

3. $x^2 - 5x + 6 = 0$ **4.** $3y^2 + 7y - 6 = 0$
    $2, 3$ $-3, \frac{2}{3}$

5. $4y^2 - 25 = 0$ $-\frac{5}{2}, \frac{5}{2}$ **6.** $8z^2 - 32 = 0$ $-2, 2$

② Solve quadratic equations by the Square Root Property.

In Exercises 7–22, solve the quadratic equation by the Square Root Property.

7. $x^2 = 49$ ± 7

8. $a^2 = 81$ ± 9

9. $x^2 - 48 = 0$ $\pm 4\sqrt{3}$

10. $x^2 - 72 = 0$ $\pm 6\sqrt{2}$

11. $y^2 - 18 = 0$ **12.** $y^2 - 8 = 0$
    $\pm 3\sqrt{2}$ $\pm 2\sqrt{2}$

13. $(x - 5)^2 = 3$ **14.** $(x + 2)^2 = 5$
    $5 \pm \sqrt{3}$ $-2 \pm \sqrt{5}$

15. $(x - 2)^2 - 6 = 0$ **16.** $(x + 1)^2 - 3 = 0$
    $2 \pm \sqrt{6}$ $-1 \pm \sqrt{3}$

17. $2(x + 4)^2 - 16 = 0$ **18.** $3(y - 1)^2 - 36 = 0$
    $-4 \pm 2\sqrt{2}$ $1 \pm 2\sqrt{3}$

19. $9(x - 7)^2 - 25 = 25$ **20.** $4(x + 3)^2 - 9 = 9$
    $\frac{21 \pm 5\sqrt{2}}{3}$ $\frac{-6 \pm 3\sqrt{2}}{2}$

21. $8(x - 4)^2 - 32 = 0$ **22.** $7(y - 6)^2 - 63 = 0$
    $2, 6$ $3, 9$

10.2 Solution by Completing the Square

① Construct perfect square trinomials.

In Exercises 23–26, what term should be added to the expression to make it a perfect square trinomial?

23. $x^2 + 12x$ 36

24. $y^2 - 16y$ 64

25. $t^2 - 15t$ $\frac{225}{4}$

26. $x^2 + 21x$ $\frac{441}{4}$

② Solve quadratic equations by completing the square.

In Exercises 27–34, solve the quadratic equation by completing the square.

27. $x^2 - 6x - 1 = 0$ $3 \pm \sqrt{10}$

28. $x^2 + 10x + 12 = 0$ $-5 \pm \sqrt{13}$

29. $x^2 - x - 1 = 0$ $\dfrac{1 \pm \sqrt{5}}{2}$

30. $t^2 + 3t + 1 = 0$ $\dfrac{-3 \pm \sqrt{5}}{2}$

31. $2y^2 + 10y + 5 = 0$ $\dfrac{-5 \pm \sqrt{15}}{2}$

32. $3x^2 - 2x - 1 = 0$ $-\frac{1}{3}, 1$

33. $4x^2 - 2x - 1 = 0$ $\dfrac{1 \pm \sqrt{5}}{4}$

34. $2y^2 + y - 3 = 0$ $-\frac{3}{2}, 1$

10.3 Solution by the Quadratic Formula

② Solve quadratic equations using the Quadratic Formula.

In Exercises 35–50, use the Quadratic Formula to solve the quadratic equation.

35. $y^2 + y - 42 = 0$ $-7, 6$

36. $x^2 - x - 20 = 0$ $-4, 5$

37. $c^2 - 6c + 6 = 0$ $3 \pm \sqrt{3}$

38. $c^2 - 6c + 5 = 0$ $1, 5$

39. $-x^2 + 3x + 70 = 0$ $-7, 10$

40. $y^2 + y - 1 = 0$ $\dfrac{-1 \pm \sqrt{5}}{2}$

41. $2y^2 + y - 42 = 0$ $\dfrac{-1 \pm \sqrt{337}}{4}$

42. $2x^2 - x - 20 = 0$ $\dfrac{1 \pm \sqrt{161}}{4}$

43. $-3x^2 - 5x + 3 = 0$ $\dfrac{-5 \pm \sqrt{61}}{6}$

44. $4x^2 + 4x + 1 = 0$ $-\frac{1}{2}$

45. $v^2 = 250$ $\pm 5\sqrt{10}$

46. $x^2 - 45x = 0$ $0, 45$

47. $0.3t^2 - 2t + 1 = 0$ $\dfrac{10 \pm \sqrt{70}}{3}$

48. $-u^2 + 3.1u + 5 = 0$ $\dfrac{31 \pm 3\sqrt{329}}{20}$

49. $0.7x^2 - 0.14x + 0.007 = 0$ 0.1

50. $0.5y^2 + 0.75y - 2 = 0$ $\dfrac{-3 \pm \sqrt{73}}{4}$

In Exercises 51–54, solve the equation.

51. $\dfrac{1}{x} + \dfrac{1}{x+1} = \dfrac{1}{2}$ $\dfrac{3 \pm \sqrt{17}}{2}$

52. $\dfrac{3}{t-1} - \dfrac{2}{t^2+t-2} = 4$ $\dfrac{-1 \pm \sqrt{193}}{8}$

53. $\sqrt{2x+5} = x - 3$ $4 + 2\sqrt{3}$

54. $x = \sqrt{4x+5}$ 5

10.4 Graphing Quadratic Equations

① Determine whether parabolas open upward or downward using the Leading Coefficient Test.

In Exercises 55–60, determine whether the parabola opens upward or downward.

55. $y = 3x^2 + 4x - 8$ Upward

56. $y = -4x^2 + 2x + 10$ Downward

57. $y = -7x^2 - 5x - 6$ Downward

58. $y = x^2 - 9x + 3$ Upward

59. $y = 3 - (x+4)^2$ Downward

60. $y = 7 + (2x-1)^2$ Upward

② Sketch graphs of quadratic equations using the point-plotting method and the vertex of a parabola.

In Exercises 61–72, sketch the parabola. Label the vertex and any intercepts.   See Additional Answers.

61. $y = x^2 - 2x + 1$

62. $y = -(x^2 - 2x + 1)$

63. $y = -x^2 + 4x - 3$

64. $y = x^2 - 6x + 9$

65. $y = -x^2 + 3x$

66. $y = x^2 - 10x$

67. $y = \frac{1}{4}(4x^2 - 4x + 3)$

68. $y = \frac{1}{3}(x^2 - 4x + 6)$

69. $y = 2x^2 + 4x + 5$

70. $y = -2x^2 + 8x - 5$

71. $y = -(3x^2 - 4x - 2)$

72. $y = 3x^2 + 2x + 3$

In Exercises 73–76, use a graphing calculator to graph the equation. Approximate the vertex from the graph.   See Additional Answers.

73. $y = 3 - x^2$

74. $y = 3 - \frac{1}{3}x^2$

75. $y = x^2 - 6x + 5$

76. $y = 20 - 11x - 3x^2$

10.5 Applications of Quadratic Equations

① Solve application problems that can be modeled by quadratic equations.

77. *Number Problem* Find two consecutive positive integers whose product is 240.   15, 16

78. *Number Problem* Find two consecutive positive integers such that the sum of their squares is 365.   13, 14

79. *Compound Interest* After 2 years, a $2000 investment, compounded annually at interest rate r, will yield an amount $2000(1 + r)^2$. If this amount is $2332.80, find the rate r.   8%

80. *Compound Interest* After 2 years, an $800 investment, compounded annually at interest rate r, will yield an amount $800(1 + r)^2$. If this amount is $882.00, find the rate r.   5%

81. *Free-Falling Object* The height h (in feet) of an object above the ground is given by

$$h = 48 - 16t^2$$

where t is time in seconds. How long does it take for the object to hit the ground? Describe the motion of this object. Was it dropped, thrown upward, or thrown downward? Explain. $\sqrt{3} \approx 1.73$ seconds. The object was dropped from a height of 48 feet.

82. *Free-Falling Object* The height h (in feet) of an object above the ground is given by

$$h = -16t^2 + 48t + 160$$

where t is time in seconds. How long does it take for the object to hit the ground? Describe the motion of the object. Was it dropped, thrown upward, or thrown downward? Explain.   5 seconds. The object was thrown upward from a height of 160 feet with a velocity of 48 feet per second. The object reaches its maximum height of 196 feet after $1\frac{1}{2}$ seconds.

Geometry **In Exercises 83–86, solve for x.**

83. Area = 32 square centimeters   6 centimeters

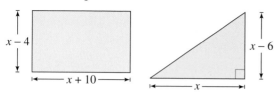

Figure for 83   Figure for 84

84. Area = 20 square feet   10 feet

85. Area = 1800 square meters   40 meters

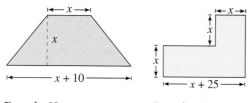

Figure for 85   Figure for 86

86. Area = 1300 square feet   20 feet

87. ▲ *Geometry*   The height of a triangle is one and one-half times its base, and its area is 60 square inches. Find the dimensions of the triangle.   Base: $4\sqrt{5} \approx 8.94$ inches; Height: $6\sqrt{5} \approx 13.42$ inches

88. ▲ *Geometry*   The height of a triangle is three times its base, and its area is 24 square inches. Find the dimensions of the triangle.   Base: 4 inches; Height: 12 inches

89. ▲ *Geometry*   The perimeter of a rectangle is 34 feet and the length of the diagonal is 13 feet. Find the dimensions of the rectangle.   5 feet × 12 feet

90. ▲ *Geometry*   The perimeter of a rectangle is 122 centimeters and the length of the diagonal is 43.6 centimeters. Find the dimensions of the rectangle.
26 centimeters × 35 centimeters

91. *Work Rate*   Working together, two people can complete a task in 10 hours. Working alone, one person takes 4 hours longer than the other to complete the task. Working alone, how long does it take each person to complete the task?   18.2 hours; 22.2 hours

92. *Work Rate*   Working together, two people can complete a task in 15 hours. Working alone, one person takes 2 hours longer than the other to complete the task. Working alone, how long does it take each person to complete the task?   29.03 hours; 31.03 hours

93. *Reduced Rates*   A Little League baseball team paid $72 for a block of tickets to a ball game. The block contained three more tickets than the team needed. By inviting three more people to attend (and share in the cost), the team lowered the price per ticket by $1.20. How many people are going to the game?
15 people

94. *Reduced Rates*   A Little League baseball team paid $120 for a block of tickets to a ball game. The block contained four more tickets than the team needed. By inviting four more people to attend (and share in the cost), the team lowered the price per ticket by $1. How many people are going to the game?   24 people

95. *Average Speed*   A train traveled the first 165 miles of a trip at one speed and the last 300 miles at an average speed that was 5 miles per hour greater. The entire trip took 8 hours. What were the two average speeds?   55 miles per hour; 60 miles per hour

96. *Average Speed*   A bus traveled the first 220 miles of a trip at one speed and the last 130 miles at an average speed that was 10 miles per hour greater. The entire trip took 6 hours. What were the two average speeds?   55 miles per hour; 65 miles per hour

97. *Path of a Ball*   The height y (in feet) of a ball thrown by a child is modeled by

$$y = -\frac{1}{10}x^2 + 3x + 3$$

where x is the horizontal distance (in feet) from where the ball is thrown.

(a) How high is the ball when it leaves the child's hand?   3 feet

(b) What is the maximum height of the ball? 25.5 feet

(c) How far from the child does the ball strike the ground?   31.0 feet

98. *Path of a Ball*   The height y (in feet) of a ball thrown by a child is modeled by

$$y = -\frac{1}{4}x^2 + x + 3$$

where x is the horizontal distance (in feet) from where the ball is thrown.

(a) How high is the ball when it leaves the child's hand?   3 feet

(b) What is the maximum height of the ball?   4 feet

(c) How far from the child does the ball strike the ground?   6 feet

99. ▲ *Geometry*   The perimeter of a rectangle of length l and width w is 40 feet.

(a) Show that $w = 20 - l$.

$2l + 2w = 40$

$l + w = 20$

$\quad w = 20 - l$

(b) Show that the area A is given by

$A = lw = l(20 - l)$.

$A = lw$

$w = 20 - l$

$A = l(20 - l)$

(c) Complete the table.

l	2	4	6	8	10	12	14	16	18
A	36	64	84	96	100	96	84	64	36

(d) Sketch the graph of the equation $A = l(20 - l)$.
   See Additional Answers.

(e) Of all rectangles with perimeters of 40 feet, which has the maximum area? How can you tell?   10 feet × 10 feet; maximum area occurs at the vertex.

100. ▦ *Data Analysis*   The revenue R (in millions of dollars) for Papa John's International for the years 1996 through 2001 can be approximated by the model

$R = -12.454t^2 + 340.25t - 1246.0,$
$6 \le t \le 11$

where t represents the year, with $t = 6$ corresponding to 1996.   (Source: Papa John's International)

(a) Use a graphing calculator to graph the model.
   See Additional Answers.

(b) Use the graph to approximate the year in which there was $911 million in revenue.   2000

(c) Use the graph to complete the table and verify your answer from part (b).

t	6	7	8	9	10	11
R	347.16	525.5	678.94	807.48	911.1	989.82

(d) Verify your answers to part (c) algebraically.
   See Additional Answers

10.6 Complex Numbers

① Write square roots of negative numbers in *i*-form and perform operations on numbers in *i*-form.

In Exercises 101–106, write the number in *i*-form.

101. $\sqrt{-48}$ $4\sqrt{3}i$

102. $\sqrt{-0.16}$ $0.4i$

103. $10 - 3\sqrt{-27}$ $10 - 9\sqrt{3}i$

104. $3 + 2\sqrt{-500}$ $3 + 20\sqrt{5}i$

105. $\frac{3}{4} - 5\sqrt{-\frac{3}{25}}$ $\frac{3}{4} - \sqrt{3}i$

106. $-0.5 + 3\sqrt{-1.21}$ $-0.5 + 3.3i$

In Exercises 107–114, perform the operation(s) and write the result in standard form.

107. $\sqrt{-81} + \sqrt{-36}$ $15i$

108. $\sqrt{-49} + \sqrt{-1}$ $8i$

109. $\sqrt{-121} - \sqrt{-84}$ $(11 - 2\sqrt{21})i$

110. $\sqrt{-169} - \sqrt{-4}$ $11i$

111. $\sqrt{-5}\sqrt{-5}$ -5

112. $\sqrt{-24}\sqrt{-6}$ -12

113. $\sqrt{-10}(\sqrt{-4} - \sqrt{-7})$ $\sqrt{70} - 2\sqrt{10}$

114. $\sqrt{-5}(\sqrt{-10} + \sqrt{-15})$ $-5\sqrt{2} - 5\sqrt{3}$

② Determine the equality of two complex numbers.

In Exercises 115–118, determine the values of a and b that satisfy the equation.

115. $12 - 5i = (a + 2) + (b - 1)i$
   $a = 10, \ b = -4$

116. $-48 + 9i = (a - 5) + (b + 10)i$
   $a = -43, \ b = -1$

117. $\sqrt{-49} + 4 = a + bi$ $a = 4, \ b = 7$

118. $-3 - \sqrt{-4} = a + bi$ $a = -3, b = -2$

③ Add, subtract, and multiply complex numbers.

In Exercises 119–126, perform the operation and write the result in standard form.

119. $(-4 + 5i) - (-12 + 8i)$ $8 - 3i$

120. $(-8 + 3i) - (6 + 7i)$ $-14 - 4i$

121. $(3 - 8i) + (5 + 12i)$ $8 + 4i$

122. $(-6 + 3i) + (-1 + i)$ $-7 + 4i$

123. $(4 - 3i)(4 + 3i)$ 25

124. $(12 - 5i)(2 + 7i)$ $59 + 74i$

125. $(6 - 5i)^2$ $11 - 60i$

126. $(2 - 9i)^2$ $-77 - 36i$

④ Use complex conjugates to write the quotient of two complex numbers in standard form.

In Exercises 127–132, write the quotient in standard form.

127. $\dfrac{7}{3i}$ $-\frac{7}{3}i$

128. $\dfrac{4}{5i}$ $-\frac{4}{5}i$

129. $\dfrac{4i}{2 - 8i}$ $-\frac{8}{17} + \frac{2}{17}i$

130. $\dfrac{5i}{2 + 9i}$ $\frac{9}{17} + \frac{2}{17}i$

131. $\dfrac{3 - 5i}{6 + i}$ $\frac{13}{37} - \frac{33}{37}i$

132. $\dfrac{2 + i}{1 - 9i}$ $-\frac{7}{82} + \frac{19}{82}i$

⑤ Solve quadratic equations with complex solutions.

In Exercises 133–138, solve the equation by using the Square Root Property.

133. $z^2 = -121$ $\pm 11i$

134. $u^2 = -25$ $\pm 5i$

135. $y^2 + 50 = 0$ $\pm 5\sqrt{2}i$

136. $x^2 + 48 = 0$ $\pm 4\sqrt{3}i$

137. $(y + 4)^2 + 18 = 0$ $-4 \pm 3\sqrt{2}i$

138. $(x - 2)^2 + 24 = 0$ $2 \pm 2\sqrt{6}i$

In Exercises 139–142, solve the equation by completing the square. Give the solutions in exact form and in decimal form rounded to two decimal places.

139. $x^2 - 2x + 26 = 0$ $1 \pm 5i$

140. $t^2 - 16t + 208 = 0$ $8 \pm 12i$

141. $x^2 - 3x + 3 = 0$
$\frac{3}{2} + \frac{\sqrt{3}}{2}i \approx 1.50 + 0.87i; \frac{3}{2} - \frac{\sqrt{3}}{2}i \approx 1.50 - 0.87i$

142. $y^2 - \frac{2}{3}y + 2 = 0$
$\frac{1}{3} + \frac{\sqrt{17}}{3}i \approx 0.33 + 1.37i; \frac{1}{3} - \frac{\sqrt{17}}{3}i \approx 0.33 - 1.37i$

In Exercises 143–146, solve the equation by using the Quadratic Formula.

143. $x^2 + 6x + 13 = 0$ $-3 \pm 2i$

144. $a^2 + 4a + 29 = 0$ $-2 \pm 5i$

145. $3z^2 - 3z + \frac{49}{64} = 0$ $\frac{1}{2} \pm \frac{\sqrt{3}}{24}i$

146. $2y^2 + y + \frac{29}{32} = 0$ $-\frac{1}{4} \pm \frac{5}{8}i$

10.7 Relations, Functions, and Graphs

① Identify the domain and range of a relation.

In Exercises 147–150, find the domain and range of the relation.

147. $\{(8, 3), (-2, 7), (5, 1), (3, 8)\}$
Domain: $\{-2, 3, 5, 8\}$; Range: $\{1, 3, 7, 8\}$

148. $\{(0, 1), (-1, 3), (4, 6), (-7, 5)\}$
Domain: $\{-7, -1, 0, 4\}$; Range: $\{1, 3, 5, 6\}$

149. $\{(2, -3), (-2, 3), (7, 0), (-4, -2)\}$
Domain: $\{-4, -2, 2, 7\}$; Range: $\{-3, -2, 0, 3\}$

150. $\{(1, 7), (-3, 4), (6, 5), (-2, -9)\}$
Domain: $\{-3, -2, 1, 6\}$; Range: $\{-9, 4, 5, 7\}$

② Determine if relations are functions by inspection or by using the Vertical Line Test.

In Exercises 151–154, determine whether the relation represents a function.

151. Function

152. Not a function

153. Not a function

Input x	Output y	(x, y)
0	0	$(0, 0)$
2	1	$(2, 1)$
4	1	$(4, 1)$
6	2	$(6, 2)$
2	3	$(2, 3)$

154. Function

Input x	Output y	(x, y)
-6	1	$(-6, 1)$
-3	0	$(-3, 0)$
0	1	$(0, 1)$
3	4	$(3, 4)$
6	2	$(6, 2)$

In Exercises 155–160, use the Vertical Line Test to determine whether *y* is a function of *x*.

155. Function

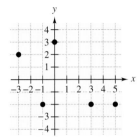

156. Not a function

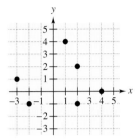

157. Not a function

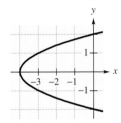

158. Function

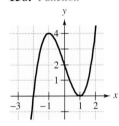

159. Function

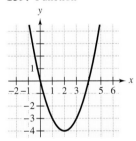

160. Not a function

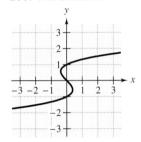

③ Use function notation and evaluate functions.

In Exercises 161–166, evaluate the function as indicated, and simplify.

161. $f(x) = 25x$ (a) $f(-1)$ (b) $f(7)$
(a) -25 (b) 175 (c) $f(10)$ (d) $f\left(-\frac{4}{3}\right)$
(c) 250 (d) $-\frac{100}{3}$

162. $f(x) = 2x - 7$ (a) $f(-1)$ (b) $f(3)$
(a) -9 (b) -1 (c) $f\left(\frac{1}{2}\right)$ (d) $f(-4)$
(c) -6 (d) -15

163. $g(t) = -16t^2 + 64$ (a) $g(0)$ (b) $g\left(\frac{1}{4}\right)$
(a) 64 (b) 63 (c) $g(1)$ (d) $g(2)$
(c) 48 (d) 0

164. $h(u) = u(u - 3)^2$ (a) $h(0)$ (b) $h(3)$
(a) 0 (b) 0 (c) $h(-1)$ (d) $h\left(\frac{3}{2}\right)$
(c) -16 (d) $\frac{27}{8}$

165. $f(x) = |2x + 3|$ (a) $f(0)$ (b) $f(5)$
(a) 3 (b) 13 (c) $f(-4)$ (d) $f\left(-\frac{3}{2}\right)$
(c) 5 (d) 0

166. $f(x) = |x| - 4$ (a) $f(-1)$ (b) $f(1)$
(a) -3 (b) -3 (c) $f(-4)$ (d) $f(2)$
(c) 0 (d) -2

167. *Demand* The demand for a product is a function of its price. Consider the demand function

$$f(p) = 40 - 0.2p$$

where *p* is the price in dollars. Find the demand for (a) $p = 10$, (b) $p = 50$, and (c) $p = 100$.
(a) 38 (b) 30 (c) 20

168. *Profit* The profit for a product is a function of the amount spent on advertising for the product. In the profit function

$$f(x) = 8000 + 2000x - 50x^2$$

x is the amount (in hundreds of dollars) spent on advertising. Find the profit for (a) $x = 5$, (b) $x = 10$, and (c) $x = 20$.
(a) $\$16{,}750$ (b) $\$23{,}000$ (c) $\$28{,}000$

④ Identify the domain of a function.

In Exercises 169 and 170, find the domain of the function.

169. $f: \{(1, 5), (2, 10), (3, 15), (4, -10), (5, -15)\}$
$D = \{1, 2, 3, 4, 5\}$

170. $g: \{(-3, 6), (-2, 4), (-1, 2), (0, 0), (1, -2)\}$
$D = \{-3, -2, -1, 0, 1\}$

Take this test as you would take a test in class. After you are done, check your work against the answers in the back of the book.

In Exercises 1–8, solve the equation. If indicated, use the specified method.

4. $\dfrac{-9 \pm \sqrt{21}}{6}$ 5. $\dfrac{1 \pm \sqrt{11}i}{2}$

6. $\dfrac{-2 \pm \sqrt{2}}{2}$ 7. $\dfrac{1 \pm \sqrt{3}i}{2}$

1. Square Root Property:

$x^2 - 144 = 0$ $-12, 12$

2. Square Root Property:

$(x + 4)^2 + 100 = 0$ $-4 \pm 10i$

3. Completing the square:

$t^2 - 6t + 11 = 0$ $3 \pm \sqrt{2}i$

4. Completing the square:

$3z^2 + 9z + 5 = 0$

5. Quadratic Formula:

$x^2 - x + 3 = 0$

6. Quadratic Formula:

$2u^2 + 4u + 1 = 0$

7. $\dfrac{1}{x + 1} - \dfrac{1}{x - 2} = 1$

8. $\sqrt{2x} = x - 1$ $2 + \sqrt{3}$

Input x	Output y	(x, y)
0	4	$(0, 4)$
1	5	$(1, 5)$
2	8	$(2, 8)$
1	-3	$(1, -3)$
0	-1	$(0, -1)$

Table for 19

In Exercises 9–11, determine whether the parabola opens upward or downward. Then find the coordinates of the vertex.

9. $y = -2x^2 + 4$ Downward; Vertex: $(0, 4)$

10. $y = 5 - 2x - x^2$ Downward; Vertex: $(-1, 6)$

11. $y = (x - 2)^2 + 3$ Upward; Vertex: $(2, 3)$

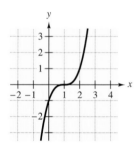

Figure for 20

In Exercises 12 and 13, sketch the parabola. Label the vertex and any intercepts.   See Additional Answers.

12. $y = x^2 - 8x + 12$

13. $y = x^2 - 4x$

In Exercises 14–17, perform the operations(s) and simplify.

14. $(2 + 3i) - \sqrt{-25}$ $2 - 2i$

15. $(2 - 3i)^2$ $-5 - 12i$

16. $\sqrt{-16}\left(1 + \sqrt{-4}\right)$ $-8 + 4i$

17. $(3 - 2i)(1 + 5i)$ $13 + 13i$

18. Write $\dfrac{5 - 2i}{3 + i}$ in standard form. $\frac{13}{10} - \frac{11}{10}i$

19. Does the table at the left represent y as a function of x? Explain.

No, because some input values, 0 and 1, have two different output values.

20. Does the graph at the left represent y as a function of x? Explain.

Yes, because it passes the Vertical Line Test.

21. Evaluate $f(x) = x^3 - 2x^2$ at the indicated values.

(a) $f(0)$ 0   (b) $f(2)$ 0   (c) $f(-2)$ -16 (d) $f\left(\frac{1}{2}\right)$ $-\frac{3}{8}$

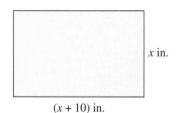

$(x + 10)$ in.

Figure for 22

22. $x(x + 10) = 96$

$x^2 + 10x = 96$

$x^2 + 10x - 96 = 0$

$(x + 16)(x - 6) = 0$

$x = -16, x = 6$

6 inches × 16 inches

22. The rectangle shown in the figure at the left has an area of 96 square inches. Use a quadratic equation to find its dimensions. Show your work.

23. Working together, two people can mow a lawn in 6 hours. When working alone, it takes one person 5 hours longer than the other. Working alone, how long does it take each person to mow the lawn?   10 hours, 15 hours

24. After 2 years, a $600 investment, compounded annually at interest rate r, will yield an amount $600(1 + r)^2$. If this amount is $655.22, find the rate r.   4.5%

Appendix A

Introduction to Graphing Calculators

Introduction   •   Using a Graphing Calculator   •   Using Special Features of a Graphing Calculator

Introduction

In Section 4.2, you studied the point-plotting method for sketching the graph of an equation. One of the disadvantages of the point-plotting method is that to get a good idea about the shape of a graph you need to plot *many* points. By plotting only a few points, you can badly misrepresent the graph.

For instance, consider the equation $y = x^3$. To graph this equation, suppose you calculated only the following three points.

x	-1	0	1
$y = x^3$	-1	0	1
Solution point	$(-1, -1)$	$(0, 0)$	$(1, 1)$

By plotting these three points, as shown in Figure A.1, you might assume that the graph of the equation is a line. This, however, is not correct. By plotting several more points, as shown in Figure A.2, you can see that the actual graph is not straight at all.

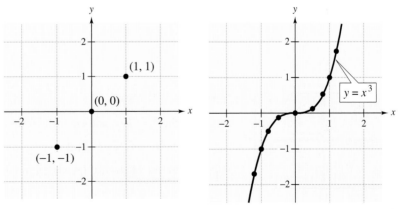

Figure A.1   Figure A.2

So, the point-plotting method leaves you with a dilemma. On the one hand, the method can be very inaccurate if only a few points are plotted. But, on the other hand, it is very time-consuming to plot a dozen (or more) points. Technology can help you solve this dilemma. Plotting several points (or even hundreds of points) on a rectangular coordinate system is something that a computer or graphing calculator can do easily.

Using a Graphing Calculator

There are many different graphing utilities: some are graphing packages for computers and some are hand-held graphing calculators. In this appendix, the steps used to graph an equation with a *TI-83* or *TI-83 Plus* graphing calculator are described. Keystroke sequences are often given for illustration; however, these may not agree precisely with the steps required by *your* calculator.*

Graphing an Equation with a *TI-83* or *TI-83 Plus* Graphing Calculator

Before performing the following steps, set your calculator so that all of the standard defaults are active. For instance, all of the options at the left of the MODE screen should be highlighted.

1. Set the viewing window for the graph. (See Example 3.) To set the standard viewing window, press ZOOM 6.

2. Rewrite the equation so that *y* is isolated on the left side of the equation.

3. Press the Y= key. Then enter the right side of the equation on the first line of the display. (The first line is labeled $Y_1 = .$)

4. Press the GRAPH key.

Example 1   Graphing a Linear Equation

Use a graphing calculator to graph $2y + x = 4$.

Solution

To begin, solve the equation for *y* in terms of *x*.

$$2y + x = 4 \qquad \text{Write original equation.}$$

$$2y = -x + 4 \qquad \text{Subtract } x \text{ from each side.}$$

$$y = -\frac{1}{2}x + 2 \qquad \text{Divide each side by 2.}$$

Press the Y= key, and enter the following keystrokes.

(−) X,T,θ,n ÷ 2 + 2

The top row of the display should now be as follows.

$$Y_1 = \text{-X/2} + 2$$

Press the GRAPH key, and the screen should look like that shown in Figure A.3.

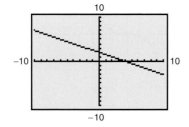

Figure A.3

---

*The graphing calculator keystrokes given in this section correspond to the *TI-83* and *TI-83 Plus* graphing calculators by Texas Instruments. For other graphing calculators, the keystrokes may differ. Consult your user's guide.

In Figure A.3, notice that the calculator screen does not label the tick marks on the *x*-axis or the *y*-axis. To see what the tick marks represent, you can press WINDOW. If you set your calculator to the standard graphing defaults before working Example 1, the screen should show the following values.

Xmin = -10   The minimum *x*-value is − 10.
Xmax = 10   The maximum *x*-value is 10.
Xscl = 1   The *x*-scale is 1 unit per tick mark.
Ymin = -10   The minimum *y*-value is − 10.
Ymax = 10   The maximum *y*-value is 10.
Yscl = 1   The *y*-scale is 1 unit per tick mark.
Xres = 1   Sets the pixel resolution

These settings are summarized visually in Figure A.4.

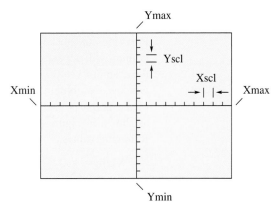

Figure A.4

Example 2   Graphing an Equation Involving Absolute Value

Use a graphing calculator to graph

$$y = |x - 3|.$$

Solution

This equation is already written so that *y* is isolated on the left side of the equation. Press the Y= key, and enter the following keystrokes.

    MATH ▶ 1 X,T,θ,n − 3)

The top row of the display should now be as follows.

$$Y_1 = \text{abs}(X - 3)$$

Press the GRAPH key, and the screen should look like that shown in Figure A.5.

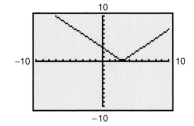

Figure A.5

Using Special Features of a Graphing Calculator

To use your graphing calculator to its best advantage, you must learn to set the viewing window, as illustrated in the next example.

Example 3   Setting the Viewing Window

Use a graphing calculator to graph

$$y = x^2 + 12.$$

Solution

Press [Y=] and enter $x^2 + 12$ on the first line.

[X,T,θ,n] [x²] [+] 12

Press the [GRAPH] key. If your calculator is set to the standard viewing window, nothing will appear on the screen. The reason for this is that the lowest point on the graph of $y = x^2 + 12$ occurs at the point (0, 12). Using the standard viewing window, you obtain a screen whose largest y-value is 10. In other words, none of the graph is visible on a screen whose y-values vary between -10 and 10, as shown in Figure A.6. To change these settings, press [WINDOW] and enter the following values.

Xmin = -10	The minimum x-value is -10.
Xmax = 10	The maximum x-value is 10.
Xscl = 1	The x-scale is 1 unit per tick mark.
Ymin = -10	The minimum y-value is -10.
Ymax = 30	The maximum y-value is 30.
Yscl = 5	The y-scale is 5 units per tick mark.
Xres = 1	Sets the pixel resolution

Press [GRAPH] and you will obtain the graph shown in Figure A.7. On this graph, note that each tick mark on the y-axis represents five units because you changed the y-scale to 5. Also note that the highest point on the y-axis is now 30 because you changed the maximum value of y to 30.

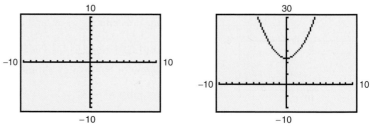

Figure A.6   Figure A.7

If you changed the y-maximum and y-scale on your calculator as indicated in Example 3, you should return to the standard setting before working Example 4. To do this, press [ZOOM] 6.

Example 4   Using a Square Setting

Use a graphing calculator to graph $y = x$. The graph of this equation is a line that makes a 45° angle with the x-axis and with the y-axis. From the graph on your calculator, does the angle appear to be 45°?

Solution

Press $\boxed{Y=}$ and enter x on the first line.

$$Y_1 = X$$

Press the $\boxed{GRAPH}$ key and you will obtain the graph shown in Figure A.8. Notice that the angle the line makes with the x-axis doesn't appear to be 45°. The reason for this is that the screen is wider than it is tall. This makes the tick marks on the x-axis farther apart than the tick marks on the y-axis. To obtain the same distance between tick marks on both axes, you can change the graphing settings from "standard" to "square." To do this, press the following keys.

$\boxed{ZOOM}$ 5   Square setting

The screen should look like that shown in Figure A.9. Note in this figure that the square setting has changed the viewing window so that the x-values vary from -15 to 15.

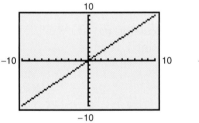

Figure A.8

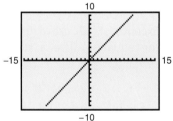

Figure A.9

There are many possible square settings on a graphing calculator. To create a square setting, you need the following ratio to be $\frac{2}{3}$.

$$\frac{Ymax - Ymin}{Xmax - Xmin}$$

For instance, the setting in Example 4 is square because $(Ymax - Ymin) = 20$ and $(Xmax - Xmin) = 30$.

Example 5   Graphing More than One Equation in the Same Viewing Window

Use a graphing calculator to graph each equation in the same viewing window.

$$y = -x + 4, \quad y = -x, \quad \text{and} \quad y = -x - 4$$

Solution

To begin, press $\boxed{Y=}$ and enter all three equations on the first three lines. The display should now be as follows.

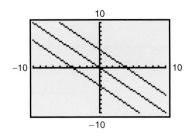

Figure A.10

$Y_1 = -X + 4$ $\boxed{(-)}\ \boxed{X,T,\theta,n}\ \boxed{+}\ 4$

$Y_2 = -X$ $\boxed{(-)}\ \boxed{X,T,\theta,n}$

$Y_3 = -X - 4$ $\boxed{(-)}\ \boxed{X,T,\theta,n}\ \boxed{-}\ 4$

Press the $\boxed{\text{GRAPH}}$ key and you will obtain the graph shown in Figure A.10. Note that the graph of each equation is a line, and that the lines are parallel to each other.

Another special feature of a graphing calculator is the *trace* feature. This feature is used to find solution points of an equation. For example, you can approximate the x- and y-intercepts of $y = 3x + 6$ by first graphing the equation, then pressing the $\boxed{\text{TRACE}}$ key, and finally pressing the $\boxed{\blacktriangleleft}\boxed{\blacktriangleright}$ keys. To get a better approximation of a solution point, you can use the following keystrokes repeatedly.

$\boxed{\text{ZOOM}}\ 2\ \boxed{\text{ENTER}}$

Check to see that you get an x-intercept of $(-2, 0)$ and a y-intercept of $(0, 6)$. Use the *trace* feature to find the x- and y-intercepts of $y = \frac{1}{2}x - 4$.

Appendix A   Exercises

▦ In Exercises 1–12, use a graphing calculator to graph the equation. (Use a standard setting.) See Examples 1 and 2.   See Additional Answers.

1. $y = -3x$

2. $y = x - 4$

3. $y = \frac{3}{4}x - 6$

4. $y = -3x + 2$

5. $y = \frac{1}{2}x^2$

6. $y = -\frac{2}{3}x^2$

7. $y = x^2 - 4x + 2$

8. $y = -0.5x^2 - 2x + 2$

9. $y = |x - 3|$

10. $y = |x + 4|$

11. $y = |x^2 - 4|$

12. $y = |x - 2| - 5$

▦ In Exercises 13–16, use a graphing calculator to graph the equation using the given window settings. See Example 3.   See Additional Answers.

13. $y = 27x + 100$

Xmin = 0
Xmax = 5
Xscl = .5
Ymin = 75
Ymax = 250
Yscl = 25
Xres = 1

14. $y = 50{,}000 - 6000x$

Xmin = 0
Xmax = 7
Xscl = .5
Ymin = 0
Ymax = 50000
Yscl = 5000
Xres = 1

15. $y = 0.001x^2 + 0.5x$

Xmin = -500
Xmax = 200
Xscl = 50
Ymin = -100
Ymax = 100
Yscl = 20
Xres = 1

16. $y = 100 - 0.5|x|$

Xmin = -300
Xmax = 300
Xscl = 60
Ymin = -100
Ymax = 100
Yscl = 20
Xres = 1

▦ In Exercises 17–20, find a viewing window that shows the important characteristics of the graph.
See Additional Answers.   Answers will vary.

17. $y = 15 + |x - 12|$

18. $y = 15 + (x - 12)^2$

19. $y = -15 + |x + 12|$

20. $y = -15 + (x + 12)^2$

▦ In Exercises 21–24, use a graphing calculator to graph both equations in the same viewing window. Are the graphs identical? If so, what basic rule of algebra is being illustrated? See Example 5.
See Additional Answers.   Yes.

21. $y_1 = 2x + (x + 1)$
    $y_2 = (2x + x) + 1$
    Associative Property
    of Addition

22. $y_1 = \frac{1}{2}(3 - 2x)$
    $y_2 = \frac{3}{2} - x$
    Distributive Property

23. $y_1 = 2\left(\frac{1}{2}\right)$

$y_2 = 1$

Multiplicative Inverse
Property

24. $y_1 = x(0.5x)$

$y_2 = (0.5x)x$

Commutative Property
of Multiplication

▦ In Exercises 25–32, use the *trace* feature of a graphing calculator to approximate the *x*- and *y*-intercepts of the graph.

25. $y = 9 - x^2$ $(-3, 0), (3, 0), (0, 9)$

26. $y = 3x^2 - 2x - 5$ $(-1, 0), \left(\frac{5}{3}, 0\right), (0, -5)$

27. $y = 6 - |x + 2|$ $(-8, 0), (4, 0), (0, 4)$

28. $y = |x - 2|^2 - 3$ $(0.268, 0), (3.732, 0), (0, 1)$

29. $y = 2x - 5$ $\left(\frac{5}{2}, 0\right), (0, -5)$

30. $y = 4 - |x|$ $(-4, 0), (4, 0), (0, 4)$

31. $y = x^2 + 1.5x - 1$ $(-2, 0), \left(\frac{1}{2}, 0\right), (0, -1)$

32. $y = x^3 - 4x$ $(-2, 0), (0, 0), (2, 0)$

▦ ▲ *Geometry*   In Exercises 33–36, use a graphing calculator to graph the equations in the same viewing window. Using a "square setting," determine the geometrical shape bounded by the graphs.
See Additional Answers.

33. $y = -4$, $y = -|x|$ Triangle

34. $y = |x|$, $y = 5$ Triangle

35. $y = |x| - 8$, $y = -|x| + 8$ Square

36. $y = -\frac{1}{2}x + 7$, $y = \frac{8}{3}(x + 5)$, $y = \frac{2}{7}(3x - 4)$
Triangle

▦ *Modeling Data*   In Exercises 37 and 38, use the following models, which give the number of pieces of first-class mail and the number of pieces of Standard A (third-class) mail handled by the U.S. Postal Service.

First Class

$y = 0.008x^2 + 1.42x + 88.7$ $0 \le x \le 10$

Standard A (Third Class)

$y = 0.246x^2 + 0.36x + 62.5$ $0 \le x \le 10$

In these models, *y* is the number of pieces handled (in billions) and *x* is the year, with $x = 0$ corresponding to 1990.   (Source: U.S. Postal Service)

37. Use the following window setting to graph both models in the same viewing window of a graphing calculator.   See Additional Answers.

$$\begin{array}{l} \text{Xmin} = 0 \\ \text{Xmax} = 10 \\ \text{Xscl} = 1 \\ \text{Ymin} = 0 \\ \text{Ymax} = 120 \\ \text{Yscl} = 10 \\ \text{Xres} = 1 \end{array}$$

38. (a) Were the numbers of pieces of first-class mail and Standard A mail increasing or decreasing over time?   Increasing

(b) Is the distance between the graphs increasing or decreasing over time? What does this mean to the U.S. Postal Service?   Decreasing. The number of pieces of third-class mail being handled is increasing more rapidly than the number of pieces of first-class mail.

Additional Answers

Chapter 1

Section 1.1 *(page 9)*

5.

6.

7.

8.

9.
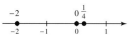

10.

11.

12.

13.

14.

15.

16.

17.

18.

23.

24.

25.

26.

27.

28.

59.

60.
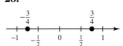

61.

62.

Section 1.2 *(page 17)*

1.

2.

3.
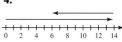

4.

5.
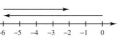

6.
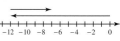

7.
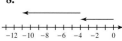

8.
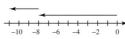

Section 1.3 *(page 29)*

111. (b)

112. (b)

Mid-Chapter Quiz *(page 33)*

1.
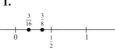

2.

3.
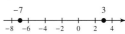

4.
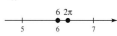

Review Exercises *(page 61)*

3.

4.

5.

6.
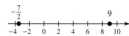

7.

8.

9.

10.
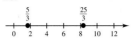

3.
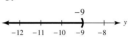

4.
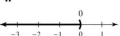

11.
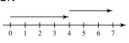

12.

5.

6.
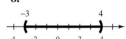

27.

28.

7.

8.

29.

30.

9.
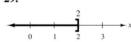

10.
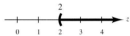

27.
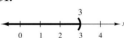

28.
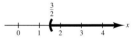

Chapter 2

Section 2.3   *(page 101)*

75.

29.

30.

31.

32.

33.
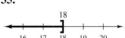

34.
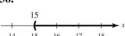

35.

36.

76.

37.
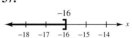

38.
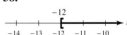

39.

40.

41.

42.
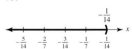

Chapter 3

Section 3.6   *(page 189)*

1.

2.
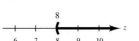

43.

44.
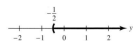

45.

46.

47.

48.

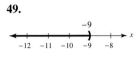

49.

50.

51.

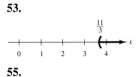

52.

53.

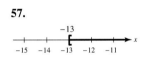

54.

55.

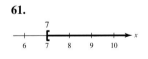

56.

57.

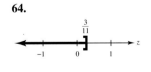

58.

59.

60.

61.

62.

63.

64.

65.

66.

67.

68.

Review Exercises   *(page 193)*

83.

84.

85.

86.

89.

90.

91.

92.

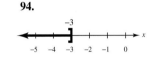

93.

94.

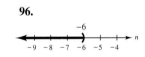

95.

96.

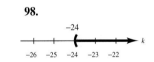

97.

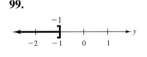

98.

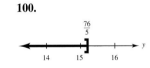

99.

100.

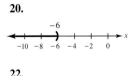

Chapter Test   *(page 196)*

19.

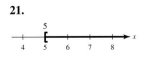

20.

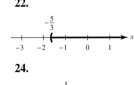

21.

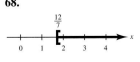

22.

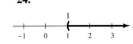

23.

24.

Chapter 4

Section 4.1   *(page 207)*

1.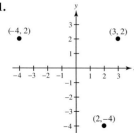

2.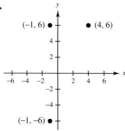

3.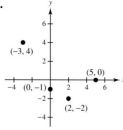

4.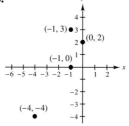

5.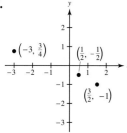

6.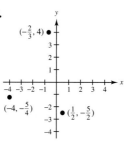

7.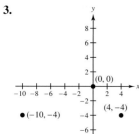

8.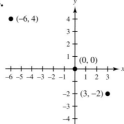

9.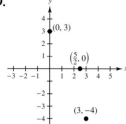

10.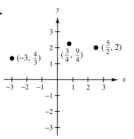

27.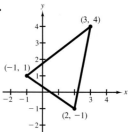

28.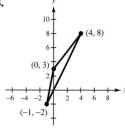

29.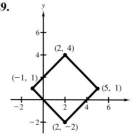

30.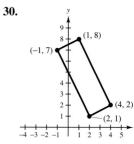

31.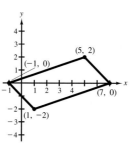

32.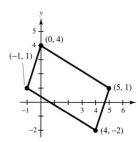

33.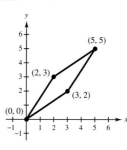

34.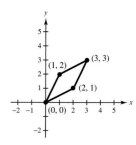

35.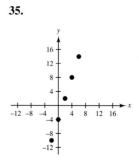

36.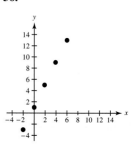

37.

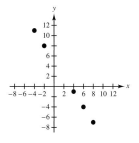

38.

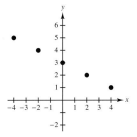

39.

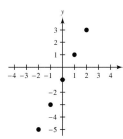

40.

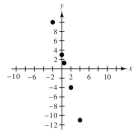

59.

x	20	40	60	80	100
$y = 0.066x$	1.32	2.64	3.96	5.28	6.60

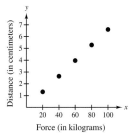

60.

x	0	2	4	6	8
$y = -800x + 9500$	9500	7900	6300	4700	3100

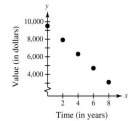

61.

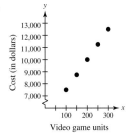

Video game units

62.

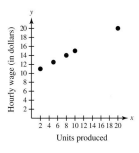

Units produced

63. (a)

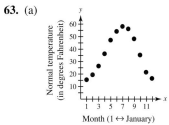

Month (1 ↔ January)

64. (a)

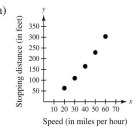

Speed (in miles per hour)

65. (a)

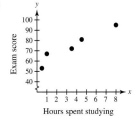

Hours spent studying

66. (a)

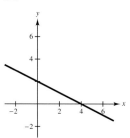

77. (b)

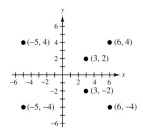

81. (a) and (b)

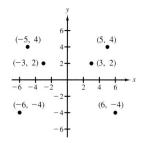

82. (a) and (b)

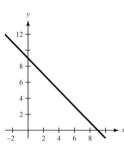

Section 4.2   *(page 218)*

9.

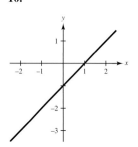

10.

11.

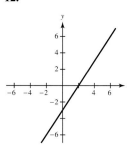

12.

13.

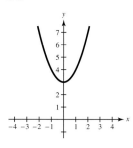

14.

15.

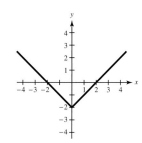

16.

37.

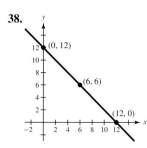

38.

39.

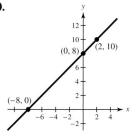

40.

41.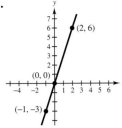

42.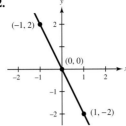

51.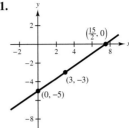

52.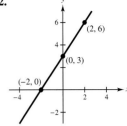

43.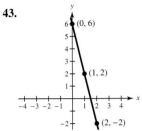

44.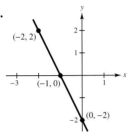

53.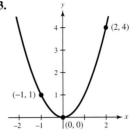

54.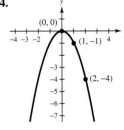

45.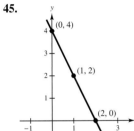

46.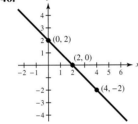

55.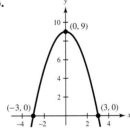

56.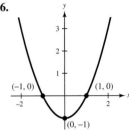

47.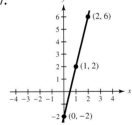

48.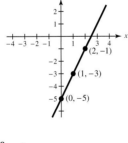

57.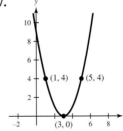

58.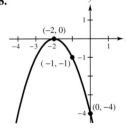

49.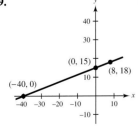

50.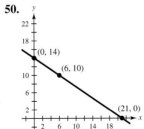

59.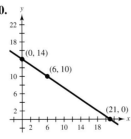

60.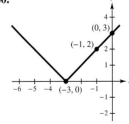

61.

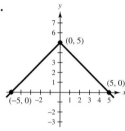

62.

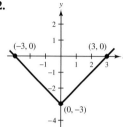

73.

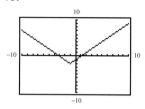

74.

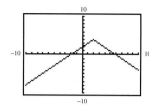

63.

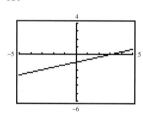

64.

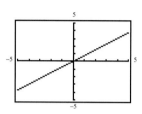

75.

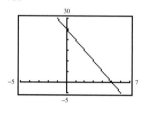

76.

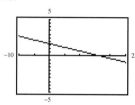

65.

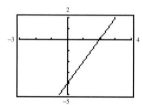

66.

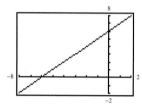

77.

Xmin = -15
Xmax = 15
Xscl = 1
Ymin = -10
Ymax = 10
Yscl = 1

78.

Xmin = -15
Xmax = 15
Xscl = 1
Ymin = -10
Ymax = 10
Yscl = 1

67.

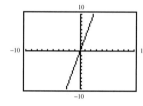

68.

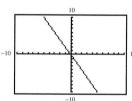

79.

Xmin = -5
Xmax = 20
Xscl = 5
Ymin = -5
Ymax = 20
Yscl = 5

80.

Xmin = -10
Xmax = 10
Xscl = 5
Ymin = -9
Ymax = 21
Yscl = 5

69.

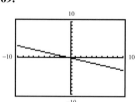

70.

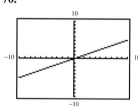

81.

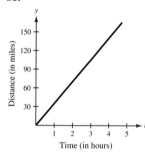

82.

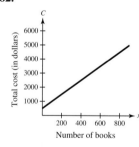

71.

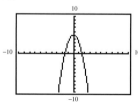

72.

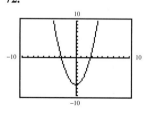

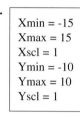
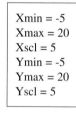

83. (a)

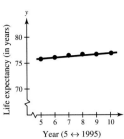

91.

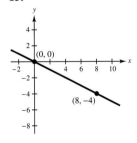

19.

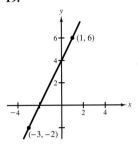

20.

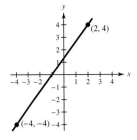

Section 4.3 *(page 231)*

13.

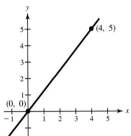

14.

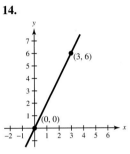

21.

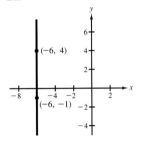

22.

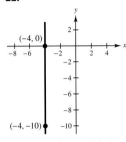

15.

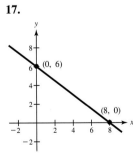

16.

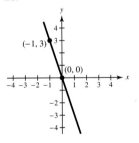

23.

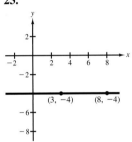

24.

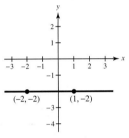

17.

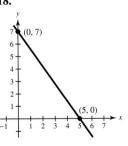

18.

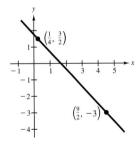

25.

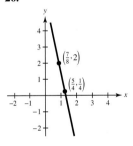

26.

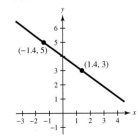

27.

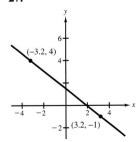

28.

29.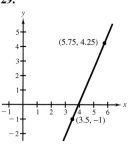

30.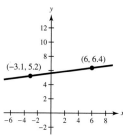

43.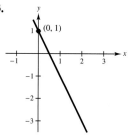

44.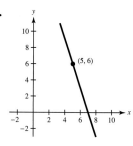

31.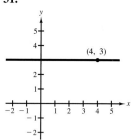

32.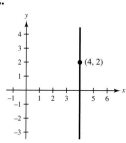

45.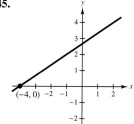

46.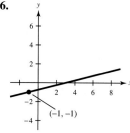

33.

x	-2	0	2	4
y	2	-2	-6	-10
Solution point	$(-2, 2)$	$(0, -2)$	$(2, -6)$	$(4, -10)$

34.

x	-2	0	2	4
y	-2	4	10	16
Solution point	$(-2, -2)$	$(0, 4)$	$(2, 10)$	$(4, 16)$

47.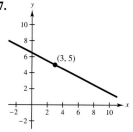

48.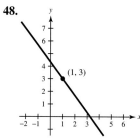

39.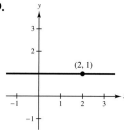

40.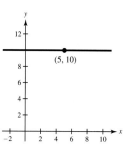

49.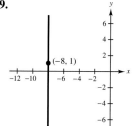

50.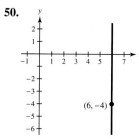

41.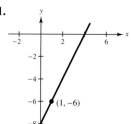

42.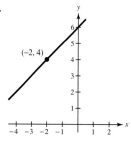

51.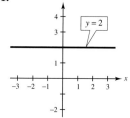

52.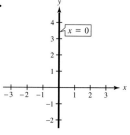

53.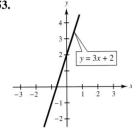

54.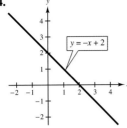

63.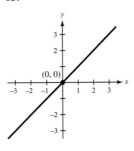

64.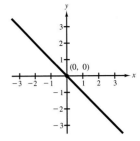

55.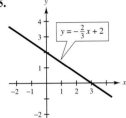

56.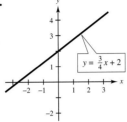

65.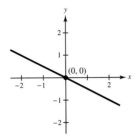

66.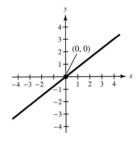

57.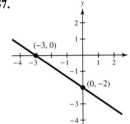

58.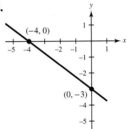

67.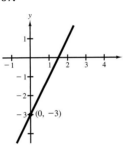

68.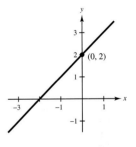

59.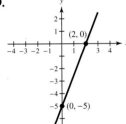

60.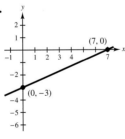

61.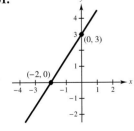

62.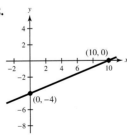

69.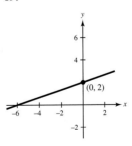

70.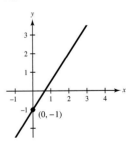

71.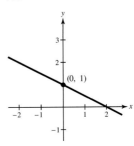

72.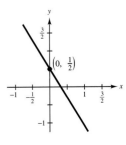

73.

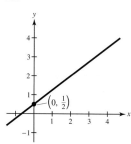

74.

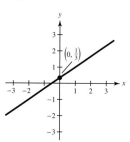

92. (d)

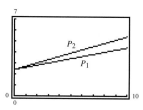

75.

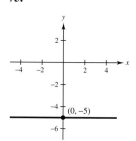

76.

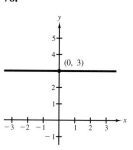

108. (a)

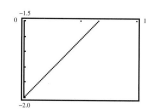

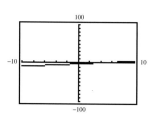

81.

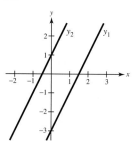

82.

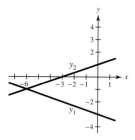

Mid-Chapter Quiz   *(page 236)*

1.

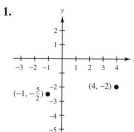

83.

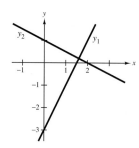

84.

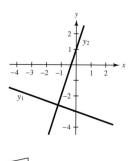

7.

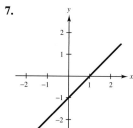

8.

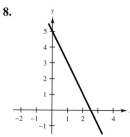

87. (a)

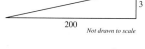

200
Not drawn to scale

88. (a)

21

9.

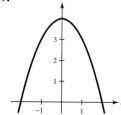

10.

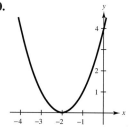

11.

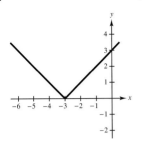

12.

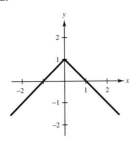

Section 4.4   *(page 244)*

1.

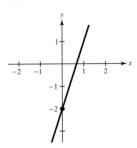

2.

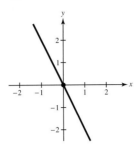

13.

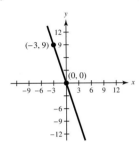

14.

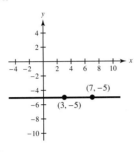

3.

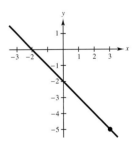

4.

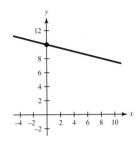

15.

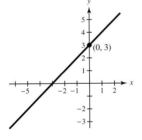

16.

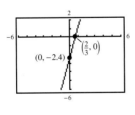

5.

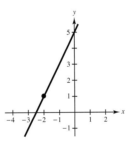

6.

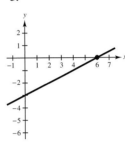

17.

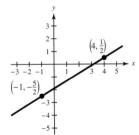

18.

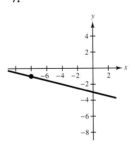

7.

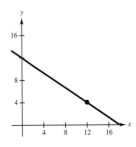

8.

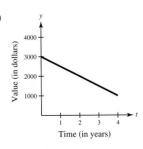

19. (b)

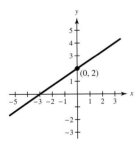

9.

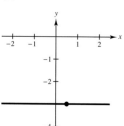

10.

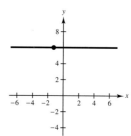

49.

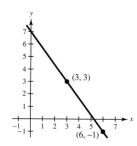

50.

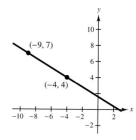

11.

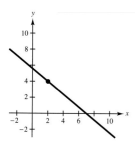

12.

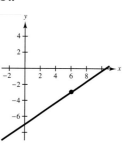

51.

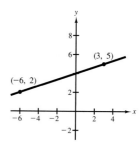

52.

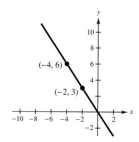

13.

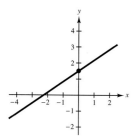

14.

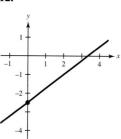

53.

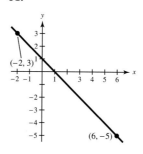

54.

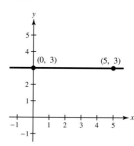

47.

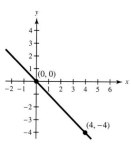

48.

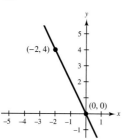

55.

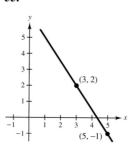

56.

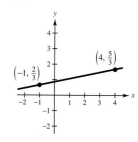

57.

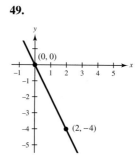

58.

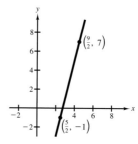

91.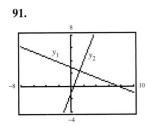

92.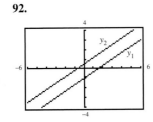

93.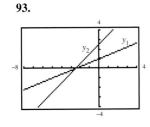

94.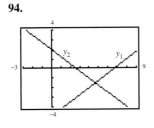

95.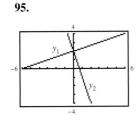

96.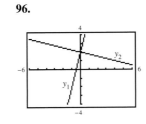

101.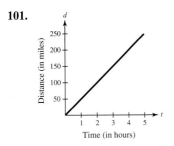

102. (b)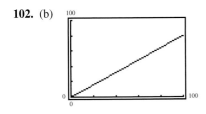

105.

(a) (b)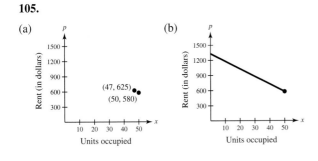

106. (a)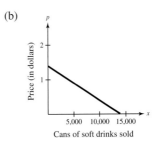

(b)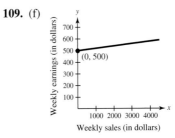

109. (f)

Section 4.5 *(page 254)*

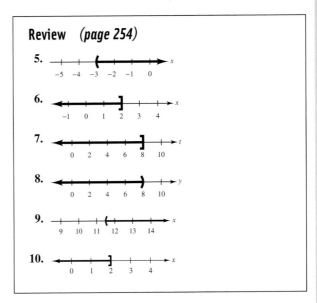

Review *(page 254)*

21.

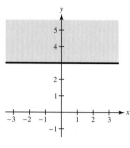

22.

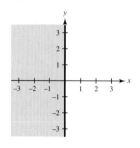

31.

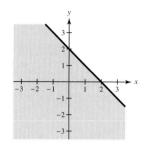

32.

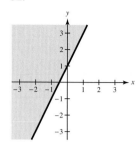

23.

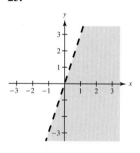

24.

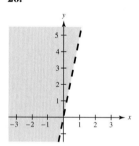

33.

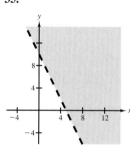

34.

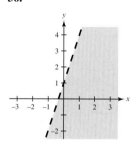

25.

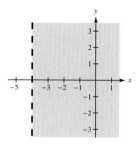

26.

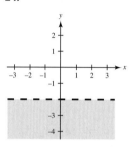

35.

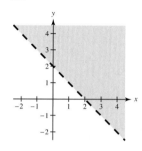

36.

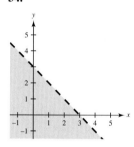

27.

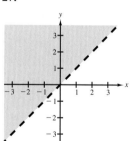

28.

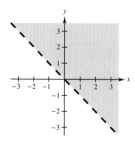

37.

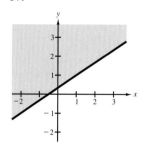

38.

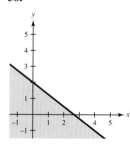

29.

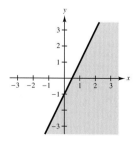

30.

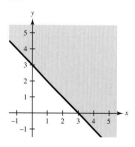

39.

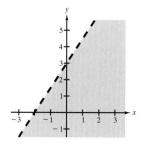

40.

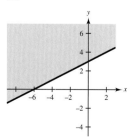

41.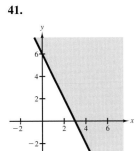

42.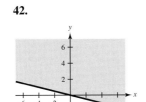

43.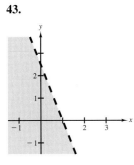

44.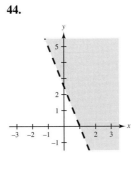

45.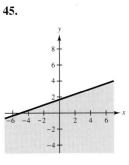

46.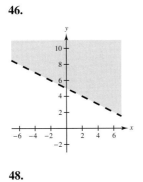

47.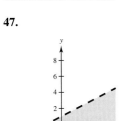

48.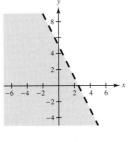

49.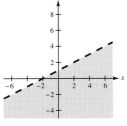

50.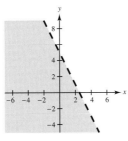

51.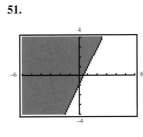

52.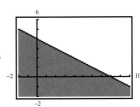

53.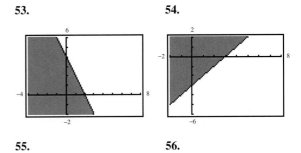

54.

55.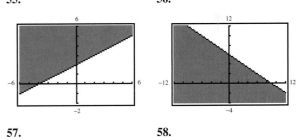

56.

57.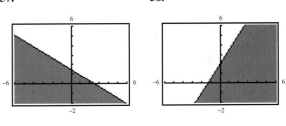

58.

65.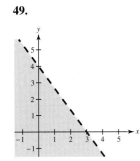

66.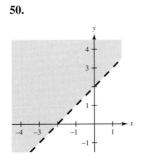

14.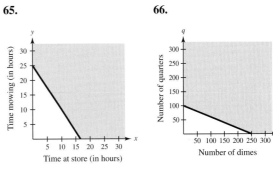

67.
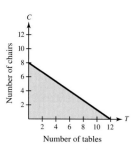

68.
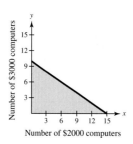

15.
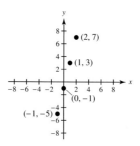

16.
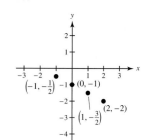

69.
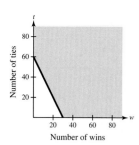

70.
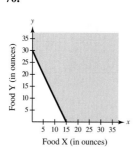

25. (a)
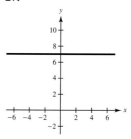

26. (a)
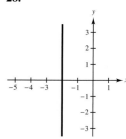

Review Exercises (*page 259*)

1.
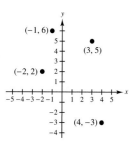

2.
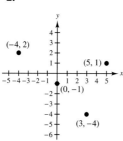

3.
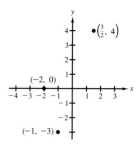

4.
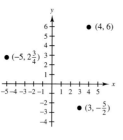

27.
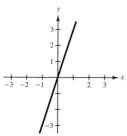

28.
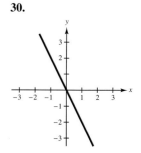

29.

30.

31.

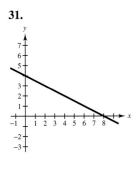

32.

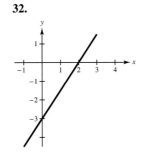

41.

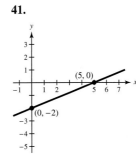

42.

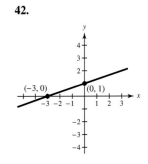

33.

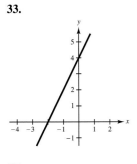

34.

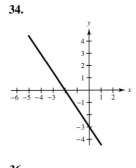

43.

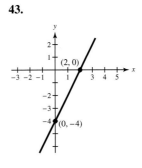

44.

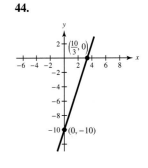

35.

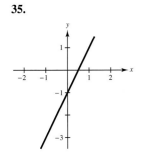

36.

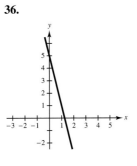

45.

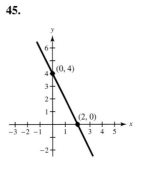

46.

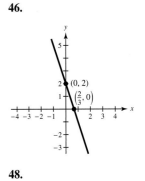

37.

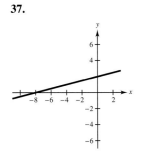

38.

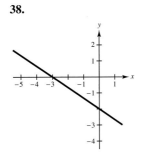

47.

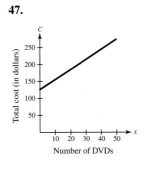

48.

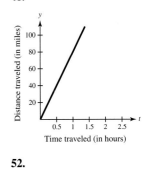

39.

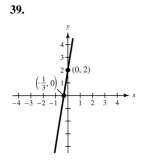

40.

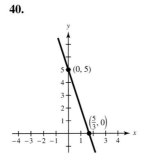

51.

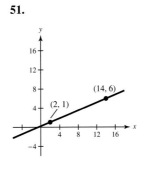

52.

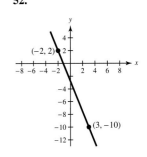

53.

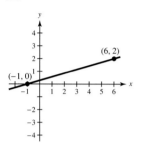

54.

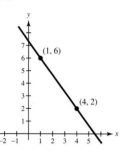

61.

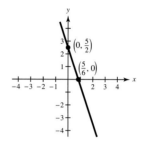

62.

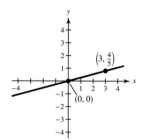

55.

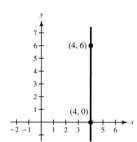

56.

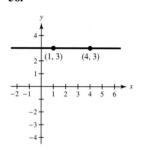

65.

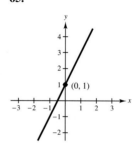

66.

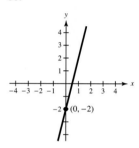

57.

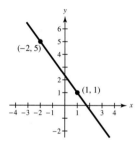

58.

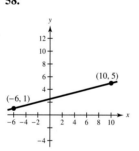

67.

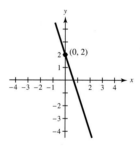

68.

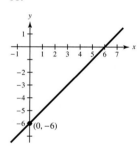

59.

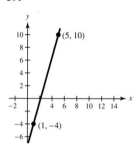

60.

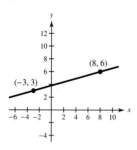

69.

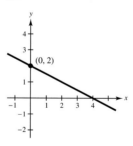

70.

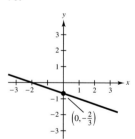

71.
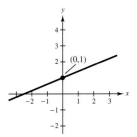

72.
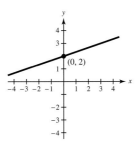

111.
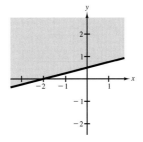

112.
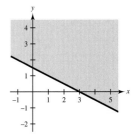

104. (a)
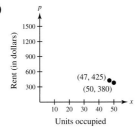

(b)
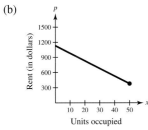

117.
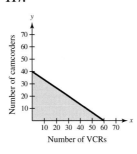

118.
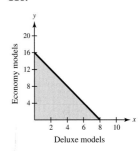

Chapter Test (*page 263*)

1.
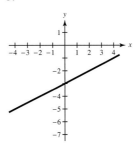

107.
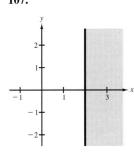

108.
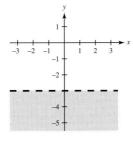

109.
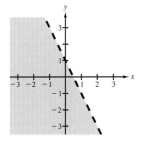

110.
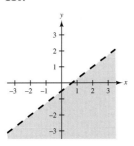

5.

6.
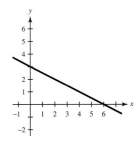

7.
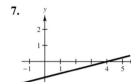

8.
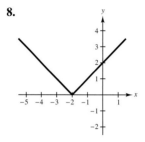

11.
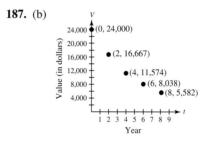

12.
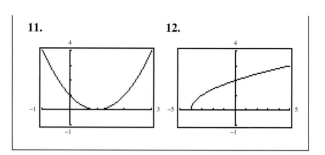

9.
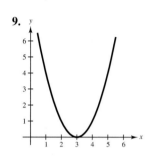

11.
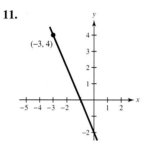

187. (b)
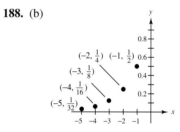

16.
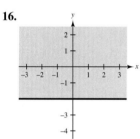

17.
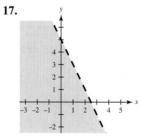

188. (b)
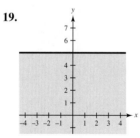

18.
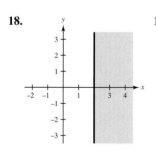

19.
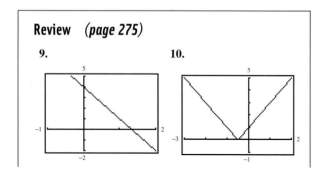

202. (b)

Expression	Value at $x = 3$	Equivalent expression	Value at $x = 3$
$\dfrac{2}{x^{-3}}$	54	$2x^3$	54
$\dfrac{1}{2x^3}$	$\dfrac{1}{54}$	$\dfrac{x^{-3}}{2}$	$\dfrac{1}{54}$
$\dfrac{1}{(2x)^{-3}}$	216	$8x^3$	216
$\dfrac{1}{8x^3}$	$\dfrac{1}{216}$	$(2x)^{-3}$	$\dfrac{1}{216}$

Chapter 5

Section 5.1   *(page 275)*

Review   *(page 275)*

9.

10.

Section 5.2   *(page 286)*

Review   *(page 286)*

11.   **12.**

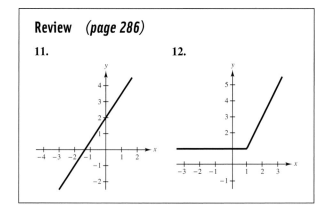

101. (b)

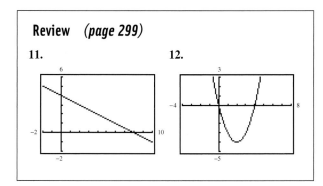

102. (b)

Section 5.3   *(page 299)*

Review   *(page 299)*

11.   **12.**

129. (b)

130. (a)

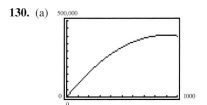

Section 5.4   *(page 309)*

71. (a)   **72.** (a)

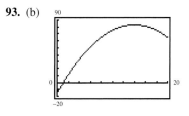

Review Exercises   *(page 313)*

93. (b)

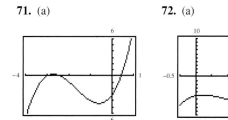

94. (a)

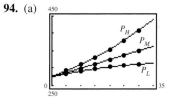

(b)

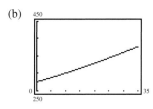

(c)

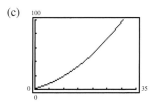

Chapter 6

Section 6.1   *(page 325)*

Review   *(page 325)*

7.

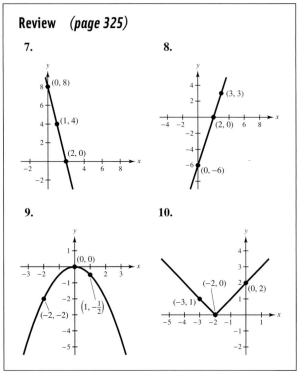

8.

9.

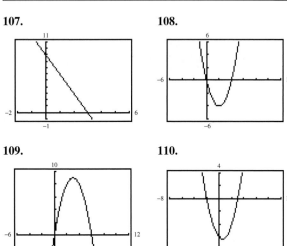

10.

107.

108.

109.

110.

Section 6.2   *(page 333)*

77.

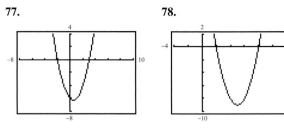

78.

79.

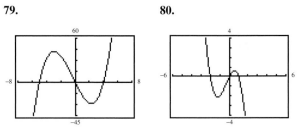

80.

81. (b)

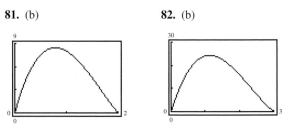

82. (b)

Section 6.3   *(page 341)*

Review   *(page 341)*

9.

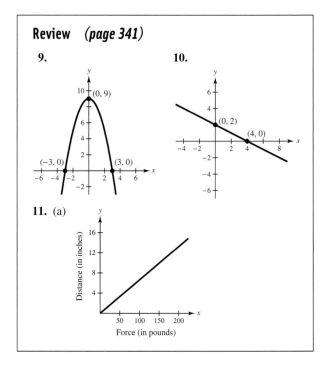

10.

11. (a)

115. (b)
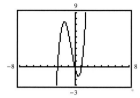

Mid-Chapter Quiz *(page 345)*

21.
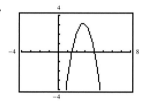

Section 6.4 *(page 352)*

119.
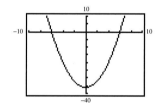

120.
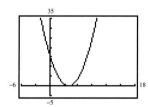

121.
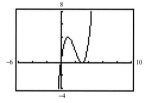

122.
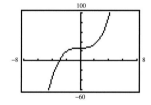

Section 6.5 *(page 362)*

77.
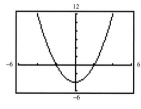

78.
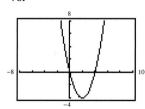

79.
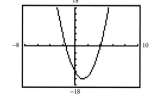

80.
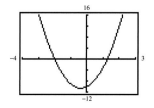

95. (a)
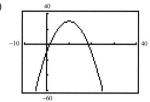

96. (a)
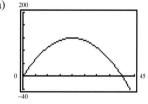

Review Exercises *(page 367)*

136. (a)
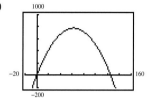

Cumulative Test: Chapters 4–6 *(page 371)*

3.
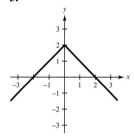

4.
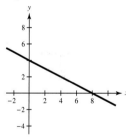

7.
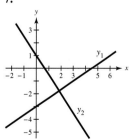

8.
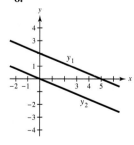

Chapter 7

Section 7.1   *(page 379)*

102. (a)

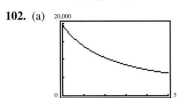

103. (b)

p	20	40	60	80
C	$20,000	$53,333	$120,000	$320,000

Section 7.2   *(page 389)*

Review   *(page 389)*

11.   **12.**

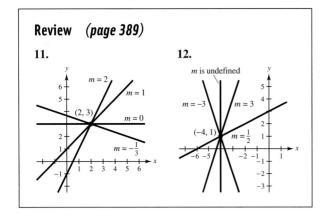

Section 7.5   *(page 417)*

61. (a)   **62.** (a)

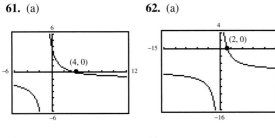

63. (a)   **64.** (a)

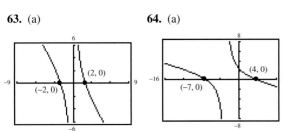

65. (a)   **66.** (a)

67. (a)   **68.** (a)

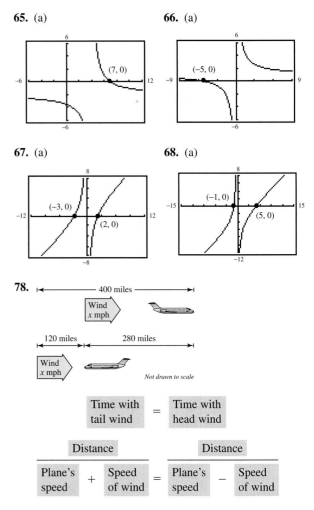

78.

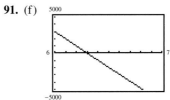

	Time with tail wind		$=$	Time with head wind	

	Distance			Distance	
Plane's speed	$+$	Speed of wind	$=$	Plane's speed	$-$	Speed of wind

91. (f)

Review Exercises   *(page 423)*

95. (a)   **96.** (a)

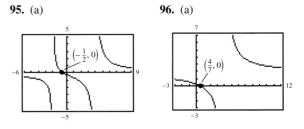

Chapter 8

Section 8.1 *(page 436)*

17.

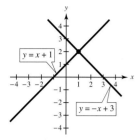

$y = x + 1$
$y = -x + 3$

18.

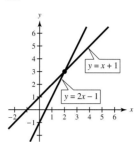

$y = x + 1$
$y = 2x - 1$

19.

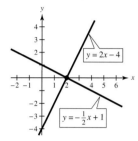

$y = 2x - 4$
$y = -\frac{1}{2}x + 1$

20.

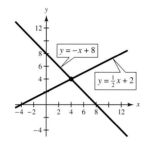

$y = -x + 8$
$y = \frac{1}{2}x + 2$

21.

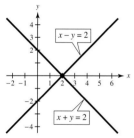

$x - y = 2$
$x + y = 2$

22.

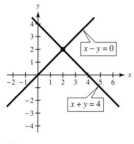

$x - y = 0$
$x + y = 4$

23.

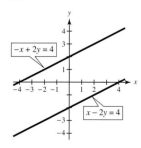

$-x + 2y = 4$
$x - 2y = 4$

24.

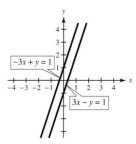

$-3x + y = 1$
$3x - y = 1$

25.

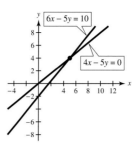

$6x - 5y = 10$
$4x - 5y = 0$

26.

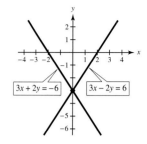

$3x + 2y = -6$
$3x - 2y = 6$

27.

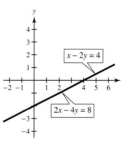

$x - 2y = 4$
$2x - 4y = 8$

28.

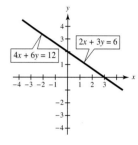

$4x + 6y = 12$
$2x + 3y = 6$

29.

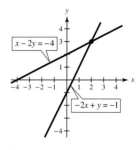

$x - 2y = -4$
$-2x + y = -1$

30.

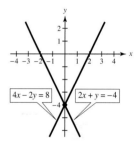

$4x - 2y = 8$
$2x + y = -4$

31.

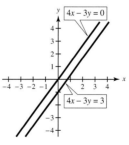

$4x - 3y = 0$
$4x - 3y = 3$

32.

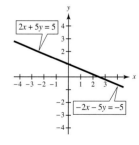

$2x + 5y = 5$
$-2x - 5y = -5$

33.

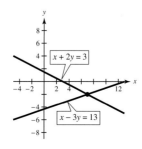

$x + 2y = 3$
$x - 3y = 13$

34.

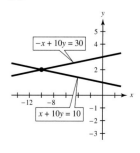

$-x + 10y = 30$
$x + 10y = 10$

35.

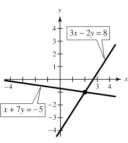

36.

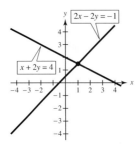

37.

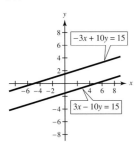

38.

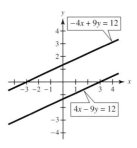

39.

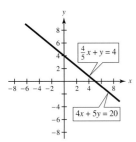

40.

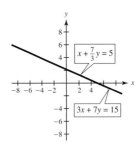

41.

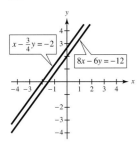

42.

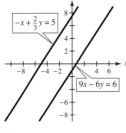

43.

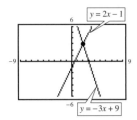

44.

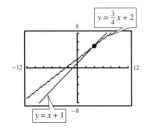

45.

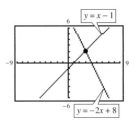

46.

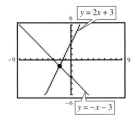

55.

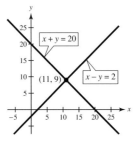

56.

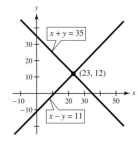

57. (a)

$$\text{Cost} = \boxed{\begin{array}{c}\text{Cost per}\\\text{unit}\end{array}} \cdot \boxed{\begin{array}{c}\text{Number}\\\text{of units}\end{array}} + \boxed{\begin{array}{c}\text{Fixed}\\\text{costs}\end{array}}$$

$$\text{Revenue} = \boxed{\begin{array}{c}\text{Price per}\\\text{unit}\end{array}} \cdot \boxed{\begin{array}{c}\text{Number}\\\text{of units}\end{array}}$$

(b)

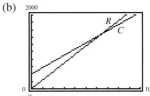

58.

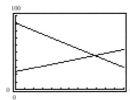

61.

Section 8.2   *(page 445)*

Review   *(page 445)*

11.   **12.**

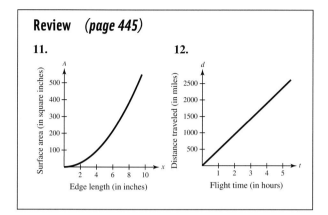

76.

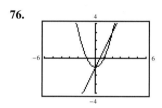

Section 8.3   *(page 455)*

Review   *(page 455)*

5.   **6.**

7.   **8.**

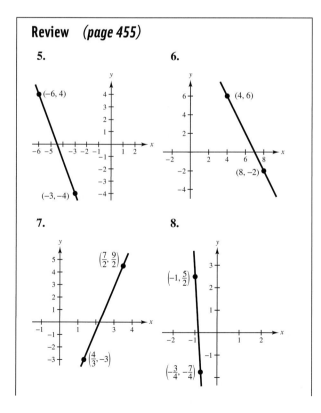

9.   **10.**

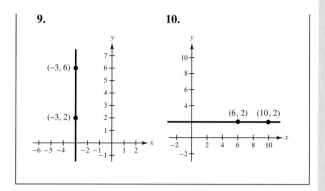

Mid-Chapter Quiz   *(page 459)*

6.   **7.**

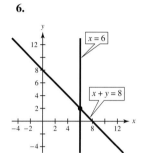

8.

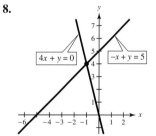

Section 8.4   *(page 467)*

Review   *(page 467)*

11. (b)

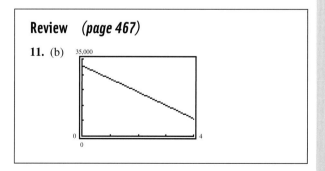

59.
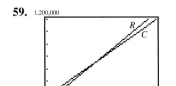

60.
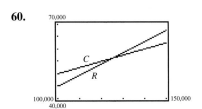

Section 8.5   *(page 476)*

Review   *(page 476)*

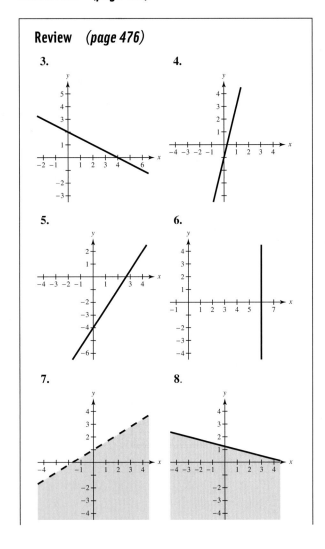

9.
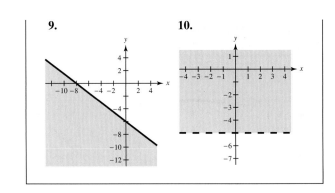

10.

7.

8.
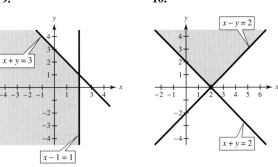

9.

10.
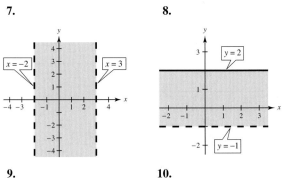

11.

12.
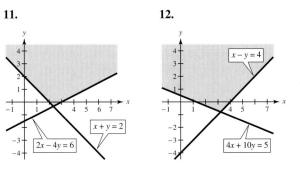

13.

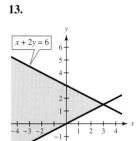

14.

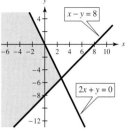

21.

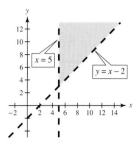

22.

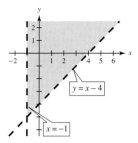

15.

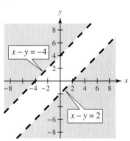

16.

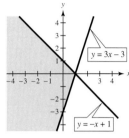

23.

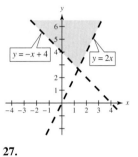

24.

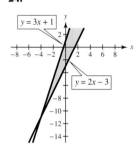

17.

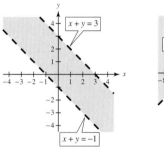

18.

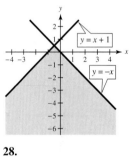

25.

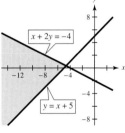

26.

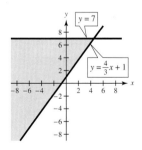

19.

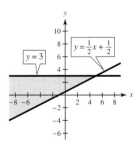

20.

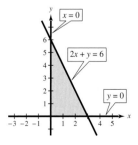

27.

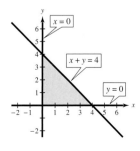

28.

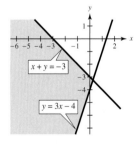

29.

30.

31.

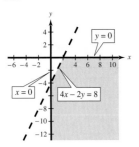

32.

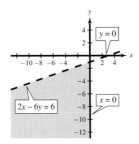

39.

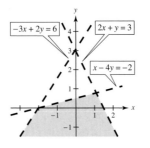

40.

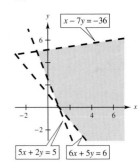

33.

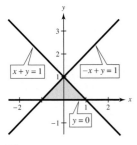

34.

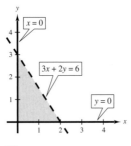

41.

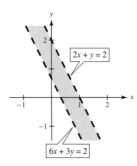

42.

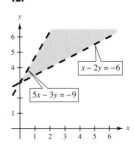

35.

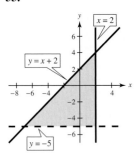

36.

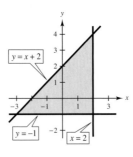

43.

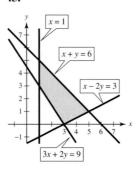

44.

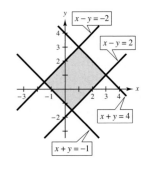

37.

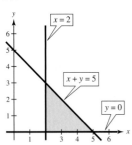

38.

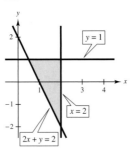

45.

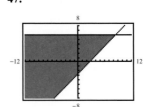

46.

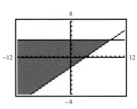

47.

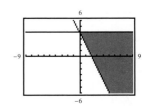

48.

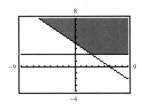

49.
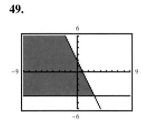

50.
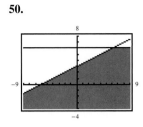

11.
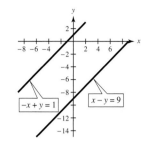

12.
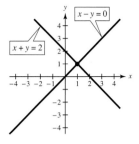

55.
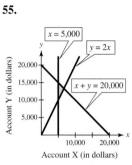

56.
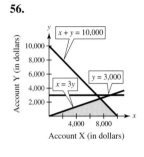

13.
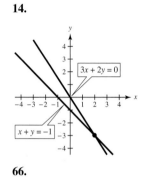

14.
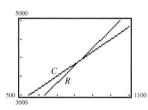

57.
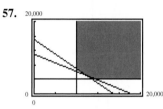

58.
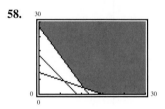

65.
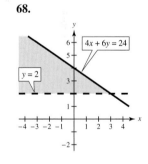

66.
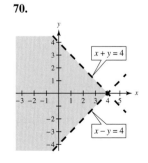

Review Exercises *(page 481)*

9.
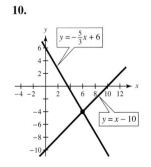

10.
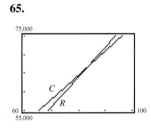

67.
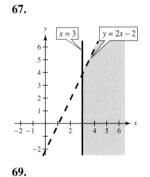

68.
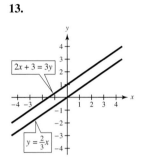

69.
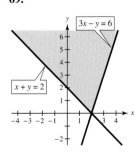

70.

71.

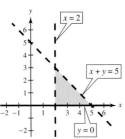

72.

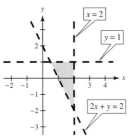

Chapter Test *(page 485)*

5.

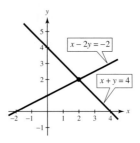

6.

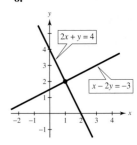

73.

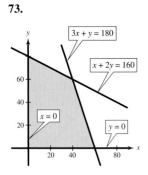

74.

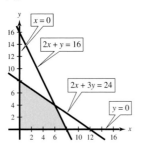

7.

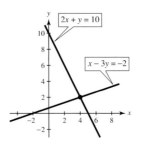

8.

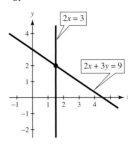

77.

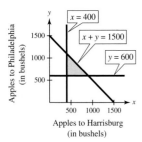

78.

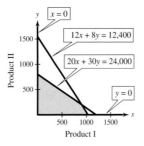

15.

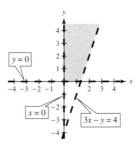

16.

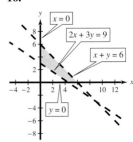

79.

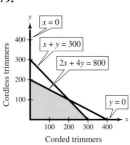

80.

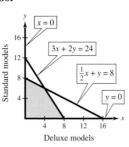

18.

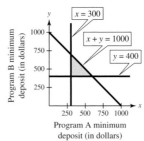

Chapter 9

Section 9.1 *(page 494)*

Review *(page 494)*

13.
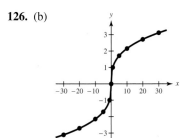

14.
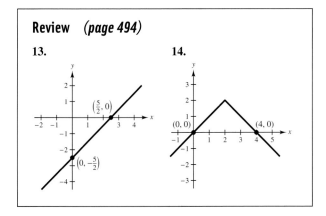

125. (b)
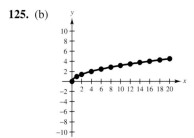

126. (b)

Section 9.2 *(page 504)*

Review *(page 504)*

3.
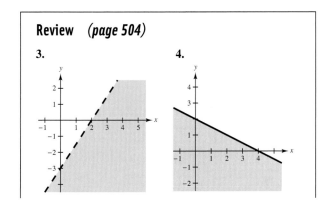

4.

5.
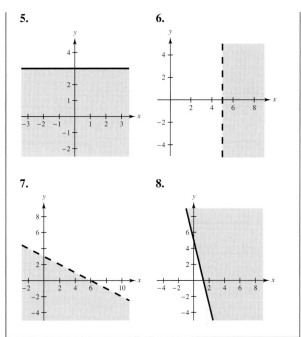

6.

7.

8.

123.
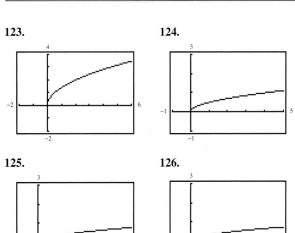

124.

125.

126.

Section 9.3 *(page 515)*

127.
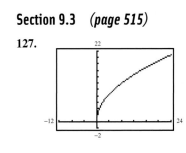

128.

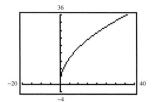

129.

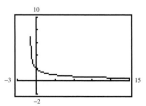

130.

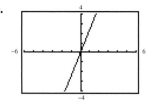

Section 9.4   *(page 525)*

Review   *(page 525)*

9.

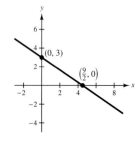

10.

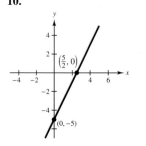

57.

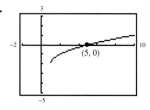

58.

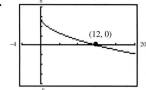

59.

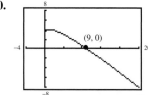

60.

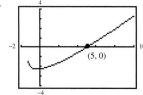

91. (a)

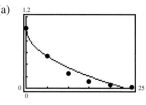

Review Exercises   *(page 530)*

115.

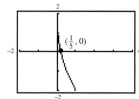

116.

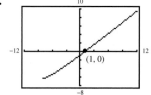

Chapter 10

Section 10.1   *(page 541)*

83.

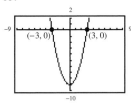

84.

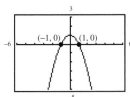

85.

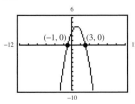

86.

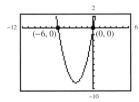

87.

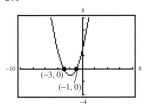

88.

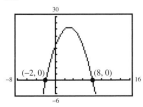

110. Isolate the variable and take the square root of each side.

$$ax^2 + c = 0$$
$$ax^2 = -c$$
$$x^2 = -\frac{c}{a}$$
$$x = \pm\sqrt{-\frac{c}{a}}$$

Suggested revision: If $a > 0$, restrict the values of c to $c \le 0$ so that the solutions will be real. If $a < 0$, restrict the values of c to $c \ge 0$ so that solutions will be real.

Section 10.2   *(page 549)*

87.

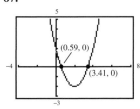

88.

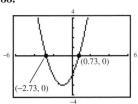

89.

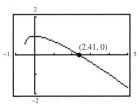

90.

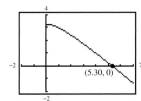

99.

(a)

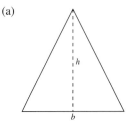

100.

(a)

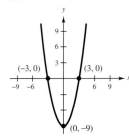

111.

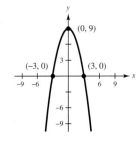

Section 10.4   *(page 569)*

41.

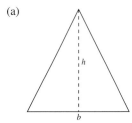

42.

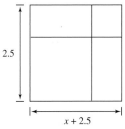

43.

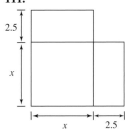

44.

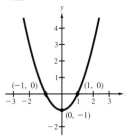

45.

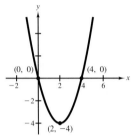

46.

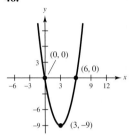

55.

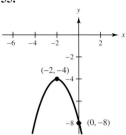

56.

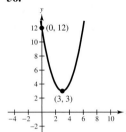

47.

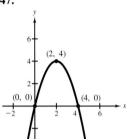

48.

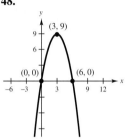

57.

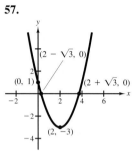

58.

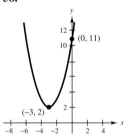

49.

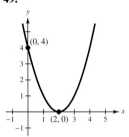

50.

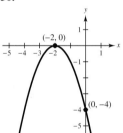

59.

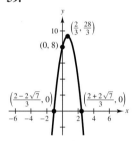

60.

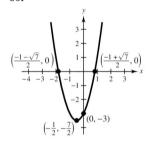

51.

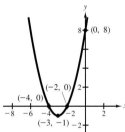

52.

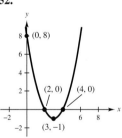

61.

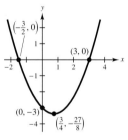

62.

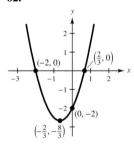

53.

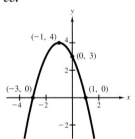

54.

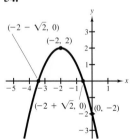

63.

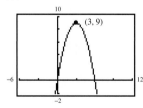

64.

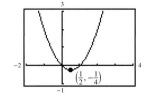

65.
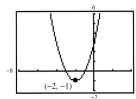

66.
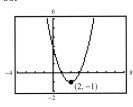

75.
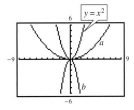

76.
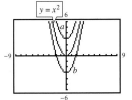

67.
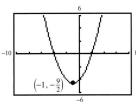

68.
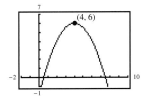

77.
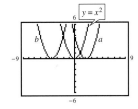

78.
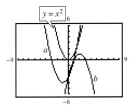

69.
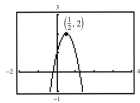

70.
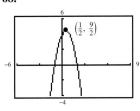

83.
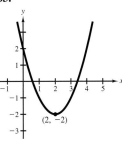

84.
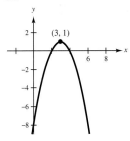

71.
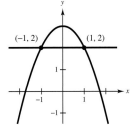

72.
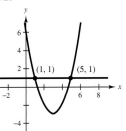

85.
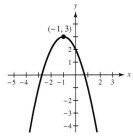

86.
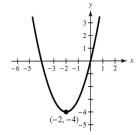

73.
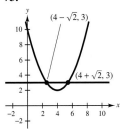

74.
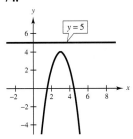

95. (c)
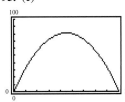

97. (a)
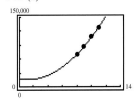

ADDITIONAL ANSWERS

98. (a)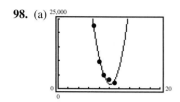

Section 10.5   *(page 577)*

Review   *(page 577)*

11.   **12.**

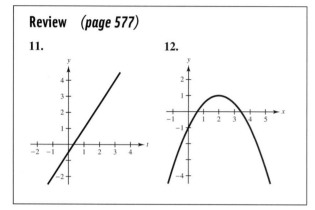

57. (a)

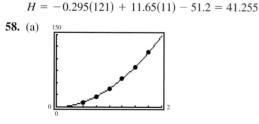

(d)

For $t = 6$, $H = -0.295(36) + 11.65(6) - 51.2 = 8.08$

For $t = 7$, $H = -0.295(49) + 11.65(7) - 51.2 = 15.895$

For $t = 8$, $H = -0.295(64) + 11.65(8) - 51.2 = 23.12$

For $t = 9$, $H = -0.295(81) + 11.65(9) - 51.2 = 29.755$

For $t = 10$,
$H = -0.295(100) + 11.65(10) - 51.2 = 35.8$

For $t = 11$,
$H = -0.295(121) + 11.65(11) - 51.2 = 41.255$

58. (a)

---

60. (i) Simplify each side by removing symbols of grouping, combining like terms, and reducing fractions on one or both sides.

   (ii) Add (or subtract) the same quantity to (from) each side of the equation.

   (iii) Multiply (or divide) each side of the equation by the same nonzero real number.

   (iv) Interchange the two sides of the equation.

63. (i) Write a verbal model that will describe what you need to know.

   (ii) Assign labels to each part of the verbal model—numbers to the known quantities and letters to the variable quantities.

   (iii) Use the labels to write an algebraic model based on the verbal model.

   (iv) Solve the resulting algebraic equation and check your solution.

64.

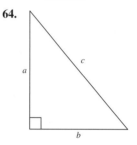

Section 10.6   *(page 589)*

Review   *(page 589)*

5. $\dfrac{x}{5}$, $x \neq 0$, $x \neq -\dfrac{3}{2}$ **6.** $\dfrac{x}{5(x + y)}$, $x \neq 0$, $x \neq y$

7. $\dfrac{9}{2(x + 3)}$, $x \neq 0$ **8.** $\dfrac{1}{x - 2}$, $x \neq 0$, $x \neq -2$

9. $\dfrac{4}{x + 3}$, $x \neq 3$ **10.** $\dfrac{2(x + 1)}{3}$, $x \neq -1$

159. (a) $1, \dfrac{-1 + \sqrt{3}i}{2}, \dfrac{-1 - \sqrt{3}i}{2}$

  (b) $2, \dfrac{-2 + 2\sqrt{3}i}{2} = -1 + \sqrt{3}i,$

     $\dfrac{-2 - 2\sqrt{3}i}{2} = -1 - \sqrt{3}i$

  (c) $4, \dfrac{-4 + 4\sqrt{3}i}{2} = -2 + 2\sqrt{3}i,$

     $\dfrac{-4 - 4\sqrt{3}i}{2} = -2 - 2\sqrt{3}i$

Section 10.7 *(page 598)*

73. Example:

Domain	Range
1	4
2	5
3	6

Review Exercises *(page 604)*

61.
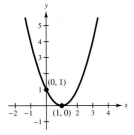

62.
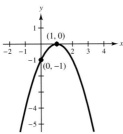

63.
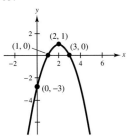

64.
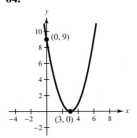

65.
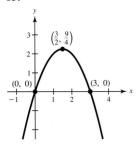

66.
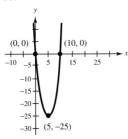

67.
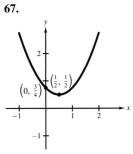

68.
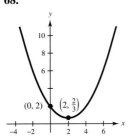

69.
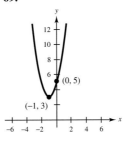

70.
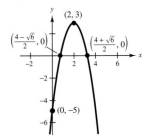

71.
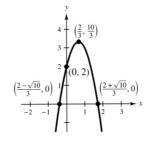

72.
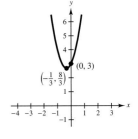

73.
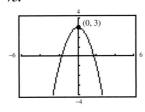

74.
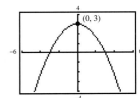

75.
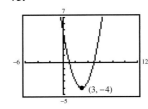

76.
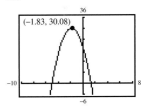

99. (d)
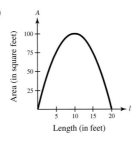

100. (a)
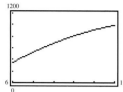

(d)
For $t = 6$,
$R(6) = -12.454(36) + 340.25(6) - 1246 = 347.156$

For $t = 7$,
$R(7) = -12.454(49) + 340.25(7) - 1246 = 525.504$

For $t = 8$,
$R(8) = -12.454(64) + 340.25(8) - 1246 = 678.944$

For $t = 9$,
$R(9) = -12.454(81) + 340.25(9) - 1246 = 807.476$

For $t = 10$,
$R(10) = -12.454(100) + 340.25(10) - 1246 = 911.1$

For $t = 11$,
$R(11) = -12.454(121) + 340.25(11) - 1246 = 989.816$

Chapter Test   *(page 610)*

12.
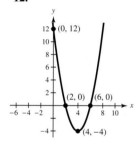

13.
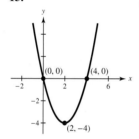

Appendix A   *(page A6)*

1.
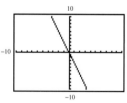

2.
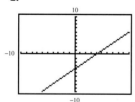

3.
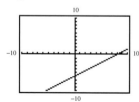

4.
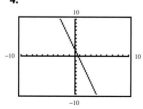

5.
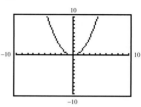

6.
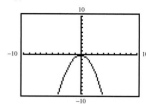

7.
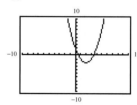

8.
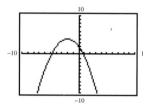

9.
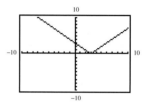

10.
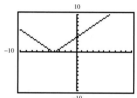

11.
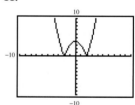

12.
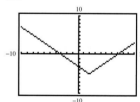

13.
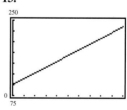

14.
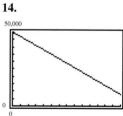

15.
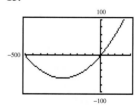

16.
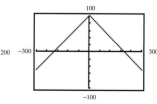

17. Sample answer:

> Xmin = 4
> Xmax = 20
> Xscl = 1
> Ymin = 14
> Ymax = 22
> Yscl = 1

18. Sample answer:

> Xmin = 8
> Xmax = 16
> Xscl = 1
> Ymin = 14
> Ymax = 22
> Yscl = 1

37.

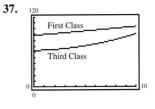

19. Sample answer:

> Xmin = -20
> Xmax = -4
> Xscl = 1
> Ymin = -16
> Ymax = -8
> Yscl = 1

20. Sample answer:

> Xmin = -18
> Xmax = -6
> Xscl = 1
> Ymin = -16
> Ymax = 10
> Yscl = 1

21.

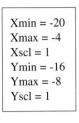

22.

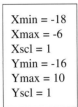

23.

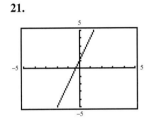

24.

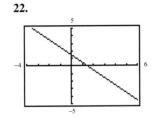

33.

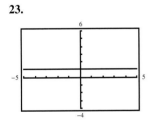

34.

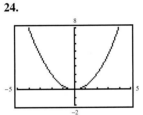

35.

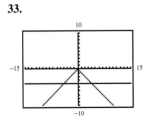

36.

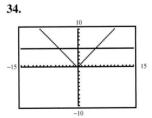

35.

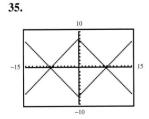

36.

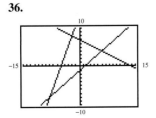

Technology Answers

Section 1.1   *(page 4)*

Between 3 and 4

Section 1.3   *(page 22)*

No, no, $1 = 0 \cdot 0$, $2 = 0 \cdot 0$. Division by zero is undefined; an error message

Section 1.4   *(page 41)*

-0.882. Rounding digit: 2; decision digit: 4

Section 1.5   *(page 48)*

$(-5)^4 = 625$, $-5^4 = -625$; $(-5)^3 = -125$, $-5^3 = -125$. If the negative sign is part of the base and the base is raised to an odd power, the result is negative. If the negative sign is part of the base and the base is raised to an even power, the result is positive. If the negative sign is not part of the base, the result is negative regardless of the power.

Section 3.5   *(page 171)*

$P = \$9580$

Section 4.2   *(page 214)*

The graph appears in the viewing window, but you cannot see where the x- and y-intercepts of the graph are located.

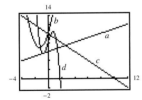

Section 5.1   *(page 271)*

≈ 170.6913258, ≈ 170.6913258. The expressions are equal.

Section 6.4   *(page 346)*

$16^2 - 9^2 = 175$ and $(16 + 9)(16 - 9) = 175$. The equation is true for all real numbers.

Section 6.5   *(page 359)*

Example 3:   General form: $x^2 - 8x + 16 = 0$
                 x-intercept: $(4, 0)$

Example 4:   General form: $x^2 + 9x + 14 = 0$
                 x-intercepts: $(-2, 0), (-7, 0)$

The number of solutions is the same as the number of x-intercepts.

Section 7.1   *(page 374)*

(a)

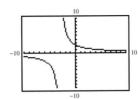

(b)

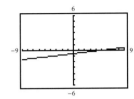

(c)

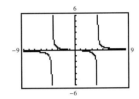

(d)

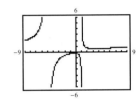

The y-value to the right of $x = -3$ is a large positive number, whereas the y-value to the left of $x = -3$ is a large negative number. When the table feature is used, at $x = -3$ the corresponding y-value displays an error. When an x-value is not in the domain of a rational expression, the graphing calculator displays a line connecting the two extremes on each side of the x-value.

Section 7.1   *(page 377)*

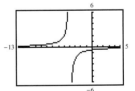

Section 7.5   *(page 414)*

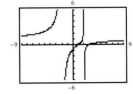 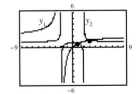

The x-intercepts and the point(s) of intersection of the graphs are the solutions in Example 5. The x-intercepts of a rational equation are the solutions of the equation.

Section 8.1 *(page 432)*

(a) $y = -\frac{3}{4}x + 3$

$y = \frac{2}{3}x + 3$

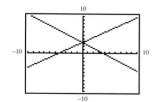

(b) $y = \frac{1}{2}x + 4$

$y = \frac{1}{2}x - \frac{5}{4}$

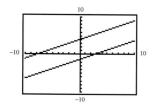

(c) $y = -x + 6$

$y = -x + 6$

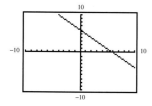

If the slopes are not equal, there is one point of intersection. If the slopes are equal, there may be no point of intersection or infinitely many points of intersection.

Section 8.1 *(page 434)*

The solution point is $\left(\frac{3}{2}, 1\right)$.

Section 8.2 *(page 444)*

The solution is (7000, 5000).

Section 9.1 *(page 488)*

When a negative number is raised to a power, the answer can be positive or negative.

(a) 16   (b) −64   (c) 16   (d) −32

When a negative number is raised to an even power, the answer is positive. When a negative number is raised to an odd power, the answer is negative.

Section 9.1 *(page 491)*

(a) 8   (b) −8

(c) Error. The square root of a negative number is not a real number.

Section 9.4 *(page 524)*

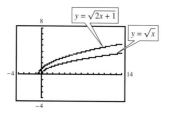

The graphs do not intersect, so there is no real solution. An equation has an extraneous solution when the graphs of the two sides of the equation do not intersect.

Section 10.2 *(page 548)*

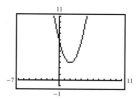

The graph does not cross the x-axis. The vertex of the graph lies on the x-axis. The graph intersects the x-axis at two locations.

Section 10.4 *(page 562)*

(a)

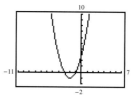

(b)

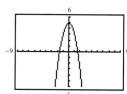

(c)

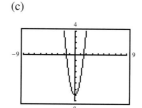

(d)

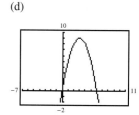

If the coefficient of the x^2-term is positive, the graph opens upward. If the coefficient of the x^2-term is negative, the graph opens downward. The graph opens downward.

Section 10.5   *(page 574)*

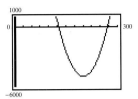

The *x*-intercepts are the same as the solutions in Example 2.

Section 10.6   *(page 581)*

(a) 11

(b) $\sqrt{-121}$ is not a real number. Some graphing utilities will give an ERROR message. Others will display the complex number as $0 + 11i$ or as $(0, 11)$.

(c) -11

Section 10.6   *(page 587)*

(a) $-\frac{5}{2}, 1$ (b) -1

(c) $\pm 2i$ (d) $-7 \pm \sqrt{2}i$

(a) and (b) have real solutions.

(c) and (d) have complex solutions.

(a) and (b) have *x*-intercepts.

(c) and (d) have no *x*-intercepts.

Quadratic equations with only complex solutions do not have *x*-intercepts, but quadratic equations with real solutions do.

Index of Applications

Index